Group Theory in Chemistry

Bonding and Molecular Spectroscopy

Asok K Mukherjee
Professor of Chemistry (retired), The University of Burdwan
Burdwan, West Bengal

Bankim Chandra Ghosh
Assistant Professor of Chemistry, Durgapur Government College
Durgapur, West Bengal

GROUP THEORY IN CHEMISTRY: Bonding and Molecular Spectroscopy

UNIVERSITIES PRESS (INDIA) PRIVATE LIMITED

Registered Office
3-6-747/1/A & 3-6-754/1, Himayatnagar, Hyderabad 500 029, Telangana, India
info@universitiespress.com: www.universitiespress.com

Distributed by
Orient Blackswan Private Limited

Registered Office
3-6-752 Himayatnagar, Hyderabad 500 029, Telangana, India

Other Offices
Bengaluru / Bhopal / Chennai / Guwahati / Hyderabad / Jaipur /
Kolkata / Lucknow / Mumbai / New Delhi / Noida / Patna / Visakhapatnam

First published 2018
Reprinted 2020

ISBN: 978-93-86235-19-0

Typeset by
Bookcraft Publishing Services (I) Pvt Ltd, Chennai

501700

Printed at
First Impression, Noida

Published by
Universities Press (India) Private Limited
3-6-747/1/A & 3-6-754/1, Himayatnagar
Hyderabad 500 029, Telangana, India

DEDICATED

To the memory of my parents,
(Late) Smt. Geeta Mukherjee and
(Late) Shri Bijaymadhab Mukherjee
(AKM)

To my parents,
Smt. Sabita Ghosh and Shri Sushil Kumar Ghosh
(BCG)

Preface

Group theory is now included in the M Sc chemistry curriculum of almost all universities. In fact, proper understanding of chemical bonding and molecular spectroscopy remains incomplete without at least a preliminary knowledge of molecular symmetry aspects which are best dealt with by the representation theory of groups. Usually chemistry students, excepting those who opt for theoretical chemistry, consider group theory as a tough subject because of the abstract way it is generally presented. The main objective of this textbook is to remove the fear of the abstract presentations of group theory from the student's mind while rigorously maintaining the correctness of the necessary mathematical statements and proofs. Towards this end, all discussions in the book start with familiar examples and then proceed to explain the concepts. A 'do it yourself' approach has been adopted for the models and drawings. Any good student with standard (10+2) level of knowledge in elementary quantum chemistry and mathematics will be able to follow the book.

The book contains 14 chapters. Chapters 1–7 develop the basic principles of group theory (GT). Chapters 8–10 demonstrate how GT is used to construct molecular orbitals and energy level diagrams and to predict spectroscopic transitions that lead to the observed UV–vis spectra of molecules. Chapter 11 discusses at length how GT principles classify the normal modes of vibration of molecules (both linear and non-linear), and predict the number of fundamental bands in the infrared and Raman spectra of molecules, analyse their nature of polarisation and justify the appearance of certain overtone bands. Examples are given to show how such analysis can establish certain structural features and choose the correct one from among a number of plausible isomeric structures of a molecule. Chapter 12 deals with spin-orbit coupling in atoms and explains the methods of obtaining atomic term symbols from electronic configurations. The method for obtaining molecular term symbols has also been briefly outlined in this chapter. A large portion of Chapter 12 dealing with term symbols, may seem to be a digression from the main focus of the book but we felt it was a necessity, since the contents of Chapter 13 cannot be developed without these. The Crystal Field Theory, its modification to Ligand Field Theory and the Molecular Orbital Theory for bonding in transition metal complexes and interpretation of their UV–vis spectra—are all based on intensive use of GT; this has been elaborately discussed in Chapter 13. The role of symmetry groups in determining the course of pericyclic reactions has been outlined in Chapter 14. The large number of examples worked out in each chapter serve to familiarise the student with the basic concepts and applications. The exercises given at the end of each chapter, aim to test the students' understanding of the concepts. Proofs of two important theorems, which may be skipped during the first reading, are given in the Appendices for the interested reader.

The authors owe their knowledge in the subject to the original papers and books which have been mentioned and recommended for further reading at the end of each chapter. The senior author (AKM) taught the subject for a long time at the M Sc level in the Department of Chemistry, The University of Burdwan (West Bengal). The style of presentation adopted in the book stems from the interactions with batch after batch of students of the chemistry department, BU.

In spite of serious attempts to keep the text free of errors, it would be presumptuous to hope that no error has crept in. We shall be grateful to teachers, students and researchers for any corrections/suggestions they send (akm_13@rediffmail.com and/or bankimghosh@yahoo.co.in).

We gratefully acknowledge our publisher, Universities Press (India) Pvt Ltd, for bringing out this book. In particular, Dr Gita Dattatri and Mr Kallol Das, helped us at various stages with valuable suggestions in preparing and editing the manuscript. It is our pleasure to acknowledge the encouragement and help received from Mr Shibshankar Ray of Digital Computer System, Burdwan, in taking up the project of writing the book. We express our respect and gratitude to the great teachers we had at the University of Burdwan whose illuminating lectures built up our confidence in pursuing the subject, and in particular, to Professor S C Rakshit from whom AKM got his first lessons in quantum mechanics. Special thanks are due to Dr Mira Mukherjee and Smt. Chhanda Ghosh, without whose inspiration and support the book could not have seen the light of the day. Finally, we are delighted to mention our lovely little Debanjana and Rishika, who frequently visited our study and innocently wondered what the two elderly people were doing.

July, 2017 A K Mukherjee

Burdwan B C Ghosh

Contents

1 Elements of Symmetry, Symmetry Operations and Point Groups

Learning objectives

- ➢ Symmetry elements and symmetry operations
- ➢ Mathematical groups, symmetry operators of an object forming a group—point group
- ➢ Symmetry multiplication table, similarity transformation, conjugate classes
- ➢ Schoenflies notation of point groups, prediction of polarity and optical activity
- ➢ Stereographic projection of a point group and its utility

1.1 INTRODUCTION

Without requiring a formal definition of the word *symmetry*, one can intuitively understand what it means. For example, a square is more symmetric than a rectangle and a circle is even more symmetric. All beautiful objects in nature have symmetry—one half of the body of a butterfly is a mirror image of the other half, a flower has symmetrically arranged petals of symmetrical shape, leaves and ice-flakes (crystals) are common examples of naturally symmetric objects. In many molecules, atoms are symmetrically arranged. Natural phenomena such as the rising and setting of the sun, high and low tides and so on, occur in a rhythmic manner which is a manifestation of symmetry in the time domain. In mathematics, symmetry is observed at many places—the curve of $\sin x$ against x remains the same if it is rotated through an angle π about an axis perpendicular to the plane of the curve and passing through the origin, the curve of $\cos x$ against x exhibits mirror symmetry with respect to the YZ plane. As a consequence of this, symmetry plays an important role in all branches of physics where the Fourier series is involved.

Symmetry in molecules can be exploited to predict properties of molecules, to derive spectroscopic selection rules, to simplify secular determinants in molecular orbital calculations, to determine which normal modes of vibration are infrared or Raman active, and so on. The mathematical tool required for utilising symmetry for such purposes is Group Theory.

Note: In a system of charged particles, the Hamiltonian operator does not change if signs of the charges are interchanged (+ to – and – to +) in the absence of an applied electric or magnetic field; this is *charge conjugation symmetry* used by particle physicists. There are other types of symmetry also. However, in this elementary book where applications are on molecular structure and spectroscopy, we shall be concerned with only geometrical symmetry.

1.2 ELEMENTS OF SYMMETRY AND SYMMETRY OPERATIONS

By *geometrical symmetry* of an object we usually mean the regularity in its geometrical shape. To be more precise, an object is said to have geometrical symmetry if it has at least one of the following features:

- a rotational axis of symmetry (*proper axis of rotation*),
- a plane of symmetry (*reflection plane*),
- a centre of symmetry (*inversion centre*), and
- an alternating axis of symmetry (*axis of improper rotation*).

These geometrical entities are called *elements of symmetry*. They exist on the basis of the fact that a given object has a *configuration*, that is, different parts of the object have definite locations in space so that one can recognise or identify the object. Even if the above four elements of symmetry are absent, a given object has its own identity; for example, a banyan tree apparently may not have such symmetry but one can indentify it. In fact, without considering *identity* as a symmetry element, the mathematics of symmetry operations (which form the most significant part of group theory applications) cannot be developed just as the entire mathematics of number systems cannot be developed without the concept of zero and one.

A *symmetry operation* is defined as an action that brings a body into a configuration which is indistinguishable from its initial configuration. Such an operation has to be done with respect to one of the symmetry elements listed above; if the symmetry element is denoted by R, the associated symmetry operator is denoted by $\hat{R}$.

1.2.1 Axis and Plane of Symmetry

A body is said to possess a proper *n-fold rotational axis of symmetry* C_n, if by an anticlockwise rotation of the body about this axis through an angle of $2\pi/n$ (n = an integer), the body is brought to a configuration equivalent to or identical with the starting configuration.

Note: Rotation about an axis may be performed clockwise or anticlockwise, but one convention should be followed throughout. In trigonometry, angles produced by anticlockwise rotation of a straight line with respect to another fixed straight line are taken to be positive. Accordingly, all rotations in this book are carried out in the anticlockwise direction.

For example, the line AB is a C_2-axis for the H_2O molecule (Fig. 1.1). The act of performing a 180° rotation about this axis is denoted by $\hat{C}_2$. Here, (I) and (II) are *equivalent* configurations and (I) and (III) are *identical*. In going from (I) to (III) directly, we have effectively done nothing to the configuration; such an action is called the *identity* operation or 'do nothing operation' and is denoted by $\hat{E}$ (from German, 'einheit' ≡ unit). Thus, $\hat{C}_2^2 = \hat{E}$ (a superscript on $\hat{C}_n$ means the number of times $\hat{C}_n$ is carried out successively).

An object is said to possess a *plane of symmetry* (reflection plane) σ, if the plane cuts the object into two equal halves such that one half is the mirror image of the other half when reflected through that plane. For example, the H_2O molecule has two planes of symmetry—one such plane is shown in Fig. 1.2(a). This plane $\sigma(1)$, is perpendicular to the one containing the three atoms of the molecule and passes through the O atom. The other plane of symmetry $\sigma(2)$, is the molecular plane itself; to understand its action as a symmetry plane, only labelling of the atoms is not sufficient as will be evident from Fig. 1.2(b).

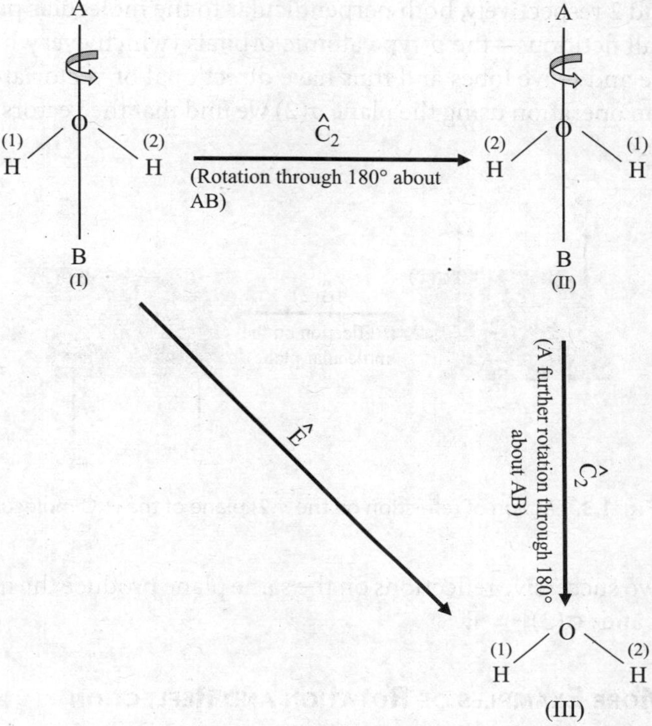

Fig. 1.1 Rotation axis of symmetry of the H_2O molecule; $\hat{C}_2^2 = \hat{E}$

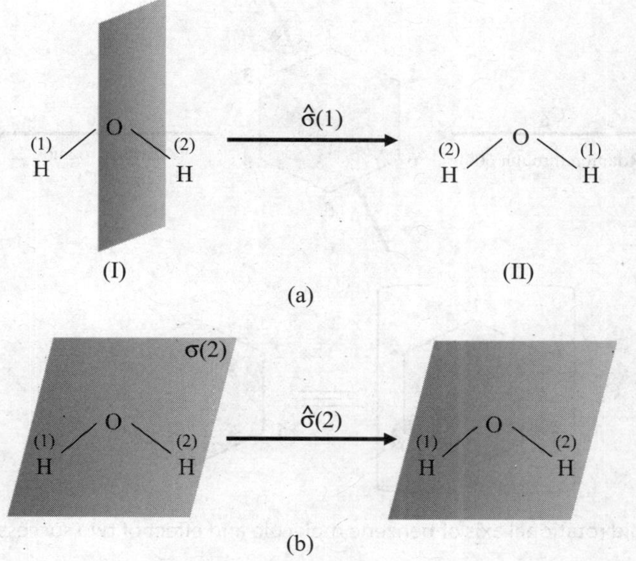

Fig. 1.2 Two planes of symmetry of the H_2O molecule

It appears from Fig. 1.2(b) that nothing has been done on the initial configuration, that is, as if $\hat{\sigma}(2) = \hat{E}$. To verify that it is not so, we imagine two vectors $\vec{v}_1$ and $\vec{v}_2$ attached to the H atoms labelled 1 and 2 respectively, both perpendicular to the molecular plane (Fig. 1.3). Such vectors are not at all fictitious—the *p*-type atomic orbitals (which every hydrogen atom does possess) have (+)ve and (–)ve lobes and thus have directional or vectorial character. Now, on applying a reflection operation using the plane $\sigma(2)$ we find that the vectors have changed their direction (Fig. 1.3).

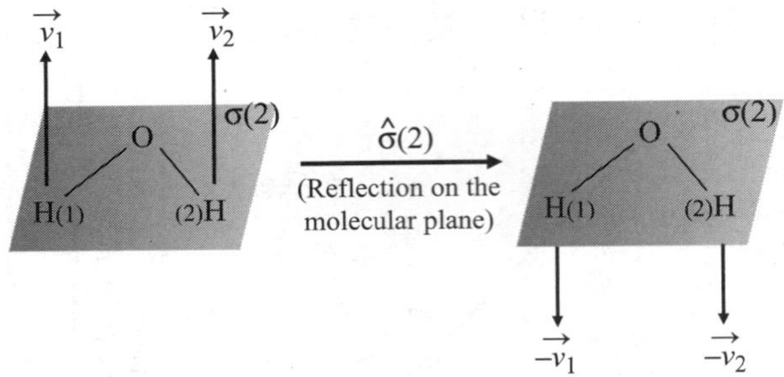

Fig. 1.3 Action of reflection on the $\sigma(2)$ plane of the H_2O molecule

Hence $\hat{\sigma}(2) \neq \hat{E}$. Two successive reflections on the same plane produce the initial configuration. Hence $[\hat{\sigma}(1)]^2 = \hat{E}$ and $[\hat{\sigma}(2)]^2 = \hat{E}$.

1.2.2 Some More Examples of Rotation and Reflection

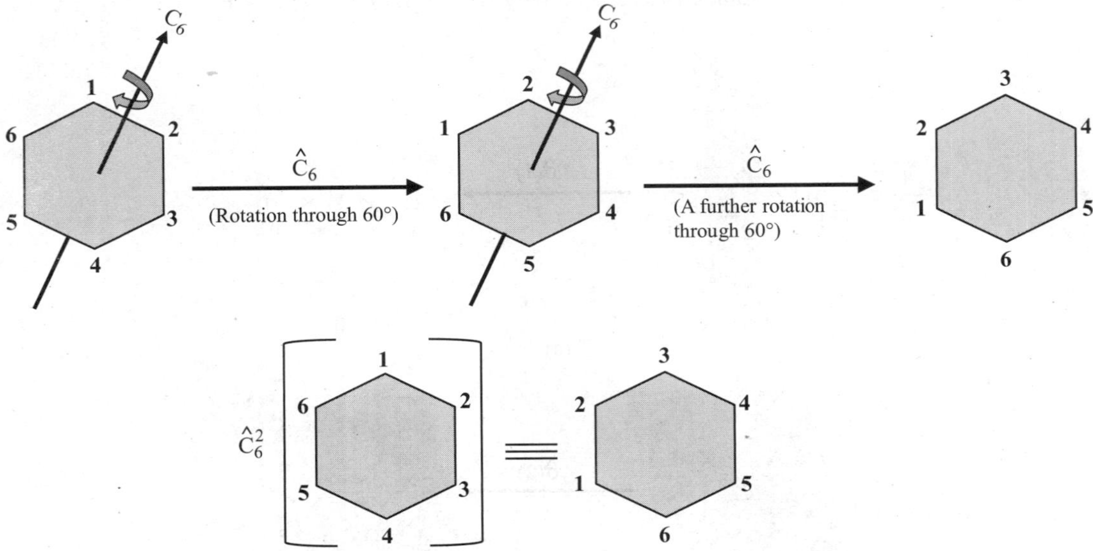

Fig. 1.4 Six-fold rotational axis of benzene molecule and effect of two successive $\hat{C}_6$ operations

Example 1.1 The benzene molecule has a C_6-axis perpendicular to the ring (Fig. 1.4). It is obvious that the effect of two successive C_6 rotations is the same as that of one C_3 rotation (that is, 120° rotation about the same axis). Hence $\hat{C}_6^2 = \hat{C}_3$.

Example 1.2 The ethane molecule in the eclipsed conformation has a C_3-axis (along the C–C single bond) as is evident from Fig. 1.5(a). The operation described in Fig. 1.5(a) is shown more conveniently in Fig. 1.5(b) where the triangle is constructed by joining the H atoms attached to the same C-atom. Two triangles are thus possible—(1, 2, 3) and (4, 5, 6). When viewed along the C–C bond, the two triangles (one in front and one on the rear side) overlap in the eclipsed conformation and so one triangle has been shown. The C_3-axis (which is actually a line along the C–C bond) is perpendicular to this triangle and passes through its centroid.

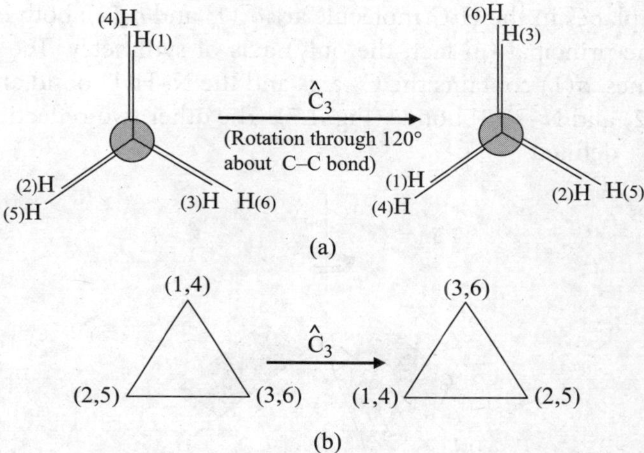

Fig. 1.5 $\hat{C}_3$ operation about the C–C bond in the eclipsed conformation of the ethane molecule

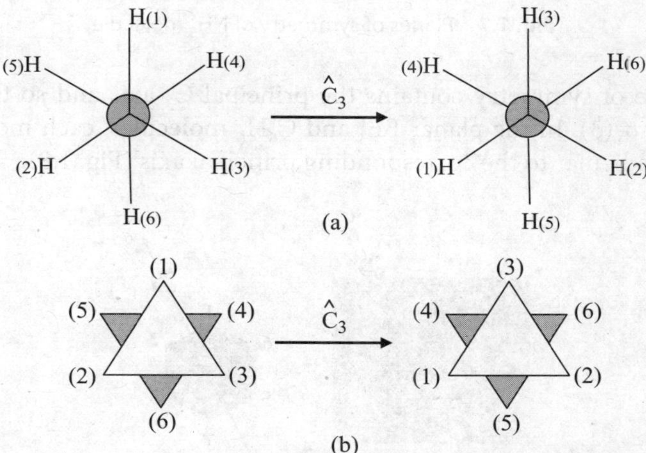

Fig. 1.6 $\hat{C}_3$ operation about the C–C bond in the staggered conformation of the ethane molecule

The staggered conformation of ethane also has a C_3-axis along the C–C bond (Fig. 1.6), where the frontal and rear-side triangles have been differentiated by shading.

Before explaining the remaining two elements of symmetry, it is necessary to classify the planes of symmetry and examine whether the result of carrying out a rotation and a reflection depends on the order in which they are carried out (*commutability*).

When the object under consideration contains a principal axis of rotation C_n (the highest-fold axis), the subscripts h (horizontal), v (vertical) and d (dihedral) are used with the notation (σ) for a reflection plane according to the following rules:

- *h*, when the reflection plane is perpendicular to C_n,
- *v*, when the reflection plane contains the C_n-axis, and
- *d*, when the object contains *n* number of C_2-axes perpendicular to C_n and the reflection plane (σ) not only contains C_n but also bisects the angle between two such successive C_2 axes.

Thus the reflection planes in the H_2O molecule are $\sigma_v(1)$ and $\sigma_v(2)$; both of them contain the C_2-axis which is the principal (in fact, the only) axis of symmetry. The NH_3 molecule has three reflection planes. $\sigma(1)$ contains the C_3-axis and the N–H(1) bond and bisects the angle between the N–H(2) and N–H(3) bonds (Fig. 1.7). The other two reflection planes, $\sigma(2)$ and $\sigma(3)$ can be similarly defined.

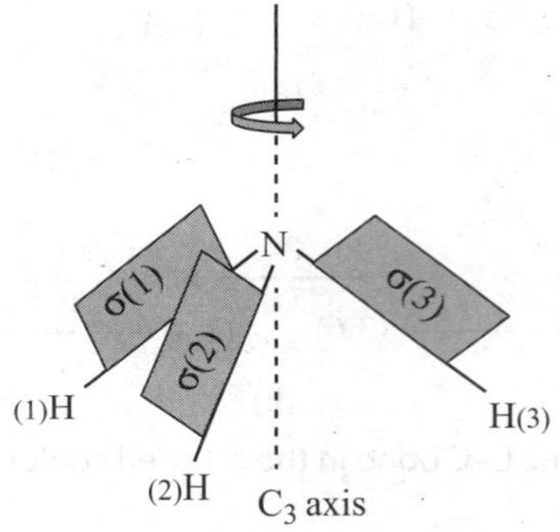

Fig. 1.7 Planes of symmetry of NH_3 molecule

In NH_3, each plane of symmetry contains the principal C_3-axis and so they are designated as $\sigma_v(1)$, $\sigma_v(2)$ and $\sigma_v(3)$. In the planar BF_3 and C_6H_6 molecules, each molecular plane is σ_h because it is perpendicular to the corresponding principal axis (Fig. 1.8).

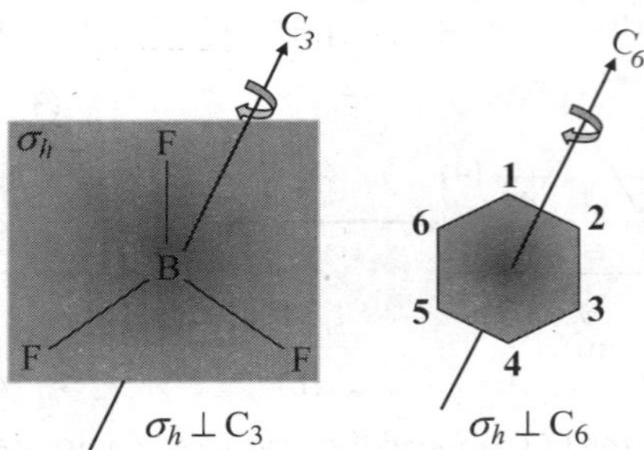

Fig. 1.8 Horizontal symmetry planes of the BF_3 and C_6H_6 molecules

A rotation and a reflection operator may or may not commute if the axis of rotation lies on the reflection plane (that is, $\hat{C}_n\hat{\sigma}_v$ and $\hat{\sigma}_v\hat{C}_n$ may or may not be the same). But they do commute if the rotation axis is perpendicular to the reflection plane. This is evident from the following example. Consider an equilateral triangle with labelled vertices [Fig. 1.9(a)]; it has a C_3-axis perpendicular to the plane of the triangle, and three C_2-axes each passing through one vertex and lying on the plane. For convenience of clearly 'seeing' the results of symmetry operations, we imagine vectors $\vec{v}_1$, $\vec{v}_2$ and $\vec{v}_3$ attached to the vertices 1, 2 and 3 respectively, each perpendicular to the plane of the triangle. In Figs 1.9(a) and (b) the configurations (I) and (II) are equivalent but not identical. Hence, in this case, $\hat{\sigma}_v \hat{C}_3 \neq \hat{C}_3 \hat{\sigma}_v$.

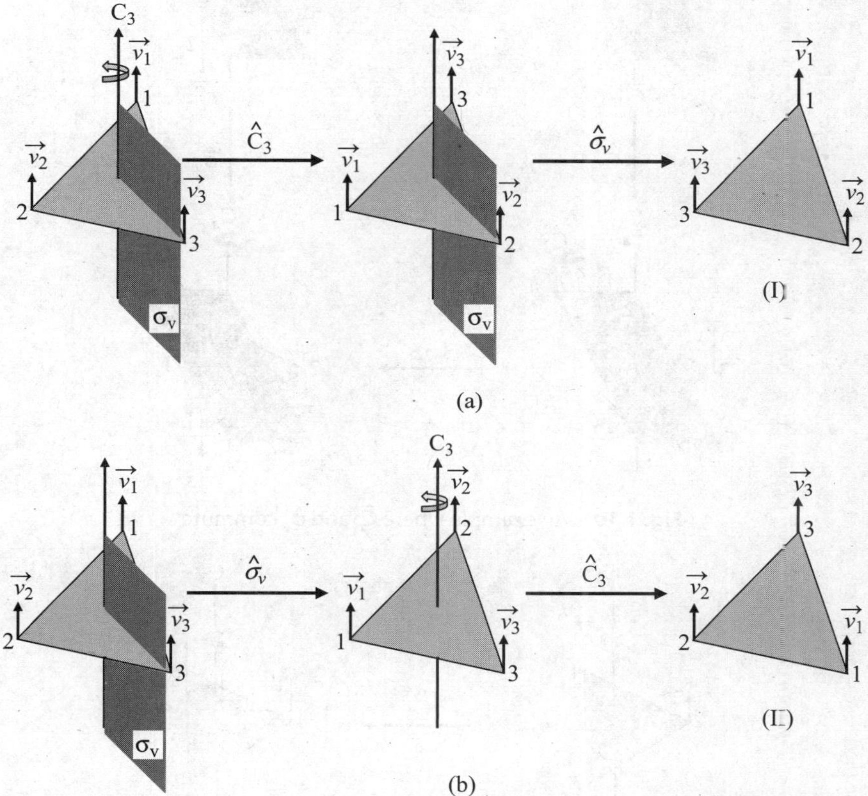

Fig. 1.9 An example showing non-commutability of rotation and reflection

Note: In carrying out successive symmetry operations, the elements of symmetry such as axes and planes do not move while the positions of the atoms or vertices change. For example, in (a) above, the σ_v plane is not shifted during the $\hat{C}_3$ operation. This has to be remembered in drawing the above and subsequent diagrams.

Now consider the sequence of operations shown in Fig. 1.10, from which it will be evident that in this case $\hat{\sigma}_v \hat{C}_2 = \hat{C}_2 \hat{\sigma}_v$.

Denoting the plane of the triangle by σ_h (since it is perpendicular to the principal axis C_3) it is possible to show that $\hat{\sigma}_h \hat{C}_3 = \hat{C}_3 \hat{\sigma}_h$ (Fig. 1.11). In general it is true that for any $C_n \perp \sigma_h$,

$$\hat{\sigma}_h \hat{C}_n^r = \hat{C}_n^r \hat{\sigma}_h$$

where r is the number of times the operation $\hat{C}_n$ has been carried out successively.

Henceforth, if an object contains n number of C_2-axes perpendicular to a C_n-axis ($n \geq 2$), then a prime will be used to denote such C_2-axes and operations.

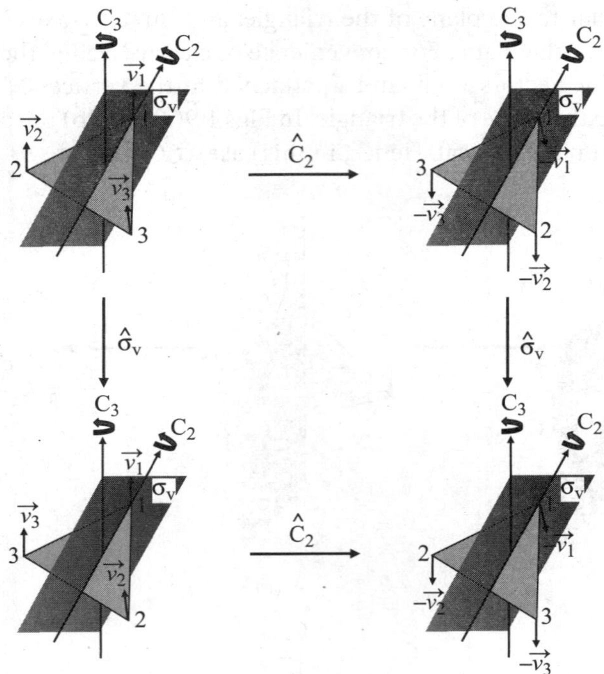

Fig. 1.10 An example where $\hat{C}_n$ and $\hat{\sigma}_v$ commute

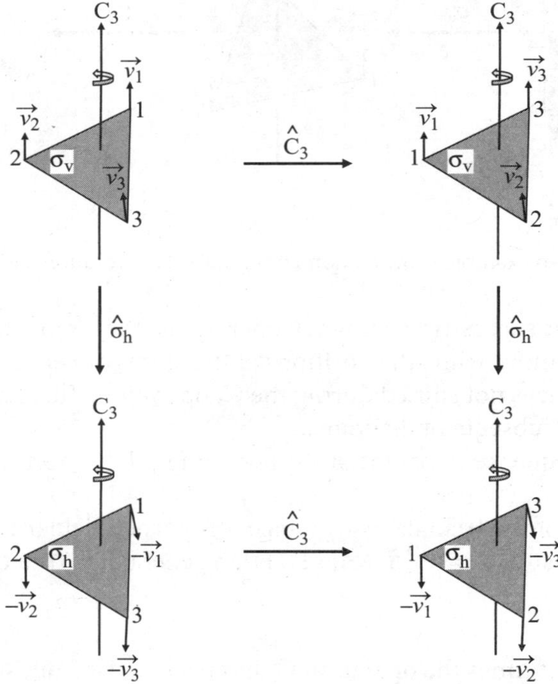

Fig. 1.11 Example showing commutation of $\hat{C}_3$ and $\hat{\sigma}_h$

1.2.3 Improper Axis of Symmetry

A body is said to have an n-fold alternating (or improper) axis of symmetry S_n, if by a rotation about that axis through an anticlockwise angle of $2\pi/n$ followed by a reflection on a plane perpendicular to that axis, one gets an equivalent configuration. It is to be noted that a proper axis (C_n) or a reflection plane (σ) sometimes may not exist independently, but the S_n axis exists. Examples are given below.

(i) The benzene molecule has a C_6-axis and a reflection plane (σ_h) (the ring plane which is perpendicular to C_6). A rotation $\hat{C}_6$ about this axis followed by a reflection on the σ_h plane leads to an equivalent configuration. The same is the result if we carry out $\hat{\sigma}_h$ first and then $\hat{C}_6$. Hence the combined operation $\hat{\sigma}_h \hat{C}_6$ (or $\hat{C}_6 \hat{\sigma}_h$) is a symmetry operation distinct from $\hat{C}_6$ or $\hat{\sigma}_h$ or from other symmetry operations possible with the benzene molecule, as can be readily verified; we call this new symmetry operation an improper rotation and denote it by $\hat{S}_6$. Here the S_6-axis coincides with the C_6-axis.

(ii) The allene molecule is an example where an S_4-axis exists but a C_4-axis or its perpendicular reflection plane does not exist independently. Consider Fig. 1.12. As (I) and (II) are not equivalent configurations, AB is not a C_4-axis. However, (I) and (III) are equivalent. Hence, the combined operation $\hat{\sigma}_\perp \hat{C}_4$ is a symmetry operation and we denote it by $\hat{S}_4$. Thus AB is an improper four-fold axis of symmetry.

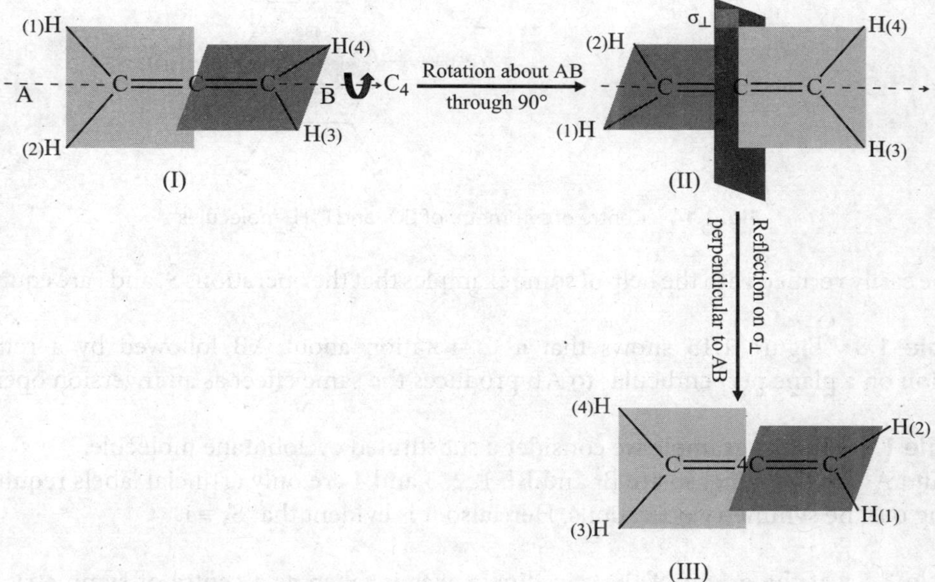

Fig. 1.12 Four-fold improper rotation axis of the allene molecule

(iii) Staggered ethane molecule has a six-fold improper axis stretched along the C–C bond (Fig. 1.13). Since IV and V are non-equivalent configurations, AB is not a proper six-fold axis of symmetry. It is an S_6-axis because IV and VI are equivalent.

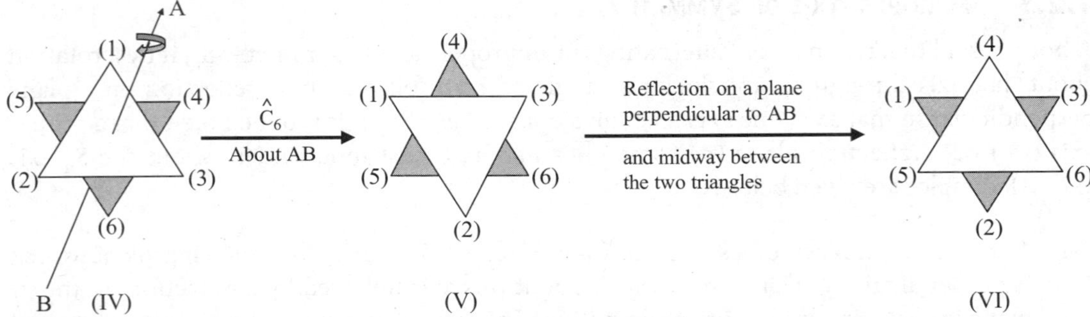

Fig. 1.13 $\hat{S}_6$ operation about the C–C bond in the staggered conformation of the ethane molecule

1.2.4 Centre of Symmetry

A body is said to possess a *centre of symmetry* (inversion centre) i, if starting from any arbitrary point P of the body, proceeding along a straight line through i and going along the same straight line an equal distance on the opposite side of i, one arrives at a point P′, which is equivalent to P. Thus CO_2 and C_6H_6 have i (Fig. 1.14).

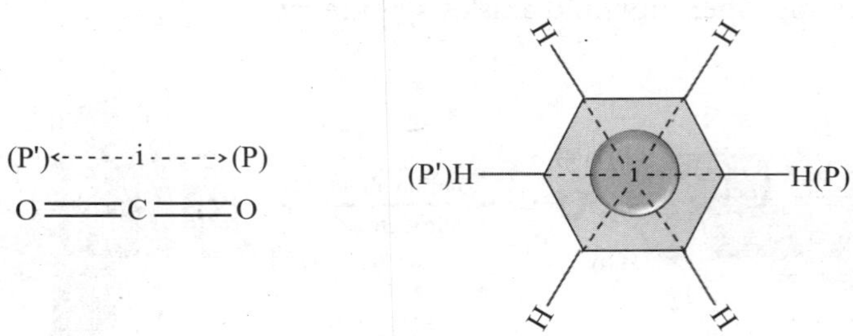

Fig. 1.14 Centre of symmetry of CO_2 and C_6H_6 molecules

It can be easily verified with the help of some examples that the operations $\hat{S}_2$ and $\hat{i}$ are equivalent.

Example 1.3 Figure 1.15 shows that a $\hat{C}_2$-rotation about AB followed by a reflection operation on a plane perpendicular to AB produces the same effect as an inversion operation.

Example 1.4 In this example we consider a substituted cyclobutane molecule.

A^1 and A^3 are the same; so are B^2 and B^4. 1, 2, 3 and 4 are only artificial labels required for carrying out the symmetry operations. Here also it is evident that $\hat{S}_2 \equiv \hat{i}$.

Example 1.5 If the origin of the coordinate axes is taken as a centre of symmetry then a general point P (x, y, z) becomes P′ $(-x, -y, -z)$ by inversion through the origin. Again P (x, y, z) after a $\hat{C}_2$ operation about the Z-axis becomes Q $(-x, -y, z)$ which after reflection on the XY-plane (perpendicular to the Z-axis) becomes P′ $(-x, -y, -z)$.

Thus in each of the above examples

$$\hat{i} = \hat{C}_2 \hat{\sigma}_\perp = \hat{\sigma}_\perp \hat{C}_2 = \hat{S}_2$$

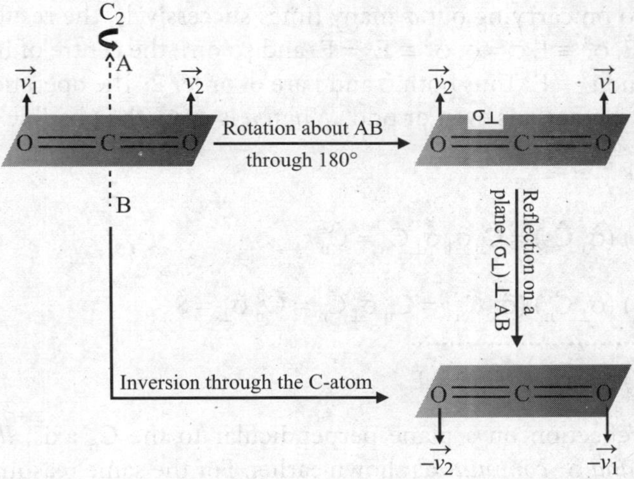

Fig. 1.15 Equivalence of inversion operation and $\hat{S}_2$ on the linear CO_2 molecule

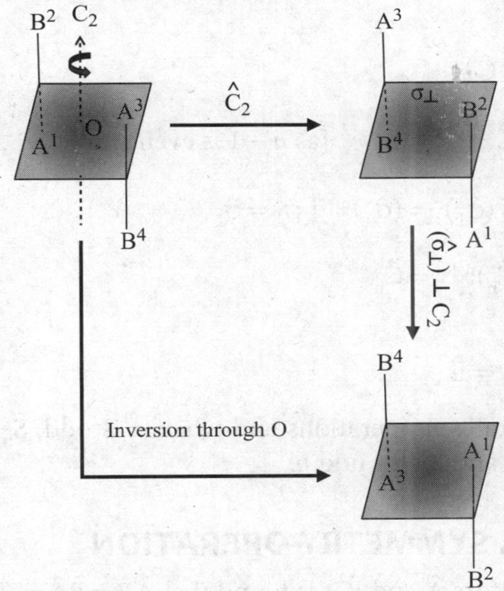

Fig. 1.16 Inversion operation on a cyclobutane derivative

1.3 REPEATED SYMMETRY OPERATIONS

If a symmetry operator $\hat{R}$ on being carried out at least *p* times successively produces the identity operator $\hat{E}$, then *p* is said to be the *order* of $\hat{R}$. Repeated operations are designated by suitable indices. Thus $\hat{C}_n^m$ means *m* successive rotations, each time by an angle of $2\pi/n$, about the axis of symmetry or a single rotation about the same axis through an angle of $2\pi m/n$.

The symmetry operations generated by a C_n-axis are $\hat{C}_n$, $\hat{C}_n^2$, $\hat{C}_n^3$, ..., $\hat{C}_n^n = \hat{E}$. Thus the operator $\hat{C}_n$ is of an order *n*. Similarly, a reflection plane generates two operations namely $\hat{\sigma}$

and $\hat{\sigma}^2 \equiv \hat{E}$ (if we go on carrying out σ many times successively, the result is the sequence of operations $\hat{\sigma}$, $\hat{\sigma}^2 \equiv \hat{E}$, $\hat{\sigma}^3 \equiv \hat{E}\,\hat{\sigma} = \hat{\sigma}$, $\hat{\sigma}^4 \equiv \hat{E}^2 = \hat{E}$ and so on); the centre of inversion i generates two operations, $\hat{i}$ and $\hat{i}^2 \equiv \hat{E}$. Thus both $\hat{\sigma}$ and $\hat{i}$ are of order 2. The operations generated by an S_n-axis depend on whether n is even or odd. When n is even, the possible operations are:

$$\hat{S}_n \equiv (\hat{\sigma}_\perp \hat{C}_n)$$

$$\hat{S}_n^2 \equiv (\hat{\sigma}_\perp \hat{C}_n)\,(\hat{\sigma}_\perp \hat{C}_n) = \hat{C}_n \hat{\sigma}_\perp \hat{\sigma}_\perp \hat{C}_n = \hat{C}_n^2$$

$$\hat{S}_n^3 \equiv (\hat{\sigma}_\perp \hat{C}_n)\,(\hat{\sigma}_\perp \hat{C}_n)\,(\hat{\sigma}_\perp \hat{C}_n) = \hat{C}_n^2 \hat{\sigma}_\perp \hat{C}_n = \hat{C}_n^3 \hat{\sigma}_\perp = \hat{S}_n^3$$

....

$$\hat{S}_n^n \equiv (\hat{\sigma}_\perp \hat{C}_n)^n = (\hat{\sigma}_\perp)^n\,(\hat{C}_n)^n = \hat{E}$$

Here $\sigma_\perp$ means a reflection on a plane perpendicular to the C_n axis; *the above results are possible because $\hat{C}_n$ and $\hat{\sigma}_\perp$ commute* as shown earlier. For the same reason, when n is odd, the operations generated are:

$$\hat{S}_n \equiv (\hat{\sigma}_\perp \hat{C}_n)$$

$$\hat{S}_n^2 \equiv (\hat{\sigma}_\perp)^2\,(\hat{C}_n)^2 = \hat{C}_n^2$$

$$\hat{S}_n^{n-1} \equiv (\hat{\sigma}_\perp)^{n-1}\,(\hat{C}_n)^{n-1} = \hat{C}_n^{n-1} \quad [\text{as } n-1 \text{ is even and } (\hat{\sigma}_\perp)^{n-1} = \hat{E}]$$

$$\hat{S}_n^n \equiv (\hat{\sigma}_\perp)^n\,(\hat{C}_n)^n = (\hat{\sigma}_\perp)^n = (\hat{\sigma}_\perp)^{n-1}\,\hat{\sigma}_\perp = \hat{\sigma}_\perp$$

$$\hat{S}_n^{n+1} \equiv (\hat{\sigma}_\perp)^{n+1}\,(\hat{C}_n)^{n+1} = \hat{C}_n$$

...

$$\hat{S}_n^{2n} \equiv (\hat{\sigma}_\perp)^{2n}\,(\hat{C}_n)^{2n} = \hat{E}$$

Thus, for even n, $\hat{S}_n$ generates n operations and when n is odd, $\hat{S}_n$ generates $2n$ operations. Order of $\hat{S}_n$ is n for even n and $2n$ for odd n.

1.4 INVERSE OF A SYMMETRY OPERATION

If for two symmetry operators $\hat{A}$ and $\hat{B}$ it is found that $\hat{A}\hat{B} = \hat{B}\hat{A} = \hat{E}$ (identity), then $\hat{A}$ is said to be the inverse of $\hat{B}$ and vice-versa. We write $\hat{A} = \hat{B}^{-1}$ or $\hat{B} = \hat{A}^{-1}$. For example

$$\hat{C}_3 \hat{C}_3^2 = \hat{C}_3^2 \hat{C}_3 = \hat{E}, \text{ and hence}$$

$$\hat{C}_3^{-1} = \hat{C}_3^2 \text{ or } (\hat{C}_3^2)^{-1} = \hat{C}_3$$

An operator of two-fold symmetry (one which on being executed twice successively produces $\hat{E}$) is the inverse of itself; thus,

$$\hat{\sigma}\hat{\sigma} = \hat{E}, \text{ that is, } \hat{\sigma}^{-1} = \hat{\sigma}$$

$$\hat{i}\,\hat{i} = \hat{E}, \text{ that is, } \hat{i}^{-1} = \hat{i}$$

1.4.1 Inverse of the Product of Several Symmetry Operations

The inverse of the product of two symmetry operators is the product of their inverses in the reverse order, that is, $(\hat{A}\hat{B})^{-1} = \hat{B}^{-1}\hat{A}^{-1}$. This can be easily proved as follows:

Let $(\hat{A}\hat{B})\hat{Z} = \hat{E}$

Multiplying both sides by $\hat{A}^{-1}$ from the left,

$$\hat{A}^{-1}\hat{A}\hat{B}\hat{Z} = \hat{A}^{-1}\hat{E}$$

or, $\hat{B}\hat{Z} = \hat{A}^{-1}$

Multiplying again by $\hat{B}^{-1}$ from the left,

$$\hat{B}^{-1}\hat{B}\hat{Z} = \hat{B}^{-1}\hat{A}^{-1} \quad \text{or} \quad \hat{Z} = \hat{B}^{-1}\hat{A}^{-1}$$

Thus, $(\hat{A}\hat{B})^{-1} = \hat{B}^{-1}\hat{A}^{-1}$ (Proved)

In general, $(\hat{A}\hat{B} \ldots \hat{Y}\hat{Z})^{-1} = \hat{Z}^{-1}\hat{Y}^{-1} \ldots \hat{B}^{-1}\hat{A}^{-1}$

1.5 CONSTRUCTION OF SYMMETRY MULTIPLICATION TABLE

The results of successive symmetry operations on an object can be written down in a tabular form which is known as a symmetry multiplication table. Let $\hat{A}$, $\hat{B}$, $\hat{C}$, $\hat{D}$, …, $\hat{P}$, … be the symmetry operations possible with an object. The symmetry product $\hat{A}\hat{B}$ means that the operation $\hat{B}$ is carried out on some initial configuration of the object and then $\hat{A}$ is carried out on the resulting configuration. In the same way, $\hat{A}\hat{B}\hat{C}\hat{D}$ means that $\hat{D}$ is the first symmetry operation to be carried out, which will be followed successively by $\hat{C}$, $\hat{B}$ and $\hat{A}$. In other words, the sequence in a symmetry product is from right to left. In fact, this sequence is followed in all mathematical operators operating on a function. For example, let $\sqrt{}$, $\left(\frac{d}{dx}\right)$, $()^3$, and 2+ be four operations, denoted respectively by $\hat{A}$, $\hat{B}$, $\hat{C}$ and $\hat{D}$. If $f(x)$ is a differentiable function (defined within some range of x where these operations can be carried out), the result of $\hat{A}\hat{B}\hat{C}\hat{D}f(x)$ is obtained as follows:

$$\hat{D}f(x) = 2 + f(x)$$

$$\hat{C}\hat{D}f(x) = \hat{C}[2 + f(x)] = [2 + f(x)]^3$$

$$\hat{B}\hat{C}\hat{D}f(x) = \hat{B}[2 + f(x)]^3 = \frac{d}{dx}[2 + f(x)]^3$$

$$\hat{A}\hat{B}\hat{C}\hat{D}f(x) = \hat{A}\left[\frac{d}{dx}\{2 + f(x)\}^3\right] = \sqrt{\frac{d}{dx}\{2 + f(x)\}^3}$$

'Multiplication of symmetry operations', therefore means successive execution of the symmetry operations from right to left. Now, if the result of $\hat{C}\hat{B}$ on some initial configuration is the same as that of carrying out the symmetry operation $\hat{P}$ directly on the initial configuration, we say

$\hat{C}\hat{B} = \hat{P}$ and this result is written down in a symmetry multiplication table in the format given in Table 1.1.

Table 1.1 Format for construction of symmetry multiplication tables

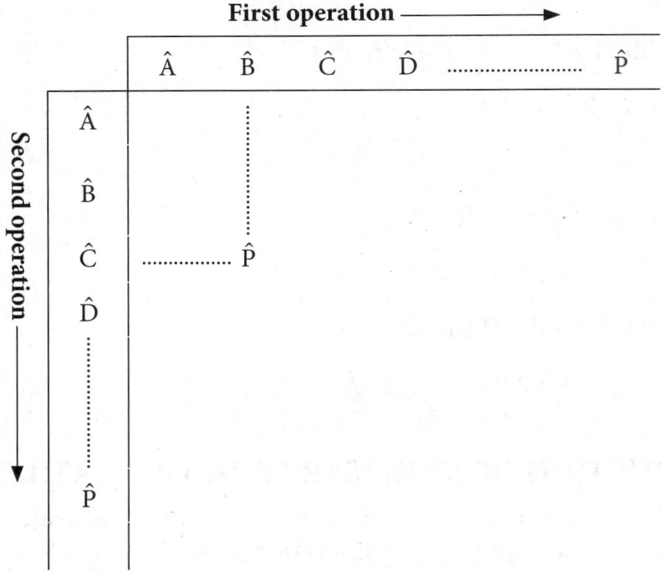

Construction of multiplication table will be made clear using the example of the pyramidal NH_3 molecule. It has the symmetry operations, $\hat{E}$, $\hat{C}_3$, $\hat{C}_3^2$, $\hat{\sigma}_v(1)$, $\hat{\sigma}_v(2)$, $\hat{\sigma}_v(3)$ (Fig. 1.17). Let us find out the product $\hat{C}_3\hat{\sigma}_v(2)$. Application of $\hat{\sigma}_v(2)$ on the initial configuration (1) interchanges the position of (a) and (c), keeping the C_3-axis and the position of (b) unchanged and thus we get configuration (2) (Fig. 1.18). When $\hat{C}_3$ is applied on (2), the changes are: (a) to (c), (c) to (b) and (b) to (a). Thus, $\{\hat{C}_3\hat{\sigma}_v(2)\}(1) \rightarrow (3)$.

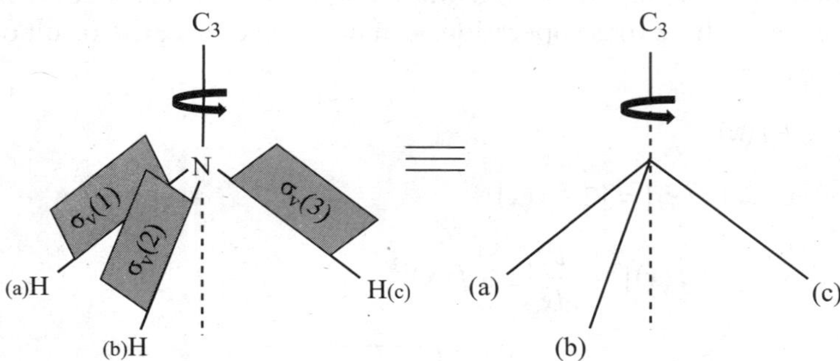

Fig. 1.17 Different symmetry elements of NH_3

Now, the configuration (3) can be obtained directly from (1) by applying $\hat{\sigma}_v(1)$, which interchanges (b) and (c), leaving (a) unchanged. Hence $\{\hat{\sigma}_v(1)\}(1) \rightarrow (3)$. Thus, $\hat{C}_3\hat{\sigma}_v(2) = \hat{\sigma}_v(1)$. Table 1.2 shows this result together with other symmetry products for NH_3.

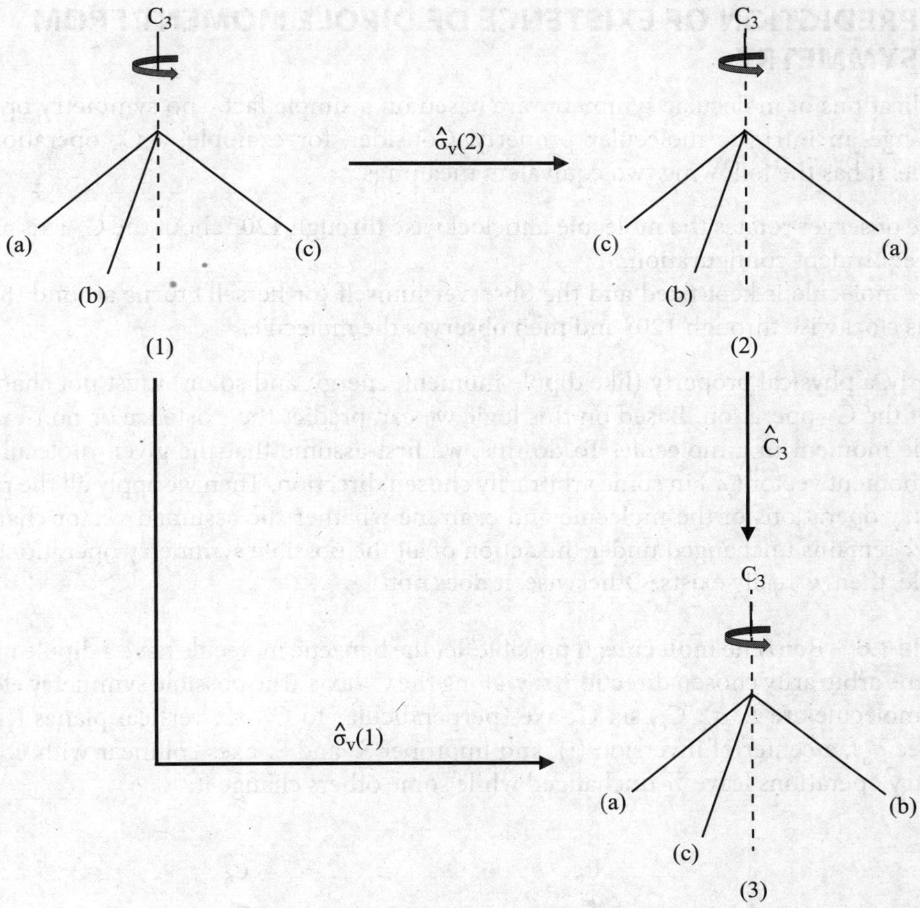

Fig. 1.18 An example of product of two symmetry operations

Table 1.2 Symmetry multiplication table for the NH_3 (pyramidal) molecule

Second operation ↓ \ First operation →	$\hat{E}$	$\hat{C}_3$	$\hat{C}_3^2$	$\hat{\sigma}_v(1)$	$\hat{\sigma}_v(2)$	$\hat{\sigma}_v(3)$
$\hat{E}$	$\hat{E}$	$\hat{C}_3$	$\hat{C}_3^2$	$\hat{\sigma}_v(1)$	$\hat{\sigma}_v(2)$	$\hat{\sigma}_v(3)$
$\hat{C}_3$	$\hat{C}_3$	$\hat{C}_3^2$	$\hat{E}$	$\hat{\sigma}_v(3)$	$\hat{\sigma}_v(1)$	$\hat{\sigma}_v(2)$
$\hat{C}_3^2$	$\hat{C}_3^2$	$\hat{E}$	$\hat{C}_3$	$\hat{\sigma}_v(2)$	$\hat{\sigma}_v(3)$	$\hat{\sigma}_v(1)$
$\hat{\sigma}_v(1)$	$\hat{\sigma}_v(1)$	$\hat{\sigma}_v(2)$	$\hat{\sigma}_v(3)$	$\hat{E}$	$\hat{C}_3$	$\hat{C}_3^2$
$\hat{\sigma}_v(2)$	$\hat{\sigma}_v(2)$	$\hat{\sigma}_v(3)$	$\hat{\sigma}_v(1)$	$\hat{C}_3^2$	$\hat{E}$	$\hat{C}_3$
$\hat{\sigma}_v(3)$	$\hat{\sigma}_v(3)$	$\hat{\sigma}_v(1)$	$\hat{\sigma}_v(2)$	$\hat{C}_3$	$\hat{C}_3^2$	$\hat{E}$

1.6 PREDICTION OF EXISTENCE OF DIPOLE MOMENT FROM SYMMETRY

All applications of molecular symmetry are based on a simple fact—no symmetry operation can change an intrinsic molecular property. Consider, for example, a $\hat{C}_3$ operation on a molecule. It has the following two equivalent meanings:

(i) The observer rotates the molecule anticlockwise through 120° about the C_3-axis and gets an equivalent configuration.
(ii) The molecule is kept fixed and the observer himself (or herself) turns around the same axis clockwise through 120° and then observes the molecule.

Obviously, a physical property (like dipole moment, energy, and so on) must not change as a result of the $\hat{C}_3$-operation. Based on this logic we can predict the existence or non-existence of dipole moment in a molecule. To do this, we first assume that the given molecule has a dipole moment vector ($\vec{\mu}$) in some arbitrarily chosen direction. Then we apply all the possible symmetry operations of the molecule and examine whether the assumed vector changes or not. If $\vec{\mu}$ remains unchanged under the action of all the possible symmetry operations of the molecule, then $\vec{\mu}$ really exists. Otherwise, it does not.

Example 1.6 **Benzene molecule**: If possible, let the benzene molecule have a dipole moment $\vec{\mu}$ in some arbitrarily chosen direction, say, along the C_6-axis. The possible symmetry elements of this molecule are E, σ_h, C_6, six C_2-axes perpendicular to C_6, six vertical planes (three σ_v and three σ_d), a center of inversion (i), and improper S_3 and S_6 axes collinear with C_6. Some symmetry operations leave $\vec{\mu}$ unchanged while some others change it:

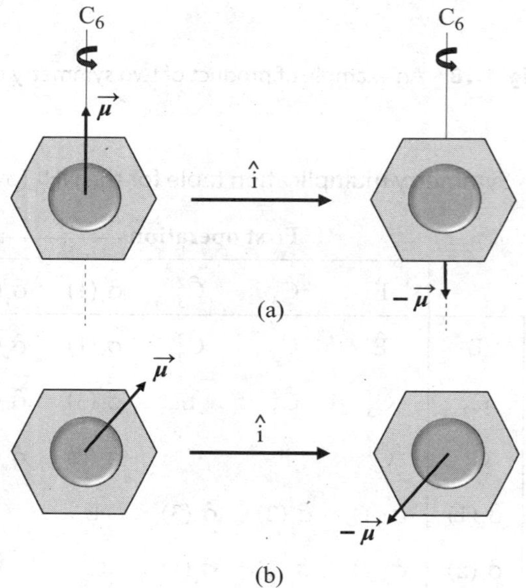

Fig. 1.19 (a) Action of operation $\hat{i}$ on the hypothetical dipole moment vector imagined along the C_6-axis. (b) Action of $\hat{i}$ on the hypothetical dipole moment vector imagined in an arbitrary direction

$$\hat{E}(\vec{\mu})=\vec{\mu},\ \hat{C}_6(\vec{\mu})=\vec{\mu},\ \hat{C}_6^5(\vec{\mu})=\vec{\mu}\ \ldots \quad \text{and} \quad \hat{i}(\vec{\mu})=-\vec{\mu},\ \hat{C}_2(\vec{\mu})=-\vec{\mu}$$

Since a symmetry operation should not change $\vec{\mu}$, $\hat{i}(\vec{\mu})$ should be equal to $\vec{\mu}$.

Hence, $\vec{\mu}=-\vec{\mu}$ that is, $\vec{\mu}=\vec{0}$ (null vector). Hence benzene cannot have a dipole moment along the C_6-axis.

If instead of lying along the C_6-axis, the assumed vector $\vec{\mu}$ has some arbitrary direction, then also $\hat{i}(\vec{\mu})=-\vec{\mu}$ and the same logic as above leads to $\vec{\mu}=\vec{0}$ (Fig. 1.19).

Example 1.7 **Water molecule**: Let the resultant dipole moment vector $\vec{\mu}$ of this molecule be directed along one O–H bond. A $\hat{C}_2$-operation changes $\vec{\mu}$ to some other vector $\vec{\mu}'$; the direction has changed (Fig. 1.20). Hence such a $\vec{\mu}$ does not exist.

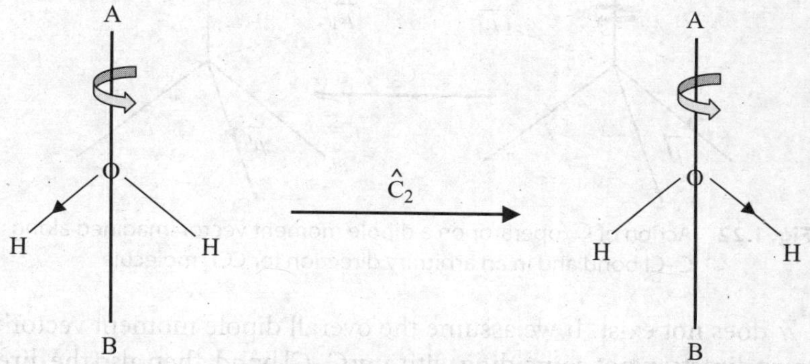

Fig. 1.20 Action of $\hat{C}_2$ on $\vec{\mu}$ imagined along a O–H bond

So the resultant dipole moment vector of water molecule cannot be directed along an O–H bond.

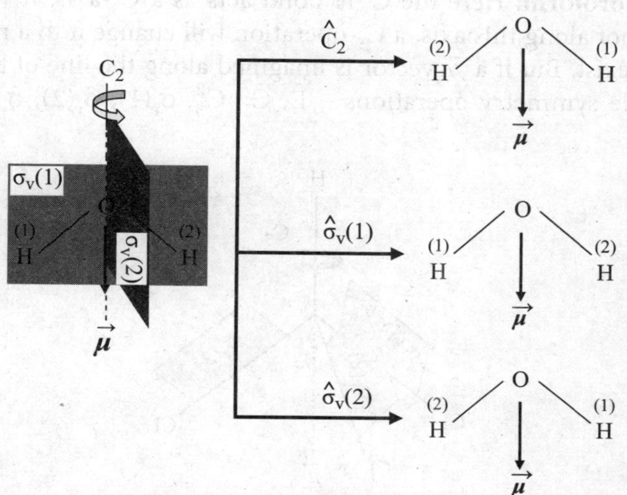

Fig. 1.21 Invariance of dipole moment vector ($\vec{\mu}$) of the H_2O molecule under the possible symmetry operations

However, if the resultant $\vec{\mu}$ is assumed to be directed along the C_2-axis of H_2O molecule, then it remains unchanged under the action of all the possible symmetry operators of the molecule (Fig. 1.21). Hence the resultant dipole moment of water must be directed along the C_2-axis.

Example 1.8 **Carbon tetrachloride**: This is a tetrahedral molecule. Each C–Cl bond acts as a C_3-axis; there are many more symmetry operations possible for a tetrahedron (discussed later). If we assume the resultant dipole moment vector $\vec{\mu}$ to be directed along a C–Cl bond, then a $\hat{C}_3$ operation with any other bond as axis changes it to a different vector $\vec{\mu}'$ (Fig. 1.22).

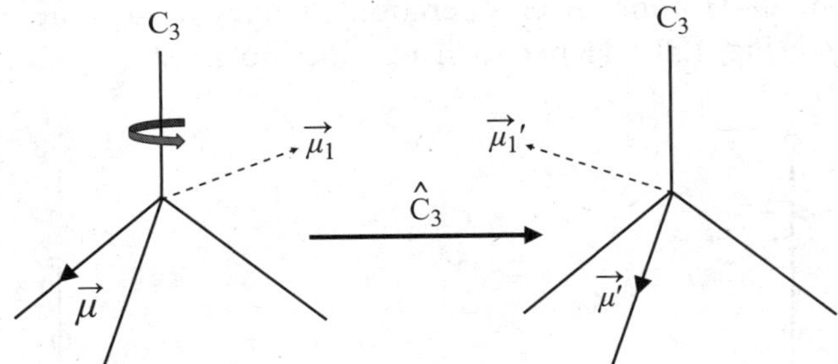

Fig. 1.22 Action of $\hat{C}_3$ operator on a dipole moment vector imagined along a C–Cl bond and in an arbitrary direction for CCl_4 molecule

Hence such a $\vec{\mu}$ does not exist. If we assume the overall dipole moment vector to be directed along an arbitrary direction not coinciding with any C–Cl bond, then also the direction changes by the action of a $\hat{C}_3$-operation (shown by $\vec{\mu}_1 \rightarrow \vec{\mu}_1'$ in Fig. 1.22). Thus, the assumed resultant $\vec{\mu}$ cannot be directed along such arbitrary directions also. Hence CCl_4 has no resultant dipole moment.

Example 1.9 **Chloroform**: Here the C–H bond acts as a C_3-axis. If we imagine a dipole moment vector $\vec{\mu}$ not along this axis, a $\hat{C}_3$-operation will change it to a new position. Hence such a $\vec{\mu}$ does not exist. But if a $\vec{\mu}$ vector is imagined along the line of the C–H bond, then none of the possible symmetry operations – $\hat{E}$, $\hat{C}_3$, $\hat{C}_3^2$, $\hat{\sigma}_v(1)$, $\hat{\sigma}_v(2)$, $\hat{\sigma}_v(3)$ – can change it (Fig. 1.23).

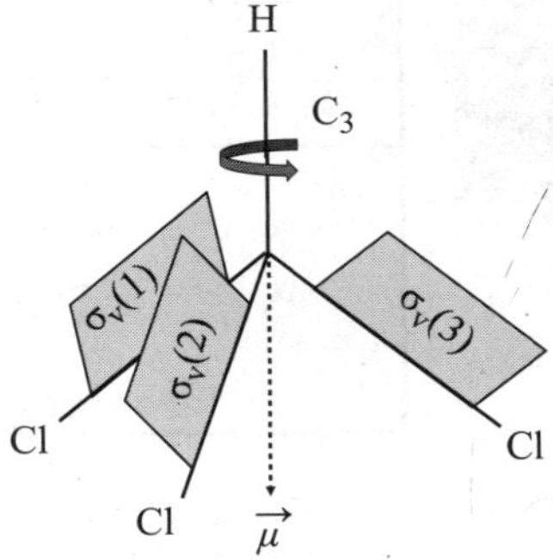

Fig. 1.23 Elements of symmetry of $CHCl_3$ molecule and its $\vec{\mu}$ along the C–H bond

$$\hat{E}(\vec{\mu}) = \hat{C}_3(\vec{\mu}) = \hat{C}_3^2(\vec{\mu}) = \hat{\sigma}_v(1)\vec{\mu} = \hat{\sigma}_v(2)\vec{\mu} = \hat{\sigma}_v(3)\vec{\mu} = \vec{\mu}$$

Hence this $\vec{\mu}$ exists. Thus, the resultant dipole moment vector of $CHCl_3$ must be directed along the line of the C–H bond.

The above illustrations will make it clear that (i) whenever a molecule has a centre of symmetry, it has no dipole moment; but the reverse statement is not true—absence of permanent dipole moment cannot ensure the presence of a centre of symmetry i, (ii) if a molecule has only one C_n-axis, it can have permanent dipole moment along this axis, (iii) if there are several C_n-axes (of the same or different order), permanent dipole moment is absent; if a symmetry plane is present, the dipole moment vector $\vec{\mu}$ must lie on that plane and if a number of symmetry planes are present, the $\vec{\mu}$ vector should be along the C_n-axis which coincides with the line of intersection of the planes; (iv) if a C_n-axis and a symmetry plane perpendicular to this axis are both present, the molecule cannot have permanent dipole moment because the C_n-axis allows $\vec{\mu}$ to be aligned along it but the $\hat{\sigma}_h$ operation inverts it.

1.7 NECESSITY OF CONSIDERING ALL THE SYMMETRY OPERATIONS OF AN OBJECT TOGETHER

From the examples cited in Section 1.6, it follows that it is not possible to come to any conclusion about some physical property or a physico-chemical phenomenon by looking at only one or two symmetry elements ignoring the others. For example, if we consider only the C_6-axis of symmetry of the benzene molecule and ignore the effects of the other elements of symmetry, it will be found that benzene has a dipole moment along the C_6-axis, which is not true. Consideration of the simultaneous effect of all the symmetry elements of an object is, therefore, a necessity and it requires a specially rich and developed branch of mathematics, known as *group theory*. The Hamiltonian operator of an atom or molecule remains unchanged (invariant) under the action of all the symmetry operators of the atom or molecule and this fact forms the basis of all applications of group theory in quantum mechanical structure calculation and spectroscopy.

1.8 MATHEMATICAL GROUP

A set *S* is called a group when a binary operation, ⊙, is defined in it (that is, a rule for combining any two elements of *S* is given) and the following conditions hold:

(i) *S* is closed under ⊙, that is, for any two elements A and B belonging to *S* we find that

$$A \odot B \in S \text{ and also } B \odot A \in S \text{ (the symbol } \in \text{ means 'belongs to')}$$

This property of *S* is called *closure property*.

(ii) *S* contains an element E such that for any element A contained in *S* we have

$$A \odot E = E \odot A = A$$

Such an element E is called the *identity element*.

(iii) For any element A in S there exists a unique element C in S such that

$$A \odot C = C \odot A = E$$

Here C is called the *inverse* of A and vice-versa, that is, $C^{-1} = A$ and $A^{-1} = C$.

Every element of S must have a unique inverse (an element must not have several inverses) which must also belong to S.

(iv) For any three elements A, B, C belonging to S, the associative law holds,

$$(A \odot B) \odot C = A \odot (B \odot C)$$

(The symbol $\odot$ for the binary operation is most often omitted for the sake of convenience.)

1.8.1 Some Simple Examples of Groups

Example 1.10 Consider the set of the fourth roots of unity:

$$(1)^{\frac{1}{4}} = (\pm 1)^{\frac{1}{2}} = \pm 1, \pm i, \text{ where } i = \sqrt{(-1)}$$

In the set $S_1 = \{1, -1, i, -i\}$ we observe the following:

(i) $i \times i = -1 \in S_1$, $i \times (-1) = -i \in S_1$,and so on. Thus the product of any two elements of S_1 belongs to S_1. The set S_1 is *closed* if the rule for combining any two elements of S_1 is *algebraic multiplication*.

(ii) $1 \times (-1) = -1 \in S_1$, $1 \times (-i) = -i \in S_1$, $1 \times i = i \in S_1$
Thus the algebraic operation '1×' has no effect on the elements of the set S_1. We say that 1 is the identity element of S_1.

(iii) The element 1 can be shown to be the product of two (same or different) elements of S_1:

$$(-i) \times i = i \times (-i) = 1, (-1) \times (-1) = 1, 1 \times 1 = 1$$

Thus, $i^{-1} = -i$, $(-1)^{-1} = -1$ and $(1)^{-1} = 1$, and we observe that the inverse of any element of S_1 also belongs to S_1.

(iv) $(-1) \times (i \times -i) = (-1 \times i) \times (-i) = i^2 = -1$

$$(1) \times (i \times -1) = (1 \times i) \times (-1) = -i$$

Thus, associativity holds in the product of any three elements of the set S_1. Hence by definition, the set S_1 is a group under the binary operation of multiplication.

Example 1.11 Consider the set I of all integers from $-\infty$ to ∞ including zero.

$$I = \{\ldots, -5, -4, -3, -2, -1, 0, 1, 2, 3, 4, 5, \ldots\}$$

Here we observe the following:

(i) The sum of any two members of I also belongs to I. Thus,

$$(-5) + (+7) = 2 \in I, (+3) + (+10) = 13 \in I, (-5) + (+5) = 0 \in I, \text{ and so on.}$$

So I is closed under *addition*.

(ii) The element 0 of I is such that when it is added to any other element of I, it leaves the latter unchanged. Thus,

$$0 + (-5) = (-5) + 0 = -5; 0 + 3 = 3 + 0 = 3, \text{ and so on.}$$

Hence 0 acts as the identity element in the set I when the rule for combining any two elements (binary operation) is addition.

(iii) Corresponding to any element (say, n) of I, there is an element $(-n)$ in I such that $n + (-n) = (-n) + n = 0$, the identity element of I. We say that under addition as the binary operation, n is the *inverse* of $(-n)$ and vice-versa:

$$(-n)^{-1} = n \text{ and } (n^{-1}) = -n \text{ under '+'.}$$

0 is the inverse of itself .We also note that the inverse of every element of I is a member of I.

(iv) Addition of any three integers obeys the associative law:

$$\{5 + (-3)\} + 4 = 5 + \{(-3) + 4\} \text{ and so on.}$$

Hence I is a group under addition.

Example 1.12 Consider the set of symmetry operations of the pyramidal NH_3 molecule:

$$S_2 = \{\hat{E}, \hat{C}_3, \hat{C}_2^2, \hat{\sigma}_v(1), \hat{\sigma}_v(2), \hat{\sigma}_v(3)\}$$

Let the rule for combining any two members of S_2 be symmetry multiplication, that is, successive execution of symmetry operations as discussed in Section 1.5. The results of all such symmetry products are listed in Table 1.2, from which we observe the following:

The product of any two members of S_2 is also a member of S_2. Thus,

$$\hat{C}_3 \cdot \hat{\sigma}_v(1) = \hat{\sigma}_v(3) \in S_2, \hat{\sigma}_v(1) \cdot \hat{\sigma}_v(2) = \hat{C}_3 \in S_2, \text{ and so on.}$$

The element $\hat{E}$ of the set, when multiplied by any other member of S_2, does not change the latter. Thus, $\hat{E}\hat{C}_3 = \hat{C}_3\hat{E} = \hat{C}_3$, $\hat{E}\,\hat{\sigma}_v(1) = \hat{\sigma}_v(1)\,\hat{E} = \hat{\sigma}_v(1)$, and so on.

So $\hat{E}$ acts as the identity element of the set S_2.

Corresponding to any element of S_2 one can find another element in S_2 such that their product is $\hat{E}$. Thus,

$$\hat{C}_3\hat{C}_3^2 = \hat{C}_3^2\,\hat{C}_3 = \hat{E} \text{ that is, } (\hat{C}_3)^{-1} = \hat{C}_3^2 \in S_2 \text{ and } (\hat{C}_3^2)^{-1} = \hat{C}_3 \in S_2$$

$$\hat{\sigma}_v(1)\,\hat{\sigma}_v(1) = \hat{E} \text{ that is, } [\hat{\sigma}_v(1)]^{-1} = \hat{\sigma}_v(1) \in S_2$$

So every element of S_2 has its inverse which also belongs to S_2.

The symmetry multiplication is associative:

$$\{\hat{C}_3\,\hat{\sigma}_v(1)\}\,\hat{\sigma}_v(2) = \hat{\sigma}_v(3)\,\hat{\sigma}_v(2) = \hat{C}_3^2; \quad \hat{C}_3\,\{\hat{\sigma}_v(1)\,\hat{\sigma}_v(2)\} = \hat{C}_3\hat{C}_3 = \hat{C}_3^2$$

Hence, $\{\hat{C}_3\,\hat{\sigma}_v(1)\}\,\hat{\sigma}_v(2) = \hat{C}_3\,\{\hat{\sigma}_v(1)\,\hat{\sigma}_v(2)\}$.

Similar results can be verified with any three members of the set S_2 using Table 1.2.

All the three sets S_1, I and S_2, together with their respective binary operations $\times$, + and symmetry multiplication, are examples of groups according to definition.

1.8.2 Some Examples of Sets (Together With Defined Binary Operations) Which are not Groups

(i) The set of integers $I = \{\ldots, -5, -4, -3, -2, -1, 0, 1, 2, 3, 4, 5, \ldots\}$ where the binary operation is '$\times$' (arithmetic multiplication) is not a group as the inverse of an integer is not an integer (excepting 1).

(ii) The set of all 3×3 matrices including the null matrix is not a group under matrix multiplication (See Chapter 3). This is because the 3×3 null matrix is singular and its inverse cannot be constructed.

Note: (a) There is only one identity element in every group. (b) A group element cannot have more than one inverse.

1.9 POINT SYMMETRY GROUPS OF MOLECULES OR IONS

The group of symmetry operators related to the four symmetry elements (proper rotation, inversion, reflection and improper rotation) is called a *point symmetry group* or simply *point group* because these symmetry operations keep at least one point, namely, the centre of gravity of the object, unmoved. (If translational, screw and glide plane symmetries, which are possible with stacks of molecules or unit cells, are also considered together with the point symmetries, the resulting group is called a *space group*. Such groups are encountered in crystallography and are beyond the scope of the present discussions.) Depending on the number and nature of the point symmetry operations, objects belong to different point groups which are designated by two types of notation—Schoenflies and Herman Mauguin. The latter is used more in crystallography. In applying group theory to quantum chemistry and spectroscopy, Schoenflies notation is used. A scheme is given below for obtaining the Schoenflies notation of the point group of an object. It should be noted that in such notations all the symmetry elements of the point group are not mentioned explicitly—only a few key elements are indicated and all the other symmetry operations can be generated from them by systematic techniques that will be discussed in a subsequent section.

1.9.1 A Scheme for Obtaining Schoenflies Notation of Point Groups of Objects

For linear objects the axis of the object acts as a C_∞-axis; rotation about this axis through any angle produces an equivalent configuration making an infinite number of equivalent configurations possible during a full rotation. Now, if a σ_h is present, the point group is $D_{\infty h}$; if no σ_h is present, the point group is $C_{\infty v}$ (this is because an infinite number of σ_v planes must be present in a linear molecule). Common examples are COS and CO_2 molecules (Fig. 1.24). For other objects, use the following scheme.

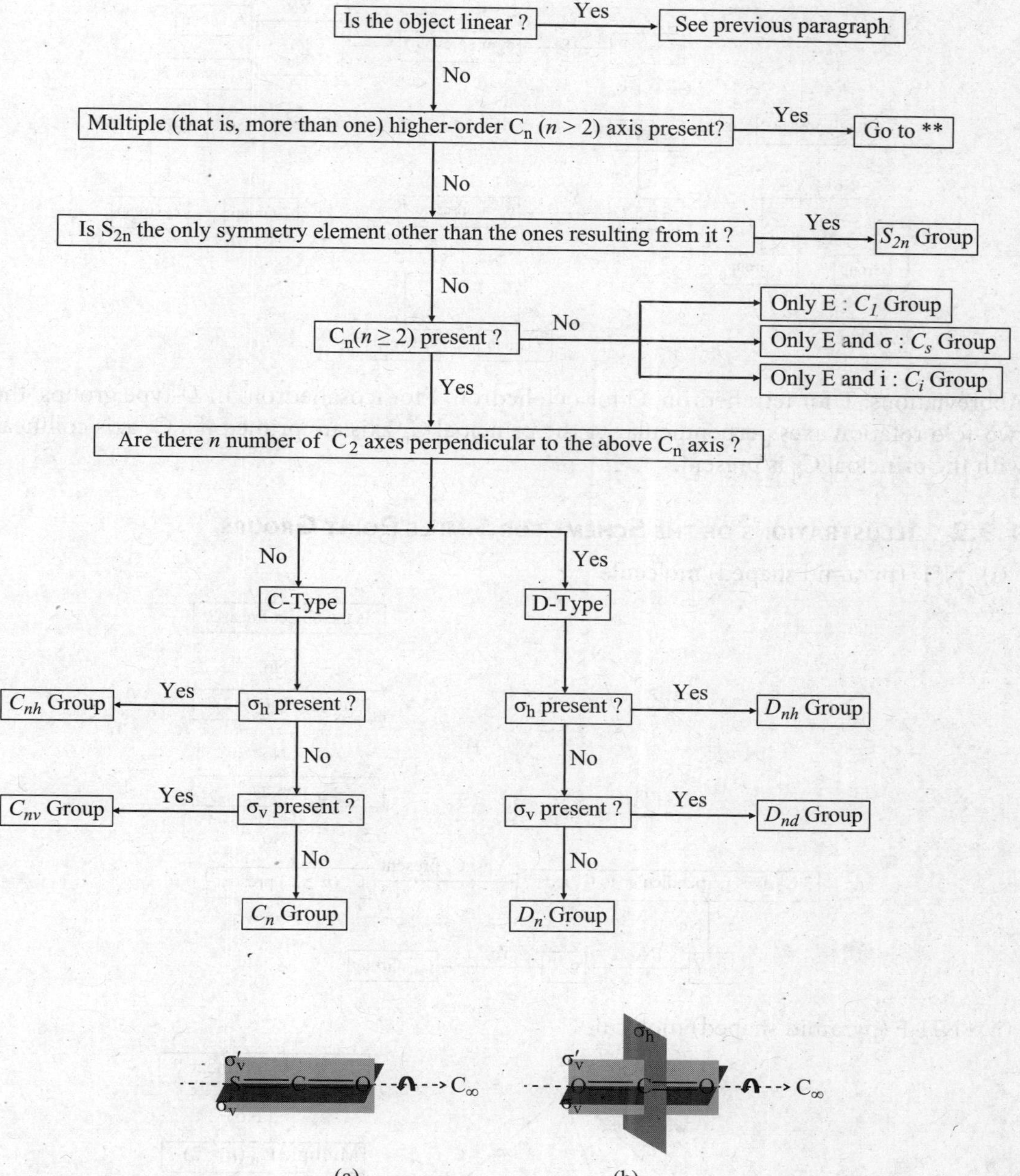

Fig. 1.24 Some symmetry elements in COS and CO_2 molecules. (a) COS; one C_∞-axis, an infinite number of σ_v planes but no σ_h, the point group is $C_{\infty v}$. (b) CO_2: one C_∞-axis, one $\sigma_h \perp C_\infty$, an infinite of σ_v planes and also an infinite number of C_2-axes $\perp C_\infty$, the point group is $D_{\infty h}$ (Only two σ_v planes have been shown in each diagram.)

** The scheme for special groups with a multiple high-order C_n axis is as follows:

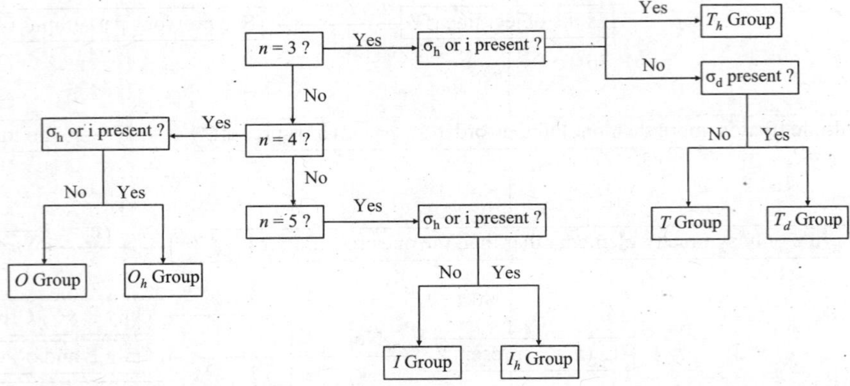

Abbreviations: T for tetrahedron, O for octahedron, I for icosahedron. In *D*-type groups, the two-fold rotation axes perpendicular to the principal C_n axis are primed if a C_2 axis collinear with the principal C_n is present.

1.9.2 Illustrations of the Scheme for Simple Point Groups

(i) NH_3 (pyramid shaped) molecule

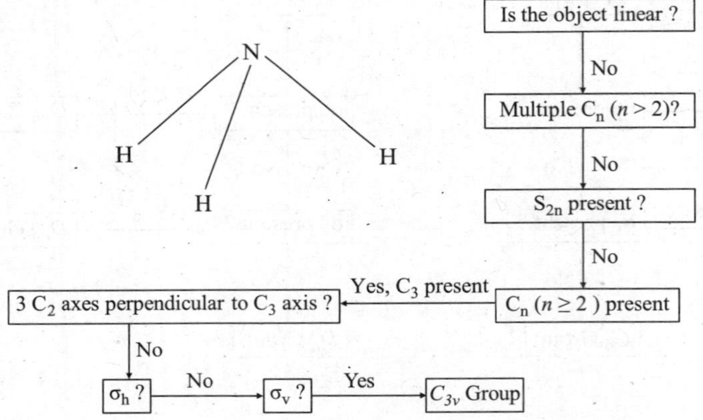

(ii) NH_2F (pyramid shaped) molecule

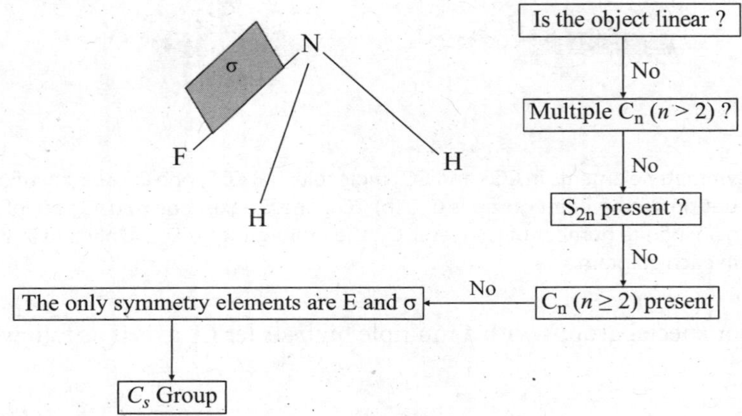

(iii) NHFCl (pyramid shaped) molecule

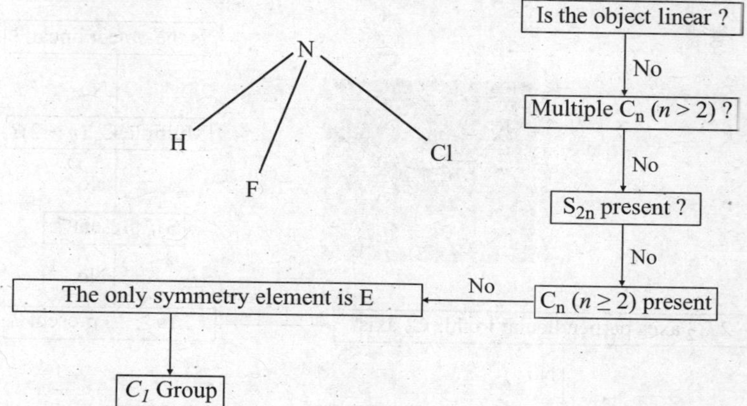

(iv) H_2O_2 in its equilibrium configuration

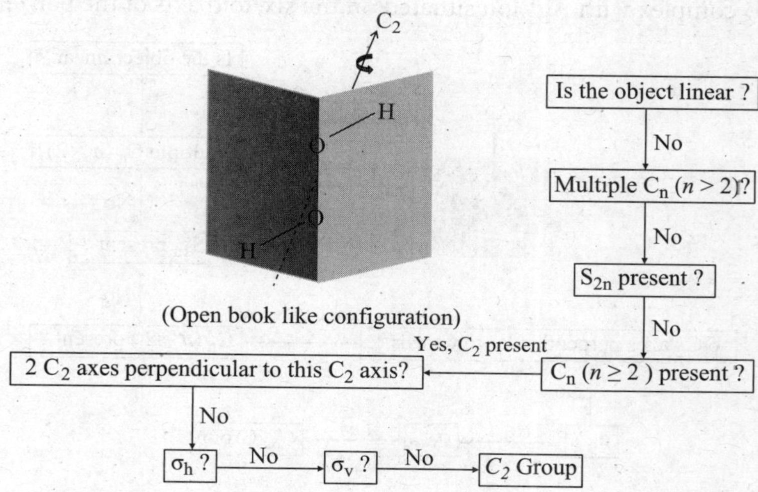

(v) *cis*-planar H_2O_2

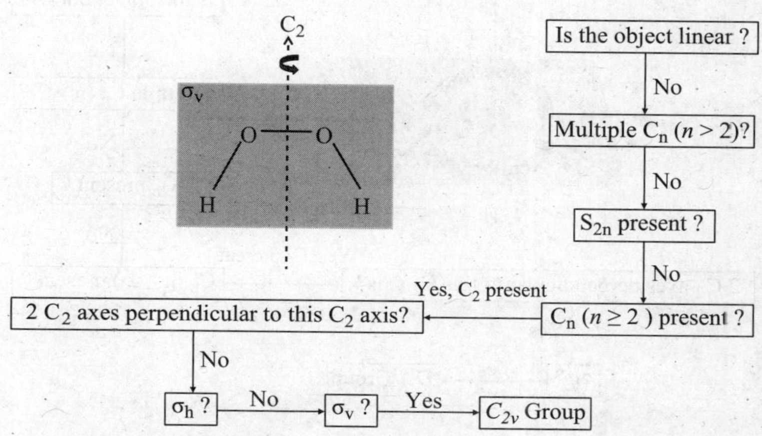

(vi) *trans*-planar H_2O_2

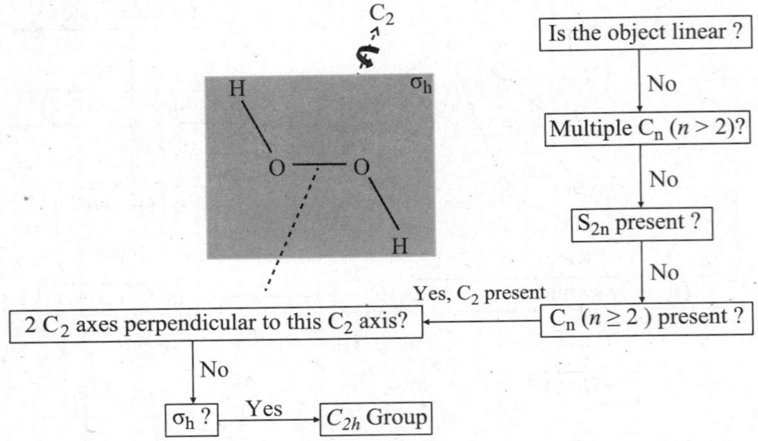

(vii) Ag^+–C_6H_6 complex with Ag^+ ion situated on the six-fold axis of the benzene ring

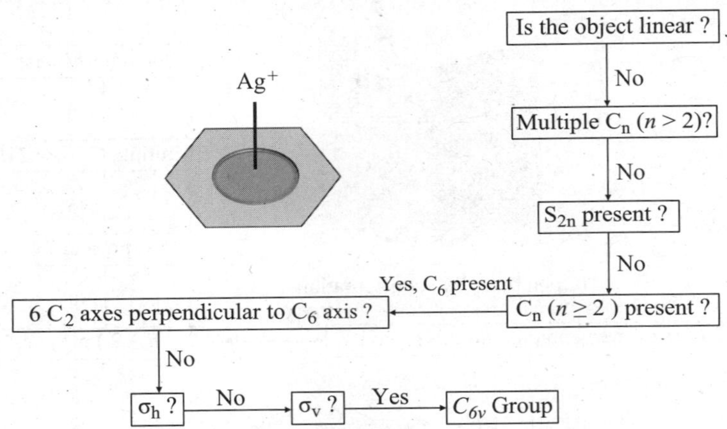

(viii) Diborane, B_2H_6

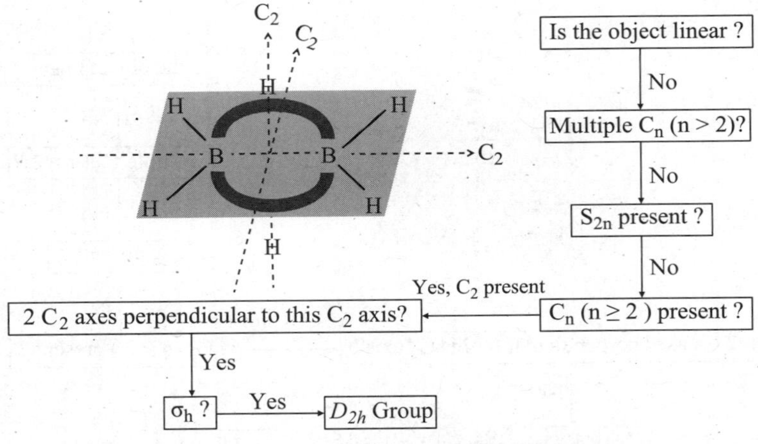

(ix) Allene, $CH_2 = C = CH_2$

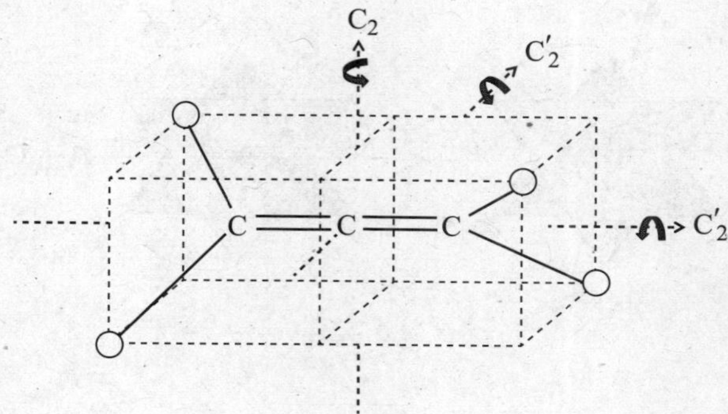

One principal C_2 axis, two other C_2'-axes $\perp$ this C_2-axis, and two vertical σ-planes (not shown in figure). Hence the point group is D_{2d}.

(x) 1, 3-dimethyl allene

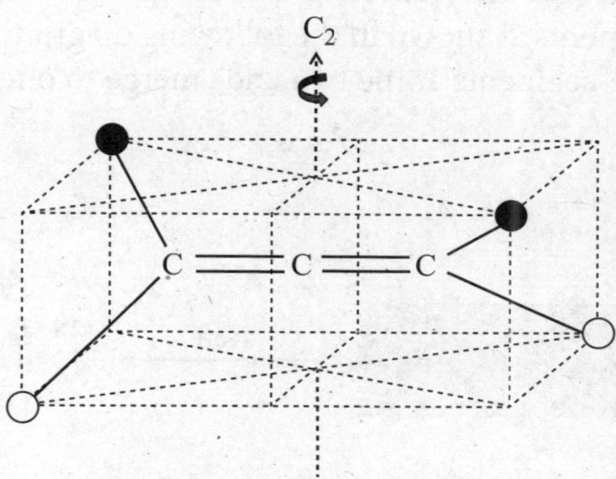

Excepting E, there is one C_2-axis and no other symmetry element. So the point group is C_2.

(xi) *cis*-$[Fe(oxalate)_3]^{3-}$ ion

The only symmetry elements besides E are one C_3-axis and three C_2-axes perpendicular to the C_3-axis. The point group is D_3.

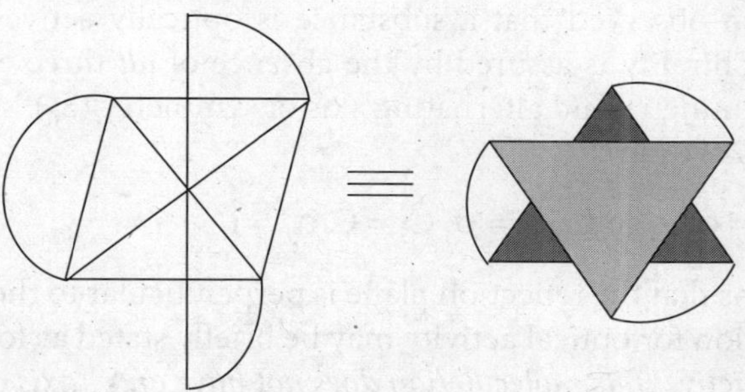

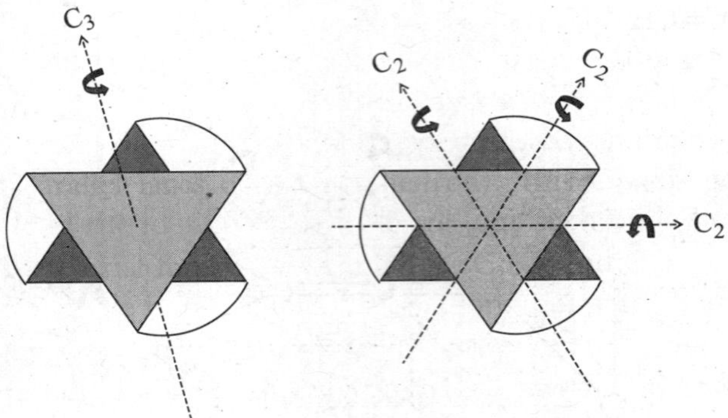

(xii) S_8 molecule (Crown shaped)

The molecule has a closed zigzag chain of eight S atoms embedded on the surface of an imaginary hollow cylinder. There are one C_4-axis, four C_2'-axes ⊥ this C_4-axis and four vertical symmetry planes. The point group is D_{4d}. (It is difficult to visualise the symmetry elements in this case without a model. A simple model can be prepared as follows: take a rectangular paper strip about 5 cm long. Place eight vertices denoting the S atoms and join them by line segments as shown in the following diagram. Roll the strip cylindrically such that the half line-segments at the two ends merge to one line.)

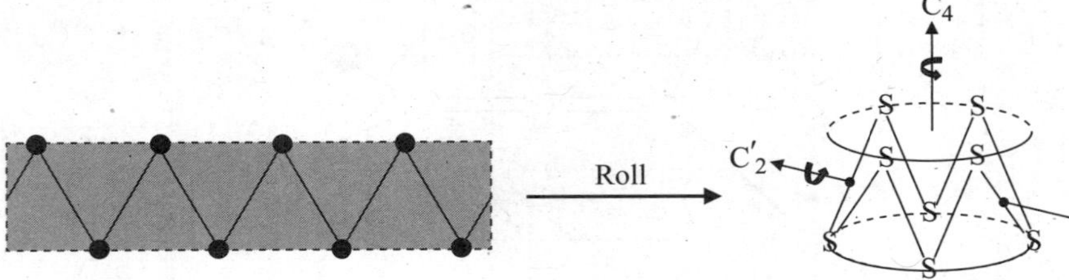

Note: The scheme for determination of point groups, S_n with n odd, has been omitted. This is because the list of symmetry operations generated by an odd order S_n axis is the same as that of the point group C_{nh} with odd n. However, S_n and C_{nh} are different point groups for even n.

1.10 OPTICAL ACTIVITY AND POINT GROUPS

Experimentally it has been observed that a substance is optically active if its molecule or constituent ion is *chiral*. Chirality is assured by the absence of *all* three elements—plane of symmetry (σ), centre of symmetry and alternating axis of symmetry (S_n).

However, for $n = 1$ and 2 we have

$$\hat{S}_1 = \hat{\sigma}_\perp \hat{C}_1 = \hat{C}_1 \hat{\sigma}_\perp = \hat{\sigma}_\perp \quad \text{and} \quad \hat{S}_2 = \hat{\sigma}_\perp \hat{C}_2 = \hat{C}_2 \hat{\sigma}_\perp = \hat{i}$$

where the notation ⊥ means that the reflection plane is perpendicular to the concerned axis of rotation. Hence the condition for optical activity may be briefly stated as follows:

A substance is optically active if its molecule/ion does not have an S_n *axis of any order.*

A molecule may have a C_n axis but no S_n axis. In such a case the substance is optically active and the molecule is called *dissymmetric*. If in addition to S_n, the C_n ($n > 1$) axis is also absent, the substance is optically active and its molecule is called *asymmetric*. Thus all asymmetric molecules are dissymmetric but the reverse is not true. It is easy to recognise that asymmetric molecules belong to point group C_1 and dissymmetric (but not asymmetric) molecules belong to C_n or D_n point groups ($n > 1$). Thus, *a molecule is optically active if its point group is C_n or D_n*. Some examples are: (i) Lactic acid (point group C_1), (ii) 1, 3-dimethyl allene (point group C_2) and (iii) *cis*-$[Fe(oxalate)_3]^{3-}$ (point group D_3).

There are, however, exceptions: a dissymmetric molecule may not show optical activity if some internal periodic motion of the molecule occurs rapidly enough, and the dextro- and levorotatory forms interconvert within the time period of oscillation of the electric vector of the incident polarised light. For example, the substituted amine, N(ABC), A ≠ B ≠ C, is dissymmetric but it does not show optical activity because of rapid oscillation of the *N* atom in an umbrella-like inversion:

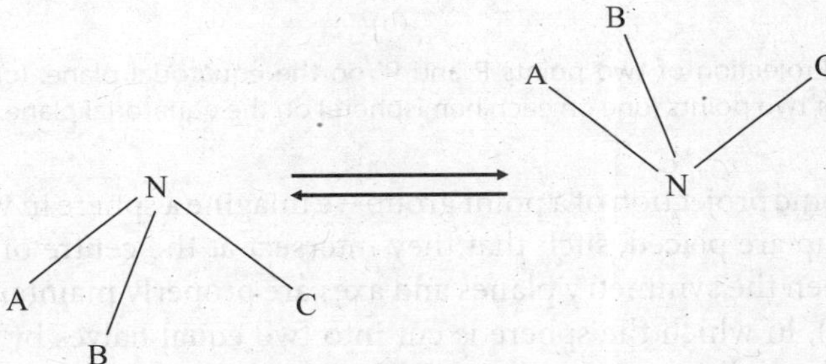

Fig. 1.25 Umbrella-like inversion of N(ABC)

A dissymmetrically substituted biphenyl is optically inactive since the structure of the molecule is convertible to that of its mirror image by free rotation about the C–C single bond that joins the phenyl rings. *cis*-1,2-dimethyl cyclohexane is dissymmetric but its optical activity cannot be observed due to rapid flipping of bonds.

1.11 STEREOGRAPHIC PROJECTION OF MOLECULAR POINT GROUPS

The Schoenflies notation of a point group indicates only the minimum symmetry elements of a point group. The full list of symmetry operations can, in principle, be generated from this minimum information by utilising the closure property. But the task is not always easy. *Stereographic projection* or *stereogram*, which generally means the projection of points of a three dimensional space on a given plane, provides a convenient way for this purpose. It also helps to construct the multiplication table of a point group *G* without considering any particular molecule or object that belongs to *G*. To explain the method of construction of such diagrams we first see how the points on the surface of a sphere can be represented on a plane. Let the sphere be divided into two equal halves (hemispheres) by an equatorial plane and let P be a point on the upper hemisphere. The extremities (N and S) of a vertical diameter are called poles. Join P to the pole (S) of the opposite (lower) hemisphere. If the line joining these meets the equatorial plane at Q [Fig. 1.26(a)], then Q is the projection of P. At the same position Q, there may be projection of another point P′ lying in the lower hemisphere [Fig. 1.26(b)]. Hence we have to use two symbols, + for points of the upper hemisphere and o for points of the lower hemisphere, to indicate their projections on the equatorial plane [Fig. 1.26(c)].

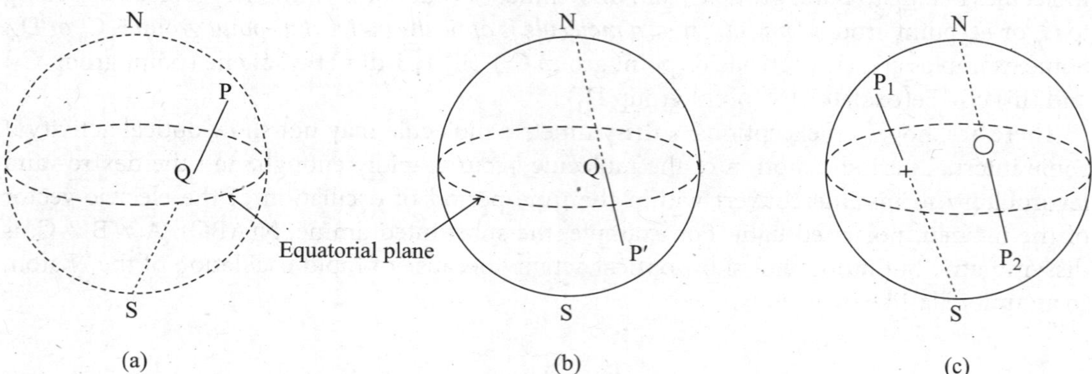

Fig. 1.26 (a) and (b) Projection of two points P and P′ on the equatorial plane. (c) Different symbols for projection of two points (one on each hemisphere) on the equatorial plane.

To draw a stereographic projection of a point group we imagine a sphere in which the symmetry elements of the group are placed, such that they intersect at the centre of the sphere and the mutual angles between the symmetry planes and axes are properly maintained. As an example consider Fig. 1.27(a), in which the sphere is cut into two equal halves by a σ_h plane, the line (NS) joining the poles acts as a C_n axis, and there is a C_2-axis (AB) on the σ_h plane. A point (P) which is not situated on any of the elements of symmetry is called a *general point*. The projection of P on the equatorial circle is denoted by + at position 1. When P is rotated anti-clockwise through $2\pi/n$ about the C_n axis (NS), it comes to P_1 in the upper hemisphere, the

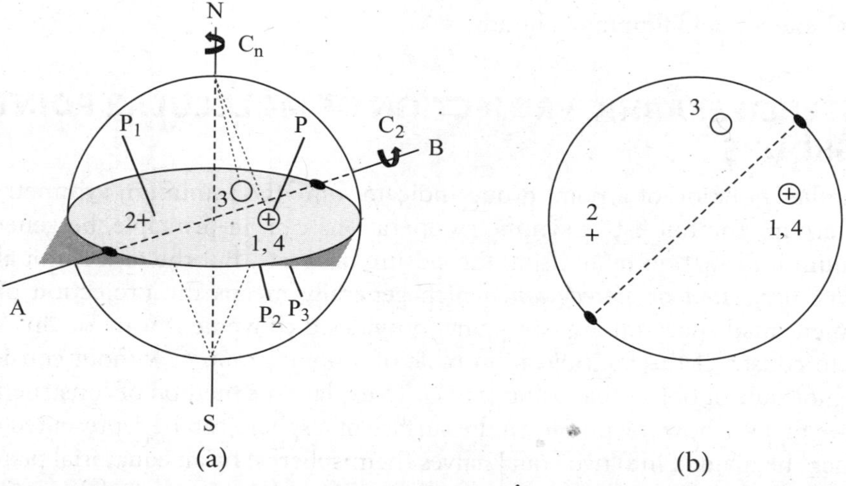

Fig. 1.27 (a) A general point P, its image P_1, P_2, P_3 produced by $\hat{C}_n$, C_2(AB) and $\hat{\sigma}_h$, and their projections on the equatorial plane (b) Stereogram showing the projections 1, 2, 3 and 4 of P, P_1, P_2 and P_3

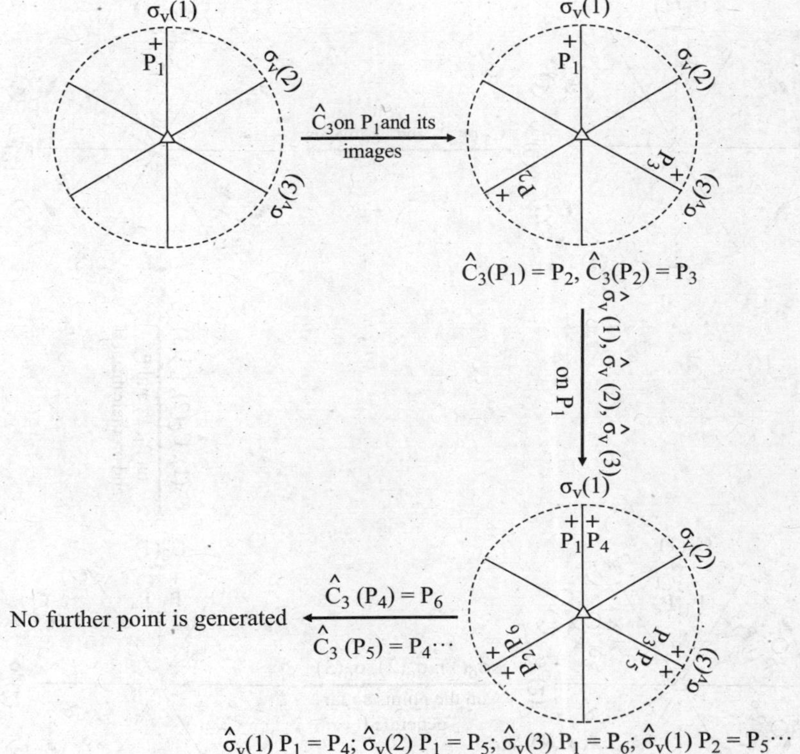

Fig. 1.29 Symmetry operations of C_{3v} carried out on a general point and its images

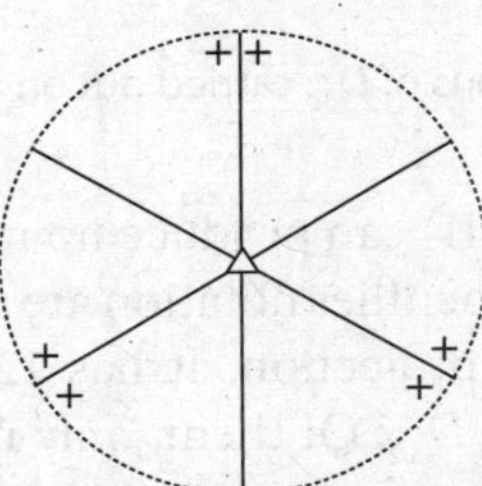

Fig. 1.30 Stereogram of the point group C_{3v}

(iii) ***D*$_{3d}$ point group**: The Schoenflies notation indicates a principal C_3-axis, three C_2 axes perpendicular to the principal axis and three dihedral planes bisecting the angles between these C_2-axes. We start with their symbols and the projection + of a general point P in the upper hemisphere.

In Fig. 1.31(II), we note that some points have been denoted by ∘. This is because a C_2 operation on a point of the upper hemisphere brings it to the lower hemisphere and the image so produced is projected on the equatorial plane as ∘. Thus

$$\hat{C}_2(1)P_1 = P_4, \hat{C}_2(2)P_1 = P_5, \hat{C}_2(3)P_1 = P_6, \hat{C}_2(1)P_2 = P_6 \text{ and so on}$$

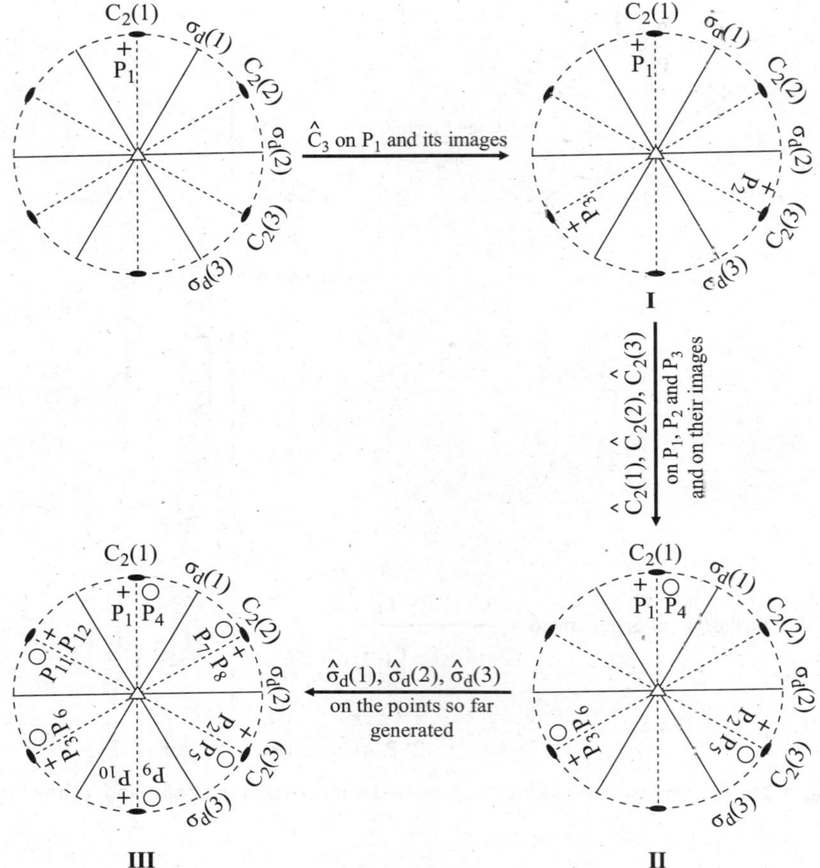

Fig. 1.31 Symmetry operations of D_{3d} carried out on a general point and its images

The points shown in Fig. 1.31(III) can produce no more new points when the symmetry operations indicated by the Schoenflies notation are applied on them. Hence Fig. 1.31(III) is the complete stereographic projection. It has 12 points and so there are a total of twelve symmetry operations in D_{3d}. Of them, only nine are indicated by the Schoenflies notation—$\hat{E}$, $\hat{C}_3$, $\hat{C}_3^2$, $\hat{C}_2(1)$, $\hat{C}_2(2)$, $\hat{C}_2(3)$, $\hat{\sigma}_d(1)$, $\hat{\sigma}_d(2)$, $\hat{\sigma}_d(3)$. Three more can be found out from Fig. 1.31(III) as follows:

The transformations $P_1 \rightarrow P_9$, $P_7 \rightarrow P_3$, $P_4 \rightarrow P_{10}$ and so on, are possible by an inversion through the centre of the auxiliary sphere. So, one symmetry operation in D_{3d} is $\hat{i}$. Similarly, the transformations $P_1 \rightarrow P_7$, $P_7 \rightarrow P_2$ ... are possible by rotation through 60° about the principal C_3-axis followed by a reflection on the equatorial plane which is perpendicular to the principal axis. Thus, there is an S_6-axis collinear with the principal axis. We, therefore, replace the symbol $\triangle$ by ⬢ at the centre (Fig. 1.32). The operations generated by this S_6-axis are

$$\hat{S}_6,\ \hat{S}_6^2 = \hat{C}_3,\ \hat{S}_6^3 = \hat{S}_2 = \hat{i},\ \hat{S}_6^4 = \hat{C}_3^2,\ \hat{S}_6^5,\ \hat{S}_6^6 = \hat{E}$$

$\uparrow$ $\qquad\qquad\qquad\qquad$ $\uparrow$ $\qquad\qquad$ $\uparrow$

The new operations (those different from the ones indicated by the Schoenflies notation D_{3d}) have been indicated by an upward arrow. The complete list of symmetry operations of D_{3d} is $\hat{E}$, $\hat{C}_3$, $\hat{C}_3^2$, $\hat{C}_2(1)$, $\hat{C}_2(2)$, $\hat{C}_2(3)$, $\hat{\sigma}_d(1)$, $\hat{\sigma}_d(2)$, $\hat{\sigma}_d(3)$, $\hat{S}_6$, $\hat{i}$, $\hat{S}_6^5$ and the full sterographic projection is Fig. 1.32.

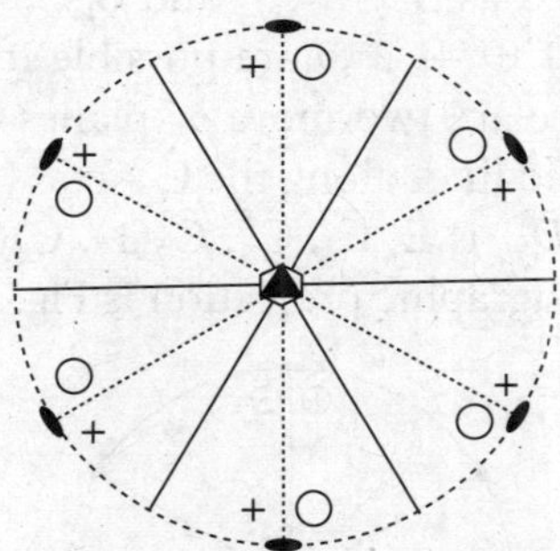

Fig. 1.32 Stereogram of the point group D_{3d}

(iv) **D_{3h} point group:** When a σ_h plane is present, it is taken as the equatorial plane and is shown by a solid circle. The Schoenflies notation D_{3h} indicates the symmetry elements—C_3 (principal axis), three $\hat{C}_2$-axes perpendicular to this C_3 and a σ_h plane. As before, we start with the symbols of these elements and the projection + of a general point in the upper hemisphere. By the $\hat{\sigma}_h$ operation, + on the equatorial plane becomes ∘ at the same position. The necessary flow chart is given in Fig. 1.33.

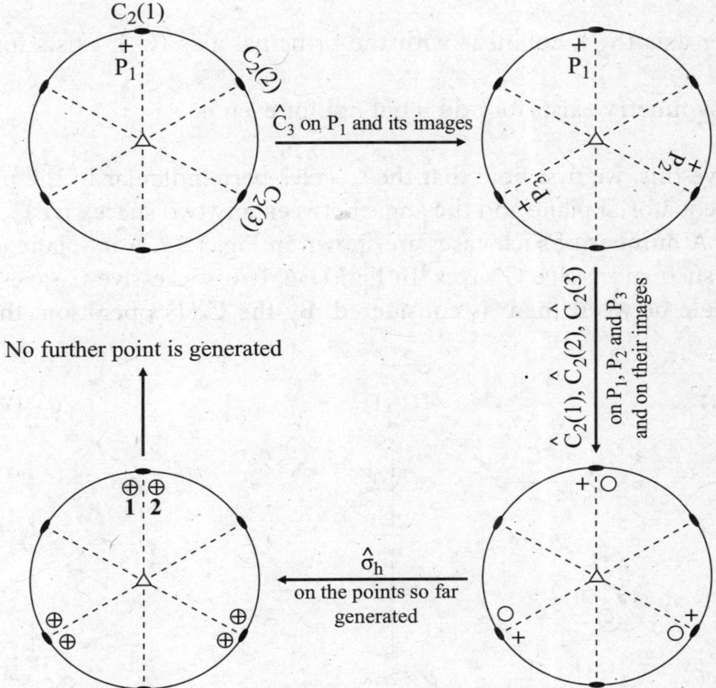

Fig. 1.33 Symmetry operations of D_{3h} carried out on a general point and its images

We thus see that there are 12 points in the stereographic projection and so 12 symmetry operations are expected in D_{3h}. The ones indicated by Schoenflies notation are:

$\hat{E}, \hat{C}_3, \hat{C}_3^2, \hat{C}_2(1), \hat{C}_2(2), \hat{C}_2(3), \hat{\sigma}_h$ (7 in number)

Two more are easily obtained as $\hat{\sigma}_h \hat{C}_3 = \hat{S}_3$ and $\hat{\sigma}_h \hat{C}_3^2 = \hat{S}_3^5$. We have to now find three more. The transformations 1 ⊕ → 2 ⊕ are possible if there is a σ_v plane placed along $\hat{C}_2(1)$. In the same way there are two more σ_v planes along $\hat{C}_2(2)$ and $\hat{C}_2(3)$. These σ_v planes are now shown by solid lines along the $\hat{C}_2$-axes (Fig. 1.34). Thus, the complete list of symmetry operations of D_{3h} is $\hat{E}, \hat{C}_3, \hat{C}_3^2, \hat{C}_2(1), \hat{C}_2(2), \hat{C}_2(3), \hat{\sigma}_h, \hat{S}_3, \hat{S}_3^5, \hat{\sigma}_v(1), \hat{\sigma}_v(2), \hat{\sigma}_v(3)$ and the complete stereographic projection is Fig. 1.34.

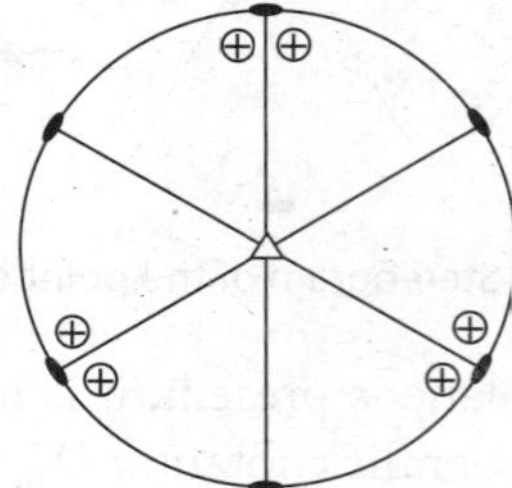

Fig. 1.34 Stereogram of D_{3h}

1.11.2 An Important Property of D_{nd} Groups

In a D_{nd} point group

(i) an improper axis (S_{2n}), collinear with the principal axis (C_n), exists for both odd and even n.

(ii) a centre of symmetry exists for odd n but not for even n.

Proof: To prove this, we first note that the C_2-axes perpendicular to the principal C_n-axis intersect on the equatorial plane and the angle between any two successive C_2-axes is $\alpha = \theta/2$, where $\theta = 2\pi/n$. A number of such cases are shown in Fig. 1.35. A σ_d-plane is symmetrically placed between such successive C_2-axes. In Fig. 1.36, two successive C_2-axes and a σ_d-plane bisecting the angle between them is considered. By the $C_2(1)$ operation, the general point

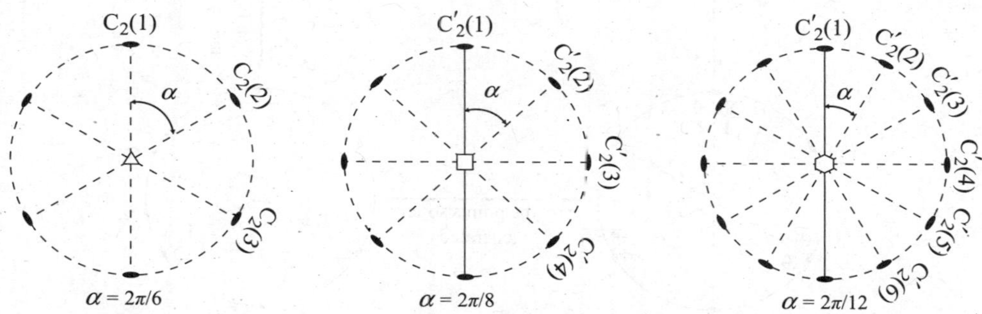

Fig. 1.35 Angle between successive two-fold axes in D_3, D_4 and D_6 groups

in the upper hemisphere whose projection on the equatorial plane is P^+, goes to the lower hemisphere and the image point has projection Q°. On being reflected on the σ_d-plane, the final projection is R°. Now, a single operation that brings P^+ to R° is an improper rotation through an angle α about an axis collinear with the C_n-axis. Hence $\hat{\sigma}_d\hat{C}_2(1) = \hat{S}_m$, where $m = 2\pi/\alpha = 2n$. This proves the statement (i).

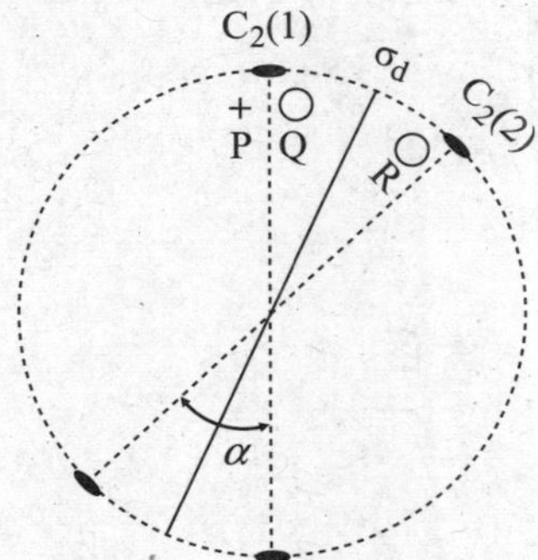

Fig. 1.36 Action of successive $\hat{C}_2(1)$ and $\hat{\sigma}_d$ operations on the projection P of a general point

Now, if n is odd,

$$\hat{S}_{2n}^{n} = (\hat{\sigma}_{\perp}\hat{C}_{2n})^{n} = \hat{\sigma}_{\perp}^{n}\hat{C}_{2n}^{n} = \hat{\sigma}_{\perp}\hat{C}_2 = \hat{i}$$

If n is even,

$$\hat{S}_{2n}^{n} = (\hat{\sigma}_{\perp}\hat{C}_{2n})^{n} = \hat{\sigma}_{\perp}^{n}\cdot\hat{C}_{2n}^{n} = \hat{E}\hat{C}_2 = \hat{C}_2 \neq \hat{i} \quad (\text{as } \hat{\sigma}_{\perp}^{n} = \hat{E})$$

This proves the statement (ii).

1.12 CONSTRUCTION OF GROUP MULTIPLICATION TABLES USING STEREOGRAPHIC PROJECTION

Having understood how to draw stereographic projections, one can easily construct the multiplication table of a point group. Let us consider Fig. 1.31(III) of the D_{3d} point group. Some parts of the multiplication table can be easily written down without the help of any diagram. For example,

$$\hat{C}_3\hat{S}_6 = \hat{C}_3\hat{\sigma}_h\hat{C}_6 = \hat{C}_6^2\hat{\sigma}_h\hat{C}_6 = \hat{\sigma}_h\hat{C}_6^3 = \hat{\sigma}_h\hat{C}_2 = \hat{i}$$

$$\hat{S}_6^2 = (\hat{\sigma}_h\hat{C}_6)^2 = \hat{E}\hat{C}_3 = \hat{C}_3; (\hat{S}_6^5)^2 = (\hat{\sigma}_h\hat{C}_6^5)^2 = \hat{E}\hat{C}_6^{10} = \hat{C}_6^4 = C_3^2$$

For the remaining part of the table, we use the stereographic projection [Fig. 1.31(III)] of D_{3d}. Thus

$$\hat{C}_3\hat{C}_2(1)(P_1) = \hat{C}_3(P_4) = P_5 = \hat{C}_2(2)(P_1), \text{ so that } \hat{C}_3\hat{C}_2(1) = \hat{C}_2(2)$$

$$\hat{\sigma}_d(1)\hat{S}_6(P_1) = \hat{\sigma}_d(1)(P_7) = P_4 = \hat{C}_2(1)(P_1), \text{ so that } \hat{\sigma}_d(1)\hat{S}_6 = \hat{C}_2(1)$$

In this way the all the blank spaces in the table can be filled up (Table 1.3).

Table 1.3 Symmetry multiplication table of D_{3d} (partial)

	$\hat{E}$	$\hat{C}_3$	$\hat{C}_3^2$	$\hat{C}_2(1)$	$\hat{C}_2(2)$	$\hat{C}_2(3)$	$\hat{\sigma}_d(1)$	$\hat{\sigma}_d(2)$	$\hat{\sigma}_d(3)$	$\hat{S}_6$	$\hat{i}$	$\hat{S}_6^5$
$\hat{E}$	$\hat{E}$	$\hat{C}_3$	$\hat{C}_3^2$	$\hat{C}_2(1)$	$\hat{C}_2(2)$	$\hat{C}_2(3)$	$\hat{\sigma}_d(1)$	$\hat{\sigma}_d(2)$	$\hat{\sigma}_d(3)$	$\hat{S}_6$	$\hat{i}$	$\hat{S}_6^5$
$\hat{C}_3$	$\hat{C}_3$	$\hat{C}_3^2$	$\hat{E}$	$\hat{C}_2(2)$						$\hat{i}$		
$\hat{C}_3^2$	$\hat{C}_3^2$	$\hat{E}$	$\hat{C}_3$									
$\hat{C}_2(1)$	$\hat{C}_2(1)$			$\hat{E}$								
$\hat{C}_2(2)$	$\hat{C}_2(2)$				$\hat{E}$							
$\hat{C}_2(3)$	$\hat{C}_2(3)$					$\hat{E}$						
$\hat{\sigma}_d(1)$	$\hat{\sigma}_d(1)$						$\hat{E}$			$\hat{C}_2(1)$		
$\hat{\sigma}_d(2)$	$\hat{\sigma}_d(2)$							$\hat{E}$				
$\hat{\sigma}_d(3)$	$\hat{\sigma}_d(3)$								$\hat{E}$			
$\hat{S}_6$	$\hat{S}_6$									$\hat{C}_3$		$\hat{E}$
$\hat{i}$	$\hat{i}$										$\hat{E}$	
$\hat{S}_6^5$	$\hat{S}_6^5$									$\hat{E}$		$\hat{C}_3^2$

1.13 CONJUGATE CLASSES

With objects having some symmetry elements, it is sometimes possible to guess intuitively which symmetry operations are similar to one another. For example, a rotation through an angle θ about an axis in the anticlockwise direction seems to be similar to the clockwise rotation through θ about the same axis. (Note that this is not always true.) However, such intuitive guesses for finding similarity is not desirable. A well defined procedure for this purpose is discussed in the following sections.

1.13.1 Similarity Transformation

Let $G = \{\hat{E}, \hat{R}_2, \hat{R}_3, \ldots\}$ be a group. (Since the identity operation $\hat{E}$ must be an element of a group, the subscript 1 is reserved for this element; this practice will be followed in all subsequent discussions). If $\hat{R}_p^{-1}\,\hat{R}_i\,\hat{R}_p = \hat{R}_j$, then $\hat{R}_i$ and $\hat{R}_j$ are said to be *similarity transforms* of one another or *conjugate* to one another. The reason for calling $\hat{R}_i$ and $\hat{R}_j$ similar when the above condition holds may be explained as follows:

The effect of $\hat{R}_p$ on some object is nullified by the operation $\hat{R}_p^{-1}$. Hence in the composite operation $\hat{R}_p^{-1}\,\hat{R}_i\,\hat{R}_p$ we expect the result to be either $\hat{R}_i$ itself or some operation whose effect is equivalent to that of $\hat{R}_i$.

The symmetry multiplication table of the NH_3 molecule (Table 1.2) will be of great use to obtain the operations conjugate to a given symmetry operation of C_{3v}. Consider $\hat{C}_3$ first. The possible similarity transformations on $\hat{C}_3$ are:

$$\hat{E}^{-1}\,\hat{C}_3\,\hat{E} = \hat{C}_3; \quad \hat{C}_3^{-1}\,\hat{C}_3\,\hat{C}_3 = \hat{C}_3; \quad (\hat{C}_3^2)^{-1}\,\hat{C}_3\,\hat{C}_3^2 = \hat{C}_3\,\hat{C}_3\,\hat{C}_3^2 = \hat{C}_3^4 = \hat{C}_3$$

$$\{\hat{\sigma}_v(1)\}^{-1}\,\hat{C}_3\,\{\hat{\sigma}_v(1)\} = \{\hat{\sigma}_v(1)\}\,\hat{C}_3\,\{\hat{\sigma}_v(1)\} = \{\hat{\sigma}_v(1)\}\,\{\hat{\sigma}_v(3)\} = \hat{C}_3^2 \text{ (see Table 1.2)}$$

$$\{\hat{\sigma}_v^{-1}(2)\}\,\hat{C}_3\,\{\hat{\sigma}_v(2)\} = \{\hat{\sigma}_v(2)\}\,\hat{C}_3\,\{\hat{\sigma}_v(2)\} = \{\hat{\sigma}_v(2)\}\,\{\hat{\sigma}_v(1)\} = \hat{C}_3^2$$

$$\{\hat{\sigma}_v(3)\}^{-1}\,\hat{C}_3\,\{\hat{\sigma}_v(3)\} = \{\hat{\sigma}_v(3)\}\,\hat{C}_3\,\{\hat{\sigma}_v(3)\} = \{\hat{\sigma}_v(3)\}\,\{\hat{\sigma}_v(2)\} = \hat{C}_3^2$$

It is thus seen that the elements conjugate to $\hat{C}_3$ are $\hat{C}_3$ itself and $\hat{C}_3^2$.

In the same way it can be shown that $\hat{\sigma}_v(1)$, $\hat{\sigma}_v(2)$ and $\hat{\sigma}_v(3)$ are conjugate to one another.

1.13.2 Definition of Conjugate Class

In a group $G = \{\hat{E}, \hat{R}_2, \hat{R}_3, \ldots\}$ a *conjugate class* (or equivalence class, or simply, class) means, a subset of group elements which are similarity transforms of one another. If $\hat{R}_p^{-1}\,\hat{R}_i\,\hat{R}_p = \hat{R}_j$, then $\hat{R}_i$ and $\hat{R}_j$ are in the same class. Thus, in the example cited in the previous section, $(\hat{C}_3, \hat{C}_3^2)$ is a conjugate class of the C_{3v} point group. It can be readily verified that in this group the conjugate classes are $(\hat{E})$, $(\hat{C}_3, \hat{C}_3^2)$ and $[\hat{\sigma}_v(1), \hat{\sigma}_v(2), \hat{\sigma}_v(3)]$.

1.13.3 Some Important Theorems on Conjugate Classes

(i) *The identity operator $\hat{E}$ forms a class by itself.*

Proof: For any symmetry operator $\hat{X}$ of the group under consideration, $\hat{X}^{-1}\,\hat{E}\,\hat{X} = \hat{X}^{-1}\,\hat{X} = \hat{E}$.

This is true for any $\hat{X}$. So, $\hat{E}$ is conjugate to itself only.

(ii) *If a symmetry operator $\hat{P}$ of a group commutes with all other elements of the group, then $\hat{P}$ forms a class by itself.*

Proof: Let $\hat{A}$ be a member of a group *G*.

Now, if $\hat{A}\,\hat{P} = \hat{P}\,\hat{A}$ then $\hat{A}^{-1}\hat{P}\,\hat{A} = \hat{A}^{-1}\,\hat{A}\,\hat{P} = \hat{P}$

As $\hat{A}$ may be any member of *G*, the above result is true for all $\hat{A}$ in *G*. Hence all similarity transformations on $\hat{P}$ will give the same result, $\hat{P}$.

Thus, $\hat{P}$ forms a class by itself.

Note:

- $\hat{E}$, $\hat{\sigma}_h$ and $\hat{i}$ are such unique symmetry operators in molecular point groups; they commute with all the other symmetry operators and individually form classes.

- A group is called *Abelian* (after the name of the famous mathematician N H Abel) if each element of a group commutes with every other element of the group. Examples are C_{2v}, C_{2h}, D_2, C_n and so on. In such a group every symmetry operator forms a class by itself.

(iii) *No symmetry operator of a group can belong simultaneously to two different classes.*

Proof: If possible, let C-I and C-II be two different classes of a group G and let $\hat{P}$ be a symmetry operator common to them. Let $\hat{A}$ and $\hat{B}$ be two different symmetry operators of G which may or may not belong to C-I and C-II.

Now, considering $\hat{P}$ as a symmetry operator of C-I we have,

$$\hat{A}^{-1}\hat{P}\hat{A} = \hat{P}_1 \tag{1.1}$$

where $\hat{P}_1$ belongs to C-I.

Left-multiplying by $\hat{A}$ and right-multiplying by $\hat{A}^{-1}$ we have

$$\hat{A}\hat{A}^{-1}\hat{P}\hat{A}\hat{A}^{-1} = \hat{A}\hat{P}_1\hat{A}^{-1} \quad \text{or,} \quad \hat{P} = \hat{A}\hat{P}_1\hat{A}^{-1} \tag{1.2}$$

Similarly, considering $\hat{P}$ as a member of C-II and carrying out similarity transformations with the help of $\hat{B}$ we have

$$\hat{P} = \hat{B}\hat{P}_2\hat{B}^{-1} \tag{1.3}$$

where $\hat{P}_2$ is a member of C-II.

Now comparing Eqs. (1.2) and (1.3), $\hat{B}\hat{P}_2\hat{B}^{-1} = \hat{A}\hat{P}_1\hat{A}^{-1}$ (1.4)

Left-multiplying both sides of Eq. (1.4) by $\hat{A}^{-1}$ and right-multiplying by $\hat{A}$, we have

$$\hat{P}_1 = \hat{A}^{-1}\hat{B}\hat{P}_2\hat{B}^{-1}\hat{A} = \hat{Z}^{-1}\hat{P}_2\hat{Z} \tag{1.5}$$

where $\hat{Z} = \hat{B}^{-1}\hat{A}$ and $\hat{Z}^{-1} = \hat{A}^{-1}\hat{B}$. Owing to closure property and existence of inverse of each group element, both $\hat{Z}$ and $\hat{Z}^{-1}$ belong to G. Hence according to Eq. (1.5), $\hat{P}_1$ and $\hat{P}_2$ are conjugate to one another, and should belong to the same class. Thus, C-I and C-II should be the same, which is contrary to the starting hypothesis. Hence $\hat{P}$ cannot be common to C-I and C-II. Thus the classes of a group form disjoint sets.

(iv) *If two operators of a group are conjugate to a third operator, then they are themselves conjugate.*

Proof: Let $\hat{A}$, $\hat{B}$ and $\hat{C}$ be three symmetry operators of a group G such that both $\hat{A}$ and $\hat{B}$ are conjugate to $\hat{C}$. We are to prove that $\hat{A}$ and $\hat{B}$ are conjugate to one another.

As $\hat{A}$ is conjugate to $\hat{C}$, $\hat{X}^{-1}\hat{A}\hat{X} = \hat{C}$ for some $\hat{X}$ in G.

As $\hat{B}$ is conjugate to $\hat{C}$, $\hat{Y}^{-1}\hat{B}\hat{Y} = \hat{C}$ for some other operator $\hat{Y}$ in G.

Hence $\hat{X}^{-1}\hat{A}\hat{X} = \hat{Y}^{-1}\hat{B}\hat{Y}$

or, $\hat{A} = \hat{X}\hat{Y}^{-1}\hat{B}\hat{Y}\hat{X}^{-1} = (\hat{Y}\hat{X}^{-1})^{-1}\hat{B}(\hat{Y}\hat{X}^{-1})$

Since $\hat{X}$ and $\hat{Y}$ belong to G, $\hat{Y}\hat{X}^{-1}$ also belongs to G. Hence $\hat{A}$ and $\hat{B}$ are conjugate to one another.

(v) *If $\hat{A}$ and $\hat{B}$ are in the same class and $\hat{A}^h = \hat{E}$ (h an integer), then $\hat{B}^h = \hat{E}$.*

Proof: $\hat{A}$ and $\hat{B}$ are conjugate.

Hence $\hat{X}^{-1}\hat{A}\hat{X} = \hat{B}$ for some $\hat{X}$ in G.

Now $\hat{A}^h = \hat{E}$

or, $\hat{A}\cdot\hat{A}\cdot\hat{A}\ldots\hat{A}$ (h factors) $= \hat{E}$

or, $\hat{A}\hat{X}\hat{X}^{-1}\hat{A}\hat{X}\hat{X}^{-1}\hat{A}\hat{X}\hat{X}^{-1}\hat{A}\ldots\hat{X}\hat{X}^{-1}\hat{A} = \hat{E}$

or, $\hat{A}\hat{X}\hat{B}\hat{B}\hat{B}\ldots\hat{B}\hat{X}^{-1}\hat{A} = \hat{E}$

or, $\hat{A}\hat{X}\hat{B}^{h-2}\hat{X}^{-1}\hat{A} = \hat{E}$

or, $\hat{X}^{-1}\hat{A}\hat{X}\hat{B}^{h-2}\hat{X}^{-1}\hat{A}\hat{X} = \hat{X}^{-1}\hat{E}\hat{X}$

or, $\hat{B}\hat{B}^{h-2}\hat{B} = \hat{E}$ that is, $\hat{B}^h = \hat{E}$

Theorem (v) means that elements of the same conjugate class of a group have the same order.
For example in C_{3v}, the classes are $(\hat{E})$, $(\hat{C}_3, \hat{C}_3^2)$, $[(\hat{\sigma}_v(1), \hat{\sigma}_v(2), \hat{\sigma}_v(3)]$. Here

$$(\hat{C}_3)^3 = \hat{E} \text{ or, } \hat{C}_3 \text{ has order 3; } (\hat{C}_3^2)^3 = \hat{E} \text{ or, } \hat{C}_3^2 \text{ has order 3.}$$

The converse of theorem (v) is not always true. For example in the C_{2v} point group, $(\hat{C}_2)^2 = \hat{E}$ and $(\hat{\sigma}_v)^2 = \hat{E}$ but $\hat{C}_2$ and $\hat{\sigma}_v$ are not in the same class.

1.14 SYMMETRY ELEMENT, GROUP ELEMENT AND ORDER OF A GROUP

Sometimes confusion arises in using the word *element* in group theory literature. A *group element* means a member of a group. A *symmetry element* means an element of symmetry, namely, an axis of rotation, a point of inversion… as defined earlier. In the case of a point group, a group element is a symmetry operator and not a symmetry element. The order of a group element is the smallest number of times it has to be combined by the composition rule of the group in order to produce the identity element. The order of a group is the total number of group elements. For example, in the point group C_{3v}, a group element is $\hat{C}_3$ (not the C_3-axis), the order of $\hat{C}_3$ is 3 (as $\hat{C}_3^3 \equiv \hat{E}$) and the order of the group C_{3v} is 6. In the mathematical group $G = \{1, -1, i, -i\}$ under multiplication, the order of the group element i is 4 and order of the group G is also 4 as it contains four elements.

1.15 SUBGROUP OF A GROUP

A subgroup of a group G means a part of a group which has all the four group properties. For example, consider the group $G = \{1, i, -i, -1\}$ where the binary operation is multiplication.

Here the subset {1, –1} exhibits all the four group properties under multiplication, as can be readily verified. Thus {1, –1} is a subgroup of *G*. Similarly in the C_{3v} point group, $\{\hat{E}, \hat{\sigma}_v(1)\}$, $\{\hat{E}, \hat{\sigma}_v(2)\}$, $\{\hat{E}, \hat{\sigma}_v(3)\}$ and $(\hat{E}, \hat{C}_3, \hat{C}_3^2)$ are subgroups.

1.15.1 Two Important Theorems on Subgroups and Groups

(i) **Lagrange's theorem:** *The order of a group is divisible by the order of its subgroup.*
For example, the point group C_{3v} is of order 6 which is divisible by the orders 2 and 3 of the subgroups $(\hat{E}, \hat{\sigma}_v)$ and $(\hat{E}, \hat{C}_3, \hat{C}_3^2)$ respectively.

(ii) **The rearrangement theorem:** *No symmetry operator appears more than once in any row or column of the multiplication table of the group.* In other words, *every row (or column) of the group multiplication table is a rearrangement of the other rows (or columns).*

Proof: Let $\{\hat{E}, \hat{R}_2, \hat{R}_3, \ldots, \hat{R}_n\}$ be a group. If possible, let the element $\hat{R}_i$ appear in the *r*th row at two different places—once at the intersection of the *r*th row and *p*th column and again at the intersection of the *r*th row and *q*th column ($p \neq q$) of the multiplication table as shown in Table 1.4.

Table 1.4 Hypothetical multiplication table of a group $\{\hat{E}, \hat{R}_2, \hat{R}_3, \ldots, \hat{R}_n\}$ where $\hat{R}_i$ occurs twice in a row.

1st operation (columns); **2nd operation** (rows)

	$\hat{E}$	$\hat{R}_2$	$\hat{R}_3$	………	$\hat{R}_p$	………	$\hat{R}_q$	………	$\hat{R}_n$
$\hat{E}$					⋮		⋮		
$\hat{R}_2$					⋮		⋮		
$\hat{R}_3$					⋮		⋮		
⋮					⋮		⋮		
$\hat{R}_r$	………	………	………	………	$\hat{R}_i$	………	$\hat{R}_i$		
⋮									
$\hat{R}_p$									
⋮									
$\hat{R}_q$									
⋮									
$\hat{R}_n$									

Then $\hat{R}_r\hat{R}_p = \hat{R}_i$ or, $\hat{R}_p = \hat{R}_r^{-1}\hat{R}_i$

$\hat{R}_r\hat{R}_q = \hat{R}_i$ or, $\hat{R}_q = \hat{R}_r^{-1}\hat{R}_i$

Thus $\hat{R}_p = \hat{R}_q$, that is, $p = q$ which is contrary to the starting hypothesis. (Proved).

1.16 CYCLIC GROUPS AND SUBGROUPS

A group of the type $(x, x^2, x^3, \ldots, x^n = E)$ is called a *cyclic group* (C_n) with generator x. For example the group of cube roots of unity, $C_3 = \{1, \omega, \omega^2\} \equiv \{\omega, \omega^2, \omega^3 = 1\}$ is a cyclic group of order 3 with generator ω. In the point group $C_{3v} = \{\hat{E}, \hat{C}_3, \hat{C}_3^2, \hat{\sigma}_v(1), \hat{\sigma}_v(2), \hat{\sigma}_v(3)\}$, we have $\{\hat{E}, \hat{C}_3, \hat{C}_3^2\}$ as a cyclic subgroup of C_{3v} with order 3 and generator $\hat{C}_3$. Similarly $\{\hat{E}, \hat{\sigma}_v(1)\}$ is a cyclic subgroup of C_{3v} with order 2 and generator $\hat{\sigma}_v(1)$. Cyclic groups are Abelian.

1.16.1 Combined Results of Lagrange's Theorem and the Rearrangement Theorem

(i) *A group of prime order must be cyclic.*

Proof: Let $G_p = \{E, A_2, A_3, \ldots, A_p\}$ be a group of order p, where p is a prime number. Let us consider a general element A_j in G_p. Since G_p is a finite group, the order of A_j must also be finite. Let s be the order of A_j. Then, $\{A_j, A_j^2, \ldots, A_j^s = E\}$ must be (a) either a part of G_p, or (b) the whole of G_p. In case (a), $\{A_j, A_j^2, \ldots, A_j^s = E\}$ is a cyclic subgroup of G_p with order s. Then according to Lagrange's theorem s is a divisor of p, which is not possible since p is prime. Thus case (a) is ruled out.

In case (b), $s = p$ and then G_p becomes a cyclic group $(A_j, A_j^2, \ldots, A_j^p)$ with generator A_j. [As cyclic groups are Abelian, every element of a group of prime order forms a class by itself.]

(ii) *There can be only two types of groups of order 4.*

Proof: Let $(\hat{E}, \hat{A}, \hat{B}, \hat{C})$ be a group of order 4. Using the results $\hat{E}\,\hat{A} = \hat{A}\,\hat{E} = \hat{A}$ ..., the first row and first column of the multiplication table can be immediately written down as in the following table. Now, 2 is a divisor of 4 and hence the group should have a subgroup of order 2. Let $(\hat{E}, \hat{A})$ be this subgroup. Then $\hat{A}^2 = \hat{E}$ (it is immaterial whether we take $\hat{B}^2 = \hat{E}$, or $\hat{C}^2 = \hat{E}$, the final result would be the same). Using the rearrangement theorem two rows and two columns can be filled up as:

	Ê	Â	B̂	Ĉ
Ê	Ê	Â	B̂	Ĉ
Â	Â	Ê	Ĉ	B̂
B̂	B̂	Ĉ	*	
Ĉ	Ĉ	B̂		

The position indicated by * can be occupied in only two ways, namely (a) * = Â or (b) * = Ê, according to the rearrangement theorem. Hence there can be only two kinds of groups of order 4. Their nature can be obtained as follows.

For case (a), * = Â and the complete multiplication table is,

	$\hat{E}$	$\hat{A}$	$\hat{B}$	$\hat{C}$
$\hat{E}$	$\hat{E}$	$\hat{A}$	$\hat{B}$	$\hat{C}$
$\hat{A}$	$\hat{A}$	$\hat{E}$	$\hat{C}$	$\hat{B}$
$\hat{B}$	$\hat{B}$	$\hat{C}$	$\hat{A}$	$\hat{E}$
$\hat{C}$	$\hat{C}$	$\hat{B}$	$\hat{E}$	$\hat{A}$

Here we find $\hat{A}^2 = \hat{E}$, $\hat{B}^2 = \hat{A}$, $\hat{C}^2 = \hat{A}$.

Hence $\hat{B}^4 = \hat{A}^2 = \hat{E}$, $\hat{B}^3 = \hat{B}\hat{A} = \hat{C}$; thus the group $(\hat{E}, \hat{A}, \hat{B}, \hat{C})$ is equivalent to $(\hat{B}^4, \hat{B}^2, \hat{B}, \hat{B}^3)$, that is, $(\hat{B}, \hat{B}^2, \hat{B}^3, \hat{B}^4 \equiv \hat{E})$. Hence the group is cyclic (with generator $\hat{B}$) and Abelian.

For case (b), * = $\hat{E}$ and the complete multiplication table is

	$\hat{E}$	$\hat{A}$	$\hat{B}$	$\hat{C}$
$\hat{E}$	$\hat{E}$	$\hat{A}$	$\hat{B}$	$\hat{C}$
$\hat{A}$	$\hat{A}$	$\hat{E}$	$\hat{C}$	$\hat{B}$
$\hat{B}$	$\hat{B}$	$\hat{C}$	$\hat{E}$	$\hat{A}$
$\hat{C}$	$\hat{C}$	$\hat{B}$	$\hat{A}$	$\hat{E}$

Here $\hat{A}\hat{B} = \hat{C} = \hat{B}\hat{A}$, $\hat{A}\hat{C} = \hat{B} = \hat{C}\hat{A}$, $\hat{B}\hat{C} = \hat{A} = \hat{C}\hat{B}$. But it cannot be shown to be cyclic as in the former case. This group is non-cyclic but Abelian, and is called *Vierergrupe* or *Klein's four-group*. The point groups C_{2v}, C_{2h} and D_2 are examples of such a group.

1.17 SOME RULES FOR DETERMINING CONJUGATE CLASSES IN A POINT GROUP

Rule 1: *Two rotations $\hat{P}$ and $\hat{Q}$ through the same angle but about different axes p and q respectively are in the same conjugate class if the group contains a third symmetry operation which brings one of the axes into the other.*

Examples: The 2-fold rotations about axes perpendicular to the principal axis are conjugate for:

(i) D_n and D_{nh} groups with odd n, and
(ii) D_{nd} groups with any integral n.

This is because any C_2-axis perpendicular to C_n can be brought to any other such C_2-axis by a suitable $\hat{C}_n$ rotation about the principal axis if n is odd. If n is even there are two sets of C_2'-axes producing two different classes; any two members of one set can be brought into coincidence by $\hat{C}_n$ rotation but no member of one set can be made to coincide with a C_2'-axis of another set by $\hat{C}_n$. The case remains the same for D_{nh}—introduction of σ_d-plane does not improve the situation. In the case of D_{nd} groups, irrespective of odd or even n, there is a σ_d-plane between successive $\hat{C}_2'$-axes. This brings any C_2'-axis of one set into coincidence with the closest C_2'-axis of the other set. Thus, for D_{nd} groups with odd n, the two sets of $\hat{C}_2'$-axes cannot be distinguished—they correspond to a single class; the prime notation on C_2 is not needed for odd n.

Proof: Let a group G contain two rotational axes of symmetry, p and q; also let rotations about them through an angle of $2\pi/n$ be denoted by the operators $\hat{P}$ and $\hat{Q}$ respectively.

By the operator $\hat{P}$, all points on the axis p remain unmoved: $\hat{P}(p) = p$ (1.6)

By the operator $\hat{Q}$, all points on the axis q remain unmoved: $\hat{Q}(q) = q$ (1.7)

Let there exist another symmetry operator $\hat{R}$ in G such that all points on p be brought to corresponding points on q by the action of $\hat{R}$:

$$\hat{R}(p) = q \tag{1.8}$$

Consequently, $\hat{R}^{-1}(q) = p$ (1.9)

Then $\hat{R}^{-1}\hat{Q}\hat{R}(p) = \hat{R}^{-1}\hat{Q}(q)$ (by Eq. 1.8)

$= \hat{R}^{-1}(q)$ (by Eq. 1.7)

$= p$ (by Eq. 1.9)

$= \hat{P}(p)$ (by Eq. 1.6)

Hence $\hat{R}^{-1}\hat{Q}\hat{R} = \hat{P}$ (1.10)

which means that $\hat{P}$ and $\hat{Q}$ are in the same conjugate class.

Rule 2: *Two reflections $\hat{\sigma}_1$ and $\hat{\sigma}_2$ on two different planes of symmetry, 1 and 2, are in the same conjugate class if there exists in the group a third symmetry operator $\hat{R}$ which brings all points of the plane 1 on to the corresponding points of the plane 2.*

Proof: The proof is exactly analogous to that of (1) above:
$\hat{\sigma}_2$ keeps all points of plane 2 unmoved, that is, $\hat{\sigma}_2(2) = 2$. Again, $\hat{R}$ brings the plane 1 to the plane 2, that is,

$$\hat{R}(1) = 2 \quad \text{or,} \quad \hat{R}^{-1}(2) = 1$$

Then $\hat{R}^{-1}\hat{\sigma}_2\hat{R}(1) = \hat{R}^{-1}\hat{\sigma}_2(2) = \hat{R}^{-1}(2) = 1 = \hat{\sigma}_1(1)$ that is, $\hat{R}^{-1}\hat{\sigma}_2\hat{R} = \hat{\sigma}_1$ (Proved)

An example is provided by the pyramidal NH_3 molecule. There are one C_3-axis and three $\hat{\sigma}_v$ planes. Any one $\hat{\sigma}_v$ plane can be brought to another by a rotation operation ($\hat{C}_3$ or $\hat{C}_3^2$) about the three-fold axis. Hence the three $\hat{\sigma}_v$ operations of the NH_3 molecule are in the same conjugate class. In a square pyramid there is a C_4-axis passing through the apex and the centre of the square base. There are four vertical planes of symmetry, but they belong to two different classes because one of the planes containing one diagonal can be brought to the other plane passing through the other diagonal by a $\hat{C}_4$ rotation but there is no symmetry operation of the square pyramid that can bring such a plane on to a symmetry plane that bisects a pair of opposite edges of the square base. Thus, with reference to Fig. 1.37, [$\hat{\sigma}_v(1)$, $\hat{\sigma}_v(2)$] and [$\hat{\sigma}_d(1)$, $\hat{\sigma}_d(2)$] are two different conjugate classes of the point group C_{4v}.

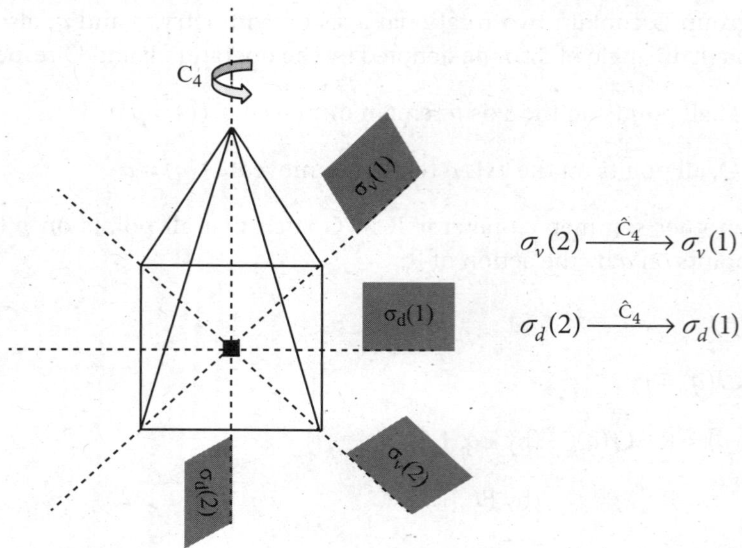

Fig. 1.37 Four planes of symmetry of a square pyramid belonging to two different conjugate classes

Rule 3: *A rotation C_n^k and its inverse C_n^{-k} [or, $(C_n^k)^{-1}$] about an axis of symmetry are in the same class if there exists in the group a symmetry plane (σ) containing that axis or a two-fold rotation axis perpendicular to the C_n-axis.*

Examples are very common. In the point group C_3, the operators $\hat{C}_3$ and $\hat{C}_3^{-1}$ are in different classes but in the point group C_{3v}, they are in the same class owing to the σ_v planes and in the point group D_3, they are in the same class owing to the presence of C_2 axes.

Proof: This rule can be proved in four parts as follows:

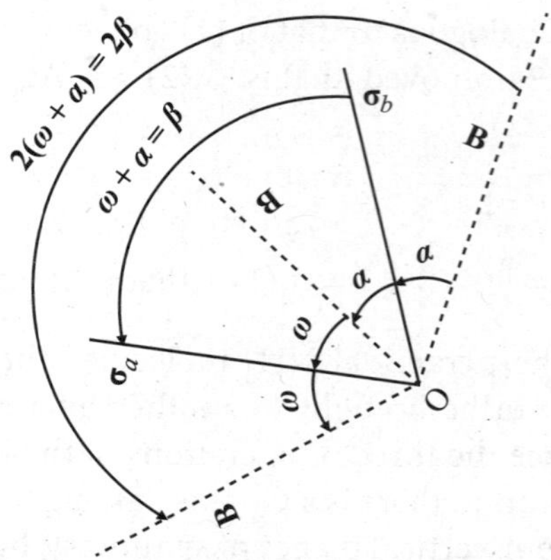

Fig. 1.38 Illustration of $\hat{\sigma}_a\hat{\sigma}_b = \hat{C}_n, n = \frac{2\pi}{2\beta}$

Part I: *The result of successive reflections on two different symmetry planes σ_a and σ_b is a rotation about their line of intersection through an angle 2β, where β is the angle between the two reflection planes.* As shown in Fig. 1.38, two reflection planes σ_a and σ_b intersect at a line perpendicular to the plane of the paper and passing through the point O. (This line has not been shown in Fig. 1.38 for the sake of clarity.) For the reflection plane σ_b, angle of incidence = angle of reflection = α and this operation produces a laterally inverted image of the object (letter B).

For the next reflection on the plane σ_a the angle of incidence and reflection are ω each and the result is the image B, which can be alternatively obtained by a rotation of B (as its initial position) about the line of intersection of σ_a and σ_b through an angle of $2(\omega + a) = 2\beta$; clearly, $(\omega + a) = \beta$ = angle between the reflection planes.

Thus $\hat{\sigma}_a \hat{\sigma}_b = \hat{C}_n$ (1.11)

where $n = \dfrac{2\pi}{2\beta} = \dfrac{\pi}{\beta}$

Part II: Taking the inverse of both sides of Eq. (1.11),

$\hat{\sigma}_b^{-1} \hat{\sigma}_a^{-1} = \hat{C}_n^{-1}$ or $\hat{\sigma}_b \hat{\sigma}_a = \hat{C}_n^{-1}$ (as each reflection operation is its own inverse)

or, $\hat{\sigma}_b = \hat{C}_n^{-1} \hat{\sigma}_a$ (1.12)

Result (1.12) shows that if a molecule has an n-fold axis of symmetry (C_n) and a reflection plane σ_a that contains this C_n-axis, then another symmetry plane σ_b can be found out by utilising relation (1.12). Continuing this process one finds $\hat{\sigma}_c = \hat{C}_n^{-1} \hat{\sigma}_b$ and so on. *Thus a total of n reflection planes can be found out if one C_n-axis and a σ-plane containing this axis are present in the object.*

Part III: From Eq. 1.12, $\hat{\sigma}_b = \hat{C}_n^{-1} \hat{\sigma}_a$ and so $\hat{\sigma}_a \hat{\sigma}_b = \hat{\sigma}_a \hat{C}_n^{-1} \hat{\sigma}_a$ (1.13)

But $\hat{\sigma}_a \hat{\sigma}_b = \hat{C}_n$ (from Eq. 1.11)

Hence from Eq. 1.13, $\hat{C}_n = \hat{\sigma}_a \hat{C}_n^{-1} \hat{\sigma}_a = (\hat{\sigma}_a)^{-1} \hat{C}_n^{-1} \hat{\sigma}_a$ [as $(\hat{\sigma}_a)^{-1} = \hat{\sigma}_a$] (1.14)

Thus $\hat{C}_n$ and $\hat{C}_n^{-1}$ are conjugate if $\hat{\sigma}_a$ exists.

To get a more generalised result we have from Eq. (1.14),

$$\hat{C}_n^k = (\hat{\sigma}_a^{-1} \hat{C}_n^{-1} \hat{\sigma}_a)(\hat{\sigma}_a^{-1} \hat{C}_n^{-1} \hat{\sigma}_a) \ldots \text{ to } k \text{ factors}$$

$$= \hat{\sigma}_a^{-1} (\hat{C}_n^{-1})^k \hat{\sigma}_a = \hat{\sigma}_a^{-1} \hat{C}_n^{-k} \hat{\sigma}_a \quad (1.15)$$

Thus $\hat{C}_n^k$ and $\hat{C}_n^{-k}$ are also conjugate if $\hat{\sigma}_a$ exists.

Part IV: If two reflection planes σ_a and σ_b are perpendicular to one another, they intersect at a line which acts as a C_2-axis (as $\frac{\pi}{\beta} = 2$). If the principal axis of rotation of the group is n-fold and σ_a contains this C_n-axis, we can call σ_a a vertical plane and denote it as σ_v; then the σ_b plane is perpendicular to the C_n-axis and we can denote it by σ_h. The C_2-axis at the intersection of σ_v and σ_h lies on the σ_h plane and is perpendicular to the C_n-axis. There are n number of σ_a, that is, σ_v planes according to result of Part (I) and each σ_v intersects the σ_h plane at one C_2-axis. Thus, there are n number of C_2-axes perpendicular to C_n, if one C_n-axis and at least one σ_v plane are found together with a σ_h plane.

Now, $\hat{\sigma}_v\hat{\sigma}_h = \hat{C}_2$; inverting both sides, $(\hat{\sigma}_h)^{-1}(\hat{\sigma}_v)^{-1} = (\hat{C}_2)^{-1}$ or, $\hat{\sigma}_h\hat{\sigma}_v = \hat{C}_2$.

Thus, $\hat{\sigma}_v\hat{\sigma}_h = \hat{\sigma}_h\hat{\sigma}_v \; (= \hat{C}_2')$ (1.16)

So $\hat{\sigma}_h$ commutes with $\hat{\sigma}_v$. The conjugate nature of $\hat{C}_n^{-1}$ and $\hat{C}_n$ by similarity transformation with $\hat{C}_2$ can now be proved:

$$(\hat{C}_2)^{-1}\hat{C}_n^{-1}\hat{C}_2 = \hat{C}_2\hat{C}_n^{-1}\hat{C}_2 \quad (\hat{C}_2 \text{ being a 2-fold operation})$$

$$= \hat{\sigma}_h\hat{\sigma}_v\hat{C}_n^{-1}\hat{\sigma}_v\hat{\sigma}_h \quad (\text{using Eq. 1.16})$$

$$= \hat{\sigma}_h(\hat{\sigma}_a\hat{C}_n^{-1}\hat{\sigma}_a)\hat{\sigma}_h \quad (\text{as } \hat{\sigma}_v = \hat{\sigma}_a)$$

$$= \hat{\sigma}_h\hat{C}_n\hat{\sigma}_h \quad (\text{using the result 1.14})$$

$$= \hat{\sigma}_h\hat{\sigma}_h\hat{C}_n \quad (\text{as } \hat{\sigma}_h \text{ commutes with } \hat{C}_n)$$

$$= \hat{C}_n$$

Extension to $(\hat{C}_2)^{-1}\hat{C}_n^{-k}\hat{C}_2 = \hat{C}_n^k$ easily follows: As $\hat{C}_n = (\hat{C}_2)^{-1}\hat{C}_n^{-1}\hat{C}_2$

$$\hat{C}_n^k = [(\hat{C}_2)^{-1}\hat{C}_n^{-1}\hat{C}_2]\,[(\hat{C}_2)^{-1}\hat{C}_n^{-1}\hat{C}_2] \ldots \text{ to } k \text{ factors}$$

$$= (\hat{C}_2)^{-1}\hat{C}_n^{-1}\hat{E}\hat{C}_n^{-1} \ldots \hat{E}\hat{C}_n^{-1}\hat{C}_2 = (\hat{C}_2)^{-1}\hat{C}_n^{-k}\hat{C}_2 \quad (\text{Proved})$$

Thus $\hat{C}_n^k$ and $\hat{C}_n^{-k}$ are also conjugate to each other if a $\hat{C}_2$-axis perpendicular to the C_n-axis exists.

SUMMARY

- The set of all symmetry operators of an object forms a *point group*.
- Two symmetry operators may or may not commute.
- In a row (or column) of the group multiplication table no symmetry operator occurs more than once (rearrangement theorem).
- For three operators $\hat{R}_a$, $\hat{R}_i$ and $\hat{R}_j$ of a group G if $\hat{R}_a^{-1}\hat{R}_i\hat{R}_a = \hat{R}_j$, then $\hat{R}_i$ and $\hat{R}_j$ are *conjugate to* or *similarity transforms of* one another. Conjugate symmetry operators of G form a *conjugate class*.
- If $\hat{R}_i$ commutes with all other members of G, then $\hat{R}_i$ forms a class by itself; $\hat{E}$ (the identity), $\hat{\sigma}_h$ and $\hat{i}$ are such operators.
- In an Abelian group all pairs of operators commute and each symmetry operator forms a separate class.
- A *cyclic group* of order n has the structure $\{\hat{R}, \hat{R}^2, \hat{R}^3, \ldots, \hat{R}^n \equiv \hat{E}\}$.
- The Schoenflies notation for a point group specifies the minimal set of symmetry operators from which all the symmetry operators can be generated by stereographic projections. Important molecular properties can be predicted from symmetry alone.

LEARNING OUTCOME

After reading this chapter the student will be able to (i) identify the point group of an object, (ii) construct the stereogram of a point group from its Schoenflies notation, (iii) utilise it to find the complete list of symmetry operations and construct the multiplication table, (iv) find the conjugate classes of symmetry operators by similarity transformation and (v) apply symmetry arguments to predict optical activity and existence of a permanent dipole moment.

EXERCISES

1. Replace each of the following by a single equivalent symmetry operator:
   (a) $\hat{S}_4^2$ (b) $\hat{S}_4^4$ (c) $\hat{C}_4^2$ (d) $\hat{i}^3$ (e) $\hat{C}_3^3$ (f) $\hat{S}_3^3$ (g) $\hat{\sigma}_h^3$ (h) $\hat{S}_6^2$ (i) $\hat{S}_6^3$

2. For the pyridine molecule, (a) find $\hat{C}_2\hat{\sigma}_v(1)$ and (b) show that $\hat{\sigma}_v(2)[\hat{C}_2\hat{\sigma}_v(1)] = [\hat{\sigma}_v(2)\hat{C}_2]\,\hat{\sigma}_v(1)$.

3. For the BF_3 molecule find out if $\hat{C}_3\hat{\sigma}_v(1) = \hat{\sigma}_v(1)\hat{C}_3$.

4. List the symmetry operations and construct symmetry multiplication tables for each of the following molecules.
   (a) Pyridine
   (b) BrF_5 (square pyramidal, with Br at the centre of the square base and F at the vertices)
   (c) BF_3
   (d) C_6H_6
   (e) *trans*-but-2-ene (assuming $-CH_3$ to be spherical).

5. Show that the sets of (a) square roots and (b) fourth roots of unity are cyclic groups under arithmetic multiplication. Point out the generator of the group in each case. Do the sets remain groups under addition?

6. Complete the unfinished work in Table 1.3.

7. How many reflection planes are there in the acetylene molecule? How many of them are σ_v planes and how many are σ_h?

8. What is the result of the improper rotation $\hat{S}_4(Z)$ on the point $(x, 0, z)$? Show that the point groups S_4 and C_{4h} are different but C_{3h} and the (hypothetical) S_3 point groups are the same.

9. Show that (a) if $Q^{-1}PQ = P$, where P and Q are members of a group, then P and Q commute; (b) if the elements P and Q of a group are conjugate, their inverses are also conjugate; (c) the set of natural numbers including 0 is a group under addition; if x and y are two different numbers in this set, they are in different conjugate classes; what is your conclusion from this result regarding the nature of the group?

10. Which of the following sets, under the binary operations stated against the sets, are not groups and why?

	Set	Operation
(a)	$\{1, \omega, \omega^2\}, \omega = e^{\frac{2\pi i}{3}}$	Arithmetic multiplication
(b)	$\{I_2, I_3, I_4, I_5\}$ where $I_m = m \times m$ unit matrix	Matrix multiplication
(c)	The set of all real numbers	Arithmetic multiplication
(d)	The set of all real numbers excluding zero	Arithmetic multiplication
(e)	The set of all possible symmetry operations of the pyridine molecule	Successive execution of symmetry operations
(f)	Set of the following matrices:	Matrix multiplication
	(i) $\begin{bmatrix} 1 & 0 \\ 0 & 1 \end{bmatrix}, \begin{bmatrix} 0 & 1 \\ -1 & 0 \end{bmatrix}, \begin{bmatrix} -1 & 0 \\ 0 & -1 \end{bmatrix}, \begin{bmatrix} 0 & -1 \\ 1 & 0 \end{bmatrix}$	
	(ii) $\begin{bmatrix} 1 & 0 & 0 & 0 \\ 0 & 1 & 0 & 0 \\ 0 & 0 & 1 & 0 \\ 0 & 0 & 0 & 1 \end{bmatrix}, \begin{bmatrix} 0 & 1 & 0 & 0 \\ 1 & 0 & 0 & 0 \\ 0 & 0 & 0 & 1 \\ 0 & 0 & 1 & 0 \end{bmatrix}, \begin{bmatrix} 0 & 0 & 0 & 1 \\ 0 & 0 & 1 & 0 \\ 0 & 1 & 0 & 0 \\ 1 & 0 & 0 & 0 \end{bmatrix}, \begin{bmatrix} 0 & 0 & 1 & 0 \\ 0 & 0 & 0 & 1 \\ 1 & 0 & 0 & 0 \\ 0 & 1 & 0 & 0 \end{bmatrix}$	

(If you are not familiar with matrix multiplication, read Chapter 3 and then attempt this problem.)

11. From the list of sets in Question 10, take out the ones which in your opinion, are groups and demonstrate all the four group properties for each of them.

12. Find whether the following pairs of symmetry operations commute:
(a) $\hat{\sigma}(XY)$ and $\hat{\sigma}(YZ)$ (b) $\hat{C}_4(Z)$ and $\hat{C}_2(X)$
(c) $\hat{C}_2(Z)$ and $\hat{C}_4(Y)$ (d) $\hat{S}_n$ (n even) and $\hat{i}$

[**Hint**: Take a Cartesian coordinate system, draw the necessary symmetry elements which intersect at the origin and examine the effect of symmetry operations on a general point (x, y, z).]

13. What are the point groups of spirane (C_5H_8), cubane (C_8H_8), octacyanomolybdate (IV) with a square antiprism structure, CH_2ClF, cyclohexane chair, dibenzene chromium (eclipsed), 1, 2-dibromo-1,2-dichloro ethane (staggered), C_{60} fullerene, IPR C_{70} fullerene. [A fullerene structure in which all the pentagons are isolated by intervening hexagons is called an IPR fullerene (IPR = isolated pentagon rule).]

14. Taking the following symmetry operators as generator, construct the point groups and verify the four group properties: (a) $\hat{S}_4$ (b) $\hat{S}_3$. Also verify that no two symmetry operators of each of the generated groups belong to the same conjugate class. What type of point group does have this feature?

15. The following is a multiplication table.

⊙	A	B	C	D
A	B	A	D	C
B	A	B	C	D
C	D	C	B	A
D	C	D	A	B

Verify that the set (A, B, C, D) forms a group under the binary operation ⊙. Obtain the results (a) ABC^{-1}, (b) $B^{-1}DB$ and (c) $A^{-1}CA$. Show that the group is Abelian.

16. Show that there can be only one group of order 3. Show also that this group is cyclic.

17. An object has a four-fold rotation axis, a two-fold rotation axis perpendicular to it and a σ_h plane. What are the possible symmetry operations that constitute the point group of the object? What is the point group?

18. Draw a stereographic projection of the C_{5h} point group and generate the complete list of symmetry operations.

19. Draw (a) a stereographic projection of the C_{4v} point group and construct the multiplication table for the group, (b) the stereograms of the C_{4h} and S_4 groups to show their difference.

20. Draw a stereographic projection of the point group of the BF_3 molecule and construct the multiplication table using it. [You will feel the ease with which you can do it now, if you have solved Question 4(c).]

21. Construct the multiplication table and then find the conjugate classes of symmetry operators in the following point groups.
(a) C_{4v} (b) the point group of BF_3
(c) the point group of benzene molecule.

22. Two groups are said to be isomorphic if they have the same multiplication table. Show that (a) the point groups C_{2v}, C_{2h} and D_2 are isomorphic. (b) the point groups S_4 and C_4 are isomorphic. (**Hint**: rename the symmetry operations of the groups as $\hat{A}$, $\hat{B}$, $\hat{C}$, $\hat{D}$; construct the multiplication tables and observe their similarity.)

23. Although the line of intersection of the reflection planes of the H_2O molecule acts as a rotation axis, the two reflection operations are not conjugate. Why?

24. In cyclopentadienyl anion all the five vertical reflection planes are in the same class but those of the benzene molecule are in two different conjugate classes. Why?

25. Explain, using symmetry argument, why BF_3 (a planar molecule) does not have a permanent dipole moment.

26. Which point groups do not allow (a) the existence of permanent dipole moment? (b) optical activity?

27. Explain, by drawing the molecular structure and showing necessary operations, that 3,4,3′,4′-tetramethyl-spiro-(1,1′) bipyrrolidinium ion is optically inactive since it belongs to the S_4 point group.

28. Draw the end-on view of each of the following molecules in a conformation which is neither staggered nor eclipsed and find the point groups:
(a) ethane (b) 1, 1, 1-trichloro ethane

29. Choose the correct option(s):

(a) $\hat{\sigma}'_v \hat{\sigma}''_v$ is equivalent to $\hat{S}^k_n/\hat{C}^k_n/\hat{\sigma}_h/\hat{i}$

(b) $\hat{C}^2_3 \hat{C}_2 \hat{C}_3$ is equivalent to $\hat{C}_2/\hat{E}/\hat{\sigma}_h/\hat{C}^{-1}_3$

(c) $\hat{i}\hat{C}^n_{2n}$ is equivalent to $\hat{\sigma}_h/\hat{i}/\hat{S}_{2n}/\hat{E}$

(d) The point groups S_4 and C_{2h} (i) are Abelian and the same/ (ii) are Abelian but different / (iii) both have i as a symmetry element/ (iv) have the same order.

30. Which of the point groups contain(s) an inversion centre: D_{3h}, D_{4h}, D_{5d}, D_{2h}?

31. Choose the correct option(s): The two-fold rotations perpendicular to the principal axis form
(a) four different classes in D_{4h}
(b) two different classes in D_{5h}
(c) one single class in D_{5h}
(d) two different classes in D_{5d}
(e) one single class in D_{5d}
(f) one single class in D_{6d}
(g) one single class in D_{4h}.

Suggested Books and References for Further Reading

1. Herstein I N, *Abstract Algebra*, Wiley, **1996.**
2. Bishop D M, *Group Theory and Chemistry*, Dover, NY, **1973.**
3. Cotton F A, *Chemical Applications of Group Theory*, John Wiley & Sons, NY, **2003.**
4. Rakshit S C, *Atomic and Molecular Symmetry Groups and Chemistry*, Levant Books, Kolkata, **2012.**

2 Symmetry of the Platonic solids

Learning objectives

- ➢ Platonic solids—reasons for their number being exactly five
- ➢ Point groups of Platonic solids and conjugate classes of their symmetry operators
- ➢ Constructing cardboard models for a tetrahedron and an icosahedron—finding the ideal tetrahedral angle
- ➢ Conjugate pairs of polyhedra

2.1 INTRODUCTION

A Platonic solid is a convex polyhedron of which all faces are equivalent, as are all edges and all vertices. By *equivalent faces* we mean that it is not possible to distinguish one face from the others; the same is true for the vertices and edges. The Greek philosopher Plato suggested that there are five such solid figures—tetrahedron, cube, octahedron, icosahedron and dodecahedron. Their relevance to the study of molecular structure is derived from the fact that tetrahedral, octahedral and pyramidal arrangements of atoms in molecules are the most common. The famous French mathematician Euler proved that for all convex polyhedra, the number of faces (F), vertices (V) and edges (E) must be related as

$$V - E + F = 2 \tag{2.1}$$

Thus, a cube has 8 vertices, 12 edges and 6 faces, that is, $V = 8$, $E = 12$ and $F = 6$ (six squares) so that $V - E + F = 8 - 12 + 6 = 2$. Equation (2.1), known as *Euler's polyhedron formula*, can similarly be verified for all the other Platonic solids. The question of why there should be exactly five Platonic solids, can be explained by a simple geometrical argument and also by symmetry-based arguments. The geometrical argument is as follows.

We start with a well known result of plane geometry—the sum of the interior angles of a polygon of s sides = $(2s - 4)$ right angles. If the polygon is regular, that is, if the s sides are all of the same length, each interior angle = $(2s - 4)/s$ right angles. Now, let the faces of a regular polyhedron be s-sided regular polygons; each vertex of the polyhedron is at the intersection of n such polygons. The sum of the interior angles meeting at a vertex must be less than 2π since a sum of 2π gives a flat surface which cannot enclose a convex polyhedron. For example, each vertex of a cube is at the intersection of three squares and sum of the interior angles at each vertex of the cube is $3 \times 90°$; for a regular tetrahedron every vertex is at the intersection of three equilateral triangles and the sum of the interior angles meeting at each vertex is $3 \times 60°$. In general, for any regular polyhedron, the sum of n interior angles meeting at a vertex $= n \times \dfrac{2s - 4}{s} \times 90°$. Hence,

$$n \times \frac{2s-4}{s} \times \frac{\pi}{2} < 2\pi \tag{2.2}$$

which leads to the condition

$$n < \frac{2s}{s-2} \tag{2.3}$$

Also, n cannot be less than 3 since two polygons by intersection cannot produce a vertex; n must at least be equal to 3.

When $s = 3$ (equilateral triangle), the permitted values of n (condition 2.3) are 3, 4 and 5. The corresponding polyhedra are respectively tetrahedron, octahedron and icosahedron. When $s = 4$ (square), $n < 4$ (condition 2.3) and so only $n = 3$ is permitted. The resulting polyhedron is a cube. When $s = 5$ (regular pentagon), condition (2.3) gives $n < 3.3$ and the permitted value of n is 3; this gives the dodecahedron. For values of $s \geq 6$, we find $n < 3$ which is not permitted. It is thus shown that only five Platonic solids are possible.

Shapes and symmetry of the Platonic solids are discussed in the following sections.

2.2 CONSTRUCTION OF A TETRAHEDRON FROM A CUBE—A CARDBOARD MODEL

A tetrahedron has four equilateral triangles as four equivalent faces, four equivalent vertices and six equivalent edges (Fig. 2.1).

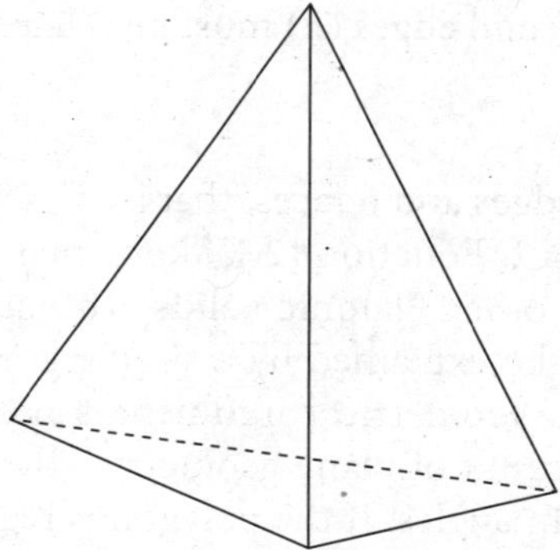

Fig. 2.1 A regular tetrahedron

Its relationship with the structure of a cube is easily understood from the following construction. On a piece of cardboard draw a square S of some edge length l (say, 4 cm) and draw four squares labelled 1, 2, 3, 4, one on each side of S (Fig. 2.2). After cutting out the resulting figure, fold the four peripheral squares vertically with respect to the plane of S to obtain a hollow (without a lid) cube. The diagonal of each face of this cube is $l\sqrt{2}$. Paste the edges with paper to hold the walls tight.

From another piece of cardboard, cut out an equilateral triangle MQR of edge length $2l\sqrt{2}$. Join the midpoints of the edges to get a smaller equilateral triangle D of edge length $l\sqrt{2}$. Fold

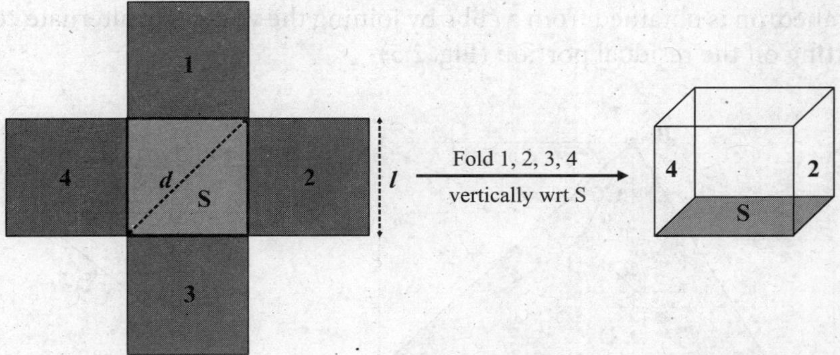

Fig. 2.2 Construction of a cube (topless) of edge length *l*

the equilateral triangles on the periphery of D such that the vertices M, Q and R meet at a common point (Fig. 2.3).

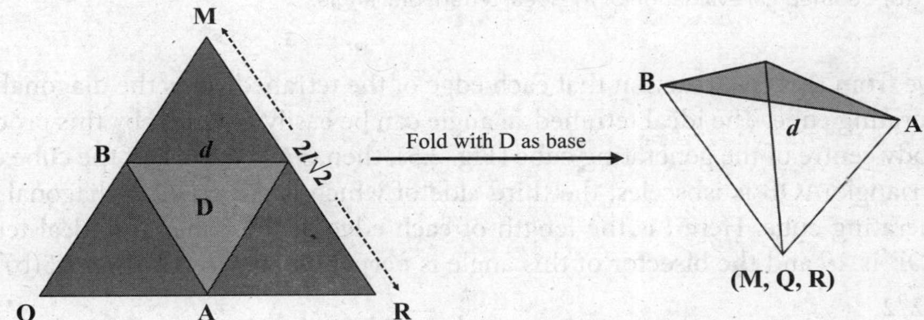

Fig. 2.3 Construction of a regular tetrahedron of edge length $l\sqrt{2}$

A regular tetrahedron is now produced, each edge of which is of length $l\sqrt{2}$ (which is equal to the length (d) of each face diagonal of the cube in Fig. 2.2. The tetrahedron thus constructed can now be placed in the hollow cube in such a way that the six edges of the tetrahedron are along the diagonals of the six faces of the cube and the four vertices of the tetrahedron are on alternate corners of the cube (Fig. 2.4).

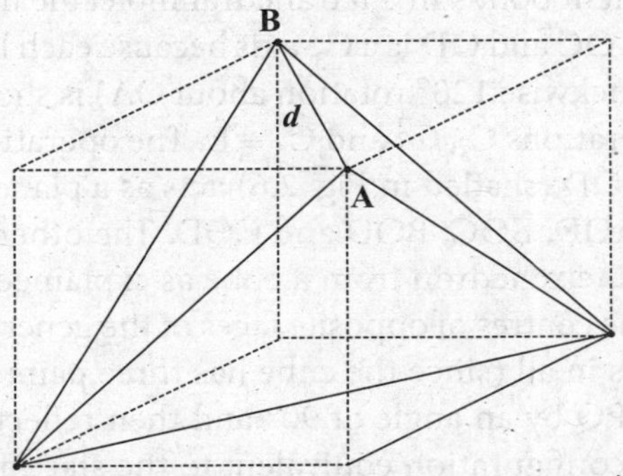

Fig. 2.4 A regular tetrahedron fitted into a cube

Hence. a tetrahedron is obtained from a cube by joining the vertices at alternate corners of the cube and cutting off the residual portion (Fig. 2.5).

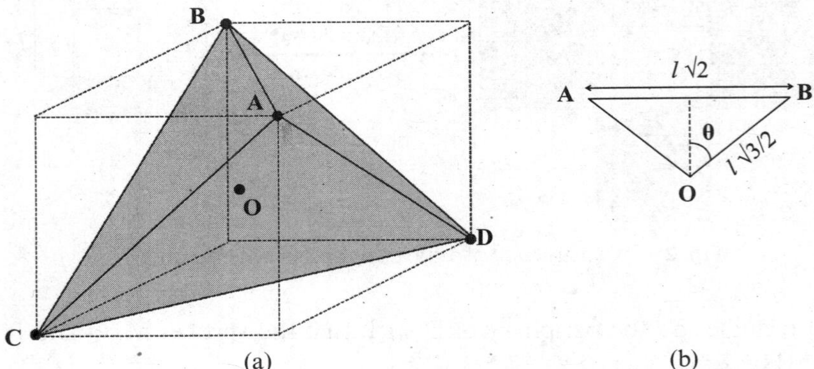

Fig. 2.5 (a) A tetrahedron obtained from a cube by taking the vertices at alternate corners of the cube. (b) Geometrical evaluation of the ideal tetrahedral angle.

We observe from this construction that each edge of the tetrahedron is the diagonal of a face of the generating cube. The ideal tetrahedral angle can be easily obtained by this procedure. If O is the body centre of the generating cube (Fig. 2.5), then OA = OB = half the cube diagonal. Thus the triangle AOB is isosceles, the third side of which is $AB = l\sqrt{2}$ = diagonal of a face of the generating cube. Here l is the length of each edge of the cube. The ideal tetrahedral angle $\angle AOB$ is 2θ and the bisector of this angle is perpendicular to AB [Fig. 2.5(b)]. Hence, $\sin\theta = \dfrac{l\sqrt{2}/2}{l\sqrt{3}/2}$ which gives $\angle AOB = 2\theta = 2\sin^{-1}\sqrt{2/3} = 109°28'$. This is the well known $H\hat{C}H$ bond angle in the tetrahedral CH_4 molecule.

2.3 SYMMETRY ELEMENTS OF A TETRAHEDRON

Drawing a tetrahedron from a cube reveals all the symmetry elements (Figs 2.6 and 2.7). By joining the vertices of the tetrahedron to its centre we get the tetrahedral skeleton (Fig. 2.6), which shows the usual covalent bonds in a tetrahedral molecule like CH_4.

Each of the lines OA, OB, OC and OD is a C_3-axis because each line is along a cube diagonal. One $\hat{C}_3$ operation (an anticlockwise 120° rotation about OA) is shown in Fig. 2.6. Each C_3-axis generates the symmetry operations $\hat{C}_3$, $\hat{C}_3^2$ and $\hat{C}_3^3 \equiv \hat{E}$. The operations $\hat{C}_3$ and $\hat{C}_3^2$ are conjugate to one another. The plane AOD (shaded in Fig. 2.6) acts as a plane of symmetry. There are six such planes—AOB, AOC, AOD, BOC, BOD and COD. The other elements of symmetry can be understood by drawing a tetrahedron from a cube as explained earlier. The line PQ of Fig. 2.7, which passes through the centres of opposite faces of the generating cube, acts as a C_2-axis. There are three such C_2-axes in all (since the cube has three pairs of opposite faces). Again, if one rotates the cube about PQ by an angle of 90° and then reflects on a plane perpendicular to PQ (Fig. 2.7), one gets a configuration equivalent to the starting one. Hence PQ acts as an S_4-axis also.

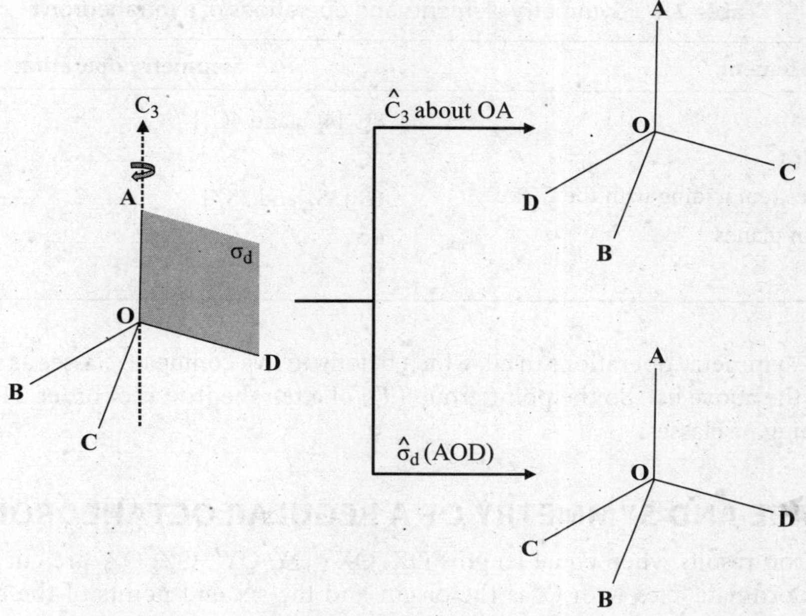

Fig. 2.6 Skeleton of a tetrahedral structure and $\hat{C}_3$ and $\hat{\sigma}_d$ operations

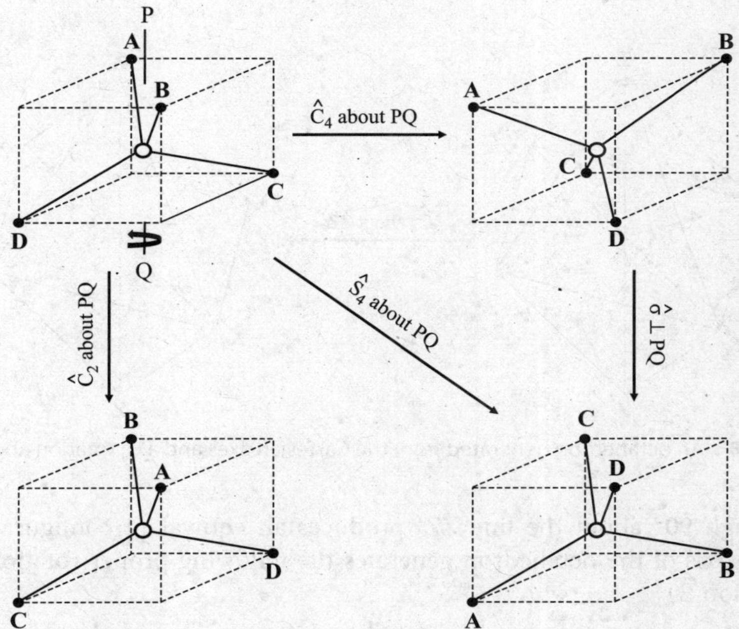

Fig. 2.7 A C_2-axis (PQ) and an S_4-axis collinear with C_2

The elements of symmetry and the corresponding symmetry operations of a regular tetrahedron are as listed in Table 2.1.

Table 2.1 Symmetry elements and operations of a tetrahedron

Symmetry element	Symmetry operation
Four C_3-axes	$8\hat{C}_3[4\hat{C}_3$ and $4\hat{C}_3^2]$
Three C_2-axes	$3\hat{C}_2$
Three S_4-axes (coinciding with the C_2-axes)	$6\hat{S}_4[3\hat{S}_4$ and $3\hat{S}_4^3]$
Six reflection planes	$6\hat{\sigma}_d$
Identity	$\hat{E}$

There are 24 symmetry operations in all, which belong to five conjugate classes as shown in the five rows of the above list. So the point group (T_d) of a tetrahedron is of order 24 and has five possible conjugate classes.

2.4 SHAPE AND SYMMETRY OF A REGULAR OCTAHEDRON

An octahedron results when equal lengths OX, OX′, OY, OY′, OZ, OZ′ are cut off from the Cartesian coordinate axes with O as the origin and the six end points of the cut off parts, namely, X, X′, Y, Y′, Z, and Z′ are joined by straight line segments as shown in Fig. 2.8. The resulting figure contains eight equilateral triangles as the external faces and hence the name *octahedron.*

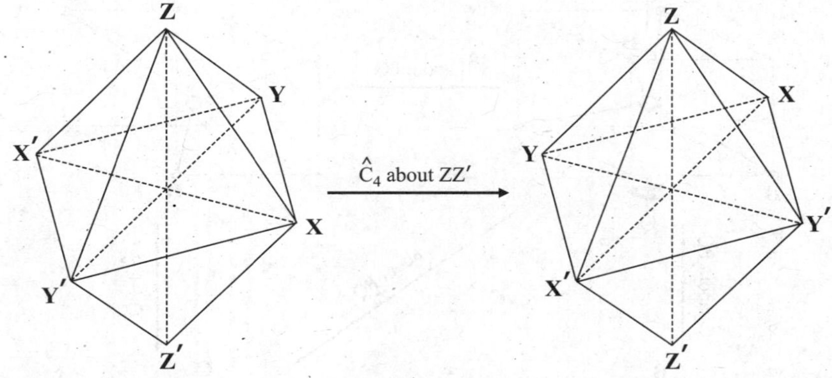

Fig. 2.8 An octahedron generated from the Cartesian axes and a $\hat{C}_4$ rotation about ZZ′

Rotation through 90° about the line ZZ′ produces an equivalent configuration (Fig. 2.8). Hence the ZZ′ axis of the octahedron generates the following proper rotations (besides the identity operation $\hat{E}$).

$$\hat{C}_4,\ \hat{C}_4^2 = \hat{C}_2,\ \hat{C}_4^3\ (\text{and } \hat{C}_4^4 = \hat{E})$$

The XX′ and YY′ lines are similar four-fold rotation axes for the octahedron. Again, with the help of the XY plane which is perpendicular to the Z-axis, the following improper rotations are also generated by the ZZ′ axis.

$\hat{S}_4 = \hat{C}_4\,\hat{\sigma}(XY),\ \hat{S}_4^2 = \hat{C}_4^2[\hat{\sigma}(XY)]^2 = \hat{C}_2$ (already considered)

$\hat{S}_4^3 = \hat{C}_4^3[\hat{\sigma}(XY)]^3 = \hat{C}_4^3[\hat{\sigma}(XY)]$ and $\hat{S}_4^4 = \hat{C}_4^4[\hat{\sigma}(XY)]^4 = \hat{E}$ (already considered).

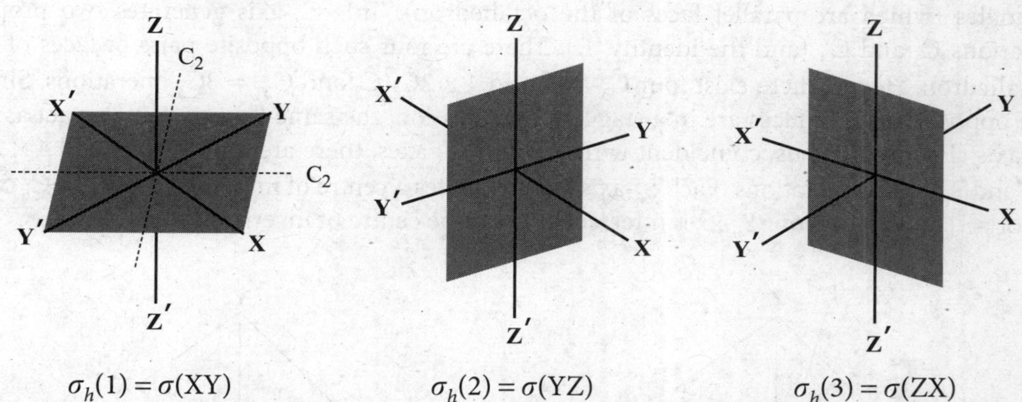

Fig. 2.9 Three σ_h planes of an octahedron. A pair of C_2 axes perpendicular to the Z-axis is shown. There are two more pairs of C_2 axes, each perpendicular to the X- and Y-axes, which are not shown in this figure for the sake of clarity.

The same operations are generated by the XX′ and YY′ axes also. Hence with all the three coordinate axes the following proper and improper rotations are possible for a regular octahedron:

$$3 \times 2\hat{C}_4(\hat{C}_4 \text{ and } \hat{C}_4^3),\ 3 \times 2\hat{S}_4(\hat{S}_4 \text{ and } \hat{S}_4^3),\ 3\hat{C}_2$$

That is, $6\hat{C}_4$, $6\hat{S}_4$ and $3\hat{C}_2$ (together with the identity operation, $\hat{E}$).

There are three σ_h planes of symmetry, each containing four vertices of the octahedron and perpendicular to the axis joining the remaining two vertices (Fig. 2.9). There are two C_2 axes perpendicular to ZZ′ (Fig. 2.9). Thus, considering the three coordinate axes there are $3 \times 2 = 6C_2$ axes not collinear with the coordinate axes. There are two reflection planes containing the ZZ′ axis bisecting the opposite pairs of edges of the square plane XYX′Y′ (Fig. 2.10).

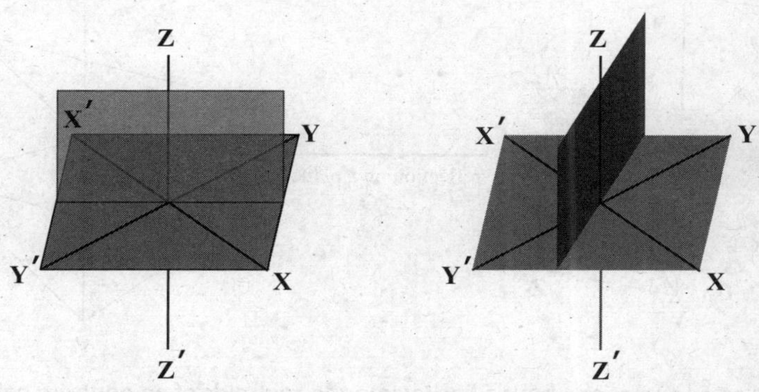

Fig. 2.10 Two σ_d planes containing the ZZ′ axis

These symmetry planes are vertical in nature as they contain the principal axis C_4 (ZZ′). But they bisect the angles between the C_2 axes, XX′ and YY′. Hence they are σ_d. From ZZ′ we get two σ_d planes. Considering all the three coordinate axes we get a total of $3 \times 2 = 6\sigma_d$ planes.

A C_3-axis can be drawn (Fig. 2.11) through the centroids of an opposite pair of equilateral triangles (which are parallel faces of the octahedron). This C_3-axis generates two proper rotations $\hat{C}_3$ and $\hat{C}_3^2$ (and the identity, $\hat{E}$). There are four such opposite pairs of faces of an octahedron. Hence, there exist four C_3-axes and $4 \times 2\hat{C}_3(\hat{C}_3$ and $\hat{C}_3^2) = 8\hat{C}_3$ operations. Since the opposite pairs of faces are in a staggered orientation, the same C_3-axis can also act as an S_6-axis (Fig. 2.11). Thus, coincident with the four C_3-axes, there are four S_6-axes and $4 \times 2\hat{S}_6$ ($\hat{S}_6$ and $\hat{S}_6^5) = 8\hat{S}_6$ operations. Each S_6-axis also generates a centre of inversion i, [as $\hat{S}_6^3 = \hat{C}_6^3\hat{\sigma}_h = \hat{C}_2\hat{\sigma}_h = \hat{i}$] and all the four S_6-axes intersect at the same centre of inversion.

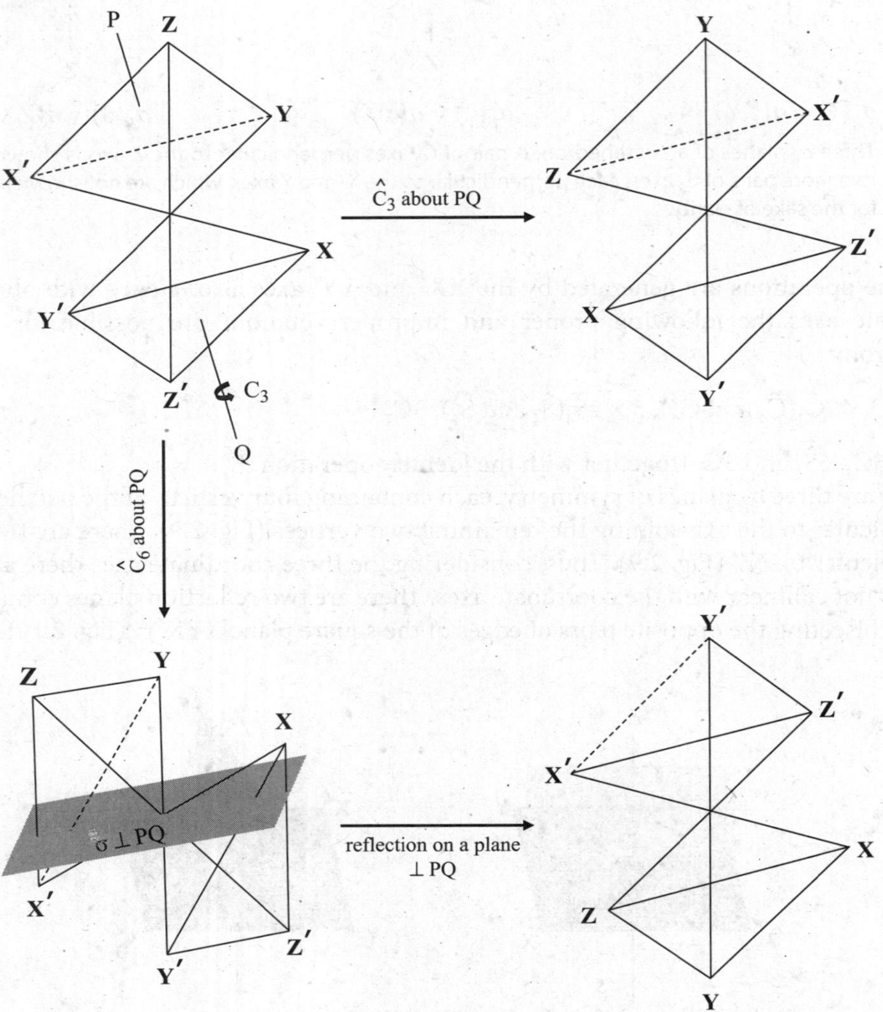

Fig. 2.11 A $\hat{C}_3$ and an $\hat{S}_6$ operation about a line joining the centroids of an opposite pair of triangular faces of an octahedron

The total list of symmetry operations of an octahedron, therefore, is as follows:

$\hat{E}$(Class 1), $8\hat{C}_3$(Class 2), $8\hat{S}_6$(Class 3), $6\hat{C}_4$(Class 4), $3\hat{C}_2 = \hat{C}_4^2$ (Class 5), $6\hat{S}_4$ (Class 6), $3\hat{\sigma}_h$(Class 7), $6\hat{\sigma}_d$(Class 8), $\hat{i}$ (Class 9) and $6\hat{C}_2$ (Class 10). So an octahedron has 48 symmetry operations belonging to 10 conjugate classes. The point group is O_h; it is of order 48 and has 10 conjugate classes.

2.5 SYMMETRY OF A REGULAR ICOSAHEDRON

To understand the structure of regular icosahedrons, it will be helpful if we have a cardboard model which can be constructed easily as follows. Draw a regular pentagon (marked 1) on a cardboard and on each side of it, draw an equilateral triangle (marked T_1).

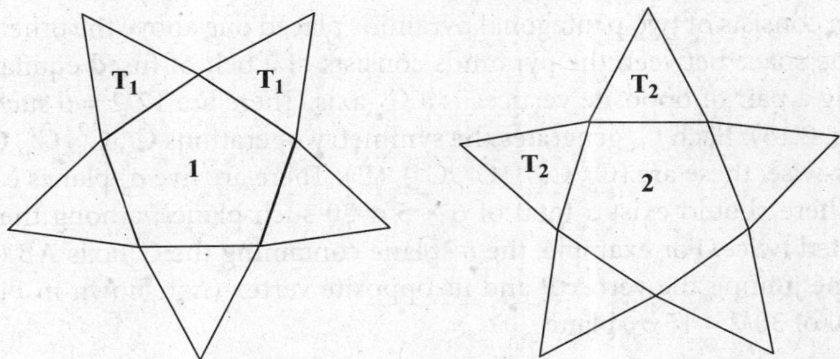

Fig. 2.12 A pair of five-vertex stars

Let us call these triangles *teeth* of pentagon 1 (Fig. 2.12). Similarly on another piece of cardboard draw another pentagon of the same size (marked 2) and its teeth (each marked T_2). These will look like a pair of five-vertex stars (Fig. 2.12). Cut out the stars and place the pentagons one above the other in a staggered orientation (Fig. 2.13).

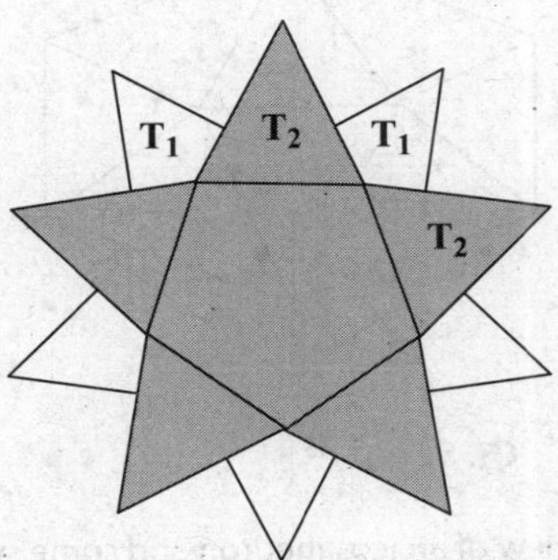

Fig. 2.13 Two five-vertex stars placed one above the other in a staggered manner

Then fold the teeth such that each T_1 points towards the plane of the pentagon 2, and each T_2 points towards the plane of the pentagon 1; join them together. This will give a pentagonal antiprism structure (Fig. 2.14).

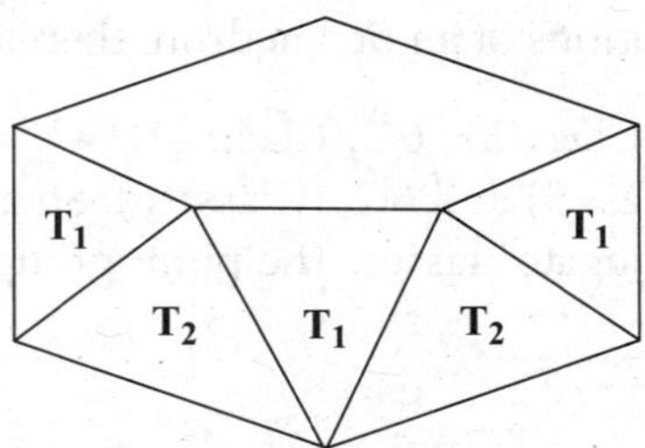

Fig. 2.14 Pentagonal antiprism

If we now construct the pentagonal pyramids above and below this antiprism, the icosahedron (I_h) is complete (Fig. 2.15). We see that I_h contains 12 equivalent vertices, 20 equilateral triangles as faces and 30 equivalent edges, consistent with Euler's polyhedron formula. In brief, an icosahedron consists of two pentagonal pyramids placed one above the other in a staggered manner and the space between the pyramids consists of a belt of fused equilateral triangles. The line joining a pair of opposite vertices is a C_5-axis. There are 12/2 = 6 such C_5-axes (one is shown in Fig. 2.15). Each C_5 generates the symmetry operations $\hat{C}_5$, $\hat{C}_5^2$, $\hat{C}_5^3$, $\hat{C}_5^4$ and $\hat{C}_5^5 = \hat{E}$. Arranged class-wise, these are ($\hat{C}_5$, $\hat{C}_5^4$), ($\hat{C}_5^2$, $\hat{C}_5^3$), ($\hat{E}$). There are five σ_v planes each containing a C_5-axis. So there should exist a total of $6 \times 5 = 30$ such planes; among them, each plane has been counted twice. For example, the σ_v plane containing the C_5-axis AB (Fig. 2.15) also contains the one joining the vertex P and its opposite vertex (not shown in Fig. 2.15). Thus, there are a total of 30/2 = 15 σ_v-planes.

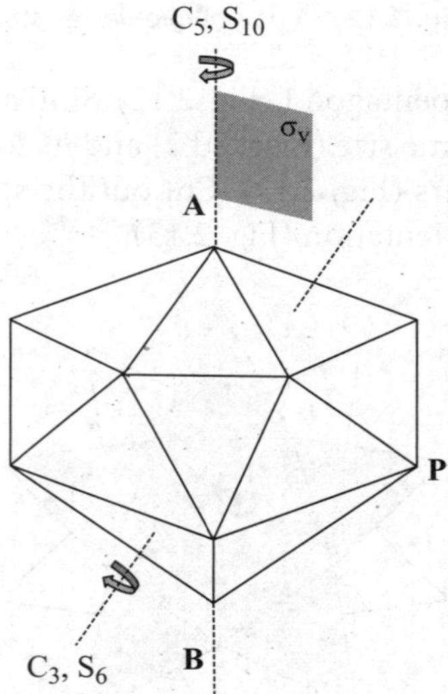

Fig. 2.15 Front view of an icosahedron and some symmetry elements

As the oppositely situated pentagons are staggered, each C_5-axis also acts as an S_{10}-axis and every such axis generates the operations: $\hat{S}_{10}$, $\hat{S}_{10}^3$, $\hat{S}_{10}^5 = \hat{S}_2 = \hat{i}$, $\hat{S}_{10}^7$ and $\hat{S}_{10}^9$. (Even powers of $\hat{S}_{10}$ result in pure $\hat{C}_5$-rotations and $\hat{E}$ which have already been considered.) Arranged class-wise, the improper rotations generated by each of the six S_{10} axes are ($\hat{S}_{10}$, $\hat{S}_{10}^9$), ($\hat{S}_{10}^3$, $\hat{S}_{10}^7$) and ($\hat{i} = \hat{S}_{10}^5$).

Joining the centroids of the opposite pairs of equilateral triangles one gets a C_3-axis. There are 20/2 = 10 such pairs of triangles and so $10C_3$ axes (one is shown in Fig. 2.15). Every C_3-axis generates the operations ($\hat{C}_3$, $\hat{C}_3^2$) and ($\hat{E}$). Since the oppositely situated triangles are staggered, the C_3-axes also act as S_6-axes; thus there are $10S_6$-axes and each of them generates the operations $\hat{S}_6$, $\hat{S}_6^3 = \hat{S}_2 = \hat{i}$, $\hat{S}_6^5$ (and $\hat{S}_6^6 = \hat{E}$). Arranged class-wise, these are ($\hat{S}_6$, $\hat{S}_6^5$) (and $\hat{i}$ which is the same as the one produced by the S_{10} axes).

Joining the centres of oppositely situated edges of the icosahedron, one gets a C_2 axis (Fig. 2.16). There are 30/2 = 15 such pairs of opposite edges and hence $15C_2$-axes. In a regular icosahedron, one can find three distinct *belt* regions, each containing five C_2 axes (Fig. 2.16). Each of the $15C_2$-axes generates a $\hat{C}_2$ operation (besides $\hat{C}_2^2 = \hat{E}$). Table 2.2 shows the complete list of symmetry operations of the point group I_h of a regular icosahedron.

Table 2.2 Symmetry elements and symmetry operations of an icosahedron

Symmetry elements (no. of such elements)	Symmetry operations	No. of conjugate classes consisting of the symmetry operations
E	$\hat{E}$	1
i	$\hat{i}$	1
C_5-axis (6 in number)	$6(\hat{C}_5, \hat{C}_5^4), 6(\hat{C}_5^2, \hat{C}_5^3)$	2
S_{10}-axis (6 in number)	$6(\hat{S}_{10}, \hat{S}_{10}^9), 6(\hat{S}_{10}^3, \hat{S}_{10}^7)$	2
C_3-axis (10 in number)	$10(\hat{C}_3, \hat{C}_3^2)$	1
S_6-axis (10 in number)	$10(\hat{S}_6, \hat{S}_6^5)$	1
C_2-axis (15 in number)	$15\hat{C}_2$	1
σ_v planes (15 in number)	$15\hat{\sigma}_v$	1
	Total number of symmetry operations = 120	Total number of conjugate classes = 10

Thus the point group (I_h) of a regular icosahedron is of order 120 and has 10 conjugate classes.

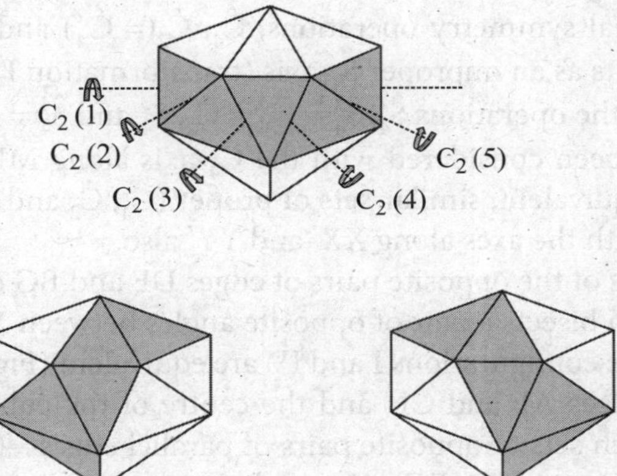

Fig. 2.16 Three belt regions of an icosahedron; five two-fold symmetry axes are shown for one belt

2.5.1 Truncation of the Icosahedron and the [60]Fullerene Structure

If the icosahedron is truncated by 12 planes at each of the 12 vertices, 12 pentagonal faces will be generated and they will be separated by 20 hexagons. Two such hexagons are shown in Fig. 2.17(a) (shaded). The resulting figure contains 60 vertices and if these 60 sites are occupied by sp^2 hybridised C atoms (one at each vertex) we get the structure of buckminsterfullerene—C_{60} [Fig. 2.17(b)]. The structure is just like that of a football. Six pairs of pentagons are arranged with opposite orientation at the ends of six diameters of the football—three of these diameters are the X, Y and Z axes and the other three are along the meridonial directions. The [60]fullerene structure, therefore, has the same symmetry operations as those of the original icosahedron. Hence the point group of [60]fullerene is I_h.

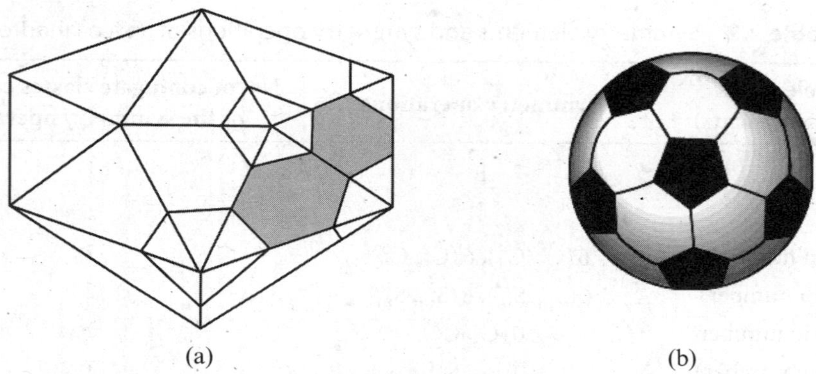

Fig. 2.17 (a) Truncated icosahedron (only three truncations are shown)
(b) Buckminsterfullerene (now called [60]fullerene)

2.6 SYMMETRY OF A CUBE

Let ABCDEFGH be a cube. We can imagine a set of mutually perpendicular line segments XX′, YY′ and ZZ′ intersecting at the centre of the cube (Fig. 2.18). There is an axis along ZZ′ which can act as a C_4 rotational axis since the configurations I and II are equivalent. A C_4-axis generates three rotational symmetry operations, $\hat{C}_4$, $\hat{C}_2(=\hat{C}_4^2)$ and $\hat{C}_4^3$ (besides the identity, $\hat{C}_4^4 = \hat{E}$). The line also acts as an improper S_4-axis (transformation I → III, Fig. 2.18).

An S_4-axis generates the operations $\hat{S}_4$, $\hat{S}_4^2 = \hat{C}_4^2 = \hat{C}_2$, S_4^3 and $\hat{S}_4^4 = (\hat{C}_4\hat{\sigma}_\perp)^4 = \hat{E}$. Out of these, $\hat{C}_2$ and $\hat{E}$ have already been considered with the C_4-axis along MN. As the coordinate axes XX′, YY′ and ZZ′ are equivalent, similar sets of proper ($\hat{C}_4$, $\hat{C}_2$ and $\hat{C}_4^3$) and improper ($\hat{S}_4$, $\hat{S}_4^3$) rotations are possible with the axes along XX′ and YY′ also.

Joining the midpoints of the opposite pairs of edges DE and BG one gets a straight line PQ (on the XY plane) which bisects a pair of opposite angles between XX′ and YY′. This line PQ acts as a C_2 axis since the configurations I and IV are equivalent (Fig. 2.19). Similarly, through the midpoints of the edges AF and CH and the centre of the cube, another C_2 axis can be drawn. There are six such sets of opposite pairs of parallel edges—(DE, BG), (AF, CH), (AD, GH), (BC, EF), (AB, EH) and (CD, FG). Hence there are six C_2 axes (other than the three C_2 axes resulting from C_4^2 about the three coordinate axes mentioned earlier).

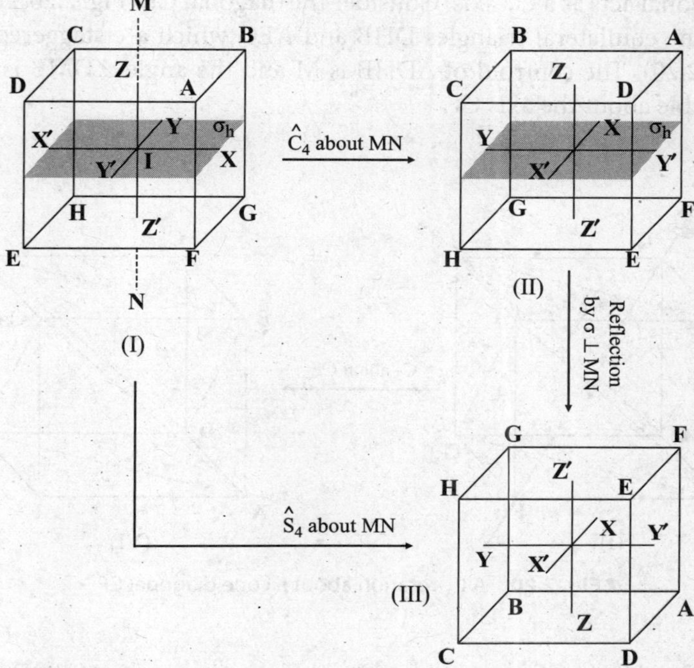

Fig. 2.18 Proper and improper rotations of a cube about the Z-axis

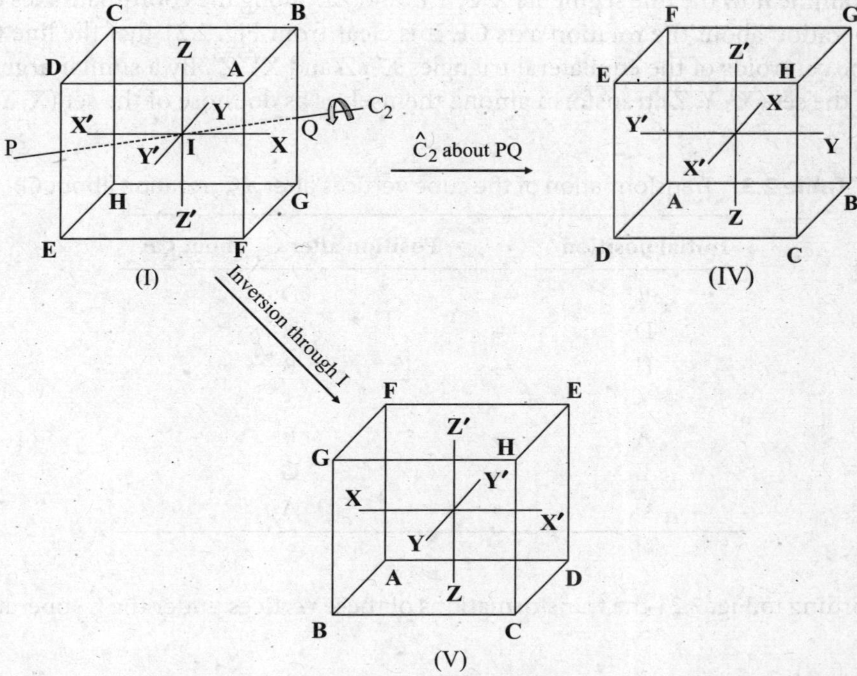

Fig. 2.19 Two-fold rotation about PQ and inversion through the centre I of the cube

Each cube diagonal acts as a C_3-axis. Consider the diagonal CF (Fig. 2.20). It passes through the centroids of the equilateral triangles DHB and AEG which are staggered with respect to each other (Fig. 2.20). The centroid of ΔDHB is M and the angle $\angle$DMB is 120°. Thus a $\hat{C}_3$ operation is possible about the axis CF.

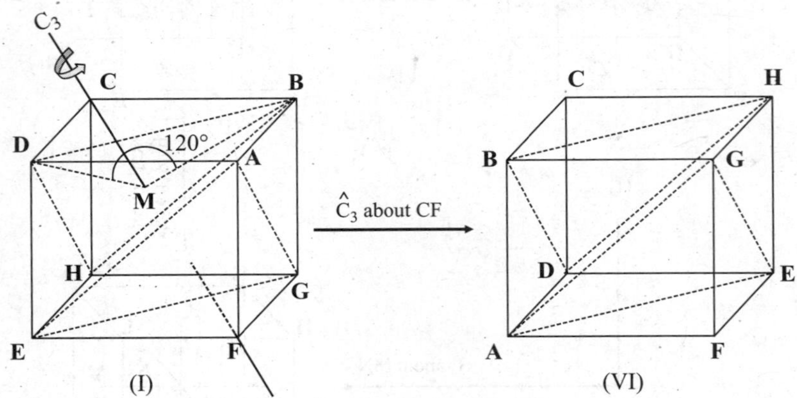

Fig. 2.20 A $\hat{C}_3$-rotation about a cube diagonal CF

During a $\hat{C}_3$ rotation about CF, the vertices C and F remain unmoved, B, D and H transform among themselves as do A, E and G (Table 2.3).

Such transformations convert configuration I to the equivalent configuration VI (Fig. 2.20). We now examine how the line segments XX′, YY′ and ZZ′ along the coordinate axes transform by a $\hat{C}_3$ operation about the rotation axis CF. It is clear from Fig. 2.21 that the line CF passes through the centroids of the equilateral triangles X′YZ and XY′Z′. By a similar argument, the vertices of the set (X′, Y, Z) transform among themselves as do those of the set (X, Y′, Z′).

Table 2.3 Transformation of the cube vertices after a $\hat{C}_3$ rotation about CF

Initial position		Position after $\hat{C}_3$ about CF
B	→	D
D	→	H
H	→	B
A	→	E
E	→	G
G	→	A

Thus, according to Fig. 2.21 the transformations of these vertices under the $\hat{C}_3$ operation about CF are:

$$Y \rightarrow Z, Z \rightarrow X', X' \rightarrow Y \quad \text{and} \quad X \rightarrow Y', Y' \rightarrow Z', Z' \rightarrow X$$

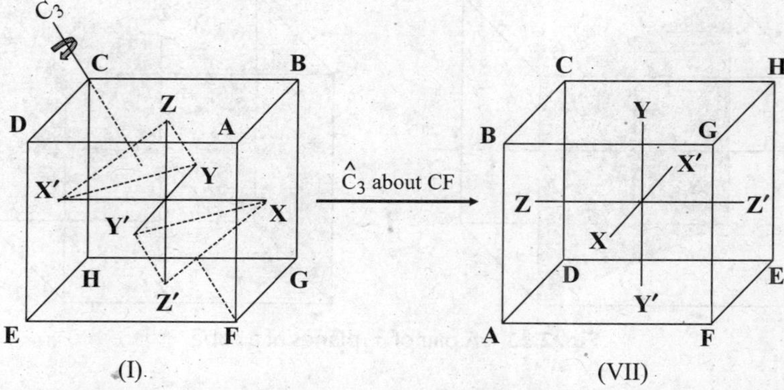

Fig. 2.21 Transformation of the coordinate axes and the cube vertices by $\hat{C}_3$ operation about CF

Then the line segment XX′ after $\hat{C}_3$ operation about CF becomes YY′, YY′ becomes ZZ′ and ZZ′ becomes X′X; the coordinate lines do not break, only their orientations change, maintaining mutual orthogonality as before. There are four cube diagonals and hence $8\hat{C}_3(4\hat{C}_3$ and $4\hat{C}_3^2)$ operations are possible.

Each cube diagonal also acts as an S_6 axis since the triangles DHB and AEG are staggered. So with four cube diagonals four $\hat{S}_6$ and four $\hat{S}_6^5$ operations are possible. The centre of the cube acts as the centre of symmetry (Fig. 2.19). There is a plane of symmetry perpendicular to the ZZ′ line (which is a C_4 axis) and containing the XX′ and YY′ lines (Fig. 2.22). This is a σ_h plane since $\hat{C}_4$ is the highest-fold rotation possible for a cube. There are two other such σ_h planes, one perpendicular to XX′ and the other to YY′.

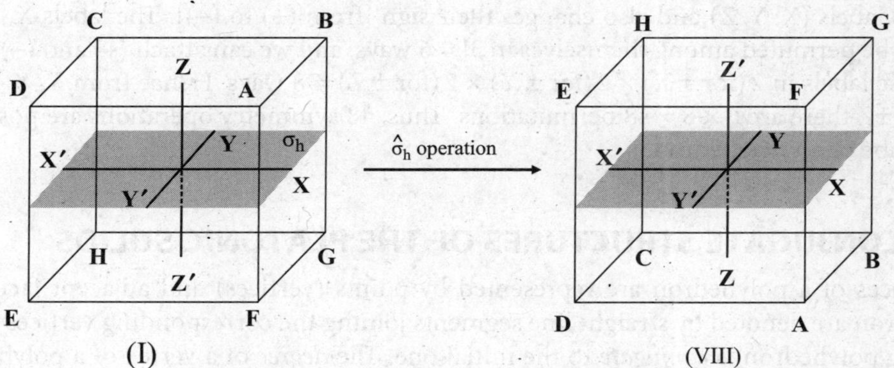

Fig. 2.22 Reflection on a σ_h [$= \sigma$(XY)] plane of a cube

There is one reflection plane, which is a σ_d plane, containing the ZZ′ line and each of the diagonals of the top and bottom faces. One pair of such σ_d planes is shown in Fig. 2.23.

There are three pairs of opposite faces and hence six σ_d planes are possible. The complete list of symmetry operations of a cube, arranged class-wise, is:

$(\hat{E})$, $(6\hat{C}_4$, that is, $3\hat{C}_4$ and $3\hat{C}_4^3)$, $(3\hat{C}_2 = 3\hat{C}_4^2)$, $(6\hat{C}_2)$, $(6\hat{S}_4$, that is, $3\hat{S}_4$ and $3\hat{S}_4^3)$, $(\hat{i})$, $(3\hat{\sigma}_h)$, $(6\hat{\sigma}_d)$, $(8\hat{C}_3$, that is, $4\hat{C}_3$ and $4\hat{C}_3^2)$, $(8\hat{S}_3$, that is, $4\hat{S}_3$ and $4\hat{S}_3^5)$ – altogether 48 symmetry operations and

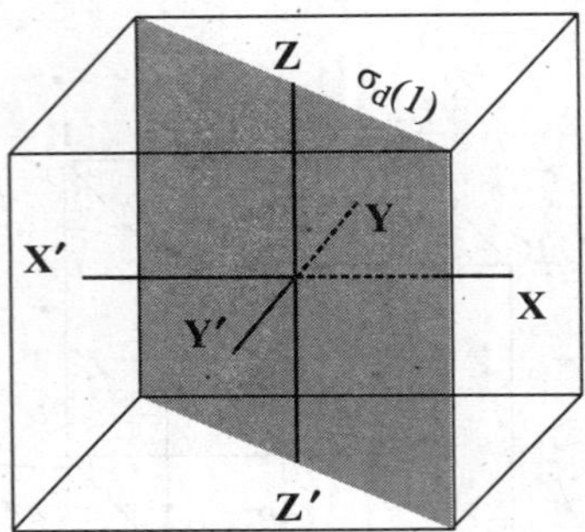

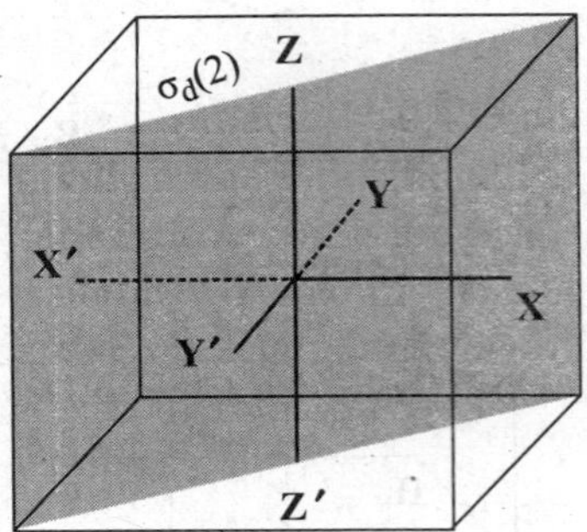

Fig. 2.23 A pair of σ_d planes of a cube

10 conjugate classes which are precisely the same as those of an octahedron. The point group of a cube is thus O_h.

Note:

(i) The symmetry operations and conjugate classes of a cube are the same as those of an octahedron. This is because a cube and an octahedron are structurally *conjugate* as explained in the next section.

(ii) It is easy to understand why there should be 48 symmetry operations for a cube or an octahedron. A system of Cartesian coordinate axes parallel to the edges of a cube and with the origin at the body centre can be imagined inside a cube (Fig. 2.18). The six points where the axes meet the faces of the cube are the six face-centres, and by joining them we get a regular octahedron (Fig. 2.24). A symmetry operation causes a permutation of the axis labels (X, Y, Z) and also changes their sign [from (+) to (–)]. The labels X, Y and Z can be permuted among themselves in $3! = 6$ ways, and we can attach (+) and (–) signs to these labels in 2 (for $\pm$ X) $\times$ 2 (for $\pm$ Y) $\times$ 2 (for $\pm$ Z) = 8 ways. Hence from X, Y, Z to $\pm$X, $\pm$Y, $\pm$Z there are $6 \times 8 = 48$ permutations. Thus, 48 symmetry operations are possible for a cube or an octahedron.

2.7 CONJUGATE STRUCTURES OF THE PLATONIC SOLIDS

If the faces of a polyhedron are represented by points (vertices) and adjacent faces of the polyhedron are denoted by straight line segments joining the corresponding vertices then the resulting polyhedron is *conjugate* to the initial one. The *degree* of a vertex of a polyhedron is the number of edges meeting at the vertex. In a Platonic solid, each face is a regular polygon of n sides and hence each is adjacent to n faces of the polyhedron; thus, the degree of each vertex in the conjugate is n. Construction of the conjugate of a Platonic solid is easy—take the face-centres and join them by noting the adjacencies of the faces (Fig. 2.24).

The conjugate of a tetrahedron is also a tetrahedron, a cube and an octahedron are conjugate pairs and the conjugate of an icosahedron is a dodecahedron as is evident from Fig. 2.24 and Table 2.4. Conjugate pairs of polyhedra have the same set of symmetry operators and belong to the same point group; the cube–octahedron pair has already been discussed in Sections 2.4 and 2.6.

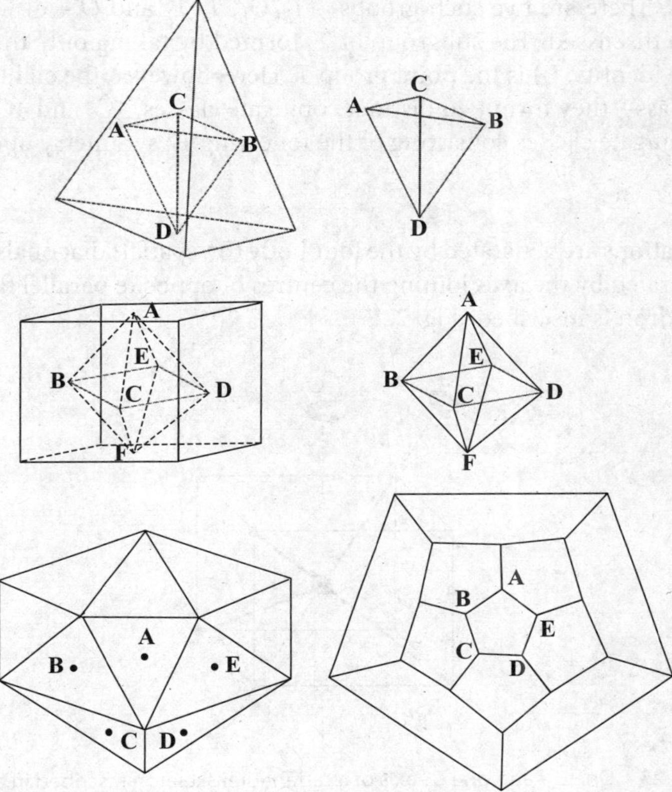

Fig. 2.24 Conjugate pairs of Platonic solids—(tetrahedron, tetrahedron), (cube, octahedron) and (icosahedron, dodecahedron); for the last pair, some face-centres of the icosahedron appearing as vertices of the conjugate dodecahedron are shown

Table 2.4 Relationship among the number and nature of faces and vertex degrees of conjugate pairs of polyhedra

Polyhedron	Conjugate
Tetrahedron: 4 faces (equilateral triangles), 4 vertices each of degree 3	Tetrahedron: 4 faces (equilateral triangles), 4 vertices each of degree 3
Cube: 6 faces (squares), 8 vertices each of degree 3	Octahedron: 8 faces (equilateral triangles), 6 vertices each of degree 4
Icosahedron: 20 faces (equilateral triangles), 12 vertices each of degree 5	Dodecahedron: 12 faces (regular pentagons), 20 vertices each of degree 3

2.8 CUBIC POINT GROUPS

We know that a regular tetrahedron and an octahedron can be inscribed in a cube. Point groups which are formed by taking all or some of the symmetry elements of a cube are called *cubic point groups*. There are five such groups – T_d, O_h, T, T_h and O – of which the first two have already been discussed. The subgroup of T_d formed by taking only the proper rotations (together with the identity, $\hat{E}$) is the point group T. Here, however, the eight $\hat{C}_3$ operations are not in the same class—they form two distinct conjugate classes, $4\hat{C}_3$ and $4\hat{C}_3^2$. Thus, the point group T has 4 conjugate classes consisting of the following 12 symmetry operations:

$$\hat{E},\ 4\hat{C}_3,\ 4\hat{C}_3^2,\ 3\hat{C}_2$$

The three-fold rotations are generated by the four body (or, spatial) diagonals, and the two-fold rotations are generated by the axes joining the centres of opposite parallel faces of the cube in which the tetrahedron is inscribed (Fig. 2.25).

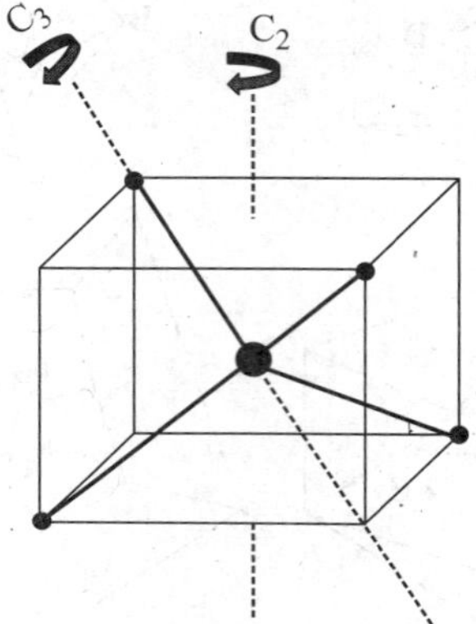

Fig. 2.25 One C_3- and one C_2-axis of a tetrahedral skeleton inscribed in a cube

T_h is the direct product group $T \times C_i$. This means that the symmetry operations of T_h are obtained by multiplying those of T by $(\hat{E}, \hat{i})$; the results of multiplication by $\hat{i}$ are as follows:

$$\hat{i}\,\hat{E} = \hat{i},\ \hat{i}\,\hat{C}_3 = \hat{C}_3\hat{i} = \hat{S}_6^5,\ \hat{i}\,\hat{C}_3^2 = \hat{C}_3^2\hat{i} = \hat{S}_6,\ \hat{i}\,\hat{C}_2 = \hat{C}_2\hat{i} = \hat{\sigma}_h$$

Hence

$$(\hat{E}, 4\hat{C}_3, 4\hat{C}_3^2, 3\hat{C}_2) \times \hat{E} = (\hat{E}, 4\hat{C}_3, 4\hat{C}_3^2, 3\hat{C}_2)$$

$$(\hat{E}, 4\hat{C}_3, 4\hat{C}_3^2, 3\hat{C}_2) \times \hat{i} = (\hat{E}, 4\hat{S}_6^5, 4\hat{S}_6, 3\hat{\sigma}_h)$$

Thus, T_h has 24 symmetry operations (like T_d) and 8 conjugate classes (unlike T_d):

$$(\hat{E}), (4\hat{C}_3), (4\hat{C}_3^2), (3\hat{C}_2), (\hat{i}), (4\hat{S}_6), (4\hat{S}_6^5), (3\hat{\sigma}_h)$$

Note: If two groups $G_a = (\hat{E}, \hat{A}_1, \hat{A}_2, \ldots)$ and $G_b = (\hat{E}, \hat{B}_1, \hat{B}_2, \ldots)$ are such that each element of G_b commutes with all elements of G_a, then the direct product of the sets $G_a \times G_b = (\hat{E}, \hat{A}_1, \hat{A}_2, \ldots, \hat{B}_1, \hat{B}_2, \ldots, \hat{A}_1\hat{B}_1, \hat{A}_2\hat{B}_1, \ldots)$ is also a group. This is called a *direct product group*. Since the operators $\hat{E}$, $\hat{i}$ and $\hat{\sigma}_h$ commute with all point symmetry operations, it is possible to construct direct product groups of C_i $\{= (\hat{E}, \hat{i})\}$ and C_s $\{= (\hat{E}, \hat{\sigma}_h)\}$ with all point groups to generate larger groups. (Details are discussed in Chapter 5.)

The point group O is a subgroup of O_h and consists of the pure rotation operations of the latter, including the identity, $\hat{E}$. Arranged class-wise, these operations are:

$$(\hat{E}), (8\hat{C}_3), (6\hat{C}_2 = 3\hat{C}_4^2), (6\hat{C}_4), (3\hat{C}_2)$$

SUMMARY

- The five regular convex polyhedra—tetrahedron, cube, octahedron, icosahedron and dodecahedron, are collectively called Platonic solids.
- The number of edges (E), faces (F) and vertices (V) of a polyhedron obey Euler's polyhedron formula, $V - E + F = 2$.
- Point groups with more than one high-order rotation axes are highly symmetric. The point group of a cube is the same (O_h) as that of an octahedron; similarly, an icosahedron and a dodecahedron have the same point group (I_h).
- A tetrahedron is obtained by joining the alternate corners of a cube. A cardboard model reveals that a tetrahedron can be inscribed in a cube and that the ideal tetrahedral angle is 109°28′.

LEARNING OUTCOME

After reading this chapter the student will be able to (i) identify the point group of a Platonic solid and prepare the list of its symmetry operations, (ii) reveal the relationships (a) among tetrahedral, octahedral and cubic structures and (b) between the [60]fullerene structure and an icosahedron by preparing simple cardboard models, (iii) carry out symmetry operations on such structures through the use of Cartesian coordinate axes and (iv) show geometrically that the ideal tetrahedral angle is 109°28′.

EXERCISES

1. Explain why a cube is a Platonic solid but a hexagonal prism is not.
2. Verify Euler's formula for the following convex polyhedra—trigonal bipyramid, trigonal prism, hexagonal prism, square pyramid, dodecahedron. Which of these are not Platonic solids and why?
3. If a convex polyhedron is truncated at a number of vertices, how are the changes ΔV, ΔE and ΔF in the number of vertices, edges and faces related? Verify the relation in case of the following figures: (a) a tetrahedron truncated at one vertex, (b) a tetrahedron truncated at two vertices, (c) an octahedron truncated at one vertex, (d) an octahedron truncated at two vertices.
4. Prove that any convex polyhedron with only pentagonal and hexagonal faces can have exactly 12 pentagons and the number of hexagons is arbitrary.
5. Fullerene structures are cages containing pentagonal and hexagonal carbon rings. Find the number of pentagons and hexagons in C_{60} and C_{70} fullerene structures using Euler's formula.

6. Draw a cube with labelled vertices and show the $\hat{C}_4$, $\hat{S}_4^3$ and $\hat{\sigma}_d$ operations using one axis/plane in each.

7. Draw an octahedron with labelled vertices and show one $\hat{C}_3$ operation.

8. For the CH_4 molecule, (a) list the pairs of symmetry operations which commute; (b) show that presence of a C_4-axis and a perpendicular symmetry plane is not necessary for the existence of an S_4-axis.

9. If one pair of opposite parallel faces of an octahedron is coloured orange and the other faces are kept white, what are the possible symmetry elements and what is the point group of the object?

10. If one face of a tetrahedron is coloured orange and the other faces are kept white, what are the possible symmetry elements and what is the point group of the object?

11. Show that the maximum possible order of a rotational symmetry axis for a Platonic solid is 5 and determine which Platonic solid it is.

12. Choose the correct option(s):
    (a) If a regular octahedron is pulled along one axis, the point group reduces from O_h to: (i) O (ii) D_{4h} (iii) D_{2d} (iv) C_{2h}.
    (b) If a regular tetrahedron is truncated at one vertex by a plane parallel to the opposite face, the point group reduces from T_d to: (i) T (ii) C_{3v} (iii) C_{3h} (iv) D_{3d}.
    (c) If a right circular cone is truncated at the vertex by a plane parallel to the base, the point group
    (i) reduces to $D_{\infty d}$ (ii) becomes $C_{\infty v}$ (iii) does not change (iv) becomes $C_{\infty h}$.
    (d) If half of a cube is chopped off by a plane through the diagonals of the top and bottom faces, the point group of the remaining portion is (i) D_{2h} (ii) C_{2h} (iii) C_{3h} (iv) C_{2v}.
    (e) A regular octahedron is truncated at a vertex by a plane perpendicular to the axis passing through that vertex. The point group of the resulting object is (i) D_{4d} (ii) C_{4h} (iii) C_{4v} (iv) D_{3d}.

Suggested books and references for further reading

1. Ladd M, *Symmetry and Group Theory in Chemistry*, Horwood Publishing Limited, Chichester, England, **1998**.
2. Euler L, *Akademia nauk Leningrad*, 4 (109) **1752**.
3. Hall L H, *Group Theory and Symmetry in Chemistry*, McGraw-Hill, NY, **1969**.
4. Kroto H W, Heath J R, O'Brien S C, Curl R F and Smalley R E, *Nature*, 318 (162) **1985**.

3 Vector Space and Matrices

Learning objectives

- ➢ Vector and function spaces—from 3-dimensional to *n*-dimensional
- ➢ Bra and ket vectors, inner product, linear independence of vectors
- ➢ Orthonormality of basis vectors/functions for spanning a space
- ➢ Matrices from linear transformation of vectors/functions in a space
- ➢ The matrix eigenvalue problem—Jacobi's method of diagonalisation

3.1 INTRODUCTION

The representation theory of groups is built up on the concept of vector spaces, which in turn is developed by a generalisation of the commonly known 3-dimensional vectors along three lines of approach: (i) extension to dimensions greater than three, (ii) allowing the vector components to be complex instead of being only real and (iii) considering functions of one or more variables to behave as vectors. Moreover, matrix algebra is an essential tool for representation of symmetry operations and for developing the principal theorems related to representation of groups. The present chapter introduces briefly the concepts of vector spaces and matrix algebra that will be required for understanding the remaining chapters of the book.

3.2 VECTORS AND VECTOR SPACES

In the ordinary 3-dimensional physical space, we specify a vector $\vec{v}$ with the help of three mutually perpendicular unit vectors $\vec{e}_1, \vec{e}_2, \vec{e}_3$ taken respectively along arbitrarily chosen (mutually perpendicular) Cartesian coordinate axes X, Y and Z as follows:

$$\vec{v} = a_x\vec{e}_1 + a_y\vec{e}_2 + a_z\vec{e}_3 \tag{3.1}$$

where a_x, a_y and a_z are the components of $\vec{v}$ along the Cartesian axes. With the same set of coordinate axes and unit vectors one can imagine another vector:

$$\vec{w} = b_x\vec{e}_1 + b_y\vec{e}_2 + b_z\vec{e}_3 \tag{3.2}$$

The sum of these two vectors is defined as

$$\vec{v} + \vec{w} = (a_x + b_x)\vec{e}_1 + (a_y + b_y)\vec{e}_2 + (a_z + b_z)\vec{e}_3 \tag{3.3}$$

This *vector addition rule* comes from the familiar law of resultant of two forces—each component of the resultant force is the sum of the corresponding components of the individual forces.

As a result of this component-wise addition rule, one can keep the unit vectors in mind and write the vectors $\vec{v}$, $\vec{w}$ and so on, in terms of ordered 3-tuples of their components:

$$\vec{v} = (a_x, a_y, a_z), \vec{w} = (b_x, b_y, b_z) \tag{3.4}$$

Equation (3.3) can then be written as

$$(a_x, a_y, a_z) + (b_x, b_y, b_z) = (a_x + b_x, a_y + b_y, a_z + b_z) \tag{3.5}$$

The unit vectors can also be written as ordered 3-tuples:

$$\vec{e}_1 = (1, 0, 0), \vec{e}_2 = (0, 1, 0) \text{ and } \vec{e}_3 = (0, 0, 1) \tag{3.6}$$

and any vector in the three dimensional space can be expressed in terms of these 3-tuples:

$$\vec{v} = a_x\,(1, 0, 0) + a_y\,(0, 1, 0) + a_z\,(0, 0, 1)$$

$$\vec{w} = b_x\,(1, 0, 0) + b_y\,(0, 1, 0) + b_z\,(0, 0, 1) \ldots \tag{3.7}$$

For this reason, the unit vectors along the Cartesian axes are called *basis vectors* of the 3-dimensional physical space.

The following are some important properties of these vectors.

(i) The sum of any two vectors is also a vector:

$$\vec{v} + \vec{w} = \vec{r} \tag{3.8}$$

If we denote the collection of all vectors by **V**, then it is closed under vector addition.

(ii) Vector addition is commutative:

$$\vec{v} + \vec{w} = \vec{w} + \vec{v} \tag{3.9}$$

and associative:

$$\vec{u} + (\vec{v} + \vec{w}) = (\vec{u} + \vec{v}) + \vec{w} \tag{3.10}$$

(iii) There exists a zero (or, null) vector $\vec{0}$, having the property that

$$\vec{0} + \vec{v} = \vec{v} \tag{3.11}$$

for every vector $\vec{v}$ in **V**. [In 3-tuple form the null vector is (0, 0, 0), that is, all the three components are zero.] The null vector $\vec{0}$ is thus the *additive identity element* in **V**.

(iv) A scalar multiple of a vector in **V** is also a vector in **V**:

$$\lambda\vec{v} = \lambda(a_x, a_y, a_z) = (\lambda a_x, \lambda a_y, \lambda a_z) \tag{3.12}$$

Thus $\lambda\vec{v}$ is a vector in **V**, since all vectors in **V** have three components.

(v) The scalar multiplication of a vector is *associative*; it is also *distributive* with respect to vector addition:

$$a(b\vec{v}) = (ab)\vec{v}\,; \quad (a + b)\vec{v} = a\vec{v} + b\vec{v} \tag{3.13}$$

(vi) Multiplication by the scalars 0 and 1 has the expected effect:

$$0\vec{v} = \vec{0} \quad \text{and} \quad 1\vec{v} = \vec{v} \tag{3.14}$$

The negative of a vector is the result of multiplication by the scalar −1:

$$(-1)\vec{v} = -\vec{v} \tag{3.15}$$

Subtraction of two vectors is defined as: $\vec{v} - \vec{w} = \vec{v} + (-1)\vec{w}$ (3.16)

Consequently $\vec{v} + (-\vec{v}) = \vec{0}$, implying that $-\vec{v}$ is the *additive inverse* of $\vec{v}$ in the collection **V**. At this point it is easy to recognise that **V** is an *additive abelian group*.

The above listing of vector properties is simply a collection of known vector manipulations written in an abstract language. But these will help us to develop higher dimensional vectors, as will be evident from the subsequent discussions. The collection **V** of vectors is called a *3-dimensional vector space*, with basis set $(\vec{e}_1, \vec{e}_2, \vec{e}_3)$ or {(1, 0, 0), (0, 1, 0), (0, 0, 1)}. Let us now consider some examples where the number of components of a vector may be different from 3.

Example 3.1 An n-degree polynomial $a_0x^n + a_1x^{n-1} + \cdots + a_{n-1}x + a_n$ can be written as an ordered $(n+1)$-tuple $(a_0, a_1, a_2, \ldots, a_{n-1}, a_n)$ and addition of such polynomials obeys the rule:

$$(a_0x^n + a_1x^{n-1} + \cdots + a_n) + (b_0x^n + b_1x^{n-1} + \cdots + b_n)$$

$$= (a_0 + b_0)x^n + (a_1 + b_1)x^{n-1} + \cdots + (a_n + b_n)$$

or, written as ordered $(n+1)$-tuples,

$$(a_0, a_1, a_2, \ldots, a_n) + (b_0, b_1, b_2, \ldots, b_n)$$

$$= (a_0 + b_0, a_1 + b_1, a_2 + b_2, \ldots, a_n + b_n) \tag{3.17}$$

which is the *component-wise addition rule*. Scalar multiplication of a polynomial amounts to multiplying each component of the $(n+1)$-tuple form by the scalar:

$$\lambda\,(a_0\,x^n + a_1\,x^{n-1} + \cdots + a_{n-1}\,x + a_n) = \lambda a_0\,x^n + \lambda a_1\,x^{n-1} + \cdots + \lambda a_{n-1}\,x + \lambda a_n$$

which means that $\lambda\,(a_0, a_1, \cdots, a_{n-1}, a_n) = (\lambda a_0, \lambda a_1, \cdots, \lambda a_{n-1}, \lambda a_n)$ (3.18)

Hence in such a representation, an n-degree polynomial behaves like a vector and the collection of all such polynomials is an '$(n+1)$-dimensional vector space' whose basis vectors are

$$\begin{aligned}
\boldsymbol{x}^n &= (1, 0, 0, \ldots 0, 0)\\
\boldsymbol{x}^{n-1} &= (0, 1, 0, \ldots 0, 0)\\
\boldsymbol{x}^{n-2} &= (0, 0, 1, \ldots 0, 0)\\
&\cdots\\
&\cdots\\
\boldsymbol{x}^0 &= (0, 0, 0, \ldots 0, 1)
\end{aligned} \tag{3.19}$$

In terms of these vectors (functions), any polynomial of degree n or less can be expressed. Thus the 4-degree polynomial $2x^4 + 3x^3 - x^2 + x + 5$ can be written as 2(1, 0, 0, 0, 0) + 3(0, 1, 0, 0, 0) − 1(0, 0, 1, 0, 0) + 1(0, 0, 0, 1, 0) + 5(0, 0, 0, 0, 1) in a collection (space) of 4-degree polynomials. It has to be mentioned here that the properties (i)–(vi) of the 3-dimensional vector space can be shown to exist for the collection of polynomials if we define the null polynomial as one in which every coefficient is zero and denote it as (0, 0, 0, ..., 0, 0).

Example 3.2 The general solution of the differential equation

$$\frac{d^2 y}{dx^2} + k^2 y = 0 \tag{3.20}$$

is $Ae^{ikx} + Be^{-ikx}$ where A and B are constants. It represents a family of solutions each of which is characterised by a particular value of A and B. In the family A and B vary parametrically. The sum of any two solutions is also a solution:

$$(A_1 e^{ikx} + B_1 e^{-ikx}) + (A_2 e^{ikx} + B_2 e^{-ikx}) = (A_1 + A_2)e^{ikx} + (B_1 + B_2)e^{-ikx} \tag{3.21}$$

It is at once evident that any solution can be written in the 2-tuple form (A, B) and the identity (3.21) can be written as

$$(A_1, B_1) + (A_2, B_2) = (A_1 + A_2, B_1 + B_2) \tag{3.22}$$

with the functions e^{ikx} and e^{-ikx} in the back of our mind. All the axioms (i)–(vi) can be shown to be valid for the collection of all functions which are solutions of Eq. (3.20). Thus, the collection of these solutions is a two-dimensional vector space with e^{ikx} and e^{-ikx} as basis.

Example 3.3 The molecular orbital theory will be discussed in detail in Chapters 8–10. A commonly used molecular orbital wave function, constructed from the atomic orbital wave functions $\phi_1, \phi_2, \ldots, \phi_n$ is the LCAO wave function of the form,

$$\psi_i = (c_{1i}\phi_1 + c_{2i}\phi_2 + \cdots + c_{ni}\phi_n) \tag{3.23}$$

ψ_i is a solution of the Schrödinger equation for the system under consideration. Linear combination $c_1\psi_1 + c_2\psi_2$ of any two solutions ψ_1 and ψ_2 is also a solution of the same Schrödinger equation. This fact gives rise to the vector properties (i)–(vi). Hence the functions like ψ_i in Eq. 3.23 may be written as vectors in the form $(c_{1i}, c_{2i}, \ldots, c_{ni})$ with $\{\phi_1, \phi_2, \ldots, \phi_n\}$ as the *basis set*. With this background we are now ready for the desired generalisation of vector spaces.

3.2.1 Definition

A collection of n-tuples of numbers or functions, $\vec{v} = (v_1, v_2, \ldots, v_n)$, is called an n-dimensional vector space if (i) the n-tuples obey a component-wise addition rule, (ii) addition of any number of members in the collection is also an n-tuple, the addition is commutative and associative, and (iii) multiplication by (real or complex) scalars is associative and distributive.

To include functions (real or complex) and vectors of higher dimension in the treatment of vector spaces, Dirac introduced the *ket* notation instead of conventional *arrow-heads*. According to this notation $|v\rangle$ (read as *ket vee*) means a vector $\vec{v}$ and $|\psi\rangle$ means a function behaving as a vector in a function space. Henceforth the ket notation will be used for vectors throughout this book except for very common ones like the position vector of a particle, linear and angular momentum vectors, and so on, which are easily recognised as directed physical quantities. The null vector will be denoted by $|0\rangle$. *Linearly independent vectors* will be discussed in the next section. If a minimum of n linearly independent vectors are required to represent any vector of a vector space we say that the vector space is n-dimensional. When the vectors in a vector space are functions, we call it a *function space*.

3.3 LINEAR DEPENDENCE OR INDEPENDENCE OF VECTORS

In an n-dimensional vector space the vectors $|v_1\rangle, |v_2\rangle, \ldots, |v_r\rangle$ are said to be linearly dependent if there exist scalars $\lambda_1, \lambda_2, \ldots, \lambda_r$ not all zero, such that

$$\lambda_1|v_1\rangle + \lambda_2|v_1\rangle + \cdots + \lambda_r|v_r\rangle = |0\rangle \tag{3.24}$$

where r may or may not be equal to n. Equation (3.24) means that if some vectors are linearly dependent, any one of them can be expressed as a linear combination of all the others.

$$|v_i\rangle = \left(\frac{-1}{\lambda_i}\right)[\lambda_1|v_1\rangle + \lambda_2|v_2\rangle + \cdots + \lambda_{i-1}|v_{i-1}\rangle + \lambda_{i+1}|v_{i+1}\rangle + \cdots + \lambda_r|v_r\rangle], \quad 1 \le i \le r \tag{3.25}$$

In the 3-dimensional vector space the vectors (1, 2, 3), (1, 0, 0), (0, 2, 0) and (0, 0, 4) are linearly dependent as shown below.

Let $\lambda_1, \lambda_2, \lambda_3$ and λ_4 be four scalars such that

$$\lambda_1(1, 0, 0) + \lambda_2(0, 2, 0) + \lambda_3(0, 0, 4) + \lambda_4(1, 2, 3) = (0, 0, 0)$$

This requires

$$\left.\begin{aligned} &\lambda_1 + \lambda_4 = 0 \quad \text{or} \quad \lambda_1 = -\lambda_4 \\ &2\lambda_2 + 2\lambda_4 = 0 \quad \text{or} \quad \lambda_2 = -\lambda_4 \\ &4\lambda_3 + 3\lambda_4 = 0 \quad \text{or} \quad \lambda_3 = -\left(\frac{3}{4}\right)\lambda_4 \end{aligned}\right\} \tag{3.26}$$

So if we choose $\lambda_1 = 1$, we find $\lambda_2 = 1$, $\lambda_4 = -1$ and $\lambda_3 = 3/4$. Thus all the scalars need not be equal to zero for condition (3.24) to hold. So the above four vectors are linearly dependent.

In the same way it can be shown that the vectors (1, 0, 0), (0, 1, 0) and (0, 0, 1) are linearly independent because in making $\lambda_1(1, 0, 0) + \lambda_2(0, 1, 0) + \lambda_3(0, 0, 1) = (0, 0, 0)$ we must have $\lambda_1 = \lambda_2 = \lambda_3 = 0$.

Note: In the familiar 2- or 3-dimensional space, vectors are linearly dependent when they are parallel or collinear so that one can be expressed in terms of the others. For example, in the 2-dimensional space $(2k, k)$ and (2, 1) are collinear (Fig. 3.1). They are also linearly dependent by definition, because we can make

$$\lambda_1(2, 1) + \lambda_2(2k, k) = (0, 0) \tag{3.27}$$

by taking $\lambda_1 = -\lambda_2 k$. For example, we may take $\lambda_2 = 1$, $\lambda_1 = -k$ or $\lambda_2 = 2$, $\lambda_1 = -2k$, and so on, which satisfy condition (3.27).

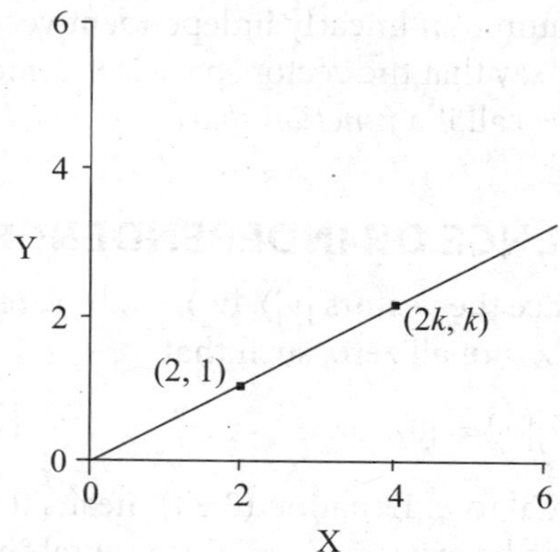

Fig. 3.1 Example of two collinear vectors in the 2-dimensional space

3.4 NECESSITY OF LINEAR INDEPENDENCE OF VECTORS

In a graphical plot, a point is the terminus of a vector with respect to the origin of coordinates. Now, if the coordinate axes are collinear or parallel, the origin cannot be chosen and therefore, points cannot be plotted. Extending this idea to n-dimensional space, we conclude that with a linearly dependent set of vectors as basis it is not possible to represent the position of a point in space or to express any vector in that space in the form of an n-tuple. But if the basis vectors $|e_1\rangle, |e_2\rangle, |e_3\rangle, \ldots, |e_n\rangle$ of an n-dimensional vector space are linearly independent, any vector $|v\rangle$ in that space can be represented in one and only one way. We can prove this as follows:

If possible, let $|v\rangle$ be represented in the following two ways:

$$|v\rangle = \lambda_1 |e_1\rangle + \lambda_2 |e_2\rangle + \cdots + \lambda_n |e_n\rangle$$

and $$|v\rangle = \beta_1 |e_1\rangle + \beta_2 |e_2\rangle + \cdots + \beta_n |e_n\rangle$$

Here λ_i and β_i are scalars (for $i = 1$ to n). Subtracting, we have

$$|0\rangle = (\lambda_1 - \beta_1)\,|e_1\rangle + (\lambda_2 - \beta_2)\,|e_2\rangle + \cdots + (\lambda_n - \beta_n)\,|e_n\rangle$$

which is possible only if each component of the vector in the RHS is zero. This gives $\lambda_i = \beta_i$ for all $i = 1$ to n, and thus the above two representations are identical.

The uniqueness of representation of vectors in a given vector space is, therefore, ensured by linear independence of the chosen basis set. The vectors in such a basis set are said to *span* the vector space.

3.5 BRA VECTORS: ADJOINTS OF KETS

In quantum mechanics wave functions describing states of a system are represented by ket vectors; they are generally complex functions. Their complex conjugates, therefore, should also form vectors in some vector space different from the space defined by the ket vectors. The complex conjugate or *adjoint* of a vector $|v\rangle$ is denoted by $\langle v|$:

$$|v\rangle^* = \langle v| \tag{3.28}$$

The notation is due to Dirac who coined the terms *bra* and *ket* (from *bracket* ≡ bra · c · ket). Corresponding to every vector $|v\rangle$ in the ket space there is a bra vector $\langle v|$ and all the conditions (i) to (vi) in Section 3.2 are valid. This collection of bra vectors is called a *bra space*.

3.6 INNER PRODUCT

Ordinary 3-dimensional vectors admit of two types of products—the cross product and the dot product. The former is a vector and the latter, a scalar. In linear algebra, the latter has been generalised in the name of *inner product*.

In a vector space an inner product of two vectors $|u\rangle$ and $|v\rangle$, denoted by $\langle u|v\rangle$, is defined as a unique *scalar* quantity satisfying the following conditions:

(i) $\langle u|v\rangle = \langle v|u\rangle^*$

(ii) $\langle au|bv\rangle = a^*b\,\langle u|v\rangle$ where a and b are scalars and a^* and b^* are their complex conjugates respectively

(iii) $\langle w|(a|u\rangle) + b|v\rangle)\rangle = a\langle w|u\rangle + b\langle w|v\rangle$

(iv) $\langle u|u\rangle \geq 0$; the equality holds only if $|u\rangle = |0\rangle$

We can define a rule for taking the inner product of two vectors in various ways, always maintaining the validity of conditions (i) to (iv). The rules that are usually followed will be clear from the examples given below.

Example 3.4 In the familiar 3-dimensional vector space, with $\vec{e}_1, \vec{e}_2, \vec{e}_3$ as unit vectors taken along the Cartesian coordinate axes, all the vectors are representable in terms of their components:

$$\left.\begin{aligned} \vec{u} &= u_1\,\vec{e}_1 + u_2\,\vec{e}_2 + u_3\,\vec{e}_3 \\ \vec{v} &= v_1\,\vec{e}_1 + v_2\,\vec{e}_2 + v_3\,\vec{e}_3 \end{aligned}\right\} \tag{3.29}$$

and their dot product is defined by

$$\vec{u}\cdot\vec{v} = |\vec{u}|\cdot|\vec{v}|\cos\theta \tag{3.30}$$

where the modulus sign on the right denotes the magnitude of the concerned vector and θ is the angle between them. With this definition we find:

$$\left.\begin{aligned} \vec{e}_1\cdot\vec{e}_2 = \vec{e}_2\cdot\vec{e}_3 = \vec{e}_3\cdot\vec{e}_1 &= 1\cdot 1\cos\frac{\pi}{2} = 0 \\ \vec{e}_1\cdot\vec{e}_1 = \vec{e}_2\cdot\vec{e}_2 = \vec{e}_3\cdot\vec{e}_3 &= 1\cdot 1\cos 0 = 1 \end{aligned}\right\} \tag{3.31}$$

and

$$\vec{u}\cdot\vec{v} = u_1\,v_1 + u_2\,v_2 + u_3\,v_3 \tag{3.32}$$

which can be verified by utilising the relations (3.29)–(3.31).

If the vector components are all real, the dot product can also be written as

$$\vec{u}\cdot\vec{v} = u_1^*\,v_1 + u_2^*\,v_2 + u_3^*\,v_3 \tag{3.33}$$

This way of the writing the dot product enables one to verify all the properties (i) to (iii) of the inner product, if the dot product is identified with the inner product. Thus,

$$(\vec{u})^* \cdot \vec{v} = \langle u \,|\, v\rangle = |\,u\rangle \cdot |\,v\rangle = u_1^* v_1 + u_2^* v_2 + u_3^* v_3$$

$$= (v_1^* u_1 + v_2^* u_2 + v_3^* u_3)^* = \langle v \,|\, u\rangle^* \qquad \text{[Condition (i) verified]}$$

The other conditions can be similarly verified.

Example 3.5 Let us consider two LCAO molecular orbital wave functions:

$$\psi_1 = c_1\phi_1 + c_2\phi_2 + c_3\phi_3 + c_4\phi_4$$

$$\psi_2 = a_1\phi_1 + a_2\phi_2 + a_3\phi_3 + a_4\phi_4$$

where c_i and a_i ($i = 1$ to 4) are constants and $\{\phi_i\}$, $i = 1$ to 4 is the set of atomic orbital wave functions that build up the molecular orbital wave functions ψ_1 and ψ_2. Here the inner product *is defined* as

$$\langle \psi_1 \,|\, \psi_2\rangle = \int \psi_1^* \psi_2 \, d\tau \tag{3.34}$$

where the integration is carried out over the entire range of variables in which the functions ϕ_i are defined. Let the basis functions ϕ_1, ϕ_2, ϕ_3 and ϕ_4 be chosen so that

$$\langle \phi_i \,|\, \phi_j\rangle = \delta_{ij} = \begin{cases} 1 \text{ for } i = j \\ 0 \text{ for } i \neq j \end{cases} \tag{3.35}$$

The symbol δ_{ij} is called *Krönecker delta*. Then,

$$\langle \psi_1 | \psi_2\rangle = c_1^* a_1 + c_2^* a_2 + c_3^* a_3 + c_4^* a_4$$

and $$\langle \psi_1 | \psi_2\rangle^* = \{c_1^* a_1 + c_2^* a_2 + c_3^* a_3 + c_4^* a_4\}^*$$

$$= a_1^* c_1 + a_2^* c_2 + a_3^* c_3 + a_4^* c_4$$

$= \langle \psi_2 \,|\, \psi_1\rangle$, which shows the validity of condition (i).

The other two conditions can be similarly shown to be valid.

The results of the above two examples lead us to the following generalisation:

If any vector in a vector space is expressible in terms of a set of n linearly independent unit vectors $\{|e_1\rangle, |e_2\rangle, |e_3\rangle, \ldots, |e_n\rangle\}$ then the inner product of any two vectors is given by

$$\langle u \,|\, v\rangle = u_1^* v_1 + u_2^* v_2 + \cdots + u_n^* v_n \tag{3.36}$$

where $|u\rangle = u_1|e_1\rangle + u_2|e_2\rangle + \cdots + u_n|e_n\rangle$

and $$\langle e_i \,|\, e_j\rangle = \delta_{ij} \tag{3.37}$$

A vector space where an inner product is defined is called an *inner product space.* Two vectors in this space are said to be *orthogonal* when their inner product is zero. Any non-null vector in this space is said to be normalised if its inner product with itself is unity. The *norm* of a vector $|u\rangle$ is defined as

$$\|u\| = \sqrt{\langle u \mid u \rangle} \tag{3.38}$$

This is analogous to the length of a vector in the 3-dimensional space. *Normalising* a vector is analogous to setting the length to unity:

$$\begin{aligned} |u\rangle\,(\text{normalised}) &= \frac{|u\rangle\,(\text{unnormalised})}{\|u\|} = \frac{|u\rangle\,(\text{unnormalised})}{\langle u \mid u \rangle} \\ &= \frac{|u\rangle\,(\text{unnormalised})}{\sqrt{u_1^2 + u_2^2 + \cdots + u_n^2}} \end{aligned} \tag{3.39}$$

Note: When a set of functions $\{\phi_1, \phi_2, \ldots, \phi_n\}$ obeys the condition $\langle \phi_i \mid \phi_j \rangle = \delta_{ij}$, we say that the set is *orthonormal*, because $\langle \phi_i \mid \phi_j \rangle = 1$ means that each function is normalised to unity and $\langle \phi_i \mid \phi_j \rangle = 0$ for $i \neq j$ means that the functions are pair-wise orthogonal.

3.7 NECESSITY OF ORTHOGONALITY OF BASIS VECTORS

Theorem: Two orthogonal vectors must be linearly independent.

Let $|u\rangle$ and $|v\rangle$ be two given mutually orthogonal non-null vectors of a given vector space. If possible, let α and β be two scalars such that

$$\alpha|u\rangle + \beta|v\rangle = |0\rangle \tag{3.40}$$

Then taking inner product with $|u\rangle$,

$$\langle u \mid (\alpha \mid u\rangle + \beta \mid v\rangle)\rangle = \langle u \mid 0\rangle = 0$$

or $\quad \alpha\langle u \mid u\rangle + \beta\langle u \mid v\rangle = 0$

or $\quad \alpha\langle u \mid u\rangle = 0 \quad$ (as $|u\rangle$ and $|v\rangle$ are orthogonal)

or $\quad \alpha = 0$

Similarly, by taking the inner product of both sides of Eq. (3.40) with $|v\rangle$, we have $\beta = 0$. Hence, if $|u\rangle$ and $|v\rangle$ are orthogonal, condition (3.24) can be satisfied only if each scalar is zero ($\alpha = \beta = 0$). So $|u\rangle$ and $|v\rangle$ are linearly independent.

This result is of great importance in choosing basis vectors of a vector space. For the unique representation of vectors in a vector space, the basis vectors must be linearly independent. Since orthogonality ensures linear independence, we always choose an orthogonal set as basis. For some other conveniences, we also choose basis functions to be individually normalised to unity. In other words, the choice of an orthonormal basis set is always preferred.

3.7.1 An Example of Orthonormal Basis Set in a Function Space

Let us consider the function space containing solutions of the differential equation,

$$\frac{d^2y}{dx^2} = -y \tag{3.41}$$

in the range $-\pi \le x \le \pi$. The general solution is either of the following:

(i) $y = Pe^{ix} + Qe^{-ix}$
(ii) $y = A \sin x + B \cos x$

where P, Q, A and B are constants.

Functions of type (i) belong to a space whose basis functions may be chosen to be e^{ix} and e^{-ix}. Let us examine whether the basis set $\{e^{ix}, e^{-ix}\}$ is orthonormal or not.

$$\langle e^{ix} | e^{-ix} \rangle = \int_{-\pi}^{\pi} (e^{ix})^* (e^{-ix})\, dx = \int_{-\pi}^{\pi} e^{-2ix}\, dx = \left[\frac{e^{-2ix}}{-2i} \right]_{-\pi}^{\pi}$$

$$= 0 \qquad [\text{as } \cos(\pm 2\pi) = 1 \text{ and } \sin(\pm 2\pi) = 0]$$

Thus, the basis functions of type (i) are orthogonal. Again,

$$\langle e^{ix} | e^{ix} \rangle = \int_{-\pi}^{\pi} (e^{ix})^* (e^{ix})\, dx = \int_{-\pi}^{\pi} dx = 2\pi \text{ and similarly } \langle e^{-ix} | e^{-ix} \rangle = 2\pi$$

Hence the two functions e^{ix} and $e^{-ix}$are not normalised to unity; they can be made so by multiplying them by $\frac{1}{\sqrt{2\pi}}$. Thus an orthonormal basis set for the collection of functions of type (i) is $\left\{ \frac{1}{\sqrt{2\pi}} e^{ix}, \frac{1}{2\pi} e^{-ix} \right\}$. In the same way, it can be shown that for the collection of functions containing solutions of type (ii), the orthonormal basis set is $\left\{ \frac{1}{\sqrt{\pi}} \cos x, \frac{1}{\sqrt{\pi}} \sin x \right\}$.

The two collections of functions constitute the same function space because they contain solutions of the same differential equation (Eq. 3.41). We thus see that the same function space may have different basis sets, the number of basis functions being always the same and equal to the dimensionality of the space.

3.8 MATRICES FROM LINEAR TRANSFORMATION OF VECTORS

Symmetry operations change a vector of a vector space into another vector in the same space without a change in magnitude. The angle between any two vectors in this space also remains unchanged by such operations. These actions are called *linear transformations* because it is found that each component of the transformed vector is a linear combination of all the components of the original vector. For a specific example, let a vector in the 3-dimensional vector space be subjected to a rotation about the Z-axis or a reflection on some plane. The result is another vector in the same space. Consider, for example, a vector $\vec{u}$ on the XY plane; let it be reflected on a plane σ which contains the Z-axis and passes through the origin. Also let σ be placed at an angle β with the X-axis (Fig. 3.3). The image of $\vec{u}$ is $\vec{u}'$. If $(u_1, u_2, 0)$ be the Cartesian components of $\vec{u}$ and $(u_1', u_2', 0)$ be those of $\vec{u}'$ with respect to the same set of coordinate axes, then from Fig. 3.2, we have:

$$u_1 = |\vec{u}|\cos\alpha,\ u_2 = |\vec{u}|\sin\alpha \text{ and } u_1' = |\vec{u}'|\cos\theta,\ u_2' = |\vec{u}'|\sin\theta \tag{3.42}$$

where α and θ are the angles made by $\vec{u}$ and $\vec{u}'$ with the X-axis respectively. Then

$$\theta = \alpha + 2(\beta - \alpha) = 2\beta - \alpha \tag{3.43}$$

$$\begin{aligned} u_1' &= |\vec{u}'|\cos(2\beta - \alpha) \\ &= |\vec{u}'|\{\cos 2\beta\cos\alpha + \sin 2\beta\sin\alpha\} \\ &= |\vec{u}|\cos\alpha\cos 2\beta + |\vec{u}|\sin\alpha\sin 2\beta \qquad (\text{as } |\vec{u}'| = |\vec{u}|) \\ &= u_1\cos 2\beta + u_2\sin 2\beta + 0 \end{aligned} \tag{3.44a}$$

Similarly, $u_2' = u_1\sin 2\beta - u_2\cos 2\beta + 0$ (3.44b)

It is thus seen that each component of the transformed vector $\vec{u}'$ is a linear combination of all the components of the original vector $\vec{u}$. Mathematically this is written as

$$\hat{T}\vec{u} = \vec{u}' \tag{3.45}$$

where $\hat{T}$ is the operator for reflection in the present case.

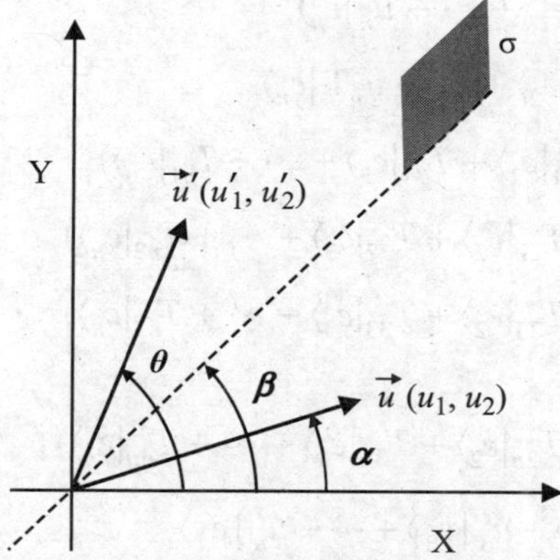

Fig. 3.2 A reflection plane σ containing the Z-axis and placed at an angle β with the X-axis; reflection of a vector $\vec{u}$ on the plane σ

The idea gained from this example can be generalised to a vector space of any dimension; the ket notation will be used instead of the arrow-head in that case. A transformation operator $\hat{T}$ is called *linear* if it keeps scalars unchanged and changes the vector $|u\rangle$ to $|u'\rangle$ such that each component of $|u'\rangle$ is a linear combination of all the components of the original vector $|u\rangle$, and the following conditions hold:

$$\left.\begin{aligned} \hat{T}(a|u\rangle) &= a\hat{T}|u\rangle = a|u'\rangle \\ \hat{T}(a|u\rangle + b|v\rangle) &= a\hat{T}|u\rangle + b\hat{T}|v\rangle = a|u'\rangle + b|v'\rangle \end{aligned}\right\} \tag{3.46}$$

for any two scalars a, b and two vectors $|u\rangle$, $|v\rangle$ in a given vector space. The action of $\hat{T}$ on any vector in the given vector space is conveniently obtained if its action on the members of the orthonormal basis set $\{|e_1\rangle, |e_2\rangle, |e_3\rangle, \ldots, |e_n\rangle\}$ is worked out first. The action of $\hat{T}$ on $|e_1\rangle$ is to convert it into a new vector which will be a linear combination of all the basis vectors:

$$\hat{T}|e_1\rangle = T_{11}|e_1\rangle + T_{21}|e_2\rangle + T_{31}|e_3\rangle + \cdots + T_{n1}|e_n\rangle \tag{3.47}$$

where T_{i1} is the coefficient of the ith basis vector in the result of operating $\hat{T}$ on $|e_1\rangle$. In the same way, transformations of the other basis vectors can be written down. The following is the list of these results:

$$\begin{aligned}
\hat{T}|e_1\rangle &= T_{11}|e_1\rangle + T_{21}|e_2\rangle + T_{31}|e_3\rangle + \cdots + T_{n1}|e_n\rangle \\
\hat{T}|e_2\rangle &= T_{12}|e_1\rangle + T_{22}|e_2\rangle + T_{32}|e_3\rangle + \cdots + T_{n2}|e_n\rangle \\
\hat{T}|e_3\rangle &= T_{13}|e_1\rangle + T_{23}|e_2\rangle + T_{33}|e_3\rangle + \cdots + T_{n3}|e_n\rangle \\
&\cdots \\
\hat{T}|e_n\rangle &= T_{1n}|e_1\rangle + T_{2n}|e_2\rangle + T_{3n}|e_3\rangle + \cdots + T_{nn}|e_n\rangle
\end{aligned} \tag{3.48}$$

The action of $\hat{T}$ on $|u\rangle$ is now obtained as follows:

$$\begin{aligned}
|u\rangle &= u_1|e_1\rangle + u_2|e_2\rangle + \cdots + u_n|e_n\rangle \\
\hat{T}|u\rangle &= \hat{T}(u_1|e_1\rangle + u_2|e_2\rangle + \cdots + u_n|e_n\rangle) \\
&= u_1\hat{T}|e_1\rangle + u_2\hat{T}|e_2\rangle + \cdots + u_n\hat{T}|e_n\rangle \\
&= u_1(T_{11}|e_1\rangle + T_{21}|e_2\rangle + T_{31}|e_3\rangle + \cdots + T_{n1}|e_n\rangle) \\
&\quad + u_2(T_{12}|e_1\rangle + T_{22}|e_2\rangle + T_{32}|e_3\rangle + \cdots + T_{n2}|e_n\rangle) \\
&\quad + u_3(T_{13}|e_1\rangle + T_{23}|e_2\rangle + T_{33}|e_3\rangle + \cdots + T_{n3}|e_n\rangle) \\
&\quad + \cdots \\
&\quad + u_n(T_{1n}|e_1\rangle + T_{2n}|e_2\rangle + T_{3n}|e_3\rangle + \cdots + T_{nn}|e_n\rangle) \\
&= u_1'|e_1\rangle + u_2'|e_2\rangle + u_3'|e_3\rangle + \cdots + u_n'|e_n\rangle
\end{aligned}$$

where $u_1' = u_1 T_{11} + u_2 T_{12} + u_3 T_{13} + \cdots + u_n T_{1n}$

$u_2' = u_1 T_{21} + u_2 T_{22} + u_3 T_{23} + \cdots + u_n T_{2n}$ and so on.

The above results can be written in a compact form as follows:

$$\hat{T}|u\rangle = \hat{T}\sum_{j=1}^{n} u_j|e_j\rangle = \sum_{j=1}^{n} u_j(\hat{T}|e_j\rangle) = \sum_{j=1}^{n} u_j \sum_{i=1}^{n}(T_{ij}|e_i\rangle) = \sum_{i=1}^{n}\left(\sum_{j=1}^{n} T_{ij}u_j\right)|e_i\rangle$$

$$= \sum_{i=1}^{n} u_i'|e_i\rangle \tag{3.49a}$$

where $$u'_i = \sum_{j=1}^{n} T_{ij}\, u_j \tag{3.49b}$$

We thus arrive at n^2 numbers, $T_{11}, T_{21}, T_{31}, \ldots, T_{nn}$ which can be defined in terms of Eqs 3.48 and condition (iii) of the definition of inner product:

$$\langle e_1|\hat{T}|e_1\rangle = \langle e_1|T_{11}\,e_1\rangle + \langle e_1|T_{21}\,e_2\rangle + \langle e_1|T_{31}\,e_3\rangle + \cdots + \langle e_1|T_{n1}\,e_n\rangle = T_{11}\langle e_1|e_1\rangle = T_{11}$$

$$\langle e_2|\hat{T}|e_1\rangle = \langle e_2|T_{11}\,e_1\rangle + \langle e_2|T_{21}\,e_2\rangle + \langle e_2|T_{31}\,e_3\rangle + \cdots + \langle e_2|T_{n1}\,e_n\rangle = T_{21}\langle e_2|e_2\rangle = T_{21}$$

and in general

$$\langle e_i|\hat{T}|e_j\rangle = T_{ij} \tag{3.50}$$

These results follow from the orthonormality of the basis kets.

For mathematical manipulations, the numbers T_{ij} are displayed in the form of an array of n rows and n columns and the array is called a *matrix*:

$$T = \begin{bmatrix} T_{11} & T_{12} & T_{13} & \cdot & \cdot & \cdot & T_{1n} \\ T_{21} & T_{22} & T_{23} & \cdot & \cdot & \cdot & T_{2n} \\ \cdot & \cdot & \cdot & \cdot & \cdot & \cdot & \cdot \\ \cdot & \cdot & \cdot & \cdot & \cdot & \cdot & \cdot \\ \cdot & \cdot & \cdot & \cdot & \cdot & \cdot & \cdot \\ \cdot & \cdot & \cdot & \cdot & \cdot & \cdot & \cdot \\ \cdot & \cdot & \cdot & \cdot & \cdot & \cdot & \cdot \\ T_{n1} & T_{n2} & T_{n3} & \cdot & \cdot & \cdot & T_{nn} \end{bmatrix} \tag{3.51}$$

Successive execution of two linear transformations $\hat{S}$ and $\hat{T}$ is called a *product* of the operations and it leads to the rule for matrix multiplication as shown below:

Let $\hat{T}\,|u\rangle = |u'\rangle$ and $\hat{S}\,|u'\rangle = |u''\rangle$. Execution of $\hat{T}$ on $|u\rangle$ first and then of $\hat{S}$ on the result gives

$$|u''\rangle = \hat{S}\,|u'\rangle = \hat{S}(\hat{T}\,|u\rangle) = \hat{S}\,\hat{T}\,|u\rangle \tag{3.52a}$$

If the matrix **U** represents the combined operation $\hat{U} = \hat{S}\,\hat{T}$, then elements of **U** can be obtained as follows:

$$|u'\rangle = \hat{T}\,|u\rangle \Rightarrow u'_i = \sum_k T_{ik} u_k$$

$$|u''\rangle = \hat{S}\,|u'\rangle \Rightarrow u''_j = \sum_i S_{ji} u'_i = \sum_i S_{ji} \sum_k T_{ik} u_k$$

$$= \sum_k \sum_i S_{ji}\, T_{ik}\, u_k \tag{3.52b}$$

The operator $\hat{U}$ $(= \hat{S}\,\hat{T})$ transforms $|u\rangle$ to $|u''\rangle$ and hence

$$u_j'' = \sum_k U_{jk}\, u_k \tag{3.52c}$$

Comparing Eqs (3.52b) and (3.52c),

$$U_{jk} = \sum_i S_{ji}\, T_{ik} \tag{3.52d}$$

which gives the elements of the matrix **U** (= **ST**). This is the rule for the matrix multiplication—to get the jk element of the product matrix **U**, the elements of the jth row of **S** are to be multiplied by the corresponding elements of the kth column of **T** and added. It immediately follows that the two matrices **S** and **T** are conformable for multiplication as **ST** only if the number of columns in **S** is equal to the number of rows in **T**.

The addition of two matrices is carried out according to the following rule:

$$\mathbf{A} = \mathbf{S} + \mathbf{T} \text{ if } A_{ij} = S_{ij} + T_{ij} \tag{3.53}$$

That is, the corresponding elements of the matrices are to be added. Two matrices are conformable for addition only if they are of the same size.

Note:

(i) The matrices discussed above have as many rows as there are columns. However, there are many other contexts in physics, economics and linear programming where matrices appear with unequal number of rows and columns. Such *rectangular matrices* can also be multiplied according to the same rule. If the matrices **A** and **B** are of size $(m \times n)$ and $(n \times p)$ respectively, then the product matrix **AB** is of size $(m \times p)$.

(ii) The matrix multiplication rule can be extended to any finite number of matrices provided that the number of columns of the left matrix is equal to number of rows in the right matrix. Thus if the matrices **A**, **B**, **C** are of size $(m \times n)$, $(n \times p)$ and $(p \times q)$ respectively, then **ABC** is a matrix of size $(m \times q)$ and its (α, ε) element is given by

$$(ABC)_{\alpha\varepsilon} = \sum_{\beta=1}^{n} \sum_{\gamma=1}^{p} A_{\alpha\beta} B_{\beta\gamma} C_{\gamma\varepsilon}, \text{ for } \alpha = 1 \text{ to } m,\ \varepsilon = 1 \text{ to } q \tag{3.54}$$

Example 3.6 Matrix addition

$$\begin{bmatrix} 1 & -1 & i \\ -1 & 2 & -1 \\ -i & -1 & 3 \end{bmatrix} + \begin{bmatrix} 0 & 1 & 1 \\ 1 & -2 & -1 \\ 1 & i & -1 \end{bmatrix} = \begin{bmatrix} 1 & 0 & i+1 \\ 0 & 0 & -2 \\ 1-i & i-1 & 2 \end{bmatrix}$$

Example 3.7 Matrix multiplication

(i) Let $\mathbf{A} = \begin{bmatrix} 1 & -1 & 2 \\ -1 & 0 & 1 \\ 2 & 1 & -1 \end{bmatrix}$, $\mathbf{B} = \begin{bmatrix} -1 & 1 & 0 \\ 2 & -1 & 1 \\ 0 & -1 & 2 \end{bmatrix}$

Then the first row elements of **AB** are:

$$(\mathbf{AB})_{11} = -1 - 2 + 0 = -3, (\mathbf{AB})_{12} = 1 + 1 - 2 = 0, (\mathbf{AB})_{13} = 0 - 1 + 4 = 3$$

The second row elements are:

$$(\mathbf{AB})_{21} = 1 + 0 + 0 = 1, (\mathbf{AB})_{22} = -1 + 0 - 1 = -2, (\mathbf{AB})_{23} = 0 + 0 + 2 = 2$$

The third row elements are:

$$(\mathbf{AB})_{31} = -2 + 2 + 0 = 0, (\mathbf{AB})_{32} = 2 - 1 + 1 = 2, (\mathbf{AB})_{33} = 0 + 1 - 2 = -1$$

$$\text{Hence} \quad \mathbf{AB} = \begin{bmatrix} -3 & 0 & 3 \\ 1 & -2 & 2 \\ 0 & 2 & -1 \end{bmatrix}$$

(ii) Let $\mathbf{A} = \begin{bmatrix} 1 & 0 & -1 \\ -1 & 2 & 1 \end{bmatrix}$, $\mathbf{B} = \begin{bmatrix} 1 & -1 \\ 0 & -2 \\ -1 & 1 \end{bmatrix}$

A is of size (2×3) and **B** is a (3×2) matrix. The product **AB** should be a (2×2) matrix and **BA** should be of size (3×3). By actual calculation it will be found that

$$\mathbf{AB} = \begin{bmatrix} 2 & -2 \\ -2 & -2 \end{bmatrix} \text{and } \mathbf{BA} = \begin{bmatrix} 2 & -2 & -2 \\ 2 & -4 & -2 \\ -2 & 2 & 2 \end{bmatrix}$$

Evidently, matrix multiplication is not always commutative.

3.8.1 Square Matrix and Determinant

A square matrix is one with equal number of rows and columns. Their multiplication is associative. A matrix by itself has no numerical value. But the *determinant* of a square matrix **A**, denoted by $|A|$ or det **A**, has a numerical value defined as follows:

If a_{ij} is the element at the intersection of the *i*th row and *j*th column of a determinant (Δ), its cofactor A_{ij} is $(-1)^{i+j}$ times the determinant obtained by deleting the entire *i*th row and *j*th column from Δ. The value of Δ is given by

$$\Delta = a_{i1} A_{i1} + a_{i2} A_{i2} + \cdots + a_{in} A_{in} = \sum_{\lambda=1}^{n} a_{i\lambda} A_{i\lambda} \tag{3.55}$$

where n is the number of columns in Δ. Any row (i) can be chosen for this expansion. For example:

$$\text{Let} \quad \Delta = \begin{vmatrix} 4 & 5 & 6 \\ 6 & 4 & 5 \\ 5 & 6 & 4 \end{vmatrix}$$

The cofactors of the elements in the first row are:

$$A_{11}=(-1)^{1+1}\begin{vmatrix}4 & 5\\ 6 & 4\end{vmatrix}=-14,\ A_{12}=(-1)^{1+2}\begin{vmatrix}6 & 5\\ 5 & 4\end{vmatrix}=1,\ A_{13}=(-1)^{1+3}\begin{vmatrix}6 & 4\\ 5 & 6\end{vmatrix}=16$$

$$\Delta=4\cdot(-14)+5\cdot(1)+6\cdot(16)=45$$

It can be easily verified that the same result is obtained if Δ is expanded in terms of any of the other two rows.

The following are some important properties of determinants:

- The value of a determinant remains unchanged if rows are changed into columns.
- A single interchange of any two rows (or columns) changes the sign of the determinant.
- For two square matrices **A** and **B** of the same size

$$\det \mathbf{AB}=\det \mathbf{A}\cdot\det \mathbf{B}$$

- If any two rows (or columns) of a determinant are identical, the value of the determinant is zero.
- If all the off-diagonal elements are zero, the value of the determinant is the product of the diagonal elements.
- Multiplying every element of any row (or column) of a determinant by a scalar λ amounts to multiplying the whole determinant by λ. Consequently, when all the elements of the $(n\times n)$ determinant, det **B**, are multiplied by λ the result is $\lambda^n\cdot\det\mathbf{B}$.

3.8.2 Unit and Singular Matrices

A square matrix is called *singular* if its determinant is zero. The *unit* or *identity* matrix $\mathbf{I}_n$ of size $n\times n$ is defined as a matrix whose diagonal elements are each unity and each off-diagonal element is zero. It can be readily verified that for any square matrix **A** of size $n\times n$, $\mathbf{AI}_n=\mathbf{I}_n\mathbf{A}=\mathbf{A}$ (which justifies the name given to $\mathbf{I}_n$). Evidently, the determinant of $\mathbf{I}_n$ is unity for any value of n.

3.8.3 The Transpose and Inverse of a Matrix

Given a matrix **A** (square or rectangular), its transpose $\mathbf{A}^T$ is defined as a matrix obtained by converting the rows of **A** into columns. Thus

$$\text{if } \mathbf{A}=[a_{ij}]_{m\times n},\ \text{then } \mathbf{A}^T=[a_{ji}]_{n\times m}$$

The transpose of the product of two matrices is the product of the transposes of the constituent matrices in the reverse order: $(\mathbf{AB})^T=\mathbf{B}^T\mathbf{A}^T$

Proof: Let $\mathbf{A}=[a_{ij}]_{m\times n}$ and $\mathbf{B}=[b_{ij}]_{n\times p}$

Now $$[(\mathbf{AB})^T]_{ij}=[\mathbf{AB}]_{ji}=\sum_{k=1}^{n}[\mathbf{A}]_{jk}[\mathbf{B}]_{ki}=\sum_{k=1}^{n}[\mathbf{B}]_{ki}[\mathbf{A}]_{jk}$$

(since matrix elements are scalars, their multiplication is commutative like the multiplication of two numbers). Hence

$$[(\mathbf{AB})^T]_{ij} = \sum_{k=1}^{n} [\mathbf{B}^T]_{ik} [\mathbf{A}^T]_{kj} = [\mathbf{B}^T \mathbf{A}^T]_{ij} \qquad \text{(Proved)}$$

The *inverse* $\mathbf{A}^{-1}$ of a matrix **A** is defined as a matrix which satisfies the condition

$$\mathbf{AA}^{-1} = \mathbf{A}^{-1}\mathbf{A} = \mathbf{I} \tag{3.56}$$

where **I** is the unit matrix of the size of **A**.

Note:

(i) The definition (3.56) ensures that $\mathbf{A}^{-1}$ can exist only if **A** has non-vanishing determinant. For, if det $\mathbf{A} = 0$, det $\mathbf{A}^{-1} = (\det \mathbf{I})/0 = 1/0$ which is undefined, making $\mathbf{A}^{-1}$ itself undefined.

(ii) The inverse of the product **AB** of two matrices **A** and **B** is the product of the inverses $\mathbf{B}^{-1}$ and $\mathbf{A}^{-1}$ in the order $\mathbf{B}^{-1}\mathbf{A}^{-1}$. The rule can be generalised to any finite number of matrices:

$$(\mathbf{ABC}\cdots)^{-1} = (\cdots\mathbf{C}^{-1}\mathbf{B}^{-1}\mathbf{A}^{-1}) \tag{3.57}$$

(See Problem 16 of Exercises)

There is a general procedure for construction of the inverse of a given matrix, but we shall bypass it because the matrices encountered in representing symmetry operators are usually *unitary* in nature and construction of their inverses is much easier, as will be evident from the next section.

3.9 SPECIAL SQUARE MATRICES

The following types of square matrices are often encountered in the development and applications of group theory.

(i) ***Hermitian matrix***: A matrix **H** is said to be *hermitian* if it is equal to its conjugate transpose, that is, $\mathbf{H} = [\mathbf{H}^T]^*$. (Change the rows of **H** into columns and take the complex conjugate of each element; if the resulting matrix is equal to the original matrix **H**, then it is hermitian.) Usually the conjugate transpose notation, (asterisk + superscript T), is replaced by the notation '†' (dagger) in the superscript: $(\mathbf{H}^*)^T = \mathbf{H}^\dagger$. Thus, a matrix **H** is hermitian if $\mathbf{H} = \mathbf{H}^\dagger$.

(ii) ***Unitary matrix***: A matrix **U** is said to be unitary if it is found that

$$\mathbf{UU}^\dagger = \mathbf{I} \text{ or, } \mathbf{U}^{-1} = \mathbf{U}^\dagger \tag{3.58}$$

where **I** is the unit matrix of the size of **U**.

3.9.1 Some Important Results about Hermitian and Unitary Matrices

(i) Given a square matrix **A**, the matrix $\mathbf{AA}^\dagger$ must be hermitian:

$$(\mathbf{AA}^\dagger)^\dagger = (\mathbf{A}^\dagger)^\dagger\mathbf{A}^\dagger = \mathbf{AA}^\dagger$$

(ii) A given matrix can be readily identified as unitary by examining whether the following two conditions hold.

$$\left.\begin{array}{l} \bullet \text{ for any column having index } j, \sum_i a_{ij}\, a_{ij}^* = 1 \\ \bullet \text{ for any two different columns having indices } j \text{ and } k, \sum_i a_{ij}\, a_{ik}^* = 0 \end{array}\right\} \tag{3.59}$$

For example, consider the matrix,

$$\mathbf{A} = \begin{bmatrix} \cos\theta & -\sin\theta & 0 \\ \sin\theta & \cos\theta & 0 \\ 0 & 0 & 1 \end{bmatrix} \tag{3.60}$$

Here, for the first column,

$$\sum_i a_{i1}\, a_{i1}^* = (\cos\theta)\,(\cos\theta)^* + (\sin\theta)\,(\sin\theta)^* + 0.0^* = 1 \tag{3.61}$$

The same is true for the other columns. Again using the first and second columns,

$$\sum_i a_{i1}\, a_{i2}^* = (\cos\theta)\,(-\sin\theta)^* + (\sin\theta)\,(\cos\theta)^* + 0.0^* = 0 \tag{3.62}$$

and the result is similar for the pair of the 2nd and 3rd columns or the pair of the 1st and 3rd columns. Thus the matrix **A** is unitary. As soon as we identify a matrix to be unitary, we can construct its inverse by taking its conjugate transpose. Thus,

$$\mathbf{A}^{-1} = \begin{bmatrix} (\cos\theta)^* & (\sin\theta)^* & 0^* \\ (-\sin\theta)^* & (\cos\theta)^* & 0^* \\ 0^* & 0^* & 1^* \end{bmatrix} \tag{3.63}$$

(iii) The product of any two unitary matrices is also unitary

Proof: Let $\mathbf{U}_1$ and $\mathbf{U}_2$ be two unitary matrices. Then

$$\mathbf{U}_1^\dagger = \mathbf{U}_1^{-1} \text{ and } \mathbf{U}_2^\dagger = \mathbf{U}_2^{-1}.$$

Hence $(\mathbf{U}_1\mathbf{U}_2)^\dagger = \mathbf{U}_2^\dagger\,\mathbf{U}_1^\dagger = \mathbf{U}_2^{-1}\,\mathbf{U}_1^{-1} = (\mathbf{U}_1\mathbf{U}_2)^{-1}$ that is, $\mathbf{U}_1\mathbf{U}_2$ is unitary.

This result can be obviously generalised to the product of any number of matrices.

3.10 EIGENVALUES AND EIGENVECTORS OF OPERATORS AND MATRICES

Let $\hat{\mathrm{T}}$ be a linear operator that transforms a vector $|u\rangle$ in the 3-dimensional space; this transformation may be, for example, a rotation about an axis or a reflection on a plane. By such an operation most vectors change in a complicated way but there are some vectors which either do not change or change to a scalar multiple of themselves. For example, if the operation is a 180° rotation about the Z-axis, vectors which are along the Z-axis do not change and those which are on the XY plane change to −1 times themselves. In general, the linear operator $\hat{T}$ has *some* vectors for which

$$\hat{T}\,|u\rangle = \lambda\,|u\rangle \tag{3.64}$$

where λ is a scalar. Such vectors are called *eigenvectors* and the number λ (which may be complex also) is the *eigenvalue* of $\hat{T}$ corresponding to the eigenvector $|u\rangle$. [The null vector $|0\rangle$ does not count because it trivially satisfies Eq. 3.64 for any value of λ.] The matrix form of Eq. 3.64 can be obtained by expressing $|u\rangle$ in terms of its components along the orthonormal vectors $|e_1\rangle$, $|e_2\rangle$ and $|e_3\rangle$ which span the 3-dimensional space:

$$|u\rangle = u_1\,|e_1\rangle + u_2\,|e_2\rangle + u_3\,|e_3\rangle \tag{3.65}$$

$$\hat{T}\,|u\rangle = u_1\,\hat{T}\,|e_1\rangle + u_2\,\hat{T}\,|e_2\rangle + u_3\,\hat{T}\,|e_3\rangle \tag{3.66}$$

or, from Eq. 3.64,

$$\begin{aligned}\lambda|u\rangle &= u_1(T_{11}|e_1\rangle + T_{21}|e_2\rangle + T_{31}|e_3\rangle) + u_2(T_{12}|e_1\rangle + T_{22}|e_2\rangle + T_{32}|e_3\rangle)\\ &\quad + u_3(T_{13}|e_1\rangle + T_{23}|e_2\rangle + T_{33}|e_3\rangle)\end{aligned} \tag{3.67}$$

But $\lambda|u\rangle = \lambda(u_1|e_1\rangle + u_2|e_2\rangle + u_3|e_3\rangle)$

$$= \lambda u_1|e_1\rangle + \lambda u_2|e_2\rangle + \lambda u_3|e_3\rangle \tag{3.68}$$

Comparing Eqs 3.67 and 3.68, we have

$$\left.\begin{aligned}\lambda u_1 &= u_1 T_{11} + u_2 T_{12} + u_3 T_{13} \text{ (from the coefficients of } |e_1\rangle)\\ \lambda u_2 &= u_1 T_{21} + u_2 T_{22} + u_3 T_{23} \text{ (from the coefficients of } |e_2\rangle)\\ \lambda u_3 &= u_1 T_{31} + u_2 T_{32} + u_3 T_{33} \text{ (from the coefficients of } |e_3\rangle)\end{aligned}\right\} \tag{3.69}$$

These can be written in matrix form as

$$\lambda\begin{bmatrix}u_1\\u_2\\u_3\end{bmatrix} = \begin{bmatrix}T_{11} & T_{12} & T_{13}\\ T_{21} & T_{22} & T_{23}\\ T_{31} & T_{32} & T_{33}\end{bmatrix}\begin{bmatrix}u_1\\u_2\\u_3\end{bmatrix}$$

or, $\lambda\mathbf{u} = \mathbf{Tu}$

or, $$(\mathbf{T} - \lambda\mathbf{I})\mathbf{u} = \mathbf{0} \tag{3.70}$$

(In Eq. 3.70 the unit matrix **I** is required because one cannot subtract the scalar λ from the matrix **T**, but the subtraction $\mathbf{T} - \lambda\mathbf{I}$ is permitted.)

If the matrix $\mathbf{T} - \lambda\mathbf{I}$ is non-singular, one can multiply both sides of Eq. 3.70 by $(\mathbf{T} - \lambda\mathbf{I})^{-1}$ and get $\mathbf{u} = \mathbf{0}$, that is, $u_1 = u_2 = u_3 = 0$ which makes $|u\rangle$ a null vector. But an eigenvector should be a non-null vector. Hence the matrix $\mathbf{T} - \lambda\mathbf{I}$ must be singular:

$$\det(\mathbf{T} - \lambda\mathbf{I}) = 0$$

or,
$$\begin{vmatrix} T_{11}-\lambda & T_{12} & T_{13} \\ T_{21} & T_{22}-\lambda & T_{23} \\ T_{31} & T_{32} & T_{33}-\lambda \end{vmatrix} = 0 \tag{3.71}$$

This is a 3-degree equation in λ. By expanding the determinant we get a 3-degree polynomial, which is called *characteristic polynomial* of the matrix **T** and when it is equated to zero, the three roots λ_1, λ_2 and λ_3 are the eigenvalues of **T** (as well as of the operator $\hat{T}$). The eigenvectors are obtained by using these values of λ in the system of linear equations (3.69) which give on simplification,

$$\left.\begin{aligned} (T_{11}-\lambda)u_1 + T_{12}u_2 + T_{13}u_3 &= 0 \\ T_{21}u_1 + (T_{22}-\lambda)u_2 + T_{23}u_3 &= 0 \\ T_{31}u_1 + T_{32}u_2 + (T_{33}-\lambda)u_3 &= 0 \end{aligned}\right\} \tag{3.72}$$

for each value of λ. On solving these equations and utilising the normalisation condition $u_1^2 + u_2^2 + u_3^2 = 1$, the eigenvector components can be obtained for each λ. The above ideas can be generalised to vector (or function) space of n-dimensions. Thus, if $\{|e_1\rangle, |e_2\rangle, |e_3\rangle, \ldots, |e_n\rangle\}$ is an orthonormal basis set, we have by the same procedure

$$\mathbf{T} = \begin{bmatrix} T_{11} & T_{12} & T_{13} & \cdot & \cdot & \cdot & T_{1n} \\ T_{21} & T_{22} & T_{23} & \cdot & \cdot & \cdot & T_{2n} \\ T_{31} & T_{32} & T_{33} & \cdot & \cdot & \cdot & T_{3n} \\ \cdot & \cdot & \cdot & \cdot & \cdot & \cdot & \cdot \\ \cdot & \cdot & \cdot & \cdot & \cdot & \cdot & \cdot \\ \cdot & \cdot & \cdot & \cdot & \cdot & \cdot & \cdot \\ T_{n1} & T_{n2} & T_{n3} & \cdot & \cdot & \cdot & T_{nn} \end{bmatrix} \tag{3.73}$$

and the eigenvalues can be obtained from

$$\det(\mathbf{T} - \lambda\mathbf{I}) = 0 \tag{3.74}$$

where **I** is now the $(n \times n)$ unit matrix; for each eigenvalue the normalised eigenvector components $(u_1, u_2, \ldots, u_n)$ can be obtained from

$$\left.\begin{aligned} (T_{11}-\lambda)u_1 + T_{12}u_2 + \cdots + T_{1n}u_n &= 0 \\ T_{21}u_1 + (T_{22}-\lambda)u_2 + \cdots + T_{2n}u_n &= 0 \\ \ldots & \\ \ldots & \\ T_{n1}u_1 + T_{n2}u_2 + \cdots + (T_{nn}-\lambda)u_n &= 0 \end{aligned}\right\} \tag{3.75}$$

The determinant on the LHS of Eq. (3.74) is called the *secular determinant* for the system under consideration.

3.10.1 An Example of Solving the Matrix Eigenvalue Equations

Consider $\mathbf{A} = \begin{bmatrix} 0 & 1 & 0 \\ 1 & 0 & 1 \\ 0 & 1 & 0 \end{bmatrix}$. Let the eigenvectors of **A** be of the form $\mathbf{X} = \begin{bmatrix} c_1 \\ c_2 \\ c_3 \end{bmatrix}$.

Then the eigenvalue equation is

$$\mathbf{AX} = \lambda \mathbf{X} \tag{3.76}$$

which gives a non-null **X** only if $\det(\mathbf{A} - \lambda \mathbf{I}) = 0$

or, $$\begin{vmatrix} 0-\lambda & 1 & 0 \\ 1 & 0-\lambda & 1 \\ 0 & 1 & 0-\lambda \end{vmatrix} = \lambda(\lambda^2 - 2) = 0$$

This has roots $0, \pm\sqrt{2}$ (eigenvalues). The full form of Eq. (3.76) is

$$\begin{bmatrix} 0 & 1 & 0 \\ 1 & 0 & 1 \\ 0 & 1 & 0 \end{bmatrix} \begin{bmatrix} c_1 \\ c_2 \\ c_3 \end{bmatrix} = \lambda \begin{bmatrix} c_1 \\ c_2 \\ c_3 \end{bmatrix}$$

or, $$\begin{bmatrix} c_2 \\ c_1 + c_3 \\ c_2 \end{bmatrix} = \begin{bmatrix} \lambda c_1 \\ \lambda c_2 \\ \lambda c_3 \end{bmatrix} \tag{3.77}$$

From the first and second rows of Eq. (3.77) we can express c_2 and c_3 in terms of c_1:

$$c_2 = \lambda c_1 \quad \text{and} \quad c_3 = \lambda c_2 - c_1 = \lambda^2 c_1 - c_1 = (\lambda^2 - 1)c_1 \tag{3.78}$$

[The third row is redundant].

From the normalisation condition $c_1^2 + c_2^2 + c_3^2 = 1$ we have

$$[1 + \lambda^2 + (\lambda^2 - 1)^2]c_1^2 = 1$$

or, $$c_1 = \frac{1}{\sqrt{1 + \lambda^2 + (\lambda^2 - 1)^2}}$$

For $\lambda = 0$, $c_1 = \frac{1}{\sqrt{2}}$, and $c_2 = \lambda c_1 = 0$, $c_3 = (\lambda^2 - 1)\, c_1 = -c_1 = -\frac{1}{\sqrt{2}}$

For $\lambda = \sqrt{2}$, $c_1 = \frac{1}{2}$ and $c_2 = \lambda c_1 = \frac{1}{\sqrt{2}}$, $c_3 = (\lambda^2 - 1)\, c_1 = c_1 = \frac{1}{2}$

For $\lambda = -\sqrt{2}$, $c_1 = \frac{1}{2}$ and $c_2 = \lambda c_1 = -\frac{1}{\sqrt{2}}$, $c_3 = (\lambda^2 - 1)c_1 = c_1 = \frac{1}{2}$

We can tabulate the results as follows:

$$\text{Eigenvalues:} \qquad \sqrt{2} \quad 0 \quad -\sqrt{2}$$

$$\text{Eigenvector matrix:} \begin{bmatrix} \frac{1}{2} & \frac{1}{\sqrt{2}} & \frac{1}{2} \\ \frac{1}{\sqrt{2}} & 0 & -\frac{1}{\sqrt{2}} \\ \frac{1}{2} & -\frac{1}{\sqrt{2}} & \frac{1}{2} \end{bmatrix}$$

We can adopt an easier alternative procedure for obtaining the eigenvectors from Eq. (3.78):

Take $c_1 = 1$. Then $c_2 = \lambda$ and $c_3 = \lambda^2 - 1$. For $\lambda = 0$ we get the unnormalised eigenvector components, $c_1 = 1$, $c_2 = 0$ and $c_3 = -1$. Upon normalisation (that is, division by the norm, $\sqrt{c_1^2 + c_2^2 + c_3^2} = \sqrt{2}$), the eigenvector is $\begin{bmatrix} \frac{1}{\sqrt{2}} & 0 & -\frac{1}{\sqrt{2}} \end{bmatrix}^T$. Similarly for $\lambda = \sqrt{2}$ and $-\sqrt{2}$ the normalised eigenvectors are found to be $\begin{bmatrix} \frac{1}{2} & \frac{1}{\sqrt{2}} & \frac{1}{2} \end{bmatrix}^T$ and $\begin{bmatrix} \frac{1}{2} & -\frac{1}{\sqrt{2}} & \frac{1}{2} \end{bmatrix}^T$ respectively.

3.10.2 Eigenvalues of the Unitary Matrix

The eigenvalues of a unitary matrix are of absolute value unity. If the unitary matrix is also real orthogonal, then the eigenvalues occur as complex conjugate pairs. For example, consider the matrix

$$\begin{bmatrix} \cos\theta & \sin\theta \\ -\sin\theta & \cos\theta \end{bmatrix}$$

Its secular determinant is

$$D = \begin{vmatrix} x - c & s \\ -s & x - c \end{vmatrix} \quad \text{where} \quad c = \cos\theta, s = \sin\theta$$

On expanding, $D = (x - c)^2 + s^2 = x^2 - 2cx + 1$

So the eigenvalues are $x = \dfrac{2c \pm \sqrt{4c^2 - 4}}{2} = c \pm \sqrt{c^2 - 1} = c \pm is$

The eigenvalues are in complex conjugate pairs, with absolute value $\sqrt{c^2 + s^2} = 1$.

3.11 TWO IMPORTANT PROPERTIES OF EIGENVALUES AND EIGENVECTORS OF HERMITIAN MATRICES

Let λ_1 and λ_2 be two distinct eigenvalues of a hermitian matrix **A** (of size $n \times n$), the corresponding eigenvectors being $\mathbf{v}_1$ and $\mathbf{v}_2$ (where each **v** is a column matrix of n rows). Then

$$\mathbf{Av}_1 = \lambda_1 \mathbf{v}_1 \tag{3.79a}$$

$$\mathbf{Av}_2 = \lambda_2 \mathbf{v}_2 \tag{3.79b}$$

Left-multiplying both sides of (3.79a) by $\mathbf{v}_2^\dagger$,

$$\mathbf{v}_2^\dagger \mathbf{Av}_1 = \mathbf{v}_2^\dagger \lambda_1 \mathbf{v}_1 = \lambda_1 \mathbf{v}_2^\dagger \mathbf{v}_1 \tag{3.80a}$$

Taking the hermitian adjoint of Eq. (3.79b) and right-multiplying by $\mathbf{v}_1$,

$$(\mathbf{Av}_2)^\dagger \mathbf{v}_1 = (\lambda_2 \mathbf{v}_2)^\dagger \mathbf{v}_1 = \lambda_2^* \mathbf{v}_2^\dagger \mathbf{v}_1 \tag{3.80b}$$

As $\mathbf{A}$ is hermitian, $\mathbf{A}^\dagger = \mathbf{A}$ and so

$$(\mathbf{Av}_2)^\dagger \mathbf{v}_1 = \mathbf{v}_2^\dagger \mathbf{A}^\dagger \mathbf{v}_1 = \mathbf{v}_2^\dagger \mathbf{Av}_1 \tag{3.81}$$

Hence from Eqs (3.80a) and (3.80b), we get

$$\lambda_2^* \mathbf{v}_2^\dagger \mathbf{v}_1 = \lambda_1 \mathbf{v}_2^\dagger \mathbf{v}_1$$

or,
$$(\lambda_2^* - \lambda_1)\mathbf{v}_2^\dagger \mathbf{v}_1 = 0 \tag{3.82}$$

Case 1: if $\mathbf{v}_1$ and $\mathbf{v}_2$ are the same eigenvectors, $\mathbf{v}_1^\dagger \mathbf{v}_1 = |\mathbf{v}_1|^2 \neq 0$ and from Eq. (3.82),

$$(\lambda_1^* - \lambda_1) = 0 \text{ or, } \lambda_1^* = \lambda_1 \Rightarrow \lambda_1 \text{ is real}$$

Case 2: if $\mathbf{v}_1$ and $\mathbf{v}_2$ are eigenvectors having different eigenvalues ($\lambda_2 \neq \lambda_1$), we have from Eq. (3.82),

$$\mathbf{v}_2^\dagger \mathbf{v}_1 = 0 \text{ that is, } \mathbf{v}_1 \text{ and } \mathbf{v}_2 \text{ are orthogonal vectors.}$$

Thus *hermitian matrices have real eigenvalues and eigenvectors corresponding to different eigenvalues are orthogonal.* Two or more eigenvalues of a hermitian matrix may be the same but the corresponding eigenvectors are different. This is a case of degeneracy. In such a case as well, the eigenvectors are orthogonal. The proof is a bit complicated and is beyond the scope of this text.

3.12 MATRIX DIAGONALISATION—AN ALTERNATIVE METHOD FOR SOLVING THE EIGENVALUE PROBLEM

An alternative and widely used procedure for solving the eigenvalue problem is matrix diagonalisation, which is based on similarity transformation by unitary matrices. The method rests on the following important theorem.

Theorem: *A similarity transformation does not change the set of eigenvalues of a matrix.*

Proof: Let $\mathbf{A}$ be a matrix of size $n \times n$. The characteristic equation for this matrix is

$$\det(\mathbf{A} - \lambda\mathbf{I}) = 0 \tag{3.83}$$

where **I** is unit matrix of size $n \times n$. In expanded from this equation is

$$\begin{vmatrix} a_{11}-\lambda & a_{12} & a_{13} & \vdots & a_{1n} \\ a_{21} & a_{22}-\lambda & a_{23} & \cdots & a_{2n} \\ a_{31} & a_{32} & a_{33}-\lambda & \cdots & a_{3n} \\ \cdots & \cdots & \cdots & \cdots & \cdots \\ a_{n1} & a_{n2} & a_{n3} & \cdots & a_{nn}-\lambda \end{vmatrix} = 0 \tag{3.84}$$

Equation 3.84 is an n-degree polynomial equation in λ and its n roots $\lambda_1, \lambda_2, \lambda_3, \ldots, \lambda_n$ are the *eigenvalues* of matrix **A**.

Let us now carry out a similarity transformation of **A** by another non-singular matix **C**:

$$\mathbf{C}^{-1}\mathbf{AC} = \mathbf{B} \tag{3.85}$$

The characteristic equation for matrix **B** is,

$$\det(\mathbf{B} - \lambda'\mathbf{I}) = 0 \tag{3.86}$$

Then $\det(\mathbf{C}^{-1}\mathbf{AC} - \lambda'\mathbf{I}) = 0$

or, $\det(\mathbf{C}^{-1}\mathbf{AC} - \lambda'\mathbf{C}^{-1}\mathbf{IC}) = 0$ (as λ' is a scalar and $\mathbf{C}^{-1}\mathbf{IC} = \mathbf{I}$)

or, $\det[\mathbf{C}^{-1}(\mathbf{A} - \lambda'\mathbf{I})\mathbf{C}] = 0$

or, $\det(\mathbf{C}^{-1}) \cdot \det(\mathbf{A} - \lambda'\mathbf{I}) \cdot \det(\mathbf{C}) = 0$

or, $\det(\mathbf{A} - \lambda'\mathbf{I}) = 0$ [as $\det(\mathbf{C}^{-1}) \cdot \det(\mathbf{C}) = \det(\mathbf{C}^{-1}\mathbf{C}) = 1$]

This is the same as Eq. (3.83). Hence

$$\{\lambda_k\} = \{\lambda'_k\} \text{ for } k = 1, 2, 3, \ldots, n. \tag{3.87}$$

Thus, the matrices **A** and **B** have the same set of eigenvalues; this set is also called *eigenspectrum.* The eigenspectrum of a matrix, therefore, is invariant under a similarity transformation.

Our next objective is to find a suitable matrix **C** which on being used in the similarity transformation $\mathbf{C}^{-1}\mathbf{AC}$ will yield a diagonal matrix containing exclusively the eigenvalues of **A** in the diagonal. The characteristic equation of **A** (Eq. 3.83) is a result of existence of non-trivial solution to the matrix eigenvalue equation,

$$\begin{bmatrix} a_{11}-\lambda & a_{12} & a_{13} & \cdots & a_{1n} \\ a_{21} & a_{22}-\lambda & a_{23} & \cdots & a_{2n} \\ a_{31} & a_{32} & a_{33}-\lambda & \cdots & a_{3n} \\ \cdots & \cdots & \cdots & \cdots & \cdots \\ a_{n1} & a_{n2} & a_{n3} & \cdots & a_{nn}-\lambda \end{bmatrix} \cdot \begin{bmatrix} v_1^k \\ v_2^k \\ v_3^k \\ \cdots \\ v_n^k \end{bmatrix} = \lambda_k \begin{bmatrix} v_1^k \\ v_2^k \\ v_3^k \\ \cdots \\ v_n^k \end{bmatrix} \tag{3.88a}$$

or, $$\mathbf{A}\mathbf{v}^k = \lambda_k \mathbf{v}^k \tag{3.88b}$$

where the column vector $\mathbf{v}^k$ is the *eigenvector* of $\mathbf{A}$ corresponding to the eigenvalues λ_k. For an $n \times n$ matrix $\mathbf{A}$, we can collect the n eigenvectors to form a matrix $\mathbf{V}$:

$$\mathbf{V} = (\mathbf{v}^1 \quad \mathbf{v}^2 \quad \mathbf{v}^3 \quad \cdots \quad \mathbf{v}^k \cdots \quad \mathbf{v}^n) \tag{3.89}$$

in which the (i, k) element of $\mathbf{V}$ is $V_{ik} = v_i^k$ (3.90)

We can also construct a diagonal matrix of the eigenvalues:

$$\Lambda = \begin{bmatrix} \lambda_1 & 0 & 0 & \cdots & 0 \\ 0 & \lambda_2 & 0 & \cdots & 0 \\ 0 & 0 & \lambda_3 & \cdots & 0 \\ \cdots & \cdots & \cdots & \cdots & \cdots \\ 0 & 0 & 0 & \cdots & \lambda_n \end{bmatrix}$$

for which the (i, k) element is $\Lambda_{ik} = \lambda_k \delta_{ik}$ (3.91)

where δ_{ik} is the Krönecker delta. Now, considering the ith row of both sides of the matrix equation (3.88a), we have

$$\begin{aligned} a_{i1} v_1^k + a_{i2} v_2^k + a_{i3} v_3^k + \cdots + a_{in} v_n^k &= \lambda_k v_i^k = \lambda_k V_{ik} \quad \text{[using Eq. (3.90)]} \\ &= \sum_j \lambda_k \delta_{jk} V_{ij} \\ &= \sum_j \Lambda_{jk} V_{ij} \quad \text{[using Eq. (3.91)]} \\ &= \sum_j V_{ij} \Lambda_{jk} \\ &= (\mathbf{V}\Lambda)_{ik} \end{aligned}$$

or, $\sum_j a_{ij} V_{jk} = (\mathbf{V}\Lambda)_{ik}$ [using the definition (3.90) in the LHS]

or, $(\mathbf{AV})_{ik} = (\mathbf{V}\Lambda)_{ik}$ for every i and k

or, $$\mathbf{AV} = \mathbf{V}\Lambda \tag{3.92}$$

If the eigenvectors of $\mathbf{A}$ are linearly independent, the matrix V is non-singular and we have from Eq. (3.92),

$$\mathbf{V}^{-1}\mathbf{AV} = \Lambda \tag{3.93}$$

Thus, *a matrix* $\mathbf{A}$ *with linearly independent eigenvectors can be transformed into a diagonal matrix* Λ *which contains the eigenvalues of* $\mathbf{A}$ *as diagonal elements, by subjecting* $\mathbf{A}$ *to a similarity transformation; the matrix* **V** *required for this similarity transformation should contain the eigenvectors of* **A** *as column vectors.*

However, both eigenvalues and eigenvectors of **A** are unknown at the beginning and the above principle cannot be applied directly. The matrix eigenvalue equation must be solved iteratively. Even if one tries to determine the eigenvalues first by solving the secular equation, an iterative procedure (such as the Newton–Raphson method) must be adopted because the secular equation is an n-degree polynomial equation in λ. A number of iterative methods for diagonalising a hermitian matrix are available (such as the Householder method and the Jacobi method). We discuss here the Jacobi method of matrix diagonalisation.

The eigenvectors of a hermitian matrix are orthogonal. Also, if they are normalised to unity then the eigenvector matrix **V** is unitary. The Jacobi method, therefore, starts with an initial guess of unitary matrix **U** to replace **V** in the similarity transformation of **A**. This procedure is discussed below.

Let a_{pq} $(= a_{qp})$ be the largest off-diagonal element (marked by double underline) in the given matrix **A** which is to be diagonalised:

$$\mathbf{A} = \begin{bmatrix} a_{11} & a_{12} & \cdots & a_{1p} & \cdots & a_{1q} & \cdots & a_{1n} \\ a_{21} & a_{22} & \cdots & a_{2p} & \cdots & a_{2q} & \cdots & a_{2n} \\ \cdots & \cdots & \cdots & \cdots & \cdots & \cdots & \cdots & \cdots \\ a_{p1} & a_{p2} & \cdots & a_{pp} & \cdots & \underline{\underline{a_{pq}}} & \cdots & a_{pn} \\ \cdots & \cdots & \cdots & \cdots & \cdots & \cdots & \cdots & \cdots \\ a_{q1} & a_{q2} & \cdots & \underline{\underline{a_{qp}}} & \cdots & a_{qq} & \cdots & a_{qn} \\ \cdots & \cdots & \cdots & \cdots & \cdots & \cdots & \cdots & \cdots \\ a_{n1} & a_{n2} & \cdots & a_{np} & \cdots & a_{nq} & \cdots & a_{nn} \end{bmatrix} \tag{3.94}$$

To start the iteration, Jacobi's method chooses a unitary matrix $\mathbf{U}_1$ such that

$$(\mathbf{U}_1)_{pp} = \cos\theta = c$$

$$(\mathbf{U}_1)_{pq} = \sin\theta = s$$

$$(\mathbf{U}_1)_{qp} = -\sin\theta = -s$$

$$(\mathbf{U}_1)_{qq} = \cos\theta = c \text{ and}$$

$$(\mathbf{U}_1)_{ii} = 1 \text{ for all other diagonal elements of } \mathbf{U}_1$$

$$(\mathbf{U}_1)_{ij} = 0 \text{ for all other off-diagonal elements of } \mathbf{U}_1$$

This means that $\mathbf{U}_1$ is the unit matrix **I** of size $n \times n$ in which the elements of the rotation matrix

$$\begin{bmatrix} \cos\theta & \sin\theta \\ -\sin\theta & \cos\theta \end{bmatrix} \text{ or, } \begin{bmatrix} c & s \\ -s & c \end{bmatrix}$$

occupy the (p, p), (p, q), (q, p) and (q, q) positions:

$$\mathbf{U}_1 = \begin{bmatrix} 1 & 0 & 0 & \cdots & 0 & \cdots & 0 & 0 & 0 & \cdots & 0 & 0 \\ 0 & 1 & 0 & \cdots & 0 & \cdots & 0 & 0 & 0 & \cdots & 0 & 0 \\ 0 & 0 & 1 & \cdots & 0 & \cdots & 0 & 0 & 0 & \cdots & 0 & 0 \\ \cdots & \cdots & \cdots & \ddots & \cdots & \cdots & \cdots & \cdots & \cdots & \cdots & \cdots & \cdots \\ 0 & 0 & 0 & \cdots & c & \cdots & 0 & s & 0 & \cdots & 0 & 0 \\ \cdots & \cdots & \cdots & \cdots & \cdots & \ddots & \cdots & \cdots & \cdots & \cdots & \cdots & \cdots \\ 0 & 0 & 0 & \cdots & 0 & \cdots & 1 & 0 & 0 & \cdots & 0 & 0 \\ 0 & 0 & 0 & \cdots & -s & \cdots & 0 & c & 0 & \cdots & 0 & 0 \\ 0 & 0 & 0 & \cdots & 0 & \cdots & 0 & 0 & 1 & \cdots & 0 & 0 \\ \cdots & \cdots & \cdots & \cdots & \cdots & \cdots & \cdots & \cdots & \cdots & \ddots & \cdots & \cdots \\ 0 & 0 & 0 & \cdots & 0 & \cdots & 0 & 0 & 0 & \cdots & 1 & 0 \\ 0 & 0 & 0 & \cdots & 0 & \cdots & 0 & 0 & 0 & \cdots & 0 & 1 \end{bmatrix} \tag{3.95}$$

In the matrix multiplications necessary for the similarity transformation $\mathbf{U}_1^{-1}\mathbf{A}\mathbf{U}_1$ $(= \mathbf{U}_1^{\dagger}\mathbf{A}\mathbf{U}_1)$, it will be convenient to consider only the 2×2 matrices consisting of the (p, p), (p, q), (q, p) and (q, q) positions because the unit column vectors for the other columns of $\mathbf{U}_1$ will effectively induce no change in the matrix elements of $\mathbf{A}$. We therefore have to consider

$$\mathbf{U}_1^{\dagger}\mathbf{A}\mathbf{U}_1 \to \begin{bmatrix} c & -s \\ s & c \end{bmatrix} \cdot \begin{bmatrix} a_{pp} & a_{pq} \\ a_{qp} & a_{qq} \end{bmatrix} \cdot \begin{bmatrix} c & s \\ -s & c \end{bmatrix}$$

$$= \begin{bmatrix} c & -s \\ s & c \end{bmatrix} \cdot \begin{bmatrix} c \cdot a_{pp} - s \cdot a_{pq} & s \cdot a_{pp} + c \cdot a_{pq} \\ c \cdot a_{qp} - s \cdot a_{qq} & s \cdot a_{qp} + c \cdot a_{qq} \end{bmatrix}$$

$$= \begin{bmatrix} c\,(c \cdot a_{pp} - s \cdot a_{pq}) - s\,(c \cdot a_{qp} - s \cdot a_{qq}) & c\,(s \cdot a_{pp} + c \cdot a_{pq}) - s\,(s \cdot a_{qp} + c \cdot a_{qq}) \\ s\,(c \cdot a_{pp} - s \cdot a_{pq}) + c\,(c \cdot a_{qp} - s \cdot a_{qq}) & s\,(s \cdot a_{pp} + c \cdot a_{pq}) + c\,(s \cdot a_{qp} + c \cdot a_{qq}) \end{bmatrix}$$

in which the off-diagonal element is

$$s\,(c \cdot a_{pp} - s \cdot a_{pq}) + c\,(c \cdot a_{qp} - s \cdot a_{qq})$$

$$= sc \cdot a_{pp} - s^2 \cdot a_{pq} + c^2 \cdot a_{qp} - cs \cdot a_{qq}$$

$$= \frac{1}{2}(2sc) \cdot (a_{pp} - a_{qq}) + (c^2 - s^2) \cdot a_{pq} \qquad (\text{as } a_{pq} = a_{qp})$$

$$= \frac{1}{2}\sin 2\theta \cdot (a_{pp} - a_{qq}) + \cos 2\theta \cdot a_{pq}$$

θ has to be so chosen that this off-diagonal element vanishes. Hence

$$\frac{1}{2}\sin 2\theta \cdot (a_{pp} - a_{qq}) + \cos 2\theta \cdot a_{pq} = 0$$

$$\text{or,} \qquad \tan 2\theta = \frac{2a_{pq}}{(a_{qq} - a_{pp})}$$

$$\text{or,} \qquad \theta = \frac{1}{2}\tan^{-1}\frac{2a_{pq}}{(a_{qq} - a_{pp})} \tag{3.96}$$

Using this value of θ, the matrices $\mathbf{U}_1$ and $\mathbf{B} = \mathbf{U}_1^\dagger \mathbf{A} \mathbf{U}_1$ are constructed. Now, if the largest off-diagonal element of $\mathbf{B}$ is b_{mn} then the above procedure is repeated by constructing a new unitary matrix $\mathbf{U}_2$ whose (m, m), (m, n), (n, m) and (n, n) elements are respectively $\cos\alpha$, $\sin\alpha$, $-\sin\alpha$ and $\cos\alpha$:

$$\mathbf{U}_2 = \begin{bmatrix} 1 & 0 & 0 & \cdots & 0 & \cdots & 0 & 0 & 0 & \cdots & 0 & 0 \\ 0 & 1 & 0 & \cdots & 0 & \cdots & 0 & 0 & 0 & \cdots & 0 & 0 \\ 0 & 0 & c & \cdots & 0 & \cdots & 0 & 0 & s & \cdots & 0 & 0 \\ \cdots & \cdots & \cdots & \ddots & \cdots & \cdots & \cdots & \cdots & \cdots & \cdots & \cdots & \cdots \\ 0 & 0 & 0 & \cdots & 1 & \cdots & 0 & s & 0 & \cdots & 0 & 0 \\ \cdots & \cdots & \cdots & \cdots & \cdots & \ddots & \cdots & \cdots & \cdots & \cdots & \cdots & \cdots \\ 0 & 0 & 0 & \cdots & 0 & \cdots & 1 & 0 & 0 & \cdots & 0 & 0 \\ 0 & 0 & 0 & \cdots & 0 & \cdots & 0 & 1 & 0 & \cdots & 0 & 0 \\ 0 & 0 & -s & \cdots & 0 & \cdots & 0 & 0 & c & \cdots & 0 & 0 \\ \cdots & \cdots & \cdots & \cdots & \cdots & \cdots & \cdots & \cdots & \cdots & \ddots & \cdots & \cdots \\ 0 & 0 & 0 & \cdots & 0 & \cdots & 0 & 0 & 0 & \cdots & 1 & 0 \\ 0 & 0 & 0 & \cdots & 0 & \cdots & 0 & 0 & 0 & \cdots & 0 & 1 \end{bmatrix} \tag{3.97}$$

As before, one finds $\alpha = \frac{1}{2}\tan^{-1}\frac{2b_{mn}}{(b_{nn} - b_{mm})}$ for making the (m, n) off-diagonal element zero.

Using this value of α, the matrices $\mathbf{U}_2$ and $\mathbf{C} = \mathbf{U}_2^\dagger \mathbf{B} \mathbf{U}_2$ are constructed. In matrix $\mathbf{C}$, the (m, n) and (n, m) elements are each 0 but now the previously nullified (p, q) and (q, p) elements may again become non-zero. The process is therefore again performed on $\mathbf{C}$. This iterative procedure is stopped when it is found that the magnitude of each off-diagonal element is less than a pre-assigned, very small quantity, say $\varepsilon = 10^{-16}$. At this stage it is said that convergence of iteration is achieved and the resulting final matrix is

$$\text{diag}\,\{\lambda\} = \begin{bmatrix} \lambda_1 & 0 & 0 & 0 & 0 \\ 0 & \lambda_2 & 0 & 0 & 0 \\ 0 & 0 & \ddots & 0 & 0 \\ 0 & 0 & 0 & \ddots & 0 \\ 0 & 0 & 0 & 0 & \lambda_n \end{bmatrix}$$

where the λ's are the eigenvalues of $\mathbf{A}$. Now,

$$\mathbf{B} = \mathbf{U}_1^{-1}\mathbf{A}\mathbf{U}_1$$

$$\mathbf{C} = \mathbf{U}_2^{-1}\mathbf{B}\mathbf{U}_2 = \mathbf{U}_2^{-1}\,\mathbf{U}_1^{-1}\mathbf{A}\mathbf{U}_1\mathbf{U}_2 = (\mathbf{U}_1\mathbf{U}_2)^{-1}\,\mathbf{A}(\mathbf{U}_1\mathbf{U}_2)$$

$$\ldots$$

$$\ldots$$

$$\text{diag}\,\{\lambda\} = (\mathbf{U}_2^{-1}\,\mathbf{U}_1^{-1}\ldots\mathbf{A}\mathbf{U}_1\mathbf{U}_2\ldots) = (\mathbf{U}_1\mathbf{U}_2\ldots)^{-1}\,\mathbf{A}(\mathbf{U}_1\mathbf{U}_2\ldots)$$

$$= \mathbf{U}^{-1}\mathbf{A}\mathbf{U} \tag{3.98}$$

where $\mathbf{U} = \mathbf{U}_1\mathbf{U}_2\ldots$ is a unitary matrix (the product of unitary matrices is also unitary). The column vectors of $\mathbf{U}$ are normalised eigenvectors of $\mathbf{A}$.

Note: All square matrices are not diagonalisable. The matrix $\mathbf{A}$ is diagonalisable only if its eigenvectors are linearly independent. Since hermitian matrices have orthogonal eigenvectors, they fulfil the condition of diagonalisability. Hence the Jacobi method can be applied safely to hermitian or real symmetric matrices.

Example 3.8

Consider the matrix $\mathbf{A} = \begin{bmatrix} 0 & 1 & 1 \\ 1 & 0 & 1 \\ 1 & 1 & 0 \end{bmatrix}$

Find the eigenvalues and the corresponding eigenvectors.

First of all we try to make the (1, 3) element zero. For this the required unitary matrix is $\mathbf{U}_1$, where

$$\mathbf{U}_1 = \begin{bmatrix} c & 0 & s \\ 0 & 1 & 0 \\ -s & 0 & c \end{bmatrix} \quad \text{(according to the Jacobi method)}$$

with $c = \cos\theta$, $s = \sin\theta$ and θ is given by the equation

$$\theta = \frac{1}{2}\tan^{-1}\frac{2a_{13}}{(a_{33} - a_{11})} = \frac{1}{2}\tan^{-1}\frac{2\cdot 1}{(0-0)} = \frac{1}{2}\tan^{-1}\infty = \frac{\pi}{4}$$

Thus $\quad c = \dfrac{1}{\sqrt{2}},\ s = \dfrac{1}{\sqrt{2}}$ and $\mathbf{A}\mathbf{U}_1 = \begin{bmatrix} 0 & 1 & 1 \\ 1 & 0 & 1 \\ 1 & 1 & 0 \end{bmatrix} \cdot \begin{bmatrix} c & 0 & s \\ 0 & 1 & 0 \\ -s & 0 & c \end{bmatrix} = \begin{bmatrix} -s & 1 & c \\ c-s & 0 & c+s \\ c & 1 & s \end{bmatrix}$

Now, $\mathbf{U}_1^{-1} = \mathbf{U}_1^{\dagger}$ (since $\mathbf{U}_1$ is unitary) and so using $\theta = \dfrac{\pi}{4}$ we have

$$\mathbf{U}_1^{-1}\mathbf{A}\mathbf{U}_1 = \begin{bmatrix} c & 0 & -s \\ 0 & 1 & 0 \\ s & 0 & c \end{bmatrix} \cdot \begin{bmatrix} -s & 1 & c \\ c-s & 0 & c+s \\ c & 1 & s \end{bmatrix} = \begin{bmatrix} -2cs & c-s & c^2 - s^2 \\ c-s & 0 & c+s \\ c^2 - s^2 & c+s & 2cs \end{bmatrix}$$

$$= \begin{bmatrix} -1 & 0 & 0 \\ 0 & 0 & \sqrt{2} \\ 0 & \sqrt{2} & 1 \end{bmatrix} = \mathbf{A}_1 \text{ (say)}$$

(As $2cs = \sin 2\theta = \sin \frac{\pi}{2} = 1, c^2 - s^2 = \cos 2\theta = \cos \frac{\pi}{2} = 0$, and so on.)

Next we make the (2, 3) off-diagonal element (of $\mathbf{A}_1$) zero. The necessary unitary matrix is

$$\mathbf{U}_2 = \begin{bmatrix} 1 & 0 & 0 \\ 0 & c & s \\ 0 & -s & c \end{bmatrix}, c = \cos\alpha, s = \sin\alpha, \text{ and}$$

$$\alpha = \frac{1}{2} \tan^{-1} \frac{2a'_{23}}{(a'_{33} - a'_{22})} \quad \text{(where prime refers to } \mathbf{A}_1\text{)}$$

Thus $\alpha = \frac{1}{2} \tan^{-1} \frac{2\sqrt{2}}{(1-0)}$ or, $\tan 2\alpha = 2\sqrt{2}$ or, $\frac{2 \tan\alpha}{1 - \tan^2\alpha} = 2\sqrt{2}$ which simplifies to

$\sqrt{2} \tan^2\alpha + \tan\alpha - \sqrt{2} = 0$, that is, $\tan\alpha = \frac{-1 \pm \sqrt{1+8}}{2\sqrt{2}} = \frac{1}{\sqrt{2}}$ (taking the positive square root only). After a simple trigonometric calculation this gives

$$c = \cos\alpha = \sqrt{\frac{2}{3}}, s = \sin\alpha = \frac{1}{\sqrt{3}}$$

$$\text{Hence } \mathbf{U}_2 = \begin{bmatrix} 1 & 0 & 0 \\ 0 & \sqrt{\frac{2}{3}} & \frac{1}{\sqrt{3}} \\ 0 & -\frac{1}{\sqrt{3}} & \sqrt{\frac{2}{3}} \end{bmatrix}$$

$$\text{This will give } \mathbf{U}_2^{-1}\mathbf{A}_1\mathbf{U}_2 = \begin{bmatrix} -1 & 0 & 0 \\ 0 & -1 & 0 \\ 0 & 0 & 2 \end{bmatrix} = \text{diag}\{\lambda\}$$

Thus the eigenvalues of **A** are −1, −1, 2. To obtain the eigenvectors of **A**, we have to calculate $\mathbf{U}_1\mathbf{U}_2$.

$$\mathbf{U}_1\mathbf{U}_2 = \begin{bmatrix} c & 0 & s \\ 0 & 1 & 0 \\ -s & 0 & c \end{bmatrix} \cdot \begin{bmatrix} 1 & 0 & 0 \\ 0 & \sqrt{\frac{2}{3}} & \frac{1}{\sqrt{3}} \\ 0 & -\frac{1}{\sqrt{3}} & \sqrt{\frac{2}{3}} \end{bmatrix} \left(c = \frac{1}{\sqrt{2}}, s = \frac{1}{\sqrt{2}} \right)$$

$$= \begin{bmatrix} \frac{1}{\sqrt{2}} & -\frac{1}{\sqrt{6}} & \frac{1}{\sqrt{3}} \\ 0 & \sqrt{\frac{2}{3}} & \frac{1}{\sqrt{3}} \\ -\frac{1}{\sqrt{2}} & -\frac{1}{\sqrt{6}} & \frac{1}{\sqrt{3}} \end{bmatrix}$$

The eigenvectors corresponding to the eigenvalues $-1, -1, 2$ are respectively

$$\begin{bmatrix} \frac{1}{\sqrt{2}} & 0 & \frac{1}{\sqrt{2}} \end{bmatrix}^T, \begin{bmatrix} -\frac{1}{\sqrt{6}} & \sqrt{\frac{2}{3}} & -\frac{1}{\sqrt{6}} \end{bmatrix}^T \text{ and } \begin{bmatrix} \frac{1}{\sqrt{3}} & \frac{1}{\sqrt{3}} & \frac{1}{\sqrt{3}} \end{bmatrix}^T$$

SUMMARY

- ❑ A vector space is a collection of n-tuples of numbers or functions obeying certain associative, distributive, closure and commutative laws with respect to addition and multiplication by scalars.
- ❑ A set of vectors in a vector space is linearly independent if no vector in the set is expressible as a linear combination of the others. Functions can behave like vectors and their collection is a function space.
- ❑ Dirac introduced the *ket* and *bra* notations for working with functions and their complex conjugates which behave like vectors in function spaces.
- ❑ The inner product of two vectors $|u\rangle$ and $|v\rangle$, denoted as $\langle u|v\rangle$, is a non-negative scalar obeying certain rules and is analogous to the scalar product of two vectors in the ordinary 3-dimensional vector space. The method of taking the inner product is to be defined on the basis of requirement of the problem under question.
- ❑ Two vectors in a vector space are orthogonal if their inner product is zero; each vector is normalised if its inner product with itself is unity.
- ❑ A basis set is an orthonormal set of vectors in terms of which all vectors in the vector space can be uniquely described. A linear transformation in a vector space converts a vector into another vector of the same space. Given a basis set in a vector space, a linear operator can be expressed as a non-singular matrix.

- A matrix **H** is called hermitian if it is equal to its conjugate transpose, $(\mathbf{H}^*)^T$ or $\mathbf{H}^\dagger$. A matrix **U** is unitary if $\mathbf{U}^{-1} = \mathbf{U}^\dagger$. A hermitian matrix **H** has real eigenvalues; eigenvectors corresponding to different eigenvalues are orthogonal; **H** is diagonalised by a similarity transformation $\mathbf{U}^{-1}\mathbf{H}\mathbf{U} = \text{diag}\{\lambda\}$ where **U** is a unitary matrix whose column vectors are eigenvectors of **H** and the diagonal elements of diag$\{\lambda\}$ are the eigenvalues.

LEARNING OUTCOME

After reading this chapter the student will (i) understand how linear transformations act on members of a vector/function space producing matrices, (ii) learn the necessary techniques of matrix algebra that will be required in the remaining chapters and (iii) apply the Jacobi method of matrix diagonalisation for obtaining eigenvalues and eigenvectors of matrices.

EXERCISES

1. Find out whether the vectors (1, 0), (0, 1) and (2, 3) are linearly dependent or not.
2. Show that the vectors (1, 0, 0), (0, 0.5, 0) and (0, 0, 2) are linearly independent.
3. Find the inner product of all possible pairs of the following vectors and examine which pairs are orthogonal:

   $|u\rangle \equiv (1, 0, 0, 1), \quad |v\rangle \equiv (0, 0, 1, 2), \quad |w\rangle \equiv (1, 1, 1, 0), \quad |a\rangle \equiv (0, 0, 1, 0),$

   $|b\rangle \equiv (-1, 0, 0, 1)$
4. Show that the following matrices are unitary:

$$\mathbf{M}_1 = \frac{1}{\sqrt{3}}\begin{bmatrix} 1 & 1 & 1 \\ 1 & \omega & \omega^2 \\ 1 & \omega^* & \omega \end{bmatrix}, \quad \mathbf{M}_2 = \begin{bmatrix} \sqrt{3}/2 & -1/2 & 0 \\ 1/2 & \sqrt{3}/2 & 0 \\ 0 & 0 & 1 \end{bmatrix} \quad \text{and} \quad \mathbf{M}_3 = \frac{1}{2}\begin{bmatrix} 1 & 1 & 1 & 1 \\ 1 & -1 & 1 & -1 \\ 1 & i & -1 & -i \\ 1 & -i & -1 & i \end{bmatrix}$$

   Here 1, ω, ω^2 are the cube roots of unity and $i = \sqrt{-1}$.
5. By a linear transformation $\hat{T}$ a vector $\vec{u}$ becomes $\vec{v}$ and $\vec{v}$ becomes $\vec{u}$. Show that the resultant of $\vec{u}$ and $\vec{v}$ remains unaffected by $\hat{T}$.
6. Using the following matrices verify that **AB** may or may not be equal to **BA** but the result det **A** · det **B** = det (**AB**) = det (**BA**) is valid:

   (a) $\mathbf{A} = \begin{bmatrix} 1 & 0 & 1 \\ -1 & -1 & 1 \\ 1 & 1 & 2 \end{bmatrix}, \mathbf{B} = \begin{bmatrix} 1 & -1 & 1 \\ 2 & 0 & -1 \\ 1 & 2 & 3 \end{bmatrix}$

   (b) $\mathbf{A} = \begin{bmatrix} 1 & 2 & 3 \\ 3 & 1 & 2 \\ 2 & 1 & 3 \end{bmatrix}, \mathbf{B} = \begin{bmatrix} 1 & 0 & -1 \\ -1 & 1 & 0 \\ 0 & -1 & 1 \end{bmatrix}$

7. Deduce the condition for the following matrix equality to hold:

$$(\mathbf{A}+\mathbf{B})\,(\mathbf{A}-\mathbf{B})=\mathbf{A}^2-\mathbf{B}^2$$

8. The differential equation $\dfrac{d^2y}{dx^2}=-y, -\pi\le x\le\pi$, has two different forms of solutions: (a) $y=\text{A}\sin x+\text{B}\cos x$ and (b) $y=Pe^{ix}+Qe^{-ix}$. Find orthonormal basis sets for these two solution spaces and verify that the matrix for transformation from one basis set to the other is unitary.

9. With the matrices **U** and **B** show that the similarity transformation $\mathbf{U}^{-1}\mathbf{B}\mathbf{U}$ block diagonalises **B**:

$$\mathbf{U}=\begin{bmatrix}1/\sqrt{2} & -1/\sqrt{2} & 0\\ 1/\sqrt{2} & 1/\sqrt{2} & 0\\ 0 & 0 & 1\end{bmatrix},\ \mathbf{B}=\begin{bmatrix}1 & 2 & 3\\ 2 & 1 & 3\\ 3 & 3 & 9\end{bmatrix}$$

Also verify that **B** and its block diagonal form have the same determinant.

10. Find the eigenvalues and eigenvectors of the following matrices by both algebraic method and Jacobi method:

$$\text{(a)}\begin{bmatrix}3 & 1\\ 1 & 3\end{bmatrix}\quad \text{(b)}\begin{bmatrix}3 & 1\\ 1 & 2\end{bmatrix}\quad \text{(c)}\begin{bmatrix}15 & \sqrt{6} & 7\\ \sqrt{6} & 18 & \sqrt{6}\\ 7 & \sqrt{6} & 15\end{bmatrix}\quad \text{(d)}\begin{bmatrix}0 & 1 & 0 & 1\\ 1 & 0 & 1 & 0\\ 0 & 1 & 0 & 1\\ 1 & 0 & 1 & 0\end{bmatrix}\quad \text{(e)}\begin{bmatrix}5 & 2\sqrt{3} & 0\\ 2\sqrt{3} & 1 & 0\\ 0 & 0 & 1\end{bmatrix}$$

[**Answer**: The eigenvalues are (c) 24, 16, 8 (d) ±2, 0, 0 (e) 7, −1, 1]

11. Verify that the following pairs of matrices (**A**, **B**) have the same set of eigenvectors and that their eigenvalues differ by a constant

$$\text{(a) }\mathbf{A}=\begin{bmatrix}1 & 2\\ 2 & 0\end{bmatrix},\ \mathbf{B}=\begin{bmatrix}3 & 2\\ 2 & 2\end{bmatrix}\qquad \text{(b) }\mathbf{A}=\begin{bmatrix}3 & 1\\ 1 & 3\end{bmatrix},\ \mathbf{B}=\begin{bmatrix}0 & 1\\ 1 & 0\end{bmatrix}$$

12. Show that if the diagonal elements of a matrix **A** are all increased by a constant c, the eigenvectors do not change and the eigenvalues are each incremented by c.

13. The matrix equation $\mathbf{AX}=\lambda\mathbf{X}$, where **A** and **X** are matrices of size $n\times n$ and $n\times 1$ respectively and λ is a constant, is correct but $(\mathbf{A}-\lambda)\mathbf{X}=\mathbf{0}$ is incorrect. Why?

14. Construct $\mathbf{A}^\dagger$ and verify that $\mathbf{AA}^\dagger$ is hermitian with the following **A** matrices

$$\text{(a)}\begin{bmatrix}2 & e^{i} & 3\\ 2 & 4 & e^{-i}\\ 1 & e^{-i} & 5\end{bmatrix}\qquad \text{(b)}\begin{bmatrix}1 & 0 & 0\\ 0 & 0 & e^{i}\\ 0 & e^{-2i} & 0\end{bmatrix}$$

15. If $\mathbf{M} = \begin{bmatrix} \cos\theta & \sin\theta \\ -\sin\theta & \cos\theta \end{bmatrix}$, show that $\mathbf{M}^2 = \begin{bmatrix} \cos 2\theta & \sin 2\theta \\ -\sin 2\theta & \cos 2\theta \end{bmatrix}$. What do you guess about the meaning of **M** from this result?

16. Show that $(\mathbf{ABC})^{-1} = \mathbf{C}^{-1}\mathbf{B}^{-1}\mathbf{A}^{-1}$ where **A**, **B** and **C** are non-singular matrices of the same size.

Suggested Books and References for Further Reading

1. Dirac P A M, *The Principles of Quantum Mechanics*, Oxford University Press, **1958.**
2. Margenau H and Murphy G M, *The Mathematics of Physics and Chemistry*, van Nostrand, **1943.**
3. Kim S K, *Group Theoretical Methods and Applications to Molecules and Crystals*, Cambridge University Press, **1999.**

4 Representation of Symmetry Operators and Point Groups by Matrices

Learning objectives

- Representing symmetry operators by matrices
- Representation of a point group by a set of matrices in a given function space
- Invariant partitioning of basis sets and reducibility of representations
- Set of characters in a representation and equivalent representations
- Reduction of representations by similarity transformation

4.1 INTRODUCTION

A geometrical symmetry operator ($\hat{R}$) transforms the configuration of an object by shifting all points in the object to symmetry-equivalent positions or by transforming vectors associated with the object. Such transformations are all results of some *movement* of the points in the object. However, $\hat{R}$ cannot directly *move* an algebraic function in this way. But in quantum mechanics, it is such *functions* (wave functions, for example) that describe the state of the system. If symmetry properties are to be exploited to extract important results from such functions, an indirect way has to be adopted. Mathematical operators ($\hat{O}_R$), one for each symmetry operator $\hat{R}$, have to be assumed to exist such that their actions on desired functions produce the same result as that of applying the symmetry operator ($\hat{R}$) on the configuration of the object or on vectors associated with it. From the examples in the following sections it will be found that matrices can adequately play the role of $\hat{O}_R$. Expressing $\hat{O}_R$ in matrix form extends the limited geometrical manipulations with $\hat{R}$ to the much wider and useful field of matrix algebra.

4.2 DERIVATION OF MATRIX FORMS OF SYMMETRY OPERATORS FROM THE CARTESIAN COORDINATES OF A POINT (REPRESENTATION IN PHYSICAL SPACE)

The action of a symmetry operator $\hat{R}$ on a point P(x, y, z) in the physical space is readily expressed in the form of a matrix as shown in the following sections.

4.2.1 Matrix for Rotation about the Z-axis

Consider the effect of an anticlockwise rotation on a point P(x, y, z) through an angle θ about the Z-axis. Such an operation will be denoted by $\hat{C}_\theta(Z)$. Let the point P(x, y, z) be shifted to Q(x', y', z') by this operation. As the operation $\hat{C}_\theta(Z)$ does not change the Z-coordinate of a point, we consider the effect of $\hat{C}_\theta(Z)$ on a point $P_1(x, y, 0)$ situated on the XY-plane (Fig. 4.1, where the Z-axis has not been shown for clarity).

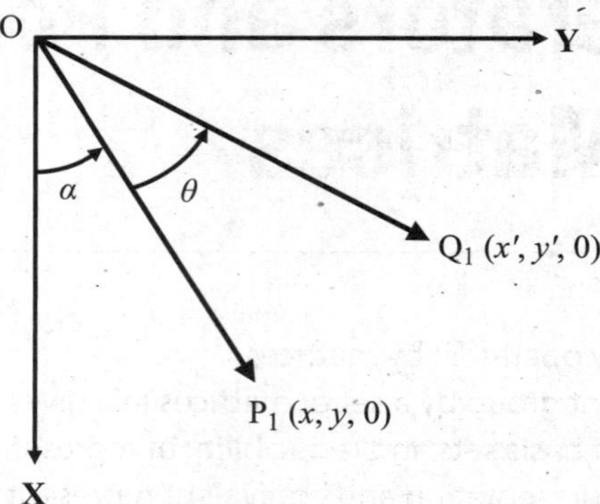

Fig. 4.1 Anticlockwise rotation of a point (x, y, 0) about the Z-axis

Let P_1 be transformed into $Q_1(x', y', 0)$ by this operation (Fig. 4.1).

Now, $$\begin{aligned} x' &= OQ_1\cos(\theta+\alpha) = OQ_1\cos\theta\cos\alpha - OQ_1\sin\theta\sin\alpha \\ &= (OP_1\cos\alpha)\cos\theta - (OP_1\sin\alpha)\sin\theta \quad (\text{as } OP_1 = OQ_1) \\ &= x\cos\theta - y\sin\theta + 0\cdot z \end{aligned}$$

Similarly, $$\begin{aligned} y' &= OQ_1\sin(\theta+\alpha) = OQ_1\sin\theta\cos\alpha + OQ_1\cos\theta\sin\alpha \\ &= (OP_1\cos\alpha)\sin\theta + (OP_1\sin\alpha)\cos\theta \\ &= x\sin\theta + y\cos\theta + 0\cdot z \end{aligned}$$

Recalling that $z' = z$ we can write the above equations in matrix form as

$$\begin{bmatrix} x' \\ y' \\ z' \end{bmatrix} = \begin{bmatrix} \cos\theta & -\sin\theta & 0 \\ \sin\theta & \cos\theta & 0 \\ 0 & 0 & 1 \end{bmatrix} \begin{bmatrix} x \\ y \\ z \end{bmatrix} \tag{4.1}$$

Thus, the matrix form for an anticlockwise rotation $\hat{C}_\theta(Z)$ acting on the Cartesian coordinates of a point is

$$\mathbf{M}(\hat{C}_\theta(Z)) = \begin{bmatrix} \cos\theta & -\sin\theta & 0 \\ \sin\theta & \cos\theta & 0 \\ 0 & 0 & 1 \end{bmatrix} \tag{4.2}$$

To get the matrix for a clockwise rotation, θ is to be replaced by $-\theta$. The correctness of this matrix form can be tested as follows: Since $\hat{C}_3\hat{C}_3^2 = \hat{E}$, one expects

$$\mathbf{M}(\hat{C}_3)\mathbf{M}(\hat{C}_3^2) = \mathbf{I}$$

where **I** is the identity matrix.

For $\hat{C}_3$, $\theta = 120°$ and

$$\mathbf{M}(\hat{C}_3) = \begin{bmatrix} \cos(\frac{2\pi}{3}) & -\sin(\frac{2\pi}{3}) & 0 \\ \sin(\frac{2\pi}{3}) & \cos(\frac{2\pi}{3}) & 0 \\ 0 & 0 & 1 \end{bmatrix} = \begin{bmatrix} -\frac{1}{2} & -\frac{\sqrt{3}}{2} & 0 \\ \frac{\sqrt{3}}{2} & -\frac{1}{2} & 0 \\ 0 & 0 & 1 \end{bmatrix} \tag{4.3}$$

For $\hat{C}_3^2$, which means two successive anticlockwise rotations, each time through 120° about the same axis, we have $\theta = 240°$. As $\cos 240° = -\frac{1}{2}$ and $\sin 240° = -\frac{\sqrt{3}}{2}$, we have

$$\mathbf{M}(\hat{C}_3^2) = \begin{bmatrix} -\frac{1}{2} & \frac{\sqrt{3}}{2} & 0 \\ -\frac{\sqrt{3}}{2} & -\frac{1}{2} & 0 \\ 0 & 0 & 1 \end{bmatrix} \tag{4.4}$$

So,
$$\mathbf{M}(\hat{C}_3)\mathbf{M}(\hat{C}_3^2) = \begin{bmatrix} -\frac{1}{2} & -\frac{\sqrt{3}}{2} & 0 \\ \frac{\sqrt{3}}{2} & -\frac{1}{2} & 0 \\ 0 & 0 & 1 \end{bmatrix} \cdot \begin{bmatrix} -\frac{1}{2} & \frac{\sqrt{3}}{2} & 0 \\ -\frac{\sqrt{3}}{2} & -\frac{1}{2} & 0 \\ 0 & 0 & 1 \end{bmatrix} = \begin{bmatrix} 1 & 0 & 0 \\ 0 & 1 & 0 \\ 0 & 0 & 1 \end{bmatrix} = \mathbf{I} \tag{4.5}$$

as can be readily verified by actually carrying out the matrix multiplication. The matrices $\mathbf{M}(\hat{C}_3)$ and $\mathbf{M}(\hat{C}_3^2)$ correspond to anticlockwise rotations about the Z-axis.

The inverse of the operation $\hat{C}_\theta(Z)$ means a rotation about the Z-axis through an angle θ in the clockwise direction. Hence the matrix for this inverse operation should be obtainable from the matrix in Eq. (4.2) by replacing θ by $-\theta$:

$$\mathbf{M}(\hat{C}_\theta^{-1}(Z)) = \begin{bmatrix} \cos(-\theta) & -\sin(-\theta) & 0 \\ \sin(-\theta) & \cos(-\theta) & 0 \\ 0 & 0 & 1 \end{bmatrix} = \begin{bmatrix} \cos\theta & \sin\theta & 0 \\ -\sin\theta & \cos\theta & 0 \\ 0 & 0 & 1 \end{bmatrix} \tag{4.6}$$

The correctness of this matrix can be tested by verifying that

(i) $\mathbf{M}(\hat{C}_\theta^{-1}(Z)) \cdot \mathbf{M}(\hat{C}_\theta(Z)) = \mathbf{I}$ and (ii) $\mathbf{M}(\hat{C}_\theta^{-1}(Z)) = \mathbf{M}(\hat{C}_{2\pi-\theta}(Z))$

4.2.2 Matrix Form for the Operator of Reflection on a Plane Which Contains the Z-axis and is Placed at an Angle β with the X-axis

Let us denote this operation by $\hat{\sigma}_z$. Then $\hat{\sigma}_z[P(x, y, z)] = P'(x', y', z')$. Since $\hat{\sigma}_z$ does not change the Z-coordinate, we may simplify the problem by investigating the effect of $\hat{\sigma}_z$ on a point situated on the XY plane (Fig. 4.2).

Then $x' = r\cos\angle P_1'OX = r\cos[2(\beta - \alpha) + \alpha]$

$$= r\cos(2\beta - \alpha)$$

$$= r\cos 2\beta \cos\alpha + r\sin 2\beta \sin\alpha$$

$$= (r\cos\alpha)\cos 2\beta + (r\sin\alpha)\sin 2\beta$$

$$= x\cos 2\beta + y\sin 2\beta + 0\cdot z$$

$$y' = r\sin(2\beta - \alpha) = x\sin 2\beta - y\cos 2\beta + 0\cdot z$$

and $$z' = z = 0\cdot x + 0\cdot y + 1\cdot z$$

Hence,

$$\begin{bmatrix} x' \\ y' \\ z' \end{bmatrix} = \begin{bmatrix} \cos 2\beta & \sin 2\beta & 0 \\ \sin 2\beta & -\cos 2\beta & 0 \\ 0 & 0 & 1 \end{bmatrix} \begin{bmatrix} x \\ y \\ z \end{bmatrix} \tag{4.7}$$

which gives

$$\mathbf{M}(\hat{\sigma}_z) = \begin{bmatrix} \cos 2\beta & \sin 2\beta & 0 \\ \sin 2\beta & -\cos 2\beta & 0 \\ 0 & 0 & 1 \end{bmatrix} \tag{4.8}$$

It can be readily verified that, whatever be the value of β, $\mathbf{M}(\hat{\sigma}_z)\cdot\mathbf{M}(\hat{\sigma}_z) = \mathbf{I}$ as required by the symmetry multiplication $\hat{\sigma}_z \cdot \hat{\sigma}_z = \hat{E}$.

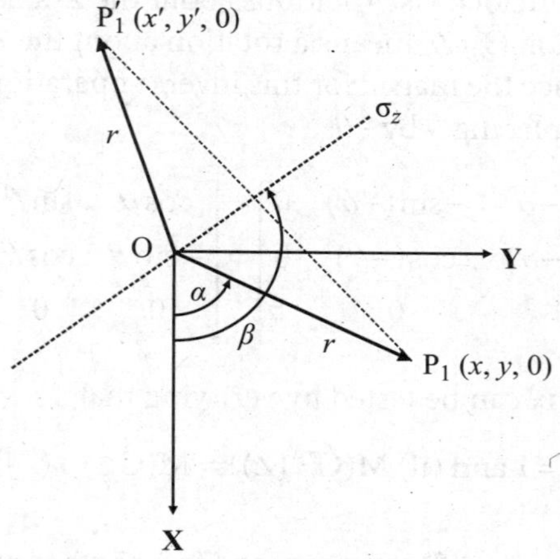

Fig. 4.2 Reflection of a point $(x, y, 0)$ on a plane σ_z containing the Z-axis and placed at an angle β with the X-axis

4.2.3 Matrix for Improper Rotation

Consider the operation $\hat{S}_\theta(Z) \equiv$ an anticlockwise rotation through an angle θ about the Z-axis followed by a reflection on the XY plane.

Let $$P(x, y, z) \xrightarrow{\hat{S}_\theta(Z)} Q'(x', y', z')$$

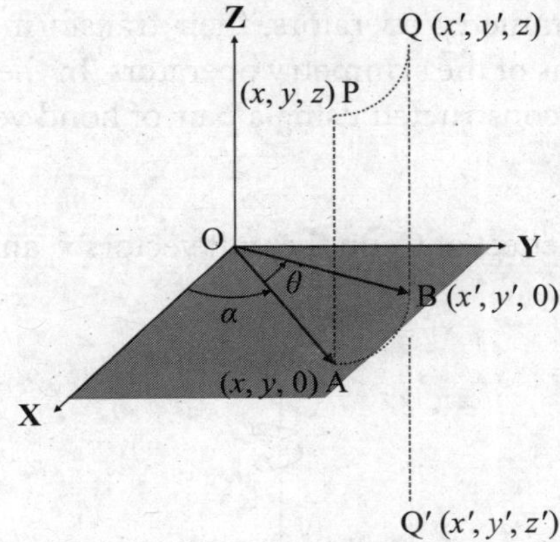

Fig. 4.3 Improper rotation of a point about the Z-axis

Rotation through an angle θ about the Z-axis brings the point P(x, y, z) to Q(x', y', z), that is, the z-coordinate does not change. When Q is reflected through the XY-plane, it comes to the final point Q′(x', y', z'). Then, obviously, $z' = -z$. Now, projections of P(x, y, z) and Q(x', y', z) on the XY plane are A(x, y, 0) and B(x', y', 0) respectively (Fig. 4.3).

Then, $x' = \text{OB} \cos(\theta + \alpha) = \text{OA} \cos\theta \cos\alpha - \text{OA} \sin\theta \sin\alpha$

$= (\text{OA} \cos\alpha) \cos\theta - (\text{OA} \sin\alpha) \sin\theta$

$= x \cos\theta - y \sin\theta \quad [\text{as OB} = \text{OA}, (\text{OA} \cos\alpha) = x, (\text{OA} \sin\alpha) = y]$

$= x \cos\theta - y \sin\theta + 0 \cdot z$

and $\quad y' = \text{OB} \sin(\theta + \alpha) = x \sin\theta + y \cos\theta + 0 \cdot z$

These, together with $z' = 0 \cdot x + 0 \cdot y + (-1) \cdot z$, give the matrix relation between the old and new coordinates:

$$\hat{S}_\theta(Z) \begin{bmatrix} x \\ y \\ z \end{bmatrix} = \begin{bmatrix} x' \\ y' \\ z' \end{bmatrix} = \begin{bmatrix} \cos\theta & -\sin\theta & 0 \\ \sin\theta & \cos\theta & 0 \\ 0 & 0 & -1 \end{bmatrix} \begin{bmatrix} x \\ y \\ z \end{bmatrix}$$

$$\text{Hence, } \mathbf{M}(\hat{S}_\theta(Z)) = \begin{bmatrix} \cos\theta & -\sin\theta & 0 \\ \sin\theta & \cos\theta & 0 \\ 0 & 0 & -1 \end{bmatrix} \tag{4.9}$$

4.3 MATRIX FORM OF SYMMETRY OPERATORS BY APPLYING THEM ON BASE VECTORS

One can choose a suitable number of vectors in the object under consideration and apply the possible symmetry operators ($\hat{R}$). Since atomic orbitals have spatial extension and orientation which can be changed by symmetry operators, their transformation properties can also be used to construct matrix forms of the symmetry operators. In the following two examples, the matrix form of $\hat{C}_2$ have been constructed using a pair of bond vectors and a set of orbitals of the atoms in H_2O molecule.

Example 4.1 Consider the effect of $\hat{C}_2$ on the unit vectors $\vec{v}_1$ and $\vec{v}_2$ along the O–H bonds of the H_2O molecule (Fig. 4.4).

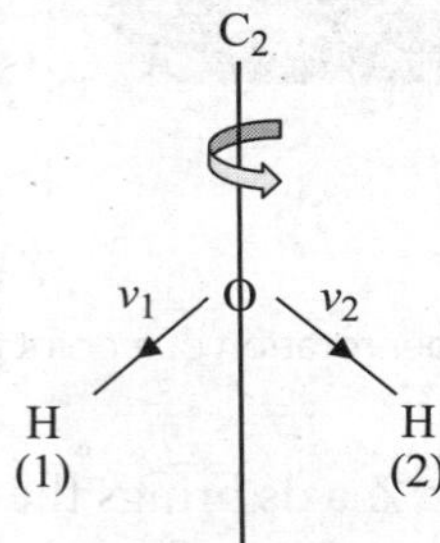

Fig. 4.4 Two unit vectors along the two O–H bonds of a H_2O molecule

Such bond vectors are required in the analysis of molecular vibrations. It is easy to verify that

$$\hat{C}_2\vec{v}_1 = \vec{v}_2 \quad \text{and} \quad \hat{C}_2\vec{v}_2 = \vec{v}_1$$

Hence, as before

$$\left.\begin{aligned} \hat{C}_2\,\vec{v}_1 &= 0\cdot\vec{v}_1 + 1\cdot\vec{v}_2 \\ \hat{C}_2\,\vec{v}_2 &= 1\cdot\vec{v}_1 + 0\cdot\vec{v}_2 \end{aligned}\right\} \tag{4.10}$$

or, in matrix form,

$$\hat{C}_2\begin{bmatrix}\vec{v}_1 & \vec{v}_2\end{bmatrix} = \begin{bmatrix}\vec{v}_1 & \vec{v}_2\end{bmatrix}\begin{bmatrix}0 & 1\\ 1 & 0\end{bmatrix} \tag{4.11}$$

Thus the required matrix for $\hat{C}_2$ is

$$\mathbf{M}^v(\hat{C}_2) = \begin{bmatrix}0 & 1\\ 1 & 0\end{bmatrix} \tag{4.12}$$

The superscript v on **M** indicates that the matrix has been constructed on the basis of transformation properties of the vectors $\vec{v}_1$ and $\vec{v}_2$. It is easy to verify that the square of the matrix $\mathbf{M}^v(\hat{C}_2)$ is the 2×2 unit matrix, which corresponds to $\hat{C}_2^2 = \hat{E}$.

Example 4.2 Representation of the $\hat{C}_2$ operator using the atomic orbitals of the H_2O molecule.

Let us consider the effect of the operator $\hat{C}_2$ on the valence shell atomic orbitals of the atoms in the H_2O molecule. With reference to Fig. 4.4, let the Z-axis be chosen along the C_2-axis and let the XY plane be the molecular plane. Also let ϕ_1, ϕ_2, ϕ_3, ϕ_4, ϕ_5 and ϕ_6 respectively denote the atomic orbitals $1s(\mathrm{H}(1))$, $1s(\mathrm{H}(2))$, $2s(\mathrm{O})$, $2p_x(\mathrm{O})$, $2p_y(\mathrm{O})$ and $2p_z(\mathrm{O})$. Since s-orbitals are spherically symmetrical,

$$\hat{C}_2\,\phi_1=\phi_2,\ \hat{C}_2\,\phi_2=\phi_1 \quad \text{and} \quad \hat{C}_2\,\phi_3=\phi_3$$

and considering the signs of the lobes of the three $2p$-orbitals of oxygen,

$$\hat{C}_2\,\phi_4=\hat{C}_2(2p_x)=-2p_x=-\phi_4;\ \hat{C}_2\,\phi_5=\hat{C}_2(2p_y)=-2p_y=-\phi_5;\ \hat{C}_2\,\phi_6=\hat{C}_2(2p_z)=2p_z=\phi_6$$

We can write these results as

$$\left.\begin{aligned}
\hat{C}_2\,\phi_1&=0\cdot\phi_1+1\cdot\phi_2+0\cdot\phi_3+0\cdot\phi_4+0\cdot\phi_5+0\cdot\phi_6\\
\hat{C}_2\,\phi_2&=1\cdot\phi_1+0\cdot\phi_2+0\cdot\phi_3+0\cdot\phi_4+0\cdot\phi_5+0\cdot\phi_6\\
\hat{C}_2\,\phi_3&=0\cdot\phi_1+0\cdot\phi_2+1\cdot\phi_3+0\cdot\phi_4+0\cdot\phi_5+0\cdot\phi_6\\
\hat{C}_2\,\phi_4&=0\cdot\phi_1+0\cdot\phi_2+0\cdot\phi_3-1\cdot\phi_4+0\cdot\phi_5+0\cdot\phi_6\\
\hat{C}_2\,\phi_5&=0\cdot\phi_1+0\cdot\phi_2+0\cdot\phi_3+0\cdot\phi_4-1\cdot\phi_5+0\cdot\phi_6\\
\hat{C}_2\,\phi_6&=0\cdot\phi_1+0\cdot\phi_2+0\cdot\phi_3+0\cdot\phi_4+0\cdot\phi_5+1\cdot\phi_6
\end{aligned}\right\} \tag{4.13}$$

In a matrix form, these equations can be written as:

$$\hat{C}_2[\phi_1\quad\phi_2\quad\phi_3\quad\phi_4\quad\phi_5\quad\phi_6]=[\phi_1\quad\phi_2\quad\phi_3\quad\phi_4\quad\phi_5\quad\phi_6]\begin{bmatrix}0&1&0&0&0&0\\1&0&0&0&0&0\\0&0&1&0&0&0\\0&0&0&-1&0&0\\0&0&0&0&-1&0\\0&0&0&0&0&1\end{bmatrix} \tag{4.14}$$

The square matrix on the RHS of Eq. (4.14) has been constructed by writing the starting orbitals ϕ_1, ϕ_2, and so on, in a row matrix and their coefficients appearing in Eq. (4.13) in columns. This matrix is a representation of the symmetry operator $\hat{C}_2$ with the set of orbitals $(\phi_1, \phi_2, \ldots, \phi_6)$ as basis.

4.4 REPRESENTATION OF SYMMETRY OPERATORS IN FUNCTION SPACE

A geometrical operator $\hat{R}$ cannot directly act on an algebraic function. Let $\hat{R}$ be replaced by a suitable mathematical operator $\hat{O}_R$ such that $\hat{O}_R$ transforms a function f to f' when $\hat{R}$ transforms the coordinates (x, y, z) of a point to (x', y', z'). Then $\hat{O}_R$ is said to be a *representation* of $\hat{R}$ if

$$f'(x', y', z') = f(x, y, z) \tag{4.15}$$

(Numerical value of the transformed function at the transformed position = numerical value of the old function at the old position.) Depending on this condition, matrices for the operators in function space can be constructed. An example is given below.

Consider an anticlockwise rotation $\hat{C}_\theta$ about the Z-axis using the five d-orbital wave functions:

$$\left.\begin{aligned} d_{x^2-y^2} &= N\left(\frac{1}{2}\right)(x^2-y^2) = d_1 \\ d_{xy} &= N(xy) = d_2 \\ d_{yz} &= N(yz) = d_3 \\ d_{zx} &= N(zx) = d_4 \\ d_{z^2} &= N\left(\frac{1}{2\sqrt{3}}\right)(3z^2-r^2) = d_5 \end{aligned}\right\} \tag{4.16}$$

where, in each case, $N = \sqrt{\frac{15}{4\pi}}\frac{1}{r^2}F(r)$ and $F(r)$ is a function of r. This r is related to the Cartesian coordinates by the relation $r^2 = x^2 + y^2 + z^2$.

If $\hat{C}_\theta$ shifts the point (x, y, z) to (x', y', z') and the mathematical operator $\hat{O}_c$ changes the function d_1 to the new function d_1' then, $\hat{O}_c$ will be a representation of $\hat{C}_\theta(Z)$ if

$$d_1'(x', y', z') = d_1(x, y, z) = N\left(\frac{1}{2}\right)(x^2-y^2) \tag{4.17}$$

It has already been found from Eq. (4.1) that

$$\begin{bmatrix} x' \\ y' \\ z' \end{bmatrix} = \begin{bmatrix} \cos\theta & -\sin\theta & 0 \\ \sin\theta & \cos\theta & 0 \\ 0 & 0 & 1 \end{bmatrix}\begin{bmatrix} x \\ y \\ z \end{bmatrix}.$$

$$\text{Hence}\quad \begin{bmatrix} x \\ y \\ z \end{bmatrix} = \begin{bmatrix} \cos\theta & -\sin\theta & 0 \\ \sin\theta & \cos\theta & 0 \\ 0 & 0 & 1 \end{bmatrix}^{-1}\begin{bmatrix} x' \\ y' \\ z' \end{bmatrix} \tag{4.18}$$

The transformation matrix on the RHS of Eq. (4.18) is unitary (as discussed in Section 3.9). Hence the inverse of this matrix is just its conjugate transpose; therefore,

$$\begin{bmatrix} x \\ y \\ z \end{bmatrix} = \begin{bmatrix} \cos\theta & \sin\theta & 0 \\ -\sin\theta & \cos\theta & 0 \\ 0 & 0 & 1 \end{bmatrix}\begin{bmatrix} x' \\ y' \\ z' \end{bmatrix} \tag{4.19}$$

This gives $x = cx' + sy'$, $y = -sx' + cy'$ and $z = z'$ where $c = \cos\theta$ and $s = \sin\theta$. Using these relations we have from Eq. (4.17),

$\hat{O}_c d_1$, that is, d_1' at the point (x', y', z')

$$= N\left(\frac{1}{2}\right)[(cx' + sy')^2 - (-sx' + cy')^2]$$

$$= N\left(\frac{1}{2}\right)(x'^2 - y'^2)\cos 2\theta + N(x'y')\sin 2\theta \text{ (on simplification)}$$

$$= d_1(x', y', z')\cos 2\theta + d_2(x', y', z')\sin 2\theta \tag{4.20}$$

The point (x, y, z) can be arbitrarily chosen; so (x', y', z') is also arbitrary. Hence the relation (4.20) is valid for any (x', y', z'). Therefore,

$$d_1' = d_1 \cos 2\theta + d_2 \sin 2\theta + d_3 \cdot 0 + d_4 \cdot 0 + d_5 \cdot 0$$

or, in matrix form, $$d_1' = \begin{bmatrix} d_1 & d_2 & d_3 & d_4 & d_5 \end{bmatrix} \begin{bmatrix} \cos 2\theta \\ \sin 2\theta \\ 0 \\ 0 \\ 0 \end{bmatrix} \tag{4.21}$$

In the same way, matrix expressions for d_2', d_3', d_4 and d_5' can be obtained. The result finally is,

$$\hat{O}_c[d_1 \quad d_2 \quad d_3 \quad d_4 \quad d_5] = [d_1' \quad d_2' \quad d_3' \quad d_4' \quad d_5']$$

$$= [d_1 \quad d_2 \quad d_3 \quad d_4 \quad d_5]\begin{bmatrix} \cos 2\theta & -\sin 2\theta & 0 & 0 & 0 \\ \sin 2\theta & \cos 2\theta & 0 & 0 & 0 \\ 0 & 0 & \cos\theta & \sin\theta & 0 \\ 0 & 0 & -\sin\theta & \cos\theta & 0 \\ 0 & 0 & 0 & 0 & 1 \end{bmatrix} \tag{4.22}$$

Thus, using the set of functions $\{d_i\}$, $i = 1$ to 5, the matrix form of $\hat{O}_c$ is

$$\mathbf{M}^d(\hat{O}_c) = \begin{bmatrix} \cos 2\theta & -\sin 2\theta & 0 & 0 & 0 \\ \sin 2\theta & \cos 2\theta & 0 & 0 & 0 \\ 0 & 0 & \cos\theta & \sin\theta & 0 \\ 0 & 0 & -\sin\theta & \cos\theta & 0 \\ 0 & 0 & 0 & 0 & 1 \end{bmatrix} \tag{4.23}$$

Since $\hat{O}_c$ is a representative of the symmetry operator $\hat{C}_\theta(\text{Z})$, the matrix $\mathbf{M}^d(\hat{O}_c)$ may be considered as a representation of $\hat{C}_\theta(\text{Z})$ itself in the function space spanned by the five d-orbital wave functions.

These examples suggest the following definition of representation of a symmetry operator $\hat{R}$: If a set of functions $(\phi_1, \phi_2, \ldots, \phi_n)$ are transformed to $(\phi_1', \phi_2', \ldots, \phi_n')$ by the operator $\hat{O}_R$

(a mathematical representative of the symmetry operator $\hat{R}$) when the arguments (x, y, z) of the functions are transformed to (x', y', z') by action of $\hat{R}$, then the matrix $\mathbf{M}^{\phi}(\hat{O}_R)$ is a representation of $\hat{R}$ subject to validity of the following relation:

$$[\phi_1' \quad \phi_2' \quad \dots \quad \phi_n'] = [(\phi_1 \quad \phi_2 \quad \dots \quad \phi_n)]\mathbf{M}^{\phi}(\hat{O}_R) \tag{4.24}$$

In the 3-dimensional physical space, if a point (x, y, z) is changed to (x', y', z') by the action of the operator $\hat{R}$, then the matrix $\mathbf{M}(\hat{R})$ is said to be a representation of $\hat{R}$ if

$$\begin{bmatrix} x' \\ y' \\ z' \end{bmatrix} = \mathbf{M}(\hat{R}) \begin{bmatrix} x \\ y \\ z \end{bmatrix} \tag{4.25}$$

We, therefore, see that for representing symmetry operators in the matrix form, we require some starting functions or vectors which are linearly independent. The set of such functions or vectors is known as the *basis set* for the representation.

Note: A symmetry operator $\hat{R}$ acts on a function f not directly as $\hat{R}f$ but as $\hat{O}_R f$ where $\hat{O}_R$ is the *mathematical operator* corresponding to $\hat{R}$. However, for the sake of brevity, the notation $\hat{R}f$ may be and will be used, keeping in mind that one $\hat{O}_R$ exists for each $\hat{R}$ and it is $\hat{O}_R$ that is acting on the function.

4.5 IMPORTANT POINTS ABOUT MATRIX REPRESENTATION OF SYMMETRY OPERATORS

(i) A given symmetry operator can have many different matrix representations depending on the basis set chosen. If there are n functions or vectors in a basis set, the representative matrix for a symmetry operator is of size $n \times n$.

(ii) Matrices representing symmetry operators must be non-singular (that is, their determinant must not be zero). This is because in a point group every symmetry operator has its own inverse, whose matrix form exists only when its determinant is non-zero (see Section 3.8.3).

(iii) From Examples 4.1 and 4.2, it is evident that if a basis function or vector remains unchanged by the action of a symmetry operator $\hat{R}$, then it contributes 1 to the diagonal of the representative matrix $\mathbf{M}(\hat{R})$; if under the action of $\hat{R}$ a basis function changes sign without changing the magnitude, or a basis vector (or, orbital) is inverted without being shifted, then it contributes -1 to the diagonal of $\mathbf{M}(\hat{R})$. If the basis function is changed to a new function or a basis vector (or orbital) is shifted to a new position by the action of $\hat{R}$, then it contributes zero to the diagonal [in this case a non-zero contribution is observed at some off-diagonal position of $\mathbf{M}(\hat{R})$]. This observation provides us with a very easy method of calculation of the trace (sum of the diagonal elements) of $\mathbf{M}(\hat{R})$ without actually constructing the matrix.

4.6 CHANGE OF BASIS AND RESULTING CHANGE IN THE MATRIX FORMS OF SYMMETRY OPERATORS

Recall Example 4.1: With the unit vectors $\vec{v}_1$ and $\vec{v}_2$ as basis, the matrix for the operator $\hat{C}_2$ of the H_2O molecule is

$$\mathbf{M}^{v}(\hat{C}_2) = \begin{bmatrix} 0 & 1 \\ 1 & 0 \end{bmatrix}$$

We may now construct a new set of vectors from $\vec{v}_1$ and $\vec{v}_2$ as

$$\vec{v}_1' = \frac{1}{\sqrt{2}}(\vec{v}_1 + \vec{v}_2) \quad \text{and} \quad \vec{v}_2' = \frac{1}{\sqrt{2}}(\vec{v}_1 - \vec{v}_2) \tag{4.26}$$

and try to get a representation of $\hat{C}_2$ in the new basis set $(\vec{v}_1', \vec{v}_2')$. (There is a well-defined method for constructing a suitable new basis set from the old ones; this will be discussed later at an appropriate place. For the time being we accept these new vectors in the way bonding and antibonding combinations of orbitals are constructed, the factor $\frac{1}{\sqrt{2}}$ being a normalisation factor.) The effect of $\hat{C}_2$ on $\vec{v}_1'$ and $\vec{v}_2'$ is as follows:

$$\hat{C}_2\vec{v}_1' = \hat{C}_2\left\{\frac{1}{\sqrt{2}}(\vec{v}_1 + \vec{v}_2)\right\} = \frac{1}{\sqrt{2}}(\vec{v}_1 + \vec{v}_2) = \vec{v}_1' = 1\cdot\vec{v}_1' + 0\cdot\vec{v}_2'$$

$$\hat{C}_2\vec{v}_2' = \hat{C}_2\left\{\frac{1}{\sqrt{2}}(\vec{v}_1 - \vec{v}_2)\right\} = -\frac{1}{\sqrt{2}}(\vec{v}_1 - \vec{v}_2) = -\vec{v}_2' = 0\cdot\vec{v}_1' - 1\cdot\vec{v}_2'$$

Thus, in matrix form,

$$\hat{C}_2\begin{bmatrix}\vec{v}_1' & \vec{v}_2'\end{bmatrix} = \begin{bmatrix}\vec{v}_1' & \vec{v}_2'\end{bmatrix}\begin{bmatrix} 1 & 0 \\ 0 & -1 \end{bmatrix} \tag{4.27}$$

and so $\mathbf{M}^{v'}(\hat{C}_2) = \begin{bmatrix} 1 & 0 \\ 0 & -1 \end{bmatrix}$ (4.28)

is the matrix representation of $\hat{C}_2$ in basis $(\vec{v}_1', \vec{v}_2')$. We now note the following two results with the above two matrix forms of $\hat{C}_2$:

(i) The relation (4.26) between the old and new basis sets can be represented in matrix form as

$$\begin{bmatrix}\vec{v}_1' & \vec{v}_2'\end{bmatrix} = \begin{bmatrix}\vec{v}_1 & \vec{v}_2\end{bmatrix}\begin{bmatrix} \frac{1}{\sqrt{2}} & \frac{1}{\sqrt{2}} \\ \frac{1}{\sqrt{2}} & -\frac{1}{\sqrt{2}} \end{bmatrix} = \begin{bmatrix}\vec{v}_1 & \vec{v}_2\end{bmatrix}\mathbf{T}\ \text{(say)} \tag{4.29}$$

The matrix T is unitary and it can be readily verified that

$$\mathbf{T}^{-1}\mathbf{M}^{v}(\hat{C}_2)\,\mathbf{T} = \mathbf{M}^{v'}(\hat{C}_2) \tag{4.30}$$

Thus, the matrix for an operator in the new basis set can be obtained from that in the old basis set by a similarity transformation using a matrix (**T**) which transforms the new basis set into the old one.

(ii) Trace of $\mathbf{M}^{v}(\hat{C}_2) = 0 =$ Trace of $\mathbf{M}^{v'}(\hat{C}_2)$, that is, a matrix and its similarity transform have equal trace.

4.6.1 An Important Theorem on Traces and its Consequence

Theorem: Similarity transformation does not change the trace of a matrix.

Proof: To prove this, we first show that for two square matrices, **A** and **B**, of the same size,

$$\text{Trace of } \mathbf{AB} = \text{Trace of } \mathbf{BA} \quad (\text{even if } \mathbf{AB} \neq \mathbf{BA}) \tag{4.31}$$

Let **A** and **B** be two square matrices of the same size, $n \times n$. Then the (i, j)-element of **AB** is given by

$$(\mathbf{AB})_{ij} = \sum_{k=1}^{n} (\mathbf{A})_{ik} (\mathbf{B})_{kj}$$

and hence, the trace of $\mathbf{AB} = \sum_{i=1}^{n} (\mathbf{AB})_{ii} = \sum_{i=1}^{n}\sum_{k=1}^{n} (\mathbf{A})_{ik} (\mathbf{B})_{ki}$

$= \sum_{i=1}^{n}\sum_{k=1}^{n} (\mathbf{B})_{ki} (\mathbf{A})_{ik}$ (as the matrix elements are numbers and their multiplication is commutative)

$= \sum_{k=1}^{n}\sum_{i=1}^{n} (\mathbf{B})_{ki} (\mathbf{A})_{ik}$ (as the order of summation is immaterial)

$= \sum_{k=1}^{n} (\mathbf{BA})_{kk} =$ trace of **BA** (proved)

Now let $\mathbf{B}^{-1}\mathbf{AB} = \mathbf{M}$ where **B** is a non-singular $n \times n$ matrix. Writing the trace of the matrix **A** as Tr **A** we have,

$\text{Tr}\,(\mathbf{M}) = \text{Tr}\,(\mathbf{B}^{-1}\mathbf{AB}) = \text{Tr}\,\mathbf{B}^{-1}(\mathbf{AB})$ (as matrix multiplication is associative)

$= \text{Tr}\,(\mathbf{AB})\mathbf{B}^{-1}$ (using Eq. 4.31)

$= \text{Tr}\,\mathbf{A}(\mathbf{BB}^{-1})$ (again using associativity)

$= \text{Tr}\,\mathbf{AI}$ [where $\mathbf{I} = (n \times n)$ unit matrix]

$= \text{Tr}\,\mathbf{A}$ (proved) (4.32)

Thus, the matrices **M** and **A**, which are similarity transforms of one another, have the same trace. This theorem has two very important implications:

Firstly, if two symmetry operators $\hat{R}_i$ and $\hat{R}_j$ of an object belong to the same class, then by definition,

$$\hat{R}_k^{-1}\hat{R}_i\hat{R}_k = \hat{R}_j \tag{4.33}$$

where $\hat{R}_k$ is a third symmetry operator of the same object. Hence, if $\mathbf{M}(\hat{R}_k)$, $\mathbf{M}(\hat{R}_i)$ and $\mathbf{M}(\hat{R}_j)$ are matrices representing $\hat{R}_k$, $\hat{R}_i$ and $\hat{R}_j$ respectively in some basis, then from Eq. (4.33), $[\mathbf{M}(\hat{R}_k)]^{-1}\,\mathbf{M}(\hat{R}_i)\,\mathbf{M}(\hat{R}_k) = \mathbf{M}(\hat{R}_j)$ and so by Eq. (4.32),

$$\text{Tr}\,\mathbf{M}(\hat{R}_i) = \text{Tr}\,\mathbf{M}(\hat{R}_j) \tag{4.34}$$

Thus *matrices for operators of the same conjugate class have the same trace.*

Secondly, the actual form of the matrix required to represent a symmetry operator $\hat{R}$, depends on the basis set chosen. But from a given basis set, one can construct another basis set by a suitable unitary transformation and the matrix representing $\hat{R}$ in this new basis has a different form; however, this new matrix is a similarity transform of the old one:

$$\mathbf{T}^{-1}\mathbf{M}^{\text{old}}(\hat{R})\,\mathbf{T} = \mathbf{M}^{\text{new}}(\hat{R})$$

where **T** is the transformation matrix for going from the old to the new basis set. Hence according to the present theorem, the old and the new matrices have different forms but the same trace. Owing to this invariance of trace under similarity transformation corresponding to a change of basis, *traces are more useful than the matrices themselves.*

For the above two reasons, traces of matrices representing symmetry operators are called *characters* (probably in analogy with the fact that change in the outer appearance of a person does not change his or her inherent character). In German, the word *spur* is also used instead of *character*.

[An alternative proof of the theorem:

Let $\mathbf{M}^{-1}\mathbf{A}\mathbf{M} = \mathbf{B}$ where **A** and **B** are two square matrices of the same size and **M** is a non-singular square matrix of the size of **A**.

Then $\quad \text{Tr}\,\mathbf{B} = \sum_i (\mathbf{B})_{ii} = \sum_i \sum_\lambda \sum_\mu (\mathbf{M}^{-1})_{i\lambda} (\mathbf{A})_{\lambda\mu} (\mathbf{M})_{\mu i}$

$$= \sum_\lambda \sum_\mu \sum_i (\mathbf{M})_{\mu i} (\mathbf{M}^{-1})_{i\lambda} (\mathbf{A})_{\lambda\mu}$$

$$= \sum_\lambda \sum_\mu (\mathbf{M}\mathbf{M}^{-1})_{\mu\lambda} (\mathbf{A})_{\lambda\mu}$$

$$= \sum_\mu \sum_\lambda (\mathbf{I})_{\mu\lambda} (\mathbf{A})_{\lambda\mu}$$

$$= \sum_\mu (\mathbf{IA})_{\mu\mu} = \sum_\mu (\mathbf{A})_{\mu\mu} = \text{Tr}\,\mathbf{A} \quad \text{(Proved)]}$$

4.6.2 An Example Showing the Validity of this Theorem

Let ϕ_1, ϕ_2, ϕ_3 denote the 1*s* orbitals of the H atoms in the NH_3 molecule labelled 1, 2 and 3 (Fig. 4.5).

The molecular point group is C_{3v} in which the operations $\hat{C}_3$ and $\hat{C}_3^2$ are in the same conjugate class and the action of these operators on these functions are as follows:

$$\hat{C}_3\phi_1 = \phi_2 = 0\cdot\phi_1 + 1\cdot\phi_2 + 0\cdot\phi_3$$
$$\hat{C}_3\phi_2 = \phi_3 = 0\cdot\phi_1 + 0\cdot\phi_2 + 1\cdot\phi_3$$
$$\hat{C}_3\phi_3 = \phi_1 = 1\cdot\phi_1 + 0\cdot\phi_2 + 0\cdot\phi_3$$

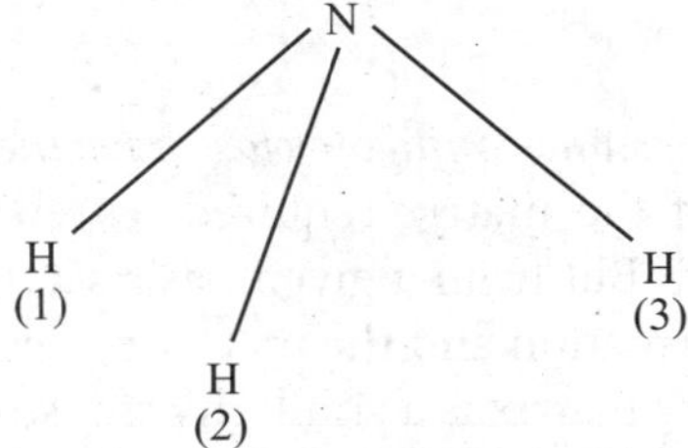

Fig. 4.5 Pyramidal NH_3 molecule

Hence the matrix representing $\hat{C}_3$ in basis (ϕ_1, ϕ_2, ϕ_3) is

$$\mathbf{M}^{\phi}(\hat{C}_3) = \begin{bmatrix} 0 & 0 & 1 \\ 1 & 0 & 0 \\ 0 & 1 & 0 \end{bmatrix}, \quad \text{Tr}\,\mathbf{M}^{\phi}(\hat{C}_3) = 0$$

Similarly,

$$\hat{C}_3^2 \phi_1 = \phi_3 = 0 \cdot \phi_1 + 0 \cdot \phi_2 + 1 \cdot \phi_3; \quad \hat{C}_3^2 \phi_2 = \phi_1 = 1 \cdot \phi_1 + 0 \cdot \phi_2 + 0 \cdot \phi_3;$$
$$\hat{C}_3^2 \phi_3 = \phi_2 = 0 \cdot \phi_1 + 1 \cdot \phi_2 + 0 \cdot \phi_3$$

and $$\mathbf{M}^{\phi}(\hat{C}_3^2) = \begin{bmatrix} 0 & 1 & 0 \\ 0 & 0 & 1 \\ 1 & 0 & 0 \end{bmatrix}, \quad \text{Tr}\,\mathbf{M}^{\phi}(\hat{C}_3^2) = 0$$

We know that $\hat{C}_3$ and $\hat{C}_3^2$ are similarity transforms of one another; this is true of their matrices as well. The matrices have equal trace, thus conjugate operators have the same character.

4.7 REPRESENTATION OF GROUPS

In the earlier sections we have seen the method of construction of matrices for representation of symmetry operators. Some well known properties of matrices are the following:

- The product of two $n \times n$ matrices is an $n \times n$ matrix.
- Matrix multiplication is associative.
- All non-singular matrices have their corresponding inverses.
- A matrix **A** does not change when it is multiplied by the unit matrix **I** of the size of **A**.

These properties hint at the four group properties. It is thus possible to represent the whole set of symmetry operators of a given point group by a set of non-singular matrices.

4.7.1 Definition and Example of Representation of a Point Group

Let $G = \{\hat{E}, \hat{R}_2, \hat{R}_3, \ldots, \hat{R}_g\}$ be the set of symmetry operators of a group of order g. Now, if there exists a second set $\Gamma = \{\alpha_1, \alpha_2, \alpha_3, \ldots, \alpha_g\}$ whose elements may be numbers or matrices such that there is a one-to-one correspondence between the sets G and Γ (that is, $\hat{E} \leftrightarrow \alpha_1$, $\hat{R}_2 \leftrightarrow \alpha_2$ and so on) and

whenever $\hat{R}_i \hat{R}_j = \hat{R}_k$ in the symmetry multiplication table,

we have $\alpha_i \alpha_j = \alpha_k$ (where the multiplication is either of numbers or of matrices),

then the set Γ is said to be a representation of the group G.

Consider the elements of C_{3v} and two sets of numbers Γ_1 and Γ_2 as tabulated below:

C_{3v}	$\hat{E}$	$\hat{C}_3$	$\hat{C}_3^2$	$\hat{\sigma}_v(1)$	$\hat{\sigma}_v(2)$	$\hat{\sigma}_v(3)$
Γ_1 :	1	1	1	1	1	1
Γ_2 :	1	1	1	−1	−1	−1

Here $\hat{C}_3 \hat{\sigma}_v(1) = \hat{\sigma}_v(2)$ (by symmetry multiplication) and, with the set Γ_1, 1 (for $\hat{C}_3$) × 1 (for $\hat{\sigma}_v(1)$) = 1, which corresponds to $\hat{\sigma}_v(2)$. Similarly with the set Γ_2, 1 (for $\hat{C}_3$) × −1 [for $\hat{\sigma}_v(1)$] = −1, which stands for $\hat{\sigma}_v(2)$. In this way it can be verified that the elements of set Γ_1 (and also those of Γ_2) multiply in accordance with the symmetry multiplication table of the C_{3v} point group. Hence Γ_1 and Γ_2 are representations of C_{3v}.

For another example, consider the representation of C_{2v} by matrices. Suppose we are dealing with the problem of finding the π-MOs of pyridine. Let (ϕ_1, ϕ_2, …, ϕ_6) be the π-type atomic orbitals of the atoms 1 to 6 taking part in π-MO formation. Let the π-conjugated atoms of this molecule be labelled as in Fig. 4.6 where the C_{2v} symmetry elements are also shown. The collection of all functions of the type $\sum_{i=1}^{6} c_i \phi_i$, where c_1, c_2, …, c_6 are appropriate numbers to be determined later, is a 6-dimensional function space with {ϕ_1, ϕ_2, …, ϕ_6} as the basis set; the MO wave functions are members of this space.

With this basis set, the matrix forms of the symmetry operators of C_{2v} are constructed as follows (taking the liberty that $\hat{R}\phi$ may be used for $\hat{O}_R\phi$):

$$\hat{E}\,\phi_1 = \phi_1 = 1\cdot\phi_1 + 0\cdot\phi_2 + 0\cdot\phi_3 + 0\cdot\phi_4 + 0\cdot\phi_5 + 0\cdot\phi_6$$

$$\hat{E}\,\phi_2 = \phi_2 = 0\cdot\phi_1 + 1\cdot\phi_2 + 0\cdot\phi_3 + 0\cdot\phi_4 + 0\cdot\phi_5 + 0\cdot\phi_6$$

…

…

$$\hat{E}\,\phi_6 = \phi_6 = 0\cdot\phi_1 + 0\cdot\phi_2 + 0\cdot\phi_3 + 0\cdot\phi_4 + 0\cdot\phi_5 + 1\cdot\phi_6$$

or,
$$\hat{E}\begin{bmatrix}\phi_1 & \phi_2 & \phi_3 & \phi_4 & \phi_5 & \phi_6\end{bmatrix} = \begin{bmatrix}\phi_1 & \phi_2 & \phi_3 & \phi_4 & \phi_5 & \phi_6\end{bmatrix}\begin{bmatrix}1 & 0 & 0 & 0 & 0 & 0\\ 0 & 1 & 0 & 0 & 0 & 0\\ 0 & 0 & 1 & 0 & 0 & 0\\ 0 & 0 & 0 & 1 & 0 & 0\\ 0 & 0 & 0 & 0 & 1 & 0\\ 0 & 0 & 0 & 0 & 0 & 1\end{bmatrix}$$

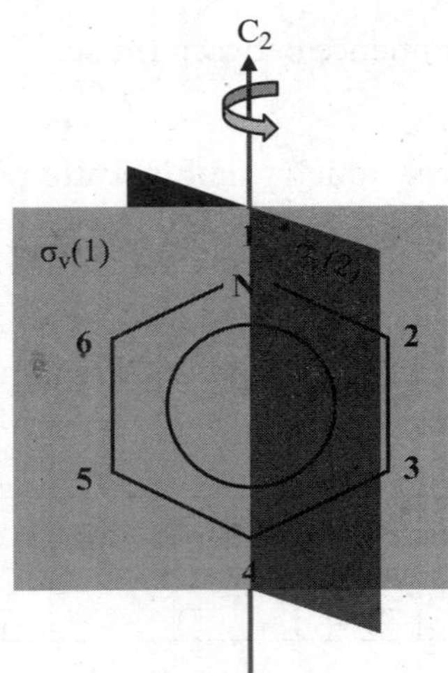

Fig. 4.6 Pyridine molecule and its symmetry elements

$$\text{that is,}\quad \mathbf{M}^{\phi}(\hat{E}) = \begin{bmatrix} 1 & 0 & 0 & 0 & 0 & 0 \\ 0 & 1 & 0 & 0 & 0 & 0 \\ 0 & 0 & 1 & 0 & 0 & 0 \\ 0 & 0 & 0 & 1 & 0 & 0 \\ 0 & 0 & 0 & 0 & 1 & 0 \\ 0 & 0 & 0 & 0 & 0 & 1 \end{bmatrix} \tag{4.35}$$

which is the matrix representation of Ê in basis $\{\phi_1, \phi_2, \phi_3, \phi_4, \phi_5, \phi_6\}$. The action of $\hat{C}_2$ on the basis functions are

$$\hat{C}_2\,\phi_1 = -\phi_1 = -1\cdot\phi_1 + 0\cdot\phi_2 + 0\cdot\phi_3 + 0\cdot\phi_4 + 0\cdot\phi_5 + 0\cdot\phi_6$$

$$\hat{C}_2\,\phi_2 = -\phi_6 = 0\cdot\phi_1 + 0\cdot\phi_2 + 0\cdot\phi_3 + 0\cdot\phi_4 + 0\cdot\phi_5 + (-1)\cdot\phi_6$$

$$\hat{C}_2\,\phi_3 = -\phi_5 = 0\cdot\phi_1 + 0\cdot\phi_2 + 0\cdot\phi_3 + 0\cdot\phi_4 + (-1)\cdot\phi_5 + 0\cdot\phi_6 \text{ and so on.}$$

$$\text{Thus,}\quad \mathbf{M}^{\phi}(\hat{C}_2) = \begin{bmatrix} -1 & 0 & 0 & 0 & 0 & 0 \\ 0 & 0 & 0 & 0 & 0 & -1 \\ 0 & 0 & 0 & 0 & -1 & 0 \\ 0 & 0 & 0 & -1 & 0 & 0 \\ 0 & 0 & -1 & 0 & 0 & 0 \\ 0 & -1 & 0 & 0 & 0 & 0 \end{bmatrix} \tag{4.36}$$

In the same way,

$$\mathbf{M}^{\phi}(\hat{\sigma}_v(1)) = \begin{bmatrix} -1 & 0 & 0 & 0 & 0 & 0 \\ 0 & -1 & 0 & 0 & 0 & 0 \\ 0 & 0 & -1 & 0 & 0 & 0 \\ 0 & 0 & 0 & -1 & 0 & 0 \\ 0 & 0 & 0 & 0 & -1 & 0 \\ 0 & 0 & 0 & 0 & 0 & -1 \end{bmatrix} \tag{4.37}$$

and $$\mathbf{M}^{\phi}(\hat{\sigma}_v(2)) = \begin{bmatrix} 1 & 0 & 0 & 0 & 0 & 0 \\ 0 & 0 & 0 & 0 & 0 & 1 \\ 0 & 0 & 0 & 0 & 1 & 0 \\ 0 & 0 & 0 & 1 & 0 & 0 \\ 0 & 0 & 1 & 0 & 0 & 0 \\ 0 & 1 & 0 & 0 & 0 & 0 \end{bmatrix} \tag{4.38}$$

These matrices multiply in accordance with the symmetry multiplication table of C_{2v}. For example, $\hat{C}_2\,\hat{\sigma}_v(1) = \hat{\sigma}_v(2)$ and multiplication reveals that

$$\mathbf{M}^{\phi}(\hat{C}_2)\mathbf{M}^{\phi}(\hat{\sigma}_v(1)) = \mathbf{M}^{\phi}(\hat{\sigma}_v(2)) \tag{4.39}$$

Thus the matrix set

$$\Gamma^{\phi} = \{\mathbf{M}^{\phi}(\hat{E}), \mathbf{M}^{\phi}(\hat{C}_2), \mathbf{M}^{\phi}(\hat{\sigma}_v(1)), \mathbf{M}^{\phi}(\hat{\sigma}_v(2))\} \tag{4.40}$$

is a representation of C_{2v}. It is a 6-dimensional representation, consisting of 6×6 matrices and requires six basis functions, ϕ_1 to ϕ_6, for their construction.

4.7.2 Representation of a Group in Function Space

One can choose the basis set for spanning a given function space in many ways. However, a representation of a point group in that space becomes meaningful only if the matrices for all the symmetry operators are constructed from the *same* set of basis functions. It can be shown that the definition of $\hat{O}_R$ given by Eq. (4.15) leads to a rule for multiplication of such operators which is paralleled by the multiplication of symmetry operators of point groups.

Let $\hat{R}_i$ and $\hat{R}_j$ be two symmetry operators of a point group G. Also let us consider a function space in which f and g are two functions. If $\hat{R}_i$ transforms the point (x, y, z) to (x', y', z') and $\hat{O}_R$ transforms f to f' then by definition,

$$f'(x', y', z') = f(x, y, z) \tag{4.41a}$$

or, $$(\hat{O}_{R_i} f)(x', y', z') = f(x, y, z) \tag{4.41b}$$

If $\hat{R}_j$ moves the point (x', y', z') to the new point (x'', y'', z''), then with the function g in the same function space,

$$(\hat{O}_{R_j}\, g)(x'', y'', z'') = g(x', y', z') \tag{4.42}$$

g may be any function and we can take it to be f'. [According to Eq. (4.41a), f' and f are in the same function space.] Hence from Eq. (4.42),

$$(\hat{O}_{R_j}\, f')\,(x'', y'', z'') = f'(x', y', z') = f(x, y, z) \quad \text{(by Eq. 4.41a)} \tag{4.43}$$

or,
$$(\hat{O}_{R_j}\,\hat{O}_{R_i}\, f)(x'', y'', z'') = f(x, y, z) \tag{4.44}$$

The transformation $(x, y, z) \rightarrow (x'', y'', z'')$ can be effected by a single symmetry operation $\hat{R}_k$. Hence $\hat{R}_k = \hat{R}_j\hat{R}_i$ and according to closure property $\hat{R}_k$ belongs to the group G and its corresponding mathematical operator $\hat{O}_{R_k}$ satisfies the condition

$$(\hat{O}_{R_k}\, f)\,(x'', y'', z'') = f\,(x, y, z) \tag{4.45}$$

Comparing Eqs (4.44) and (4.45), we have

$$\hat{O}_{R_k} = \hat{O}_{R_j}\,\hat{O}_{R_i} \quad \text{whenever} \quad \hat{R}_k = \hat{R}_j\hat{R}_i$$

Thus the mathematical operators $\hat{O}_R$ multiply in accordance with the symmetry multiplication table of the group.

4.8 REDUCIBLE AND IRREDUCIBLE REPRESENTATIONS

Before going to the actual definitions, we continue our discussion with the example used in Section 4.7.1. There the basis set was $\mathbf{b} = \{\phi_1, \phi_2, \ldots, \phi_6\}$. Let us now choose a new basis set $\mathbf{b}' = \{\psi_1, \psi_2, \ldots, \psi_6\}$ such that, with reference to Fig. 4.6,

$$\psi_1 = \phi_1,\ \psi_2 = \phi_2,\ \psi_3 = \phi_6,\ \psi_4 = \phi_3,\ \psi_5 = \phi_5 \text{ and } \psi_6 = \phi_4$$

This change of basis functions is based on the fact that ψ_2 and ψ_3 (ϕ_2 and ϕ_6 of the old basis set) are symmetry partners of one another. That is, under the action of all possible symmetry operations of the C_{2v} point group, ψ_2 and ψ_3 transform among themselves and do not mix with other functions of the basis set; similarly ψ_4 and ψ_5 are symmetry partners. In other words, the basis set $\{\psi_i\}$, $i = 1$ to 6, has the following *invariant partitioning*:

$$\{\psi_1 \,|\, \psi_2, \psi_3 \,|\,|\, \psi_4, \psi_5 \,|\, \psi_6\}$$

In the context of group theory, an *invariant subset* of a set means that the action of all symmetry operators of the group on the members of that subset does not produce results which require members from other subsets of the set.

With this new basis set, it is easy to verify that the matrices representing the symmetry operators of C_{2v} are,

$$\mathbf{M}^{\psi}(\hat{E}) = \begin{bmatrix} 1 & 0 & 0 & 0 & 0 & 0 \\ 0 & 1 & 0 & 0 & 0 & 0 \\ 0 & 0 & 1 & 0 & 0 & 0 \\ 0 & 0 & 0 & 1 & 0 & 0 \\ 0 & 0 & 0 & 0 & 1 & 0 \\ 0 & 0 & 0 & 0 & 0 & 1 \end{bmatrix} \qquad \mathbf{M}^{\psi}(\hat{C}_2) = \begin{bmatrix} -1 & 0 & 0 & 0 & 0 & 0 \\ 0 & 0 & -1 & 0 & 0 & 0 \\ 0 & -1 & 0 & 0 & 0 & 0 \\ 0 & 0 & 0 & 0 & -1 & 0 \\ 0 & 0 & 0 & -1 & 0 & 0 \\ 0 & 0 & 0 & 0 & 0 & -1 \end{bmatrix}$$

$$\mathbf{M}^{\psi}(\hat{\sigma}_v(1)) = \begin{bmatrix} -1 & 0 & 0 & 0 & 0 & 0 \\ 0 & -1 & 0 & 0 & 0 & 0 \\ 0 & 0 & -1 & 0 & 0 & 0 \\ 0 & 0 & 0 & -1 & 0 & 0 \\ 0 & 0 & 0 & 0 & -1 & 0 \\ 0 & 0 & 0 & 0 & 0 & -1 \end{bmatrix} \quad \mathbf{M}^{\psi}(\hat{\sigma}_v(2)) = \begin{bmatrix} 1 & 0 & 0 & 0 & 0 & 0 \\ 0 & 0 & 1 & 0 & 0 & 0 \\ 0 & 1 & 0 & 0 & 0 & 0 \\ 0 & 0 & 0 & 0 & 1 & 0 \\ 0 & 0 & 0 & 1 & 0 & 0 \\ 0 & 0 & 0 & 0 & 0 & 1 \end{bmatrix} \tag{4.46}$$

By actual multiplication it may be verified that these matrices multiply among themselves in accordance with the symmetry multiplication table of C_{2v}. Hence the matrix set

$$\Gamma^{\psi} = \{\mathbf{M}^{\psi}(\hat{\mathrm{E}}), \mathbf{M}^{\psi}(\hat{\mathrm{C}}_2), \mathbf{M}^{\psi}(\hat{\sigma}_v(1)), \mathbf{M}^{\psi}(\hat{\sigma}_v(2))\} \tag{4.47}$$

is also a representation of C_{2v} in the basis set $\{\psi_i\}$.

It may now be observed that all the matrices in Γ^{ψ} are block-factored in the same fashion:

$$\mathbf{M}^{\psi}(\hat{\mathrm{E}}) = \begin{bmatrix} 1 & 0 & 0 & 0 & 0 & 0 \\ 0 & 1 & 0 & 0 & 0 & 0 \\ 0 & 0 & 1 & 0 & 0 & 0 \\ 0 & 0 & 0 & 1 & 0 & 0 \\ 0 & 0 & 0 & 0 & 1 & 0 \\ 0 & 0 & 0 & 0 & 0 & 1 \end{bmatrix} \quad \mathbf{M}^{\psi}(\hat{\mathrm{C}}_2) = \begin{bmatrix} -1 & 0 & 0 & 0 & 0 & 0 \\ 0 & 0 & -1 & 0 & 0 & 0 \\ 0 & -1 & 0 & 0 & 0 & 0 \\ 0 & 0 & 0 & 0 & -1 & 0 \\ 0 & 0 & 0 & -1 & 0 & 0 \\ 0 & 0 & 0 & 0 & 0 & -1 \end{bmatrix}$$

$$\mathbf{M}^{\psi}(\hat{\sigma}_v(1)) = \begin{bmatrix} -1 & 0 & 0 & 0 & 0 & 0 \\ 0 & -1 & 0 & 0 & 0 & 0 \\ 0 & 0 & -1 & 0 & 0 & 0 \\ 0 & 0 & 0 & -1 & 0 & 0 \\ 0 & 0 & 0 & 0 & -1 & 0 \\ 0 & 0 & 0 & 0 & 0 & -1 \end{bmatrix} \quad \mathbf{M}^{\psi}(\hat{\sigma}_v(2)) = \begin{bmatrix} 1 & 0 & 0 & 0 & 0 & 0 \\ 0 & 0 & 1 & 0 & 0 & 0 \\ 0 & 1 & 0 & 0 & 0 & 0 \\ 0 & 0 & 0 & 0 & 1 & 0 \\ 0 & 0 & 0 & 1 & 0 & 0 \\ 0 & 0 & 0 & 0 & 0 & 1 \end{bmatrix} \tag{4.48}$$

Now if **A** and **B** are two matrices block-factored in the same manner as shown below,

$$\mathbf{A} = \begin{bmatrix} \mathbf{A}_1 & \mathbf{0} & \mathbf{0} \\ \mathbf{0} & \mathbf{A}_2 & \mathbf{0} \\ \mathbf{0} & \mathbf{0} & \mathbf{A}_3 \end{bmatrix} \quad \text{and} \quad \mathbf{B} = \begin{bmatrix} \mathbf{B}_1 & \mathbf{0} & \mathbf{0} \\ \mathbf{0} & \mathbf{B}_2 & \mathbf{0} \\ \mathbf{0} & \mathbf{0} & \mathbf{B}_3 \end{bmatrix}$$

then their multiplication occurs as

$$\mathbf{AB} = \begin{bmatrix} \mathbf{A}_1\mathbf{B}_1 & \mathbf{0} & \mathbf{0} \\ \mathbf{0} & \mathbf{A}_2\mathbf{B}_2 & \mathbf{0} \\ \mathbf{0} & \mathbf{0} & \mathbf{A}_3\mathbf{B}_3 \end{bmatrix}$$

[Note: If a matrix is block-factored, it is alternatively called a *direct sum* of the blocks. For example, the three matrices in Eq. (4.49) can be written as follows

$$\mathbf{A} = \mathbf{A}_1 \oplus \mathbf{A}_2 \oplus \mathbf{A}_3, \quad \mathbf{B} = \mathbf{B}_1 \oplus \mathbf{B}_2 \oplus \mathbf{B}_3, \quad \mathbf{AB} = \mathbf{A}_1\mathbf{B}_1 \oplus \mathbf{A}_2\mathbf{B}_2 \oplus \mathbf{A}_3\mathbf{B}_3]$$

With the symmetry multiplication table of C_{2v}, the result

$$\hat{C}_2\hat{\sigma}_v(1) = \hat{\sigma}_v(2) \tag{4.50}$$

corresponds to

$$\mathbf{M}^{\psi}(\hat{C}_2)\,\mathbf{M}^{\psi}(\hat{\sigma}_v(1)) = \mathbf{M}^{\psi}(\hat{\sigma}_v(2)) \tag{4.51}$$

By the rule for multiplication of block-factored matrices, we can write Eq. (4.51) in the following form:

$$\begin{bmatrix} \boxed{\mathbf{M}_1(\hat{C}_2)\,\mathbf{M}_1(\hat{\sigma}_v(1))} & 0 & 0 & 0 & 0 & 0 \\ 0 & \multicolumn{2}{c}{\multirow{2}{*}{$\boxed{\mathbf{M}_2(\hat{C}_2)\,\mathbf{M}_2(\hat{\sigma}_v(1))}$}} & 0 & 0 & 0 \\ 0 & & & 0 & 0 & 0 \\ 0 & 0 & 0 & \multicolumn{2}{c}{\multirow{2}{*}{$\boxed{\mathbf{M}_3(\hat{C}_2)\,\mathbf{M}_3(\hat{\sigma}_v(1))}$}} & 0 \\ 0 & 0 & 0 & & & 0 \\ 0 & 0 & 0 & 0 & 0 & \boxed{\mathbf{M}_4(\hat{C}_2)\,\mathbf{M}_4(\hat{\sigma}_v(1))} \end{bmatrix}$$

$$= \begin{bmatrix} \boxed{\mathbf{M}_1(\hat{\sigma}_v(2))} & 0 & 0 & 0 & 0 & 0 \\ 0 & \multicolumn{2}{c}{\multirow{2}{*}{$\boxed{\mathbf{M}_2(\hat{\sigma}_v(2))}$}} & 0 & 0 & 0 \\ 0 & & & 0 & 0 & 0 \\ 0 & 0 & 0 & \multicolumn{2}{c}{\multirow{2}{*}{$\boxed{\mathbf{M}_3(\hat{\sigma}_v(2))}$}} & 0 \\ 0 & 0 & 0 & & & 0 \\ 0 & 0 & 0 & 0 & 0 & \boxed{\mathbf{M}_4(\hat{\sigma}_v(2))} \end{bmatrix} \tag{4.52}$$

Here $\mathbf{M}_j(\hat{R})$ means the jth block of the matrix $\mathbf{M}^{\psi}(\hat{R})$, $j = 1, 2, 3, 4$. Equating the corresponding blocks we have,

$$\begin{aligned} &\mathbf{M}_1(\hat{C}_2)\,\mathbf{M}_1(\hat{\sigma}_v(1)) = \mathbf{M}_1(\hat{\sigma}_v(2)), \text{ (which is readily verified as } [-1]\times[-1] = [1]) \\ &\mathbf{M}_2(\hat{C}_2)\,\mathbf{M}_2(\hat{\sigma}_v(1)) = \mathbf{M}_2(\hat{\sigma}_v(2)) \\ &\mathbf{M}_3(\hat{C}_2)\,\mathbf{M}_3(\hat{\sigma}_v(1)) = \mathbf{M}_3(\hat{\sigma}_v(2)) \\ &\mathbf{M}_4(\hat{C}_2)\,\mathbf{M}_4(\hat{\sigma}_v(1)) = \mathbf{M}_4(\hat{\sigma}_v(2)) \end{aligned} \tag{4.53}$$

All these are in correspondence with the symmetry product $\hat{C}_2\hat{\sigma}_v(1) = \hat{\sigma}_v(2)$. Hence we conclude that each of the matrix sets listed below is a representation of C_{2v}:

$$\Gamma_1 = \{\mathbf{M}_1(\hat{E}), \mathbf{M}_1(\hat{C}_2), \mathbf{M}_1(\hat{\sigma}_v(1)), \mathbf{M}_1(\hat{\sigma}_v(2))\}$$
$$= \{[1], [-1], [-1], [1]\} \tag{4.54}$$

$$\Gamma_2 = \{\mathbf{M}_2(\hat{E}), \mathbf{M}_2(\hat{C}_2), \mathbf{M}_2(\hat{\sigma}_v(1)), \mathbf{M}_2(\hat{\sigma}_v(2))\}$$
$$= \left\{\begin{bmatrix}1 & 0\\0 & 1\end{bmatrix}, \begin{bmatrix}0 & -1\\-1 & 0\end{bmatrix}, \begin{bmatrix}-1 & 0\\0 & -1\end{bmatrix}, \begin{bmatrix}0 & 1\\1 & 0\end{bmatrix}\right\} \tag{4.55}$$

$$\Gamma_3 = \{\mathbf{M}_3(\hat{E}), \mathbf{M}_3(\hat{C}_2), \mathbf{M}_3(\hat{\sigma}_v(1)), \mathbf{M}_3(\hat{\sigma}_v(2))\}$$
$$= \left\{\begin{bmatrix}1 & 0\\0 & 1\end{bmatrix}, \begin{bmatrix}0 & -1\\-1 & 0\end{bmatrix}, \begin{bmatrix}-1 & 0\\0 & -1\end{bmatrix}, \begin{bmatrix}0 & 1\\1 & 0\end{bmatrix}\right\} \tag{4.56}$$

and
$$\Gamma_4 = \{\mathbf{M}_4(\hat{E}), \mathbf{M}_4(\hat{C}_2), \mathbf{M}_4(\hat{\sigma}_v(1)), \mathbf{M}_4(\hat{\sigma}_v(2))\}$$
$$= \{[1], [-1], [-1], [1]\} \tag{4.57}$$

Hence the 6-dimensional representation Γ^ψ, consisting of the matrices $\mathbf{M}^\psi$, is *reducible* as it contains within it the representations Γ_1, Γ_2, Γ_3 and Γ_4 of smaller dimensions.

The singleton matrices of the set Γ_1 are constructed by operating the symmetry operators of C_{2v} on the function ψ_1 only. Hence ψ_1 is the basis function of the one-dimensional representation Γ_1. In the same way, (ψ_2, ψ_3) is the set of basis functions of the two-dimensional representation Γ_2, (ψ_4, ψ_5) is the basis set for Γ_3 and ψ_6 is the basis function for Γ_4. It is readily seen from Eqs (4.54) to (4.57) that the matrix sets Γ_1 and Γ_4 are the same and so are Γ_2, Γ_3. Being one dimensional, Γ_1 (or Γ_4) is *irreducible*. To examine whether Γ_2 (or Γ_3) can be reduced further, we have to proceed a bit further.

4.8.1 Equivalence of a Number of Representations of a Group

Two representations Γ^ϕ and Γ^ψ of a group, whose basis sets are $\phi = \{\phi_1, \phi_2, \ldots, \phi_n\}$ and $\psi = \{\psi_1, \psi_2, \ldots, \psi_n\}$ respectively, are said to be *equivalent* if (i) the matrices $\mathbf{M}^\phi(\hat{R})$ and $\mathbf{M}^\psi(\hat{R})$ are similarity transforms of one another for every symmetry operator $\hat{R}$ of the group and (ii) the basis sets of the two such representations span the same function space, that is, the collection of all linear combinations of the basis functions of one representation is the same as that of the basis functions of the other. Obviously, equivalent representations have the same set of characters. The representations Γ^ϕ and Γ^ψ of C_{2v} as given by Eqs (4.40) and (4.47) are equivalent because of the following reasons:

(i) It can be easily verified that

$$\mathbf{M}^\psi(\hat{C}_2) = \mathbf{T}^{-1}\mathbf{M}^\phi(\hat{C}_2)\mathbf{T} \tag{4.58}$$

where $\mathbf{T}$ is the transformation matrix for going to the basis set ψ from the set ϕ; its construction follows from the relations

$$
\begin{aligned}
\psi_1 &= \phi_1 = 1\cdot\phi_1 + 0\cdot\phi_2 + 0\cdot\phi_3 + 0\cdot\phi_4 + 0\cdot\phi_5 + 0\cdot\phi_6 \\
\psi_2 &= \phi_2 = 0\cdot\phi_1 + 1\cdot\phi_2 + 0\cdot\phi_3 + 0\cdot\phi_4 + 0\cdot\phi_5 + 0\cdot\phi_6 \\
\psi_3 &= \phi_6 = 0\cdot\phi_1 + 0\cdot\phi_2 + 0\cdot\phi_3 + 0\cdot\phi_4 + 0\cdot\phi_5 + 1\cdot\phi_6 \\
\psi_4 &= \phi_3 = 0\cdot\phi_1 + 0\cdot\phi_2 + 1\cdot\phi_3 + 0\cdot\phi_4 + 0\cdot\phi_5 + 0\cdot\phi_6 \\
\psi_5 &= \phi_5 = 0\cdot\phi_1 + 0\cdot\phi_2 + 0\cdot\phi_3 + 0\cdot\phi_4 + 1\cdot\phi_5 + 0\cdot\phi_6 \\
\psi_6 &= \phi_4 = 0\cdot\phi_1 + 0\cdot\phi_2 + 0\cdot\phi_3 + 1\cdot\phi_4 + 0\cdot\phi_5 + 0\cdot\phi_6
\end{aligned} \tag{4.59}
$$

Thus, collecting the coefficients of the ϕ functions,

$$
\mathbf{T} = \begin{bmatrix} 1 & 0 & 0 & 0 & 0 & 0 \\ 0 & 1 & 0 & 0 & 0 & 0 \\ 0 & 0 & 0 & 1 & 0 & 0 \\ 0 & 0 & 0 & 0 & 0 & 1 \\ 0 & 0 & 0 & 0 & 1 & 0 \\ 0 & 0 & 1 & 0 & 0 & 0 \end{bmatrix} \tag{4.60}
$$

(ii) Tr $\mathbf{M}^{\psi}(\hat{C}_2) = $ Tr $\mathbf{M}^{\phi}(\hat{C}_2) = -2$ and a similar result holds for each of the C_{2v} operators.
(iii) The basis sets ϕ and ψ span the same function space—the same collection of all functions in which the π-MO wave functions of pyridine are only some special members.

4.8.2 An Example of Irreducible Representation of Dimension Greater than One

Consider a set of Cartesian coordinate axes with the origin at the centre of the N atom of the NH_3 molecule. Let $\vec{e}_1, \vec{e}_2, \vec{e}_3$ be three unit vectors starting from the origin and stretched along the X-, Y- and Z-axes respectively. Then, under the action of all the symmetry operators of the molecular point group C_{3v}, the vector $\vec{e}_3$ remains unchanged (Fig. 4.7) but $\vec{e}_1$ and $\vec{e}_2$ get mixed up and give linear combinations:

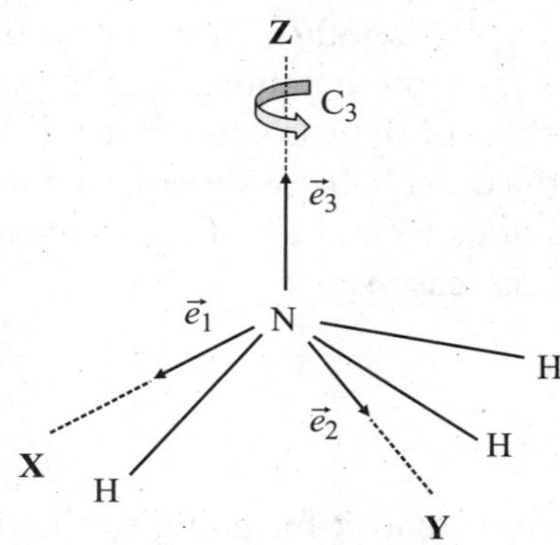

Fig. 4.7 Pyramidal NH_3 molecule and basis vectors $\vec{e}_1, \vec{e}_2, \vec{e}_3$ along the Cartesian axes

$$\hat{C}_3(\vec{e}_1) = \cos 120° (\vec{e}_1) - \sin 120° (\vec{e}_2), \ \ldots$$

Hence $\vec{e}_1$ and $\vec{e}_2$ together are required to construct the matrices for representing the C_{3v} symmetry operators. The set of these 2×2 matrices is a 2-dimensional representation of C_{3v}. By no means can this be reduced to a representation with dimension less than 2. In other words, the basis set is invariantly partitioned as $\{\vec{e}_1, \vec{e}_2 | \vec{e}_3\}$ under the action of the C_{3v} symmetry operators and hence from this basis set we get one 1-dimensional and one 2-dimensional irreducible representations.

Note: Irreducible representations with dimension greater than one are called *degenerate representations*. The reason for such a nomenclature will be understood when we study the applications of group theory.

4.8.3 Block-diagonalisation by Similarity Transformation and Reducibility

The matrices in Γ_2 of Eq. (4.55) are each 2-dimensional. We shall now investigate whether there are representations equivalent to it but block-diagonal in form. The basis set of Γ_2 is (ψ_2, ψ_3) with respect to Fig. 4.6. Let us construct new basis functions (recall Section 4.6) as follows:

$$\psi_2' = \frac{1}{\sqrt{2}}(\psi_2 + \psi_3) \quad \text{and} \quad \psi_3' = \frac{1}{\sqrt{2}}(\psi_2 - \psi_3)$$

Then, $\hat{C}_2 \psi_2' = \frac{1}{\sqrt{2}} \hat{C}_2 (\phi_2 + \phi_6) = \frac{1}{\sqrt{2}}(-\phi_2 - \phi_6) = -\psi_2'$

$$\hat{C}_2 \psi_3' = \frac{1}{\sqrt{2}} \hat{C}_2 (\phi_2 - \phi_6) = \frac{1}{\sqrt{2}}(-\phi_6 + \phi_2) = \psi_3'$$

And so the matrix representation of C_2 with basis set (ψ_2', ψ_3') is

$$M'(\hat{C}_2) = \begin{bmatrix} -1 & 0 \\ 0 & 1 \end{bmatrix}$$

In this way we can construct a representation of the C_{2v} point group with (ψ_2', ψ_3') as the basis set:

$$\begin{array}{cccccc} & \hat{E} & \hat{C}_2 & \hat{\sigma}_v(1) & \hat{\sigma}_v(2) & \\ \Gamma_2' = & \left\{ \begin{bmatrix} 1 & 0 \\ 0 & 1 \end{bmatrix} \right. & \begin{bmatrix} -1 & 0 \\ 0 & 1 \end{bmatrix} & \begin{bmatrix} -1 & 0 \\ 0 & -1 \end{bmatrix} & \left. \begin{bmatrix} 1 & 0 \\ 0 & -1 \end{bmatrix} \right\} & \end{array}$$

We note that the character of each operator of C_{2v} in Γ_2' is the same as that in Γ_2. Hence Γ_2 and Γ_2' are equivalent representations of C_{2v}. But the latter consists of matrices which are conformably block-factored. Hence Γ_2' and consequently, its equivalent representation Γ_2 are reducible, the component representations being

$$\begin{array}{llllll} & \hat{E} & \hat{C}_2 & \hat{\sigma}_v(1) & \hat{\sigma}_v(2) & \\ \Gamma_2'(1) = & \{\ [1] & [-1] & [-1] & [1]\ \} & \text{with basis } \psi_2' \end{array}$$

and

$$\begin{array}{lcccc} & \hat{E} & \hat{C}_2 & \hat{\sigma}_v(1) & \hat{\sigma}_v(2) \\ \Gamma_2'(2) = \{ & [1] & [1] & [-1] & [-1]\,\} \end{array} \quad \text{with basis } \psi_3'$$

Note here, that the matrices of Γ_2' result from similarity transformation on those of Γ_2 by the matrix

$$\mathbf{T} = \begin{bmatrix} \frac{1}{\sqrt{2}} & \frac{1}{\sqrt{2}} \\ \frac{1}{\sqrt{2}} & -\frac{1}{\sqrt{2}} \end{bmatrix}$$

which is unitary and is required for a change of basis from the set (ψ_2, ψ_3) to (ψ_2', ψ_3'). Thus,

$$\mathbf{T}^{-1}\mathbf{M}_2(\hat{\sigma}_v(2))\,\mathbf{T} = \begin{bmatrix} \frac{1}{\sqrt{2}} & \frac{1}{\sqrt{2}} \\ \frac{1}{\sqrt{2}} & -\frac{1}{\sqrt{2}} \end{bmatrix}^{-1} \begin{bmatrix} 0 & 1 \\ 1 & 0 \end{bmatrix} \begin{bmatrix} \frac{1}{\sqrt{2}} & \frac{1}{\sqrt{2}} \\ \frac{1}{\sqrt{2}} & -\frac{1}{\sqrt{2}} \end{bmatrix}$$

$$= \begin{bmatrix} \frac{1}{\sqrt{2}} & \frac{1}{\sqrt{2}} \\ \frac{1}{\sqrt{2}} & -\frac{1}{\sqrt{2}} \end{bmatrix} \begin{bmatrix} \frac{1}{\sqrt{2}} & -\frac{1}{\sqrt{2}} \\ \frac{1}{\sqrt{2}} & \frac{1}{\sqrt{2}} \end{bmatrix} = \begin{bmatrix} 1 & 0 \\ 0 & -1 \end{bmatrix} = \mathbf{M}_2'(\hat{\sigma}_v(2))$$

Similarly, $\mathbf{T}^{-1}\mathbf{M}_2(\hat{C}_2)\ \mathbf{T} = \mathbf{M}_2'(\hat{C}_2)$ = the matrix for $\hat{C}_2$ in Γ_2', and so on. Note further that the new (primed) basis set is invariantly partitioned.

In a similar way Γ_3 can also be shown to be reducible.

4.8.4 Definition of Reducible and Irreducible Representations

From the above discussions we now arrive at the following definition of reducible and irreducible representations. A representation of a group is reducible if all the matrices of the representation are either block-diagonalised or can be made so by suitable transformation of the basis set, such that the corresponding blocks for all the group elements (symmetry operators) have the same size. In other words, a representation is reducible if its basis set is invariantly partitioned into smaller subsets (or can be made so by suitable similarity transformation with unitary matrices).

4.9 UNITARY REPRESENTATION OF POINT GROUPS

A representation of a point group is said to be *unitary* if all the operators of the group are represented by unitary matrices; for such a representation, the basis set must be orthonormal (pair-wise orthogonal and individually normalised to unity). It is possible to construct a set of orthonormal functions (or vectors) from a given set of non-orthogonal but linearly independent functions (or vectors) by suitable mathematical techniques (for example, the Schmidt orthogonalisation method). Hence, from any given representation with a non-orthogonal basis set, it is possible to construct an equivalent unitary representation. It is always convenient to work with unitary representations because unitary matrices are easy to handle, particularly, for obtaining the inverse of matrices.

4.10 CONCLUDING REMARKS

After going through the preceding sections, several questions might arise in the reader's mind:

(i) How many representations are possible for a group? The answer is, obviously: as many as one likes, depending on the choice of basis set.
(ii) How many *irreducible* representations are possible for a group?
(iii) What are the sets of characters of the possible irreducible representations of a group?
(iv) Given the matrices of an irreducible representation in a particular vector or function space, how to get the basis vectors?

The Great Orthogonality Theorem discussed in Chapter 5 addresses all these questions.

SUMMARY

- Symmetry operators ($\hat{R}$) are linear transformations; in the 3-dimensional physical space they are represented by 3×3 matrices consisting of the coefficients of transformation of the coordinates (x, y, z) of a general point.
- $\hat{R}$ does not directly act on mathematical functions $\{f(x, y, z)\}$ but transforms the coordinates (x, y, z) in their arguments to (x', y', z').
- Corresponding to each $\hat{R}$, a mathematical operator $\hat{O}_R$ exists which transforms the function f to $f'(= \hat{O}_R f)$ such that, numerically, $f'(x', y', z') = f(x, y, z)$. This definition of $\hat{O}_R$ ensures closure property of the group formed by symmetry operators. Square non-singular matrices perform the task of $\hat{O}_R$ for $\hat{R}$.
- In an n-dimensional function space with n linearly independent basis functions, the operators of a point group (G) are represented by non-singular $n \times n$ matrices which multiply in accordance with the symmetry multiplication table of G. The whole set of matrices, one for each $\hat{R}$, is called a *representation* of G in the given basis. With orthonormal basis sets the representations are unitary.
- In a representation of G the matrix traces are called *characters*. Symmetry operators in the same conjugate class have the same character.
- With different choices of basis sets in a given function space, a point group may have many representations; any two such representations are equivalent if they have the same character set.
- If the basis set of a representation is invariantly partitioned, the matrices are block-factored in the same fashion. Such a representation is called *reducible*, each set of corresponding blocks being a representation of smaller dimension. If the matrices in a representation are not initially block-factored but can be made so by some similarity transformation, then also the representation is reducible.
- A representation is *irreducible* if its matrices cannot be further block-factored; its basis set cannot be invariantly partitioned.

LEARNING OUTCOME

After reading this chapter the student will be able to (i) construct representations from given basis sets, (ii) judge whether two given representations are equivalent or not and (iii) test the reducibility of a given representation of a point group.

EXERCISES

1. Construct the matrix for (a) reflection of a point (x, y, z) through a plane which contains the Z axis and is at an angle of 45° to the X and Y axes, (b) inversion of the same point through the origin.

2. Obtain the matrix representation of the operator $\hat{\sigma}$ for a reflection on a plane which is placed at an angle β with the X-axis and contains the Z-axis, using the five *d*-orbital functions given in Eq. (4.16). Find whether the trace of the matrix is 1, irrespective of the angle β.

3. Find the result of operating $\hat{C}_3(Z)$ on the functions: (a) xyz (b) $x^2 + y^2 + z^2$ (c) $x^2 - y^2$.

4. (a) Verify the matrix Eq. (4.58) using the transformation matrix **T** as given in Eq. (4.60) and the matrices $\mathbf{M}^{\phi}(\hat{C}_2)$ and $\mathbf{M}^{\psi}(\hat{C}_2)$ from Eqs (4.36) and (4.46).
   (b) Do the same for the operators $\hat{\sigma}_v(1)$ and $\hat{\sigma}_v(2)$.
   [This exercise will convince you that suitable similarity transformation block-factorises matrices.]

5. Imagine vertical planes $\hat{\sigma}_v(1)$ and $\hat{\sigma}_v(2)$ bisecting the angles between the XZ and YZ planes of a coordinate system. Determine the matrices that represent the effect of reflections of a general point (x, y, z) by these planes. By matrix product determine:
   (a) what operation results when the two reflections are carried out successively.
   (b) what operation results when (i) $\hat{\sigma}_v(1)$ is followed by a reflection on the XY plane and (ii) $\hat{\sigma}_v(2)$ is followed by a reflection on the XY plane.

6. Without constructing the actual matrices, construct the set of characters of a representation of the point groups of the following molecules in the basis set given against each.

Molecule	Basis set
Pyridine	The set of *p*-orbitals of the conjugated atoms which form the π-MOs
Ammonia	The set of unit vectors stretched along the N–H bonds
Benzene	The set of *p*-orbitals of the conjugated atoms which form the π-MOs

7. Using the set of B–F stretching (in-plane) vectors of BF_3 as the basis set construct the matrices for the symmetry operators of the molecular point group and verify that the matrices representing symmetry operators of the same conjugate class have equal trace.

8. Construct the matrices for the symmetry operators of the point group of NH_3 using (a) the set of 1*s* orbitals of H and (b) the set of N–H stretching vectors as basis. In each case verify that the matrices representing symmetry operators of the same conjugate class have the same trace.

9. Using the set of unit vectors stretched along the Pt–Cl bonds of the square planar $[PtCl_4]^{2-}$ ion construct the matrices representing the $\hat{C}_4$ and $\hat{C}_4^3$ operators. Verify that their characters are the same and explain why they are so. Write down the sets of characters of all the operators of the point group of the ion with the given basis set without actually constructing the matrices.

10. Verify that the coordinate set (x, y, z) of a general point forms a basis set for a representation of the O_h point group.

11. Find the invariant partitions of the set of Cartesian coordinates (x, y, z) of a general point for the following point groups: C_{2v}, C_{3v}, C_{4h}, D_{2d}, D_{3d}, T_d, O_h. In each case, state the number and dimension of the possible irreducible representations with the invariant subsets as bases.
[**Hint**: take the highest-fold axis of rotation as the Z-axis for the axial groups; for cubic groups draw the coordinate axes parallel to the edges of the generating cube.]

12. Short questions
(a) Explain why the matrices **A** and **B** cannot represent symmetry operators:

$$\mathbf{A} = \begin{bmatrix} \sqrt{3}/2 & \sqrt{3}/2 & 0 \\ 1/2 & 1/2 & 0 \\ 0 & 0 & 1 \end{bmatrix}, \mathbf{B} = \begin{bmatrix} e^{i\theta} & 1 \\ 1 & e^{-i\theta} \end{bmatrix}$$

(b) The set of basis functions of an irreducible representation of a group is $\{d_{xy}, d_{yz}, d_{zx}\}$. What is the character of the identity operator in this representation and why?
(c) What is the trace of the commutator [**P**, **Q**] of two square matrices **P** and **Q**?
(d) With the dipole moment vector of NH_3 as basis, write the character set of for the operators of C_{3v}.
(e) Support or contradict the statement: The characters of a representation multiply in accordance with the symmetry multiplication table of a point group.

Suggested Books and References for Further Reading

1. Hammermesh M, *Group Theory*, Addison-Wesley Publishing Company, Inc., Reading, Mass., **1962**.
2. Hiene V, *Group Theory in Quantum Mechanics*, Pergamon Press, NY, **1960**.

5 The Great Orthogonality Theorem and its Consequences

Learning objectives

- ➢ The Great Orthogonality Theorem (GOT)—statement and meaning
- ➢ Important properties of irreducible representations from GOT
- ➢ Orthogonality of character vectors from GOT
- ➢ Totally symmetric irreducible representation of a point group
- ➢ Construction of character tables of finite and infinite point groups
- ➢ Polar and axial vectors as bases of irreducible representations
- ➢ Nomenclature of irreducible representations—Mulliken symbols
- ➢ Reduction formula
- ➢ Construction of character tables of cyclic groups

5.1 INTRODUCTION

In the previous chapter several questions were raised regarding the possible number of irreducible representations of a group and sets of characters in such representations. The central theorem of group theory which addresses these questions is known as the *Great Orthogonality Theorem* (GOT). It summarises the principal properties of irreducible representations of point groups. In this chapter we shall state and explain the GOT without proof and discuss some of its important consequences that lead to the construction of character tables of point groups. (A proof of the GOT is given in Appendix A1.)

5.2 STATEMENT OF THE THEOREM

The GOT states that for a group of order *g*,

$$\sum_{\hat{R}} [\mathbf{D}^i(\hat{R})]^*_{mn} [\mathbf{D}^j(\hat{R})]_{pq} = \frac{g}{\sqrt{l_i l_j}} \delta_{ij}\delta_{mp}\delta_{nq} \tag{5.1}$$

where $[\mathbf{D}^i(\hat{R})]_{mn}$ means the element at the *m*th row and *n*th column of the matrix $\mathbf{D}^i(\hat{R})$ that represents the symmetry operator $\hat{R}$ of the group in the *i*th irreducible representation, the asterisk stands for complex conjugate, l_i and l_j are the dimensions of the *i*th and *j*th irreducible representations respectively and each δ is a Krönecker delta. The sum in Eq. (5.1) is taken over all the operators $\hat{R}$ of the group.

The meaning of the theorem is clarified with an illustration discussed in the following paragraphs. Let D^1, D^2, D^3 and D^4 be four irreducible representations of a group $G = \{\hat{E}, \hat{R}_2, \hat{R}_3, \ldots, \hat{R}_g\}$ whose matrices are as shown in Table 5.1.

Table 5.1 Some matrix sets representing different operators of a group of order g in irreducible representations of different dimensions

	$\hat{E}$	$\hat{R}_2$	$\hat{R}_3$	…	$\hat{R}_g$	Dimension
D^1	$[1]$	$[r_2]$	$[r_3]$	…	$[r_g]$	1
D^2	$[1]$	$[\alpha_2]$	$[\alpha_3]$	…	$[\alpha_g]$	1
D^3	$\begin{bmatrix} 1 & 0 \\ 0 & 1 \end{bmatrix}$	$\begin{bmatrix} a_2 & b_2 \\ c_2 & d_2 \end{bmatrix}$	$\begin{bmatrix} a_3 & b_3 \\ c_3 & d_3 \end{bmatrix}$	…	$\begin{bmatrix} a_g & b_g \\ c_g & d_g \end{bmatrix}$	2
D^4	$\begin{bmatrix} 1 & 0 & 0 \\ 0 & 1 & 0 \\ 0 & 0 & 1 \end{bmatrix}$	$\begin{bmatrix} p_2 & q_2 & s_2 \\ t_2 & u_2 & v_2 \\ w_2 & x_2 & y_2 \end{bmatrix}$	$\begin{bmatrix} p_3 & q_3 & s_3 \\ t_3 & u_3 & v_3 \\ w_3 & x_3 & y_3 \end{bmatrix}$	…	$\begin{bmatrix} p_g & q_g & s_g \\ t_g & u_g & v_g \\ w_g & x_g & y_g \end{bmatrix}$	3

Let us consider the irreducible representation, D^1. Here the matrices are all of size 1×1. So $m = 1, n = 1, p = 1, q = 1$. Using $i = j = 1$ for D^1, we have from the GOT (Eq. 5.1),

$$1^* \cdot 1 + r_2^* r_2 + r_3^* r_3 + \cdots + r_g^* r_g = \frac{g}{\sqrt{1.1}} \delta_{11} \delta_{11} \delta_{11} = g \tag{5.2}$$

or, written in a matrix form,

$$[1^* \quad r_2^* \quad r_3^* \quad \ldots \quad r_g^*] \begin{bmatrix} 1 \\ r_2 \\ r_3 \\ \vdots \\ r_g \end{bmatrix} = g \tag{5.3}$$

This means that the collection of matrix elements of D^1 behaves as a vector like the column vector in Eq. (5.3). Denoting this vector by $|u\rangle$, Eq. (5.3) can be written in the inner product form,

$$\langle u \,|\, u \rangle = g \tag{5.4}$$

Similarly for the irreducible representation D^2, we have $\langle v \,|\, v \rangle = g$, where

$$|v\rangle = [1 \quad \alpha_2 \quad \alpha_3 \quad \ldots \quad \alpha_g]^T \tag{5.5}$$

Here the superscript T denotes transpose. Henceforth we shall write the column vectors containing the matrix elements of the representations in this transposed form in order to minimise space requirement.

Next let us consider the irreducible representation, D^3. To collect the (1, 2) elements of the matrices $\mathbf{D}^3(\hat{R})$ of D^3, we consider $i = j = 3$, and $m = 1, n = 2$. Then from the GOT,

$$0^* \cdot 0 + b_2^* \, b_2 + b_3^* \, b_3 + \cdots + b_g^* \, b_g = \frac{g}{\sqrt{2.2}} \delta_{33} \delta_{11} \delta_{22} = \frac{g}{2}$$

or, $$\langle w|w\rangle = \frac{g}{2} = \frac{g}{l_3} \tag{5.6}$$

where $|w\rangle$ in matrix form is $|w\rangle = [0 \quad b_2 \quad b_3 \quad \ldots \quad b_g]^T$

and $l_3 = 2 =$ dimension of D^3.

Similarly, collecting the (2, 3) elements of the matrices in D^4, we have

$$0^* \cdot 0 + v_2^* \, v_2 + v_3^* \, v_3 + \cdots + v_g^* \, v_g = \frac{g}{\sqrt{3.3}} \delta_{44} \, \delta_{22} \, \delta_{33}$$

or, $$\langle p|p\rangle = \frac{g}{3} = \frac{g}{l_4} \tag{5.7}$$

where $|p\rangle = [0 \quad v_2 \quad v_3 \quad \ldots \quad v_g]^T$ and $l_4 = 3$ is the dimension of D^4.

Thus, in a particular l_i-dimensional irreducible representation D^i, of a group of order g, the g-tuple of matrix elements collected from a particular position of all the matrices of D^i behaves like a g-dimensional vector normalised to $\frac{g}{l_i}$.

Let us now consider two vectors by collecting matrix elements from two non-equivalent positions (1, 2) and (2, 2) of a particular irreducible representation, say D^3:

$$|w\rangle = [0 \quad b_2 \quad b_3 \quad \ldots \quad b_g]^T \quad \text{and} \quad |q\rangle = [0 \quad d_2 \quad d_3 \quad \ldots \quad d_g]^T$$

Then from the GOT we have

$$\langle w|q\rangle = 0^* \cdot 0 + b_2^* \, d_2 + b_3^* \, d_3 + \cdots + b_g^* \, d_g = \frac{g}{3} \delta_{33} \, \delta_{12} \, \delta_{22} = 0 \tag{5.8}$$

Thus the vectors $|w\rangle$ and $|q\rangle$ are orthogonal.

In the same way, vectors constructed from two different irreducible representations D^1 and D^2 are orthogonal:

$$\langle u|v\rangle = \frac{g}{\sqrt{1.1}} \delta_{12} \, \delta_{11} \, \delta_{11} = 0 \quad \text{(according to the GOT)} \tag{5.9}$$

The results discussed above can now be summarised as follows:

For a group of order g we can construct g-tuples (or, g-dimensional vectors) by collecting one element from a particular position of each of the matrices representing the operators $\hat{E}$, $\hat{R}_2$, $\hat{R}_3$, …, $\hat{R}_g$ in various irreducible representations of the group. Every such vector is normalised to $\frac{g}{l_i}$ if the vector components are elements collected from the matrices in the ith irreducible representation. The vectors are mutually orthogonal if

- they correspond to two different irreducible representations, and/or
- they correspond to the same irreducible representation but their elements are collected from non-equivalent positions of the matrices.

Note that there are three Krönecker deltas in the GOT. The first of these is required for orthogonality of two vectors corresponding to different irreducible representations; the second and third are required for orthogonality resulting from non-equality of the row-indices and column-indices respectively.

5.2.1 An Example Showing the Validity of the GOT

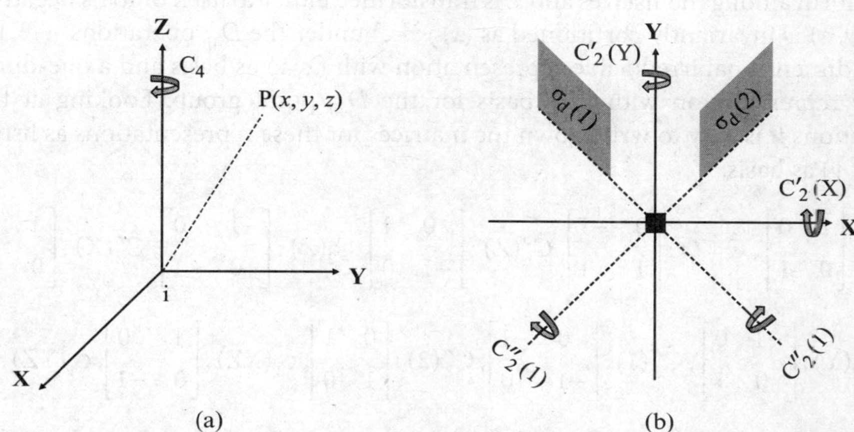

Fig. 5.1 (a) Cartesian axes, C_4 axis along the Z direction and a general point P. (b) Four symmetry axes of D_{4h} perpendicular to the principal C_4(Z) axis. Two σ_d planes are shown: the other two vertical planes of symmetry are σ_v(XZ) and σ_v(YZ).

Consider the matrix representation of D_{4h} with the Cartesian coordinates (x, y, z) of a general point P as basis. The coordinate axes are mutually perpendicular, the centre of inversion is taken at the origin and the Z-axis is taken as the principal C_4-axis (Fig. 5.1a).

The anticlockwise rotation operations change the coordinates as follows:

$$\hat{C}_4(Z)(x, y, z) \rightarrow (y, -x, z);\quad \hat{C}_4^3(Z)(x, y, z) \rightarrow (-y, x, z);\quad \hat{C}_2(Z)(x, y, z) \rightarrow (-x, -y, z)$$

There are four C_2' axes; two of them are along the coordinate axes and two are on the XY-plane bisecting the angles between these axes (Fig. 5.1b).

By these operations, the transformations of (x, y, z) are as follows

$$\hat{C}_2'(X)(x, y, z) \rightarrow (x, -y, -z);\quad \hat{C}_2'(Y)(x, y, z) \rightarrow (-x, y, -z);$$

$$\hat{C}_2''(1)(x, y, z) \rightarrow (-y, -x, -z);\quad \hat{C}_2''(2)(x, y, z) \rightarrow (y, x, -z)$$

Out of four vertical symmetry planes, two are XZ and YZ planes and the other two are dihedral, $\sigma_d(1)$ and $\sigma_d(2)$, containing the $\hat{C}_2''(1)$ and $\hat{C}_2''(2)$ axes respectively. The corresponding transformations are:

$$\hat{\sigma}_v(XZ)(x, y, z) \rightarrow (x, -y, z);\quad \hat{\sigma}_v(YZ)(x, y, z) \rightarrow (-x, y, z);$$

$$\hat{\sigma}_d(1)(x, y, z) \rightarrow (-y, -x, z)\quad \hat{\sigma}_d(2)(x, y, z) \rightarrow (y, x, z)$$

The Z-axis acts also as an S_4-axis and the XY-plane is the σ_h plane. The associated transformations are as follows:

$$\hat{S}_4(Z)(x, y, z) \rightarrow (y, -x, -z); \quad \hat{S}_4^3(Z)(x, y, z) \rightarrow (-y, x, -z);$$

$$\hat{\sigma}_h(x, y, z) \rightarrow (x, y, -z) \quad \hat{i}\,(x, y, z) \rightarrow (-x, -y, -z)$$

The results of transformation show that under the action of all symmetry operators of D_{4h}, x and y transform among themselves and Z is transformed either to itself or to its negative. Hence the set (x, y, z) is invariantly partitioned as $(x, y, | \, z)$ under the D_{4h} operations. We, therefore, expect a 2-dimensional irreducible representation with (x, y) as basis and a one-dimensional irreducible representation with z as basis for the D_{4h} point group. Looking at the above transformations it is easy to write down the matrices for these representations as listed below.

With (x, y) as basis:

$$\hat{E} : \begin{bmatrix} 1 & 0 \\ 0 & 1 \end{bmatrix}, \hat{C}_4(Z) : \begin{bmatrix} 0 & -1 \\ 1 & 1 \end{bmatrix}, \hat{C}_4^3(Z) : \begin{bmatrix} 0 & 1 \\ -1 & 0 \end{bmatrix}, \hat{C}_2(Z) : \begin{bmatrix} -1 & 0 \\ 0 & -1 \end{bmatrix}, \hat{C}_2'(X) : \begin{bmatrix} 1 & 0 \\ 0 & -1 \end{bmatrix},$$

$$\hat{C}_2'(Y) : \begin{bmatrix} -1 & 0 \\ 0 & 1 \end{bmatrix}, \hat{C}_2''(1) : \begin{bmatrix} 0 & -1 \\ -1 & 0 \end{bmatrix}, \hat{C}_2''(2) : \begin{bmatrix} 0 & 1 \\ 1 & 0 \end{bmatrix}, \hat{\sigma}_v(XZ) : \begin{bmatrix} 1 & 0 \\ 0 & -1 \end{bmatrix}, \hat{\sigma}_v(YZ) : \begin{bmatrix} -1 & 0 \\ 0 & 1 \end{bmatrix},$$

$$\hat{\sigma}_d(1) : \begin{bmatrix} 0 & -1 \\ -1 & 0 \end{bmatrix}, \hat{\sigma}_d(2) : \begin{bmatrix} 0 & 1 \\ 1 & 0 \end{bmatrix}, \hat{S}_4(Z) : \begin{bmatrix} 0 & -1 \\ 1 & 0 \end{bmatrix}, \hat{S}_4^3(Z) : \begin{bmatrix} 0 & 1 \\ -1 & 0 \end{bmatrix}, \hat{\sigma}_h : \begin{bmatrix} 1 & 0 \\ 0 & 1 \end{bmatrix},$$

$$\hat{i} : \begin{bmatrix} -1 & 0 \\ 0 & -1 \end{bmatrix}$$

With z as basis:

$$\hat{E} : [1], \hat{C}_4(Z) : [1], \hat{C}_4^3(Z) : [1], \hat{C}_2(Z) : [1], \hat{C}_2'(X) : [-1], \hat{C}_2'(Y) : [-1], \hat{C}_2''(1) : [-1],$$

$$\hat{C}_2''(2) : [-1], \hat{\sigma}_v(XZ) : [1], \hat{\sigma}_v(YZ) : [1], \hat{\sigma}_d(1) : [1], \hat{\sigma}_d(2) : [1], \hat{S}_4(Z) : [-1], \hat{S}_4^3(Z) : [-1],$$

$$\hat{\sigma}_h : [-1], \hat{i} : [-1].$$

The order of the group D_{4h} is 16. By collecting the (1, 1) elements from the matrices of the 2-dimensional representation, one can construct the following 16-component vector (written in a transposed form for economy of space):

$$|v_1\rangle = [1 \quad 0 \quad 0 \quad -1 \quad 1 \quad -1 \quad 0 \quad 0 \quad 1 \quad -1 \quad 0 \quad 0 \quad 0 \quad 0 \quad 1 \quad -1]^T$$

Similarly, from the other three positions we get the following three vectors:

From (1, 2) positions

$$|v_2\rangle = [0 \quad -1 \quad 1 \quad 0 \quad 0 \quad 0 \quad -1 \quad 1 \quad 0 \quad 0 \quad -1 \quad 1 \quad -1 \quad 1 \quad 0 \quad 0]^T$$

From (2, 1) positions

$$|v_3\rangle = [0 \quad 1 \quad -1 \quad 0 \quad 0 \quad 0 \quad -1 \quad 1 \quad 0 \quad 0 \quad -1 \quad 1 \quad 1 \quad -1 \quad 0 \quad 0]^T$$

From (2, 2) positions

$$|v_4\rangle = [1 \quad 0 \quad 0 \quad -1 \quad -1 \quad 1 \quad 0 \quad 0 \quad -1 \quad 1 \quad 0 \quad 0 \quad 0 \quad 0 \quad 1 \quad -1]^T$$

By collecting the elements of the 1×1 matrices of the representation with z as basis we get the vector,

$$|v_5\rangle = [1 \quad 1 \quad 1 \quad 1 \quad -1 \quad -1 \quad -1 \quad -1 \quad 1 \quad 1 \quad 1 \quad 1 \quad -1 \quad -1 \quad -1 \quad -1]^T$$

It can now be readily verified that all these vectors are pair-wise orthogonal:

$$\langle v_1 | v_2\rangle = \langle v_1 | v_3\rangle = \ldots = \langle v_4 | v_5\rangle = 0$$

It is also observed that the individual vectors $|v_1\rangle$ to $|v_4\rangle$ are normalised to 8 and the vector $|v_5\rangle$ is normalised to 16. The norms are equal to the order (g) of the group divided by the dimension (l_i) of the representation for each vector.

5.3 SOME PROPERTIES OF IRREDUCIBLE REPRESENTATIONS DERIVABLE FROM THE GOT

5.3.1 Property 1

Sum of the squares of the dimensions of the possible irreducible representations of a group is equal to the order of the group.

Proof (partial): Let a group of order g have c possible irreducible representations, namely, $D^1, D^2, D^3, \ldots, D^c$ whose dimensions are respectively $l_1, l_2, l_3, \ldots, l_c$. D^1 means a set of matrices of size $l_1 \times l_1$ and there are g such matrices in the set, each matrix representing a symmetry operator. Hence, by collecting elements from the matrices of D^1 one can construct l_1^2 number of g-tuples and the GOT states that these g-tuples behave like g-dimensional orthogonal vectors. Considering all the irreducible representations one can construct $l_1^2 + l_2^2 + l_3^2 + \cdots + l_c^2$ vectors (each g-dimensional) which are mutually orthogonal and hence linearly independent. But in a g-dimensional vector space there can be a maximum of g linearly independent vectors. Hence

$$l_1^2 + l_2^2 + l_3^2 + \cdots + l_c^2 \le g \tag{5.10}$$

[The complete proof requires the concept of regular representations and is given in Appendix A2.]

5.3.2 Property 2

The characters of an irreducible representation behave like the components of a g-dimensional vector normalised to g (where g is the order of the group).

Proof: The character $\chi^i(\hat{R})$ of the operator $\hat{R}$ in the ith irreducible representation of a group is given by

$$\chi^i(\hat{R}) = \sum_{m=1}^{l_i} [\mathbf{D}^i(\hat{R})]_{mm} \tag{5.11}$$

where $\mathbf{D}^i(\hat{R})$ is the $l_i \times l_i$ matrix representing $\hat{R}$ in the ith irreducible representation of the group.

Now,

$$\sum_R [\chi^i(\hat{R})]^*[\chi^i(\hat{R})] = \sum_R \sum_{m=1}^{l_i} [\mathbf{D}^i(\hat{R})]^*_{mm} \sum_{n=1}^{l_i} [\mathbf{D}^i(\hat{R})]_{nn}$$

$$= \sum_m \sum_n \sum_R [\mathbf{D}^i(\hat{R})]^*_{mm}[\mathbf{D}^i(\hat{R})]_{nn} \qquad \text{(as the order of taking the sum is immaterial)}$$

$$= \sum_m \sum_n \frac{g}{l_i} \delta_{mn}\delta_{mn} \qquad \text{(using the GOT and the fact that } \delta_{ii} = 1\text{)}$$

$$= \frac{g}{l_i} \sum_m (\delta_{m1}\delta_{m1} + \delta_{m2}\delta_{m2} + \delta_{m3}\delta_{m3} + \cdots + \delta_{mm}\delta_{mm} + \cdots)$$

$$= \frac{g}{l_i} \sum_{m=1}^{l_i} \delta_{mm}\delta_{mm} \qquad (\text{as } \delta_{mk} = 0 \text{ for } m \neq k)$$

$$= \frac{g}{l_i} (\delta_{11}\delta_{11} + \delta_{22}\delta_{22} + \delta_{33}\delta_{33} + \cdots \text{ to } l_i \text{ terms})$$

$$= \frac{g}{l_i} \times l_i = g \tag{5.12}$$

This result can be written in the form of inner product of a g-dimensional vector with itself:

$$[\chi^i(\hat{E}) \quad \chi^i(\hat{R}_2) \quad \chi^i(\hat{R}_3) \quad \ldots \quad \chi^i(\hat{R}_g)]^* \begin{bmatrix} \chi^i(\hat{E}) \\ \chi^i(\hat{R}_2) \\ \chi^i(\hat{R}_3) \\ \vdots \\ \chi^i(\hat{R}_g) \end{bmatrix} = g \tag{5.13}$$

where $\hat{E}, \hat{R}_2, \hat{R}_3, \ldots, \hat{R}_g$ are the symmetry operators of the group. This completes the proof.

Note:

(i) The result of Eq. (5.12) means that the sum of the modulus-squares of the characters must be equal to g if the representation is irreducible. If for a given representation (Γ) this sum exceeds g then Γ must be reducible.

(ii) Vectors of the type $|v^{(i)}\rangle = [\chi^i(\hat{E}) \quad \chi^i(\hat{R}_2) \quad \chi^i(\hat{R}_3) \quad \ldots \quad \chi^i(\hat{R}_g)]^T$ will henceforth be called *character vectors*.

5.3.3 Property 3

Character vectors corresponding to different irreducible representations of a group are orthogonal. That is, $\langle v^{(i)} | v^{(j)} \rangle = 0$ for $i \neq j$.

Proof: $\sum_R [\chi^i(\hat{R})]^*[\chi^j(\hat{R})]$

$$= \sum_R \sum_{m=1}^{l_i} [\mathbf{D}^i(\hat{R})]^*_{mm} \sum_{n=1}^{l_j} [\mathbf{D}^j(\hat{R})]_{nn}$$

$$= \sum_{m=1}^{l_i} \sum_{n=1}^{l_j} \sum_R [\mathbf{D}^i(\hat{R})]^*_{mm} [\mathbf{D}^j(\hat{R})]_{nn} \text{ (as the order of taking the sum is immaterial)}$$

$$= \sum_{m=1}^{l_i} \sum_{n=1}^{lj} \frac{g}{\sqrt{l_i l_j}} \delta_{ij}\, \delta_{mn}\, \delta_{mn}$$

$$= 0 \text{ for } i \neq j \tag{5.14}$$

or,
$$[\chi^i(\hat{E}) \quad \chi^i(\hat{R}_2) \quad \chi^i(\hat{R}_3) \quad \ldots \quad \chi^i(\hat{R}_g)]^* \begin{bmatrix} \chi^j(\hat{E}) \\ \chi^j(\hat{R}_2) \\ \chi^j(\hat{R}_3) \\ \vdots \\ \chi^j(\hat{R}_g) \end{bmatrix} = 0$$

or,
$$\langle v^{(i)} | v^{(j)} \rangle = 0 \text{ for } i \neq j \text{ (proved)} \tag{5.15}$$

Equations (5.12) and (5.14) can be combined into the form

$$\sum_R [\chi^i(\hat{R})]^*[\chi^j(\hat{R})] = g\delta_{ij} \tag{5.16}$$

This result is known as the *character orthogonality theorem*.

Note: The result (5.12) can be written as $\sum_R |\chi^i(\hat{R})|^2 = g$.

5.3.4 Property 4

The number of irreducible representations of a group is equal to the number of conjugate classes in it.

Proof (partial): Let there be c classes in the group such that v_1 symmetry operators belong to class 1, v_2 operators belong to class 2 and so on. Then in the ith irreducible representation, v_1 symmetry operators are represented by v_1 number of matrices which are different in form but have the same trace, $\chi_1^i(\hat{R})$ (say). (Traces of all matrices of the same conjugate class are equal.) Similarly, v_2 operators of the same class are represented by matrices of the same character $\chi_2^i(\hat{R})$, and so on. Collecting these characters we can construct a c-tuple, $[\chi_1^i(\hat{R}) \quad \chi_2^i(\hat{R})$

$\chi_3^i(\hat{R})$ ... $\chi_c^i(\hat{R})]$ for the ith irreducible representation. Now, the character orthogonality theorem

$$\sum_R [\chi^i(\hat{R})]^*[\chi^j(\hat{R})] = g\delta_{ij}$$

can be expanded as

$$\begin{aligned}
&[\chi_1^i(\hat{R})]^* \chi_1^i(\hat{R}) + [\chi_1^i(\hat{R})]^* \chi_1^i(\hat{R}) + \cdots \text{ to } v_1 \text{ terms for class 1} \\
&\quad + [\chi_2^i(\hat{R})]^* \chi_2^i(\hat{R}) + [\chi_2^i(\hat{R})]^* \chi_2^i(\hat{R}) + \cdots \text{ to } v_2 \text{ terms for class 2} \\
&\quad + \ldots \\
&\quad + [\chi_c^i(\hat{R})]^* \chi_c^i(\hat{R}) + [\chi_c^i(\hat{R})]^* \chi_c^i(\hat{R}) + \cdots \text{ to } v_c \text{ terms for class c} \\
&= g\delta_{ij}
\end{aligned}$$

or, $$v_1[\chi_1^i(\hat{R})]^* \chi_1^i(\hat{R}) + v_2[\chi_2^i(\hat{R})]^* \chi_2^i(\hat{R}) + \cdots v_c[\chi_c^i(\hat{R})]^* \chi_c^i(\hat{R}) = g\delta_{ij}$$

or, written in matrix form,

$$[\sqrt{v_1}\chi_1^i(\hat{R}) \quad \sqrt{v_2}\chi_2^i(\hat{R}) \quad \sqrt{v_3}\chi_3^i(\hat{R}) \quad \ldots \quad \sqrt{v_c}\chi_c^i(\hat{R})]^* \begin{bmatrix} \sqrt{v_1}\chi_1^i(\hat{R}) \\ \sqrt{v_2}\chi_2^i(\hat{R}) \\ \sqrt{v_3}\chi_3^i(\hat{R}) \\ \vdots \\ \sqrt{v_c}\chi_c^i(\hat{R}) \end{bmatrix} = g\delta_{ij} \tag{5.17}$$

For each irreducible representation we thus have a c-tupple, written in a transposed matrix form,

$$[\sqrt{v_1}\chi_1^i(\hat{R}) \quad \sqrt{v_2}\chi_2^i(\hat{R}) \quad \sqrt{v_3}\chi_3^i(\hat{R}) \quad \ldots \quad \sqrt{v_c}\chi_c^i(\hat{R})]^T$$

and the collection of all such c-tuples from all the possible irreducible representations of a group is a set of orthogonal c-dimensional vectors. If n is the number of possible irreducible representations of the group, we have n number of c-dimensional vectors which are mutually orthogonal according to the above result. But in a c-dimensional vector space there cannot be more than c orthogonal (linearly independent) vectors. Hence

$$n \le c \tag{5.18}$$

5.3.5 Property 5

Sum of the characters of an irreducible representation of a group is equal to the order of the group if it is the totally symmetric irreducible representation; for the other irreducible representations the sum is zero.

Proof: For any two irreducible representations $D^{(i)}$ and $D^{(j)}$ of a group of order g, we have from the character orthogonality theorem,

$$\sum_R [\chi^i(\hat{R})]^* [\chi^j(\hat{R})] = g\delta_{ij}$$

Now, if $i = 1$ means the totally symmetric irreducible representation, then each $\chi^i(\hat{R}) = 1$ and we have from the above,

$$\sum_R [\chi^j(\hat{R})] = g\delta_{1j} \tag{5.19}$$

If j is not the totally symmetric irreducible representation (that is, $j \neq 1$) , we have from Eq. (5.19),

$$\sum_R [\chi^j(\hat{R})] = 0 \quad (\text{as } \delta_{1j} = 0 \text{ for } j \neq 1\,) \tag{5.20}$$

And if $j = 1$, the irreducible representation is totally symmetric; then Eq. (5.19) yields

$$\sum_R [\chi^1(\hat{R})] = g\delta_{11} = g \tag{5.21}$$

Note:

(i) This result may be used to test whether a given set of matrices is an irreducible representation of a group or not—the sum of the characters must be either 0 or g.

(ii) Multiplying both sides of Eq. (5.20) by the dimension l_j of the jth representation we have

$$\sum_R l_j[\chi^j(\hat{R})] = 0 \quad (\text{as } l_j \text{ is constant for the } j\text{th representation})$$

and noting that l_1 is 1 we can rewrite Eq. (5.21) as

$$\sum_R l_1[\chi^1(R)] = l_1 g = g$$

Combining these two we have the result:

$$\sum_R l_j[\chi^j(R)] = g \text{ for the totally symmetric irreducible representation}$$

$$= 0 \text{ for the other irreducible representations}$$

5.3.6 Property 6

The reduction formula: A reducible representation $\Gamma^{(\text{red})}$ of a group can be expressed as a sum of a number of irreducible representations and in such an expression the number of times (ν_i) that a particular irreducible representation ($D^{(i)}$) appears is given by

$$\nu_i = \frac{1}{g}\sum_R [\chi^i(\hat{R})]^* [\chi^{\text{red}}(\hat{R})] \tag{5.22}$$

where g is the order of the group and the summation is taken over all the symmetry operators ($\hat{R}$) of the group.

Proof: Let $\Gamma^{(\text{red})}$ be reducible into component irreducible representations as

$$\Gamma^{(\text{red})} = \nu_1 \text{D}^{(1)} + \nu_2 \text{D}^{(2)} + \nu_3 \text{D}^{(3)} + \cdots + \nu_i \text{D}^{(i)} + \cdots + \nu_c \text{D}^{(c)} \tag{5.23}$$

where $\text{D}^{(1)}, \text{D}^{(2)}, \ldots, \text{D}^{(c)}$ are the possible irreducible representations of the group. This means the following:

(i) The matrix for $\hat{\text{R}}$ in $\Gamma^{(\text{red})}$ is a direct sum of the matrices for $\hat{\text{R}}$ in $\text{D}^{(1)}$ (occurring ν_1 times), $\text{D}^{(2)}$ (occurring ν_2 times) and so on. That is,

$$\Gamma^{(\text{red})}(\hat{\text{R}}) = [\mathbf{D}^{(1)}(\hat{\text{R}}) \oplus \mathbf{D}^{(1)}(\hat{\text{R}}) \oplus \ldots \nu_1 \text{ times}] \oplus [\mathbf{D}^{(2)}(\hat{\text{R}}) \oplus \mathbf{D}^{(2)}(\hat{\text{R}}) \oplus \ldots \nu_2 \text{ times}] \oplus \ldots \tag{5.24}$$

or, more explicitly,

$$\Gamma^{(\text{red})}(\hat{\text{R}}) = \begin{bmatrix} \boxed{\mathbf{D}^{(1)}(\hat{\text{R}})} & \mathbf{0} & \mathbf{0} & \mathbf{0} & \mathbf{0} & \mathbf{0} & \mathbf{0} & \mathbf{0} & \mathbf{0} \\ \mathbf{0} & \boxed{\mathbf{D}^{(1)}(\hat{\text{R}})} & \mathbf{0} & \mathbf{0} & \mathbf{0} & \mathbf{0} & \mathbf{0} & \mathbf{0} & \mathbf{0} \\ \mathbf{0} & \mathbf{0} & \ddots & \mathbf{0} & \mathbf{0} & \mathbf{0} & \mathbf{0} & \mathbf{0} & \mathbf{0} \\ \mathbf{0} & \mathbf{0} & \mathbf{0} & \nu_1 \text{ times} & \mathbf{0} & \mathbf{0} & \mathbf{0} & \mathbf{0} & \mathbf{0} \\ \mathbf{0} & \mathbf{0} & \mathbf{0} & \mathbf{0} & \boxed{\mathbf{D}^{(2)}(\hat{\text{R}})} & \mathbf{0} & \mathbf{0} & \mathbf{0} & \mathbf{0} \\ \mathbf{0} & \mathbf{0} & \mathbf{0} & \mathbf{0} & \mathbf{0} & \boxed{\mathbf{D}^{(2)}(\hat{\text{R}})} & \mathbf{0} & \mathbf{0} & \mathbf{0} \\ \mathbf{0} & \mathbf{0} & \mathbf{0} & \mathbf{0} & \mathbf{0} & \mathbf{0} & \ddots & \mathbf{0} & \mathbf{0} \\ \mathbf{0} & \mathbf{0} & \mathbf{0} & \mathbf{0} & \mathbf{0} & \mathbf{0} & \mathbf{0} & \nu_2 \text{ times} & \mathbf{0} \\ \mathbf{0} & \mathbf{0} & \mathbf{0} & \mathbf{0} & \mathbf{0} & \mathbf{0} & \mathbf{0} & \mathbf{0} & \ldots \end{bmatrix} \tag{5.25}$$

where, **0** means null matrix of appropriate size.

(ii) The set of basis functions of $\Gamma^{(\text{red})}$ can be invariantly partitioned as $\{f_1^{(1)} | f_2^{(1)} | \ldots | f_{\nu 1}^{(1)} | f_1^{(2)} | f_2^{(2)} | \ldots | f_{\nu 2}^{(2)} | \ldots\}$, where $f_1^{(1)}$ is a basis set for $\text{D}^{(1)}$, $f_2^{(1)}$ is another possible basis set for $\text{D}^{(1)}$, and so on. Such a partitioning means that under the action of all symmetry operators of the group, functions of any subset, say $f_1^{(i)}$, remain linear combinations of the functions of the same subset $f_1^{(i)}$, and do not mix with functions of other subsets. In other words, each subset corresponds to an invariant subspace of the function space spanned by the basis set of $\Gamma^{(\text{red})}$. Note that if $\text{D}^{(i)}$ is a one-dimensional irreducible representation, each $f^{(i)}$ consists of only one function; if $\text{D}^{(i)}$ is a k-dimensional irreducible representation, each $f^{(i)}$ is a set of k functions.

Now from Eq. (5.24) we have, for any operator $\hat{\text{R}}$,

$$\chi^{(\text{red})}(\hat{\text{R}}) = \nu_1[\chi^{(1)}(\hat{\text{R}})] + \nu_2[\chi^{(2)}(\hat{\text{R}})] + \cdots + \nu_i[\chi^{(i)}(\hat{\text{R}})] + \cdots + \nu_c[\chi^{(c)}(\hat{\text{R}})] \tag{5.26}$$

Multiplying both sides of Eq. (5.25) by $[\chi^{(i)}(\hat{\text{R}})]^*$ and summing the result over all $\hat{\text{R}}$,

$$\sum_R [\chi^i(\hat{\text{R}})]^* [\chi^{\text{red}}(\hat{\text{R}})]$$

$$= \nu_1 \sum_R [\chi^{(i)}(\hat{R})]^* [\chi^{(1)}(\hat{R})] + \nu_2 \sum_R [\chi^{(i)}(\hat{R})]^* [\chi^{(2)}(\hat{R})] + \cdots$$

$$+ \nu_i \sum_R [\chi^{(i)}(\hat{R})]^* [\chi^{(i)}(\hat{R})] + \cdots + \nu_c \sum_R [\chi^{(i)}(\hat{R})]^* [\chi^{(c)}(\hat{R})]$$

$$= \nu_i g \delta_{ii} = \nu_i g$$

or, $$\nu_i = \frac{1}{g} \sum_R [\chi^j(\hat{R})]^* \chi^{\text{red}}(\hat{R}) \quad \text{[Eq. (5.22) proved]}$$

5.4 THE TOTALLY SYMMETRIC IRREDUCIBLE REPRESENTATION

For any point group, it is generally true that there are some functions or vectors which do not change under the action of all symmetry operators of the group. For example, the z-coordinate of a point in an object of C_{2v} symmetry does not change by the operations $\hat{C}_2(Z)$, $\hat{\sigma}_v(XZ)$ and $\hat{\sigma}_v(YZ)$. So if we take z or any function $\phi(z)$ of z as basis, the transformations associated with the C_{2v} symmetry operators are: $\hat{E}\,\phi(z) = [1]\phi(z)$, $\hat{C}_2(Z)\phi(z) = [1]\phi(z)$, $\hat{\sigma}_v(XZ)\phi(z) = [1]\phi(z)$ and $\hat{\sigma}_v(YZ)\phi(z) = [1]\phi(z)$. Hence the characters ($\chi$) of a representation with $\phi(z)$ as basis are:

χ (character)			
$\hat{E}$	$\hat{C}_2(Z)$	$\hat{\sigma}_v(XZ)$	$\hat{\sigma}_v(YZ)$
1	1	1	1

This is called *the totally symmetric representation* of the C_{2v} point group. In the same way it can be argued that every point group has a *totally symmetric irreducible representation*, the characters in which are 1, 1, 1, 1,... The basis function (or vector) for this irreducible representation, however, should be suitably chosen for the point group under consideration.

5.5 CONSTRUCTION OF CHARACTER TABLES OF SOME SIMPLE FINITE POINT GROUPS

Character table of a point group (G) means a table containing the characters (χ) of all the possible irreducible representations of G, together with some common basis functions or vectors which transform according to these irreducible representations. Such tables are constructed on the basis of the results obtained from the GOT discussed in Sections 5.1–5.3. Some simple point groups will be considered here.

5.5.1 CHARACTER TABLE OF THE C_{2v} POINT GROUP

This group is Abelian and has four different conjugate classes, namely, ($\hat{E}$), ($\hat{C}_2(Z)$), ($\hat{\sigma}_v(XZ)$) and ($\hat{\sigma}_v(YZ)$). So it has four irreducible representations. Let us denote them by D^1, D^2, D^3, D^4 and let their dimensions be l_1, l_2, l_3 and l_4 respectively. Then, from the result in Section 5.3.1, $l_1^2 + l_2^2 + l_3^2 + l_4^2$ = order of the group = 4. Since the dimensions are non-zero positive integers, this condition is satisfied only if $l_1 = l_2 = l_3 = l_4 = 1$. Again, for any irreducible representation, the character of the identity operator (Ê) is equal to its dimension (since the matrix representation of Ê in a d-dimensional unitary representation is a $d \times d$ unit matrix). Hence each entry in the column under Ê is 1. Again, every point group has a totally symmetric irreducible representation. Let it be D^1 in this case. The character table of C_{2v} then takes the form:

χ **(character)**

	Ê	$\hat{C}_2(Z)$	$\hat{\sigma}_v(XZ)$	$\hat{\sigma}_v(YZ)$
D^1	1	1	1	1
D^2	1			
D^3	1			
D^4	1			

To determine the remaining characters we proceed as follows. D^2 being one dimensional requires only one basis function, say ϕ. Since a symmetry operator $\hat{R}$ does not change the magnitude of ϕ, the action of any $\hat{R}$ on ϕ must give ϕ or $-\phi$. So the characters are either 1 or −1. Again, as the character-vector of D^2 must be orthogonal to that of D^1, we require two 1's and two −1's in D^2. Out of these, the entry under Ê has already been found to be 1. Thus, we may take D^2 as follows:

	Ê	$\hat{C}_2(Z)$	$\hat{\sigma}_v(XZ)$	$\hat{\sigma}_v(YZ)$	
$D^1 = \{$	1	1	1	1	$\}$
$D^2 = \{$	1	1	−1	−1	$\}$

The irreducible representations D^3 and D^4 can be produced by simply rearranging the 1 and two −1's in the blocked portion of D^2; the vector corresponding to each arrangement will be orthogonal to D^1. Hence the character table of C_{2v} is Table 5.2.

Table 5.2 Character table of C_{2v} (without basis sets)

χ **(characters)**

	Ê	$\hat{C}_2(Z)$	$\hat{\sigma}_v(XZ)$	$\hat{\sigma}_v(YZ)$
D^1	1	1	1	1
D^2	1	1	−1	−1
D^3	1	−1	−1	1
D^4	1	−1	1	−1

It may readily be verified that D^2, D^3 and D^4 are pair-wise orthogonal and that in each of the irreducible representations the sum of the squares of the characters = 4 = the order of the group C_{2v}. Table 5.2 is not the full form of the character table of C_{2v} because the basis functions which can generate these characters are not mentioned in it. We now proceed to find some common basis functions. It is customary to take the principal axis of symmetry as the Z-axis. In the present case it is the C_2-axis and so we write the corresponding operation as $\hat{C}_2(Z)$. If the plane of the paper is taken as the YZ plane, the *x*-axis is perpendicular to it. Thus, a right-handed coordinate system is as shown in Fig. 5.2.

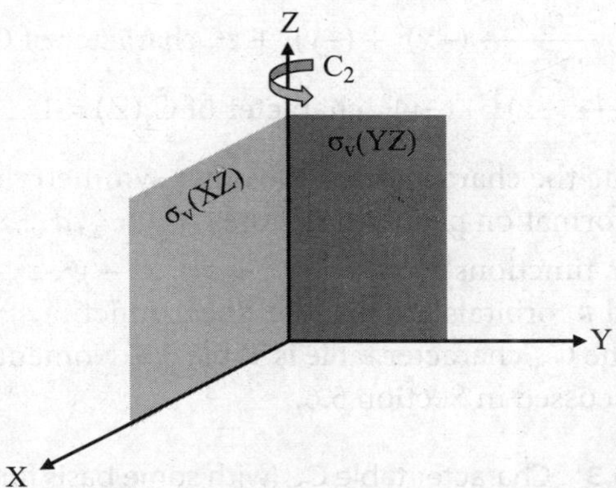

Fig. 5.2 Symmetry elements of C_{2v} together with the Cartesian axes

The two planes of symmetry in C_{2v} are perpendicular to each other. Hence if the YZ plane is taken as one σ_v plane then the other σ_v is the XZ plane. We, therefore, write $\hat{\sigma}_v(1)$ and $\hat{\sigma}_v(2)$ as $\hat{\sigma}_v(XZ)$ and $\hat{\sigma}_v(YZ)$ respectively. After setting the symmetry axes and planes in this way, we can arrive at some very common and often-used basis functions which are linear and quadratic functions of the coordinates. Such functions are needed in testing the vanishing or survival of many integrals encountered in quantum mechanical calculation of molecular orbitals and spectroscopic transitions. Action of the C_{2v} symmetry operators on linear functions of *x*, *y* and *z* produces the following results:

$$\hat{E}(x)=x=[1](x), \hat{C}_2(Z)(x)=-x=[-1](x), \hat{\sigma}_v(XZ)(x)=x=[1](x), \hat{\sigma}_v(YZ)(x)=-x=[-1](x)$$

$$\hat{E}(y)=y=[1](y), \hat{C}_2(Z)(y)=-y=[-1](y), \hat{\sigma}_v(XZ)(y)=-y=[-1](y), \hat{\sigma}_v(YZ)(y)=y=[1](y)$$

$$\hat{E}(z)=z=[1](z), \hat{C}_2(Z)(z)=z=[1](z), \hat{\sigma}_v(XZ)(z)=z=[1](z), \hat{\sigma}_v(YZ)(z)=z=[1](z)$$

We thus see that none of the *x*, *y* and *z* functions mix with other functions by the action of the C_{2v} symmetry operators. Hence they can independently act as the basis of different irreducible representations of C_{2v}. By comparing with the entries in Table 5.2 we see that the characters of the matrices generated by *x* are those of D^4. Similarly *y* generates the characters of D^3 and *z*, the characters of D^1. Note that *z* remains unchanged by the action of the symmetry operators of C_{2v}; thus *z* acts as a basis of the totally symmetric irreducible representation of C_{2v}.

The quadratic functions of coordinates like x^2, y^2, z^2, xy, yz, zx (and some of their linear combinations) transform by the action of the C_{2v} symmetry operators as follows:

$$\hat{C}_2(Z)x^2 \longrightarrow (-x)(-x) = x^2, \text{ character of } \hat{C}_2(Z) = 1$$

$$\hat{C}_2(Z)y^2 \longrightarrow (-y)(-y) = y^2, \text{ character of } \hat{C}_2(Z) = 1$$

$$\hat{C}_2(Z)z^2 \longrightarrow (z)(z) = z^2, \text{character of } \hat{C}_2(Z) = 1$$

$$\hat{C}_2(Z)xy \longrightarrow (-x)(-y) = xy, \text{character of } \hat{C}_2(Z) = 1$$

$$\hat{C}_2(Z)yz \longrightarrow (-y)(z) = -yz, \text{character of } \hat{C}_2(Z) = -1$$

$$\hat{C}_2(Z)zx \longrightarrow (z)(-x) = -zx, \text{character of } \hat{C}_2(Z) = -1$$

$$\hat{C}_2(Z)(x^2 + y^2 + z^2) \xrightarrow{\hat{C}_2(z)} (-x)^2 + (-y)^2 + z^2, \text{character of } \hat{C}_2(Z) = 1$$

$$\hat{C}_2(Z)(x^2 - y^2) \longrightarrow (-x)^2 - (-y)^2, \text{character of } \hat{C}_2(Z) = 1$$

In this way, we can generate the characters for the other symmetry operators of C_{2v}. We recall at this stage that the transformation properties of the s, $d_{x^2-y^2}$, d_{z^2}, d_{xy}, d_{yz} and d_{zx} orbitals are like those of the quadratic functions $r^2 = (x^2 + y^2 + z^2)$, $x^2 - y^2$, z^2, xy, yz and zx respectively and those of the p_x, p_y and p_z orbitals are like the linear functions x, y and z respectively. An almost complete form of the C_{2v} character table is Table 5.3. Nomenclature of these irreducible representations will be discussed in Section 5.6.

Table 5.3 Character table C_{2v} (with some basis functions)

	χ (characters)				Basis functions	
	$\hat{E}$	$\hat{C}_2(Z)$	$\hat{\sigma}_v(XZ)$	$\hat{\sigma}_v(YZ)$	Linear	Quadratic
D^1	1	1	1	1	z, p_z	$z^2, s, d_{z^2}, d_{x^2-y^2}$
D^2	1	1	−1	−1		xy, d_{xy}
D^3	1	−1	−1	1	y, p_y	yz, d_{yz}
D^4	1	−1	1	−1	x, p_x	zx, d_{zx}

5.5.2 Character Tables of D_2 and C_{2h} Point Groups

It may be recalled that there are two types of groups of order 4. Of them one is cyclic, namely, the C_4 point group with the $\hat{C}_4$ operator as the generator:

$$C_4 \text{ point group} = \{\hat{C}_4, \hat{C}_4^2, \hat{C}_4^3, \hat{C}_4^4 \equiv \hat{E}\}$$

The other is non-cyclic but Abelian and is called *Klein's four group* or *Vierergruppe*. The point groups C_{2v}, C_{2h} and D_2 belong to this category. Each of these groups being Abelian has four conjugate classes and four possible irreducible representations. Hence according to the rules just stated, the set of characters will be the same for C_{2v}, C_{2h} and D_2—only the symmetry operations and basis functions will be different.

For D_2, the symmetry operators, arranged class-wise, are $(\hat{E})$, $(\hat{C}_2(Z))$, $(\hat{C}_2(X))$, $(\hat{C}_2(Y))$. The character sets for the four possible irreducible representations (owing to four conjugate classes) of D_2 are the same as in Table 5.2. The only task remaining is to determine the basis functions involving linear and quadratic functions of Cartesian coordinates. With the Cartesian X, Y, and Z axes along the C_2 axes, we have

$$\hat{E}(x) = x,\ \hat{C}_2(Z)(x) = -x,\ \hat{C}_2(Y)(x) = -x,\ \hat{C}_2(X)(x) = x$$

Thus x acts as basis function of some irreducible representation of D_2 and the characters are as follows:

χ (character)			
$\hat{E}$	$\hat{C}_2(Z)$	$\hat{C}_2(Y)$	$\hat{C}_2(X)$
1	−1	−1	1

This set corresponds to D³ of Table 5.2. In the same way, the full character table of D_2, with basis functions, can be written down by taking the character sets from Table 5.2. This has been shown in Table 5.4.

Table 5.4 Character table of D_2

	χ (characters)				Basis functions	
	$\hat{E}$	$\hat{C}_2(Z)$	$\hat{C}_2(Y)$	$\hat{C}_2(X)$	**Linear**	**Quadratic**
D¹	1	1	1	1		$x^2, y^2, z^2, s, d_{z^2}, d_{x^2-y^2}$
D²	1	1	−1	−1	z	xy
D³	1	−1	−1	1	x	yz
D⁴	1	−1	1	−1	y	zx

Table 5.5 Character table of C_{2h}

	χ (characters)				Basis functions	
	$\hat{E}$	$\hat{C}_2(Z)$	$\hat{i}$	$\hat{\sigma}_h$	**Linear**	**Quadratic**
D¹	1	1	1	1		$xy, x^2, y^2, z^2, s, d_{z^2}, d_{x^2-y^2}$
D²	1	1	−1	−1	z	
D³	1	−1	−1	1	x, y	
D⁴	1	−1	1	−1		yz, zx, d_{yz}, d_{zx}

For C_{2h}, the symmetry operators are ($\hat{E}$), ($\hat{C}_2(Z)$), ($\hat{\sigma}_h$) or ($\hat{\sigma}(XY)$) and ($\hat{i}$) arranged class-wise. The Z-axis is taken as the C_2 axis and the σ_h plane as the XY plane. The sets of characters for the four possible irreducible representations are the same as those of Tables 5.4 or 5.2. But since the symmetry operators produce different results by acting on the linear and quadratic functions of coordinates, the basis sets will be different from those of C_{2v} or D_2. Thus, the results

$$\hat{E}(x) = x, \hat{C}_2(Z)(x) = -x, \hat{i}(x) = -x, \hat{\sigma}_h(x) = x; \hat{E}(y) = y, \hat{C}_2(Z)(y) = -y, \hat{i}(y) = -y, \hat{\sigma}_h(y) = y$$

give the character set: 1 (for $\hat{E}$), −1 (for $\hat{C}_2(Z)$), −1 (for $\hat{i}$) and 1 for $\hat{\sigma}_h$ and these characters correspond to a representation of C_{2h} where x and y individually act as bases. In this way the character table with proper basis sets for C_{2h} is as shown in Table 5.5.

5.5.3 Character Table of the C_{3v} Point Group

This group has three conjugate classes—($\hat{E}$), ($\hat{C}_3$, $\hat{C}_3^2$) and ($\hat{\sigma}_v(1)$, $\hat{\sigma}_v(2)$, $\hat{\sigma}_v(3)$). Hence it has three irreducible representations, D^1, D^2 and D^3 (say), with dimensions l_1, l_2 and l_3 respectively. Then $l_1^2 + l_2^2 + l_3^2 = 6$ (= order of C_{3v}). With non-zero positive integral values of the dimensions, this condition is satisfied by $l_1 = l_2 = 1$, $l_3 = 2$. [We could have taken $l_1 = 2$ and $l_2 = l_3 = 1$ but such a permutation of numbers 2, 1, 1 among the symbols l_1, l_2 and l_3 is immaterial; it produces no new irreducible representation.] Since these dimensions are the characters of $\hat{E}$, and since one irreducible representation must be the totally symmetric one, the character table takes the following form:

χ (characters)

	$\hat{E}$	$2\hat{C}_3$ ($\hat{C}_3$, $\hat{C}_3^2$)	$3\hat{\sigma}_v$($\hat{\sigma}_v(1)$, $\hat{\sigma}_v(2)$, $\hat{\sigma}_v(3)$)
D^1	1	1	1
D^2	1	a	b
D^3	2	c	d

Here, since $\hat{C}_3$ and $\hat{C}_3^2$ are in the same class, they have the same characters, and so at the top of the second column we write $2\hat{C}_3$. Similarly $3\hat{\sigma}_v$ means that all the three $\hat{\sigma}_v$ operators have the same character.

The unknown characters a to d can be obtained from the results of the GOT as follows:

As D^1 and D^2 are orthogonal, $1 \cdot 1^* + 2 \cdot a \cdot 1^* + 3 \cdot b \cdot 1^* = 0$

or, $\quad 2a + 3b = -1 \qquad (5.27)$

As D^2 is normalised to 6 (= the order of C_{3v}), $1 + 2a^2 + 3b^2 = 6$ (considering a and b to be real)

or, $\quad 2a^2 + 3b^2 = 5 \qquad (5.28)$

Solving Eqs (5.27) and (5.28) for a and b, we get $b = -1$, $a = 1$ and $b = \frac{3}{5}$, $a = -\frac{7}{5}$.

The fractional value of b is not acceptable because D^2, being one dimensional, requires one basis function, say ϕ, and if $b = \frac{3}{5}$, $\hat{\sigma}_v\phi = (\frac{3}{5})\ \phi$; this means a reduction in the magnitude of ϕ by a reflection; this is not allowed. Thus the irreducible representation D^2 becomes

χ (characters)

	$\hat{E}$	$2\hat{C}_3$	$3\hat{\sigma}_v$
D^2	1	1	−1

By similar arguments one can find that the acceptable solution set for c and d is $c = -1$, $d = 0$. Thus the character table for C_{3v} is as shown in Table 5.6, where some basis sets are given.

Table 5.6 Character table of C_{3v}

	χ (characters)			Basis functions	
	$\hat{E}$	$2\hat{C}_3$	$3\hat{\sigma}_v$	**Linear**	**Quadratic**
D^1	1	1	1	z	$x^2 + y^2, z^2$
D^2	1	1	−1		
D^3	2	−1	0	(x, y)	$(xz, yz)\ (x^2 - y^2, xy)$

The basis functions for these irreducible representations in terms of Cartesian coordinates (Table 5.6) can be easily obtained. The principal axis, that is the $\hat{C}_3$-axis, is the Z-axis. Also, the σ_v planes contain the Z-axis. Hence the Z-coordinate of a point remains unchanged by the action of all the C_{3v} symmetry operators. Thus z acts as a basis of the totally symmetric irreducible representation irreducible representation, D^1 of Table 5.6.

By the $\hat{C}_3$ operation, x becomes $x' = x \cos 120° - y \sin 120°$ = a linear combination of x and y; y becomes $y' = x \sin 120° + y \cos 120°$ = another linear combination of x and y. Thus x and y transform 'together' and do not mix with z under the action of $\hat{C}_3$. Writing in a matrix form, the trace of the 2×2 transformation matrix is $2 \cos 120° = -1$. So the character of $\hat{C}_3$ is −1 with (x, y) as the basis set. With the same basis set, the character of $\hat{C}_3^2$ (which is conjugate to $\hat{C}_3$) is also −1.

Each σ_v-plane contains the Z-axis. Let one σ_v contain the X-axis also. The other two σ_v-planes are at angles of 120° and 240° respectively (Fig. 5.3).

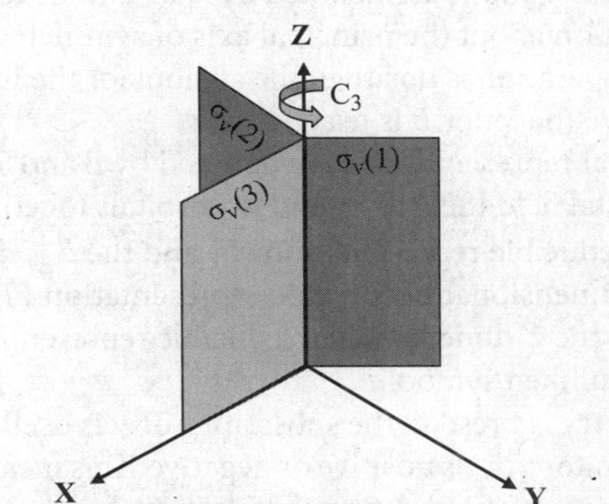

Fig. 5.3 Reflection planes of the C_{3v} point group

If a σ_v-plane contains the Z-axis and is placed at an angle of β with the positive direction of x-axis, then (as discussed in Chapter 4),

$$\hat{\sigma}_v(x) = x \cos 2\beta + y \sin 2\beta$$

$$\hat{\sigma}_v(y) = x \sin 2\beta - y \cos 2\beta$$

$$\hat{\sigma}_v(z) = z$$

Thus x and y transform together and do not mix with z. Hence (x, y) is a basis set for $\hat{\sigma}_v$ and the representative matrix will have a trace = $\cos 2\beta - \cos 2\beta = 0$. This is true for all the three $\hat{\sigma}_v$ operators of C_{3v} (as these reflection operators are in the same class).

By the identity operator $\hat{E}$, both x and y remain unchanged; hence with (x, y) as the basis set $\hat{E}$ is represented by the 2×2 unit matrix whose trace is 2. Therefore, the set (x, y) acts as a basis set of a 2-dimensional irreducible representation with characters $\chi(\hat{E}) = 2$, $\chi(\hat{C}_3 \text{ or } \hat{C}_3^2) = -1$, $\chi(\hat{\sigma}_v) = 0$ (for all the three $\hat{\sigma}_v$ operations); this corresponds to the irreducible representation, D^3 of Table 5.6.

Again, the operation of $\hat{C}_3$ on the quadratic function $x^2 + y^2$ is given by

$$(x')^2 + (y')^2 = (xc - ys)^2 + (xs + yc)^2 = (x^2 + y^2)c^2 + (x^2 + y^2)s^2 = (x^2 + y^2)$$

where $c = \cos 120°$, $s = \sin 120°$. Similarly it can be shown that $\hat{\sigma}_v$ (on any of the three reflection planes) also does not change $x^2 + y^2$; thus $\chi(\hat{C}_3) = 1 = \chi(\hat{\sigma}_v)$ with $x^2 + y^2$ as a basis. Such invariance of z^2 under the C_{3v} operations is also obvious. Thus the quadratic functions z^2 and $x^2 + y^2$ individually act as bases of D^1 (Table 5.6). The pair of quadratic functions xz and yz behave as the pair (x, y) since z remains unchanged by the C_{3v} symmetry operators. Hence (xz, yz) is a basis set for D^3 (Table 5.6). Similarly it can be verified that $(x^2 - y^2, xy)$ is a basis set of the D^3 representation of C_{3v}.

5.6 NOMENCLATURE OF IRREDUCIBLE REPRESENTATIONS: MULLIKEN SYMBOLS

(i) 1-dimensional representations are denoted by *A* or *B*. If the basis function is symmetric with respect to rotation about the principal axis of symmetry (that is, if the character χ, of the principal $\hat{C}_n$ operation is positive) the symbol for the irreducible representation is *A*; if $\chi(\hat{C}_n)$ is negative the letter *B* is used.

(ii) 2- and 3-dimensional representations are denoted by *E* and *T* respectively. For example in a tetrahedral crystal field the $d_{x^2-y^2}$ and d_{z^2} orbitals together act as basis functions of a 2-dimensional irreducible representation (*E*) and the d_{xy}, d_{yz} and d_{zx} orbitals together form bases of a 3-dimensional irreducible representation (*T*) of the T_d point group. In the C_{3v} point group, the 2-dimensional irreducible representation D^3 with (x, y) as basis is denoted by the Mulliken symbol *E*.

(iii) If a centre of symmetry is present, the subscript *g* or *u* is used according as the character of the inversion operator $\chi(\hat{i})$, is positive or negative. This means that if the basis function changes sign by the action of $\hat{i}$, the representation is of *u*-type [*u* = ungerade (German) = odd] and if it does not change sign, the representation is of *g*-type (*g* = gerade = even). For example, an octahedron has a centre of symmetry and in octahedral crystal field the *d*-orbitals, which do not change sign under the operation of inversion, are of *g*-type; in an octahedral crystal field they split into two sets ($d_{x^2-y^2}, d_{z^2}$) and (d_{xy}, d_{yz}, d_{zx}); they are respectively basis sets of the irreducible representations E_g and T_{2g} of the O_h point group (discussed in Chapter 13).

(iv) Primes and double primes are attached to all letter symbols, when appropriate, to indicate irreducible representations which are, respectively, symmetric and antisymmetric with respect to $\hat{\sigma}_h$. When both $\hat{i}$ and $\hat{\sigma}_h$ are present (for example, in C_{nh} groups with even n), the character of $\hat{i}$ is to be considered first and then, *if necessary*, $\hat{\sigma}_h$ is considered.

(v) Subscripts 1 and 2 are attached to *A*'s and *B*'s to designate irreducible representations which are, respectively, symmetric and antisymmetric with respect to a C_2-axis perpendicular to the principal C_n, or, if such a C_2 axis is lacking, to a vertical σ_v or σ_d.

(vi) The point groups D_2 and D_{2h}, each having three perpendicular C_2 axes, have representations B_1, B_2 and B_3 each of which is symmetric with respect to rotation about one C_2 axis but antisymmetric with respect to the remaining two; hence the notation *B*. Thus B_1 is antisymmetric with respect to the operations $\hat{C}_2(X)$ and $\hat{C}_2(Y)$, B_2 is antisymmetric with respect to $\hat{C}_2(X)$ and $\hat{C}_2(Z)$, and B_3 is antisymmetric with respect to $\hat{C}_2(Y)$ and $\hat{C}_2(Z)$.

(vii) The 3-dimensional representations of the cubic groups T_d, O and O_h are divided into T_1 and T_2 depending on their behaviour under the operations $\hat{S}_4$ and $\hat{\sigma}_d$ of T_d and under the operations $\hat{C}_2'$ and $\hat{C}_4$ of O.

(viii) Irreducible representations of dimension 4 and 5, denoted by *U* and *V* respectively, occur in the I_h point group.

To illustrate these rules, we now assign Mulliken symbols to the irreducible representations of the character tables that have so far been constructed. First consider the character table of C_{2v} (Section 5.5.1, Table 5.3). Here both D^1 and D^2 have positive χ for the principal $\hat{C}_2(Z)$ and so both are of *A* type. D^3 and D^4, with negative $\chi(\hat{C}_2)$, are both of *B* type [Rule (i)]. There is no inversion centre (i) nor a symmetry plane (σ_h). Hence neither *u*/*g* subscript nor primes can be used [Rules (iii) and (iv)]. D^1 and D^2 have respectively positive and negative characters for $\hat{\sigma}_v(XZ)$; hence they are designated as A_1 and A_2 respectively [Rule (v)]. For D^3 and D^4, the characters for $\hat{\sigma}_v(XZ)$ are respectively negative and positive. Hence D^4 is designated as B_1 and D^3 as B_2. Table 5.2 is rewritten as Table 5.7 with Mulliken symbols; the rows are rearranged in the order A_1, A_2, B_1, B_2.

Table 5.7 Character table of C_{2v} with Mulliken symbols of irreducible representations

	χ (characters)				Basis functions	
	$\hat{E}$	$\hat{C}_2(Z)$	$\hat{\sigma}_v(XZ)$	$\hat{\sigma}_v(YZ)$	Linear	Quadratic
$D^1 = A_1$	1	1	1	1	z, p_z	$z^2, s, d_{z^2}, d_{x^2-y^2}$
$D^2 = A_2$	1	1	−1	−1		xy, d_{xy}
$D^4 = B_1$	1	−1	1	−1	x, p_x	zx, d_{zx}
$D^3 = B_2$	1	−1	−1	1	y, p_y	yz, d_{yz}

We next consider Table 5.5, the character table of C_{2h}. According to Rule (i) D^1 and D^2 are of *A* type and D^3 and D^4 are of *B* type because of symmetry and antisymmetry with respect to the principal axis, $C_2(Z)$. There is a centre of inversion and according to Rule (iii), D^1 and D^4 are of *g* type [$\chi(\hat{i})$ positive] and D^2 and D^3 are of *u* type [$\chi(\hat{i})$ negative]. Thus for Table 5.4, $D^1 = A_g$, $D^2 = A_u$, $D^3 = B_u$ and $D^4 = B_g$. In this way all four irreducible representations of C_{2h} are assigned four different symbols and we do not need primes or numerical subscripts. In the same way, it is easy to verify that for Table 5.6 (character table for C_{3v}), $D^1 = A_1$, $D^2 = A_2$ and $D^3 = E$. The reader may check the results with the character tables given in Appendix A3.

5.7 POLAR AND AXIAL VECTORS AS BASES FOR GROUP REPRESENTATIONS

Displacement vectors corresponding to translation, rotation and vibration of molecules can act as basis vectors for representation of molecular point groups. A translational displacement ($\vec{T}$) can be resolved into three components along the Cartesian X, Y and Z axes:

$$\vec{T} = T_x\vec{e}_1 + T_y\vec{e}_2 + T_z\vec{e}_3 \tag{5.29}$$

But a rotation is best described in terms of the angular momentum ($\vec{L}$) which is a vector product of two vectors ($\vec{L} = \vec{r} \times \vec{p}$, where $\vec{r}$ is the position vector of a point on the rotating body and $\vec{p}$ is its linear momentum). For the sake of generalisation, let us consider the transformation properties of a general vector which is expressible as the vector product of two vectors with given set of Cartesian coordinates. Action of symmetry operators on the position vector (x, y, z) [that is, $\vec{r} = x\vec{e}_1 + y\vec{e}_2 + z\vec{e}_3$] has already been discussed. Let us now consider the action of symmetry operators on the vector product $\vec{R}$ of two vectors (x_1, y_1, z_1) and (x_2, y_2, z_2):

$$\vec{R} = (x_1\vec{e}_1 + y_1\vec{e}_2 + z_1\vec{e}_3) \times (x_2\vec{e}_1 + y_2\vec{e}_2 + z_2\vec{e}_3) \tag{5.30}$$

$$= \begin{vmatrix} \vec{e}_1 & \vec{e}_2 & \vec{e}_3 \\ x_1 & y_1 & z_1 \\ x_2 & y_2 & z_2 \end{vmatrix} \tag{5.31}$$

The components of $\vec{R}$ are

$$R_x = (y_1z_2 - y_2z_1),\ R_y = (x_2z_1 - x_1z_2) \quad \text{and} \quad R_z = (x_1y_2 - x_2y_1) \tag{5.32}$$

We now consider, for example, the action of C_{3v} symmetry operators on R_x, R_y and R_z. The action of $\hat{C}_3$ rotation about the *z*-axis on $\vec{r} = (x, y, z)$ is known:

$$\left.\begin{aligned} \hat{C}_3(Z)[x] &= x' = x\cos 120° + y\sin 120° = -\tfrac{1}{2}x + \tfrac{\sqrt{3}}{2}y \\ \hat{C}_3(Z)[y] &= y' = -x\sin 120° + y\cos 120° = -\tfrac{\sqrt{3}}{2}x - \tfrac{1}{2}y \\ \hat{C}_3(Z)[z] &= z' = z \end{aligned}\right\} \tag{5.33}$$

Then

$$\left.\begin{aligned}
\hat{C}_3(Z)\,[R_x] = R'_x &= y'_1z'_2 - y'_2z'_1 = (-\tfrac{\sqrt{3}}{2}x_1 - \tfrac{1}{2}y_1)z_2 - (-\tfrac{\sqrt{3}}{2}x_2 - \tfrac{1}{2}y_2)z_1 \\
&= -\tfrac{1}{2}(y_1z_2 - y_2z_1) + \tfrac{\sqrt{3}}{2}(x_2z_1 - x_1z_2) = -\tfrac{1}{2}R_x + \tfrac{\sqrt{3}}{2}R_y \\
\hat{C}_3(Z)\,[R_y] = R'_y &= x'_2z'_1 - x'_1z'_2 = (-\tfrac{1}{2}x_2 + \tfrac{\sqrt{3}}{2}y_2)z_1 - (-\tfrac{1}{2}x_1 + \tfrac{\sqrt{3}}{2}y_1)z_2 \\
&= -\tfrac{1}{2}(x_2z_1 - x_1z_2) + \tfrac{\sqrt{3}}{2}(y_2z_1 - y_1z_2) = -\tfrac{\sqrt{3}}{2}R_x - \tfrac{1}{2}R_y \\
\hat{C}_3(Z)\,[R_z] = R'_z &= x'_1y'_2 - x'_2y'_1 \\
&= (-\tfrac{1}{2}x_1 + \tfrac{\sqrt{3}}{2}y_1)\cdot(-\tfrac{\sqrt{3}}{2}x_2 - \tfrac{1}{2}y_2) - (-\tfrac{1}{2}x_2 + \tfrac{\sqrt{3}}{2}y_2)\cdot(-\tfrac{\sqrt{3}}{2}x_1 - \tfrac{1}{2}y_1) \\
&= \tfrac{1}{4}x_1y_2 - \tfrac{3}{4}y_1x_2 - \tfrac{\sqrt{3}}{4}y_1y_2 - \tfrac{1}{4}x_2y_1 + \tfrac{3}{4}y_2x_1 + \tfrac{\sqrt{3}}{4}y_1y_2 \\
&= x_1y_2 - x_2y_1 = R_z
\end{aligned}\right\} \quad (5.34)$$

Let one of the three σ_v planes be chosen as the σ_v(XZ) plane. This keeps x and z unchanged and transforms y to $-y$. Hence

$$\left.\begin{aligned}
\hat{\sigma}_v(XZ)[R_x] &= y'_1z'_2 - y'_2z'_1 = (-y_1)z_2 - (-y_2)z_1 = -(y_1z_2 - y_2z_1) = -R_x \\
\hat{\sigma}_v(XZ)[R_y] &= x'_2z'_1 - x'_1z'_2 = x_2z_1 - x_1z_2 = R_y \\
\hat{\sigma}_v(XZ)[R_z] &= x'_1y'_2 - x'_2y'_1 = x_1(-y_2) - x_2(-y_1) = -(x_1y_2 - x_2y_1) = -R_z
\end{aligned}\right\} \quad (5.35)$$

We thus find that under the $\hat{C}_3$ operation, R_x and R_y transform among themselves, the transformation matrix being obtained by expressing the relations (5.34) as

$$[R'_x \quad R'_y] = [R_x \quad R_y]\begin{bmatrix} -\frac{1}{2} & -\frac{\sqrt{3}}{2} \\ \frac{\sqrt{3}}{2} & -\frac{1}{2} \end{bmatrix} \quad (5.36)$$

and R_z remains invariant under $\hat{C}_3$. The transformations of R_x, R_y and R_z by the $\hat{\sigma}_v$(XZ) operation [given by Eq. (5.35)] can be similarly written in matrix forms:

$$[R'_x \quad R'_y] = [R_x \quad R_y]\begin{bmatrix} -1 & 0 \\ 0 & 1 \end{bmatrix}, \; R'_z = [-1]R_z \quad (5.37)$$

Since the $\hat{C}_3^2$ operation is conjugate to $\hat{C}_3$ and the $\hat{\sigma}_v$(XZ) operation is conjugate to the other two $\hat{\sigma}_v$ operations of C_{3v}, we need not find the matrix forms of these operations and the characters of the irreducible representations with R_x, R_y and R_z as bases can be immediately written down by noting the traces of the above matrices:

	$\hat{E}$	$2\hat{C}_3$	$3\hat{\sigma}_v$	Basis
D^3	2	−1	0	(R_x, R_y)
D^2	1	1	−1	R_z

(Recall that character of the identity operation = number of basis functions required.) The difference of the vector $\vec{R}$ from the position vector $\vec{r}$ [$=(x, y, z)$] is that the component z of $\vec{r}$ remains invariant under all symmetry operations of C_{3v} (if the principal axis C_3 is taken along the z-axis), but R_z changes sign on reflection by $\hat{\sigma}_v(XZ)$. Moreover, the inversion operator $\hat{i}$ (if present) transforms $\vec{r}$ to $-\vec{r}$ but $\vec{R}$ is not changed by $\hat{i}$ (since each component of $\vec{R}$ contains products of two components of $\vec{r}$). For such reasons, vectors having the properties of $\vec{r}$ are called *polar vectors* and the ones having the properties of $\vec{R}$ are called *axial vectors*. The cross product of two polar vectors is an axial vector.

The quantum mechanical angular momentum operator $\hat{L}$ transforms as an axial vector since it is the vector product of the position and momentum operators. Hence in some character tables R_x, R_y and R_z are not given as basis vectors; instead, the components of $\hat{L}$, that is, $\hat{L}_x$, $\hat{L}_y$, $\hat{L}_z$ or those of the spin operators $\hat{S}_x$, $\hat{S}_y$, $\hat{S}_z$ which have similar properties as R_x, R_y, R_z are given. For another example, let us consider the transformation properties of $\vec{R}=(R_x, R_y, R_z)$ under C_{2v} symmetry. With the C_2-axis as the Z-axis, we have

$$\left.\begin{aligned}
&\hat{C}_2(Z)[R_x] = R'_x = y'_1z'_2 - y'_2z'_1 = (-y_1)z_2 - (-y_2)z_1 = -(y_1z_2 - y_2z_1) = -R_x \\
&\hat{C}_2(Z)[R_y] = R'_y = x'_2z'_1 - x'_1z'_2 = (-x_2)z_1 - (-x_1)z_2 = -(x_2z_1 - x_1z_2) = -R_y \\
&\hat{C}_2(Z)[R_z] = R'_z = x'_1y'_2 - x'_2y'_1 = (-x_1)(-y_2) - (-x_2)(-y_1) = (x_1y_2 - x_2z_1) = R_z \\
&\hat{\sigma}_v(XZ)[R_x] = \hat{\sigma}_v(xz)[y_1z_2 - y_2z_1] = -y_1z_2 + y_2z_1 = -R_x \\
&\hat{\sigma}_v(XZ)[R_y] = \hat{\sigma}_v(xz)[x_2z_1 - x_1z_2] = x_2z_1 - x_1z_2 = R_y \\
&\hat{\sigma}_v(XZ)[R_z] = \hat{\sigma}_v(xz)[x_1y_2 - x_2y_1] = -x_1y_2 + x_2y_1 = -R_z \\
&\hat{\sigma}_v(YZ)[R_x] = \hat{\sigma}_v(yz)[y_1z_2 - y_2z_1] = y_1z_2 - y_2z_1 = R_x \\
&\hat{\sigma}_v(YZ)[R_y] = \hat{\sigma}_v(yz)[x_2z_1 - x_1z_2] = -x_2z_1 + x_1z_2 = -R_y \\
&\hat{\sigma}_v(YZ)[R_z] = \hat{\sigma}_v(yz)[x_1y_2 - x_2y_1] = -x_1y_2 + x_2y_1 = -R_z
\end{aligned}\right\} \tag{5.38}$$

It is thus found that the Cartesian components of $\vec{R}$ can separately act as bases of one-dimensional representations of C_{2v}. The sets of characters with R_x, R_y and R_z as basis, obtainable from Eq. (5.38), are found to be those of B_2, B_1 and A_2 respectively of C_{2v}. (See Appendix A3 where the character tables are given with R_x, R_y and R_z included in the columns for bases.)

5.8 CONSTRUCTION OF CHARACTER TABLES OF CYCLIC POINT GROUPS

This is based on the following facts:

(i) The set of nth roots of unity, for example, $G_1 = \{\varepsilon, \varepsilon^2, \varepsilon^3, \ldots, \varepsilon^n\}$, where $\varepsilon = e^{\frac{2\pi i}{n}}$ and n is a positive integer, is a cyclic group under multiplication, with ε as the generator. $[(1)^{\frac{1}{n}} = e^{\frac{2\pi i r}{n}}, r = 1, 2, \ldots, n]$.

(ii) The sum (s) of the nth roots of unity = 0

$$\text{That is, } s = \varepsilon + \varepsilon^2 + \varepsilon^3 + \cdots + \varepsilon^n = 0 \text{ [because } s = \frac{\varepsilon(\varepsilon^n - 1)}{(\varepsilon - 1)} = 0, \text{ as } \varepsilon^n = 1] \tag{5.39}$$

(iii) The cyclic point group $C_n = \{\hat{C}_n, \hat{C}_n^2, \hat{C}_n^3, \ldots, \hat{C}_n^n \equiv \hat{E}\}$ is in one-to-one correspondence with the group G_1.

(iv) If we define a vector $|V_r\rangle$ as

$$|V_r\rangle = (\varepsilon^r, \varepsilon^{2r}, \varepsilon^{3r}, \ldots, \varepsilon^{nr}), \text{ with } r = \text{an integer,}$$

$$\begin{aligned} \text{Then} \quad \langle V_r | V_r \rangle &= \varepsilon^r(\varepsilon^r)^* + \varepsilon^{2r}(\varepsilon^{2r})^* + \cdots + \varepsilon^{nr}(\varepsilon^{nr})^* \\ &= e^{\frac{2\pi i r}{n}} e^{-\frac{2\pi i r}{n}} + e^{\frac{4\pi i r}{n}} e^{-\frac{4\pi i r}{n}} + \cdots \\ &= 1 + 1 + \cdots \text{ etc.} = n, \text{ which is equal to the order of the group } G_1 \end{aligned} \tag{5.40}$$

(v) If two different vectors $|V_m\rangle$ and $|V_p\rangle$ are defined as

$$|V_m\rangle = (\varepsilon^m \quad \varepsilon^{2m} \quad \varepsilon^{3m} \quad \ldots \quad \varepsilon^{nm})$$

$$|V_p\rangle = (\varepsilon^p \quad \varepsilon^{2p} \quad \varepsilon^{3p} \quad \varepsilon^{4p} \quad \ldots \quad \varepsilon^{np}) \quad (m \neq p,\ m \text{ and } p \text{ are positive integers})$$

$$\begin{aligned} \text{then} \quad \langle V_m | V_p \rangle &= \varepsilon^p(\varepsilon^m)^* + \varepsilon^{2p}(\varepsilon^{2m})^* + \cdots + \varepsilon^{np}(\varepsilon^{nm})^* \\ &= \varepsilon^p \cdot \varepsilon^{n-m} + \varepsilon^{2p} \cdot \varepsilon^{n-2m} + \cdots + \varepsilon^{np} \cdot \varepsilon^{n-nm} \\ &= \varepsilon^n \cdot \varepsilon^{p-m} + \varepsilon^n \cdot \varepsilon^{2(p-m)} + \cdots + \varepsilon^n \cdot \varepsilon^{n(p-m)} \\ &= \varepsilon^{p-m} + \varepsilon^{2(p-m)} + \cdots + \varepsilon^{n(p-m)} \quad [\text{as } \varepsilon^n = 1] \\ &= \varepsilon^r + \varepsilon^{2r} + \cdots + \varepsilon^n \text{ (where } r = p - m = \text{an integer)} \\ &= \frac{\varepsilon^r\{(\varepsilon^r)^n - 1\}}{(\varepsilon^r - 1)} = \frac{\varepsilon^r\{(\varepsilon^n)^r - 1\}}{(\varepsilon^r - 1)} = \frac{\varepsilon^r\{1 - 1\}}{(\varepsilon^r - 1)} = 0 \end{aligned} \tag{5.41}$$

Thus $|V_m\rangle$ and $|V_p\rangle$ are orthogonal to one another whenever $m \neq p$.

The points (iv) and (v) suggest that vectors of the form mentioned above have the characteristics of the character-vectors of the irreducible representations of a point group as demanded by the great orthogonality theorem. Again, the groups C_n and G_1 are in one-to-one correspondence. Hence each vector of the form $|V_r\rangle$ may be regarded as the character-vector of some irreducible representation of a cyclic group C_n.

5.8.1 Example: Construction of Character Table of C_5

The cyclic group $C_5 = \{\hat{C}_5, \hat{C}_5^2, \hat{C}_5^3, \hat{C}_5^4, \hat{C}_5^5 \equiv \hat{E}\}$ is Abelian and so each element forms a class by itself. Hence there are 5 possible irreducible representations, say, $\Gamma_1, \Gamma_2, \ldots, \Gamma_5$. Taking $r = 1, 2, 3, 4$ and 5 in $|V_r\rangle$ we get the characters of these irreducible representations (Table 5.7).

Table 5.7 Characters obtained from the vectors $|V_r\rangle$, $r = 1$ to 5, for the cyclic group C_5

	χ (characters)					
	$\hat{C}_5$	$\hat{C}_5^2$	$\hat{C}_5^3$	$\hat{C}_5^4$	$\hat{C}_5^5 \equiv \hat{E}$	
Γ_1	ε	ε^2	ε^3	ε^4	$\varepsilon^5 = 1$	from $\|V_1\rangle$
Γ_2	ε^2	ε^4	ε^6	ε^8	$\varepsilon^{10} = 1$	from $\|V_2\rangle$
Γ_3	ε^3	ε^6	ε^9	ε^{12}	$\varepsilon^{15} = 1$	from $\|V_3\rangle$
Γ_4	ε^4	ε^8	ε^{12}	ε^{16}	$\varepsilon^{20} = 1$	from $\|V_4\rangle$
Γ_5	ε^5	ε^{10}	ε^{15}	ε^{20}	$\varepsilon^{25} = 1$	from $\|V_5\rangle$

As expected, we have found a totally symmetric irreducible representation, for example, Γ_5 (since $\varepsilon^5 = 1$). We can simplify the above table further by noting that $\varepsilon^* = \varepsilon^4$, $(\varepsilon^2)^* = \varepsilon^3$ and so on, and rewrite the table as:

Table 5.8 Simplified form of the previous table

	χ (characters)				
	$\hat{E}$	$\hat{C}_5$	$\hat{C}_5^2$	$\hat{C}_5^3$	$\hat{C}_5^4$
Γ_5	1	1	1	1	1
Γ_1	1	ε	ε^2	ε^{2*}	ε^*
Γ_2	1	ε^2	ε^*	ε	ε^{2*}
Γ_3	1	ε^{2*}	ε	ε^*	ε^2
Γ_4	1	ε^*	ε^{2*}	ε^2	ε

It is found that Γ_1 and Γ_4 are complex conjugate pairs and so are Γ_2 and Γ_3. Hence in a more organised form, the character table of C_5 is Table 5.9.

Table 5.9 Character table of the cyclic C_5 point group

	χ (characters)					
	$\hat{E}$	$\hat{C}_5$	$\hat{C}_5^2$	$\hat{C}_5^3$	$\hat{C}_5^4$	
Γ_5	1	1	1	1	1	
Γ_1	1	ε	ε^2	ε^{2*}	ε^*	
Γ_4	1	ε^*	ε^{2*}	ε^2	ε	$\varepsilon = e^{\frac{2\pi i}{5}}$
Γ_2	1	ε^2	ε^*	ε	ε^{2*}	
Γ_3	1	ε^{2*}	ε	ε^*	ε^2	

Note: The point groups C_n and S_n (even n) have the same order and both are cyclic. Hence they both have the same character table; the basis sets and Mulliken symbols of the irreducible representations however, differ (see Appendix A3).

5.9 POINT GROUPS OF LINEAR OBJECTS AND THEIR CHARACTER TABLES

Linear objects may be of two types: (i) those with dissimilar ends (such as heteronuclear diatomic molecules, thiocyanate ion, and so on) and (ii) those having both ends similar (such as homonuclear diatomic molecules, the CO_2 molecule in equilibrium configuration, and so on). For each type, the line joining the atoms is a C_∞-axis because rotation about this axis through any angle brings the object to an equivalent configuration, and there are an infinite number of σ_v planes intersecting at this axis. For type (ii) molecules, a σ_h plane passing through the middle of the object and perpendicular to the C_∞-axis is also present. There are an infinite number of two-fold rotation axes ($\hat{C}_2'$) on this σ_h plane and perpendicular to the principal (C_∞)-axis. Hence the point group of type (ii) objects is $D_{\infty h}$ while that of type (i) objects is $C_{\infty v}$ (Figs 5.4 and 5.5).

Fig. 5.4 The C_∞-axis, two (out of many) σ_v planes of an object of point group $C_{\infty v}$

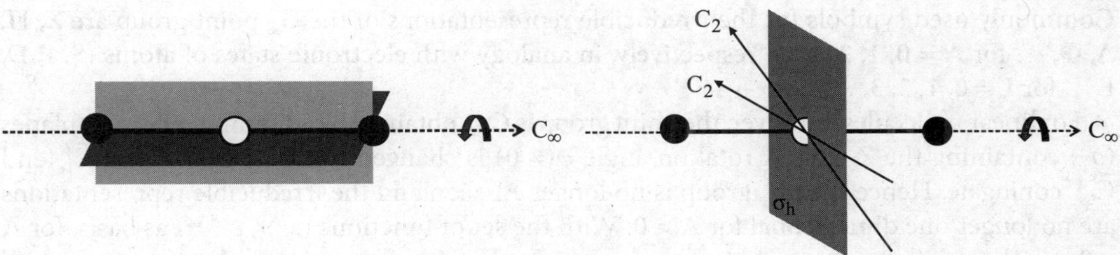

Fig. 5.5 The C_∞-axis, two σ_v planes (out of many), the σ_h plane and two C_2 (out of many possible) axes of an object of point group $D_{\infty h}$

5.9.1 CHARACTER TABLE OF THE $C_{\infty v}$ POINT GROUP

The symmetry operations of $C_{\infty v}$ are the identity ($\hat{E}$), rotation ($\hat{C}_\infty^\phi$) about the axis of the linear object (taken as the Z-axis) through any angle ϕ, and reflection ($\hat{\sigma}_v$) on a plane containing the C_∞-axis. The clockwise and anticlockwise rotations through angle ϕ (that is, $\hat{C}_\infty^\phi$ and $\hat{C}_\infty^{-\phi}$) are conjugate to each other and form one conjugate class. The infinite number of rotation operations are, therefore, written class-wise as $2\hat{C}_\infty^\phi$, $2\hat{C}_\infty^{2\phi}$, $2\hat{C}_\infty^{3\phi}$ ... All the reflection operations are in a single class because any one reflection plane can be brought into another such plane by rotation about the C_∞-axis. Owing to an infinite number of conjugate classes (for the $\hat{C}_\infty^\phi$ operations), the point group $C_{\infty v}$ has an infinite number of irreducible representations. The column headings for the character table of $C_{\infty v}$ are written as

$$\hat{E} \quad 2\hat{C}_\infty^\phi \quad 2\hat{C}_\infty^{2\phi} \quad 2\hat{C}_\infty^{3\phi} \quad \ldots \quad \infty\hat{\sigma}_v$$

The characters of the possible irreducible representations of $C_{\infty v}$ are obtained from those of the pure rotation group C_∞ which consists of the set of all possible pure rotations $\hat{C}_\infty^\phi$ about the Z-axis. The irreducible representations of this group (C_∞) have basis functions which are eigenfunctions of the z-component of the angular momentum operator, $J_z = x\frac{\partial}{\partial y} - y\frac{\partial}{\partial x}$ or, $\frac{\partial}{\partial \phi}$ (the quantum mechanical factor $\frac{h}{2\pi i}$ has been omitted for convenience). The group C_∞ is Abelian because

$$\hat{C}_\infty^{\phi_1} \cdot \hat{C}_\infty^{\phi_2} = \hat{C}_\infty^{\phi_2}\,\hat{C}_\infty^{\phi_1} = \hat{C}_\infty^{\phi_1+\phi_2}$$

That is, successive rotations about the Z-axis through two different angles ϕ_1 and ϕ_2 can be carried out in any order to obtain the same result (= a rotation equivalent to the sum of the two rotational angles). Hence all operations in C_∞ are in different classes and all of its irreducible representations are one-dimensional. It can be readily verified that the functions $e^{\pm i\Lambda\phi}$ are eigenfunctions of $\frac{\partial}{\partial \varphi}$ with eigenvalue $\pm i\Lambda$. Again, for the eigenfunctions to be single-valued, one requires

$$e^{\pm i\Lambda\phi} = e^{\pm i\Lambda(2\pi+\phi)} \quad \text{that is,} \quad e^{\pm i2\pi\Lambda} = 1$$

which is possible only if Λ is an integer. The permitted values of Λ are, therefore, 0, 1, 2, 3, … Commonly used symbols for the irreducible representations of the C_∞ point group are Σ, Π, Δ, Φ, … for $\Lambda = 0, 1, 2, 3, \ldots$ respectively, in analogy with electronic states of atoms (S, P, D, F … for L = 0, 1, 2, 3, …).

For linear molecules, however, the point group is $C_{\infty v}$, obtained by adjoining reflection planes (σ_v) containing the Z-axis. A rotation angle $\phi (\neq 0)$ is changed to $-\phi$ by $\hat{\sigma}_v$, making $\hat{C}_\infty^{\phi}$ and $\hat{C}_\infty^{-\phi}$ conjugate. Hence the $C_{\infty v}$ group is no longer Abelian and the irreducible representations are no longer one dimensional for $\Lambda > 0$. With the set of functions ($e^{i\Lambda\phi}$, $e^{-i\Lambda\phi}$) as basis, for $\Lambda \neq 0$, we observe that

$$\hat{C}_\infty^{\phi}(e^{i\Lambda\phi}) = e^{i\Lambda(\phi+\phi)} = e^{i\Lambda\phi}\cdot e^{i\Lambda\phi} = e^{i\Lambda\phi}\cdot e^{i\Lambda\phi} + 0\cdot e^{-i\Lambda\phi}$$

$$\hat{C}_\infty^{\phi}(e^{-i\Lambda\phi}) = e^{-i\Lambda(\phi+\phi)} = e^{-i\Lambda\phi}\cdot e^{-i\Lambda\phi} = 0\cdot e^{i\Lambda\phi} + e^{-i\Lambda\phi}\cdot e^{-i\Lambda\phi}$$

$$\hat{\sigma}_v(e^{i\Lambda\phi}) = e^{-i\Lambda\phi} = 0\cdot e^{i\Lambda\phi} + 1\cdot e^{-i\Lambda\phi}$$

$$\hat{\sigma}_v(e^{-i\Lambda\phi}) = e^{i\Lambda\phi} = 1\cdot e^{i\Lambda\phi} + 0\cdot e^{-i\Lambda\phi}$$

Hence we get two dimensional matrix representations for $\Lambda > 0$:

$$\mathbf{M}(\hat{C}_\infty^{\phi}) = \begin{bmatrix} e^{i\Lambda\phi} & 0 \\ 0 & e^{-i\Lambda\phi} \end{bmatrix}, \quad \text{Trace} = 2\cos\Lambda\phi$$

$$\mathbf{M}(\hat{\sigma}_v) = \begin{bmatrix} 0 & 1 \\ 1 & 0 \end{bmatrix}, \quad \text{Trace} = 0$$

All σ_v reflections are equivalent (conjugate) as mentioned earlier and all the $\hat{\sigma}_v$ operations have the same character, 0, for the two dimensional irreducible representations.

For $\Lambda = 0$, a reflection $\hat{\sigma}_v$ does not change the basis function but since $\hat{\sigma}_v^2 = \hat{E}$, a $\hat{\sigma}_v$ operation can only multiply the basis function by ±1. [In other words, both $\hat{\sigma}_v f = f$ and $\hat{\sigma}_v f = -f$ satisfy the condition $\hat{\sigma}_v^2 f = f$. The first case is obvious and in the second case $\hat{\sigma}_v^2 f = \hat{\sigma}_v \hat{\sigma}_v f = \hat{\sigma}_v(-f) = -(-f) = f$]. Thus the one dimensional Σ representation of C_∞ gives rise to two different one dimensional representations for $C_{\infty v}$, namely, Σ^+ and Σ^-, which are invariant under rotations but are symmetric and antisymmetric with respect to reflection ($\hat{\sigma}_v$). We thus arrive at the character table of $C_{\infty v}$ (Table 5.10, where Mulliken symbols are shown alongside the common symbols).

Table 5.10 Character table of $C_{\infty v}$

Λ		$\hat{E}$	$2\hat{C}_\infty^\phi$	$2\hat{C}_\infty^{2\phi}$	$2\hat{C}_\infty^{3\phi}$	…	$\infty\hat{\sigma}_v$	**Basis**	
0	$A_1 \equiv \Sigma^+$	1	1	1	1	…	1	(z)	$x^2 + y^2, z^2$
	$A_2 \equiv \Sigma^-$	1	1	1	1	…	−1	(R_z)	
1	$E_1 \equiv \Pi$	2	$2\cos\phi$	$2\cos 2\phi$	$2\cos 3\phi$	…	0	$(x, y), (R_x, R_y)$	(yz, zx)
2	$E_2 \equiv \Delta$	2	$2\cos 2\phi$	$2\cos 4\phi$	$2\cos 6\phi$	…	0		$(x^2 - y^2, xy)$
3	$E_3 \equiv \Phi$	2	$2\cos 3\phi$	$2\cos 6\phi$	$2\cos 9\phi$	…	0		
4	$E_4 \equiv \Gamma$	2	$2\cos 4\phi$	$2\cos 8\phi$	$2\cos 12\phi$	…	0		
…	…	…	…	…	…	…	…		
…	…	…	…	…	…	…	…		

(In using the table, the columns under $2\hat{C}_\infty^{2\phi}$, $2\hat{C}_\infty^{3\phi}$, … may be ignored as they yield no new information other than that given by $2\hat{C}_\infty^{\phi}$.)

Some common basis functions of these irreducible representations are easily obtained by applying the $C_{\infty v}$ symmetry operators on the components of the position vector $\vec{r} = (x, y, z)$ of a general point and on the rotation vector (R_x, R_y, R_z). Thus,

$$\hat{E}z = z,\ \hat{C}_\infty^\phi z = z \text{ (for any } \phi),\ \hat{\sigma}_v z = z \text{ (for all } \sigma_v \text{ planes)}$$

Hence z is a basis function for the totally symmetric irreducible representation, A_1 or Σ^+ (each character = 1). Again, the transformations of x and y under $\hat{C}_\infty^\phi$ and $\hat{\sigma}_v$ are as follows:

$$\begin{aligned} \hat{C}_\infty^\phi x &= x\cos\phi - y\sin\phi \\ \hat{C}_\infty^\phi y &= x\sin\phi + y\cos\phi \end{aligned} \quad \Rightarrow \hat{C}_\infty^\phi: \begin{bmatrix} \cos\phi & -\sin\phi \\ \sin\phi & \cos\phi \end{bmatrix}, \chi(\hat{C}_\infty^\phi) = 2\cos\phi$$

$$\begin{aligned} \hat{\sigma}_v(XZ)\,x &= x = 1\cdot x + 0\cdot y \\ \hat{\sigma}_v(XZ)\,y &= -y = 0\cdot x - 1\cdot y \end{aligned} \quad \Rightarrow \hat{\sigma}_v(XZ): \begin{bmatrix} 1 & 0 \\ 0 & -1 \end{bmatrix}, \chi(\hat{\sigma}_v) = 0$$

Also, with $\Lambda = 1$, $(e^{i\phi} + e^{-i\phi}) = 2\cos\phi = 2\frac{x}{r}$ which transforms as x and $(e^{i\phi} - e^{-i\phi}) = 2i\sin\phi = 2i\frac{y}{r}$ which transforms as y. Thus (x, y) is a basis set for the two dimensional irreducible representation

E_1 which is also written as Π for its correspondence with $\Lambda = 1$. The remaining two dimensional irreducible representations have Mulliken symbols E_2, E_3 and so on.

The z-component of the rotation vector $R_z \approx \left(x\frac{\partial}{\partial y} - y\frac{\partial}{\partial x} \right)$ transforms as follows:

With $\phi = \frac{\pi}{2}$, $\hat{C}_\infty^\phi R_z = \hat{C}_\infty^\phi \left(x\frac{\partial}{\partial y} - y\frac{\partial}{\partial x} \right) = \left(-y\frac{\partial}{\partial x} - x\frac{\partial}{\partial(-y)} \right) = \left(x\frac{\partial}{\partial y} - y\frac{\partial}{\partial x} \right) = R_z$

Thus, with R_z as basis, $\chi(\hat{C}_\infty^\phi) = 1$ for $\phi = \frac{\pi}{2}$. One can choose two mutually perpendicular X and Y axes at any orientation with respect to the Z axis. Hence for any ϕ, $\chi(\hat{C}_\infty^\phi) = 1$ with R_z as basis. Again, with the $\hat{\sigma}_v(XZ)$ plane,

$$\hat{\sigma}_v(XZ)R_z = \hat{\sigma}_v(XZ)\left(x\frac{\partial}{\partial y} - y\frac{\partial}{\partial x} \right) = \left(x\frac{\partial}{\partial(-y)} - (-y)\frac{\partial}{\partial x} \right) = -\left(x\frac{\partial}{\partial y} - y\frac{\partial}{\partial x} \right) = -R_z$$

which gives $\chi(\hat{\sigma}_v) = -1$. Thus R_z is a basis of the irreducible representation, Σ^- of $C_{\infty v}$.

As z belongs to A_1(or Σ^+) and (x, y) to E_1(or Π), the set (xz, yz) also belongs to E_1. Again, $x^2 + y^2$ transforms to $(x \cos\phi - y \sin\phi)^2 + (x \cos\phi + y \sin\phi)^2 = x^2 + y^2$ under $\hat{C}_\infty^\phi$ and under $\hat{\sigma}_v$, $x^2 + y^2$ transforms to $x^2 + (-y)^2 = x^2 + y^2$. Similarly, z^2 remains unchanged by $\hat{C}_\infty^\phi$ and $\hat{\sigma}_v$. Thus the quadratic functions z^2 and $x^2 + y^2$ and their linear combination $x^2 + y^2 + z^2$ are other bases of A_1(or Σ^+). The columns for basis functions as indicated in Table 5.9 may thus be verified. The character table of $D_{\infty h}$ can be easily constructed from that of $C_{\infty v}$ by utilising the concept of direct product groups. This has been done in Chapter 6.

SUMMARY

- Properties of irreducible representations of a point group are consequences of the *Great Orthogonality Theorem* which is central to all developments and applications of group theory. The theorem means that matrix elements of an n-dimensional irreducible representation behave like the components of an n-dimensional vector; two such vectors are orthogonal if the matrix elements are collected from two different irreducible representations or from different rows or columns of the same representation.
- The following fundamental questions are addressed: (i) How many irreducible representations are possible for a group? (ii) What are their dimensions and set of characters? How to determine whether a given representation is reducible and if so, how to get the component irreducible representations present in it?

LEARNING OUTCOME

After reading this chapter the student will be able to (i) construct the character tables of finite and infinite point groups, (ii) recognise both polar and axial vectors as bases of irreducible representations, (iii) assign Mulliken symbols to the possible irreducible representations of point groups, (iv) apply the reduction formula to find the component irreducible representations of a given reducible representation and (v) use the analogy between the set of symmetry operators of a cyclic point group (C_n) and the set of nth roots of unity to construct character tables for C_n.

EXERCISES

1. Using the coordinates (x, y, z) of a general point as basis construct the matrices for the symmetry operators of the groups (a) C_{3v}, (b) D_{3h}, (c) C_{2v}, (d) C_{3h}, (e) T_d (f) O_h. Hence find the matrices of the possible irreducible representations by inspection. Verify the orthogonality theorem for each group.

2. Verify, by drawing symmetry elements and coordinate axes, that $(x^2 - y^2, xy)$ is a basis set of the E representation of the C_{3v} point group.

3. Supply the missing characters in the character table of the C_{3v} point group giving reasons:

χ (characters)

C_{3v}	$\hat{E}$	$2\hat{C}_3$	$3\hat{\sigma}_v$
D^1	1	1	1
D^2	1	1	*
D^3	*	−1	0

4. Determine the characters of the irreducible representations of C_{2h} using the components R_x, R_y and R_z of an axial vector $\vec{R}$ as basis.

5. Show that the components R_x, R_y and R_z of an axial vector $\vec{R}$ cannot act as bases of the u-type irreducible representations of a centrosymmetric point group.

6. A representation Γ of C_{3v} has the character set {6, 0, 0} for $\hat{E}$, $(\hat{C}_3, \hat{C}_3^2)$ and the three $\hat{\sigma}_v$ operators. Show that Γ contains all the irreducible representations A_1, A_2 and E of C_{3v}.

7. The characters of representations (Γ) of some point groups are given. Reduce them.

C_{2v}	$\hat{E}$	$\hat{C}_2(Z)$	$\hat{\sigma}_v(XZ)$	$\hat{\sigma}_v(YZ)$
Γ	4	−2	0	−2

C_{3h}	$\hat{E}$	$\hat{C}_3$	$\hat{C}_3^2$	$\hat{\sigma}_h$	$\hat{S}_3$	$\hat{S}_3^5$
Γ	4	1	1	2	−1	−1

D_{4h}	$\hat{E}$	$2\hat{C}_4$	$\hat{C}_2$	$2\hat{C}_2'$	$2\hat{C}_2''$	$\hat{i}$	$2\hat{S}_4$	$\hat{\sigma}_h$	$2\hat{\sigma}_v$	$2\hat{\sigma}_d$
Γ	4	2	0	−2	−2	2	0	−2	0	0

D_{4d}	$\hat{E}$	$2\hat{S}_8$	$2\hat{C}_4$	$2\hat{S}_3^5$	$\hat{C}_2$	$4\hat{C}_2'$	$4\hat{\sigma}_d$
Γ	6	0	−2	0	−2	0	0

T_d	$\hat{E}$	$8\hat{C}_3$	$3\hat{C}_2$	$6\hat{S}_4$	$6\hat{\sigma}_d$
Γ	9	0	1	−1	3

8. The effect of symmetry operations of D_4 on the orthonormal basis functions ϕ_1, ϕ_2, ϕ_3, ϕ_4 of a four-dimensional representation Γ of D_4 are given in the following table:

D_4	$\hat{E}$	$\hat{C}_4$	$\hat{C}_4^3$	$\hat{C}_2$	$\hat{C}_2'(1)$	$\hat{C}_2'(2)$	$\hat{C}_2''(1)$	$\hat{C}_2''(2)$
ϕ_1	ϕ_1	ϕ_2	ϕ_4	ϕ_3	$-\phi_4$	$-\phi_2$	$-\phi_1$	$-\phi_3$
ϕ_2	ϕ_2	ϕ_3	ϕ_1	ϕ_4	$-\phi_3$	$-\phi_1$	$-\phi_4$	$-\phi_2$
ϕ_3	ϕ_3	ϕ_4	ϕ_2	ϕ_1	$-\phi_2$	$-\phi_4$	$-\phi_3$	$-\phi_1$
ϕ_4	ϕ_4	ϕ_1	ϕ_3	ϕ_2	$-\phi_1$	$-\phi_3$	$-\phi_2$	$-\phi_4$

Show that $\Gamma = A_2 + B_1 + E$.

9. The effects of the symmetry operations of C_{3v} on the orthonormal basis functions ϕ_1, ϕ_2, ϕ_3, ϕ_4, ϕ_5, ϕ_6 of a six-dimensional representation Γ of C_{3v} are given in the table below.

	$\hat{E}$	$\hat{C}_3$	$\hat{C}_3^2$	$\hat{\sigma}_v(1)$	$\hat{\sigma}_v(2)$	$\hat{\sigma}_v(3)$
ϕ_1	ϕ_1	$-\frac{1}{2}(\phi_3 - \sqrt{3}\phi_4)$	$-\frac{1}{2}(\phi_5 + \sqrt{3}\phi_6)$	$-\phi_1$	$\frac{1}{2}(\phi_5 + \sqrt{3}\phi_6)$	$\frac{1}{2}(\phi_3 - \sqrt{3}\phi_4)$
ϕ_2	ϕ_2	$-\frac{1}{2}(\sqrt{3}\phi_3 + \phi_4)$	$\frac{1}{2}(\sqrt{3}\phi_5 - \phi_6)$	ϕ_2	$\frac{1}{2}(\sqrt{3}\phi_5 - \phi_6)$	$-\frac{1}{2}(\sqrt{3}\phi_3 + \phi_4)$
ϕ_3	ϕ_3	$-\frac{1}{2}(\phi_5 - \sqrt{3}\phi_6)$	$-\frac{1}{2}(\phi_1 + \sqrt{3}\phi_2)$	$-\phi_5$	$\frac{1}{2}(\phi_3 + \sqrt{3}\phi_4)$	$\frac{1}{2}(\phi_1 - \sqrt{3}\phi_2)$
ϕ_4	ϕ_4	$-\frac{1}{2}(\sqrt{3}\phi_5 + \phi_6)$	$\frac{1}{2}(\sqrt{3}\phi_1 - \phi_2)$	ϕ_6	$\frac{1}{2}(\sqrt{3}\phi_3 - \phi_4)$	$-\frac{1}{2}(\sqrt{3}\phi_1 + \phi_2)$
ϕ_5	ϕ_5	$-\frac{1}{2}(\phi_1 - \sqrt{3}\phi_2)$	$-\frac{1}{2}(\phi_3 + \sqrt{3}\phi_4)$	$-\phi_3$	$\frac{1}{2}(\phi_1 + \sqrt{3}\phi_2)$	$\frac{1}{2}(\phi_5 - \sqrt{3}\phi_6)$
ϕ_6	ϕ_6	$-\frac{1}{2}(\sqrt{3}\phi_1 + \phi_2)$	$\frac{1}{2}(\sqrt{3}\phi_3 - \phi_4)$	ϕ_4	$\frac{1}{2}(\sqrt{3}\phi_1 - \phi_2)$	$-\frac{1}{2}(\sqrt{3}\phi_5 + \phi_6)$

Show that $\Gamma = A_1 + A_2 + 2E$.

10. Explain why the following pairs of point groups have the same character tables (excepting Mulliken symbols and basis sets):
(a) S_6 and C_6 (b) C_{3v} and D_3 (c) C_{4v} and D_4
[**Note:** Even Mulliken symbols of the irreducible representations are the same for C_{nv} and D_n.]

11. What information is obtained from the Mulliken symbols of the following irreducible representations?
(a) A_g in C_{2h} (b) B_2 in C_{4v} (c) E in D_3 (d) A_1'' in D_{3h}
(e) E' in D_{3h} (f) E_u in D_{4h} (g) B_{1g} in D_{4h} (h) T_g in T_h.

12. Supply the missing characters in the character table given below for a certain point group (where Ê is the identity operator). Can you identify the point group?

	Ê	2N̂	P̂	2Q̂	2R̂
Γ_1	*	*	*	*	*
Γ_2	1	*	1	−1	−1
Γ_3	*	−1	1	*	−1
Γ_4	*	−1	1	*	1
Γ_5	*	0	*	0	0

13. The *p*-orbitals p_x, p_y and p_z of an isolated atom M are degenerate. By consulting the appropriate character table show how the degeneracy is lifted when the following environments are created by X atoms surrounding M.
(a) MX_2 (linear) (b) MX_2 (bent) (c) MX_3 (trigonal planar)
(d) MX_3 (pyramidal) (e) MX_4 (tetrahedral) (f) MX_4 (square planar)
(g) MX_6 (octahedral).

14. Justify or refute the statement: All irreducible representations of Abelian groups are one dimensional.

15. Construct the character table of the D_3 point group by using transformations of a general point (*x*, *y*, *z*); take the Z-axis as the C_3-axis and the X-axis as one of the C_2 axes. Show the linear and quadratic functions of the Cartesian coordinates (*x*, *y*, *z*) in the basis sets of the irreducible representations. Also show the polar vector components R_x, R_y and R_z at adequate positions in the basis set column(s) of the character table that you have developed. Verify the character orthogonality theorem using this table.

16. Construct the matrices for one symmetry operator of each class of the pure rotation group *T* using the coordinates (*x*, *y*, *z*) of a general point as basis. By comparing the traces with those given in the character table of *T* (Appendix A3), show that (a) (*x*, *y*, *z*) and (R_x, R_y, R_z) are separately basis sets for the irreducible representation T, (b) $x^2 + y^2 + z^2$ is a basis function for the totally symmetric representation.

17. A point group (G) has order *g* and the characters of a representation (Γ) of G are as follows:

χ (characters)

G	Ê	$\nu_2\hat{R}_2$	$\nu_3\hat{R}_3$	…	$\nu_c\hat{R}_c$
Γ	*g*	0	0	…	0

where ν_i is the number of operators in the *i*th class. Show that in Γ each irreducible representation ($D^{(i)}$) of *G* occurs with a frequency equal to its dimension l_i.

18. Construct the matrices for the symmetry operators of point group *O* by examining the effect of these operators on a general point (*x*, *y*, *z*). By a comparison of the traces with those listed in the character table of *O* (Appendix A3), show that:

(a) the functions $f_1 = 2z^2 - x^2 - y^2$ and $f_2 = x^2 - y^2$ transform according to the E representation of the point group O.
(b) f_1, f_2, f_3 transform as (i) T_1, if $f_1 = x, f_2 = y, f_3 = z$ (ii) T_2, if $f_1 = xy, f_2 = yz, f_3 = xz$.

19. Short questions:
(a) In an octahedral field the set of *d*-orbitals splits into t_{2g} and e_g but in a tetrahedral field the *g* subscripts are not needed. Why?
(b) Show that the dipole moment vector of H_2O molecule can act as a basis of the totally symmetric irreducible representation of the molecular point group.
(c) A representation of the C_{2v} point group has $\chi(\hat{E}) = 6$. Is it reducible? Give reasons.
(d) In a point group the operators $\hat{C}_3$ and $\hat{C}_3^2$ belong to different conjugate classes. What may be the point group?
(e) The character table of C_{3v} is smaller than that of C_{2h} although the former is of higher order. Why?

Suggested Books and References for Further Reading

1. Murnaghan F D, *The theory of Group Representations*, The Johns Hopkins Press, Baltimore, **1938.**
2. Weyl H, *Gruppentheorie und Quantenmechanik*, S Hirzel Verlag, Leipzig, **1928**; translated version *Theory of Groups and Quantum Mechanics*, Dover Publications, Inc., NY, **1950.**
3. Eyring H, Walter J and Kimball G E, *Quantum Chemistry*, John Wiely & Sons, Inc., NY **1944.**
4. Wigner E P, *Gruppentheorie und ihre Anwendung auf die Quantenmechanik der Atomspektren*, Friedr. Vieweg & Sohn, Brunswick, **1931**; revised and translated edition, Academic Press Inc., NY, **1959.**

6 Direct Product Groups and Enumeration of Crystallographic Point Groups

Learning objectives

- ➢ Direct product groups—large groups from smaller ones
- ➢ Construction of character tables of large groups using direct product relationships
- ➢ Non-compatibility of rotation and translation, enumeration of crystallographic point groups

6.1 INTRODUCTION

A method for generating large point groups by utilising the closure property of groups and commutation of certain pairs of symmetry operators will be discussed in this chapter. This method not only leads to easy construction of character tables of many large point groups but also makes it possible to explain why there are exactly 32 crystallographic point groups in nature.

6.2 DIRECT PRODUCT GROUPS

If there are two point groups, $G_a = \{\hat{E}, \hat{A}_2, \hat{A}_3, \ldots, \hat{A}_h\}$ and $G_b = \{\hat{E}, \hat{B}_2, \hat{B}_3, \ldots, \hat{B}_k\}$ such that

- no operator except $\hat{E}$ is common to G_a and G_b, and
- each operator of G_b commutes with all operators of G_a,

then the set S, denoted by $G_a \times G_b$ and consisting of the operators

$$\{\hat{E}, \hat{A}_2, \hat{A}_3, \ldots, \hat{A}_h, \hat{B}_2, \hat{A}_2\hat{B}_2, \hat{A}_3\hat{B}_2, \ldots, \hat{A}_h\hat{B}_2, \ldots, \hat{B}_k, \hat{A}_2\hat{B}_k, \hat{A}_3\hat{B}_k, \ldots, \hat{A}_h\hat{B}_k\}$$

is called the direct product group of G_a and G_b. It is a group of order hk. To prove that this is so, we proceed as follows:
The product of any two operators of $G_a \times G_b$

$= (\hat{A}_i\hat{B}_j)(\hat{A}_m\hat{B}_n)$

$= \hat{A}_i\hat{B}_j\hat{B}_n\hat{A}_m$ (as $\hat{A}_m$ and $\hat{B}_n$ commute)

$= \hat{A}_i\hat{B}_p\hat{A}_m$ (as $\hat{B}_j, \hat{B}_n \in G_b$, $\hat{B}_j\hat{B}_n = \hat{B}_p \in G_b$ owing to closure property)

$= \hat{A}_i\hat{A}_m\hat{B}_p$ (as $\hat{A}_m$ and $\hat{B}_p$ commute)

$= \hat{A}_q\hat{B}_p$ (where $\hat{A}_q = \hat{A}_i\hat{A}_m \in G_a$)

= an operator of $G_a \times G_b$ according to construction of its members.

Hence $G_a \times G_b$ is closed, as required. The other group properties can be similarly shown.

Example 6.1

$$C_3 = \{\hat{E}, \hat{C}_3, \hat{C}_3^2\}, C_s = \{\hat{E}, \hat{\sigma}\} \text{ and } C_i = \{\hat{E}, \hat{i}\}$$

If the reflection plane σ in C_s is perpendicular to the C_3-axis, then σ behaves as σ_h and

$$C_3 \times C_s = \{\hat{E}, \hat{C}_3, \hat{C}_3^2, \hat{\sigma}_h, \hat{S}_3, \hat{S}_3^5\} = C_{3h}$$

But $C_3 \times C_i = \{\hat{E}, \hat{C}_3, \hat{C}_3^2, \hat{i}, \hat{i}\,\hat{C}_3, \hat{i}\,\hat{C}_3^2\} \neq C_{3h}$ because C_{3h} does not contain $\hat{i}$ which can be readily verified by drawing the stereographic projection of C_{3h}. Also, $C_3 \times C_i$ cannot be D_{3d} because the latter contains more than 6 elements. It can be shown that $C_3 \times C_i = S_6$ (see Section 6.2.2).

Example 6.2

$$C_4 = \{\hat{E}, \hat{C}_4, \hat{C}_4^2 = \hat{C}_2, \hat{C}_4^3\}, C_s = \{\hat{E}, \hat{\sigma}_h\} \text{ and } C_i = \{\hat{E}, \hat{i}\}$$

As before, taking σ to be perpendicular to the C_4-axis, we have:

$$C_4 \times C_s = \{\hat{E}, \hat{C}_4, \hat{C}_2, \hat{C}_4^3, \hat{\sigma}_h, \hat{S}_4, \hat{i}, \hat{S}_4^3\} = C_{4h}$$

and $$C_4 \times C_i = \{\hat{E}, \hat{C}_4, \hat{C}_2, \hat{C}_4^3, \hat{i}, \hat{i}\,\hat{C}_4, \hat{i}\,\hat{C}_2, \hat{i}\,\hat{C}_4^3\}$$

$$= \{\hat{E}, \hat{C}_4, \hat{C}_2, \hat{C}_4^3, \hat{\sigma}_h, \hat{S}_4, \hat{i}, \hat{S}_4^3\} = C_{4h}$$

($\hat{i} = \hat{\sigma}_h\hat{C}_2$, $\hat{i}\,\hat{C}_4 = \hat{\sigma}_h\hat{C}_2\hat{C}_4 = \hat{\sigma}_h\hat{C}_4^2\hat{C}_4 = \hat{\sigma}_h\hat{C}_4^3 = \hat{S}_4^3$, $\hat{i}\,\hat{C}_2 = \hat{\sigma}_h\hat{C}_2\hat{C}_2 = \hat{\sigma}_h$, $\hat{i}\hat{C}_4^3 = \hat{i}\,\hat{C}_2\hat{C}_4 = \hat{\sigma}_h\hat{C}_2^2\hat{C}_4 = \hat{\sigma}_h\hat{C}_4 = \hat{S}_4$)

In general, $C_n \times C_s = C_{nh}$ for odd n. In case of even n, C_{nh} is either $C_n \times C_s$ or $C_n \times C_i$. For the sake of convenience we shall take

$$C_{nh} = C_n \times C_s \text{ (for odd } n) \quad \text{and} \quad C_{nh} = C_n \times C_i \text{ (for even } n)$$

It can be shown that for any operator $\hat{R}$ of $G_a \times G_b$, the character of $\hat{R}$ is the product of the characters of the corresponding operators in the component groups G_a and G_b, that is,

$$\chi(\hat{R}_{a\times b}) = \chi(\hat{R}_a) \cdot \chi(\hat{R}_b) \tag{6.1}$$

where $\hat{R}_{a\times b}$ belongs to $(G_a \times G_b)$, $\hat{R}_a$ belongs to G_a, $\hat{R}_b$ belongs to G_b and $\hat{R}_{a\times b} = \hat{R}_a \cdot \hat{R}_b$.

Using this rule we can construct the character table for C_{nh} type of groups as shown in the following section. [The rule (6.1) is based on the construction of direct products of matrices discussed in Section 7.5.]

6.2.1 Construction of Character Tables of Some Direct Product Groups

Example 6.3 Character table of C_{4h}

Table 6.1 Character table of C_i

	χ (character)	
	$\hat{E}$	$\hat{i}$
D^+	1	1
D^-	1	–1

Since $C_{4h} = C_4 \times C_i$ we first require the individual character tables of C_4 and C_i. With only two operators ($\hat{E}$ and $\hat{i}$), the character table of C_i (Table 6.1) can be easily written down using the method discussed in Section 5.5. Again, the fourth roots of unity, $\{\exp(2\pi ir/4), r = 1, 2, 3 \text{ and } 4\}$ form a cyclic group $\{i, i^2, i^3, i^4\}$ which is isomorphic to the cyclic point group C_4. Hence the character table of C_4 is Table 6.2. On simplification and rearrangement of rows and columns this becomes Table 6.3.

Table 6.2 Character table of C_4 isomorphic to the group of square roots of unity

	χ (character)			
	$\hat{C}_4$	$\hat{C}_4^2 \equiv \hat{C}_2$	$\hat{C}_4^3$	$\hat{C}_4^4 \equiv \hat{E}$
D_1	i	i^2	i^3	$i^4 = 1$
D_2	i^2	i^4	i^6	$i^8 = 1$
D_3	i^3	i^6	i^9	$i^{12} = 1$
D_4	i^4	i^8	i^{12}	$i^{16} = 1$

Table 6.3 Character table of C_4 (rearranged)

	χ (character)			
	$\hat{C}_4^4 \equiv \hat{E}$	$\hat{C}_4$	$\hat{C}_4^2 \equiv \hat{C}_2$	$\hat{C}_4^3$
D_4	1	1	1	1
D_2	1	–1	1	–1
D_3	1	$-i$	–1	i
D_1	1	i	–1	$-i$

Now using Eq. (6.1), the character table of C_{4h} is as shown in Table 6.4. We observe that D_3^+ and D_1^+ are complex conjugate pairs. So are D_3^- and D_1^-. Moreover, $\hat{i}\hat{C}_4 = \hat{S}_4^3$, $\hat{i}\,\hat{C}_2 = \hat{\sigma}_h$ and $\hat{i}\hat{C}_4^3 = \hat{S}_4$ and so we can write the character table in a more organised form as shown in Table 6.5.

6.2.2 Some More Examples of Direct Product Groups and their Character Tables

Example 6.4

$$C_3 \times C_i = S_6$$

Proof: The stereographic projection of the point group C_3 is (1) of Fig. 6.1 and the C_i point group consists of the set of symmetry operators $\{\hat{E}, \hat{i}\}$. A general starting point is imagined in the upper hemisphere and its projection on the equatorial plane is P_1^+ as shown in (1) of Fig. 6.1. The result of operating $\hat{i}$ on P_1^+ is P_2° and then the operations $\hat{C}_3$ and $\hat{C}_3^2$ on P_2° give P_4° and P_3° respectively as shown in (2) and (3) of Fig. 6.1. From these it is evident that if an S_6-axis collinear with the C_3-axis exists, the transformations $P_1^+ \rightarrow P_3^\circ$ and $P_1^+ \rightarrow P_4^\circ$ are equivalent to the result of the single operations $\hat{S}_6$ and $\hat{S}_6^5$ respectively (Fig. 6.2). Thus, $\hat{i}\hat{C}_3^2 = \hat{S}_6$ and $\hat{i}\hat{C}_3 = \hat{S}_6^5$. Hence, the direct product group $C_3 \times C_i$ contains the set of symmetry operations, $\{\hat{E}, \hat{C}_3, \hat{C}_3^2\} \times \{\hat{E}, \hat{i}\} = \{\hat{E}, \hat{C}_3, \hat{C}_3^2, \hat{i}, \hat{S}_6, \hat{S}_6^5\}$ which comprises the S_6 point group.

Table 6.4 Character table of C_{4h} according to Eq. (6.1)

	χ (character)							
	$\hat{E} \times$ sym. op. of C_4				$\hat{i} \times$ sym. op. of C_4			
	$\hat{E}$	$\hat{C}_4$	$\hat{C}_2$	$\hat{C}_4^3$	$\hat{i}$	$\hat{i}\hat{C}_4$	$\hat{i}\hat{C}_2$	$\hat{i}\hat{C}_4^3$
D_4^+	1	1	1	1	1	1	1	1
D_4^-	1	1	1	1	−1	−1	−1	−1
D_2^+	1	−1	1	−1	1	−1	1	−1
D_2^-	1	−1	1	−1	−1	1	−1	1
D_3^+	1	$-i$	−1	i	1	$-i$	−1	i
D_3^-	1	$-i$	−1	i	−1	i	1	$-i$
D_1^+	1	i	−1	$-i$	1	i	−1	$-i$
D_1^-	1	i	−1	$-i$	−1	$-i$	1	i

Table 6.5 Organised form of the character table of C_{4h} (without basis set)

	χ (character)							
	$\hat{E} \times$ sym. op. of C_4				$\hat{i} \times$ sym. op. of C_4			
	$\hat{E}$	$\hat{C}_4$	$\hat{C}_2$	$\hat{C}_4^3$	$\hat{i}$	$\hat{S}_4^3$	$\hat{\sigma}_h$	$\hat{S}_4$
D_4^+	1	1	1	1	1	1	1	1
D_4^-	1	1	1	1	−1	−1	−1	−1
D_2^+	1	−1	1	−1	1	−1	1	−1
D_2^-	1	−1	1	−1	−1	1	−1	1
D_3^+	1	$-i$	−1	i	1	$-i$	−1	i
D_1^+	1	i	−1	$-i$	1	i	−1	$-i$
D_1^-	1	i	−1	$-i$	−1	$-i$	1	i
D_3^-	1	$-i$	−1	i	−1	i	1	$-i$

Example 6.5

$$D_3 \times C_i = D_{3d}$$

Proof: The stereographic projection of the point group $D_3 = \{\hat{E}, \hat{C}_3, \hat{C}_3^2, C_2(1), C_2(2), C_2(3)\}$ is (1) of Fig. 6.3. The result of operating $\hat{i}$ on the projection P_1^+ of a general point is shown in the same figure where some D_3 operations on the resulting point P_2° are also shown. It is evident from these figures that $\hat{i}(P_1^+) \to P_2^\circ$, $\hat{C}_3(P_2^\circ) \to P_3^\circ$ and the transformation $P_1^+ \to P_3^\circ$ is caused by $\hat{S}_6^5$; hence $\hat{i}\hat{C}_3 = \hat{S}_6^5$. Similarly, $\hat{i}\hat{C}_3^2 = \hat{S}_6$. Also, $C_2(1)(P_2^\circ) \to P_4^+$ and the transformation $P_1^+ \to P_4^+$ is caused by $\hat{\sigma}_d(1)$; thus $C_2(1)\hat{i} = \hat{\sigma}_d(1)$. Similarly $C_2(2)\hat{i} = \hat{\sigma}_d(2)$ and $C_2(3)\hat{i} = \hat{\sigma}_d(3)$.

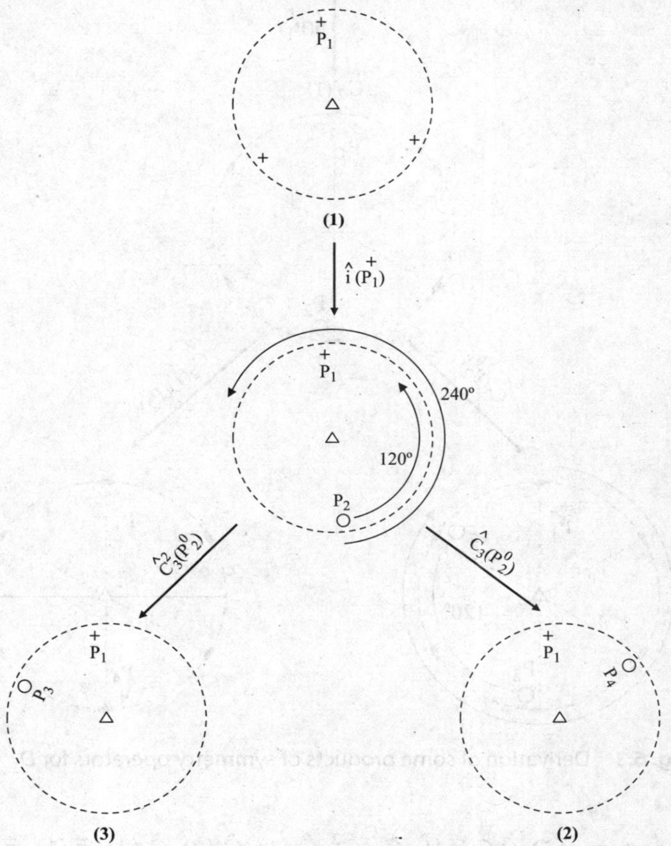

Fig. 6.1 Stereograms for obtaining the products of $\hat{i}$ with the symmetry operators of the point group C_3

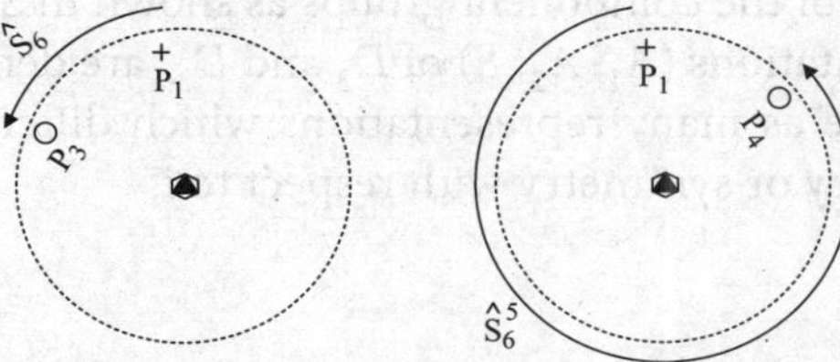

Fig. 6.2 Stereograms showing that the transformations $P_1^+ \to P_3^\circ$ and $P_1^+ \to P_4^\circ$ are equivalent to $\hat{S}_6$ and $\hat{S}_6^5$ operations respectively

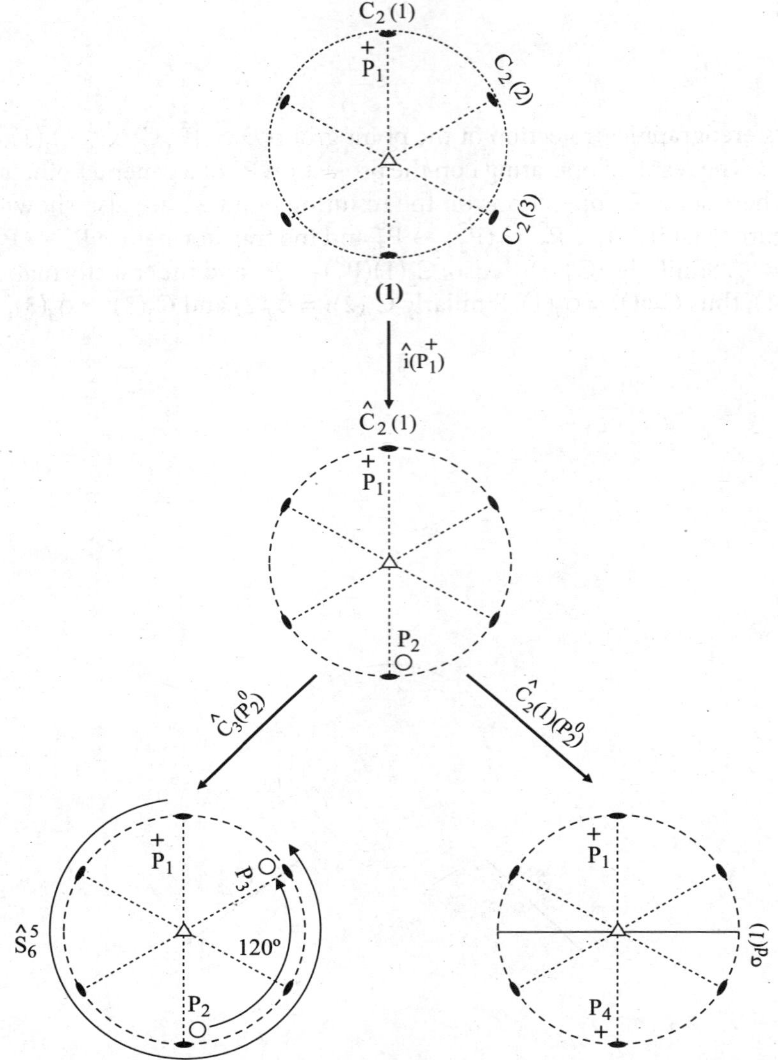

Fig. 6.3 Derivation of some products of symmetry operators for $D_3 \times C_i$

Thus, the set of operators of $D_3 \times C_i$ is $\{\hat{E}, \hat{C}_3, \hat{C}_3^2, \hat{C}_2(1), \hat{C}_2(2), \hat{C}_2(3), \hat{\sigma}_d(1), \hat{\sigma}_d(2), \hat{\sigma}_d(3), \hat{i}, \hat{S}_6, \hat{S}_6^5\}$ which is the D_{3d} set of operators. Hence $D_3 \times C_i = D_{3d}$. The character table of D_{3d} can therefore be constructed from those of the component groups as shown in Scheme 6.1.

The irreducible representations (A_1, A_2, E) of D_3 and D_{3d} are denoted by the same Mulliken symbols; but D_{3d} has twice as many representations which differ by the subscripts u and g depending on antisymmetry or symmetry with respect to $\hat{i}$.

Example 6.6

$D_3 \times C_s = D_{3h}$

Proof: Starting from the projection P_1^+ of a general point in the stereogram of D_3 and considering the σ-plane of C_s to be perpendicular to the C_3-axis of D_3, the necessary products of symmetry operations for $D_3 \times C_s$ can be obtained from the flow chart given in Fig. 6.4. The σ-plane being perpendicular to the C_3-axis now behaves like σ_h (hence drawn as a solid circle on the equatorial plane from (2) onwards in Fig. 6.4). It is found that $\hat{\sigma}_h \hat{C}_3 = \hat{S}_3^5$, $\hat{\sigma}_h \hat{C}_3^2 = \hat{S}_3$, $\hat{\sigma}_h \hat{C}_2(1) = \hat{\sigma}_v(1)$, $\hat{\sigma}_h \hat{C}_2(2) = \hat{\sigma}_v(2)$ and $\hat{\sigma}_h \hat{C}_2(3) = \hat{\sigma}_v(3)$.

Hence $D_3 \times C_s = \{\hat{E}, 2\hat{C}_3, 3\hat{C}_2\} \times \{\hat{E}, \hat{\sigma}_h\}$

$$= \hat{E}, 2\hat{C}_3, 3\hat{C}_2, \hat{\sigma}_h, 2\hat{S}_3, 3\hat{\sigma}_v\} = D_{3h}$$

The character table of D_{3h} can therefore be constructed in a manner analogous to Scheme 6.1.

Scheme 6.1 Tabular scheme for construction of character table of $D_3 \times C_i = D_{3d}$

D_3	χ (character) $\hat{E}$	$2\hat{C}_3$	$3\hat{C}_2$
A_1	1	1	1
A_2	1	1	−1
E	2	−1	0

×

C_i	χ (character) $\hat{E}$	$\hat{i}$
A_g	1	1
A_u	1	−1

↓

$\hat{E} \times$ (operator of D_3)	$\hat{i} \times$ (operator of D_3)
χ	χ
χ	$-\chi$
(where χ = body of the character table of D_3)	

↓

D_{3d}	χ (character) $\hat{E}$	$2\hat{C}_3$	$3\hat{C}_2$	$\hat{i}$	$2\hat{S}_6$	$3\hat{\sigma}_d$
A_{1g}	1	1	1	1	1	1
A_{2g}	1	1	−1	1	1	−1
E_g	2	−1	0	2	−1	0
A_{1u}	1	1	1	−1	−1	−1
A_{2u}	1	1	−1	−1	−1	1
E_u	2	−1	0	−2	1	0

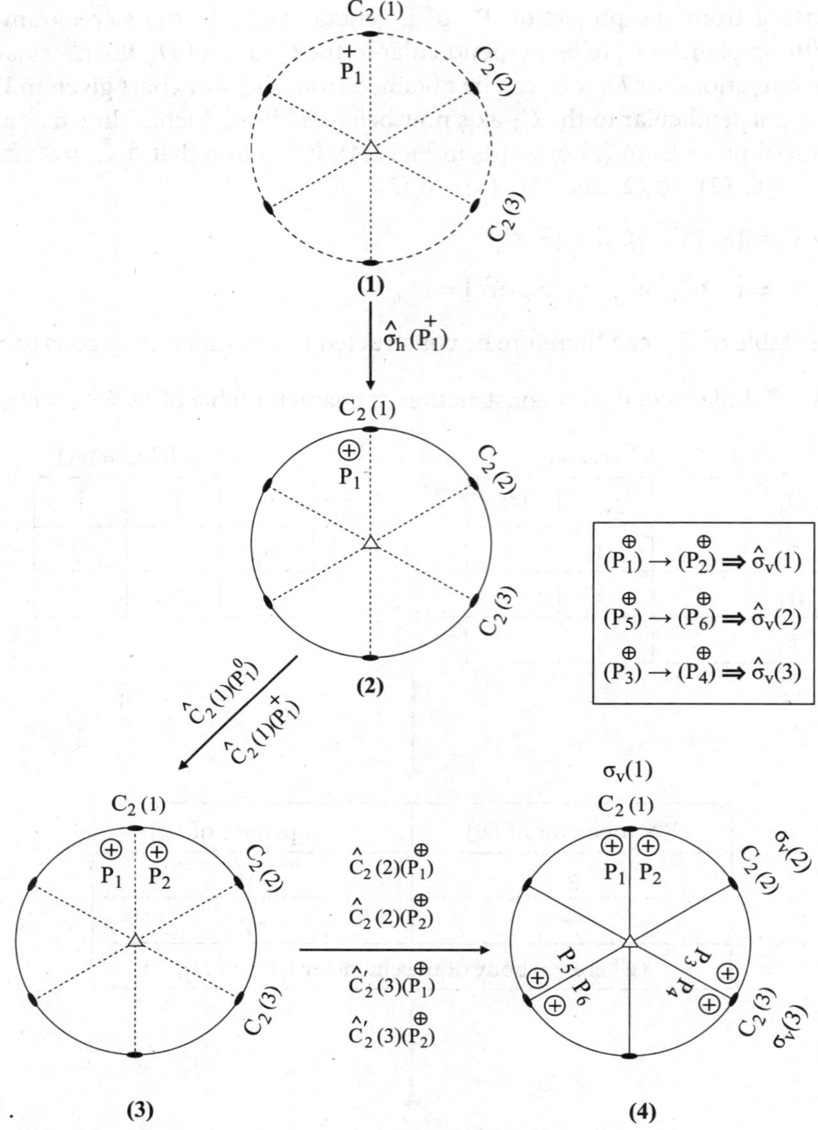

Fig. 6.4 Necessary stereograms for the construction of symmetry products for $D_3 \times C_s$

Example 6.7

$$D_{\infty h} = C_{\infty v} \times C_i$$

The symmetry elements of $C_{\infty v}$ and $D_{\infty h}$ were discussed in Section 5.9. For a linear object, one can imagine a 180° rotation about the C_∞-axis and then it is found that $\hat{i}\hat{C}_\infty^\phi = \hat{\sigma}_h\hat{C}_2\hat{C}_\infty^\phi$ (where the C_2-axis is taken collinear with the C_∞^ϕ-axis) $= \hat{\sigma}_h\hat{C}_\infty^{(\phi+\pi)} = \hat{S}_\infty^\Phi$ ($\Phi = \phi + \pi$ being another ϕ with respect to the C_∞^ϕ-axis) and $\hat{i}\hat{\sigma}_v = \hat{\sigma}_h\hat{C}_\infty^\pi\hat{\sigma}_v = \hat{\sigma}_h\hat{\sigma}_v'$ (where $\hat{\sigma}_v'$ is another vertical reflection plane) $= \hat{C}_2$ (a two-fold rotation about the line of intersection of $\hat{\sigma}_h$ and $\hat{\sigma}_v'$).

Hence, the list of symmetry operations of $D_{\infty h}$ may be written as

$$\hat{E}, 2\hat{C}_{\infty}^{\phi}, 2\hat{C}_{\infty}^{2\phi}, \ldots, \infty\hat{\sigma}_v, \hat{i}, 2\hat{S}_{\infty}^{\Phi}(\equiv \hat{i}\hat{C}_{\infty}^{\phi}), 2\hat{S}_{\infty}^{2\Phi}(\equiv \hat{i}\hat{C}_{\infty}^{2\phi}), \ldots, \infty\hat{C}_2'((\equiv \hat{i}\hat{\sigma}_v)$$

The $\hat{\sigma}_h$ operation of $D_{\infty h}$ is included in $\hat{S}_{\infty}^{\Phi}$ because $\hat{S}_{\infty}^{\Phi=2\pi}$ gives $\hat{\sigma}_h$. Hence $D_{\infty h} = C_{\infty v} \times C_i$ and the character table of $D_{\infty h}$ can be easily constructed by the procedure for direct product groups. The general structure of this character table is, therefore, as follows:

$\hat{E}$	$2\hat{C}_{\infty}^{\phi}$	$2\hat{C}_{\infty}^{2\phi}$	…	$\infty\hat{\sigma}_v$	$\hat{i}$	$2\hat{S}_{\infty}^{\Phi}$	$2\hat{S}_{\infty}^{2\Phi}$	…..	$\infty\hat{C}_2$
		χ				χ			
		χ				$-\chi$			

where χ is the array of characters of $C_{\infty v}$. This follows from the character table of C_i:

C_i	$\hat{E}$	$\hat{i}$
A_g	1	1
A_u	1	−1

In a compact form, omitting the redundant symmetry operations $\hat{C}_{\infty}^{2\phi}$, $\hat{C}_{\infty}^{3\phi}$, …, the character table of $D_{\infty h}$ is as shown in Table 6.6. In this table the superscript on $\hat{S}_{\infty}$ is Φ that is, $\phi + \pi$. The angles in the column under $\hat{S}_{\infty}^{\Phi}$ are each Φ. Since $\cos\Phi = \cos(\pi + \phi) = -\cos\phi$ and $\cos 2\Phi = \cos(2\pi + 2\phi) = \cos 2\phi$, and so on, Table 6.6 is rewritten in the standard form of Table 6.7. [The prime on C_2 is unnecessary as the C_2 about the principal axis need not be explicitly stated.]

Table 6.6 Character table of $D_{\infty h}$ obtained from the direct product $C_{\infty v} \times C_i$

$D_{\infty h}$	$\hat{E}$	$2\hat{C}_{\infty}^{\phi}$	…	$\infty\hat{\sigma}_v$	$\hat{i}$	$2\hat{S}_{\infty}^{\Phi}$	…	$\infty\hat{C}_2$		**basis**
Σ_g^+	1	1	…	1	1	1	…	1		x^2+y^2, z^2
Σ_g^-	1	1	…	−1	1	1	…	−1	(R_z)	
Π_g	2	$2\cos\phi$	…	0	2	$2\cos\Phi$	…	0	(R_x, R_y)	(yz, zx)
Δ_g	2	$2\cos 2\phi$	…	0	2	$2\cos 2\Phi$	…	0		(x^2-y^2, xy)
…	…	…	…	…	…	…	…	…		
Σ_u^+	1	1	…	1	−1	−1	…	−1	(z)	
Σ_u^-	1	1	…	−1	−1	−1	…	1		
Π_u	2	$2\cos\phi$	…	0	−2	$-2\cos\Phi$	…	0	(x, y)	
Δ_u	2	$2\cos 2\phi$	…	0	−2	$-2\cos 2\Phi$	…	0		
…	…	…	…	…	…	…	…	…		

Table 6.7 Character table of $D_{\infty h}$ in standard form

$D_{\infty h}$	$\hat{E}$	$2\hat{C}_{\infty}^{\phi}$	...	$\infty\hat{\sigma}_v$	$\hat{i}$	$2\hat{S}_{\infty}^{\phi}$	...	$\infty\hat{C}_2$		basis
Σ_g^+	1	1	...	1	1	1	...	1		x^2+y^2, z^2
Σ_g^-	1	1	...	−1	1	1	...	−1	(R_z)	
Π_g	2	$2\cos\phi$	...	0	2	$-2\cos\phi$	...	0	(R_x, R_y)	(yz, zx)
Δ_g	2	$2\cos 2\phi$	...	0	2	$2\cos 2\phi$	...	0		(x^2-y^2, xy)
...	...	...	...	...	...	...	...	...		
Σ_u^+	1	1	...	1	−1	−1	...	−1	(z)	
Σ_u^-	1	1	...	−1	−1	−1	...	1		
Π_u	2	$2\cos\phi$	...	0	−2	$2\cos\phi$	...	0	(x, y)	
Δ_u	2	$2\cos 2\phi$	...	0	−2	$-2\cos 2\phi$	...	0		
...	...	...	...	...	...	...	...	...		

6.3 ENUMERATION OF CRYSTALLOGRAPHIC POINT GROUPS

The number of molecular point groups is infinite but crystals have an important constraint—they do not permit C_5-axis and C_n-axis with $n > 6$. For this reason the number of crystallographic point groups is limited to *thirty two* (32). In this section the enumeration of these point groups has been shown using the concept of direct product groups.

6.3.1 Non-compatibility of $\hat{C}_5$ and $\hat{C}_n$ ($n > 6$) Rotations with Translational Symmetry

It is a common experience that a plane surface (such as the floor of a room) can be covered with isosceles triangles, rhombuses or rectangles (which have a C_2-axis), equilateral triangles (C_3-axis), squares (C_4-axis) and regular hexagons (C_6-axis). But it is not possible to cover the plane with regular pentagons and regular heptagons or higher polygons without leaving gaps. The vertices of these figures covering a plane form 2-dimensional arrays or lattices. Hence 2-dimensional lattices do not permit a C_5-axis and a C_n-axis with $n > 6$. A mathematical proof is given below.

Crystals are formed by arranging unit cells side by side and stacking the layers so formed one above the other. If we consider only the base of such a stack, we get a two-dimensional lattice; there is a definite distance, say l, between two successive points along a line on which similar lattice points are situated (Fig. 6.5).

In other words, the lattice exhibits *translational symmetry* such that a translation operation on a lattice point shifts it to the next lattice point at a distance l. Let O and O′ be two lattice points on a line and let $OO' = pl$, where p is an integer. If the lattice has an n-fold rotational axis of symmetry, then by rotating about O clockwise through an angle $\theta = \dfrac{2\pi}{n}$ we get from

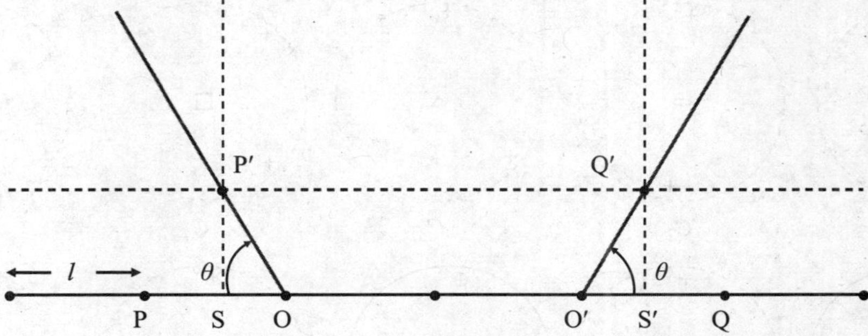

Fig. 6.5 A geometrical construction to show non-compatibility of $\hat{C}_n$ rotation with translational symmetry for $n = 5$ and $n > 6$

the point P another lattice point P′ such that OP = OP′. Similarly, anticlockwise rotation through angle θ about the point O′ takes the point Q to the position Q′ such that O′Q′ = O′Q. Because of rotational symmetry, the points P′ and Q′ must also be lattice points and the line P′Q′ must be another line through the lattice parallel to PQ and hence the distance between P′ and Q′ must be an integral multiple of length l.

Thus $\quad$ P′Q′ = OO′ + OS + O′S′

or, $\quad ql = pl + 2l\cos\theta \qquad$ (where q is an integer)

or, $\quad 2\cos\theta = q - p =$ an integer $\quad$ [(+), (–) or zero]

For $\quad n = 1, \theta = 2\pi \quad$ and $\quad 2\cos\theta = 2$

For $\quad n = 2, \theta = \pi \quad$ and $\quad 2\cos\theta = -2$

For $\quad n = 3, \theta = 120° \quad$ and $\quad 2\cos\theta = -1$

For $\quad n = 4, \theta = 90° \quad$ and $\quad 2\cos\theta = 0$

For $\quad n = 5, \theta = 72° \quad$ and $\quad 2\cos\theta =$ a non-integer

For $\quad n = 6, \theta = 60° \quad$ and $\quad 2\cos\theta = 1$

For $\quad n = 7, \theta = \dfrac{2\pi}{7} \quad$ and $\quad 2\cos\theta =$ a non-integer

For higher values of n also, $2\cos\theta$ has a non-integral value. Thus, $\hat{C}_n$ rotation with $n = 5$ or $n > 6$ is not compatible with translational symmetry. Hence the only permitted values of n are 1, 2, 3, 4 and 6.

6.3.2 Crystallographic Point Groups with One Principal Axis of Symmetry

For crystals, the permitted point groups with only one C_n-axis are the cyclic groups:

$$C_1, C_2, C_3, C_4, C_6$$

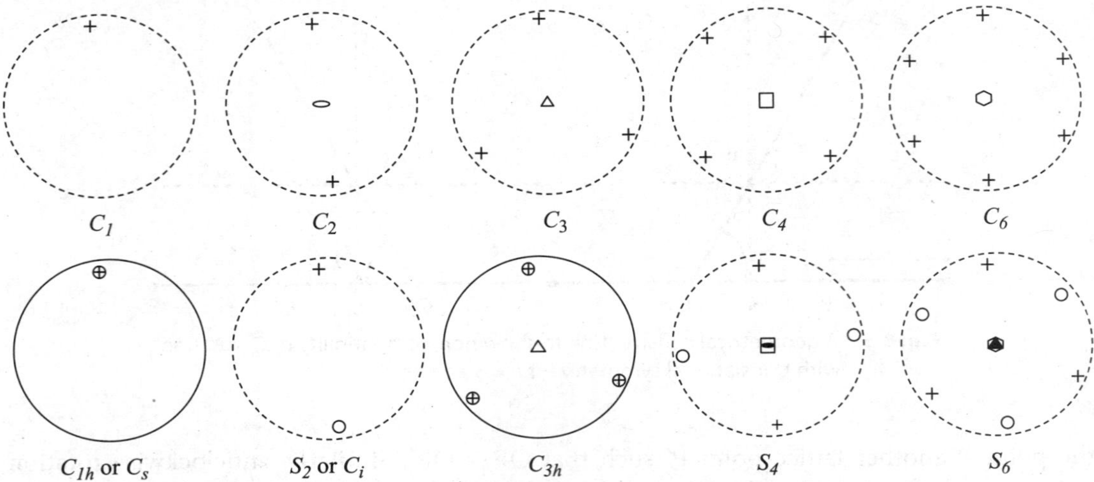

Fig. 6.6 Point groups of crystals obtained from the permitted pure rotation and roto-reflection operations

Considering the roto-reflection operations using these permitted axes we have the following point groups:

$$\{\hat{S}_1 = \hat{\sigma}_h \hat{C}_1 = \hat{\sigma}_h,\ \hat{S}_1^2 = \hat{\sigma}_h^2 \hat{C}_1^2 = \hat{E}\} = \{\hat{E}, \hat{\sigma}_h\} \equiv C_s \text{ or } C_{1h}$$

$$\{\hat{S}_2 = \hat{\sigma}_h \hat{C}_2 = \hat{i},\ \hat{S}_2^2 = \hat{i}^2 = \hat{E}\} = \{\hat{E}, \hat{i}\} \equiv C_i \text{ or } S_2$$

$$\{\hat{S}_3 = \hat{\sigma}_h \hat{C}_3,\ \hat{S}_3^2 = \hat{\sigma}_h^2 \hat{C}_3^2 = \hat{C}_3^2,\ \hat{S}_3^3 = \hat{\sigma}_h^3 \hat{C}_3^3 = \hat{\sigma}_h,\ \hat{S}_3^4 = \hat{\sigma}_h^4 \hat{C}_3^4 = \hat{C}_3,\ \hat{S}_3^5 = \hat{\sigma}_h^5 \hat{C}_3^5 = \hat{\sigma}_h \hat{C}_3^2,$$
$$\hat{S}_3^6 = \hat{\sigma}_h^6 \hat{C}_3^6 = \hat{E}\} \equiv C_{3h}$$

$$\{\hat{S}_4 = \hat{\sigma}_h \hat{C}_4,\ \hat{S}_4^2 = \hat{\sigma}_h^2 \hat{C}_4^2 = \hat{C}_2,\ \hat{S}_4^3 = \hat{\sigma}_h^3 \hat{C}_4^3 = \hat{\sigma}_h \hat{C}_4^3,\ \hat{S}_4^4 = \hat{\sigma}_h^4 \hat{C}_4^4 = \hat{E}\} \equiv S_4$$

$$\{\hat{S}_6 = \hat{\sigma}_h \hat{C}_6,\ \hat{S}_6^2 = \hat{\sigma}_h^2 \hat{C}_6^2 = \hat{C}_3,\ \hat{S}_6^3 = \hat{\sigma}_h^3 \hat{C}_6^3 = \hat{i},\ \hat{S}_6^4 = \hat{\sigma}_h^4 \hat{C}_6^4 = \hat{C}_3^2,\ \hat{S}_6^5 = \hat{\sigma}_h^5 \hat{C}_6^5 = \hat{\sigma}_h \hat{C}_6^5,$$
$$\hat{S}_6^6 = \hat{\sigma}_h^6 \hat{C}_6^6 = \hat{E}\} \equiv S_6$$

Stereographic projections of the point groups so far obtained are as in Fig. 6.6. Crystallographers use a different improper rotation, namely, roto-inversion (rotation followed by inversion)

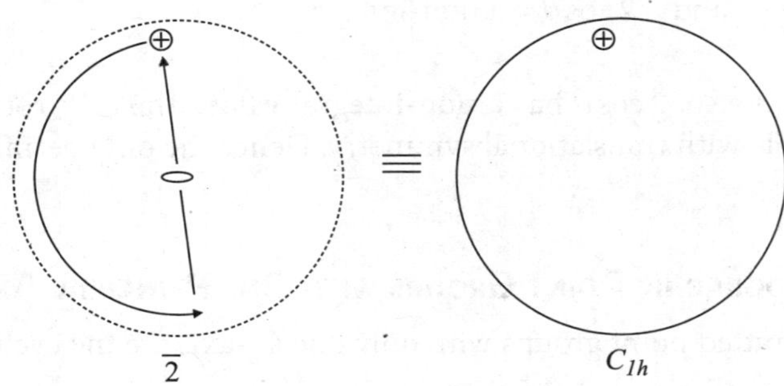

Fig. 6.7 Roto-inversion group, $\bar{2} = C_{1h}$

instead of roto-reflection; this is denoted by $\bar{n}$. However, this operation produces the same set of point groups as does roto-reflection. One typical case is illustrated in Fig. 6.7.

In the same way, $\bar{1} = S_2$, $\bar{3} = S_6$, $\bar{4} = S_4$, $\bar{6} = C_{3h}$. All these point groups have already been obtained as roto-reflection groups (Fig. 6.6).

Since the operators $\hat{\mathrm{i}}$ and $\hat{\sigma}_h$ commute with all other point symmetry operators, we can construct direct products of each of the above groups with the groups $C_s (= \hat{E}, \hat{\sigma}_h)$ and $C_i (= \hat{E}, \hat{\mathrm{i}})$. By symmetry multiplication it is easy to verify that

$$C_n \times C_s = C_n \times C_i = C_{nh} \text{ for even } n,$$

and $$C_n \times C_s = C_{nh},\ C_n \times C_i = S_{2n} \text{ for odd } n.$$

For example,

$$C_1 \times C_s = \{\hat{E}\} \times \{\hat{E}, \hat{\sigma}_h\} = \{\hat{E}, \hat{\sigma}_h\} = C_s \text{ or } C_{1h}$$

$$C_1 \times C_i = \{\hat{E}\} \times \{\hat{E}, \hat{\mathrm{i}}\} = \{\hat{E}, \hat{\mathrm{i}}\} = C_i \text{ or } S_2$$

$$C_3 \times C_s = \{\hat{E}, \hat{C}_3, \hat{C}_3^2\} \times \{\hat{E}, \hat{\sigma}_h\} = \{\hat{E}, \hat{C}_3, \hat{C}_3^2, \hat{\sigma}_h, \hat{S}_3, \hat{S}_3^5\} = C_{3h}$$

$$C_3 \times C_i = \{\hat{E}, \hat{C}_3, \hat{C}_3^2\} \times \{\hat{E}, \hat{\mathrm{i}}\} = \{\hat{E}, \hat{C}_3, \hat{C}_3^2, \hat{\sigma}_h, \hat{S}_6, \hat{S}_6^5\} = S_6$$

$$C_2 \times C_s = \{\hat{E}, \hat{C}_2\} \times \{\hat{E}, \hat{\sigma}_h\} = \{\hat{E}, \hat{C}_2, \hat{\sigma}_h, \hat{\mathrm{i}}\} = C_{2h}$$

$$C_2 \times C_i = \{\hat{E}, \hat{C}_2\} \times \{\hat{E}, \hat{\mathrm{i}}\} = \{\hat{E}, \hat{C}_2, \hat{\sigma}_h, \hat{\mathrm{i}}\} = C_{2h}$$

$$C_4 \times C_s = \{\hat{E}, \hat{C}_4, \hat{C}_2, \hat{C}_4^3\} \times \{\hat{E}, \hat{\sigma}_h\} = \{\hat{E}, \hat{C}_4, \hat{C}_2, \hat{C}_4^3, \hat{\sigma}_h, \hat{S}_4, \hat{\mathrm{i}}, \hat{S}_4^3\} = C_{4h}$$

$$C_4 \times C_i = \{\hat{E}, \hat{C}_4, \hat{C}_2, \hat{C}_4^3\} \times \{\hat{E}, \hat{\mathrm{i}}\} = \{\hat{E}, \hat{C}_4, \hat{C}_2, \hat{C}_4^3, \hat{\sigma}_h, \hat{S}_4, \hat{\mathrm{i}}, \hat{S}_4^3\} = C_{4h}$$

$$C_6 \times C_s = \{\hat{E}, \hat{C}_6, \hat{C}_3, \hat{C}_2, \hat{C}_3^2, \hat{C}_6^5\} \times \{\hat{E}, \hat{\sigma}_h\} = \{\hat{E}, \hat{C}_6, \hat{C}_3, \hat{C}_2, \hat{C}_3^2, \hat{C}_6^5, \hat{\sigma}_h, \hat{S}_6, \hat{S}_3, \hat{\mathrm{i}}, \hat{S}_3^5, \hat{S}_6^5\} = C_{6h}$$

$$C_6 \times C_i = \{\hat{E}, \hat{C}_6, \hat{C}_3, \hat{C}_2, \hat{C}_3^2, \hat{C}_6^5\} \times \{\hat{E}, \hat{\mathrm{i}}\} = \{\hat{E}, \hat{C}_6, \hat{C}_3, \hat{C}_2, \hat{C}_3^2, \hat{C}_6^5, \hat{\sigma}_h, \hat{S}_6, \hat{S}_3, \hat{\mathrm{i}}, \hat{S}_3^5, \hat{S}_6^5\} = C_{6h}$$

New point groups that are obtained from these direct products are C_{2h}, C_{4h} and C_{6h} (Fig. 6.8):

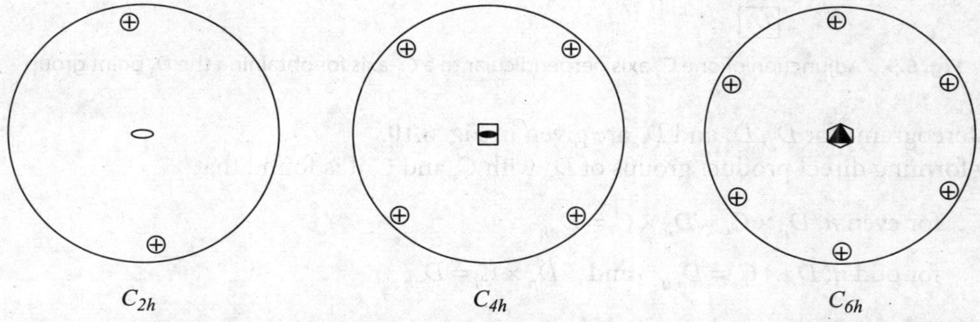

Fig. 6.8 Point groups obtained from direct products (excluding the ones already obtained)

Groups of higher order with only one C_n axis as the principal axis can be obtained by adjoining a C_2-axis perpendicular to this C_n. This is so because a rotation through an angle π about this

C_2-axis brings the C_n-axis to itself (only with a reversal of orientation). Other axes of higher order attached to the C_n-axis will transform it to a different C_n-axis and hence such cases are to be discarded because we are dealing with groups having only one C_n-axis.

Again, attachment of one C_2 axis perpendicular to the principal C_n axis generates $(n - 1)$ other C_2 axes, each perpendicular to the C_n axis. This gives the D_n groups, namely, D_2, D_3, D_4 and D_6. (D_1 is not possible because n for C_1 is less than n for C_2 and so C_1 cannot be the principal axis.) Adjunction of one C_2 to a principal C_3-axis is shown in Fig. 6.9.

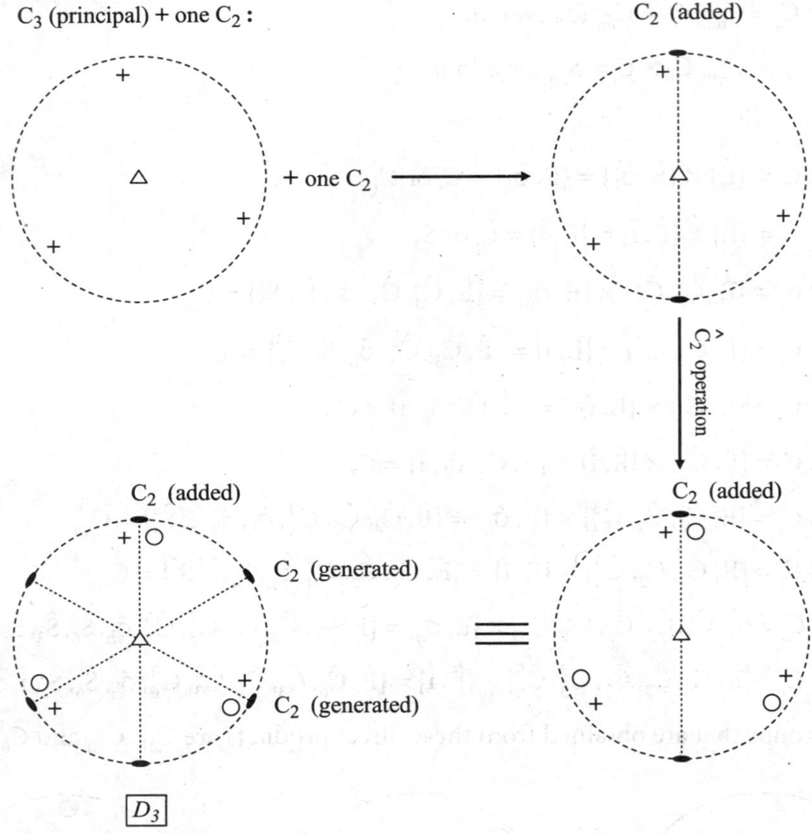

Fig. 6.9 Adjunction of one C_2 axis perpendicular to a C_3-axis for obtaining the D_3 point group

The stereograms for D_2, D_4 and D_6 are given in Fig. 6.10.

By forming direct product groups of D_n with C_s and C_i it is found that

for even n, $D_n \times C_s = D_n \times C_i = D_{nh}$

and for odd n, $D_n \times C_s = D_{nh}$ and $D_n \times C_i = D_{nh}$

For example,

$$D_2 \times C_s = \{\hat{E}, \hat{C}_2(z), \hat{C}_2(y), \hat{C}_2(x)\} \times \{\hat{E}, \hat{\sigma}_h = xy \text{ plane}\}$$
$$= \{\hat{E}, \hat{C}_2(z), \hat{C}_2(y), \hat{C}_2(x), \hat{\sigma}(xy), \hat{i}, \hat{\sigma}_v(xz), \hat{\sigma}_v(yz)\} = D_{2h}$$

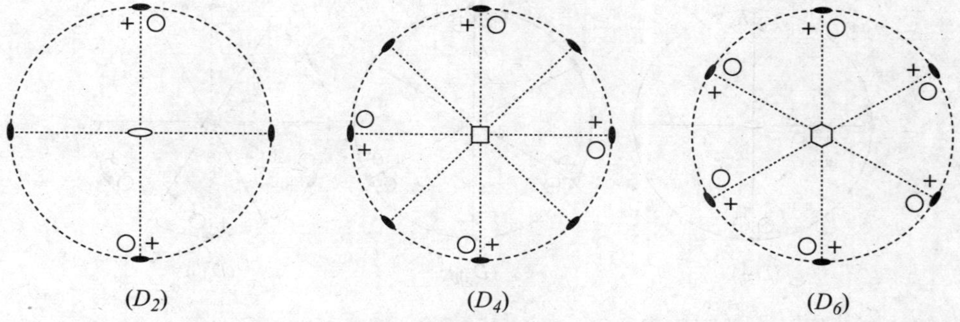

Fig. 6.10 Stereographic projections for D_2, D_4 and D_6

$$D_2 \times C_i = \{\hat{E}, \hat{C}_2(z), \hat{C}_2(y), \hat{C}_2(x)\} \times \{\hat{E}, \hat{i}\}$$

$$= \{\hat{E}, \hat{C}_2(z), \hat{C}_2(y), \hat{C}_2(x), \hat{\sigma}(xy), \hat{i}, \hat{\sigma}_v(xz), \hat{\sigma}_v(yz)\} = D_{2h}$$

$$D_3 \times C_s = \{\hat{E}, \hat{C}_3, \hat{C}_3^2, C_2(1), C_2(2), C_2(3)\} \times \{\hat{E}, \hat{\sigma}_h\}$$

$$= \{\hat{E}, \hat{C}_3, \hat{C}_3^2, C_2(1), C_2(2), \hat{C}_2(3), \hat{\sigma}_h, \hat{S}_3, \hat{S}_3^5, \hat{\sigma}_v(1), \hat{\sigma}_v(2), \hat{\sigma}_v(3)\} = D_{3h}$$

$$D_3 \times C_i = \{\hat{E}, \hat{C}_3, \hat{C}_3^2, C_2(1), \hat{C}_2(2), \hat{C}_2(3)\} \times \{\hat{E}, \hat{i}\}$$

$$= \{\hat{E}, \hat{C}_3, \hat{C}_3^2, \hat{C}_2(1), \hat{C}_2(2), \hat{C}_2(3), \hat{i}, \hat{S}_6, \hat{S}_6^5, \hat{\sigma}_d(1), \hat{\sigma}_d(2), \hat{\sigma}_d(3)\} = D_{3d}$$

[$\hat{i}\hat{C}_3 = \hat{\sigma}_h \hat{C}_6^3 \hat{C}_6^2 = \hat{\sigma}_h \hat{C}_6^5 = \hat{S}_6^5$, $\hat{i}\hat{C}_3^2 = \hat{\sigma}_h \hat{C}_6^3 \hat{C}_6^4 = \hat{\sigma}_h \hat{C}_6 = \hat{S}_6$, $\hat{i}\hat{C}_2(1) = \hat{\sigma}_d(2)$ as shown in the partial stereogram for D_{3d} (Fig. 6.11).]

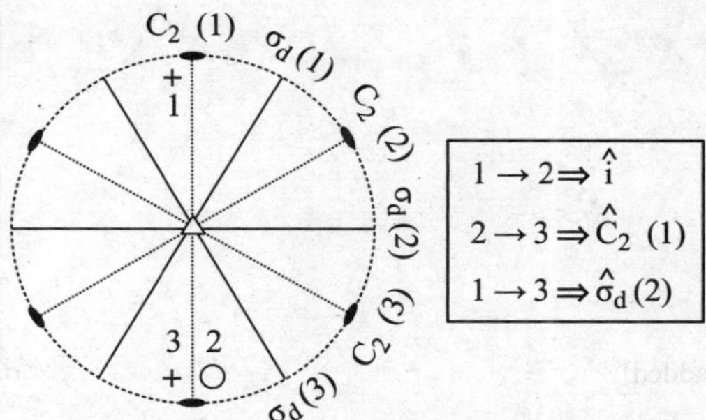

Fig. 6.11 Partial stereogram for D_{3d} showing that $\hat{i}\hat{C}_2(1) = \hat{\sigma}_d(2)$

In this way we get the following five direct product groups from the four $D_n (n = 2, 3, 4, 6)$ groups:

$$D_{2h} = D_2 \times C_s = D_2 \times C_i;\ D_{3h} = D_3 \times C_s;\ D_{3d} = D_3 \times C_i;\ D_{4h} = D_4 \times C_s = D_4 \times C_i;$$

$$D_{6h} = D_6 \times C_s = D_6 \times C_i$$

Stereograms for these five different point groups are given in Fig. 6.12.

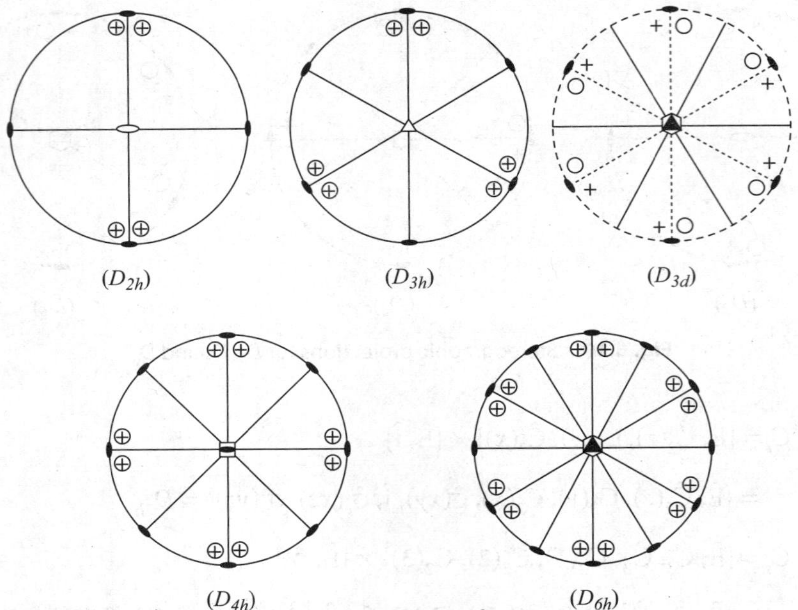

Fig. 6.12 Stereograms for the five different direct product groups obtained from D_n

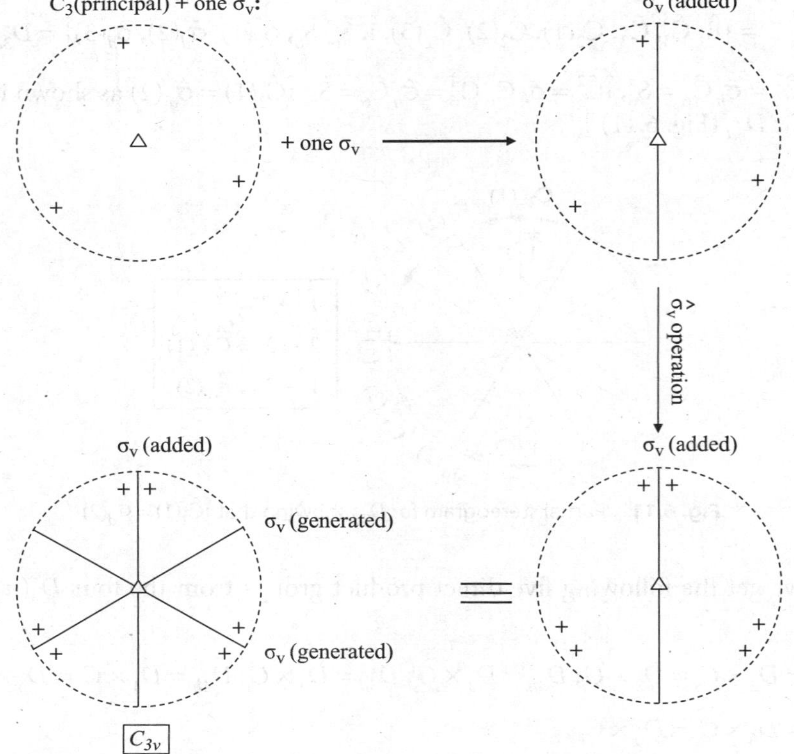

Fig. 6.13 Scheme for obtaining C_{3v} by adjunction of a σ_v plane to the principal C_3-axis

Adjunction of one σ_v plane to a C_n group gives C_{nv}. Addition of one σ_v generates $(n - 1)$ other σ_v planes. The case for C_3 is shown in Fig. 6.13.

In this way, C_{2v}, C_{4v} and C_{6v} point groups (Fig. 6.14) are obtained from C_2, C_4 and C_6.

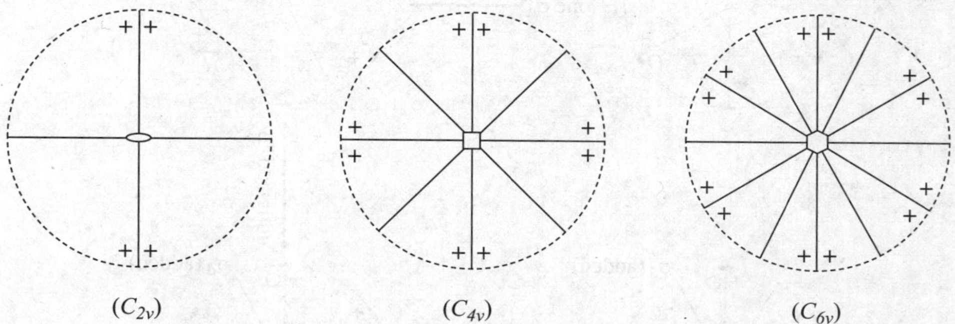

Fig. 6.14 Full stereograms of C_{2v}, C_{4v} and C_{6v} point groups

Adjunction of one σ_v plane to a D_n group produces a D_{nh} group. The case of D_{3h} is shown in Fig. 6.15.

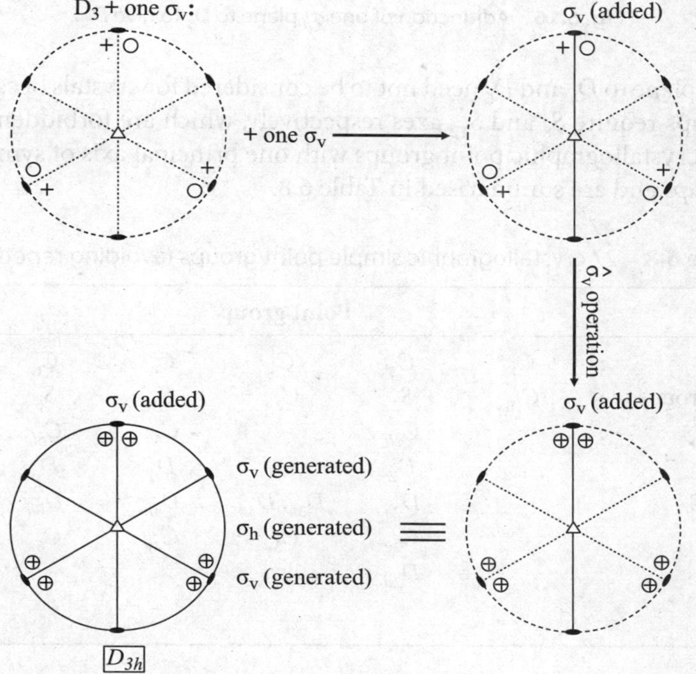

Fig. 6.15 Adjunction of one σ_v plane to D_3 to give D_{3h}

Thus adjunction of one σ_v to D_n produces no new point group. Adjunction of one σ_d plane to a D_n group produces a D_{nd} group for $n = 2$ and 3. This is shown in Fig. 6.16 for the case of D_{3d}.

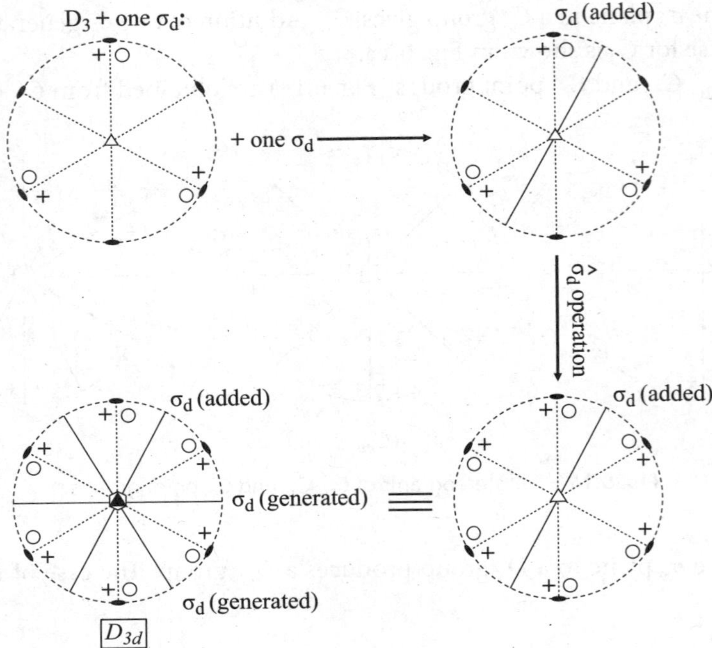

Fig. 6.16 Adjunction of one σ_d plane to D_3 to give D_{3d}

Adjunction of σ_d plane to D_4 and D_6 need not to be considered for crystals because the resulting D_{4d} and D_{6d} groups require S_8 and S_{12} axes respectively, which are forbidden for crystals. We thus arrive at 27 crystallographic point groups with one principal axis of symmetry. These are called *simple groups* and are summarised in Table 6.8.

Table 6.8 27 crystallographic simple point groups (avoiding repetition)

Origin		Point group					Total number
Cyclic groups	:	C_1	C_2	C_3	C_4	C_6	5
Roto-inversion groups	:	C_{1h}	S_2	C_{3h}	S_4	S_6	5
$C_n \times C_s$ or $C_n \times C_i$	:		C_{2h}		C_{4h}	C_{6h}	3
$C_n + C_2 \perp C_n$	:		D_2	D_3	D_4	D_6	4
$D_n \times C_s$ or $D_n \times C_i$	:		D_{2h}	D_{3h}, D_{3d}	D_{4h}	D_{6h}	5
$C_n + \sigma_v$	:		C_{2v}	C_{3v}	C_{4v}	C_{6v}	4
$D_n + \sigma_d$	:		D_{2d}				1
							27

6.3.3 Enumeration of Crystallographic Point Groups with More than One n-fold Axis of Symmetry (C_n, $n > 2$)

Platonic solids are examples of crystals having more than one C_n-axis with $n > 2$. Among the five regular polyhedra, the dodecahedron and icosahedron have C_5-rotational axes which are not compatible with translational symmetry. Hence they are excluded from crystallographic point groups. So the pure rotation groups having more than one high order rotation axis (C_n, $n > 2$) are T (for tetrahedron) and O (for octahedron).

The full symmetry group of a cube is the same as that of an octahedron and it is denoted by O_h (see Sections 2.4 and 2.6). The subgroup of O_h containing only the pure rotation operations and the identity ($\hat{E}$) is the point group O. In fact, O_h is the direct product group, $O \times C_i$. The groups O and O_h are called *the pure rotational octahedral group* and *the full octahedral group* respectively.

The symmetry group of a regular tetrahedron is denoted by T_d. The pure rotation operations of T_d, together with $\hat{E}$, constitute the *pure rotational tetrahedral group*, T. Finally, a group of order 24 is obtained by taking the direct product, $T \times C_i$, whose resulting symmetry operations are:

Symmetry operations of T	**Symmetry operations of C_i**	**Symmetry operations of $T \times C_i$**
$(\hat{E}, 4\hat{C}_3, 4\hat{C}_3^2, 3\hat{C}_2)$	$(\hat{E}, \hat{i})$	$(\hat{E}, 4\hat{C}_3, 4\hat{C}_3^2, 3\hat{C}_2, \hat{i}, 4\hat{S}_6^5, 4\hat{S}_6, 3\hat{\sigma}_h)$

The point group thus obtained is denoted by T_h and it may be verified that it is a subgroup of O_h. Further, T_h is not same as T_d, because T_h has three $\hat{\sigma}_h$ operations and also an $\hat{i}$. Thus, five crystallographic point groups (namely, O, O_h, T, T_d and T_h) with more than one C_n-axis are possible. The total number of crystallographic point groups is, therefore, $27 + 5 = 32$.

SUMMARY

- ❑ *Direct product group* ($G_a \times G_b$) is defined. As the symmetry operators $\hat{\sigma}_h$ and $\hat{i}$ commute with all other symmetry operators, direct product groups can be constructed with the point groups C_s and C_i as one component group.
- ❑ Character tables of direct products of two groups can be obtained by multiplying the characters of the corresponding operators in the component groups.
- ❑ Non-compatibility of translational symmetry with rotation $\hat{C}_n$, ($n = 5$ and $n > 6$), in crystal lattices restricts the number of crystallographic point groups to 32; of them, 27 are generated by adjunction of one perpendicular C_2-axis or one σ_v/σ_d plane to C_n and D_n groups.

LEARNING OUTCOME

After reading this chapter the student will be able to (i) generate the full list of symmetry operators and construct character tables of many large groups using the principle of direct products and (ii) understand why there are only 32 crystallographic point groups although the number of molecular point groups is not limited.

EXERCISES

1. By drawing a stereogram show that (a) $\hat{i}\hat{C}_3 = \hat{S}_6^5$ and $\hat{i}\hat{C}_3^2 = \hat{S}_6$ (b) point group $\bar{6} \equiv$ point group C_{3h}.
2. Establish the direct product relationships:

$D_{3d} = D_3 \times C_i$ or $C_{3v} \times C_i$, $D_{2h} = D_2 \times C_s$ or $D_2 \times C_i$.

3. Construct the character tables of C_{3h} and D_{3h} by the direct product method.

4. Explain why the body of the character table of D_{3h} and D_{3d} should be the same and only basis sets and Mulliken symbols differ.

5. Choose the correct answer(s):
   (a) (i) $C_5 \times C_i = C_{5h}$ (ii) $C_5 \times C_i = D_{5d}$ (iii) $C_5 \times C_s = C_{5h}$ (iv) $C_3 \times C_i = S_6$ (v) $C_4 \times C_i = C_{4d}$.
   (b) The direct product group $C_i \times C_s$ is (i) C_4 (ii) C_{2h} (iii) D_2 (iv) C_{2v}.
   (c) The result of taking direct product of a group G_a with itself is
   (i) not a group, (ii) a group of order twice that of G_a, (iii) the group G_a itself, (iv) an Abelian group.

6. Show that (a) adjunction of one σ_v plane to C_2 gives the point group C_{2v}, (b) adjunction of one C_2'-axis to C_2 gives the point group D_2 and (c) adjunction of one σ_d plane to D_2 gives the point group D_{2d}.

Suggested Books and References for Further Reading

1. Tinkham M, *Group Theory and Quantum mechanics*, McGraw-Hill, NY, **1964.**
2. Joshi A W, *Elements of Group Theory for Physicists*, Wiley Eastern Limited, New Delhi, **1982.**

7 Link between Group Theory and Quantum Mechanics

Learning objectives

- Invariance of the Hamiltonian and its commutation with symmetry operators
- Eigenfunctions of $\hat{H}$ as bases of irreducible representations
- Projection operators, symmetry adapted linear combinations
- Direct product representations, vanishing of integrals

7.1 INTRODUCTION

The basic fact that makes group theory a powerful tool in quantum mechanics and spectroscopy is that the Hamiltonian operator ($\hat{H}$) of a system remains unchanged by the action of its point symmetry operators. As a consequence of this invariance of $\hat{H}$, solutions of the Schrödinger equation become the bases of irreducible representations of the point group of the system. This fact enables the molecular orbitals and possible spectroscopic states of a system to be labelled in terms of Mulliken symbols of the irreducible representations. Such labelling is essential in simplifying secular equations in MO calculations and in deriving spectroscopic selection rules with the help of a theorem of vanishing integrals which is central to all applications of group theory. The results obtained or predictions made from such applications are completely rigorous as they are derived solely on the basis of symmetry.

7.2 INVARIANCE OF THE HAMILTONIAN OPERATOR

By a symmetry operation, one brings an object from a given configuration to another which is indistinguishable from the former. A symmetry operation, therefore, effectively means a permutation of the labels of the various particles of the system without actually changing the configuration. For example, if (x_1, y_1, z_1), (x_2, y_2, z_2) and (x_3, y_3, z_3) are the coordinates of the vertices of an equilateral triangle and, after a reflection through a vertical plane of symmetry through the vertex labelled 1, if the coordinates become (X_1, Y_1, Z_1), (X_2, Y_2, Z_2), and (X_3, Y_3, Z_3) respectively, then

$$(X_1, Y_1, Z_1) = (x_1, y_1, z_1)$$

$$(X_2, Y_2, Z_2) = (x_3, y_3, z_3)$$

$$(X_3, Y_3, Z_3) = (x_2, y_2, z_2)$$

Thus the set of coordinates of the vertices are the same as before, only their labels have been subjected to the permutation: $1 \leftrightarrow 1$, $2 \leftrightarrow 3$. Similar argument holds for all symmetry operations. Hence the kinetic energy part of the Hamiltonian operator remains the same as before:

$$\sum_{i=1}^{n} -\frac{\hbar^2}{2m_i}\left(\frac{\partial^2}{\partial x_i^2}+\frac{\partial^2}{\partial y_i^2}+\frac{\partial^2}{\partial z_i^2}\right)=\sum_{j=1}^{n} -\frac{\hbar^2}{2m_j}\left(\frac{\partial^2}{\partial X_j^2}+\frac{\partial^2}{\partial Y_j^2}+\frac{\partial^2}{\partial Z_j^2}\right)$$

(before symmetry operation) (after symmetry operation)

Also, the Coulombic part of potential energy, which depends on the various inter-particle distances in a charge distribution remains unchanged because

$$|r_{pq}|=\sqrt{(x_p-x_q)^2+(y_p-y_q)^2+(z_p-z_q)^2}=\sqrt{(X_p-X_q)^2+(Y_p-Y_q)^2+(Z_p-Z_q)^2}$$

Potential energy due to spin–orbit interaction depends on the magnitude of the resultant orbital angular momentum and the resultant spin angular momentum and the angle between these vectors. But neither the magnitude of a vector nor any inter-vector angle changes by a point symmetry operation. Hence, under all symmetry operations of the point group of an object, its Hamiltonian operator ($\hat{H}$) does not change:

$$\hat{R}\hat{H}=\hat{H} \tag{7.1}$$

This result can be used to prove the following important theorem.

Theorem: *$\hat{H}$ commutes with all the symmetry operators of the point group of the system.*

Proof: Let f be a function of coordinates (x, y, z) and by the action of a symmetry operator $\hat{R}$ let it change to the function ϕ (that is, $\hat{R}f=\phi$). This means that

if $\quad \hat{R}(x, y, z)=(X, Y, Z)$, then $f(x, y, z)=\phi(X, Y, Z)$

Now, the action of $\hat{R}$ on a function $\hat{H}f$ means the action of the transformed operator $\hat{R}\hat{H}$ on the transformed function $\hat{R}f$:

$$\hat{R}\hat{H}f=(\hat{R}\hat{H})(\hat{R}f)=\hat{H}\phi \text{ [since } \hat{R}\hat{H}=\hat{H} \text{ and } \hat{R}f=\phi]$$

Replacing f by $1\cdot f$ or $\hat{R}^{-1}\hat{R}f$ (as $\hat{R}^{-1}\hat{R}=$ identity) we have

$$\hat{R}\hat{H}[\hat{R}^{-1}(\hat{R}f)]=\hat{H}\phi$$

or, $\quad (\hat{R}\hat{H}\hat{R}^{-1})\phi=\hat{H}\phi$

or, $\quad \hat{R}\hat{H}\hat{R}^{-1}=\hat{H}$

or, right-multiplying both sides by $\hat{R}$,

$$\hat{R}\hat{H}=\hat{H}\hat{R} \tag{7.2}$$

This result leads to an important theorem stated in the next section.

7.3 EIGENFUNCTIONS OF THE HAMILTONIAN AS BASES OF IRREDUCIBLE REPRESENTATIONS

Theorem: *Non-degenerate eigenfunctions of the Hamiltonian operator of a system act as bases of one-dimensional irreducible representations of the system's point group. A set of k-fold degenerate* *eingenfunctions behaves like a basis set of some k-dimensional irreducible representation of the point group.*

Proof: Case (A): *When the eigenfunctions are non-degenerate*

Let E_i be a non-degenerate eigenvalue of a system satisfying the Schrödinger equation

$$\hat{H}\psi_i = E_i\psi_i \tag{7.3}$$

Since a symmetry operator $\hat{R}$ of the point group of the system commutes with the Hamiltonian $\hat{H}$, we have

$$\hat{H}\hat{R}\psi_i = \hat{R}\hat{H}\psi_i = \hat{R}(E_i\psi_i) = E_i(\hat{R}\psi_i) \tag{7.4}$$

(since $\hat{R}$ commutes with the numerical factor E_i). Comparing Eqs (7.3) and (7.4), we see that $\hat{R}\psi_i$ is also an eigenfunction of $\hat{H}$. So $\hat{R}\psi_i$ is either ψ_i itself or a constant multiple of ψ_i.

Let $$\hat{R}\psi_i = C\psi_i \tag{7.5}$$

where C is a constant. Let ψ_i be initially normalised. Then for $\hat{R}\psi_i$ to be a normalised eigenfunction we must have

$$\langle C\psi_i | C\psi_i\rangle = 1 \text{ or, } \quad C^2 = 1 \qquad (\text{as } \langle\psi_i|\psi_i\rangle = 1)$$

or, $$C = \pm 1$$

Hence $$\hat{R}\psi_i = \pm\psi_i \tag{7.6}$$

which is true for any symmetry operator of the point group. Let $(\hat{E}, \hat{R}_2, \hat{R}_3, \ldots, \hat{R}_g)$ be the point group. Then $\hat{E}\psi_i = \psi_i = [1]\psi_i$, $\hat{R}_2\psi_i = [1]\psi_i$ or $[-1]\psi_i$ and so on. Hence, the symmetry operators can be represented by the matrices [1] or [−1] with ψ_i as basis. So each of the matrix sets D^1, D^2, D^3, … as tabulated below, is a representation of the group provided the matrices multiply in accordance with the symmetry multiplication table of the group.

	$\hat{E}$	$\hat{R}_2$	$\hat{R}_3$	…	$\hat{R}_g$
D^1	[1]	[1]	[1]	…	[1]
D^2	[1]	[−1]	[−1]	…	[1]
D^3	[1]	[1]	[−1]	…	[−1]
…	…	…	…	…	…

Such representations are irreducible since they are all one-dimensional.

Case (B): *When the eigenvalue is of k-fold degeneracy*

Let E_k be some energy eigenvalue of k-fold degeneracy, the corresponding eigenfunctions being $\psi_{1k}, \psi_{2k}, \psi_{3k}, \ldots, \psi_{kk}$. Then from the Schrödinger equation,

$$\hat{H}\psi_{1k} = E_k\psi_{1k}$$

$$\hat{H}\psi_{2k} = E_k\psi_{2k} \tag{7.7}$$

$$\cdots$$

$$\hat{H}\psi_{kk} = E_k\psi_{kk}$$

Applying the symmetry operator $\hat{R}$ on both sides of any of these equations we have

$$\hat{R}\hat{H}\psi_{ik} = \hat{R}(E_k\,\psi_{ik}) \quad \text{or} \quad \hat{H}\hat{R}\psi_{ik} = E_k(\hat{R}\psi_{ik}) \tag{7.8}$$

(as $\hat{R}$ commutes with $\hat{H}$ and also with the constant factor E_k). Thus $\hat{R}\psi_{ik}$ is also an eigenfunction of $\hat{H}$ with the same eigenvalue E_k. Hence comparing with the set of Eqs (7.7), it may be stated that $\hat{R}\psi_{ik}$ is any of the functions $\psi_{1k}, \psi_{2k}, \psi_{3k}, \ldots, \psi_{kk}$ and most generally, a linear combination of them.

$$\text{Hence}\quad \hat{R}\psi_{ik} = a_{1i}\psi_{1k} + a_{2i}\psi_{2k} + a_{3i}\psi_{3k} + \cdots + a_{ki}\psi_{kk}$$

$$= [\psi_{1k} \quad \psi_{2k} \quad \psi_{3k} \quad \cdots \quad \psi_{kk}][a_{1i} \quad a_{2i} \quad a_{3i} \quad \cdots \quad a_{ki}]^T \ (T \text{ for transpose}) \tag{7.9}$$

Putting $i = 1, 2, 3, \ldots, k$ we get respectively the first, second, third, ... columns of a matrix representation for $\hat{R}$:

$$\mathbf{M}(\hat{R}) = \begin{bmatrix} a_{11} & a_{12} & a_{13} & \cdot & \cdot & \cdot & \cdot & \cdot & \cdot & a_{1k} \\ a_{21} & a_{22} & a_{23} & \cdot & \cdot & \cdot & \cdot & \cdot & \cdot & a_{2k} \\ a_{31} & a_{32} & a_{3k} & \cdot & \cdot & \cdot & \cdot & \cdot & \cdot & a_{3k} \\ \cdot & \cdot & \cdot & \cdot & \cdot & \cdot & \cdot & \cdot & \cdot & \cdot \\ \cdot & \cdot & \cdot & \cdot & \cdot & \cdot & \cdot & \cdot & \cdot & \cdot \\ \cdot & \cdot & \cdot & \cdot & \cdot & \cdot & \cdot & \cdot & \cdot & \cdot \\ \cdot & \cdot & \cdot & \cdot & \cdot & \cdot & \cdot & \cdot & \cdot & \cdot \\ \cdot & \cdot & \cdot & \cdot & \cdot & \cdot & \cdot & \cdot & \cdot & \cdot \\ a_{k1} & a_{k2} & a_{k3} & \cdot & \cdot & \cdot & \cdot & \cdot & \cdot & a_{kk} \end{bmatrix} \tag{7.10}$$

We now prove that the matrices for the symmetry operators constructed in this way, with (ψ_{1k}, $\psi_{2k}, \psi_{3k}, \ldots, \psi_{kk}$) as the basis set, multiply in accordance with the multiplication table of the group.

Let $\hat{S}$ and $\hat{T}$ be two other symmetry operators of the group under consideration such that

$$\hat{R}\hat{S} = \hat{T} \tag{7.11}$$

It is to be proved that

$$\mathbf{M}(\hat{R}) \cdot \mathbf{M}(\hat{S}) = \mathbf{M}(\hat{T}) \tag{7.12}$$

We have already obtained

$$\hat{R}\psi_{jk} = \sum_r a_{rj}\,\psi_{rk} \tag{7.13}$$

Let $\hat{S}\psi_{rk} = \sum_s b_{sr}\psi_{sk}$ (7.14)

and $\hat{T}\psi_{sk} = \sum_t c_{ts}\psi_{tk}$ (7.15)

Then $\hat{R}\hat{S}\psi_{rk} = \hat{R}\sum_s b_{sr}\psi_{sk}$

$$= \sum_s b_{sr}\hat{R}\psi_{sk} = \sum_s b_{sr}\sum_t a_{ts}\psi_{tk} = \sum_t\sum_s a_{ts}b_{sr}\psi_{tk} \quad (7.16)$$

And from Eq. (7.15), changing the dummy index s by r,

$$\hat{T}\psi_{rk} = \sum_t c_{tr}\psi_{tk} \quad (7.17)$$

Thus, comparing Eqs (7.11), (7.16) and (7.17),

$$c_{tr} = \sum_s a_{ts}b_{sr} \quad (7.18)$$

which is just the rule for matrix multiplication.

Hence $\mathbf{M}(\hat{T}) = \mathbf{M}(\hat{R})\cdot\mathbf{M}(\hat{S})$ whenever $\hat{T} = \hat{R}\hat{S}$ (7.19)

Thus the set of matrices $D^{\psi} = \{\mathbf{M}(\hat{E}), \mathbf{M}(\hat{R}), \mathbf{M}(\hat{S}), \mathbf{M}(\hat{T}), \ldots\}$ for the symmetry operators of the point group $\{\hat{E}, \hat{R}, \hat{S}, \hat{T}\ldots\}$ is a representation of the group with $(\psi_{1k} \quad \psi_{2k} \quad \psi_{3k} \quad \cdots \quad \psi_{kk})$ as the basis set. It now remains to be proved that this representation is irreducible. For the sake of argument let the representation D^{ψ} be reducible. Then we can construct matrices for the symmetry operators $\hat{R}, \hat{S}, \hat{T}\ldots$ with a lesser number of eigenfunctions as basis. This means that the eigenfunctions $\psi_{1k} \quad \psi_{2k} \quad \psi_{3k} \quad \cdots \quad \psi_{kk}$ can be suitably combined linearly to generate a new set of functions which consists of subsets I, II and so on, such that on application of $\hat{R}$, any function of subset I transforms into a linear combination of subset I functions only, any function of subset II transforms into a linear combination of subset II functions only, and so on. One such partitioning is shown below.

$$\{\underbrace{\Phi_{1k}, \Phi_{2k}, \Phi_{3k}, \ldots, \Phi_{\lambda k}}_{\text{Subset I}} | \underbrace{\Phi_{\lambda+1,k}, \Phi_{\lambda+2,k}, \ldots, \Phi_{\mu k}}_{\text{Subset II}} | \underbrace{\Phi_{\mu+1,k}, \Phi_{\mu+2,k}, \Phi_{\mu+3,k}, \ldots, \Phi_{kk}}_{\text{Subset III}}\}$$

Then the matrix representing the action of $\hat{R}$ on the new set of functions $\{\Phi\}$ has the block factored form (7.20) as shown below. Hence with the functions of subset I as basis we have the matrix set $D^{I} = \{\mathbf{M}^{I}(\hat{E}), \mathbf{M}^{I}(\hat{R}), \mathbf{M}^{I}(\hat{S}), \mathbf{M}^{I}(\hat{T}), \ldots\}$ as a λ-dimensional representation of the group. Similarly the set II functions produce another representation $D^{II} = \{\mathbf{M}^{II}(\hat{E}), \mathbf{M}^{II}(\hat{R}), \mathbf{M}^{II}(\hat{S}), \mathbf{M}^{II}(\hat{T}), \ldots\}$ of dimensionality less than k, and so on. The basis functions of a λ-dimensional representation must be λ-fold degenerate (as we have found that k-fold degenerate eigenfunctions form a basis for a k-dimensional representation). Hence the eigenfunctions of subset I must have the same eigenvalue, E (say).

$$\begin{bmatrix} - & - & - & & & & & \\ - & - & - & & & & & \\ - & - & - & & & & & \\ & & & - & - & & & \\ & & & - & - & & & \\ & & & & & - & - & - \\ & & & & & - & - & - \\ & & & & & - & - & - \end{bmatrix} \qquad (7.20)$$

By the same argument, if the number of functions in subset II is p then they must be p-fold degenerate with eigenvalue E'(say). There is no loss of mathematical rigor in the discussions so far made if $E \neq E'$. But this contradicts our assumption that the original basis set is of k-fold degeneracy. Hence D^{ψ} cannot be reducible.

7.3.1 An Example Showing the Validity of the Above Theorem (Non-degenerate Case)

[Molecular orbitals result from an approximate solution of the Schrödinger equation and have been discussed in detail in the next three chapters. Here we adopt an elementary approach for the simple case of ethylene π-MOs.] The point group of the ethylene molecule is D_{2h} (see Fig. 7.1, where the molecular plane is the XY plane). The π-MOs may be considered separately from the σ-MOs and they are formed by a linear combination of the carbon $2p$-orbitals which are perpendicular to the molecular plane. Because of symmetry each of them should contribute equally (that is, 50%) to form a π-MO; hence if c_1 and c_2 are the combination coefficients, then $c_1^2 = c_2^2 = \frac{1}{2}$. For the bonding and antibonding π-MOs respectively, $c_1 = c_2 = \frac{1}{\sqrt{2}}$ and $c_1 = \frac{1}{\sqrt{2}}, c_2 = -\frac{1}{\sqrt{2}}$. Thus the π-MO wave functions are,

$$\text{Bonding: } \frac{1}{\sqrt{2}}(\phi_1 + \phi_2) \qquad \text{Antibonding: } \frac{1}{\sqrt{2}}(\phi_1 - \phi_2)$$

(The atomic orbitals represented by ϕ_1 and ϕ_2 are perpendicular to the XY plane, their positive lobes being directed towards the positive z-direction). The action of the various symmetry operators of D_{2h} on the bonding π-MO wave function is as follows.

$$\hat{E}\left[\frac{1}{\sqrt{2}}(\phi_1 + \phi_2)\right] = \frac{1}{\sqrt{2}}(\phi_1 + \phi_2) \Rightarrow \chi(\hat{E}) = 1$$

$$\hat{C}_2(Z)\left[\frac{1}{\sqrt{2}}(\phi_1 + \phi_2)\right] = \frac{1}{\sqrt{2}}(\phi_2 + \phi_1) \Rightarrow \chi(\hat{C}_2(Z)) = 1$$

$$\hat{C}_2(Y)\left[\frac{1}{\sqrt{2}}(\phi_1 + \phi_2)\right] = \frac{1}{\sqrt{2}}(-\phi_1 - \phi_2) \Rightarrow \chi(\hat{C}_2(Y)) = -1$$

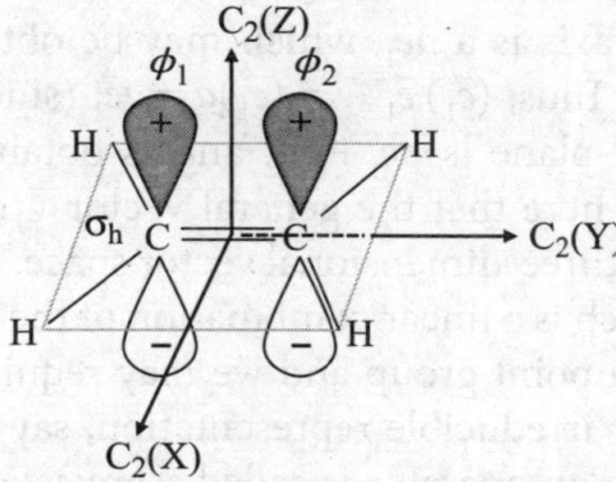

Fig. 7.1 *p*-orbitals of C in the C_2H_4 molecule forming π-MOs, some symmetry elements of D_{2h} and orientation of Cartesian axes

$$\hat{C}_2(X)\left[\frac{1}{\sqrt{2}}(\phi_1+\phi_2)\right]=\frac{1}{\sqrt{2}}(-\phi_2-\phi_1)\Rightarrow \chi(\hat{C}_2(X))=-1$$

$$\hat{i}\left[\frac{1}{\sqrt{2}}(\phi_1+\phi_2)\right]=\frac{1}{\sqrt{2}}(-\phi_2-\phi_1)\Rightarrow \chi(\hat{i})=-1$$

$$\hat{\sigma}(XY)\left[\frac{1}{\sqrt{2}}(\phi_1+\phi_2)\right]=\frac{1}{\sqrt{2}}(-\phi_1-\phi_2)\Rightarrow \chi(\hat{\sigma}(XY))=-1$$

$$\hat{\sigma}(YZ)\left[\frac{1}{\sqrt{2}}(\phi_1+\phi_2)\right]=\frac{1}{\sqrt{2}}(\phi_1+\phi_2)\Rightarrow \chi(\hat{\sigma}(YZ))=1$$

$$\hat{\sigma}(XZ)\left[\frac{1}{\sqrt{2}}(\phi_1+\phi_2)\right]=\frac{1}{\sqrt{2}}(\phi_2+\phi_1)\Rightarrow \chi(\hat{\sigma}(XZ))=1$$

From the character table of D_{2h} we find that these characters correspond to the irreducible representation B_{1u}. Thus the bonding π-MO wave function of ethylene acts as a basis of B_{1u}. In the same way, operation of the D_{2h} symmetry operators on the antibonding π^*- MO wave function yields the following characters.

			χ **(character)**				
$\hat{E}$	$\hat{C}_2(Z)$	$\hat{C}_2(Y)$	$\hat{C}_2(X)$	$\hat{i}$	$\hat{\sigma}(XY)$	$\hat{\sigma}(YZ)$	$\hat{\sigma}(XZ)$
1	−1	−1	1	1	−1	1	−1

These are the characters of B_{3g}. Hence the π^*-MO wave function acts as a basis function of B_{3g}.

7.4 PROJECTION OPERATORS

In vector algebra we are familiar with projection of a vector along a specified line or on a plane. For example, in the three-dimensional vector space the projection of a general vector

$\vec{v} = a\vec{e}_1 + b\vec{e}_2 + c\vec{e}_3$ along the X-axis is a $a\vec{e}_1$, which may be obtained by operating the scalar product operator $(\vec{e}_1)\,\vec{e}_1\cdot$ on $\vec{v}$. Thus, $(\vec{e}_1)\,\vec{e}_1\cdot\vec{v} = (\vec{e}_1)a = a\vec{e}_1$ (since $\vec{e}_1\cdot\vec{e}_2 = \vec{e}_1\cdot\vec{e}_3 = 0$). Similarly the projection of $\vec{v}$ on the XY-plane is $a\vec{e}_1 + b\vec{e}_2$ and is obtained by applying the operator $(\vec{e}_1)\vec{e}_1\cdot + (\vec{e}_2)\vec{e}_2\cdot$ on $\vec{v}$. We note here that the general vector $\vec{v}$ is a linear combination of the basis vectors $\vec{e}_1$, $\vec{e}_2$ and $\vec{e}_3$ of the three-dimensional vector space. In the same way, we may come across a general function F which is a linear combination of the basis functions of the possible irreducible representations of a point group and we may require a part of F which is a basis function of only one particular irreducible representation, say D^i. An operator $\hat{P}^{(i)}$ has to be designed for this purpose. Such an operator is called a *projection operator* for the irreducible representation D^i.

7.4.1 Construction of Projection Operators

Let D^j be an irreducible representation of a group of order g. Suppose that the matrices $\mathbf{D}^j(\hat{R})$ of this representation have been constructed by using the set of basis functions $\{\phi_1^{(j)}, \phi_2^{(j)}, \ldots, \phi_d^{(j)}\}$, where d is the dimension of D^j. Let the matrices be given and only one basis function $\phi_k^{(j)}$ is known and we want to find the other basis functions of D^j (called the 'partners' of $\phi_k^{(j)}$ in D^j).

The result of operating the symmetry operator $\hat{R}$ on $\phi_k^{(j)}$ is a linear combination of all the functions in the basis set of D^j and the coefficients of the combination are the elements of the kth column of the matrix $\mathbf{D}^j(\hat{R})$ representing $\hat{R}$ in D^j. Thus,

$$\begin{aligned}\hat{R}\phi_k^{(j)} &= a_{1k}\phi_1^{(j)} + a_{2k}\phi_2^{(j)} + a_{3k}\phi_3^{(j)} + \cdots + a_{dk}\phi_d^{(j)}\\ &= \phi_1^{(j)}[\mathbf{D}^{(j)}(\hat{R})]_{1k} + \phi_2^{(j)}[\mathbf{D}^{(j)}(\hat{R})]_{2k} + \phi_3^{(j)}[\mathbf{D}^{(j)}(\hat{R})]_{3k} + \cdots + \phi_d^{(j)}[\mathbf{D}^{(j)}(\hat{R})]_{dk}\\ &= \sum_{\lambda=1}^{d}\phi_\lambda^{(j)}\,[\mathbf{D}^{(j)}(\hat{R})]_{\lambda k} \qquad (7.21)\end{aligned}$$

(where $a_{\lambda k} = [\mathbf{D}^{(j)}(\hat{R})]_{\lambda k}$, $\lambda = 1$ to d)

As the kth column elements of the matrix $\mathbf{D}^j(\hat{R})$ appear as coefficients in this expression, we say that the basis function $\phi_k^{(j)}$ 'belongs to the kth columm of the jth irreducible representation'. Multiplying both sides of Eq. (7.21) by $[\mathbf{D}^j(\hat{R})_{\mu b}]^*$ and summing over all $\hat{R}$ we have,

$$\begin{aligned}\sum_R[\mathbf{D}^{(i)}(\hat{R})]^*_{\mu b}\,\hat{R}\,\phi_k^{(j)} &= \sum_R\sum_{\lambda=1}^{d}\phi_\lambda^{(j)}[\mathbf{D}^{(j)}(\hat{R})]_{\lambda k}[\mathbf{D}^{(i)}(\hat{R})]^*_{\mu b}\\ &= \sum_{\lambda=1}^{d}\phi_\lambda^{(j)}\sum_R[\mathbf{D}^{(j)}(\hat{R})]_{\lambda k}[\mathbf{D}^{(i)}(\hat{R})]^*_{\mu b}\\ &= \sum_{\lambda=1}^{d}\phi_\lambda^{(j)}\frac{g}{\sqrt{dd'}}\delta_{ij}\delta_{\lambda\mu}\delta_{kb} \qquad \text{(by the GOT)}\\ &= \phi_\mu^{(j)}\frac{g}{\sqrt{dd'}}\delta_{ij}\delta_{kb} \qquad [\text{as } \delta_{\mu\mu} = 1] \qquad (7.22)\end{aligned}$$

where d' is the dimension of the ith irreducible representation, D^i.

The identity (7.22) covers the following cases:

(i) If $i = j$, we have $d = d'$ and

$$\frac{d'}{g}\sum_{\hat{R}}[\mathbf{D}^{(i)}(\hat{R})]^*_{\mu b}\,\hat{R}\,\phi_k^{(i)} = \phi_\mu^{(i)}\delta_{kb} \tag{7.23a}$$

(ii) If $i \neq j$, we have from Eq. (7.22),

$$\sum_{\hat{R}}[\mathbf{D}^{(i)}(\hat{R})]^*_{\mu b}\,\hat{R}\,\phi_k^{(j)} = 0$$

and to make it appear similar to Eq. (7.23a), we write

$$\frac{d'}{g}\sum_{\hat{R}}[\mathbf{D}^{(i)}(\hat{R})]^*_{\mu b}\,\hat{R}\,\phi_k^{(j)} = 0 \tag{7.23b}$$

Generalising Eqs (7.23a) and (7.23b) we have the form of a new operator,

$$\hat{P}_{\mu b}^{(i)} = \frac{d'}{g}\sum_{\hat{R}}[\mathbf{D}^{(i)}(\hat{R})]^*_{\mu b}\,\hat{R} \tag{7.24}$$

such that $\hat{P}_{\mu b}^{(i)}\,\phi_k^{(j)} = \phi_\mu^{(i)}\delta_{kb}\delta_{ij}$ (7.25)

This operator has the following properties:

$\hat{P}_{\mu b}^{(j)}\,\phi_b^{(j)} = \phi_\mu^{(j)}$ (generation of a partner basis function in the same irreducible representation) (7.26a)

$\hat{P}_{\mu b}^{(i)}\,\phi_b^{(j)} = 0$ (annihilation of basis functions of other irreducible representations) (7.26b)

$\hat{P}_{\mu b}^{(j)}.\,\phi_m^{(j)} = 0$ [annihilation of a basis function belonging to the same irreducible representation but to a different column ($b \neq m$) of the basis row-matrix] (7.26c)

The operator is called a projection operator, the name being justified as follows: Let us consider a general function (F) expressed as a linear combination of all the basis functions of all the irreducible representations of a group G, that is,

$$F = \sum_{j=1}^{c}\sum_{\lambda=1}^{d} a_\lambda^{(j)}\ \phi_\lambda^{(j)} \tag{7.27}$$

where d is the dimension of the irreducible representation D^j and c is number of classes, that is, the number of possible irreducible representations of G; $a_\lambda^{(j)}$ are numerical coefficients in the linear combination producing F from the ϕ's. Then using Eqs (7.25) and (7.27),

$$\hat{P}^{(i)}_{\mu k}\ F = \hat{P}^{(i)}_{\mu k} \sum_{j=1}^{c}\sum_{\lambda=1}^{d} a^{(j)}_{\lambda}\ \phi^{(j)}_{\lambda} = \sum_{j=1}^{c}\sum_{\lambda=1}^{d} a^{(j)}_{\lambda}\ \phi^{(j)}_{\mu}\ \delta_{ij}\ \delta_{k\lambda} = \sum_{\lambda=1}^{d} a^{(i)}_{\lambda}\ \phi^{(i)}_{\mu}\ \delta_{k\lambda} = a^{(i)}_{k}\ \phi^{(i)}_{\mu} \quad (7.28)$$

Thus, from the general function *F*, the operator $\hat{P}^{(i)}_{\mu k}$ projects out only one basis function belonging to the μth column of the basis row-matrix of the *i*th irreducible representation and annihilates the others. Equations (7.26a) and (7.28) also suggest that from a given general function we can find all the basis functions of a given irreducible representation D^i, by repeatedly using $\hat{P}^{(i)}_{\mu k}$, changing the values of μ and *k* every time.

Thus, $\hat{P}^{(i)}_{\mu k}\ F = a^{(i)}_{k}\ \phi^{(i)}_{\mu}$ and ignoring the numerical factor $a^{(i)}_{k}$, we find

$$\hat{P}^{(i)}_{\lambda\mu}\ \phi^{(i)}_{\mu} = \phi^{(i)}_{\lambda},\ \hat{P}^{(i)}_{\nu\lambda}\ \phi^{(i)}_{\lambda} = \phi^{(i)}_{\nu},\ \text{and so on.} \quad (7.29)$$

We shall illustrate this procedure afterwards.

An important property of the operator $\hat{P}^{(i)}_{\lambda\lambda}$ is:

$$\hat{P}^{(i)}_{\lambda\lambda}\ \phi^{(i)}_{\lambda} = \phi^{(i)}_{\lambda}\ \delta_{\lambda\lambda} = \phi^{(i)}_{\lambda} \quad (7.30a)$$

$$\hat{P}^{(i)}_{\lambda\lambda}\ \hat{P}^{(i)}_{\lambda\lambda}\ \phi^{(i)}_{\lambda} = \hat{P}^{(i)}_{\lambda\lambda}\ \phi^{(i)}_{\lambda} = \phi^{(i)}_{\lambda} \quad (7.30b)$$

Thus $\quad \hat{P}^{(i)}_{\lambda\lambda}\ \hat{P}^{(i)}_{\lambda\lambda}\ \hat{P}^{(i)}_{\lambda\lambda} \ldots \hat{P}^{(i)}_{\lambda\lambda} = \hat{P}^{(i)}_{\lambda\lambda} \quad (7.31)$

which means the operator $\hat{P}^{(i)}_{\lambda\lambda}$ is *idempotent*. Also, $\phi^{(i)}_{\lambda}$ is an eigenfunction of $\hat{P}^{(i)}_{\lambda\lambda}$ with eigenvalue unity as evident from Eq. (7.30a).

Some authors call $\hat{P}^{(i)}_{\lambda\lambda}$ the *projection operator* and $\hat{P}^{(i)}_{\mu\lambda}$ the *transfer operator* because the former projects $\phi^{(i)}_{\lambda}$ from an arbitrary function *F* containing this basis and the latter transforms $\phi^{(i)}_{\lambda}$ into its partner $\phi^{(i)}_{\mu}$, in the same irreducible representation, D^i.

When the matrices $\{\mathbf{D}^{(i)}(\hat{R})\}$ of the irreducible representation D^i are given, its basis functions can be generated from an arbitrary function *F* in the function space of the representation using the following scheme:

(i) Operate $\hat{P}^{(i)}_{\lambda\lambda}$ on *F* and normalise to generate $\phi^{(i)}_{\lambda}$.
(ii) Operate $\hat{P}^{(i)}_{\mu\lambda}$ on $\phi^{(i)}_{\lambda}$ to obtain $\phi^{(i)}_{\mu}$, and so on.

Van Vleck has appropriately called this operator the *basis function generating machine*. One problem in applying the projection operator of the type $\hat{P}^{(i)}_{\lambda\mu}$ is that the matrices $\mathbf{D}^{(i)}(\hat{R})$ are to be known for the particular irreducible representation D^i. Hence, a projection operator which involves traces (characters) and not the individual matrix elements, is expected to be more convenient. We construct this as follows:

The result of operating $\hat{P}^{(i)}_{\lambda\lambda}$ on the general function *F* defined in Eq. (7.27) is,

$$\hat{P}^{(i)}_{\lambda\lambda} F = a^{(i)}_{\lambda}\ \phi^{(i)}_{\lambda} \qquad (\text{as } \hat{P}^{(i)}_{\lambda\lambda}\ \phi^{(j)}_{\lambda} = 0 \text{ when } i \neq j)$$

and $\quad \sum_{\lambda} \hat{P}^{(i)}_{\lambda\lambda} F = a^{(i)}_{1}\ \phi^{(i)}_{1} + a^{(i)}_{2}\ \phi^{(i)}_{2} + \cdots + a^{(i)}_{d'}\ \phi^{(i)}_{d'}$

or, $$\hat{P}^{(i)} F = a_1^{(i)} \phi_1^{(i)} + a_2^{(i)} \phi_2^{(i)} + \cdots + a_{d'}^{(i)} \phi_{d'}^{(i)} \tag{7.32}$$

We thus get a new projection operator:

$$\hat{P}^{(i)} \equiv \sum_{\lambda=1}^{d'} \hat{P}_{\lambda\lambda}^{(i)} \equiv \sum_{\lambda=1}^{d'} \sum_{\hat{R}} \frac{d'}{g} [\mathbf{D}^{(i)}(\hat{R})]_{\lambda\lambda}^{*} \hat{R} \quad \text{(using the definition of } \hat{P}_{\lambda\lambda}^{(i)}\text{)}$$

$$\equiv \sum_{\hat{R}} \frac{d'}{g} \sum_{\lambda=1}^{d'} [\mathbf{D}^{(i)}(\hat{R})]_{\lambda\lambda}^{*} \hat{R}$$

$$= \frac{d'}{g} \sum_{\hat{R}} [\chi^{(i)}(\hat{R})]^{*} \hat{R} \tag{7.33}$$

where $\chi^{(i)}(\hat{R}) = \mathrm{Tr}\ [\mathbf{D}^{(i)}(\hat{R})]$ in the ith irreducible representation. This $\hat{P}^{(i)}$ projects out of a general function F only that part of the function which is composed of the basis functions of the ith irreducible representation and annihilates the residual part.

7.4.2 Examples of Using Projection Operators

Problem 7.1 Let $\Phi = x + y + z$ be a general function in a function space of the C_{3v} point group. We illustrate here the use of the projection operator of the types $\hat{P}_{\lambda\lambda}^{(E)}$ and $\hat{P}_{\mu\lambda}^{(E)}$ to obtain the basis functions of the two dimensional irreducible representation E, of C_{3v}.

The matrices for the different operators of C_{3v} in the E-type irreducible representation are listed below:

$$\hat{E}: \begin{bmatrix} 1 & 0 \\ 0 & 1 \end{bmatrix}, \hat{C}_3: \begin{bmatrix} -\frac{1}{2} & -\frac{\sqrt{3}}{2} \\ \frac{\sqrt{3}}{2} & -\frac{1}{2} \end{bmatrix}, \hat{C}_3^2: \begin{bmatrix} -\frac{1}{2} & \frac{\sqrt{3}}{2} \\ -\frac{\sqrt{3}}{2} & -\frac{1}{2} \end{bmatrix}, \hat{\sigma}_v(1): \begin{bmatrix} 1 & 0 \\ 0 & -1 \end{bmatrix}, \hat{\sigma}_v(2): \begin{bmatrix} -\frac{1}{2} & \frac{\sqrt{3}}{2} \\ \frac{\sqrt{3}}{2} & \frac{1}{2} \end{bmatrix}$$

$$\hat{\sigma}_v(3): \begin{bmatrix} -\frac{1}{2} & -\frac{\sqrt{3}}{2} \\ -\frac{\sqrt{3}}{2} & \frac{1}{2} \end{bmatrix}$$

We want to project out of Φ the two basis functions of the representation E using the projection operator–transfer operator technique.

$$\hat{P}_{11}^{(E)} \Phi = \frac{2}{6}\Big[1 \cdot \hat{E}(\Phi) + \left(-\frac{1}{2}\right)\hat{C}_3(\Phi) + \left(-\frac{1}{2}\right)\hat{C}_3^2(\Phi) + 1 \cdot \hat{\sigma}_v(1)(\Phi) + \left(-\frac{1}{2}\right)\hat{\sigma}_v(2)(\Phi)$$
$$+ \left(-\frac{1}{2}\right)\hat{\sigma}_v(3)(\Phi)\Big]$$

To get the results of the C_{3v} symmetry operations on Φ, we require the results of these operations separately on x, y and z.

$$\hat{E}(x)=x,\ \hat{E}(y)=y,\ \hat{E}(z)=z\cdot\hat{C}_3(x)=[x\quad y]\begin{bmatrix}-\frac{1}{2}\\ \frac{\sqrt{3}}{2}\end{bmatrix}=-\frac{x}{2}+\frac{\sqrt{3}y}{2}=\frac{-x+\sqrt{3}y}{2}$$

$$\hat{C}_3(y)=[x\quad y]\begin{bmatrix}-\frac{\sqrt{3}}{2}\\ -\frac{1}{2}\end{bmatrix}=-\frac{\sqrt{3}x}{2}-\frac{y}{2}=\frac{-\sqrt{3}x-y}{2},\ \hat{C}_3(z)=z$$

$$\hat{C}_3^2(x)=[x\quad y]\begin{bmatrix}-\frac{1}{2}\\ -\frac{\sqrt{3}}{2}\end{bmatrix}=-\frac{x}{2}-\frac{\sqrt{3}y}{2}=\frac{-x-\sqrt{3}y}{2}$$

$$\hat{C}_3^2(y)=[x\quad y]\begin{bmatrix}\frac{\sqrt{3}}{2}\\ -\frac{1}{2}\end{bmatrix}=\frac{\sqrt{3}x}{2}-\frac{y}{2}=\frac{\sqrt{3}x-y}{2},\ \hat{C}_3^2(z)=z,\ \hat{\sigma}_v(1)(x)=[x\quad y]\begin{bmatrix}1\\ 0\end{bmatrix}=x$$

$$\hat{\sigma}_v(1)(y)=[x\quad y]\begin{bmatrix}0\\ -1\end{bmatrix}=-y,\ \hat{\sigma}_v(1)(z)=z$$

$$\hat{\sigma}_v(2)(x)=[x\quad y]\begin{bmatrix}-\frac{1}{2}\\ \frac{\sqrt{3}}{2}\end{bmatrix}=-\frac{x}{2}+\frac{\sqrt{3}y}{2}=\frac{-x+\sqrt{3}y}{2}$$

$$\hat{\sigma}_v(2)(y)=[x\quad y]\begin{bmatrix}\frac{\sqrt{3}}{2}\\ \frac{1}{2}\end{bmatrix}=\frac{\sqrt{3}x}{2}+\frac{y}{2}=\frac{\sqrt{3}x+y}{2},\ \hat{\sigma}_v(2)(z)=z$$

$$\hat{\sigma}_v(3)(x)=[x\quad y]\begin{bmatrix}-\frac{1}{2}\\ -\frac{\sqrt{3}}{2}\end{bmatrix}=-\frac{x}{2}-\frac{\sqrt{3}y}{2}=\frac{-x-\sqrt{3}y}{2}$$

$$\hat{\sigma}_v(3)(y)=[x\quad y]\begin{bmatrix}-\frac{\sqrt{3}}{2}\\ \frac{1}{2}\end{bmatrix}=-\frac{\sqrt{3}x}{2}+\frac{y}{2}=\frac{-\sqrt{3}x+y}{2},\ \hat{\sigma}_v(3)(z)=z$$

Then by the symmetry operators of C_{3v}, the function Φ transforms as follows:

$$\hat{E}(\Phi)=x+y+z$$

$$\hat{C}_3(\Phi)=\frac{(-x+\sqrt{3}y)}{2}+\frac{(-\sqrt{3}x-y)}{2}+z=\frac{(-1-\sqrt{3})}{2}x+\frac{(-1+\sqrt{3})}{2}y+z$$

$$\hat{C}_3^2(\Phi)=\frac{(-x-\sqrt{3}y)}{2}+\frac{(\sqrt{3}x-y)}{2}+z=\frac{(-1+\sqrt{3})}{2}x+\frac{(-1-\sqrt{3})}{2}y+z$$

$$\hat{\sigma}_v(1)(\Phi)=x-y+z$$

$$\hat{\sigma}_v(2)(\Phi) = \frac{(-x+\sqrt{3}y)}{2} + \frac{(\sqrt{3}x+y)}{2} + z = \frac{(-1+\sqrt{3})}{2}x + \frac{(1+\sqrt{3})}{2}y + z$$

$$\hat{\sigma}_v(3)(\Phi) = \frac{(-x-\sqrt{3}y)}{2} + \frac{(-\sqrt{3}x+y)}{2} + z = \frac{(-1-\sqrt{3})}{2}x + \frac{(1-\sqrt{3})}{2}y + z$$

Now,

$$\hat{P}_{11}^{(E)}(\Phi) = \frac{2}{6}\Bigg[1\cdot(x+y+z) - \frac{1}{2}\left\{\frac{(-1-\sqrt{3})}{2}x + \frac{(-1+\sqrt{3})}{2}y + z\right\}$$

$$-\frac{1}{2}\left\{\frac{(-1+\sqrt{3})}{2}x + \frac{(-1-\sqrt{3})}{2}y + z\right\} + 1\cdot(x-y+z)$$

$$-\frac{1}{2}\left\{\frac{(-1+\sqrt{3})}{2}x + \frac{(1+\sqrt{3})}{2}y + z\right\} - \frac{1}{2}\left\{\frac{(-1-\sqrt{3})}{2}x + \frac{(1-\sqrt{3})}{2}y + z\right\}\Bigg]$$

$$= x$$

Thus, one basis function of the E representation has been projected out of Φ. The other (partner) basis function of E can be obtained by applying the transfer operator $\hat{P}_{21}^{(E)}$ on the basis function already obtained:

$$\hat{P}_{21}^{(E)}(x) = \frac{2}{6}\Bigg[(0\cdot x) + \frac{\sqrt{3}}{2}\left\{\frac{(-x+\sqrt{3}y)}{2}\right\} - \frac{\sqrt{3}}{2}\left\{\frac{(-x-\sqrt{3}y)}{2}\right\}$$

$$+0\cdot\{x\} + \frac{\sqrt{3}}{2}\left\{\frac{(-x+\sqrt{3}y)}{2}\right\} - \frac{\sqrt{3}}{2}\left\{\frac{(-x-\sqrt{3}y)}{2}\right\}\Bigg]$$

$$= y$$

We thus get the full basis set (x, y) of the E representation from the given function Φ.

It can now be verified that if we use all the projection and transfer operators $\hat{P}_{11}^{(E)}, \hat{P}_{12}^{(E)}, \hat{P}_{22}^{(E)}$ and $\hat{P}_{21}^{(E)}$, no function other than x and y can be projected out of the function $\Phi = x+y+z$. Similarly by applying the projection operator $\hat{P}_{11}^{(A_1)}$ for the A_1 representation, one should get z from $\Phi = x+y+z$.

Problem 7.2 Consider the functions $\Phi = x^2 + y^2 + 2xy$ and $\Psi = x^2 - y^2 + 2xy$. The standard character table of C_{3v} (Appendix A3) shows that $x^2 + y^2$ is a basis function of A_1 while $(x^2 - y^2, xy)$ is a basis set of the 2-dimensional irreducible representation E. Hence Φ contains only the second member of the basis set of E and Ψ contains both the members. We expect the operator $\hat{P}_{11}^{(E)}$ to project out nothing from Φ and $x^2 - y^2$ (the first member) from Ψ. The second member xy, should be generated by applying $\hat{P}_{21}^{(E)}$ on $x^2 - y^2$ which is obtained from Ψ. These can be readily verified. In terms of the C_{3v} symmetry operators, the projection operator $\hat{P}_{11}^{(E)}$ has the expression,

$$\hat{P}_{11}^{(E)} = \frac{2}{6}\left[1\cdot\hat{E} - \frac{1}{2}\cdot\hat{C}_3 - \frac{1}{2}\cdot\hat{C}_3^2 + 1\cdot\hat{\sigma}_v(1) - \frac{1}{2}\cdot\hat{\sigma}_v(2) - \frac{1}{2}\cdot\hat{\sigma}_v(3)\right]$$

$$\hat{P}_{11}^{(E)}(x^2) = \frac{2}{6}\left[x^2 - \frac{1}{2}\cdot\frac{(-x+\sqrt{3}y)^2}{4} - \frac{1}{2}\cdot\frac{(-x-\sqrt{3}y)^2}{4} + x^2 - \frac{1}{2}\cdot\frac{(-x+\sqrt{3}y)^2}{4} - \frac{1}{2}\cdot\frac{(-x-\sqrt{3}y)^2}{4}\right]$$

$$= \frac{2}{6}\left[x^2 - \frac{1}{8}\cdot(x^2+3y^2-2\sqrt{3}xy) - \frac{1}{8}\cdot(x^2+3y^2+2\sqrt{3}xy) + x^2 - \frac{1}{8}\cdot(x^2+3y^2-2\sqrt{3}xy) - \frac{1}{8}\cdot(x^2+3y^2+2\sqrt{3}xy)\right]$$

$$= \frac{2}{6}\left[2x^2 - \frac{1}{2}\cdot(x^2+3y^2)\right]$$

Similarly, $\hat{P}_{11}^{(E)}(y^2) = \frac{2}{6}\left[2y^2 - \frac{1}{2}\cdot(3x^2+y^2)\right]$

$$\hat{P}_{11}^{(E)}(xy) = \frac{2}{6}\left[xy - \frac{1}{2}\cdot\frac{(-x+\sqrt{3}y)}{2}\cdot\frac{(-\sqrt{3}x-y)}{2} - \frac{1}{2}\cdot\frac{(-x-\sqrt{3}y)}{2}\cdot\frac{(\sqrt{3}x-y)}{2} + 1x(-y) - \frac{1}{2}\cdot\frac{(-x+\sqrt{3}y)}{2}\cdot\frac{(\sqrt{3}x+y)}{2} - \frac{1}{2}\cdot\frac{(-x-\sqrt{3}y)}{2}\cdot\frac{(-\sqrt{3}x+y)}{2}\right]$$

$= 0$ (on simplification)

Hence $\hat{P}_{11}^{(E)}(\Phi) = \hat{P}_{11}^{(E)}(x^2) + \hat{P}_{11}^{(E)}(y^2) + \hat{P}_{11}^{(E)}(2xy)$

$$= \frac{2}{6}\left[2x^2 - \frac{1}{2}\cdot(x^2+3y^2) + 2y^2 - \frac{1}{2}\cdot(3x^2+y^2) + 0\right]$$

$= 0$, as expected

$$\hat{P}_{11}^{(E)}(\Psi) = \hat{P}_{11}^{(E)}(x^2) - \hat{P}_{11}^{(E)}(y^2) + \hat{P}_{11}^{(E)}(2xy)$$

$$= \frac{2}{6}\left[2x^2 - \frac{1}{2}\cdot(x^2+3y^2) - 2y^2 + \frac{1}{2}\cdot(3x^2+y^2) + 0\right]$$

$$= \frac{2}{6}[3(x^2-y^2)]$$

$= (x^2 - y^2)$, the first member of the basis set $(x^2 - y^2, xy)$ of E.

The full form of the operator $\hat{P}_{21}^{(E)}$ is

$$\hat{P}_{21}^{(E)} = \frac{2}{6}\left[0\cdot\hat{E} + \frac{\sqrt{3}}{2}\cdot\hat{C}_3 - \frac{\sqrt{3}}{2}\cdot\hat{C}_3^2 + 0\cdot\hat{\sigma}_v(1) + \frac{\sqrt{3}}{2}\cdot\hat{\sigma}_v(2) - \frac{\sqrt{3}}{2}\cdot\hat{\sigma}_v(3)\right]$$

$$= \frac{2}{6}\left[\frac{\sqrt{3}}{2}\cdot\hat{C}_3 - \frac{\sqrt{3}}{2}\cdot\hat{C}_3^2 + \frac{\sqrt{3}}{2}\cdot\hat{\sigma}_v(2) - \frac{\sqrt{3}}{2}\cdot\hat{\sigma}_v(3)\right]$$

$$\hat{P}_{21}^{(E)}(x^2) = \frac{2}{6}\left[\frac{\sqrt{3}}{2}\cdot\frac{(-x+\sqrt{3}y)^2}{4} - \frac{\sqrt{3}}{2}\cdot\frac{(-x-\sqrt{3}y)^2}{4} + \frac{\sqrt{3}}{2}\cdot\frac{(-x+\sqrt{3}y)^2}{4} - \frac{\sqrt{3}}{2}\cdot\frac{(-x-\sqrt{3}y)^2}{4}\right]$$

$$= \frac{2}{6}\left[\frac{\sqrt{3}}{8}\cdot(x^2+3y^2-2\sqrt{3}xy) - \frac{\sqrt{3}}{8}\cdot(x^2+3y^2+2\sqrt{3}xy) + \frac{\sqrt{3}}{8}\cdot(x^2+3y^2-2\sqrt{3}xy) - \frac{\sqrt{3}}{8}\cdot(x^2+3y^2+2\sqrt{3}xy)\right]$$

$$= \frac{2}{6}\left[\frac{\sqrt{3}}{8}\cdot(-8\sqrt{3})\,xy\right]$$

$$= -xy$$

Similarly, $\hat{P}_{21}^{(E)}(y^2) = xy$

Hence $\hat{P}_{21}^{(E)}(x^2-y^2) = \hat{P}_{21}^{(E)}(x^2) - \hat{P}_{21}^{(E)}(y^2)$

$$= -2xy$$

$$\approx xy \text{ (ignoring the numerical factor)}$$

which is the second member of the basis set $(x^2 - y^2, xy)$ of E.

Problem 7.3 In this example we will see that the total projection operator $\hat{P}^{(E)} = \hat{P}_{11}^{(E)} + \hat{P}_{22}^{(E)}$ annihilates all the redundant parts of a function and keeps only that part which transforms according to the irreducible representation E of C_{3v}. Consider the function $\Omega = x^2 + y^2 + z^2 + xyz$. Here $x^2 + y^2$ and z^2 are separately bases of A_1 and xyz transforms as E (for xy) $\times A_1$ (for z) $= E$ (see character table of C_{3v}). Hence we expect $\hat{P}^{(E)}\Omega = xyz$.

The action of $\hat{P}_{11}^{(E)}$ on x^2 and y^2 has already been found in the earlier example.

$$\hat{P}_{11}^{(E)}(z^2) = \frac{2}{6}\left[1\cdot\hat{E}(z^2) - \frac{1}{2}\cdot\hat{C}_3(z^2) - \frac{1}{2}\cdot\hat{C}_3^2(z^2) + 1\cdot\hat{\sigma}_v(1)(z^2) - \frac{1}{2}\cdot\hat{\sigma}_v(2)(z^2) - \frac{1}{2}\cdot\hat{\sigma}_v(3)(z^2)\right]$$

$$= \frac{2}{6}\left[z^2 - \frac{1}{2}z^2 - \frac{1}{2}z^2 + z^2 - \frac{1}{2}z^2 - \frac{1}{2}z^2\right] \text{ (as } z \text{ remains unchanged by all } C_{3v} \text{ operators)}$$

$$= 0$$

$$\hat{P}_{11}^{(E)}(xyz) = \frac{2}{6}\left[xyz - \frac{1}{2}\cdot\frac{(-x+\sqrt{3}y)}{2}\cdot\frac{(-\sqrt{3}x-y)}{2}\cdot z - \frac{1}{2}\cdot\frac{(-x-\sqrt{3}y)}{2}\cdot\frac{(\sqrt{3}x-y)}{2}\cdot z + 1x(-y)z - \frac{1}{2}\cdot\frac{(-x+\sqrt{3}y)}{2}\cdot\frac{(\sqrt{3}x+y)}{2}\cdot z - \frac{1}{2}\cdot\frac{(-x-\sqrt{3}y)}{2}\cdot\frac{(-\sqrt{3}x+y)}{2}\cdot z\right]$$

$$= 0 \text{ (on simplification)}$$

The full form of $\hat{P}_{22}^{(E)}$ is

$$\hat{P}_{22}^{(E)} = \frac{2}{6}\left[1\cdot\hat{E} - \frac{1}{2}\cdot\hat{C}_3 - \frac{1}{2}\cdot\hat{C}_3^2 - 1\cdot\hat{\sigma}_v(1) + \frac{1}{2}\cdot\hat{\sigma}_v(2) + \frac{1}{2}\cdot\hat{\sigma}_v(3)\right]$$

$$\hat{P}_{22}^{(E)}(x^2) = \frac{2}{6}\left[x^2 - \frac{1}{2}\cdot\frac{(-x+\sqrt{3}y)^2}{4} - \frac{1}{2}\cdot\frac{(-x-\sqrt{3}y)^2}{4} - x^2 + \frac{1}{2}\cdot\frac{(-x+\sqrt{3}y)^2}{4} + \frac{1}{2}\cdot\frac{(-x-\sqrt{3}y)^2}{4}\right]$$

$$= \frac{2}{6}\left[x^2 - \frac{1}{8}\cdot(x^2+3y^2-2\sqrt{3}xy) - \frac{1}{8}\cdot(x^2+3y^2+2\sqrt{3}xy) - x^2 + \frac{1}{8}\cdot(x^2+3y^2-2\sqrt{3}xy) + \frac{1}{8}\cdot(x^2+3y^2+2\sqrt{3}xy)\right]$$

$$= 0$$

Similarly, $\hat{P}_{22}^{(E)}(y^2) = 0$

$$\hat{P}_{22}^{(E)}(xyz) = \frac{2}{6}\left[xyz - \frac{1}{2}\cdot\frac{(-x+\sqrt{3}y)}{2}\cdot\frac{(-\sqrt{3}x-y)}{2}\cdot z - \frac{1}{2}\cdot\frac{(-x-\sqrt{3}y)}{2}\cdot\frac{(\sqrt{3}x-y)}{2}\right.$$

$$\cdot z - x(-y)z + \frac{1}{2} \cdot \frac{(-x+\sqrt{3}y)}{2} \cdot \frac{(\sqrt{3}x+y)}{2} \cdot z + \frac{1}{2} \cdot \frac{(-x-\sqrt{3}y)}{2} \cdot \frac{(-\sqrt{3}x+y)}{2} \cdot z \Bigg]$$

$$= xyz \text{ (on simplification)}$$

$$\hat{P}^{(E)}\Omega = 0 + xyz = xyz$$

Problem 7.4 Suppose that the basis set for some representation of a group is given, together with the effect of the group operators on the basis functions. We show here how to construct the reducible representation Γ of the group using the given basis set and then find the irreducible representations that are present in Γ (by using the reduction formula). Then by using the projection operator of the second type, $\hat{P}^{(i)} = \sum_{\lambda} \hat{P}^{i}_{\lambda\lambda}$, we illustrate how to construct the basis functions for the irreducible representations that are components of Γ.

Let an orthonormal set (g_1, g_2, g_3) of functions be the basis set for some representation of the C_{3v} point group. Also let the results of applying the group operations on these functions be as given in the following table:

	$\hat{E}$	$\hat{C}_3$	$\hat{C}_3^2$	$\hat{\sigma}_v(1)$	$\hat{\sigma}_v(2)$	$\hat{\sigma}_v(3)$	
g_1	g_1	g_2	g_3	g_1	g_3	g_2	
g_2	g_2	g_3	g_1	g_3	g_2	g_1	$\langle g_i \mid g_j \rangle = \delta_{ij}$
g_3	g_3	g_1	g_2	g_2	g_1	g_3	

According to this table, each function remains unchanged by the action of $\hat{E}$, giving $\chi(\hat{E}) = 3$; no function remains unchanged by $\hat{C}_3$ or $\hat{C}_3^2$ and for each reflection operation there is one function in the set that remains unchanged. Hence the characters of a representation Γ with the basis set (g_1, g_2, g_3) are as follows:

χ (character)

	$\hat{E}$	$2\hat{C}_3$	$3\hat{\sigma}_v$
Γ	3	0	1

This is a reducible representation because the sum of squares of the characters is $1 \cdot 3^2 + 2 \cdot 0^2 + 3 \cdot 1^2 = 12$, which exceeds the order of the group, 6. By applying the reduction formula it can be easily shown that Γ consists of one A_1 type and one E type irreducible representation, that is, $\Gamma = A_1 + E$. Using the C_{3v} character table (Appendix A3), the basis functions belonging to these irreducible representations are constructed as follows:

$$\hat{P}^{(A_1)}\, g_1 = \frac{1}{6}[1 \cdot \hat{E}\, g_1 + 1 \cdot \hat{C}_3\, g_1 + 1 \cdot \hat{C}_3^2\, g_1 + 1 \cdot \hat{\sigma}_v(1) g_1 + 1 \cdot \hat{\sigma}_v(2) g_1 + 1 \cdot \hat{\sigma}_v(3) g_1]$$

$$= \frac{1}{6}[1 \cdot g_1 + 1 \cdot g_2 + 1 \cdot g_3 + 1 \cdot g_1 + + 1 \cdot g_2 + 1 \cdot g_3]$$

$$= \frac{1}{3}(g_1 + g_2 + g_3)$$

Similarly, $P^{(A_1)} g_2 = P^{(A_1)} g_3 = \frac{1}{3}(g_1 + g_2 + g_3)$

$$\hat{P}^{(A_2)} g_1 = \frac{1}{6}[g_1 + g_2 + g_3 - g_1 - g_2 - g_3] = 0$$

$$\hat{P}^{(A_2)} g_2 = \frac{1}{6}[g_2 + g_3 + g_1 - g_3 - g_2 - g_1] = 0. \text{ Similarly, } \hat{P}^{(A_2)} g_3 = 0$$

$$\hat{P}^{(E)} g_1 = \frac{2}{6}[2g_1 + (-g_2) + (-g_3) + 0.(g_1 + g_3 + g_2)] = \frac{2}{6}(2g_1 - g_2 - g_3)$$

$$\hat{P}^{(E)} g_2 = \frac{2}{6}(2g_2 - g_3 - g_1)$$

$$\hat{P}^{(E)} g_3 = \frac{2}{6}(2g_3 - g_1 - g_2)$$

It seems that there are three possible basis functions for the two dimensional *E* representation, namely, $\hat{P}^{(E)} g_1$, $\hat{P}^{(E)} g_2$ and $\hat{P}^{(E)} g_3$. But they are not linearly independent, as

$$\hat{P}^{(E)} g_1 + \hat{P}^{(E)} g_2 + \hat{P}^{(E)} g_3 = 0$$

It is possible to find two linearly independent functions from these three in many ways. One possible and simple way is to take, $\hat{P}^{(E)} g_1$ and $(\hat{P}^{(E)} g_2 - \hat{P}^{(E)} g_3)$. Thus two possible basis functions for the *E*-type representation are

$$f_1 = \hat{P}^{(E)} g_1 = \frac{2}{6}(2g_1 - g_2 - g_3) \approx \frac{1}{\sqrt{6}}(2g_1 - g_2 - g_3) \text{ (after normalisation)}$$

$$f_2 = \hat{P}^{(E)} g_2 - \hat{P}^{(E)} g_3 = \frac{2}{6}(3g_2 - 3g_3) \approx \frac{1}{\sqrt{2}}(g_2 - g_3) \text{ (after normalisation)}$$

It can be readily verified that f_1 and f_2 are orthogonal. So the basis functions belonging to the component irreducible representations of Γ are:

A_1: $\frac{1}{\sqrt{3}}(g_1 + g_2 + g_3)$ (after normalisation)

E_1: $\begin{cases} \frac{1}{\sqrt{6}}(2g_1 - g_2 - g_3) & \text{(after normalisation)} \\ \frac{1}{\sqrt{2}}(g_2 - g_3) & \text{(after normalisation)} \end{cases}$

Note: This technique will be followed in the construction of *symmetry adapted linear combinations* in connection with *Molecular Orbital Theory* and other applications of group theory.

7.5 DIRECT PRODUCT REPRESENTATIONS OF A GROUP AND THEIR REDUCTION

We begin with a discussion of direct products of matrices. The direct product of two matrices $\mathbf{A} = [a_{ij}]_{m\times n}$ and $\mathbf{B} = [b_{ij}]_{p\times q}$ is defined as follows:

$$\mathbf{A}\otimes\mathbf{B} = \begin{bmatrix} a_{11} & a_{12} & \cdot & \cdot & \cdot & a_{1n} \\ a_{21} & a_{22} & \cdot & \cdot & \cdot & a_{2n} \\ \cdot & \cdot & \cdot & \cdot & \cdot & \cdot \\ \cdot & \cdot & \cdot & \cdot & \cdot & \cdot \\ a_{m-1,1} & a_{m-1,2} & \cdot & \cdot & \cdot & a_{m-1,n} \\ a_{m1} & a_{m2} & \cdot & \cdot & \cdot & a_{mn} \end{bmatrix} \otimes \begin{bmatrix} b_{11} & b_{12} & \cdot & \cdot & \cdot & b_{1q} \\ b_{21} & b_{22} & \cdot & \cdot & \cdot & b_{2q} \\ \cdot & \cdot & \cdot & \cdot & \cdot & \cdot \\ \cdot & \cdot & \cdot & \cdot & \cdot & \cdot \\ b_{p-1,1} & b_{p-1,2} & \cdot & \cdot & \cdot & b_{p-1,q} \\ b_{p1} & b_{p2} & \cdot & \cdot & \cdot & b_{pq} \end{bmatrix}$$

$$= \begin{bmatrix} a_{11}\mathbf{B} & a_{12}\mathbf{B} & \cdot & \cdot & \cdot & a_{1n}\mathbf{B} \\ a_{21}\mathbf{B} & a_{22}\mathbf{B} & \cdot & \cdot & \cdot & a_{2n}\mathbf{B} \\ \cdot & \cdot & \cdot & \cdot & \cdot & \cdot \\ \cdot & \cdot & \cdot & \cdot & \cdot & \cdot \\ \cdot & \cdot & \cdot & \cdot & \cdot & \cdot \\ a_{m1}\mathbf{B} & a_{m2}\mathbf{B} & \cdot & \cdot & \cdot & a_{mn}\mathbf{B} \end{bmatrix} \text{(in a compact form)} \quad (7.34)$$

If $\mathbf{A}$ and $\mathbf{B}$ are square matrices of size $m\times m$ and $p\times p$ respectively, then the diagonal elements of $\mathbf{A}\otimes\mathbf{B}$ are clearly,

$$a_{11}b_{11}, a_{11}b_{22}, \ldots a_{11}b_{pp}, a_{22}b_{11}, a_{22}b_{22}, \ldots, a_{22}b_{pp}, \ldots a_{mm}b_{pp}$$

$$\begin{aligned} \text{Hence, } \operatorname{Tr}(\mathbf{A}\otimes\mathbf{B}) &= a_{11}(b_{11}+b_{22}+\cdots+b_{pp}) + a_{22}(b_{11}+b_{22}+\cdots+b_{pp}) + \cdots \\ &= a_{11}\operatorname{Tr}\mathbf{B} + a_{22}\operatorname{Tr}\mathbf{B} + \cdots + a_{mm}\operatorname{Tr}\mathbf{B} \\ &= \operatorname{Tr}\mathbf{A}\cdot\operatorname{Tr}\mathbf{B} \end{aligned} \quad (7.35)$$

Equation (7.35) is of fundamental importance in forming the characters of direct product representations as will be evident from the following paragraphs.

Let the matrices of two representations Γ_1 and Γ_2 of a group be as listed below.

	$\hat{E}$	$\hat{R}_2$	$\hat{R}_3$	...	$\hat{R}_g$
Γ_1:	$\mathbf{M}_1$	$\mathbf{M}_2$	$\mathbf{M}_3$	...	$\mathbf{M}_g$
Γ_2:	$\mathbf{B}_1$	$\mathbf{B}_2$	$\mathbf{B}_3$	...	$\mathbf{B}_g$

Then the set of direct product matrices as shown below is also a representation of the group:

	$\hat{E}$	$\hat{R}_2$	$\hat{R}_3$	...	$\hat{R}_g$
$\Gamma_1\otimes\Gamma_2$:	$\mathbf{M}_1\otimes\mathbf{B}_1$	$\mathbf{M}_2\otimes\mathbf{B}_2$	$\mathbf{M}_3\otimes\mathbf{B}_3$	...	$\mathbf{M}_g\otimes\mathbf{B}_g$

The validity of this statement is shown with the help of an example. Let $\phi = \{\phi_1, \phi_2, \phi_3\}$ and $\psi = \{\psi_1, \psi_2, \psi_3, \psi_4\}$ be the basis sets of two irreducible representations of a group. Let the operator $\hat{R}$ be represented by the matrices **A** and **B** in ϕ and ψ basis respectively, that is,

$$(\phi_1' \; \phi_2' \; \phi_3') = \hat{R}(\phi_1 \; \phi_2 \; \phi_3) = [\phi_1 \quad \phi_2 \quad \phi_3]\begin{bmatrix} a_{11} & a_{12} & a_{13} \\ a_{21} & a_{22} & a_{23} \\ a_{31} & a_{32} & a_{33} \end{bmatrix}$$

$$= [\phi_1 \quad \phi_2 \quad \phi_3]\mathbf{A}$$

and $$(\psi_1' \; \psi_2' \; \psi_3' \; \psi_4') = \hat{R}(\psi_1 \; \psi_2 \; \psi_3 \; \psi_4) = [\psi_1 \quad \psi_2 \quad \psi_3 \quad \psi_4]\begin{bmatrix} b_{11} & b_{12} & b_{13} & b_{14} \\ b_{21} & b_{22} & b_{23} & b_{24} \\ b_{31} & b_{32} & b_{33} & b_{34} \\ b_{41} & b_{42} & b_{43} & b_{44} \end{bmatrix}$$

$$= [\psi_1 \quad \psi_2 \quad \psi_3 \quad \psi_4]\mathbf{B}$$

Let us now see how to represent the symmetry operator $\hat{R}$ by a matrix with the basis set $\phi \times \psi$, that is, $\{\phi_1\psi_1, \phi_1\psi_2, \phi_1\psi_3, \phi_1\psi_4, \phi_2\psi_1, \phi_2\psi_2, \phi_2\psi_3, \ldots, \phi_2\psi_4\}$. We can examine the effect of $\hat{R}$ on the product functions of the new set. For example, on the seventh member $\phi_2\psi_3$ the effect of $\hat{R}$ is as follows:

$$\hat{R}(\phi_2\psi_3) = \hat{R}(\phi_2)\hat{R}(\psi_3) = (\phi_1 a_{12} + \phi_2 a_{22} + \phi_3 a_{32})(\psi_1 b_{13} + \psi_2 b_{23} + \psi_3 b_{33} + \psi_4 b_{43})$$

$$= (\phi_1\psi_1 a_{12}b_{13} + \phi_1\psi_2 a_{12}b_{23} + \phi_1\psi_3 a_{12}b_{33} + \cdots \text{12 such terms}) \tag{7.36}$$

In terms of a row matrix $[\phi_1\psi_1 \quad \phi_1\psi_2 \quad \phi_1\psi_3 \quad \ldots \quad \phi_3\psi_4]$ the result Eq. (7.36) can be expressed as,

$$\hat{R}(\phi_2\psi_3) = [\phi_1\psi_1 \quad \phi_1\psi_2 \quad \phi_1\psi_3 \quad \ldots \quad \phi_3\psi_4]\begin{bmatrix} a_{12}b_{13} \\ a_{12}b_{23} \\ \cdot \\ \cdot \\ \cdot \end{bmatrix} \tag{7.37}$$

By constructing the direct product $\mathbf{A} \otimes \mathbf{B}$ it can be readily verified that the column matrix in the RHS of Eq. (7.37) is the seventh column of $\mathbf{A} \otimes \mathbf{B}$ and it may also be noted that the function $\phi_2\psi_3$ is the 7th column member of the row-matrix whose elements are those of the ordered *cartesian product* of the sets ϕ and ψ. Similarly by operating $\hat{R}$ on the first member $\phi_1\psi_1$ of $\phi \times \psi$ one gets the first member of $\mathbf{A} \otimes \mathbf{B}$. This is true for all operators $\hat{R}$ of the group. Hence if the set of matrices {**A**} is a representation Γ^ϕ of the group in ϕ basis and the set {**B**} is a representation Γ^ψ of the same group in ψ basis then the set $\{\mathbf{A} \otimes \mathbf{B}\}$ of the direct products of matrices **A** and **B** also acts as a representation (written as $\Gamma^{\phi \times \psi}$) of the group in basis $\{\phi_1\psi_1, \phi_1\psi_2, \phi_1\psi_3, \ldots, \phi_3\psi_4\}$. In this example, Γ^ϕ is 3-dimensional, Γ^ψ is 4-dimensional and $\Gamma^{\phi \times \psi}$ is of dimension 12. In general, from any two representations (reducible or irreducible) of dimensions p and q, a direct product representation of dimension pq can be formed. We need not be afraid of the large size of the direct product matrices because in practice we require only the characters and not the actual forms of the matrices. According to Eq. (7.35), the characters of the various symmetry operators of a group in a direct product representation $\Gamma^{\phi\times\psi}$ are just the products of the corresponding characters in the component representations Γ^{ϕ} and Γ^{ψ}. Thus

$$\chi^{\phi\times\psi}(\hat{R}) = \chi^{\phi}(\hat{R})\cdot\chi^{\psi}(\hat{R}) \tag{7.38}$$

To illustrate, we show the direct products of some irreducible representations of the C_{4v} point group as follows:

C_{4v}	$\hat{E}$	$\hat{C}_2$	$2\hat{C}_4$	$2\hat{\sigma}_v$	$2\hat{\sigma}_d$
A_1	1	1	1	1	1
A_2	1	1	1	−1	−1
B_1	1	1	−1	1	−1
B_2	1	1	−1	−1	1
E	2	−2	0	0	0
$A_1\times A_2$	1	1	1	−1	−1
$B_1\times E$	2	−2	0	0	0
$E\times E$	4	4	0	0	0
$B_1\times B_2\times E$	2	−2	0	0	0

The direct product of one irreducible representation D, with the totally symmetric representation (A_1 in the present case) of the group is D itself, because each character in the latter is 1. Hence, $D\times A_1$ is irreducible; excepting this special case, all direct product representations are, in general, expected to be reducible and we can reduce them using the working formula,

$$a_i = \frac{d}{g}\sum[\chi^{j}(\hat{R})]^{*}[\chi^{\text{red}}(\hat{R})] \tag{7.39}$$

where a_i is the number of times the irreducible representations D^i occurs in the reducible representation Γ^{red}, $\chi^i(\hat{R})$ and $\chi^{\text{red}}(\hat{R})$ are respectively the characters of the operator $\hat{R}$ in D^i and Γ^{red}, d is the dimension of D^i and g is the order of the group. Thus, we can reduce $E\times E$ as follows:

If $E\times E = a_1A_1 + a_2A_2 + a_3B_1 + a_4B_2 + a_5E$ then

$$a_1 = \frac{1}{8}[1\cdot 4 + 1\cdot 4 + 2\times(1\cdot 0 + 1\cdot 0 + 1\cdot 0)] = 1$$

$$a_2 = \frac{1}{8}[1\cdot 4 + 1\cdot 4 + 2\times\{1\cdot 0 + (-1)\cdot 0 + (-1)\cdot 0\}] = 1$$

$$a_3 = \frac{1}{8}[1\cdot 4 + 1\cdot 4 + 2\times\{(-1)\cdot 0 + 1\cdot 0 + (-1)\cdot 0\}] = 1$$

$$a_4 = \frac{1}{8}[1\cdot 4 + 1\cdot 4 + 2\times\{(-1)\cdot 0 + (-1)\cdot 0 + 1\cdot 0\}] = 1$$

$$a_5 = \frac{1}{8}[2\cdot 4 + (-2)\cdot 4 + 2\times(0\cdot 0 + 0\cdot 0 + 0\cdot 0)] = 0$$

So $$E\times E = A_1 + A_2 + B_1 + B_2 \quad (7.40)$$

A check at this point is that the dimension of the representation $E\times E$ is 4 and the sum of the dimensions of the component irreducible representations, A_1, A_2, B_1 and B_2 is also 4.

Let us now consider a useful theorem on direct product representations.

7.5.1 Direct Product of Two Representations

Theorem: *In the direct product of an irreducible representation of a group with itself, the totally symmetric representation occurs exactly once; in the direct product of two different irreducible representations of a group the totally symmetric representation never occurs.* The result is valid strictly for representations involving real characters.

Proof: Let $D^1, D^2, \ldots, D^c$ be the possible irreducible representations of a group of which D^1 is the totally symmetric one. Now, if the direct product representation $D^i\times D^i$ is decomposable as,

$$D^i\times D^i = a_1D^1 + a_2D_2 + \cdots + a_cD^c$$

then $$a_i = \frac{1}{g}\sum_{\hat{R}}[\chi^1(\hat{R})]^*[\chi^{i\times i}(\hat{R})]$$

$$= \frac{1}{g}\sum_{\hat{R}}[\chi^i(\hat{R})][\chi^i(\hat{R})] \quad [\text{as } \chi^1(\hat{R}) = 1 \text{ for all } \hat{R} \text{ and } \chi^{i\times i}(\hat{R}) = \chi^i(\hat{R})\times\chi^i(\hat{R})]$$

$$= \frac{1}{g}\times g = 1$$

[Since the sum of the squares of the real characters of an irreducible representation is equal to the order (g) of the group.]

Similarly, if D^1 occurs a_1 times in $\Gamma^{i\times j}$, $(i\neq j)$, then

$$a_1 = \frac{1}{g}\sum_{\hat{R}}1^*\cdot[\chi^i(\hat{R})][\chi^i(\hat{R})] = 0 \quad \text{(because of othogonality of character vectors)}$$

These results can be easily verified from the list of direct products given below the character table of C_{4v}. Thus, $E\times E = A_1 + A_2 + B_1 + B_2$ contains A_1 exactly once. Similarly, $B_1\times E = E$, which can be shown easily, and it does not contain A_1.

7.5.2 The Theorem of Vanishing Integrals

Consider the integrals

$$I_1 = \int_{-a}^{+a} x^2\, dx,\ I_2 = \int_{-a}^{+a} x^3\, dx \quad \text{and} \quad I_3 = \int_{-a}^{+a} (bx^2 + cx^3)\, dx$$

where a, b and c are constants. Without carrying out the integrations it can be said that $I_2 = 0$ since x^3 is an odd function and $I_1 \neq 0$ as x^2 is even; also $I_3 \neq 0$ since $bx^2 + cx^3$ contains an even part (bx^2) in the range $-a \leq x \leq a$. The transformation $x \rightarrow -x$ is caused by a reflection on the YZ plane, a $\hat{C}_2$ operation about the Z-axis or an inversion through the origin. Even functions are symmetric and odd functions are antisymmetric with respect to such operations. The above results of integration have been generalised in the group theoretical vanishing integral theorem as explained below.

Consider an object of point group G. Let X be a set of coordinates [(x, y, z) or (r, θ, ϕ)] and $F_\lambda(X)$ and $\phi_\nu(X)$ be two functions associated with some physical properties of the object. Also let the functions transform respectively according to the irreducible representations D^λ and D^ν of the group G. In quantum mechanics we often come across integrals of the type

$$M = \int \phi_\nu^*(X)\hat{O}_\mu(X)F_\lambda(X)dX \tag{7.41}$$

where $\hat{O}_\mu(X)$ is an operator which transforms according to the irreducible representation D^μ of the group. Group theoretical methods can be applied to ascertain whether such integrals will survive or vanish without actually carrying out the integration. The theorem of vanishing integrals states that *if the direct product representation $D^\nu \times D^\mu \times D^\lambda$ on reduction is found* ***not*** *to contain the totally symmetric representation $D^{(1)}$ of the group G then the integral* M *vanishes.*

Proof: If $D^\nu \times D^\mu \times D^\lambda$ does not contain $D^{(1)}$ of G, then the projection operator $\hat{P}^{(1)}$, when applied to the integrand function $\phi_\nu^*(X)\hat{O}_\mu(X)F_\lambda(X)$, projects out zero:

$$\hat{P}^{(1)}Q(X) = 0, \text{ where } Q(X) = \phi_\nu^*(X)\hat{O}_\mu(X)F_\lambda(X) \tag{7.42}$$

Let $\quad \hat{R}Q = Q' \quad$ and $\quad \hat{R}X = X' \quad$ [that is, $\hat{R}(x, y, z) = (x', y', z')$]

If $\hat{R}$ is a symmetry operator, functional values must not change by the action of $\hat{R}$, that is,

$$Q'(X') = Q(X) \tag{7.43}$$

Integrating over the whole space,

$$\int_{\text{whole space}} Q'(X')dX' = \int_{\text{whole space}} Q(X)dX$$

or,

$$\int_{\text{whole space}} (\hat{R}Q)(X')dX' = \int_{\text{whole space}} Q(X)dX$$

Since change in variable name is immaterial if the range of integration is fixed, we have

$$\int_{\text{whole space}} (\hat{R}Q)(X)dX = \int_{\text{whole space}} Q(X)dX \tag{7.44}$$

Summing over all R̂ of the group G,

$$\sum_R \int (\hat{R}Q)(X)dX = \sum_R \int Q(X)dX$$

$$= \int Q(X)dX + \int Q(X)dX + \int Q(X)dX + \cdots \text{ to } g \text{ terms}$$

$$= g\int Q(X)dX \text{ where } g \text{ is the order of the group}$$

or, $$\int \frac{1}{g}\sum_R 1^* \cdot (\hat{R}Q)(X)dX = \int Q(X)dX$$

or, $$\int \hat{P}^{(1)}Q(X)dX = \int Q(X)dX \tag{7.45}$$

Now, if $D^\nu \times D^\mu \times D^\lambda$ does not contain $D^{(1)}$ of G, then according to Eq. (7.42), $\hat{P}^{(1)}Q(X) = 0$ and hence $\int Q(X)dX = 0$ (proved).

Alternatively it may be stated that the direct product of any two irreducible representations, say, D^λ and D^ν, must contain the third irreducible representation D^μ, in order that the integral may not vanish (according to Theorem 7.5.1). We shall come across many applications of the vanishing integral theorem in the remaining part of the book; here are some simple examples.

Problem 7.5 Consider the point group C_s where the reflection plane σ is the YZ plane. Let the group be represented by functions of x in $-a \le x \le a$ and let the function space consist of only even functions $f_{\text{even}}(x)$ and odd functions $f_{\text{odd}}(x)$; it can be shown group theoretically that

$$I = \int_{-a}^{a} f_{\text{even}}(x) f_{\text{odd}}(x)dx = 0$$

The character table of C_s is

C_s	Ê	σ̂
A'	1	1
A''	1	−1

$\hat{\sigma}(x) = -x$. Hence $\hat{\sigma}[f_{\text{even}}(x)] = f_{\text{even}}(-x) = f_{\text{even}}(x)$

and $\hat{\sigma}[f_{\text{odd}}(x) = f_{\text{odd}}(-x)] = -f_{\text{odd}}(x)$

Thus $f_{\text{even}}(x)$ transforms as A' and f_{odd} transforms as A''. So the integrand in the given integral (I) corresponds to the direct product representation $A' \times A'' = A''$ which is not the totally symmetric representation. Hence $I = 0$.

Problem 7.6 Consider a system of point group C_{3v}. Let ϕ_1 and ϕ_2 be two functions related to this system and let Ô be an operator. If ϕ_1, Ô and ϕ_2 transform according to the representations

A_1, A_2 and E respectively, we can test whether $\langle \phi_1 | \hat{O} | \phi_2 \rangle$ vanishes or survives. The required direct product representation for this integral is as tabulated below.

	χ (character)		
	$\hat{E}$	$2\,\hat{C}_3$	$3\hat{\sigma}_v$
$A_1 \times A_2 \times E$:	$(1\times1\times2)=2$	$(1\times1\times-1)=-1$	$(1\times-1\times0)=0$

This is E itself. Hence the direct product does not contain A_1 and therefore, $\langle \phi_1 | \hat{O} | \phi_2 \rangle = 0$.

7.6 APPLICATION OF THE RULE FOR VANISHING INTEGRALS IN OBTAINING SYMMETRY-DETERMINED SELECTION RULES

The intensity of an electronic transition from the ground state of wave function Ψ_{gr} to an excited state of wave function Ψ_{ex} is proportional to the square of the transition moment integral,

$$I = \int \Psi_{\text{ex}}^{*} \left(-\sum_i q_i \vec{r}_i\right) \Psi_{\text{gr}}\, d\tau \tag{7.46}$$

where q_i is the charge of the ith particle (electron and nucleus), and $\vec{r}_i = x_i \vec{e}_1 + y_i \vec{e}_2 + z_i \vec{e}_3$ is the position vector of the ith particle in a suitable frame of coordinates. Let Ψ_{gr} and Ψ_{ex} transform respectively according to the irreducible representations D_{gr} and D_{ex}. Also, let the dipole moment operator $\hat{M} = \sum_i q_i \vec{r}_i$ transform according to the irreducible representation D_M. Obviously, D_M is composed of the irreducible representations whose bases are x and/or y and/or z. Hence if the direct product of the representations D_{ex} and D_{gr} contains some irreducible representation with x and/or y and/or z as basis then the direct product $D_{ex} \times D_M \times D_{gr}$ must contain the totally symmetric irreducible representation at least once and the integral I does not vanish. Therefore, the group theoretical selection rule for an electronic transition is as follows:

If the direct product $D_{ex} \times D_{gr}$ contains some irreducible representation with x and/or y and/or z as basis, the transition $\Psi_{gr} \rightarrow \Psi_{ex}$ is electric dipole allowed. Otherwise such a transition is forbidden.

Note:

(i) A transition which is electric dipole forbidden may be magnetic dipole allowed. The operator for the latter quantity belongs to the symmetry species of R_x, R_y and R_z (rotation about x, y and z axes). However, the strength of the interaction of the electronic charges with the fluctuating electric vectors in radiation is about 10^5 times as high as that with the magnetic vectors. Hence in practice, magnetically allowed transitions are extremely difficult to detect and we are primarily concerned with electric dipole allowed transitions.

(ii) ***The Laporté selection rule***: For centrosymmetric molecules/ions $g \rightarrow g$ or $u \rightarrow u$ type transitions are forbidden. The dipole moment vector $\vec{\mu}$ is of the u type since it changes sign by inversion. For a $g \rightarrow g$ or $u \rightarrow u$ type transition, the intensity integral I corresponds to the direct product representation $g \times u \times g$ or, $u \times u \times u$ which is u in any case; it does not contain the totally symmetric representation which is of g-type, Hence, $I = 0$. The Laporté rule can be simplified without strictly adhering to the above direct product rule, as explained below.

Let $I = \left\langle \Psi_{ex} \middle| -\sum_i q_i \vec{r}_i \middle| \Psi_{gr} \right\rangle$

Then the x-component of this integral is,

$$I_x = \left\langle \Psi_{ex} \middle| -\sum_i q_i x_i \middle| \Psi_{gr} \right\rangle$$

Under the inversion operation $\hat{i}$,

$$\hat{i} I_x = \hat{i} \left\langle \Psi_{ex} \middle| -\sum_i q_i x_i \middle| \Psi_{gr} \right\rangle = \left\langle \hat{i} \Psi_{ex} \middle| \sum_i q_i x_i \middle| \hat{i} \Psi_{gr} \right\rangle \quad (\text{as } \hat{i} x_k = -x_k)$$

If both Ψ_{ex} and Ψ_{gr} are of g-type in a centrosymmetric molecule/ion, we have

$\hat{i}\Psi_{ex} = \Psi_{ex}$ and $\hat{i}\Psi_{gr} = \Psi_{gr}$, so that

$$\hat{i} I_x = \left\langle \Psi_{ex} \middle| \sum_i q_i x_i \middle| \Psi_{gr} \right\rangle = -I_x$$

But as I_x means a real physical quantity, it should not change under a symmetry operation, that is, $\hat{i} I_x$ should be equal to I_x.

Thus $I_x = -I_x$ that is, $I_x = 0$.

Again if both Ψ_{ex} and Ψ_{gr} are u-type functions, $\hat{i}\Psi_{ex} = -\Psi_{ex}$ and $\hat{i}\Psi_{gr} = -\Psi_{gr}$, again making $\hat{i} I_x = -I_x$, and by similar argument, $I_x = 0$ in this case also. The same is true for I_y and I_z. Hence the intensity integral I is zero for $g \rightarrow g$ and $u \rightarrow u$ transitions.

The $d_{xy}, d_{yz}, d_{zx}, d_{x^2-y^2}$ and d_{z^2} wave functions are all symmetric with respect to the inversion operation (a d_{xy} wave function remains d_{xy} when the signs of both x and y are changed). So in an octahedral field, the d orbitals of a transition metal ion are of g character. Hence $d \rightarrow d$ transition in an octahedral field means a $g \rightarrow g$ transition which is forbidden. The forbidden-ness, however, is sometimes removed as a result of vibronic coupling (which is discussed in Chapter 13).

7.7 A COROLLARY TO THE VANISHING INTEGRAL THEOREM: THE GROUP THEORETICAL MATRIX ELEMENT THEOREM

The integral $\langle \psi_i | \hat{O} | \psi_j \rangle$ of an operator $\hat{O}$ between the functions ψ_i and ψ_j is called a *matrix element* for the operator since they occur in many secular matrices of quantum mechanical calculations. *For an operator* $\hat{O}$ *which remains invariant under all symmetry operations of the point group of a system, the matrix element* $O_{ij} = \langle \psi_i | \hat{O} | \psi_j \rangle$ *vanishes if* ψ_i *and* ψ_j *transform according to* ***different*** *irreducible representations of the group.*

Proof: $\hat{O}$ transforms according to the totally symmetric representation D^1, because it remains invariant under all the symmetry operations. Let ψ_i and ψ_j transform according to the irreducible representations D^i and D^j, $i \neq j$, of the group. Then the matrix element $O_{ij} = \langle \psi_i | \hat{O} | \psi_j \rangle$ transforms according to the direct product representation

$$D^i \times D^1 \times D^j = D^i \times D^j \quad \text{(since in } D^1 \text{ each character} = 1)$$

which does not contain D^1 if $i \neq j$. Hence $O_{ij} = 0$.

Note: If the irreducible representation is unitary, its basis set is orthogonal. Hence if ψ_i and ψ_j are *different* functions of the basis set of the same multi-dimensional representation, then also $O_{ij} = 0$ for a totally symmetric operator $\hat{O}$ (like $\hat{H}$).

SUMMARY

- The link between group theory and quantum mechanics is the fact that eigenfunctions of the Hamiltonian operator ($\hat{H}$) are bases of irreducible representations of the system's point group (G). This results from the invariance of $\hat{H}$ under all symmetry operations ($\hat{R}$) of G.
- Projection operators are designed to filter out of a general function (F), that part which acts as a basis of an irreducible representation of G. The direct product of a number of irreducible representations of G is also a representation and its character set is obtained by multiplying the corresponding characters in the component representations. An integral vanishes if the direct product representation for the integrand does not contain the totally symmetric representation of G.

LEARNING OUTCOME

After reading this chapter the student will be able to (i) construct group theoretical projection operators, (ii) generate basis functions of a desired irreducible representation from a given function which contains any basis function as a component part, (iii) construct symmetry adapted linear combinations by using projection operators and (iv) examine whether a given integral vanishes or survives by constructing direct product representations.

EXERCISES

1. Show that the molecular orbital wave functions ψ_1 and ψ_2 of H_2^+ ion are bases of irreducible representations: $\psi_1 = \frac{1}{\sqrt{2}}(\phi_1 + \phi_2)$ and $\psi_2 = \frac{1}{\sqrt{2}}(\phi_1 - \phi_2)$ where each $\phi_i = 1s$ orbital of H.

2. For the point group C_{5v}, show that (a) $E_1 \times E_1 = A_1 + A_2 + E_2$ (b) $E_1 \times E_2 = E_1 + E_2$.

3. The basis set $\{x, y, z\}$ of an irreducible representation $D^{(i)}$ of a group is written as $\{\Phi_\mu^{(i)}\}$, $\mu = 1, 2, 3$. If $\hat{P}_{\mu k}^{(i)}$ represents a projection operator, find $\hat{P}_{12}^{(i)}(x)$, $\hat{P}_{12}^{(i)}(y)$, $\hat{P}_{31}^{(i)}(x)$, $\hat{P}_{22}^{(i)}(x^2 + y^2 + z^2)$ and $\hat{P}_{32}^{(i)}(y)$.
[Hint: $x = \Phi_1^{(i)}$, $y = \Phi_2^{(i)}$ and $z = \Phi_3^{(i)}$]

4. The functions Φ_1 and Φ_2 transform according to the irreducible representation A_2 and the operator $\hat{O}$ transforms according to E of the point group D_3. Examine, using the D_3 character table, whether the integral $\langle \Phi_1 | \hat{O} | \Phi_2 \rangle$ vanishes or not.

5. Find the direct product $E \times E$ in the point group C_3 and reduce it.

6. The matrices of the symmetry operators of point group C_{3v} for the irreducible representation E with the cartesian coordinates (x, y) as basis are given in Section 7.4.2. The z coordinate remains unchanged by the C_{3v} operators. Use the projection operator method to find the basis functions of E which are contained in the function $\Phi = xz + yz + z^2$.

Suggested Books and References for Further Reading

1. Schonland D S, *Molecular Symmetry*, D Van Nostrand Company Ltd., London, **1965**.
2. Carter R L, *Molecular Symmetry and Group Theory*, Wiley, **1997**.
3. Vincent A, *Molecular Symmetry and Group Theory: A Programmed Introduction to Chemical Applications*, Wiley, **2000**.

8 Chemical Bonding I: Molecular Orbital Theory

Learning objectives

- Definition of MOs, LCAO–MO method, necessary conditions for atomic orbitals to form MOs
- Diatomic molecules: symmetry type and energy order of MOs from united atom–separated atom model
- Ligand group orbitals and symmetry matching for MOs of polyatomic molecules
- Allowedness of electronic transitions from symmetry types of MOs
- Walsh correlation diagram: ascertaining linear or bent structures of XH_2 molecules

8.1 INTRODUCTION

High speed computers were not available in the early years of quantum mechanics. Attempts to explain how atoms are held together in molecules led to two different approaches—(i) the valence bond (VB) theory (initiated by Heitler and London and developed extensively by Pauling and Slater) and (ii) the molecular orbital (MO) theory (initiated by Hund and Mulliken and developed by Hückel, Coulson, Longuet-Higgins and others). Both involve approximate methods for solving the Schrödinger equation for multi-electron systems. The VB method, according to Pauling, is a quantitative formulation of the Lewis concept of electron pair bond on quantum mechanical grounds. In the MO theory, valence shell orbitals of the constituent atoms are combined linearly and the Schrodinger equation is solved approximately to get a set of molecular orbitals which are filled by electrons according to the *Aufbau* principle just like atomic orbitals in atoms. Not only chemical bonding but also spectroscopic results are well explained by the MO theory; the MO wave functions are amenable to symmetry-based classification, and group theoretical selection rules can be applied to predict electronic transitions from an occupied to an unoccupied MO. This chapter deals essentially with the application of group theoretical methods to MO calculations.

8.2 THE BORN–OPPENHEIMER APPROXIMATION

Whatever be the approach, VB or MO, the starting point is an approximation suggested by Born and Oppenheimer. In a molecule, the electrons and nuclei move simultaneously (the latter resulting in rotation, vibration and translation of the molecule). It is not possible to exactly solve the Schrodinger equation of the molecule considering all these motions together. However, electrons are much lighter than nuclei (about 1836 times lighter than the lightest nucleus, the proton); so electrons move much faster than the nuclei. It is therefore reasonable to assume that the electronic motion in a molecule is the same as it would be if the nuclei were at rest. Born and Oppenheimer utilised this idea in quantum mechanical terms by expressing the total Hamiltonian operator ($\hat{H}$) as a sum of the electronic and nuclear Hamiltonians

$$\hat{H} = \hat{H}_{el} + \hat{H}_{nu} \tag{8.1}$$

and the total wave function of the system as a product function

$$\Psi(r, R) = \psi_R(r) \cdot \phi(R) \tag{8.2}$$

where r is the set of electronic coordinates (x, y, z) and R is the set of nuclear coordinates (X, Y, Z), $\psi_R(r)$ is the electronic wave function at the positions of the nuclei given by R and $\phi(R)$ is the nuclear wave function. It was further assumed that, at least to a first approximation, the electronic Schrödinger equation

$$\hat{H}_{el}\psi_R(r) = E_{el}\psi_R(r) \tag{8.3}$$

can be solved separately for a particular set of nuclear positions (R). This means that the electronic energy and the wave function depend *functionally* on the electronic coordinates and *parametrically* on R. At each R one can calculate the total energy (E) of the system as

$$E = \text{electronic energy, } E_{el} \text{ [from Eq. (8.3)]} + \text{nucleus–nucleus repulsion energy}$$

This energy is termed 'potential energy of the molecular system at nuclear configuration R' because the atoms are held fixed in their positions. By plotting E against R a potential energy surface (PES), most often a hypersurface, is obtained. The set of nuclear coordinates (R) at the minimum of the PES gives the *molecular structure*.

8.3 MOLECULAR ORBITAL (MO) THEORY

The MO theory starts by assuming that just as an electron in an atom occupies one atomic orbital centred about the nucleus, an electron in a molecule occupies one *molecular orbital* which, in principle, is spread over all the nuclei of the molecule. Wave functions of these orbitals are eigenfunctions of the Hamiltonian operator of the system. Hence a molecular orbital may be defined as *a one-electron multicentre orbital whose wave function acts as a basis of some irreducible representation of the point group of the system.*

The exact solution of the electronic Schrödinger equation (Eq. 8.3) for a multielectron system is not possible owing to the non-separability of the variables in the inter-electron repulsion terms. Although the Hatree–Fock self-consistent field (SCF) method for atoms was known at the time the MO theory was being developed, it took about twenty years to formulate the SCF method for molecules (C C J Roothaan, 1951). However, in the mean time, simple approaches based largely on symmetry and chemical intuition had been developed. One such approach is the LCAO–MO method, which is so rich in conceptual contents that it explains a wide variety of chemical phenomena and spectroscopic observations at least qualitatively.

8.3.1 The Linear Combination of Atomic Orbitals–Molecular Orbital (LCAO–MO) Method

One method of approximation generally applied to solve the Schrödinger equation of a system is the variation method. In this method, a trial solution (ψ_g) is guessed such that it contains some adjustable parameters. The guessed function is used in the Schrödinger equation for the system and an expression for energy is obtained in the form $E = \dfrac{\langle \psi_g | \hat{\mathrm{H}} | \psi_g \rangle}{\langle \psi_g | \psi_g \rangle}$ which contains the parameters. Then by varying the parameters, the energy E is minimised to get a value that is close to the actual ground state energy E_0 (but still greater than E_0 unless the guess, by chance, happens to be the actual wave function). A proof of the *variation theorem* and the elaborate procedure for applying the variation method may be found in any standard text book of quantum chemistry. The LCAO–MO method is the simplest one to apply the variation method on molecules.

Before going to more complex molecules, let us begin with the simplest possible molecular system, the H_2^+ ion. It consists of two nuclei and a single electron.

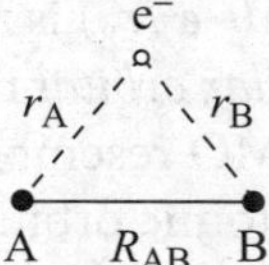

Fig. 8.1 Relative positions of the electron and nuclei in the H_2^+ ion

The Hamiltonian operator for this system is

$$\hat{\mathrm{H}} = -\frac{\hbar^2}{2m}\nabla_e^2 - \frac{\hbar^2}{2M}\nabla_{\mathrm{A}}^2 - \frac{\hbar^2}{2M}\nabla_{\mathrm{B}}^2 - \frac{e^2}{r_{\mathrm{A}}} - \frac{e^2}{r_{\mathrm{B}}} + \frac{e^2}{R_{\mathrm{AB}}} \tag{8.4}$$

where m and M are masses of the electron and proton respectively, r_{A} and r_{B} are distances of the electrons from the nuclei (that is, protons) A and B respectively, R_{AB} is the internuclear distance (Fig. 8.1) and ∇_e^2, ∇_{A}^2 and ∇_{B}^2 are the Laplacian operators for the electron and the nuclei A and B. Under the Born–Oppenheimer approximation, this $\hat{\mathrm{H}}$ [of Eq. (8.4)] can be simplified to

$$\hat{\mathrm{H}} = -\frac{\hbar^2}{2m}\nabla_e^2 - \frac{e^2}{r_{\mathrm{A}}} - \frac{e^2}{r_{\mathrm{B}}} + \frac{e^2}{R_{\mathrm{AB}}} \tag{8.5}$$

because the nuclei are assumed to be held fixed. The potential energy part e^2/R_{AB} acts as an additive constant (for a particular value of R_{AB}) and does not affect the eigenfunction of the Hamiltonian, $\hat{\mathrm{H}}$; only the potential energy e^2/R_{AB} due to internuclear repulsion should be added to the energy of the electron to get the total energy of the molecular system with the positions of the nuclei fixed at A and B. Omitting the e^2/R_{AB} part of the Hamiltonian, its required form now simplifies to

$$\hat{\mathrm{H}} = -\frac{\hbar^2}{2m}\nabla_e^2 - \frac{e^2}{r_{\mathrm{A}}} - \frac{e^2}{r_{\mathrm{B}}} \tag{8.6}$$

Note that for the isolated atom A, the Schrödinger equation is

$$\hat{\mathrm{H}}_{\mathrm{A}}\phi_{\mathrm{A}} = E_{\mathrm{A}}\phi_{\mathrm{A}} \tag{8.7}$$

$$\text{where}\quad \hat{H}_A = -\frac{\hbar^2}{2m}\nabla_e^2 - \frac{e^2}{r_A} \tag{8.8}$$

and ϕ_A is the wave function of an atomic orbital of A. Similarly for the isolated atom B,

$$\hat{H}_B\phi_B = E_B\phi_B \tag{8.9}$$

$$\text{where}\quad \hat{H}_B = -\frac{\hbar^2}{2m}\nabla_e^2 - \frac{e^2}{r_B} \tag{8.10}$$

and ϕ_B is the wave function of an atomic orbital of B.

In the region of nucleus A, the most significant part of the total Hamiltonian $\hat{H}$ of Eq. (8.6) contains the term $\hat{H}_A$; the remaining part $(-e^2/r_B)$ is not exactly zero but is small. Hence in the neighbourhood of nucleus A, the *molecular orbital* resembles the atomic orbital ϕ_A. Similarly in the neighbourhood of nucleus B, the MO resembles mainly ϕ_B. The complete MO picture, therefore, has characteristics which the atomic orbitals ϕ_A and ϕ_B separately possess. Hence it is natural to express the molecular orbital wave function as a linear combination of ϕ_A and ϕ_B:

$$\Psi_{MO} = c_1\phi_A + c_2\phi_B \tag{8.11}$$

The combining coefficients c_1 and c_2 are to be determined variationally.

One may combine any atomic orbital of A with any one of B, but all such combinations may not yield meaningful results. Before arriving at the guiding principles, let us carry out a variational calculation with the functions of Eq. (8.11) in brief. For convenience, Ψ_{MO} will henceforth be written simply as Ψ, and the constants c_1 and c_2 are assumed to be real. The following integrals will be necessary for the calculation

$$\begin{aligned}\langle\Psi|\hat{H}|\Psi\rangle &= \langle c_1\phi_A + c_2\phi_B|\hat{H}|c_1\phi_A + c_2\phi_B\rangle \\ &= c_1^2H_{AA} + 2c_1c_2H_{AB} + c_2^2H_{BB}\end{aligned} \tag{8.12}$$

where the following symbols have been used:

$$\left.\begin{aligned}H_{AA} &= \langle\phi_A|\hat{H}|\phi_A\rangle \\ H_{BB} &= \langle\phi_B|\hat{H}|\phi_B\rangle \\ H_{AB} &= \langle\phi_A|\hat{H}|\phi_B\rangle = \langle\phi_B|\hat{H}|\phi_A\rangle = H_{BA}\end{aligned}\right\} \tag{8.13}$$

$$\begin{aligned}\langle\Psi|\Psi\rangle &= \langle c_1\phi_A + c_2\phi_B|c_1\phi_A + c_2\phi_B\rangle \\ &= c_1^2 + c_2^2 + 2c_1c_2S_{AB}\end{aligned} \tag{8.14}$$

Here, the atomic orbitals ϕ_A and ϕ_B are assumed to be initially normalised and

$$S_{AB} = \langle\phi_A|\phi_B\rangle \tag{8.15}$$

Using these expressions in $E = \dfrac{\langle\Psi|\hat{H}|\Psi\rangle}{\langle\Psi|\Psi\rangle}$ and rearranging, one gets

$$c_1^2(H_{AA} - E) + 2c_1c_2(H_{AB} - ES_{AB}) + c_2^2(H_{BB} - E) = 0 \tag{8.16}$$

To minimise E, $\frac{\partial E}{\partial c_1}$ and $\frac{\partial E}{\partial c_2}$ must each be zero. We then get two linear equations:

$$c_1(H_{AA} - E) + c_2(H_{AB} - ES_{AB}) = 0 \tag{8.17a}$$

$$c_1(H_{AB} - ES_{AB}) + c_2(H_{BB} - E) = 0 \tag{8.17b}$$

These equations have non-trivial solutions for c_1 and c_2 only if

$$\begin{vmatrix} H_{AA} - E & H_{AB} - ES_{AB} \\ H_{AB} - ES_{AB} & H_{BB} - E \end{vmatrix} = 0 \tag{8.18}$$

The determinant on the left hand side of Eq. (8.18) is called the *secular determinant* for the system. Such determinants are a common occurrence in quantum mechanics, wherever the variation method is applied.

Equation (8.18) is quadratic in E and the two roots correspond to two molecular orbitals (that is, two possible linear combinations of the atomic orbitals ϕ_A and ϕ_B). Thus, two AOs on linear combination give two MOs. (The result may be generalised—n atomic orbitals on linear combination yield n molecular orbitals.) The integral H_{AA}, defined in Eq. (8.13), is

$$E_A = H_{AA} = \int_{r_A=0}^{\infty} \phi_A^* \hat{H} \phi_A d\tau \tag{8.19}$$

This E_A is the energy of the electron when it is in orbital ϕ_A but is feeling the influence of both the nuclei of the H_2^+ ion, since $\hat{H}$ is the Hamiltonian operator of the whole system and not of the isolated atom A. Thus, E_A is different from the electronic energy of the isolated atom A. The quantity H_{BB} (= E_B) has a similar meaning. Such integrals are called *Coulomb integrals.*

The integral $S_{AB} = \int \phi_A^* \phi_B \, d\tau$ is called the *overlap integral* between the atomic orbitals ϕ_A and ϕ_B and, physically, it is a measure of the volume in which the probability density of the electron from the orbitals ϕ_A and ϕ_B is appreciable. The maximum value of the overlap integral is unity when two orbitals completely coincide. The integral

$$H_{AB} = H_{BA} = \int \phi_A^* \hat{H} \phi_B \, d\tau \tag{8.20}$$

is called *exchange* or *resonance* integral. Note that even if the atoms A and B are the same, $H_{AB} \neq H_{AA}$(or, H_{BB}) because the atomic orbitals ϕ_A and ϕ_B are centred about nuclei at two different atoms. However, for homonuclear diatomic systems like the H_2^+ ion, if the same types of orbitals are used (say, 1s of each H) then $H_{AA} = H_{BB}$ and Eq. (8.18) gives

$$(H_{AA} - E)^2 = (H_{AB} - ES_{AB})^2 \tag{8.21}$$

from which it can be readily verified that two values of E are,

$$E = \frac{H_{AA} - H_{AB}}{1 - S_{AB}}, \frac{H_{AA} + H_{AB}}{1 + S_{AB}}$$

Putting $E = \dfrac{H_{AA} + H_{AB}}{1 + S_{AB}} = E_1$ (say) in Eq. (8.17a) or (8.17b), it is found that $c_1 = c_2$ and for normalisation of the MO wave function Ψ we require

$$1 = \int \Psi^* \Psi d\tau = c_1^2 + c_2^2 + 2c_1 c_2 S_{AB} \quad \text{[from Eq. (8.14)]}$$

or, $$2c_1^2 + 2c_1^2 S_{AB} = 1$$

or, $$c_1 = c_2 = \frac{1}{\sqrt{2 + 2S_{AB}}}$$

Thus, corresponding to the MO of energy E_1 the normalised MO wave function is

$$\Psi_1 = c_1(\phi_A + \phi_B) \quad (\text{as } c_1 = c_2)$$

$$= \frac{1}{\sqrt{2 + 2S_{AB}}}(\phi_A + \phi_B) \tag{8.22}$$

Similarly for the other value of E, that is, $E = \dfrac{H_{AA} - H_{AB}}{1 - S_{AB}} = E_2$ (say), the Eq. (8.17a) or (8.17b) gives $c_1 = -c_2$ and after normalisation the MO wave function is found to be

$$\Psi_2 = \frac{1}{\sqrt{2 - 2S_{AB}}}(\phi_A - \phi_B) \tag{8.23}$$

Equation (8.22) shows that Ψ_1 is a result of an in-phase combination of ϕ_A and ϕ_B. Just as two waves on being superimposed *in phase* result in an increase in amplitude (constructive interference), the linear combination [Eq. (8.22)] results in an increase in probability of finding the electron in a region between A and B. This MO, therefore, is called the *bonding* MO (the increased electron density between the nuclei 'binds' the nuclei). Similarly, Eq. (8.23) shows that Ψ_2 is a result of the out-of-phase combination of the atomic orbitals ϕ_A and ϕ_B; such a

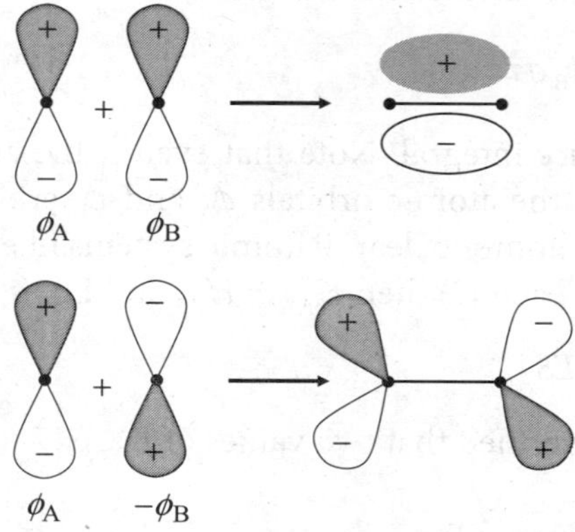

Fig. 8.2 Bonding and antibonding combinations of ϕ_A and ϕ_B

superposition of two waves decreases the resultant amplitude (destructive interference) and hence the probability of finding the electron is much reduced in the region between the two nuclei; Ψ_2 is called an *antibonding* MO. If ϕ_A and ϕ_B denote *p*-type orbitals of the atoms A and B, then the two modes of combination are as shown in Fig. 8.2.

The integral H_{AB} is generally negative (as can be revealed by using the actual expressions for the atomic orbital wave functions). Hence with respect to the energy $E_A(= H_{AA})$, the bonding molecular orbital (BMO) is stabilised (owing to the expression $H_{AA} + H_{AB}$ in the numerator) and the antibonding molecular orbital (ABMO) is destabilised (owing to the term $H_{AA} - H_{AB}$ in the numerator). However, the extent of stabilisation of the BMO is somewhat less than the extent of destabilisation of the ABMO because the denominator of the former is higher than that of the latter, S_{AB} being positive [Fig. 8.3(a)].

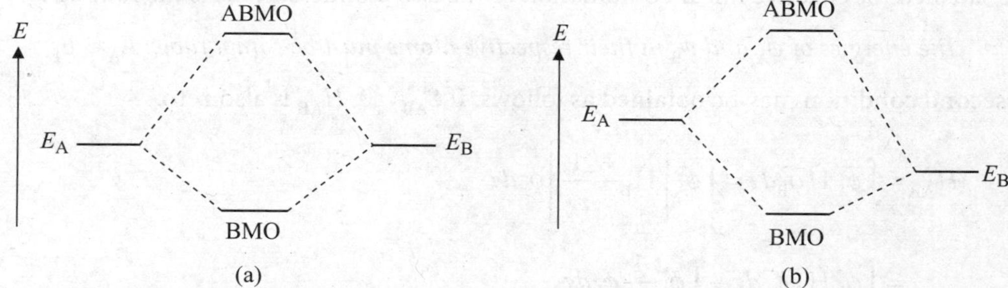

Fig. 8.3 Stabilisation of BMO and destabilisation of ABMO: case (a) $E_A = E_B$, case (b) $E_A > E_B$

Bonding and antibonding MOs result from the combination of two different atomic orbitals as well: for example, if the 2*s* of atom A and one 2*p* orbital of atom B are used (that is, $\phi_A = \phi_{2s(A)}$ and $\phi_B = \phi_{2p_x(B)}$) then $H_{AA} \neq H_{BB}$ (that is, $E_A \neq E_B$) and Eq. (8.18) gives

$$(H_{AA} - E)(H_{BB} - E) = (H_{AB} - ES_{AB})^2 = \text{a positive quantity} \tag{8.24}$$

Since H_{AA} and H_{BB} are different, one of them must be lower. Let $H_{AA} < H_{BB}$. Then if $E < H_{AA}$, both the factors $H_{AA} - E$ and $H_{BB} - E$ are positive and condition (8.24) holds; if $E > H_{BB}$, both the factors are negative and condition (8.24) again holds. But if $H_{AA} < E < H_{BB}$ then $H_{AA} - E = -\text{ve}$ while $H_{BB} - E = +\text{ve}$ and condition (8.24) does not hold. Thus, out of the two possible values of *E*, one must be lower than the lower of H_{AA} and H_{BB} and the other must be higher than the higher of the two. The former case gives a bonding MO and the latter, an antibonding MO. The stabilisation of the BMO is less than the destabilisation of the ABMO for the different denominators in the energy expressions as discussed earlier [Fig. 8.3(b)].

8.3.2 Conditions to be Satisfied by the Atomic Orbitals to Form MOs

We are now in a position to ascertain which atomic orbitals of A will combine linearly with which ones of B to produce MOs.

The energy E_A determines the size of the atomic orbital ϕ_A. A larger value of E_A implies (grossly) that the Coulombic attraction by the nucleus on the electron is less and the electron cloud spreads more widely about the atomic nucleus, making the ϕ_A orbital larger. Thus if

$E_A >> E_B$, the orbital ϕ_A is much larger than ϕ_B and the overlap integral S_{AB} is small, as will be the exchange integral H_{AB}. Under this condition, Eq. (8.18) gives

$$(H_{AA} - E)(H_{BB} - E) = (H_{AB} - ES_{AB})^2 \approx 0 \tag{8.25}$$

Equation (8.25) is satisfied if either of the two factors on the LHS is zero. When $(H_{AA} - E) = 0$, we have from Eq. (8.17a),

$$c_1 \cdot 0 + c_2(H_{AB} - ES_{AB}) = 0 \quad \text{or,} \quad c_2 = 0$$

That is, $\Psi_{MO} = c_1\phi_A \equiv \phi_A$ (on normalisation). Thus, in this case the MO is effectively the pure atomic orbital, ϕ_A. Similarly when $(H_{BB} - E) = 0$, the MO is effectively the atomic orbital ϕ_B. In neither case is there any linear combination of the two atomic orbitals. Hence the first condition to be satisfied for effective linear combination of the two atomic orbitals is the following:

The energies of ϕ_A and ϕ_B in their respective atoms must be comparable: $E_A \approx E_B$.

The second condition may be obtained as follows. If $S_{AB} = 0$, H_{AB} is also zero.

$$H_{AB} = \int \phi_A^* \hat{H} \phi_B d\tau = \int \phi_A^* \left[\hat{H}_B - \frac{1}{r_A}\right] \phi_B d\tau$$

$$= \int \phi_A^* \hat{H}_B \phi_B d\tau - \int \phi_A^* \frac{1}{r_A} \phi_B d\tau$$

$$= \int \phi_A^* E_B \phi_B \, d\tau - \int \frac{1}{r_A} (\phi_A^* \phi_B) \, d\tau$$

$$= E_B \int \phi_A^* \phi_B \, d\tau - \left[\frac{1}{r_A} \int (\phi_A^* \phi_B) d\tau - \int \left(-\frac{1}{r_A^2} \int (\phi_A^* \phi_B) d\tau \right) d\tau \right] \text{(by partial integration)}$$

$$= E_B \, S_{AB} - \left[\frac{1}{r_A} S_{AB} + \int \frac{1}{r_A^2} S_{AB} \, d\tau \right]$$

$$= 0 \text{ (for } S_{AB} = 0)$$

Under such a condition, it has been shown that E is equal either to E_A or to E_B; Ψ_{MO} is then either ϕ_A or ϕ_B and no effective combination results. The second condition is, therefore, stated as follows:

There must be non-zero overlap of the orbitals to be combined linearly.

[This condition corresponds to the maximum overlap criterion of VB theory.]

The third condition is found by considering the symmetry of the orbitals ϕ_A and ϕ_B with respect to the internuclear axis AB. In brief, their symmetry must ensure that $S_{AB} \neq 0$. If the symmetry is such that the overlap region consists of parts of '+' and '–' signs, making the sum zero, then S_{AB} becomes zero and the combination of ϕ_A and ϕ_B is not allowed. For example, consider the overlap of an *s*-orbital of A and p_x orbital of B (Fig. 8.4).

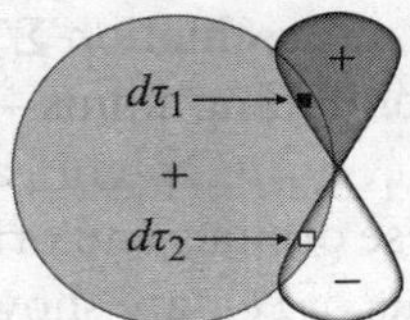

Fig. 8.4 s–p_x overlap of two atomic orbitals, $S_{AB} = 0$

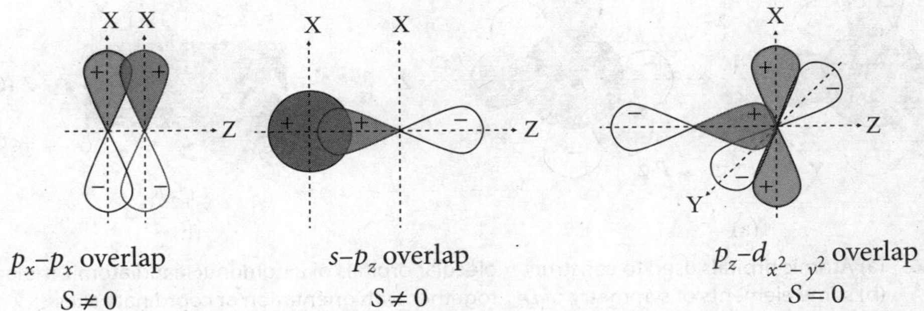

Fig. 8.5 Some combinations of s, p and d orbitals with zero or non-zero overlap

Corresponding to every volume element $d\tau_1$ where ϕ_A and ϕ_B overlap with the same sign, there is a volume element $d\tau_2$ where ϕ_A and ϕ_B overlap with the opposite sign. This makes the overlap integral S_{AB} zero, which makes the s–p_x combination not allowed for making molecular orbitals. Some more combinations with 'allowed' and 'not allowed' nature are shown in Fig. 8.5. We thus arrive at the third condition:

> *The atomic orbitals to be combined must be of the same symmetry type with respect to the internuclear axis.*

Finally, the most rigorous condition for molecular orbital formation is that *any MO wave function must form a basis of some irreducible representation of the molecular point group* (see Section 7.3).

8.4 HOMONUCLEAR DIATOMIC MOLECULES

The LCAO–MO method may now be extended to homonuclear diatomic molecules. Let us consider the diatomic molecules of the second row elements of the periodic table. According to the first condition ($E_A \approx E_B$) of Section 8.3.2, the 1s orbital of atom A should not combine with the 2s orbital of atom B owing to high energy difference. In other words, the orbital of the nth quantum shell of atom A should combine with orbitals of the same (that is, the nth) shell of atom B; it is to be remembered here that A and B refer to the same element for a homonuclear diatomic. The case of combining 1s(A) with 1s(B) has already been considered in Section 8.3.1; put $\phi_A = 1s(A)$, $\phi_B = 1s(B)$ and the result is the formation of one bonding and one antibonding MO with wave functions $1s(A) + 1s(B)$ and $1s(A) - 1s(B)$ respectively (apart from the normalising factors). The point group of the molecule is $D_{\infty h}$ and taking the line joining the two nuclei as the Z-axis, it is easy to verify that the bonding and antibonding wave functions are bases of the irreducible representation Σ_g^+ and Σ_u^+ respectively. We are now left with the valence shell orbitals of the combining atoms—2s, 2p_x, 2p_y and 2p_z of each atom; they are denoted for convenience as $s_1, s_2, p_{x1}, p_{x2}, p_{y1}, p_{y2}, p_{z1}$ and p_{z2} (that is, the atoms are labelled 1 and 2 instead of A and B). Some of these orbitals and orientation of coordinate axes are shown in Fig. 8.6(a). Some symmetry elements of $D_{\infty h}$ are shown in Fig. 8.6(b).

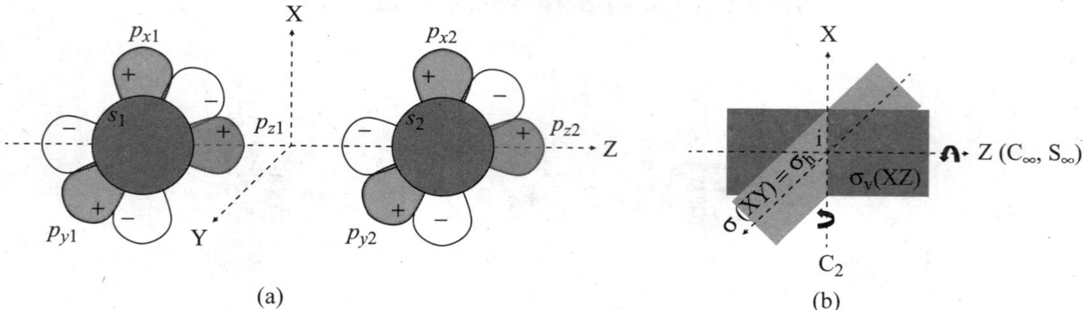

Fig. 8.6 (a) Atomic orbitals used to construct molecular orbitals of a homonuclear diatomic molecule, (b) some elements of symmetry of $D_{\infty h}$ together with orientation of coordinate axes

As the point group is of infinite order, application of the usual procedure for construction of SALCs by projection operators is not easy. We adopt here a somewhat direct and simple procedure for constructing the LCAO–MOs based on symmetry arguments. The action of the different symmetry operators on the orbitals $s_1, s_2, p_{x1}, p_{x2}, p_{y1}, p_{y2}, p_{z1}$ and p_{z2} are as follows:

- $\hat{E}$ keeps all the eight orbitals unchanged. Hence the character of $\hat{E}$ for a representation with all the above orbitals in the basis set is 8 $[\chi(\hat{E}) = 8]$.
- The operator $\hat{C}_\infty^\phi$ keeps the s_1, s_2, p_{z1} and p_{z2} orbitals unchanged; they contribute 4 to $\chi(\hat{C}_\infty^\phi)$. Again, on each atom, x changes to $x \cos\phi - y \sin\phi$ and y to $x \sin\phi + y \cos\phi$; hence the p_{x1}, p_{x2}, p_{y1} and p_{y2} orbitals contribute $2 \times 2 \cos\phi$ to the character of $\hat{C}_\infty^\phi$. Thus, total $\chi(\hat{C}_\infty^\phi) = 4 + 4\cos\phi$.
- Out of the infinite number of σ_v-planes, it is convenient to take the σ_v(XZ) plane because all the $\hat{\sigma}_v$ operators are of the same conjugate class. The $\hat{\sigma}_v$(XZ) operator keeps s_1, s_2, p_{z1} and p_{z2} unchanged, contributing 4 to the character $\chi(\hat{\sigma}_v)$. The orbitals p_{x1} and p_{x2} remain unchanged by $\hat{\sigma}_v$(XZ) and p_{y1} and p_{y2} are changed to their respective negatives. Hence $\chi(\hat{\sigma}_v) = 4 + 2 - 2 = 4$.
- The other operators $\hat{i}$, $\hat{S}_\infty^\phi$ and $\hat{C}_2$ shift the atoms from their positions. So $\chi(\hat{R}) = 0$ for $\hat{R} = \hat{i}$, $\hat{S}_\infty^\phi$ and $\hat{C}_2$. Hence the characters of a representation Γ_{AO} of $D_{\infty h}$, with the set of atomic orbitals ($s_1, s_2, \ldots, p_{z2}$) as the basis set, are as follows:

χ **(character)**

	$\hat{E}$	$2\hat{C}_\infty^\phi$	…	$\infty\hat{\sigma}_v$	$\hat{i}$	$\hat{S}_\infty^\phi$	…	$\infty\hat{C}_2$
Γ_{AO}	8	$4 + 4\cos\phi$	…	4	0	0	…	0

By inspection of the character table of $D_{\infty h}$ it can be readily verified that

$$\Gamma_{AO} = 2\Sigma_g^+ + 2\Sigma_u^+ + \Pi_g + \Pi_u \tag{8.26}$$

Thus,

χ **(character)**

	$\hat{E}$	$2\hat{C}_\infty^\phi$	...	$\infty\hat{\sigma}_v$	$\hat{i}$	$\hat{S}_\infty^\phi$	...	$\infty\hat{C}_2$
$2\Sigma_g^+$	2×1	2×1	...	2×1	2×1	2×1	...	2×1
$2\Sigma_u^+$	2×1	2×1	...	2×1	$2\times(-1)$	$2\times(-1)$	...	$2\times(-1)$
Π_g	2	$2\cos\phi$	...	0	2	$-2\cos\phi$	...	0
Π_u	2	$2\cos\phi$	...	0	−2	$2\cos\phi$	...	0
Γ_{AO}	8	$4+4\cos\phi$	...	4	0	0	...	0

The required MO wave functions must form bases of the irreducible representations of the point group. Hence the reduction (8.26) of Γ_{AO} means that out of the eight MOs obtained from the eight AOs, two MOs of different energy will be of σ_g^+ type, another two of different energy will be of σ_u^+ type, two degenerate MOs of equal energy will be of the π_g type and two other degenerate MOs will be of the π_u type. Henceforth, symmetry types of MOs (or orbital energy levels) will be denoted by lower case Greek letters (σ, π, δ ...) and the overall electronic states of the molecule by upper case Greek letters (Σ, Π, Δ ...). The linear combinations of atomic orbitals adapted to the symmetries of the component irreducible representations will be suitable for the final construction of the molecular orbitals.

The atomic orbitals s_1 and s_2 (the 2s orbitals of atoms 1 and 2) are of equal energy. Also, when the atoms are close enough, they overlap with non-zero value of the overlap integral S_{12}. Hence these two orbitals may be combined linearly to yield bonding and antibonding MOs:

$$\phi_1 = N_1(s_1 + s_2) \quad \text{and} \quad \phi_2 = N_2(s_1 - s_2) \tag{8.27}$$

They are of different energy (as we have seen earlier) and hence they should be orthogonal (since eigenfunctions of $\hat{H}$ with different eigenvalues must be orthogonal). These can be readily verified:

The individual functions s_1 and s_2, being centred on two different atoms, are not orthogonal, but their linear combinations given by Eqs (8.27) are necessarily orthogonal as shown below:

$$\int \phi_1^*\phi_2 d\tau = N_1^*N_2\left[\int s_1^*s_1 d\tau - \int s_1^*s_2 d\tau + \int s_2^*s_1 d\tau - \int s_2^*s_2 d\tau\right] = 0 \tag{8.28}$$

The factors N_1 and N_2 used for normalisation are not of much use at the present moment. By applying the symmetry operators of $D_{\infty h}$ we find

$$\hat{E}\phi_1 = \phi_1,\ \hat{C}_\infty^\phi\phi_1 = \phi_1,\ \hat{\sigma}_v(XZ)\phi_1 = \phi_1,\ \hat{i}\phi_1 = \phi_1 \text{ and so on}$$

Hence with ϕ_1 as basis, the character of each operator is 1. Thus ϕ_1 is a basis of the totally symmetric irreducible representation Σ_g^+. Similarly,

$$\hat{E}\phi_2 = \phi_2,\ \hat{C}_\infty^\phi\phi_2 = \phi_2,\ \hat{\sigma}_v(XZ)\phi_2 = \phi_2,\ \hat{i}\phi_2 = N_2(s_2 - s_1) = -\phi_2,\ \hat{S}_\infty^\phi\phi_2 = N_2(s_2 - s_1) = -\phi_2 \ ...$$

Hence ϕ_2 acts as a basis of the irreducible representation Σ_u^+ having the set of characters,

$$\chi(\hat{E}) = 1, \chi(\hat{C}_\infty) = 1, \chi(\hat{\sigma}_v) = 1, \chi(\hat{i}) = -1, \chi(\hat{S}_\infty^\phi) = -1 \ldots \chi(\hat{C}_2') = -1$$

The orbitals p_{z1} and p_{z2} behave in the same way as s_1 and s_2. It can be verified that their linear combinations,

$$\phi_3 = N_3(p_{z1} + p_{z2}) \quad \text{and} \quad \phi_4 = N_4(p_{z1} - p_{z2}) \tag{8.29}$$

are LCAO–MO wave functions which act as bases of Σ_u^+ and Σ_g^+ respectively. The p_{x1} and p_{x2} orbitals on the two atoms are drawn parallel to the X-axis. Similarly the p_{y1} and p_{y2} are parallel to the Y-axis. We now examine the effect on the linear combinations,

$$\phi_5 = N_5(p_{x1} + p_{x2}) \quad \text{and} \quad \phi_6 = N_6(p_{y1} + p_{y2}) \tag{8.30}$$

$$\hat{E}\phi_5 = \phi_5$$

$$\begin{aligned}
\hat{C}_\infty^\phi \phi_5 &= N_5[\hat{C}_\infty^\phi p_{x1} + \hat{C}_\infty^\phi p_{x2}] \\
&= N_5[p_{x1} \cos\phi + p_{y1} \sin\phi + p_{x2} \cos\phi + p_{y2} \sin\phi] \\
&= N_5[(p_{x1} + p_{x2})\cos\phi + (p_{y1} + p_{y2})\sin\phi] \\
&= \phi_5 \cos\phi + \phi_6 \sin\phi
\end{aligned}$$

Thus, ϕ_5 and ϕ_6 do not independently form a basis of any irreducible representation of $D_{\infty h}$; they get mixed up on application of $\hat{C}_\infty^\phi$.

$$\begin{aligned}
\hat{C}_\infty^\phi \phi_6 &= N_6[\hat{C}_\infty^\phi p_{y1} + \hat{C}_\infty^\phi p_{y2}] \\
&= N_6[-p_{x1} \sin\phi + p_{y1} \cos\phi - p_{x2} \sin\phi + p_{y2} \cos\phi] \\
&= N_6[-(p_{x1} + p_{x2})\sin\phi + (p_{y1} + p_{y2})\cos\phi] \\
&= -\phi_5 \sin\phi + \phi_6 \cos\phi
\end{aligned}$$

In matrix form, these results are

$$\hat{C}_\infty^\phi [\phi_5, \phi_6] = [\phi_5 \quad \phi_6] \begin{bmatrix} \cos\phi & -\sin\phi \\ \sin\phi & \cos\phi \end{bmatrix}, \text{Trace} = 2\cos\phi$$

$$\hat{E} [\phi_5, \phi_6] = [\phi_5 \quad \phi_6] \begin{bmatrix} 1 & 0 \\ 0 & 1 \end{bmatrix}, \text{Trace} = 2$$

Considering all the symmetry operators of $D_{\infty h}$, it can be shown that the pair of functions (ϕ_5, ϕ_6) acts as a basis set of a 2-dimensional irreducible representation of $D_{\infty h}$, with the character set given below:

χ **(character)**

$\hat{E}$	$2\hat{C}_\infty^\phi$	$\cdots$	$\infty\hat{\sigma}_v$	$\hat{i}$	$\hat{S}_\infty^\phi$	$\cdots$	$\infty\hat{C}_2$
2	$2\cos\phi$	$\cdots$	0	-2	$2\cos\phi$	$\cdots$	0

These are the characters of Π_u. The u subscript results from the fact that $\hat{\imath}\,p_{x1} = -p_{x2}$ and $\hat{\imath}\,p_{y1} = -p_{y2}$, which make $\hat{\imath}\,\phi_5 = -\phi_5$ and $\hat{\imath}\,\phi_6 = -\phi_6$ and yields a negative trace −2, for the matrix representation of $\hat{\imath}$. In the same way, it can be shown that $\phi_7 = (p_{x1} - p_{x2})$ and $\phi_8 = (p_{y1} - p_{y2})$ act together as a basis set of the 2-dimensional irreducible representation, Π_g of $D_{\infty h}$. [$\hat{\imath}\,p_{x1} = -p_{x2}$, $\hat{\imath}\,p_{x2} = -p_{x1}$ so that $\hat{\imath}(p_{x1} - p_{x2}) = (p_{x1} - p_{x2})$, …; this results in the g subscript.]

Using only symmetry arguments and without carrying out any quantitative calculations we thus arrive at the LCAO–MOs of the homonuclear diatomic molecules as listed in Table 8.1.

Table 8.1 Expressions for molecular orbitals of homonuclear diatomics in terms of constituent atomic orbitals

Atomic orbital	LCAO–MO (without normalisation factor)	Symmetry label of MO
$1s_1, 1s_2$	$1s_1 + 1s_2$	$\sigma_g^+(1s)$
	$1s_1 - 1s_2$	$\sigma_u^+(1s)$
$2s_1, 2s_2$	$2s_1 + 2s_2$	$\sigma_g^+(2s)$
	$2s_1 - 2s_2$	$\sigma_u^+(2s)$
$2p_{z1}, 2p_{z2}$	$2p_{z1} + 2p_{z2}$	$\sigma_g^+(2p)$
	$2p_{z1} - 2p_{z2}$	$\sigma_u^+(2p)$
$2p_{x1}, 2p_{x2}$	$2p_{x1} + 2p_{x2}$, $2p_{y1} + 2p_{y2}$	$\pi_u(2p)$
$2p_{y1}, 2p_{y2}$	$2p_{x1} - 2p_{x2}$, $2p_{y1} - 2p_{y2}$	$\pi_g(2p)$

8.4.1 Energy Order of the MOs of Homonuclear Diatomic Molecules: 'United Atom'–'Separated Atom' Viewpoints

Two extreme possibilities can be considered to obtain the order of energy of the MOs of a homonuclear diatomic molecule. In one case, the two atoms are separated by an infinite distance and the energy levels of 1s, 2s, 2p, … orbitals are shown at the extreme right of a diagram. As the atoms approach one another, MOs are formed by overlap of atomic orbitals of suitable (comparable) energy and proper symmetry. When the atoms are brought to zero separation, the *united atom limit* is achieved and the result is a single atom with nuclear charge $2Z_A$ (where Z_A is the nuclear charge of atom A in the diatomic molecule A_2); the resulting orbitals of the united atom can be qualitatively obtained by a simple drawing (Fig. 8.7). The group theoretical notations correspond to irreducible representations of the $D_{\infty h}$ point group and some of them have been explained earlier. As a short-cut guide it may be noted that (i) throughout the journey from separated to united atom state, the straight line segment joining the nuclei is along the Z-axis and its centre acts as a centre of inversion; (ii) the g or u subscripts are used by noting the signs of the orbital lobes on inversion through this centre, the symbol σ or π is used depending on whether the lobes are symmetric or antisymmetric with respect to a 180° rotation about the internuclear axis (or a reflection on the YZ plane), and finally, (iii) the symbols (σ_g, π_u …) do not depend on the internuclear distance (R). The transformations can occur both ways—the united atoms may be brought to infinite separation ($R \rightarrow \infty$) or the separated atoms may be brought to $R = 0$. Hence, in Fig. 8.7 doubled-headed arrows have been used. Moreover, an asterisk is used for the opposite phase type of linear combinations, that is, when a '+' lobe combines with '–' lobe.

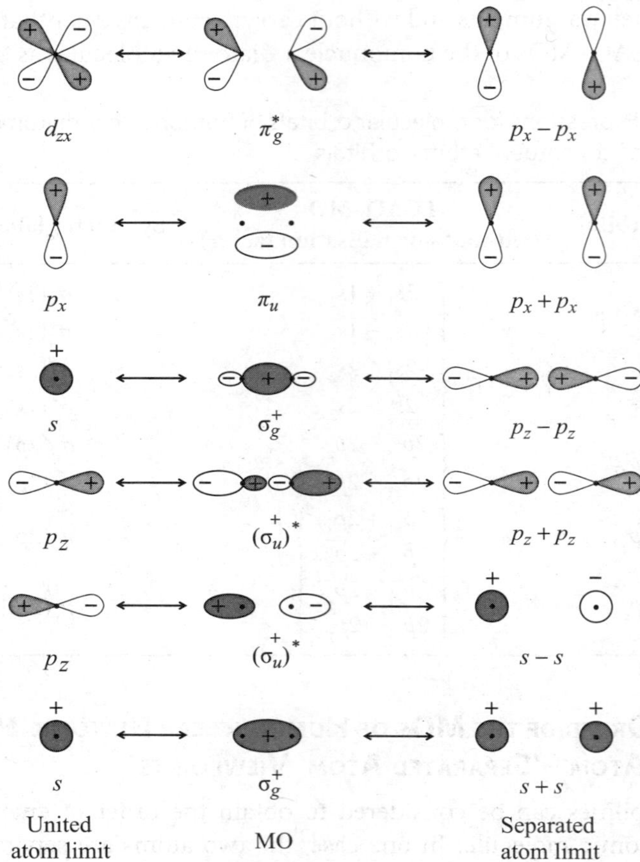

Fig. 8.7 Schematic representation of changes of orbitals from separated atom to the united atom limit, forming molecular orbitals somewhere midway. Conventionally, an asterisk is used for an antibonding MO.

A *correlation diagram* can now be drawn. It is a single diagram that not only indicates which atomic orbitals of the two atoms of a homonuclear diatomic molecule are to be combined to form the MOs but also gives, in a qualitative but correct way, the energy-order of the MOs. In the correlation diagram (Fig. 8.8) the energy-levels at the right have been drawn by thick lines to indicate that they are degenerate. Those on the left are drawn somewhat lower than the corresponding level on the right because the united atom has a higher nuclear charge which causes the electronic energy to be lower. For example, the 1*s* of the united atom is drawn at a level lower than the thick 1*s* line of the separated atoms. A slight shortening of the internuclear distance (R) results in splitting of the degenerate levels on the right; similarly, a slight increase in R (from zero) results in splitting of the degenerate levels on the left, and in both cases the splitting occurs according to the symmetry of the $D_{\infty h}$ point group. Every thick line for one *ns* level on the right is two-fold degenerate as *ns* orbitals of the separated atoms are of equal energy. Hence, the approach of the atoms towards one another results in splitting of the 1*s* level into σ_g^+ and σ_u^+ type, the former being lower in energy. But an *ns* level on the left is non-degenerate as it corresponds to one (united) atom—it does not split. Every np_z orbital on the right is doubly degenerate due to two similar atoms (separated) and it splits into σ_g^+ and σ_u^+ under the $D_{\infty h}$ environment; each of the np_x and np_y levels on the right, for a similar reason splits into π_u and π_g levels. Two points are to be noted here: (i) π_u from p_x and that from p_y have the same energy because extent of lateral overlap of two p_x orbitals is the same as that of two p_y (in other words, p_x and p_y are interconvertible by rotation about the *Z*-axis); thus π_u is doubly degenerate, and so is π_g; (ii) π_u is of lower energy than π_g (just the opposite of σ_g and σ_u). The second point results from the fact that the $p_x + p_x$ lateral overlap (bonding interaction) changes sign by the operation of $\hat{\text{i}}$ while the $p_x - p_x$ lateral overlap (antibonding interaction) remains unchanged by $\hat{\text{i}}$. The case with p_y is similar (Fig. 8.7). The 3-fold degenerate *np* level of the united atom (left side of Fig. 8.8), on slight separation in the $D_{\infty h}$ environment, splits into one non-degenerate σ_u-type and a two-fold degenerate π_u-type level; σ_u (resulting from direct overlap) is of lower energy than π_u (which results from lateral overlap).

The next task is to correlate the two sides by joining the levels with lines such that (i) one level on the right is correlated to the nearest level of the same symmetry on the left and (ii) correlation lines joining states of the same symmetry do not cross. Condition (i) is required to avoid combination of atomic orbitals with high energy-difference and condition (ii) is called the *non-crossing rule*. The correlation diagram shown in Fig. 8.8 is applicable to homonuclear diatomic molecules of hydrogen, helium and the first row elements of the periodic table. The

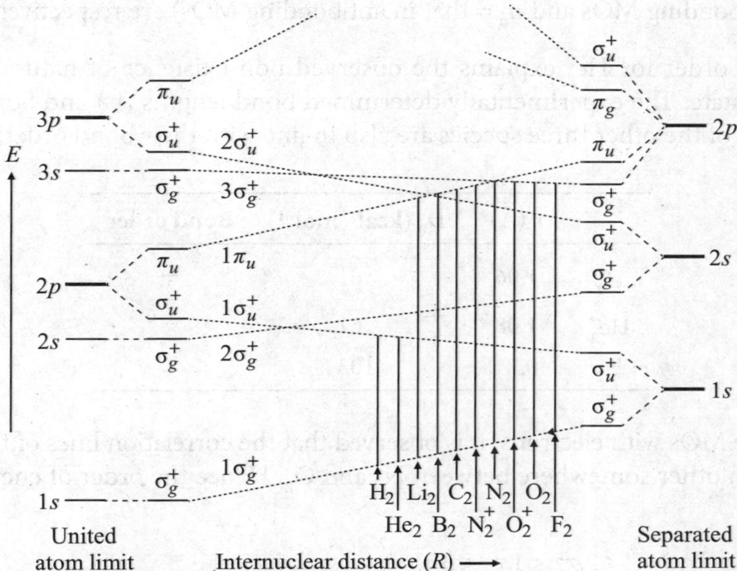

Fig. 8.8 Correlation diagram for homonuclear diatomic molecules. The positions of the diatomic molecules are indicated approximately by vertical lines. The diagram indicates only energy order and no quantitative measure of energy.

position of a diatomic molecule X_2 in this diagram is indicated by a vertical line; it depends on the equilibrium bond length of X_2, that is, on the atomic size of X.

In the increasing order of energy, the *g*-type MOs are labelled consecutively as $1\sigma_g^+$, $2\sigma_g^+$, $3\sigma_g^+$, ... The π-MOs are also similarly prefixed with successive integers 1, 2, 3, ... in the increasing order of energy. The MOs are filled with electrons according to the *Aufbau* principle. This means that the lower energy MOs are filled with electrons before those of higher energy, each MO can accommodate a maximum of two electrons of opposite spin and, when a degenerate set of MOs is encountered (like a π-MO), each individual member of the set is occupied singly before any electron pairing occurs (Hund's rule). It is now possible to use the correlation diagram to predict some important properties of homonuclear diatomic molecules. Thus, for H_2 and He_2 the available low energy MOs are $1\sigma_g^+$ and $1\sigma_u^+$, which are respectively bonding and antibonding. Electron occupancies of these two MOs for H_2^+, H_2, He_2^+ and He_2 are as follows:

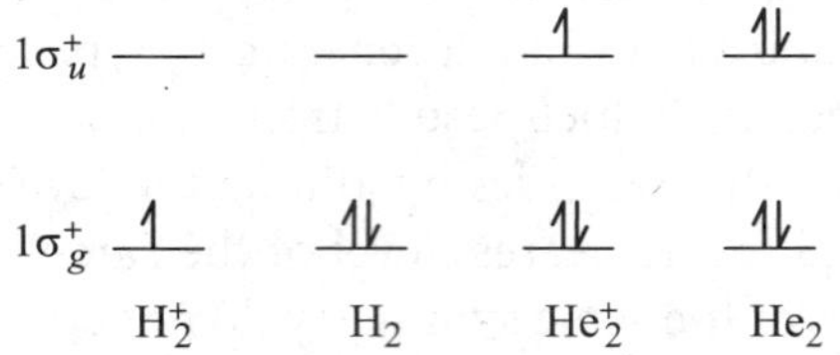

	H_2^+	H_2	He_2^+	He_2
$1\sigma_u^+$	—	—	↑	↑↓
$1\sigma_g^+$	↑	↑↓	↑↓	↑↓

H_2 has the highest excess of bonding electrons. Hence in this series of diatomics, H_2 should have the shortest bond length. Also, the bond orders (defined as $\frac{(n_b - n_a)}{2}$, where n_b = number of electrons in bonding MOs and n_a = that in antibonding MOs) are respectively $\frac{1}{2}$, 1, $\frac{1}{2}$ and 0. The zero bond order for He_2 explains the observed non-existence of natural diatomic He_2 in the ground state. The experimentally determined bond lengths (r_e) and bond dissociation enthalpies (D_0) of the other three species are also in line with their bond orders:

	r_e(Å)	D_0 (kcal · mol^{-1})	Bond order
H_2^+	1.06	61	$\frac{1}{2}$
He_2^+	1.08	61	$\frac{1}{2}$
H_2	0.74	103	1

In filling up the MOs with electrons, it is observed that the correlation lines of the $1\pi_u$ and $3\sigma_g^+$ MOs cross each other somewhere between N_2 and O_2. Hence the order of energy from Li_2 to N_2 is

$$1\sigma_g^+ < 1\sigma_u^+ < 2\sigma_g^+ < 2\sigma_u^+ < 1\pi_u < 3\sigma_g^+$$

and for O_2 to F_2, the energy order at the end part is inverted, that is, $3\sigma_g^+ < 1\pi_u$. These are shown in the two schemes given in Fig. 8.9.

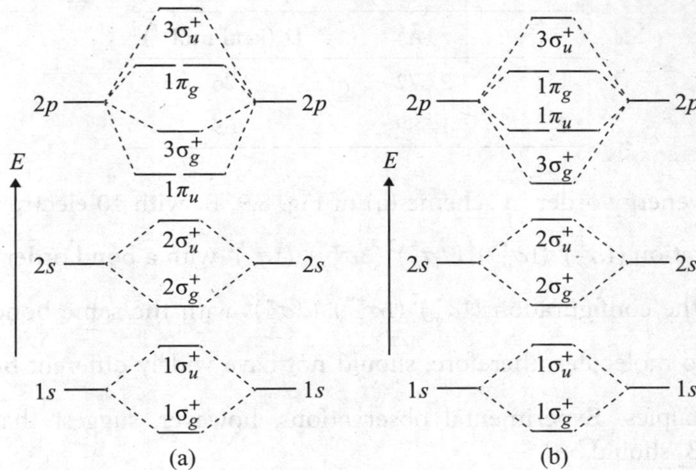

Fig. 8.9 Energy level ordering of the MOs of the first row diatomics: (a) applicable to Li_2–N_2 and (b) applicable to O_2, F_2

Electronic configurations of some homonuclear diatomics are given below as illustration of using the schemes in Fig. 8.9. Antibonding MOs are indicated by an asterisk.

Be_2 (8 electrons): $(1\sigma_g^+)^2\,(1\sigma_u^{+*})^2\,(2\sigma_g^+)^2\,(2\sigma_u^{+*})^2$; BO = 0

N_2 (14 electrons): $(1\sigma_g^+)^2\,(1\sigma_u^{+*})^2\,(2\sigma_g^+)^2\,(2\sigma_u^{+*})^2\,(1\pi_u)^4\,(3\sigma_g^+)^2$; BO = 3

O_2 (16 electrons): $(1\sigma_g^+)^2\,(1\sigma_u^{+*})^2\,(2\sigma_g^+)^2\,(2\sigma_u^{+*})^2\,(1\pi_u)^4\,(3\sigma_g^+)^2\,(1\pi_g^{\ *})^2$; BO = 2

The bond orders are in accordance with the classical triple and double bonds, N ≡ N and O = O. In the $(1\pi_g^*)$ MO of O_2, each individual MO of the doubly degenerate level is singly occupied with two electron spins parallel (Hund's rule of maximum multiplicity):

↿ ↿

$1\pi_g^*$

This explains the observed paramagnetism of O_2. The various oxygen species have the following electronic configurations in the $(1\pi_g^*)$ level:

$1\pi_g^*$ ↿ __ ↿ ↿ ⇅ ↿ ⇅ ⇅

O_2^+ O_2 O_2^- $O_2^=$

Thus the peroxide ion (O_2^{-2}) is diamagnetic while each of the other three oxygen species is paramagnetic.

The case of diatomic boron B_2, requires special consideration. Experiment shows that B_2 has shorter bond length (r_e) and higher bond dissociation enthalpy (D_e) than Li_2:

	r_e(Å)	D_0(kcal·mol^{-1})
Li_2	2.672	26
B_2	1.589	69

According to the energy order in scheme (a) of Fig. 8.9, B_2 with 10 electrons is expected to have the configuration $(1\sigma_g^+)^2\,(1\sigma_u^{+*})^2\,(2\sigma_g^+)^2\,(2\sigma_u^{+*})^2\,(1\pi_u)^2$ with a bond order of $\frac{6-4}{2}=1$ and Li_2 should have the configuration $(1\sigma_g^+)^2\,(1\sigma_u^{+*})^2\,(2\sigma_g^+)^2$ with the same bond order, namely, $\frac{4-2}{2}=1$. The two molecules, therefore, should not have widely different bond lengths and dissociation enthalpies. Experimental observations, however, suggest that the electronic configuration of B_2 should be

$$(1\sigma_g^+)^2\,(1\sigma_u^{+*})^2\,(2\sigma_g^+)^2\,(2\sigma_u^{+*})^1\,(1\pi_u)^2\,(3\sigma_g^+)^1$$

This is possible because the three energy levels $(2\sigma_u^{+*})$, $(1\pi_u)$ and $(3\sigma_g^+)$ of B_2 are so close that the energy required to promote an electron from $(2\sigma_u^{+*})$ to $(3\sigma_g^+)$ is much less than the spin pairing energy required to keep the electrons paired in the $(2\sigma_u^{+*})$ MO. The pairing energy [Fig. 8.10(a)] is essentially the Coulombic repulsion energy between two electrons placed in the same orbital. The alternative configuration (b) gives a bond order of 2 [instead of 1 as obtained from (a)] and this explains the observed shorter r_e and higher D_e of B_2 relative to Li_2; the excess number of bonding electrons in the core configuration, $(1\sigma_g^+)^2\,(1\sigma_u^{+*})^2\,(2\sigma_g^+)^2$ is denoted by [2]:

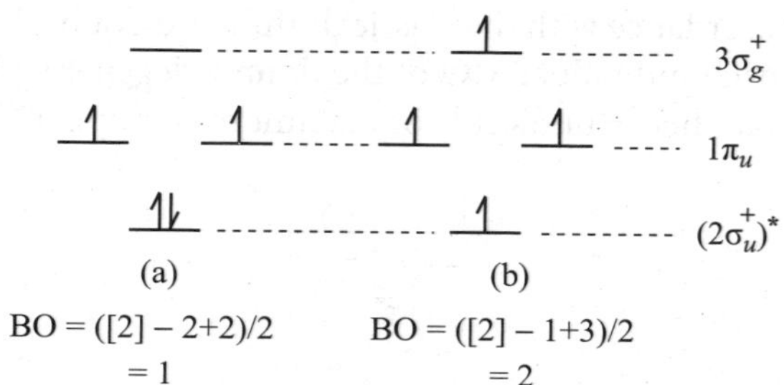

Fig. 8.10 Two alternative configurations for B_2 of which (b) corresponds to the experimental observation

8.5 LCAO–MO METHOD APPLIED TO HETERONUCLEAR DIATOMIC MOLECULES

For a heteronuclear diatomic molecule AB, the point group is $C_{\infty v}$ (instead of $D_{\infty h}$ for homonuclear diatomics). Hence, the u and g subscripts are not needed. In increasing order of energy, the σ-type MOs (formed by direct overlap of atomic orbitals) are designated consecutively as $1\sigma^+$, $2\sigma^+$, ... and the π-type MOs as 1π, 2π, ... (Fig. 8.11). The energy levels of the more electronegative atom are lower. Electronic configurations, bond orders and magnetic

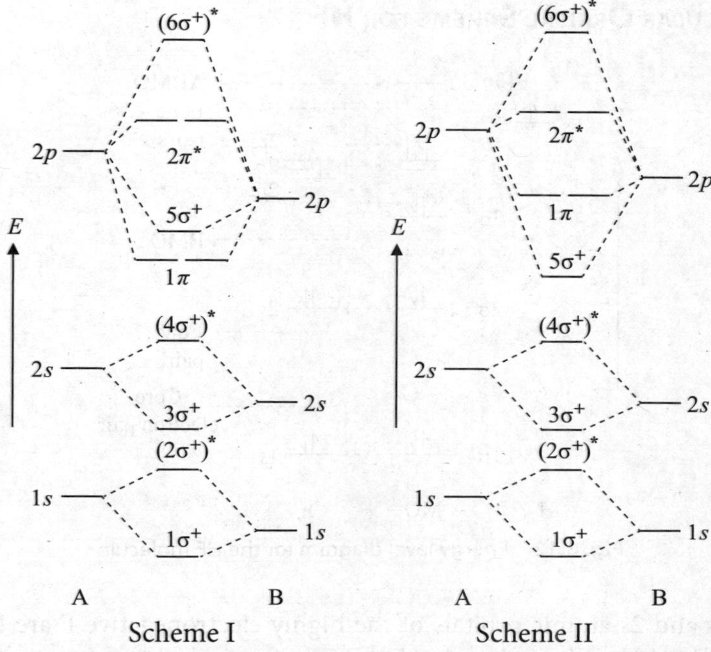

Fig. 8.11 MO energy levels for heteronuclear diatomic molecules; two alternative schemes are shown, as in the homonuclear diatomic case. Without actual computation it is not possible to say which scheme is operating; however, in most of the cases both yield the same bond order and the same number of unpaired electrons responsible for paramagnetism.

behaviour of some common heteronuclear diatomics are given below, which may be obtained from any of the schemes in Fig. 8.11. Here, we have followed Scheme 1.

Compound	Electronic configuration	Bond order	Magnetism
BeO (12e):	$(1\sigma^+)^2 (2\sigma^{+*})^2 (3\sigma^+)^2 (4\sigma^{+*})^2 (1\pi)^4$	2	dia-
BeF (13e):	$(1\sigma^+)^2 (2\sigma^{+*})^2 (3\sigma^+)^2 (4\sigma^{+*})^2 (1\pi)^4 (5\sigma^+)^1$	2.5	para-
BO (13e):	$(1\sigma^+)^2 (2\sigma^{+*})^2 (3\sigma^+)^2 (4\sigma^{+*})^2 (1\pi)^4 (5\sigma^+)^1$	2.5	para-
BF (14e):	$(1\sigma^+)^2 (2\sigma^{+*})^2 (3\sigma^+)^2 (4\sigma^{+*})^2 (1\pi)^4 (5\sigma^+)^2$	3	dia-
CN (13e):	$(1\sigma^+)^2 (2\sigma^{+*})^2 (3\sigma^+)^2 (4\sigma^{+*})^2 (1\pi)^4 (5\sigma^+)^1$	2.5	para-
CN^- (14e):	$(1\sigma^+)^2 (2\sigma^{+*})^2 (3\sigma^+)^2 (4\sigma^{+*})^2 (1\pi)^4 (5\sigma^+)^2$	3	dia-
CO (14e):	$(1\sigma^+)^2 (2\sigma^{+*})^2 (3\sigma^+)^2 (4\sigma^{+*})^2 (1\pi)^4 (5\sigma^+)^2$	3	dia-
NO^+ (14e):	$(1\sigma^+)^2 (2\sigma^{+*})^2 (3\sigma^+)^2 (4\sigma^{+*})^2 (1\pi)^4 (5\sigma^+)^2$	3	dia-
NO (15e):	$(1\sigma^+)^2 (2\sigma^{+*})^2 (3\sigma^+)^2 (4\sigma^{+*})^2 (1\pi)^4 (5\sigma^+)^2 (2\pi^*)^1$	2.5	para-

The easy ionisation of NO to NO^+ (nitrosyl ion) is because the outermost electron of NO occupies an antibonding orbital.

8.5.1 Molecular Orbital Scheme for HF

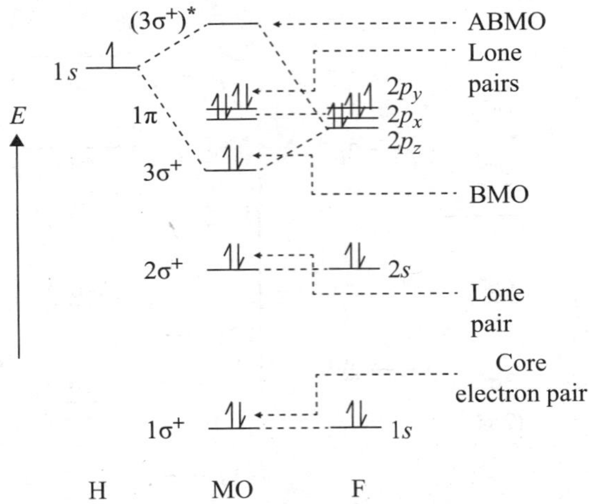

Fig. 8.12 Energy level diagram for the HF molecule

In energy, the 1*s* and 2*s* atomic orbitals of the highly electronegative F are far below the 1*s* orbital of H. The 1*s* of H and any 2*p* orbital of F are of comparable energy but only $2p_z$ of F is of the right symmetry to overlap with the 1*s* of H for linear combination. Hence $2p_x$ and $2p_y$ of F remain almost unaffected by formation of HF (they remain *non-bonding MOs* in HF). The scheme for the formation of MOs of HF is shown in Fig. 8.12.

8.6 POLYATOMIC AB$_n$ TYPE OF MOLECULES ($n = 2, 3, ...$)

If the B atoms are not bonded to one another but are each bonded to A (the central atom) and the shape of the molecule is known, then a group theoretical treatment for the construction of molecular orbitals is relatively easy. The necessary steps are the following.

(i) Using the set of valence shell atomic orbitals of the B atoms (called ligand atoms) construct a reducible representation (Γ) of the point group of the molecule and reduce Γ to find the component irreducible representations.
(ii) Combine the valence shell atomic orbitals of the B atoms to produce SALCs corresponding to the irreducible representations found in Step (i). Such SALCs are called *ligand group orbitals* (LGO).
(iii) Find the irreducible representations to which the atomic orbitals of the central atom belong.
(iv) Get the required MOs by combining the ligand group orbitals with central atom orbitals of the same symmetry and comparable energy.

A *correlation diagram* here means an energy level diagram where the LGOs and the central atom orbitals of proper symmetry are shown on two sides and are connected to the resulting MOs which they form; the MOs are shown in the middle in increasing order of energy. The procedure is best explained with the following examples.

8.6.1 MOLECULAR ORBITALS OF H_2O

The point group of the H_2O molecule (known to be V-shaped) is C_{2v}. The H-atoms are labelled A and B, and the molecule is imagined to be placed on the YZ plane. Orientation of the coordinate axes is shown in Fig. 8.13.

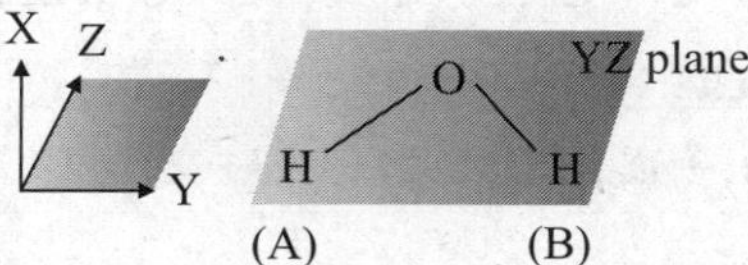

Fig. 8.13 The H_2O molecule on the YZ plane and orientation of the coordinate axes

The valence shell orbitals of the central atom O and the ligand atoms H_A and H_B are $1s\,(H_A)$, $1s\,(H_B)$, $2s\,(O)$, $2p_x\,(O)$, $2p_y\,(O)$ and $2p_z\,(O)$. These are denoted respectively as ϕ_1, ϕ_2, ϕ_3, ϕ_4, ϕ_5 and ϕ_6. The action of the C_{2v} symmetry operators on these orbitals will produce results as listed in Table 8.2.

Table 8.2 Result of operating the C_{2v} symmetry operators on $\phi_i\,(i = 1$ to $6)$

	$\hat{E}$	$\hat{C}_2(Z)$	$\hat{\sigma}_v(XZ)$	$\hat{\sigma}_v(YZ)$
ϕ_1	ϕ_1	ϕ_2	ϕ_2	ϕ_1
ϕ_2	ϕ_2	ϕ_1	ϕ_1	ϕ_2
ϕ_3	ϕ_3	ϕ_3	ϕ_3	ϕ_3
ϕ_4	ϕ_4	$-\phi_4$	ϕ_4	$-\phi_4$
ϕ_5	ϕ_5	$-\phi_5$	$-\phi_5$	ϕ_5
ϕ_6	ϕ_6	ϕ_6	ϕ_6	ϕ_6

Hence the characters of a representation Γ, with ligand orbitals ϕ_1 and ϕ_2 as bases, are as follows:

χ (character)

	$\hat{E}$	$\hat{C}_2(Z)$	$\hat{\sigma}_v(XZ)$	$\hat{\sigma}_v(YZ)$
Γ	2	0	0	2

Using the character table of C_{2v} it can be readily shown that $\Gamma = A_1 + B_2$. Applying projection operator on ϕ_1 we now find the required SALCs:

$$\hat{P}^{(A_1)}\,\phi_1 = \frac{1}{4}[1^* \cdot \hat{E} \cdot \phi_1 + 1^* \cdot \hat{C}_2(Z) \cdot \phi_1 + 1^* \cdot \hat{\sigma}_v(XZ) \cdot \phi_1 + 1^* \cdot \hat{\sigma}_v(YZ) \cdot \phi_1]$$

$$= \frac{1}{4}[\phi_1 + \phi_2 + \phi_2 + \phi_1]$$

$$\approx \frac{1}{\sqrt{2}}(\phi_1 + \phi_2) \text{ [normalising, by taking the overlap of } 1s(\text{H}_\text{A}) \text{ and } 1s(\text{H}_\text{B}) \text{ to be zero]}$$

$$= s_1 \text{ (say)}$$

$$\hat{\text{P}}^{(B_2)}\,\phi_1 = \frac{1}{4}[1^* \cdot \hat{\text{E}} \cdot \phi_1 + (-1)^* \cdot \hat{\text{C}}_2(\text{Z}) \cdot \phi_1 + (-1)^* \cdot \hat{\sigma}_v(\text{XZ}) \cdot \phi_1 + 1^* \cdot \hat{\sigma}_v(\text{YZ}) \cdot \phi_1]$$

$$= \frac{1}{4}[\phi_1 - \phi_2 - \phi_2 + \phi_1]$$

$$\approx \frac{1}{\sqrt{2}}(\phi_1 - \phi_2) \quad \text{(after normalisation)}$$

$$= s_2 \text{ (say)}$$

($\hat{\text{P}}^{(A_1)}$ and $\hat{\text{P}}^{(B_2)}$ on ϕ_2 produce the same SALCs.)

These two are the required ligand group orbitals. Table 8.2 also reveals that the central atom orbitals ϕ_3, ϕ_4, ϕ_5 and ϕ_6 transform respectively as A_1, B_1, B_2 and A_1 under the action of the C_{2v} symmetry operators.

The B_1-type atomic orbital (ϕ_4) of the central atom has no partner of the same symmetry in the LGO set and hence it remains a non-bonding orbital in the H_2O molecule.

The two B_2-type orbitals, one from the LGO set and the other from the central atom, combine linearly to produce bonding and antibonding MOs (Fig. 8.14). The combining coefficients are to be determined variationally by actual computation. It is clear from Fig. 8.14 that the in-phase combination,

$$c_1\phi_5 + c_2\frac{1}{\sqrt{2}}(\phi_1 - \phi_2) \text{ or simply, } 2p_y(\text{O}) + [1s(\text{H}_\text{A}) - 1s(\text{H}_\text{B})]$$

gives an antibonding MO with three nodes while the out-of-phase combination,

$$c_1\phi_5 - c_2\frac{1}{\sqrt{2}}(\phi_1 - \phi_2) \text{ or simply, } 2p_y(\text{O}) - [1s(\text{H}_\text{A}) - 1s(\text{H}_\text{B})]$$

gives a bonding MO with one node for the water molecule.

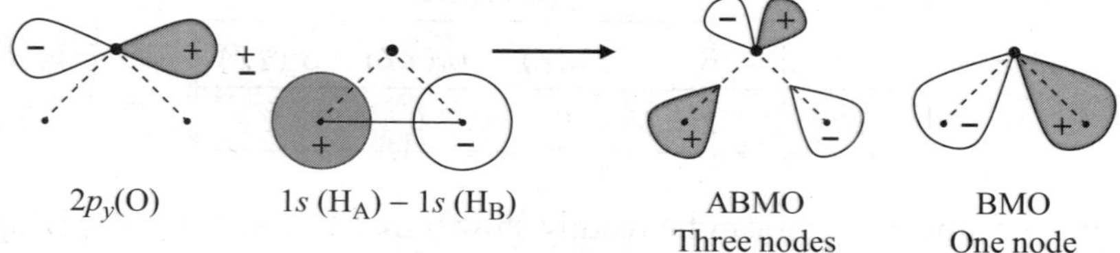

Fig. 8.14 Combination of the oxygen $2p_y$ orbital with the B_2-type LGO of the H atoms to form bonding and antibonding MOs (the figures on the RHS are easily obtained by properly superposing the figures of the LHS)

We are now left with two A_1 type central atom orbitals, $2s$(O) and $2p_z$(O), and one A_1-type LGO. The most general way of combining them is as follows:

$$\psi_i(A_1) = c_{1i} \cdot 2s(\text{O}) + c_{2i} \cdot 2p_z(\text{O}) + c_{3i} \cdot [1s(\text{H}_\text{A}) + 1s(\text{H}_\text{B})],\ i = 1, 2, 3$$

The coefficients can be determined only by actual computation but here we adopt a conceptually reasonable qualitative procedure.

$2s$(O) and the LGO, $1s(\text{H}_\text{A}) + 1s(\text{H}_\text{B})$, being of the same symmetry may combine to give bonding MOs, keeping the $2p_z$(O) orbital as a non-bonding MO. But overlap of $2s$(O) and A_1-type LGO does not depend on the HOH angle and this fact makes a linear H_2O quite feasible (Fig. 8.15).

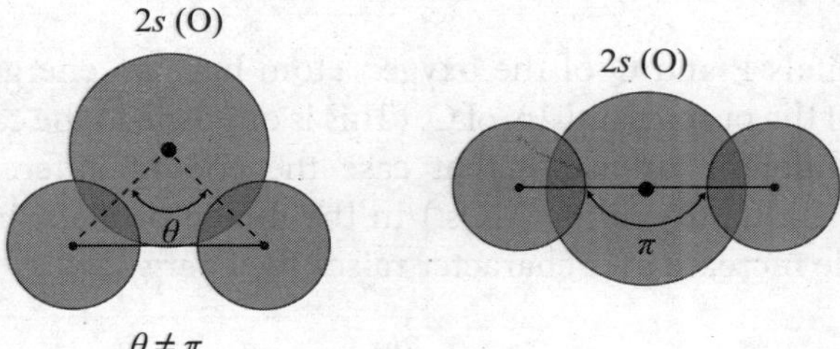

Fig. 8.15 Overlap of $2s(\text{O})$ and $1s(\text{H}_\text{A}) + 1s(\text{H}_\text{B})$ does not depend on θ

However, the experimentally determined bond angle (≈ 104°) points to the fact that the $2p_z$(O) orbital also takes part in bonding so that a bent molecule results. This is possible if the two A_1 orbitals of O, that is, $2s$ and $2p_z$ mix in both the in-phase and out-of-phase manner and the resulting orbitals combine with the remaining A_1 type LGO. This scheme, as depicted in Figs 8.16 and 8.17 shows that finally one bonding, one antibonding and one non-bonding MO result, and the NBMO is not a pure $2p_z$ but a mixture of $2s$ and $2p_z$ of O.

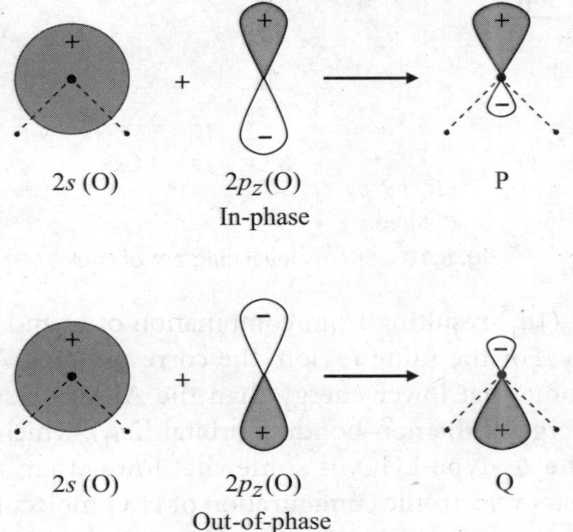

Fig. 8.16 Mixing of $2s$ and $2p_z$ orbitals of O in in-phase and out-of-phase manner

The in-phase combination (P) has electron density away from the sites for H in the O–H bonds (shown by dotted lines); hence it combines with the A_1 type LGO $[1s(\text{H}_\text{A}) + 1s(\text{H}_\text{B})]$ only very weakly. Thus P remains as a non-bonding MO. The out-of-phase combination (Q) combines appreciably with the A_1 type LGO to produce bonding and antibonding MOs (Fig. 8.17).

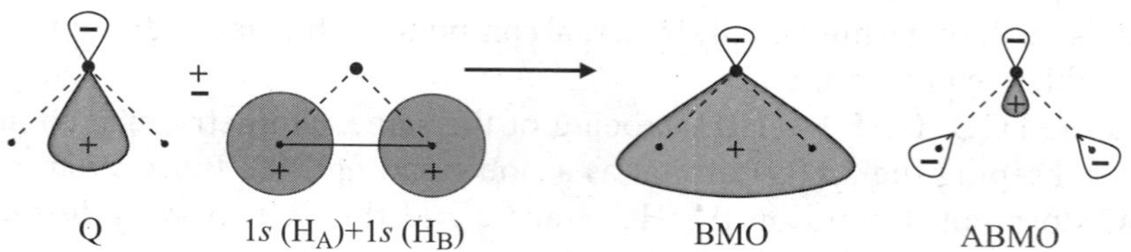

Fig. 8.17 Bonding and antibonding combinations of Q with the A_1-type LGO

The $2s$–$2p_z$ mixed orbitals P and Q of the oxygen atom have an energy that is intermediate between the energies of the pure $2s$ and $2p_z$ of O. (This is opposite to the case of same-symmetry orbitals situated on different atoms; in that case the energy difference increases during formation of bonding and antibonding MOs.) In the mixed orbitals, increase in s-character lowers the energy while increase in p-character raises the energy.

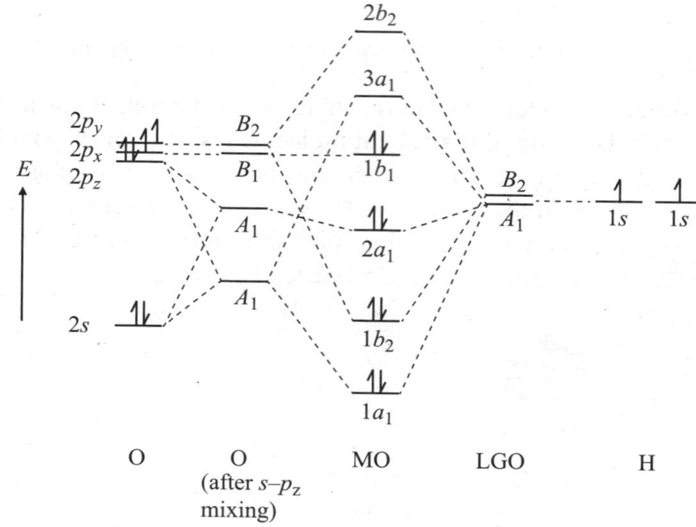

Fig. 8.18 Energy level diagram of H_2O

Hence the bonding MO ($1a_1$) resulting from combination of Q and $1s(H_A) + 1s(H_B)$ will be the lowermost in energy. For the same reason, the corresponding ABMO ($3a_1$), which has some s-character, has somewhat lower energy than the ABMO resulting from the B_2 type orbital. Similarly, the energy of the non-bonding orbital ($2a_1$), which results from very weak interaction of P with the A_1-type LGO, is somewhat lower than the pure $2p_z$(O) orbital (Fig. 8.18). The ground state electronic configuration of H_2O molecule is

$$(1a_1)^2\,(1b_2)^2\,(2a_1)^2\,(1b_1)^2$$

where the core electrons of oxygen have not been shown.

Photoelectron spectroscopy of water gives four peaks which reveal four orbital energies of the H_2O molecule. The peaks appear at 12.62, 13.78, 17.02 and 32.2 eV. These are assignable to ionisation from the $1b_1$, $2a_1$, $1b_2$ and $1a_1$ levels respectively. The highest ionisation energy corresponds to the $1a_1$ orbital which has dominant s character.

8.6.2 Molecular Orbitals of NH_3

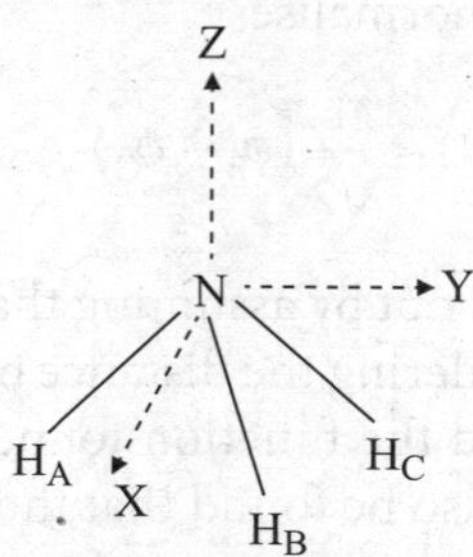

Fig. 8.19 Orientation of Cartesian axes for the NH_3 molecule

The pyramidal structure of the NH_3 molecule (of point group C_{3v}) is shown in Fig. 8.19 along with the orientation of the coordinate axes. The valence shell orbitals of NH_3 are the 1*s* orbitals of H-atoms (labelled by subscripts A, B and C) and the 2*s*, $2p_x$, $2p_y$ and $2p_z$ orbitals of the central atom, N. Let ϕ_A, ϕ_B, ϕ_C denote the 1*s* orbitals of H_A, H_B and H_C respectively. With these orbitals as basis, a representation Γ of the C_{3v} point group is as follows.

χ (character)

	$\hat{E}$	$\hat{C}_3$	$3\hat{\sigma}_v$
Γ	3	0	1

This results from the fact that all the 1s orbitals of H are shifted from their positions by the $\hat{C}_3$ and $\hat{C}_3^2$ operations, and one of them remains in its own position without change of sign by each $\hat{\sigma}_v$ operation.

The reduction formula easily shows that $\Gamma = A_1 + E$ (8.31)

The task is now to construct LGOs corresponding to the component irreducible representations, A_1 and E, of Γ.

Application of the projection operator method on ϕ_A, ϕ_B and ϕ_C yields the following results:

$$\hat{P}^{(A_1)}\phi_A = \frac{1}{\sqrt{3}}(\phi_A + \phi_B + \phi_C) \tag{8.32}$$

$$\left.\begin{aligned} \hat{P}^E\phi_A &= \frac{1}{\sqrt{6}}(2\phi_A - \phi_B - \phi_C) \\ \hat{P}^E\phi_B &= \frac{1}{\sqrt{6}}(2\phi_B - \phi_C - \phi_A) \\ \hat{P}^E\phi_C &= \frac{1}{\sqrt{6}}(2\phi_C - \phi_A - \phi_B) \end{aligned}\right\} \tag{8.33}$$

The functions in Eq. (8.33) are not linearly independent as their sum is zero. But two orthogonal functions are required as bases for the 2-dimensional irreducible representation,

E. For constructing two such functions, we retain the first function of Eq. (8.33), take the difference of the remaining two and normalise:

$$\hat{P}^E \phi_B - \hat{P}^E \phi_C = \frac{1}{\sqrt{6}}(3\phi_B - 3\phi_C) \approx \frac{1}{\sqrt{2}}(\phi_B - \phi_C) \qquad (8.34)$$

Here, normalisation has been carried out by assuming that the pair-wise overlap of ϕ_A, ϕ_B and ϕ_C is zero (which is reasonable, considering the distance between the H atoms). Orthogonality of the first function of Eq. (8.33) and the function found in Eq. (8.34) can be readily tested. From the C_{3v} character table, it can also be found that the central atom orbitals transform as

$$\text{N}(2s) : A_1,\ \text{N}(2p_z) : A_1,\ \text{N}(2p_x, 2p_y) : E$$

Matching the symmetry of the N orbitals and the LGOs, the molecular orbitals can be constructed. However, linear combination of 2*s* (N) and the A_1-type LGO (Fig. 8.20), may produce a planar NH_3 molecule because *s*–*s* overlap does not depend on the angle (θ) between the N–H bonds. Hence the $2p_z$(N) orbital, which is also of A_1 symmetry, must also participate in order to give a pyramidal shape to the NH_3 molecule. In other words, we need to consider the s–p_z mixing of N-orbitals before constructing the LCAO–MOs of the NH_3 molecule. (This is analogous to *s*–*p* mixing in H_2O.)

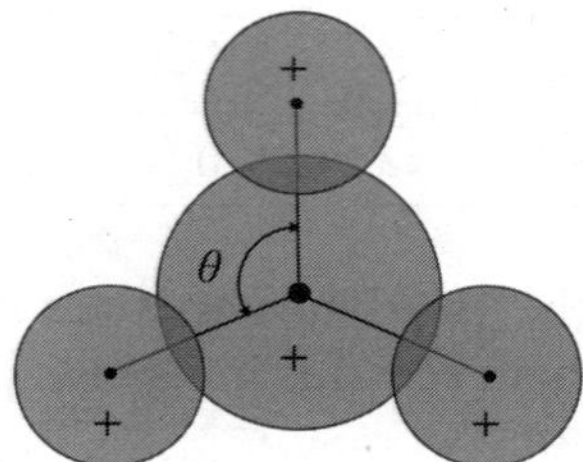

Fig. 8.20 Overlap of 2*s* (N) and the A_1-type LGO; extent of overlap does not depend on θ

By s–p_z mixing, two orbitals of A_1 symmetry are produced on N; one of them, $2s(\text{N}) + 2p_z(\text{N})$, has sufficient electron density towards the A_1-type hydrogen LGO while the other, $2s(\text{N}) - 2p_z(\text{N})$, has no such density for combining with the hydrogen LGOs; it therefore remains practically a non-bonding orbital centred on N in NH_3. The former combines with the A_1-type hydrogen LGO by direct overlap to form bonding and antibonding MOs ($1a_1$ and $3a_1^*$, Figs 8.21 and 8.22).

The *E*-type orbitals ($2p_x$, $2p_y$) of N combine with the *E*-type LGOs of H to form doubly degenerate π-bonding and antibonding MOs. The energy level diagram is shown in Fig. 8.22 where the distribution of eight valence electrons (five from N and three from H) is also shown. The electronic configuration of NH_3 is

$$(1a_1)^2\ (1e)^4\ (2a_1)^2$$

where two electrons have occupied the non-bonding ($2a_1$) orbital. This MO is essentially concentrated on N and hence the $(2a_1)^2$ electron pair is the lone pair of N in NH_3. With the remaining six electrons in the bonding MOs (and none in the antibonding MOs), the total bond order is $(6 - 0)/2 = 3$, distributed over three N–H linkages. Hence each N–H linkage is a single bond. This corresponds to the Lewis structure of NH_3.

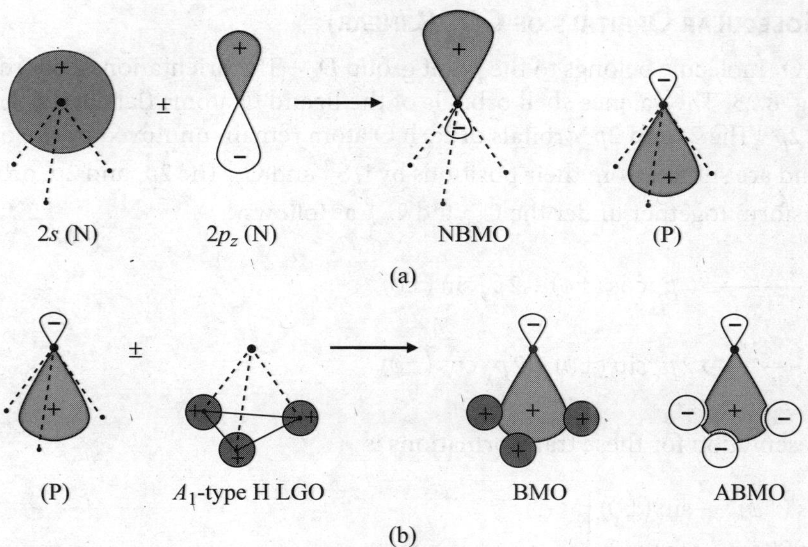

Fig. 8.21 (a) In-phase and out-of-phase sp_z mixing of N orbitals (b) combination of suitable sp_z mixed orbital of N with the hydrogen A_1-type LGO in in-phase and out-of-phase manner

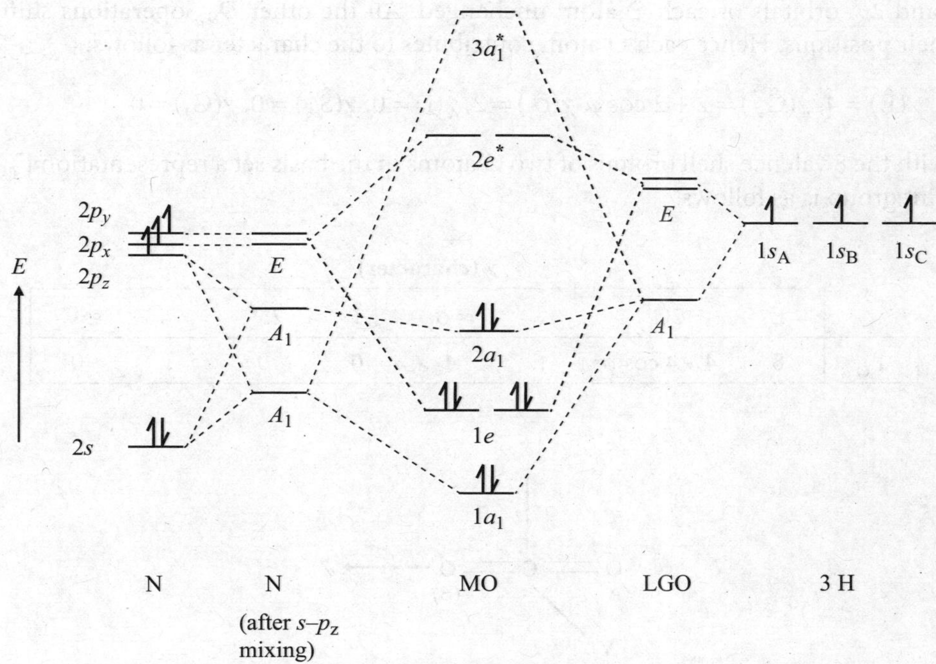

Fig. 8.22 Energy level diagram for NH_3

Photoelectron spectroscopy of ammonia gives three peaks at 10.2, 15.0 and 27.0 eV which correspond respectively to ionisation from the $2a_1$, le and $1a_1$ levels.

8.6.3 Molecular Orbitals of CO_2 (Linear)

The linear CO_2 molecule belongs to the point group $D_{\infty h}$. The orientation of coordinate axes is shown in Fig. 8.23. The valence shell orbitals of the ligand O atoms (labelled A and B) are 2*s*, $2p_x$, $2p_y$ and $2p_z$. The 2*s* and $2p_z$ orbitals of each O atom remain unmoved by the operations $\hat{E}$, $\hat{C}_\infty^\phi$ and $\hat{\sigma}_v$ and are shifted from their positions by $\hat{i}$, $\hat{S}_\infty^\phi$ and C_2. The $2p_x$ and $2p_y$ orbitals of each O atom transform together under the $\hat{C}_\infty^\phi$ and $\hat{C}_\infty^{-\phi}$ as follows:

$$2p_x \xrightarrow{\hat{C}_\infty^{\pm\phi}} 2p_x \cos(\pm\phi) + 2p_y \sin(\pm\phi)$$

$$2p_y \xrightarrow{\hat{C}_\infty^{\pm\phi}} 2p_x \sin(\pm\phi) + 2p_y \cos(\pm\phi)$$

Matrix representation for these transformations is

$$\begin{bmatrix} \cos(\pm\phi) & -\sin(\pm\phi) \\ \sin(\pm\phi) & \cos(\pm\phi) \end{bmatrix}$$

where $+\phi$ and $-\phi$ correspond respectively to anticlockwise and clockwise rotations about the Z-axis. Trace of the matrix is 2 cos ϕ in each rotation for every O atom. Any $\hat{\sigma}_v$ operation keeps the 2*s* and $2p_z$ orbitals of each O atom unchanged. All the other $D_{\infty h}$ operations shift them from their positions. Hence each O atom contributes to the character as follows:

$$\chi(\hat{E}) = 4, \chi(\hat{C}_\infty^\phi) = 2 + 2\cos\phi, \chi(\hat{\sigma}_v) = 2, \chi(\hat{i}) = 0, \chi(\hat{S}_\infty^\phi) = 0, \chi(C_2) = 0$$

Thus, with the 8 valence shell orbitals of two O atoms in the basis set a representation Γ_O of the $D_{\infty h}$ point group is as follows:

	χ (character)							
	$\hat{E}$	$2\hat{C}_\infty^\phi$	…	$\infty\hat{\sigma}_v$	$\hat{i}$	$2\hat{S}_\infty^\phi$	…	∞C_2
Γ_O	8	$4 + 4\cos\phi$	…	4	0	0	…	0

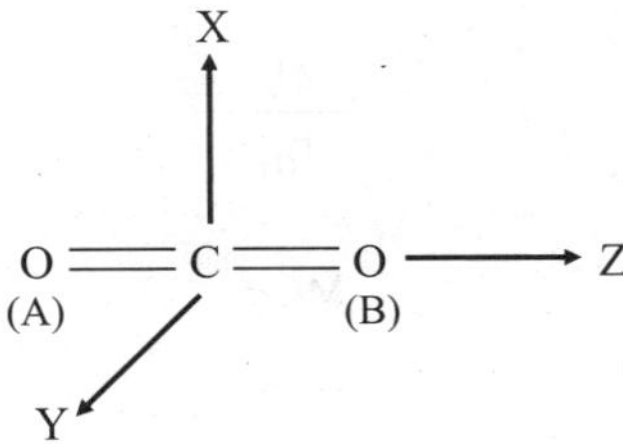

Fig. 8.23 Orientation of Cartesian axes for the CO_2 molecule

By inspection of the character table of $D_{\infty h}$ it can be readily found that

$$\Gamma_O = 2\Sigma_g^+ + 2\Sigma_u^+ + \Pi_u + \Pi_g \tag{8.35}$$

The character table also shows that the valence shell orbitals of the central carbon atom transform as follows:

$$2s : \Sigma_g^+, \quad (2p_x, 2p_y) : \Pi_u, \quad 2p_z : \Sigma_u^+$$

By making in-phase and out-of-phase combinations of the oxygen 2*s* orbitals, which are of the same symmetry, we can construct two LGOs:

$\psi_1 = 2s(\mathrm{O_A}) + 2s(\mathrm{O_B})$ (Symmetry label: Σ_g^+, because $\hat{\imath}$ does not change the sign of ψ_1)

$\psi_2 = 2s(\mathrm{O_A}) - 2s(\mathrm{O_B})$ (Symmetry label: Σ_u^+, because $\hat{\imath}$ changes the sign of ψ_2)

The normalisation factors have been omitted for clarity.

Similarly, the Π_g and Π_u-type LGOs are formed by combining the $2p_x$ and $2p_y$ orbitals of O separately (Fig. 8.24) as follows:

$$\left.\begin{aligned}\psi_3 &= 2p_x(\mathrm{O_A}) + 2p_x(\mathrm{O_B})\\ \psi_4 &= 2p_y(\mathrm{O_A}) + 2p_y(\mathrm{O_B})\end{aligned}\right\} : \Pi_u \qquad \left.\begin{aligned}\psi_5 &= 2p_x(\mathrm{O_A}) - 2p_x(\mathrm{O_B})\\ \psi_6 &= 2p_y(\mathrm{O_A}) - 2p_y(\mathrm{O_B})\end{aligned}\right\} : \Pi_g$$

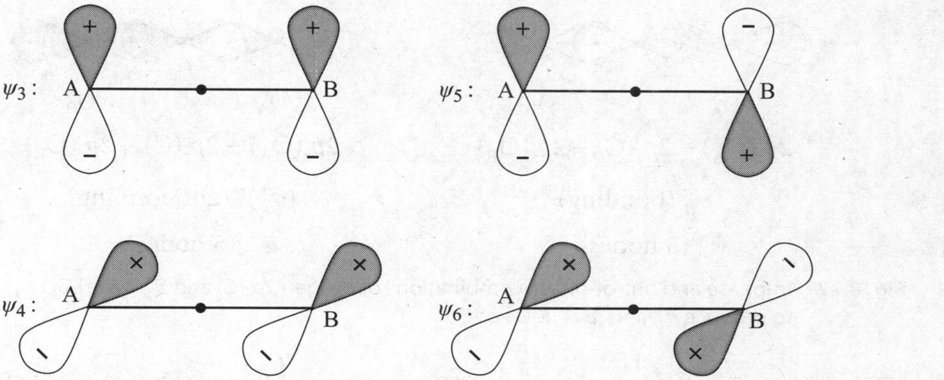

Fig. 8.24 In-phase and out-of-phase combinations of oxygen 2*p*-orbitals to form Π_g and Π_u-type LGOs

The $2p_z$ orbitals of $\mathrm{O_A}$ and $\mathrm{O_B}$ form Σ_g^+ and Σ_u^+ LGOs by in-phase and out-of-phase combinations (Fig. 8.25).

$$\psi_7 = 2p_z(\mathrm{O_A}) - 2p_z(\mathrm{O_B}) : \Sigma_g^+ \qquad \psi_8 = 2p_z(\mathrm{O_A}) + 2p_z(\mathrm{O_B}) : \Sigma_u^+$$

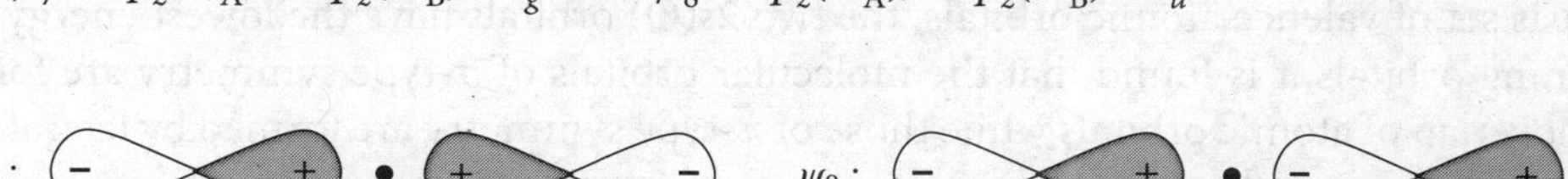

Fig. 8.25 Formation of Σ_g^+ and Σ_u^+ LGOs from the $2p_z$ orbitals of oxygen

The LCAO–MOs of CO_2 are now constructed by matching the symmetry types and energy of the carbon orbitals and LGOs. Oxygen being much more electronegative than carbon, the

$2s$(O) orbitals are the lowest in energy. Hence the LGOs ψ_1 and ψ_2 are not expected to combine with any of the carbon Σ_g^+ orbitals; these two orbitals therefore remain as non-bonding MOs in CO_2. The 2s (C) and the LGO $2p_z(O_A) - 2p_z(O_B)$ (that is, ψ_7), both of Σ_g^+ symmetry, combine to form bonding and antibonding MOs of σ_g^+ symmetry.

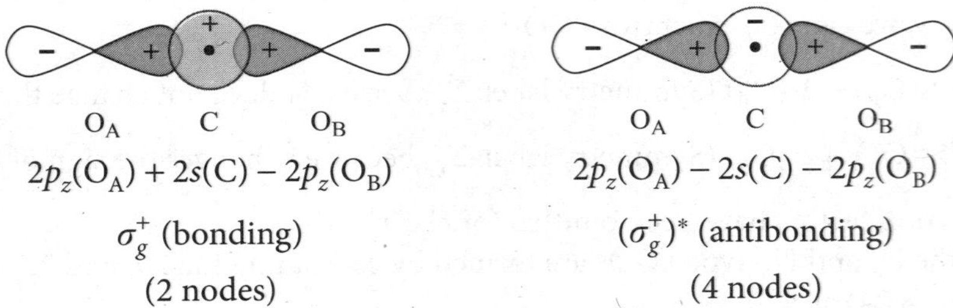

Fig. 8.26 In-phase and out-of-phase combinations of 2s(C) and Σ_g^+ type LGO, ψ_7 to produce σ_g^+ and $(\sigma_g^+)^*$ MOs of CO_2

Similarly the carbon $2p_z$ orbital of Σ_u^+ symmetry combines with the Σ_u^+ LGO (ψ_8) to give bonding σ_u^+ and antibonding $(\sigma_u^+)^*$ MOs (Fig. 8.27).

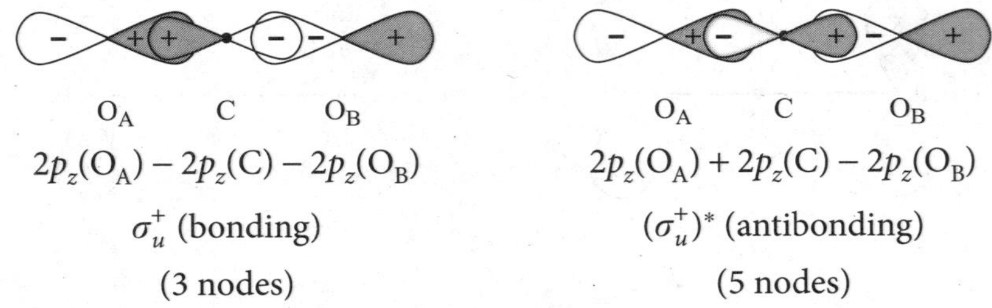

Fig. 8.27 In-phase and out-of-phase combinations of oxygen $2p_z$(C) and Σ_u^+-type LGO ψ_8, to produce σ_u^+ and $(\sigma_u^+)^*$ MOs of CO_2

The Π_u-type bonding and antibonding MOs are formed by combination of LGOs with the $2p_x$(C) and $2p_y$(C) orbitals as follows (Fig. 8.28):

The Π_g-type LGO has no partner in the set of carbon atom orbitals and hence it remains a pair of π-non-bonding MOs in the CO_2 molecule (Fig. 8.29).

The energy level diagram for the MOs of CO_2 is shown in Fig. 8.30.

The non-bonding MOs formed by the oxygen 2s orbitals are the lowest in energy because in the basis set of valence atomic orbitals, the two $2s$(O) orbitals have the lowest energy. For the remaining orbitals it is found that the molecular orbitals of σ-type symmetry are formed by direct overlap of atomic orbitals while those of π-type symmetry are formed by lateral overlap; the generally expected energy order is

$$\sigma(\text{BMO}) < \pi(\text{BMO}) < \text{NBMO} < \pi^*(\text{ABMO}) < \sigma^*(\text{ABMO})$$

Among two MOs of the same type, the one with less number of nodes has the lower energy. Thus, $2\sigma_g^+$ and $2\sigma_u^+$ are both of sigma type but the former has two nodes while the latter has three (Figs 8.26 and 8.27). Hence $2\sigma_g^+$ is of lower energy than $2\sigma_u^+$. All these are depicted in

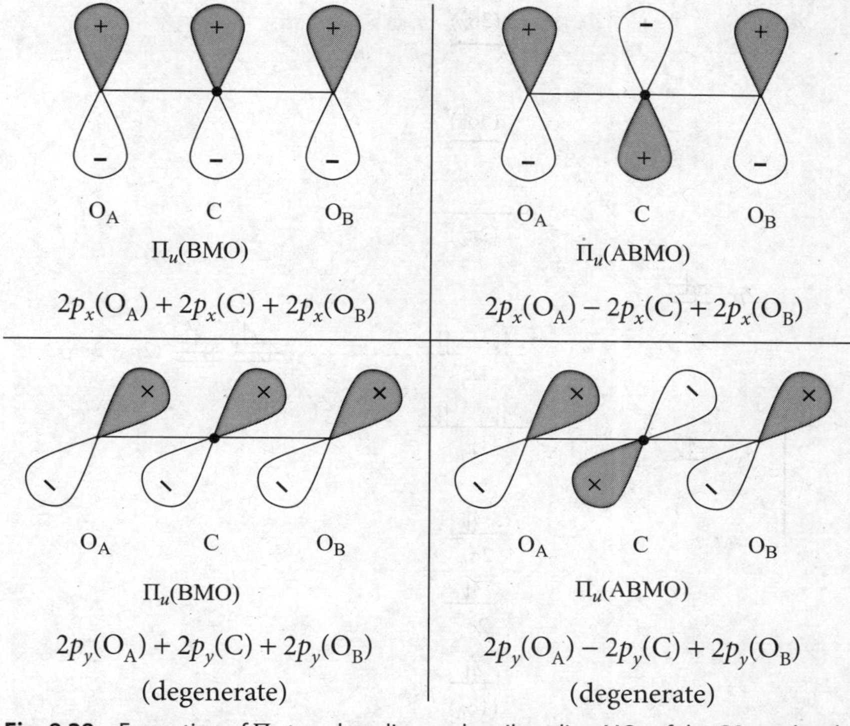

Fig. 8.28 Formation of Π_u-type bonding and antibonding MOs of the CO_2 molecule

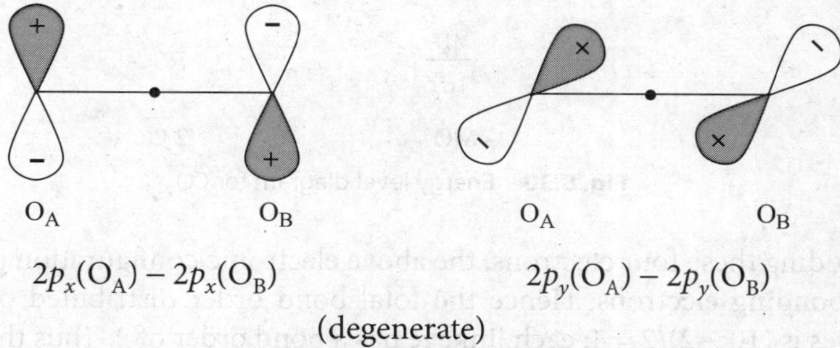

Fig. 8.29 π-type non-bonding MOs of the CO_2 molecule

Fig. 8.30 with $(2 \times 6) + 4 = 16$ valence electrons occupying the MOs according to the *Aufbau* principle.

The electronic configuration of CO_2 (with 16 valence electrons) is

$$(1\sigma_g^+)^2\,(1\sigma_u^+)^{*2}\,(2\sigma_g^+)^2\,(2\sigma_u^+)^2\,(1\pi_u)^4\,(1\pi_g)^4$$

Consistent with this electronic configuration, the photoelectron spectrum of CO_2 shows four bands in the range 13 eV to 20 eV. These correspond to ionisation from $1\pi_g$, $1\pi_u$, $2\sigma_u^+$ and $2\sigma_g^+$ levels. The first band does not show any fine structure owing to non-bonding character of the $1\pi_g$ MO; the second band, which results from $1\pi_u$, shows pronounced fine structure. Four electrons in the non-bonding MO, $1\pi_g$, form the lone pairs of electrons (one pair on each

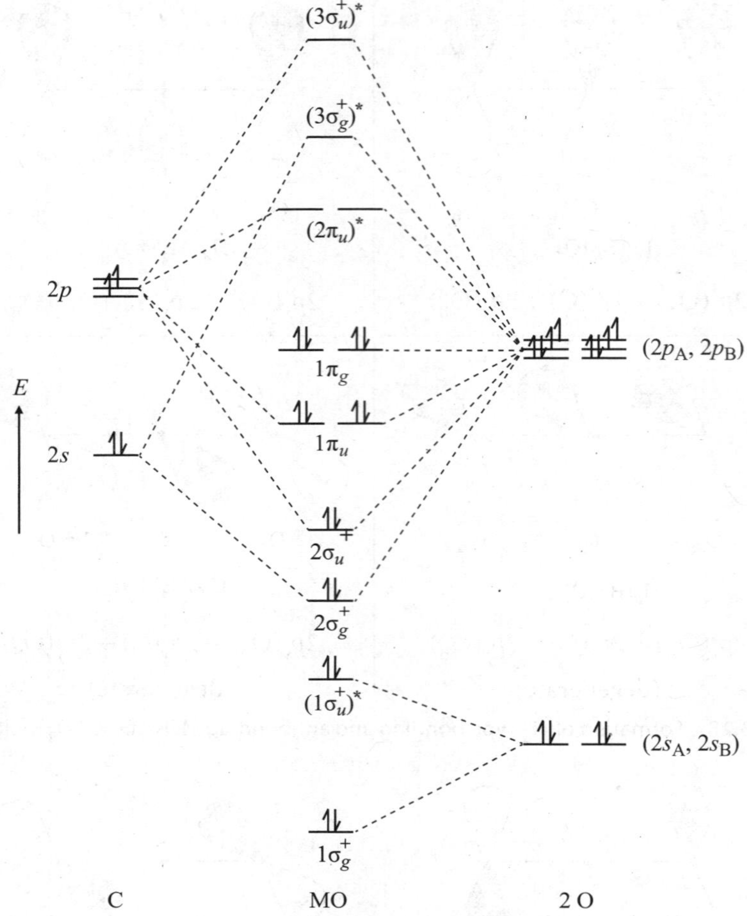

Fig. 8.30 Energy level diagram for CO_2

O atom); excluding these four electrons, the above electronic configuration gives ten bonding and two antibonding electrons. Hence the total bond order distributed over two carbon–oxygen linkages is $(10 - 2)/2 = 4$; each linkage has a bond order of 2. Thus the Lewis structure $:\ddot{O} = C = \ddot{O}:$ is justified by the MO theory.

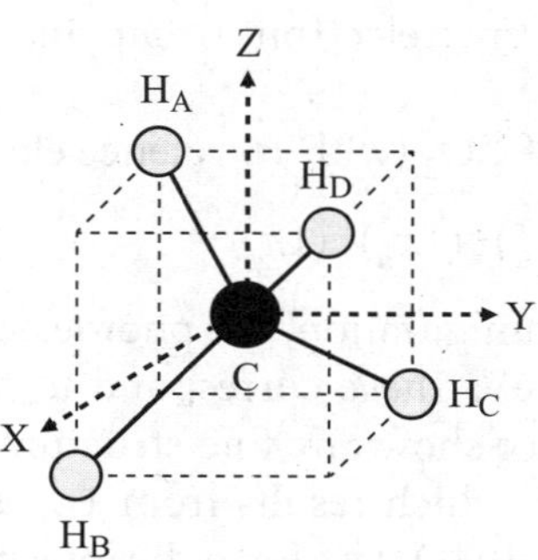

Fig. 8.31 Orientation of Cartesian axes for CH_4

8.6.4 Molecular Orbitals of CH_4

The four H atoms (labelled by subscripts A, B, C and D) of the CH_4 molecule (tetrahedral) are at alternate corners of a cube and the C-atom is at the body centre. These, together with axial orientation, are shown in Fig. 8.31. The molecular point group is T_d. Each C–H bond acts as a C_3-axis and each symmetry plane containing two of the four C–H bonds acts as a σ_d plane. Let ϕ_A, ϕ_B, ϕ_C, ϕ_D denote the 1*s* orbitals of H_A, H_B, H_C and H_D respectively. A representation Γ of T_d, with the 1*s*-orbitals of the four H atoms as basis, is as in the next page (top).

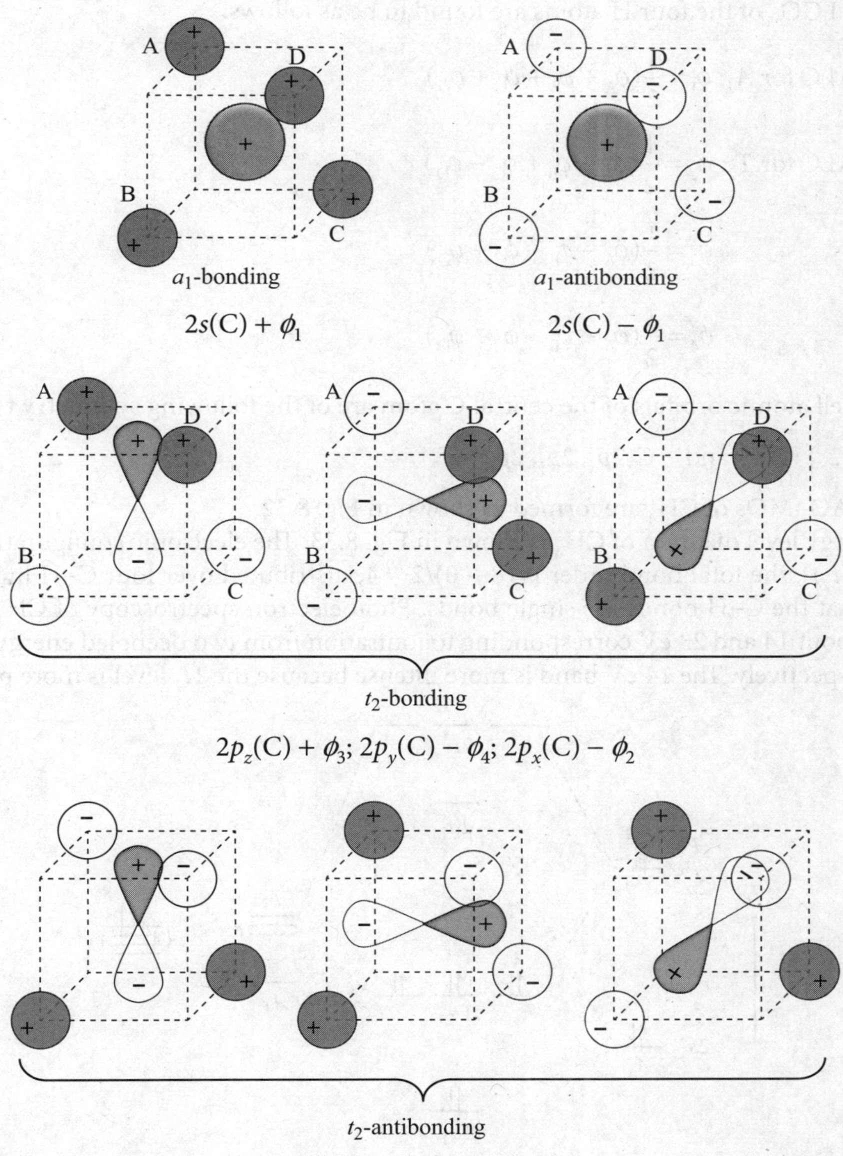

Fig. 8.32 Formation of CH_4 molecular orbitals by combination of carbon atomic orbitals and LGOs of four H atoms

	χ (character)				
	$\hat{E}$	$8\hat{C}_3$	$3\hat{C}_2$	$6\hat{S}_4$	$6\hat{\sigma}_d$
Γ	4	1	0	0	2

By inspection of the character table of T_d it is readily found that $\Gamma = A_1 + T_2$. The results of the $\hat{C}_3$, $\hat{C}_3^2$, $\hat{C}_2$, $\hat{S}_4$, $\hat{S}_4^3$ and $\hat{\sigma}_d$ operations on the 1*s* orbitals of the four H-atoms can be easily obtained by utilising Figs 2.6 and 2.7 of Chapter 2. Finally, the SALCs of the A_1 and T_2 representations, that is, the LGOs of the four H atoms are found to be as follows:

$$\text{SALC for } A_1\text{: } \phi_1 = \frac{1}{2}(\phi_A + \phi_B + \phi_C + \phi_D)$$

$$\text{SALC for } T_2\text{: } \phi_2 = \frac{1}{2}(\phi_A - \phi_B + \phi_C - \phi_D)$$

$$\phi_3 = \frac{1}{2}(\phi_A - \phi_B - \phi_C + \phi_D)$$

$$\phi_4 = \frac{1}{2}(\phi_A + \phi_B - \phi_C - \phi_D)$$

Valence shell atomic orbitals of the central C atom are of the following symmetry types:

$$\text{C}(2s) : A_1 \quad \text{and} \quad \text{C}(2p_x, 2p_y, 2p_z) : T_2$$

Hence LCAO–MOs of CH_4 are formed as shown in Fig. 8.32.

The energy level diagram of CH_4 is shown in Fig. 8.33. The electronic configuration of CH_4 is $(1a_1)^2\ (1t_2)^6$, the total bond order is $(8 - 0)/2 = 4$, distributed over four C–H linkages; this suggests that the C–H bonds are single bonds. Photoelectron spectroscopy of CH_4 gives two bands at about 14 and 23 eV corresponding to ionisation from two occupied energy levels, $1t_2$ and $1a_1$ respectively. The 14 eV band is more intense because the $1t_2$ level is more populated.

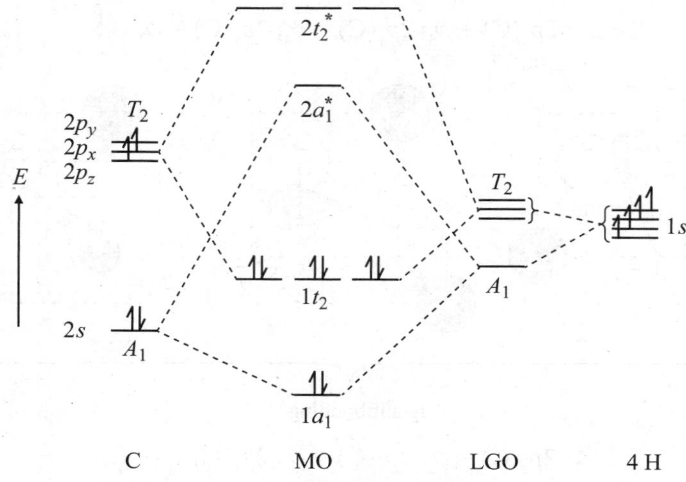

Fig. 8.33 Energy level diagram for methane.

8.6.5 Molecular Orbitals of Ethylene

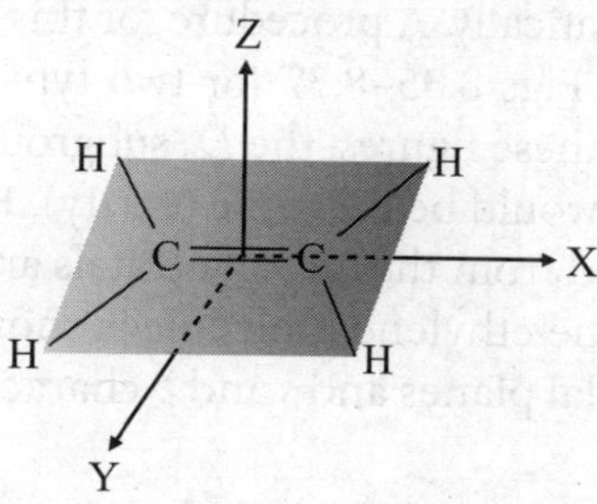

Fig. 8.34 Orientation of Cartesian axes for C_2H_4

Let the planar ethylene molecule (of point group D_{2h}) be placed on the XY plane with an orientation of the coordinate axes as shown in Fig. 8.34. The σ_h plane is the XY plane. The valence shell atomic orbitals are $1s$ (H) for each of the four H atoms, $2s$ (C), $2p_x$ (C), $2p_y$ (C), and $2p_z$ (C) for each of the two carbon atoms. The characters of representations of D_{2h} generated by these atomic orbitals as basis, are as in Table 8.3.

Table 8.3 Characters of representations of D_{2h} generated by the transformation matrices of the valence shell atomic orbitals of ethylene

Basis	$\hat{E}$	$\hat{C}_2(Z)$	$\hat{C}_2(Y)$	$\hat{C}_2(X)$	$\hat{i}$	$\hat{\sigma}(XY)$	$\hat{\sigma}(XZ)$	$\hat{\sigma}(YZ)$	Result of reduction
$1s$(4H)	4	0	0	0	0	4	0	0	$= A_g + B_{1g} + B_{2u} + B_{3u}$
$2s$(2C)	2	0	0	2	0	2	2	0	$= A_g + B_{3u}$
$2p_x$(2C)	2	0	0	2	0	2	2	0	$= A_g + B_{3u}$
$2p_y$(2C)	2	0	0	−2	0	2	−2	0	$= B_{1g} + B_{2u}$
$2p_z$(2C)	2	0	0	−2	0	−2	2	0	$= B_{2g} + B_{1u}$

Each representation can be reduced by applying the reduction formula and the results are shown in the extreme right column of Table 8.3. It is found that the group orbitals of the four $1s$(H) orbitals act as bases of the irreducible representations A_g, B_{1g}, B_{2u} and B_{3u} of the point group D_{2h}. In Table 8.3, there are three A_g-type of irreducible representations—one due to the four $1s$(H) orbitals, one due to the two $2s$(C) orbitals and the third due to the $2p_x$(C) orbitals. Hence the a_g-type molecular orbitals of ethylene are expected to be stretched on the XY plane, covering all the nuclei of C_2H_4 molecule. Similarly, two b_{1g}, two b_{2u} and three b_{3u}-type MOs may be constructed by matching the symmetries of the C and H group orbitals. It is also evident from Table 8.3 that the B_{2g} and B_{1u} representations resulting from the $2p_z$(C) orbitals have no partner of the same symmetry. Hence the two $2p_z$(C) orbitals will combine among themselves to produce b_{2g} and b_{1u} MOs of C_2H_4; these two MOs will cover only the two C atoms and exclude the H atoms; moreover, they will have the XY plane as a nodal plane, that is, they will correspond to the π-bond in C=C of C_2H_4. The energy order of the MOs can be obtained by considering the number of nodes. For MOs with the same number of nodes, the one having the greater s character will have lower energy than the one having the greater p character. For qualitative construction of an energy level diagram, it is therefore necessary to combine the matched SALCs (group orbitals) diagrammatically. A procedure for this purpose, based on the projection operator method, is illustrated in Figs 8.35–8.37 for two typical MOs. The other MOs can be similarly constructed. In drawing these figures, the D_2 subgroup of D_{2h} is sufficient; if the whole group were considered, the result would be the same (verify). Figure 8.38 shows that the SALCs of B_{2g} and B_{1u} symmetries obtained from the $2p_z$(C) orbitals are the π-type MOs of ethylene.

The energy level diagram for the ethylene molecule is shown in Fig. 8.39. It is constructed by considering the number of nodal planes and *s* and *p* character. For example, the $1a_g$ MO has

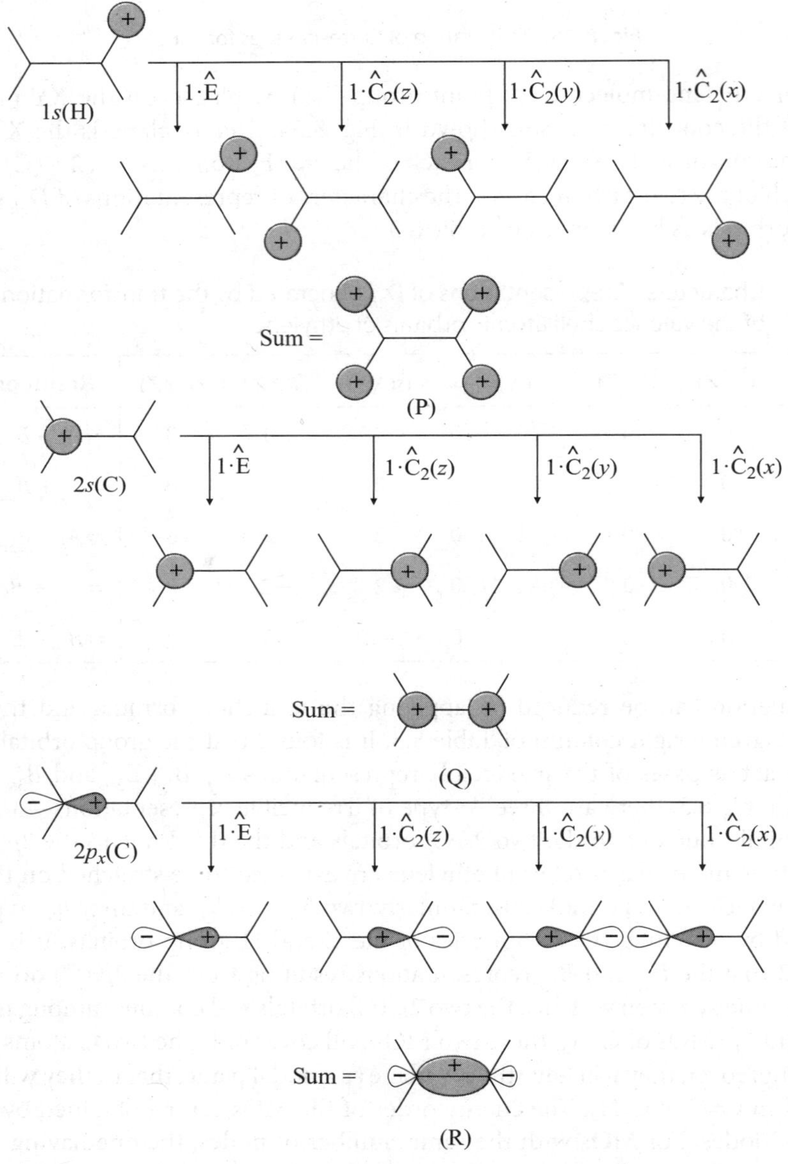

Fig. 8.35 Construction of A_{1g} SALCs for C_2H_4 using the projection operator method

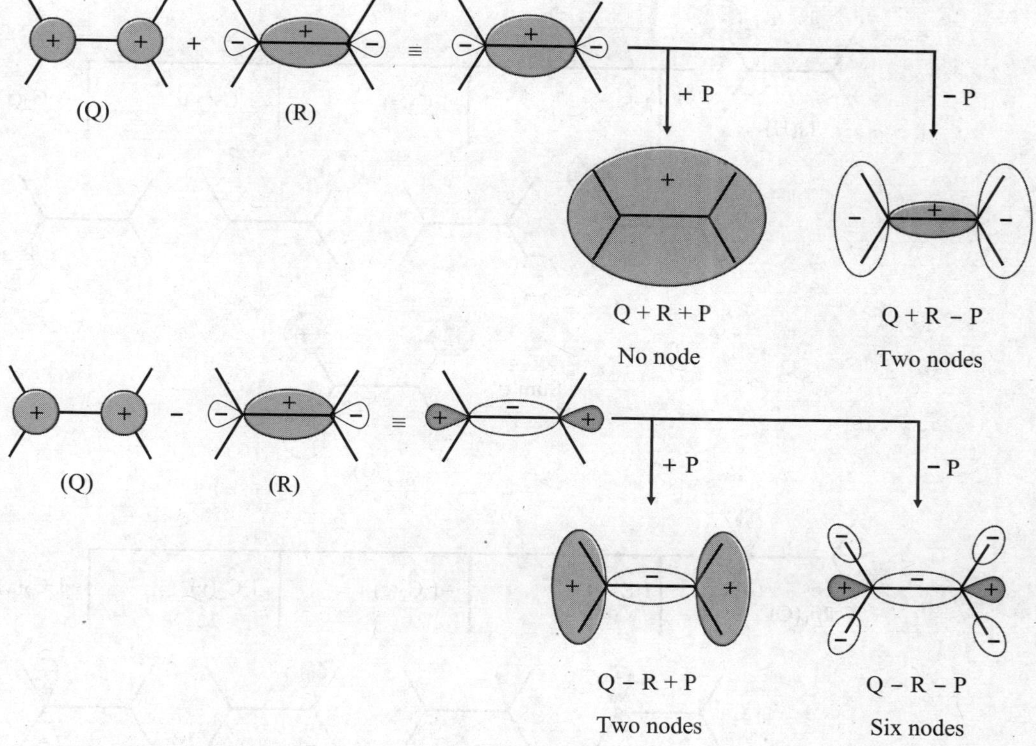

Fig. 8.36 Linear combination of A_{1g} SALCs to produce a_{1g} type MOs of C_2H_4

no node, and $1b_{3u}$ and $1b_{2u}$ have one node each; the $1a_g$ MO is lowermost in energy and among the other two, $1b_{3u}$ is made up of all s orbitals [$1s$(4H) and $2s$(2C)] while $1b_{2u}$ is composed of both s and p orbitals [$1s$(4H) and $2p_x$(2C)]. Hence $1b_{3u}$ is lower in energy than $1b_{2u}$. The electronic configuration of ethylene is

$$(1a_g)^2\,(1b_{3u})^2\,(1b_{2u})^2\,(2a_g)^2\,(1b_{1g})^2\,(1b_{1u})^2$$

The lowest energy electronic transition in ethylene is $\pi \rightarrow \pi^*$, that is, $(1b_{1u})^2 \rightarrow (1b_{1u})^1 (1b_{2g})^1$. Since the inner occupied orbitals are paired, the overall electronic ground state is of symmetry type $b_{1u} \times b_{1u} = A_g$ and the excited state has symmetry $b_{1u} \times b_{2g} = B_{3u}$ (these can be verified by using the D_{2h} character table). The transition moment integral $\langle \psi_{gr} | \hat{\mu} | \psi_{ex} \rangle$ corresponds to three direct product representations owing to the three Cartesian components of the dipole moment operator, namely, $\hat{\mu}_x$, $\hat{\mu}_y$, $\hat{\mu}_z$ which transform according to x, y and z respectively. Using the D_{2h} character table (Appendix A3), the necessary direct products are found to be:

$$\underset{\text{(ground state)}}{A_g \times} \begin{Bmatrix} B_{3u}\ (\text{for } \hat{\mu}_x) \\ B_{2u}\ (\text{for } \hat{\mu}_y) \\ B_{1u}\ (\text{for } \hat{\mu}_z) \end{Bmatrix} \underset{\text{(excited state)}}{\times B_{3u}} = \begin{Bmatrix} A_g \\ B_{1g} \\ B_{2g} \end{Bmatrix}$$

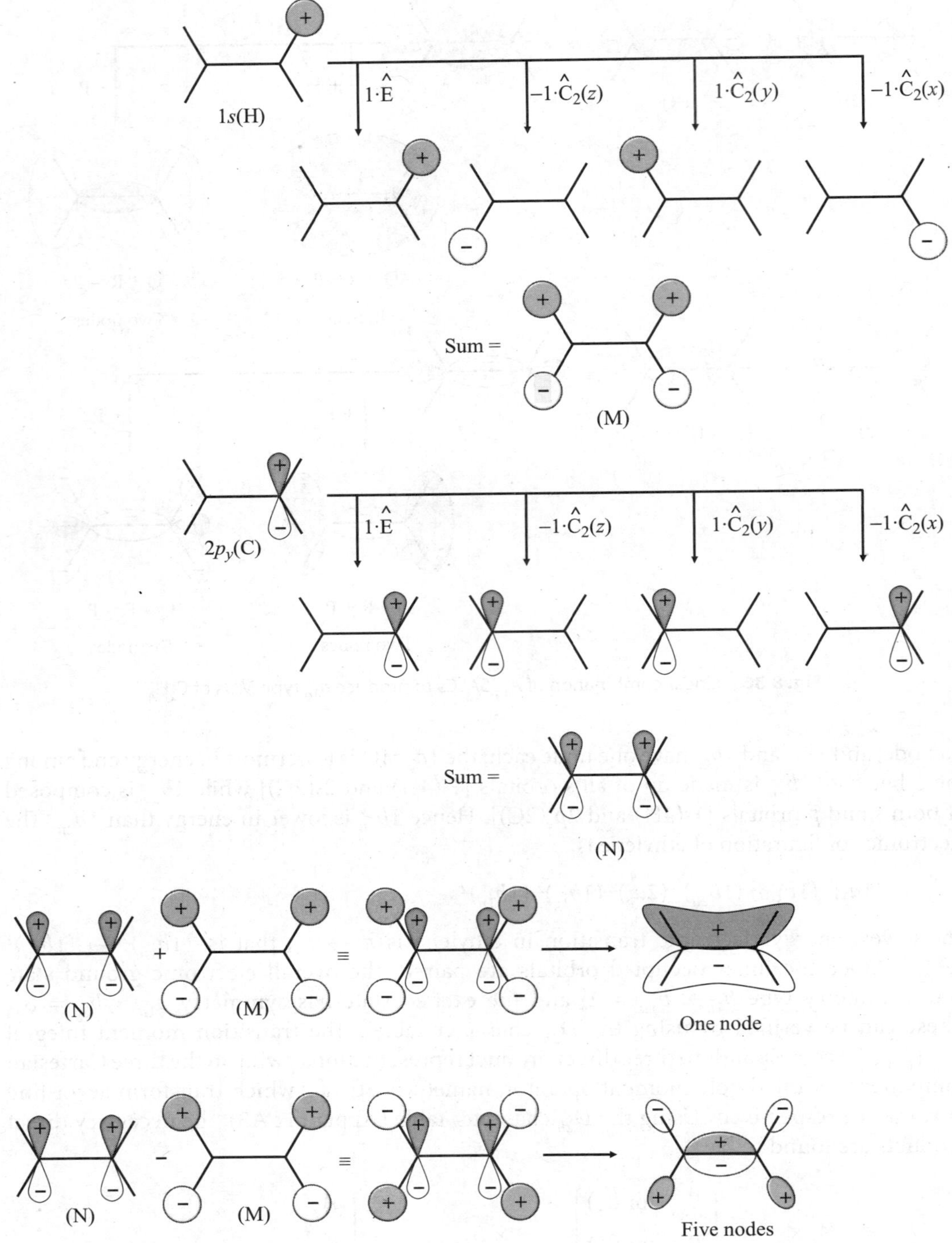

Fig. 8.37 Construction of B_{2u} SALCs using the projection operator method and construction of b_{2u} MOs of C_2H_4

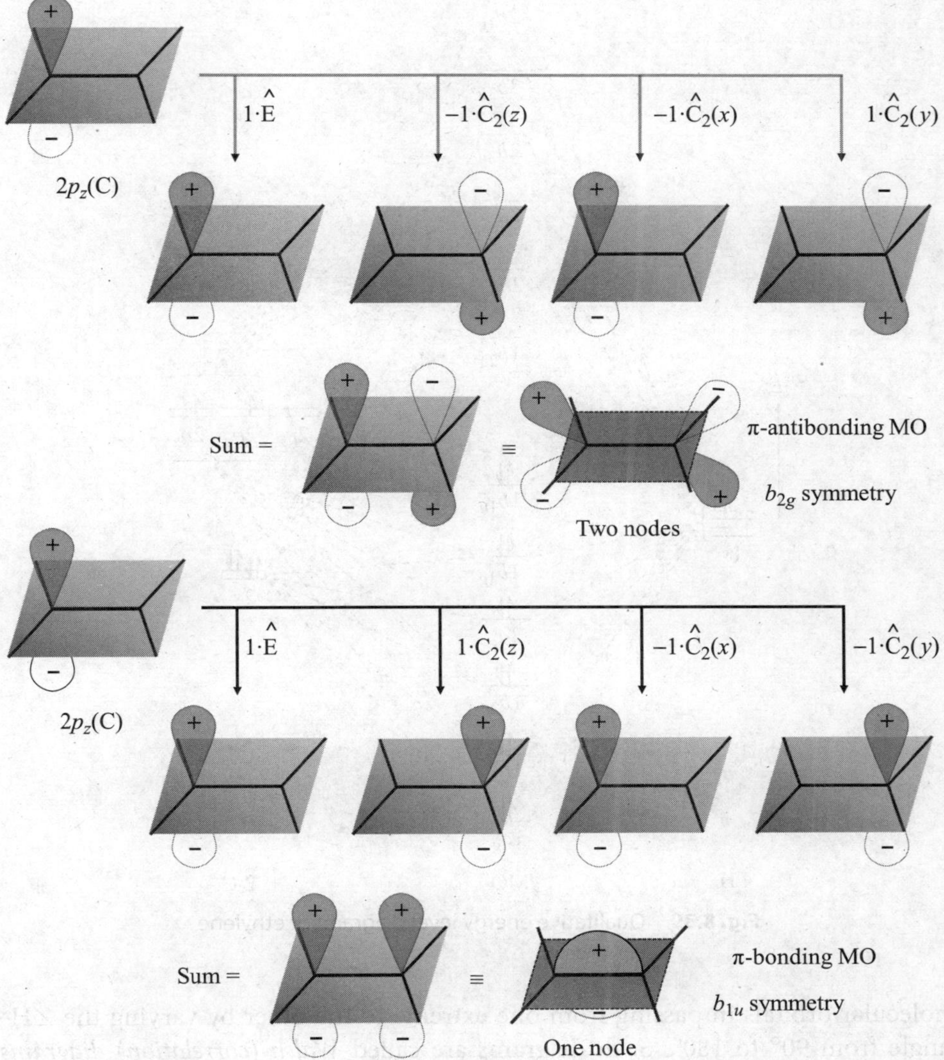

Fig. 8.38 Construction of B_{2g} and B_{1u} SALCs from $2p_z$(C) atomic orbitals and formation of the b_{2g} and b_{1u} MOs which are the π-bonding and antibonding MOs of ethylene

At least the x-component of the transition moment integral gives the totally symmetric irreducible representation A_g. Hence the integral is non-vanishing, that is, the transition is allowed, and it is polarised in the x-direction.

8.7 WALSH DIAGRAM: PREDICTION OF GEOMETRY OF XH_2 TYPE OF MOLECULES

Molecules of the type XH_2 can have two extremes geometries, linear and bent, belonging to point groups $D_{\infty h}$ and C_{2v} respectively. Their structure can be predicted from a correlation of

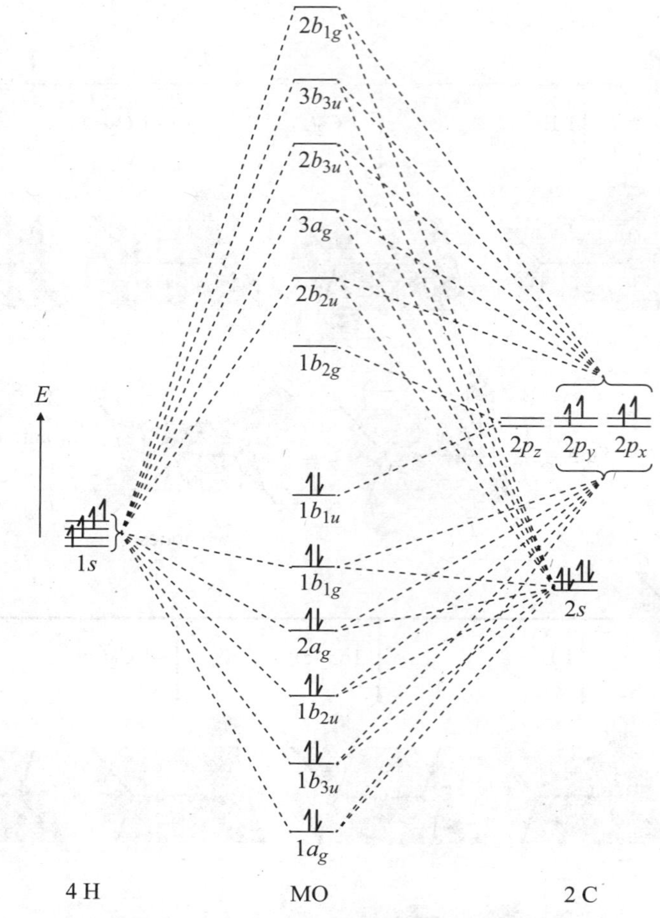

Fig. 8.39 Qualitative energy level diagram for ethylene

their molecular orbitals in passing from one extreme to the other by varying the ∠H–X–H bond angle from 90° to 180°. Such diagrams are called *Walsh (correlation) diagrams* after A D Walsh who devised such diagrams for these and many other types of molecules.

Molecular orbitals of the linear system are considered first. The 1*s* orbitals of two H atoms give the LGOs, $1s(H_A) + 1s(H_B)$ and $1s(H_A) - 1s(H_B)$ of symmetry type Σ_g^+ and Σ_u^+ respectively in the $D_{\infty h}$ point group. The internuclear axis is taken as the Z-axis. The spherically symmetrical 2*s* orbital of X belongs to Σ_g^+ and its $2p_z$ orbital is of symmetry Σ_u^+. The $2p_x$ and $2p_y$ orbitals of X transform as Π_u. By matching symmetry, the following MOs are constructed:

$$[1s(H_A) + 1s(H_B)] + 2s(X) \longrightarrow 2\sigma_g^+$$

$$[1s(H_A) + 1s(H_B)] - 2s(X) \longrightarrow 3\sigma_g^+$$

$$[1s(H_A) - 1s(H_B)] + 2p_z(X) \longrightarrow 1\sigma_u^+$$

$$[1s(H_A) - 1s(H_B)] - 2p_z(X) \longrightarrow 2\sigma_u^+$$

The $2p_x$ and $2p_y$ orbitals of X do not match the symmetry of the LGOs and hence they remain as π_u-type degenerate non-bonding MOs on X. Moreover, the 1*s* orbital of X which is of Σ_g^+ symmetry, because of its low energy, does not combine with the hydrogen LGOs; this atomic orbital is taken as the $1\sigma_g^+$ orbital in the MO picture. The energy order is found by considering the number of nodal planes and amount of *s*-character.

Next, we construct the MOs of the bent XH_2 molecules. Here the Z-axis is taken to be the C_2-axis, the YZ-plane being the molecular plane. The LGOs $1s(H_A) + 1s(H_B)$ and $1s(H_A) - 1s(H_B)$ are now bases of the irreducible representations A_1 and B_2 respectively in the C_{2v} point group. The valence shell atomic orbitals of X are of symmetry types:

$$2s \text{ and } 2p_z : A_1 \qquad 2p_x : B_1 \qquad 2p_y : B_2$$

The linear combination of three A_1-type atomic orbitals give the MOs, $2a_1$, $3a_1$ and $4a_1$. The B_2-symmetry orbitals combine as follows:

$$[1s(H_A) - 1s(H_B) + 2p_y(X)] \longrightarrow 1b_2$$

$$[1s(H_A) - 1s(H_B) - 2p_y(X)] \longrightarrow 2b_2$$

The B_1 symmetry orbital ($2p_x$) of X remains as the non-bonding MO, $1b_1$.

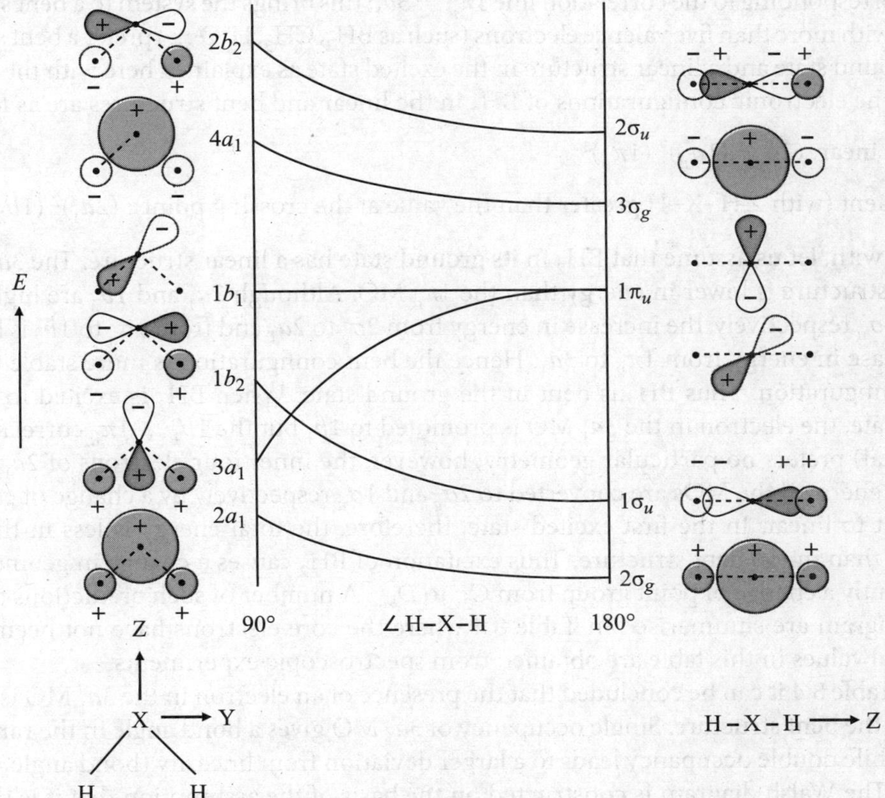

Fig. 8.40 Walsh (correlation) diagrams for XH_2 molecules

Figure 8.40 shows the correlation of the MOs as the angle changes from 90° to 180°. The MOs of the two extreme geometries are shown on two sides of the diagram in increasing order of energy. The correlation lines are drawn on the basis of shapes and signs of lobes on the two extremes. The Y-axis in the bent case becomes Z-axis in the linear case. The '+' sign in the superscripts of the MO symbols have been omitted in the diagram as there are no MOs with '–' superscript.

The $2a_1$ orbital is higher in energy than the $2\sigma_g$ MO because of the stronger electrostatic interactions between the 1*s* (H) atomic orbitals which are closer in the bent configuration than in the linear structure. The $1\sigma_u$ MO is lower in energy than $1b_2$ since the atomic orbital overlap is direct in the linear structure but lateral in the bent structure. Of particular importance in this diagram is the correlation between the MOs $3a_1$ and $1\pi_u$. When the structure bends, $1\pi_u$ (in $D_{\infty h}$) changes to $3a_1$ (in C_{2v}); the former is of purely *p*-character but the latter has appreciable *s*-character and consequently has remarkably lower energy. This causes a steep decrease in energy of the correlation line from $1\pi_u$ to $3a_1$.

Shapes of molecules of the type XH_2 in their ground and excited states can be predicted by feeding the MOs with electrons of XH_2 according to the *Aufbau* principle.

Systems with three or four valence electrons (such as LiH_2, BeH_2, BH_2^+...) prefer the linear structure in the ground state because the $2\sigma_g$ and $1\sigma_u$ orbitals are lower in energy than the other MOs listed in the Walsh diagram. On excitation, one electron is promoted to the next higher orbital corresponding to the correlation line $1\pi_u \ldots 3a_1$; this brings the system to a bent structure. Systems with more than five valence electrons (such as BH_2, CH_2, H_2O ...) prefer a bent structure in the ground state and a linear structure in the excited state as explained here with the example of BH_2. The electronic configurations of BH_2 in the linear and bent structures are as follows.

Linear: $(2\sigma_g)^2\ (1\sigma_u)^2\ (1\pi_u)^1$

Bent (with ∠H–X–H greater than the value at the crossing point): $(2a_1)^2\ (1b_2)^2\ (3a_1)^1$

To begin with, let us assume that BH_2 in its ground state has a linear structure. The $3a_1$ MO of the bent structure is lower in energy than the $1\pi_u$ MO. Although $2a_1$ and $1b_2$ are higher than $2\sigma_g$ and $1\sigma_u$ respectively, the increase in energy from $2\sigma_g$ to $2a_1$ and from $1\sigma_u$ to $1b_2$ is less than the decrease in energy from $1\pi_u$ to $3a_1$. Hence the bent configuration is more stable than the linear configuration. Thus BH_2 is bent in the ground state. When BH_2 is excited to the first excited state, the electron in the $3a_1$ MO is promoted to $1b_1$ but the $1b_1 \ldots 1\pi_u$ correlation line (horizontal) prefers no particular geometry; however, the inner four electrons of $2a_1$ and $1b_2$ get lower energy if the MOs are converted to $2\sigma_g$ and $1\sigma_u$ respectively by a change of geometry from bent to linear. In the first excited state, therefore, the total energy is less in the linear structure than in the bent structure. Thus excitation of BH_2 causes a change in geometry and consequently a change of point group from C_{2v} to $D_{\infty h}$. A number of such predictions from the Walsh diagram are summarised in Table 8.4 where the core electrons have not been shown. Numerical values in this table are obtained from spectroscopic experiments.

From Table 8.4 it can be concluded that the presence of an electron in the $3a_1$ MO is crucial; it favours the bent structure. Single occupancy of $3a_1$ MO gives a bond angle in the range 131° to 144° while double occupancy leads to a larger deviation from linearity (bond angle = 102.4° to 110°). The Walsh diagram is constructed on the basis of the assumption that it is the same for all X. Still the predictions made from it agree quite well with experiments.

Table 8.4 Electronic configurations of some XH_2 molecules and prediction of structures from the Walsh diagram

System	Ground state		Excited state	
	Configuration	Geometry	Configuration	Geometry
LiH_2	$(2\sigma_g)^2 (1\sigma_u)^1$	Linear	$(2a_1)^2 (3a_1)^1$	Bent
BeH_2	$(2\sigma_g)^2 (1\sigma_u)^2$	Linear	$(2a_1)^2 (1b_2)^1 (3a_1)^1$	Bent
BH_2	$(2a_1)^2 (1b_2)^2 (3a_1)^1$	Bent (131°)	$(2\sigma_g)^2 (1\sigma_u)^2 (1\pi_u)^1$	Linear
CH_2	$(2a_1)^2 (1b_2)^2 (3a_1)^2$	Bent (102.4°)	$(2a_1)^2 (1b_2)^2 (3a_1)^1 (1b_1)^1$	Bent (134°)
NH_2	$(2a_1)^2 (1b_2)^2 (3a_1)^2 (1b_1)^1$	Bent (103.4°)	$(2a_1)^2 (1b_2)^2 (3a_1)^1 (1b_1)^2$	Bent (144°)
OH_2	$(2a_1)^2 (1b_2)^2 (3a_1)^2 (1b_1)^2$	Bent (104.5°)		

SUMMARY

- ❑ Like atomic orbitals in atoms, there are molecular orbitals (MO) in molecules; MO is defined and the role of the Born–Oppenheimer approximation in constructing MOs by the method of linear combination of atomic orbitals (LCAO) is explained.
- ❑ Only atomic orbitals of comparable energy and proper symmetry can combine to form MOs.
- ❑ Symmetry adapted linear combinations (SALCs) of the surrounding (ligand) atoms with respect to the molecular point group are first constructed using projection operators. These SALCs are then combined linearly with the central atom orbitals by matching symmetry.
- ❑ The variation method gives the combining coefficients and MO energies.
- ❑ Energy level diagrams with symmetry labels for the MOs are useful in explaining molecular structure and chemical and spectroscopic properties.
- ❑ Walsh correlation diagram helps to choose between linear and bent structures of XH_2 type of molecules.

LEARNING OUTCOME

After reading this chapter the student will be able to (i) derive energy level diagrams for the MOs of diatomic molecules from the united atom-separated atom viewpoint and assign $D_{\infty h}$ point group symbols to the MOs, (ii) construct SALCs for obtaining ligand group orbitals by the projection operator method, (iii) apply the LCAO–MO method to polyatomic molecules using group theory, (iv) calculate bond orders and spin-only magnetic moments, (v) interpret photoelectron spectra, (vi) explain UV–vis spectra of molecules in terms of electronic transitions between MOs and (vii) apply Walsh correlation diagrams to choose between linear and bent structures for XH_2 type of molecules.

EXERCISES

1. Show the positions of Li_2 and O_2 in the united atom–separated atom correlation diagram and from this draw the approximate energy level diagram for these two diatomic molecules.

2. Construct the LGOs from 1*s*(H) orbitals in the following species:
(a) H_2O (b) BH_3 (planar) (c) AH_4 (planar) (d) C_2H_4.
In each case write the approximate energy order of the LGOs.

3. From the valence shell atomic orbitals construct molecular orbitals and approximate energy level diagrams for the following species:
(a) CH_3^+ (planar) (b) BeH_2 (linear) (c) acetylene molecule (d) vinyl cation.

4. Apply the LCAO–MO method with proper symmetry considerations to show that
(a) O_2 is paramagnetic with a bond order of 2.
(b) N_2 is diamagnetic with a bond order of 3.
(c) CO can act as a suitable ligand for metals in low oxidation states.
(d) B_2 is paramagnetic.

5. Find the bond order and spin-only magnetic moment in the following cases: (a) BeF (b) CN^- (c) NO^+ (d) BF.

6. Construct the energy level diagram for the system AH_3, where A is an element of the second row of the periodic table, assuming both planar and pyramidal geometry. Using it explain why (a) CH_3^+ is planar while CH_3^- is pyramidal (b) NH_3 in a planar structure is unstable.

7. Sketch the Walsh diagram and predict the geometry of the following in the ground and first excited states: BH_2^+, HgH_2, BeH_2.

8. Explain why BeH_2 is linear while H_2O is bent using the Walsh diagram.

Suggested Books and References for Further Reading

1. Born M and Oppenheimer J R, *Ann. der Phys.*, 84 (457) **1927.**
2. Mulliken R S, *Rev. Mod. Phys.*, 4 (1) **1932.**
3. Roothaan C C J, *Rev. Mod. Phys.*, 23 (69) **1951.**
4. Coulson C A, *Valence*, Oxford University Press, London, **1961.**
5. Kettle S F A, *Symmetry and Structure: Readable Group Theory for Chemists*, John Wiley & Sons Inc., Chichester, England **2007.**
6. Bishop D M, *Group Theory and Chemistry*, Dover, NY, **1973.**
7. Walsh A D, *J. Chem. Soc.* (2260) **1953.**
8. Baird N C, *J. Chem. Educ.*, 55 (412) **1978.**

9 Chemical Bonding II: Localised Molecular Orbitals

Learning objectives

- ➢ Hybridisation of atomic orbitals
- ➢ Construction of MOs using σ- and π- orbitals

9.1 INTRODUCTION

In the examples discussed in Chapter 8, the pristine atomic orbitals of the constituent atoms were used for construction of the LCAO–MOs. However, as the number of atomic orbitals increases, the task becomes more and more difficult. As the MOs extend over all the atoms in the molecule, it is not possible to recognise the different chemical bonds and the lone pair orbitals. But electron pair bonds and lone pairs of electrons are so important in explaining reaction mechanisms and formation of coordination compounds that the concept of such bonds cannot be abandoned. Moreover, such bonds retain their identity in a variety of molecules; for example, the C=O double bond can be detected by its infrared stretching frequency (at about 1800 cm^{-1}) in all ketone molecules irrespective of the terminal atoms or groups; the infrared stretching frequency (3000 cm^{-1}) of the C–H bond is not only used to detect such bonds in an unknown compound but also to estimate their number in the molecule. It is therefore reasonable to assume that electrons in MOs are localised in regions covering pairs of nuclei, that is, forming A–B types of bonds in molecules. Such bonds are classified as *sigma-*, *pi-*, *delta-* and so on. A compromise has to be made at this stage between the fully delocalised MO theory and the Valence Bond theory; the method that arises in this way is the method of localised molecular orbitals. It assumes that in molecules, the atoms have σ and π orbitals. An orbital with lobes symmetric with respect to a two-fold rotation about the bond axis is a σ-orbital and one which is antisymmetric is a π-orbital; MOs are formed by linear combination of such orbitals. Using this approach, MO calculations for polyatomic molecules are greatly simplified. The concept of atomic orbital hybridisation (which originally belonged to the VB theory) is retained in the localised MO theory. The hybridisation schemes and some illustrative calculations of MOs of polyatomic molecules using the localised MO method are discussed in this chapter.

9.2 HYBRIDISATION OF ATOMIC ORBITALS

In the LCAO–MO method applied to a molecule of type AB_n, the MO wave function may be obtained by any one of the following two procedures:

(i) Construct intermediate linear combinations of the central atom orbitals such that the resulting orbitals, called *hybrid orbitals*, may be directed along the bonds in the molecule AB_n, and then combine these hybrid orbitals with the ligand atomic orbitals to form MOs.

(ii) Construct intermediate linear combinations of the ligand atomic orbitals such that the resulting orbitals, called *ligand group orbitals*, are directed towards the central atom, and then combine them with suitable atomic orbitals of the central atom.

The first part of the scheme (i) is known as *hybridisation of atomic orbitals* and is discussed with some examples in the following sections.

9.2.1 Tetrahedral Hybridisation

Let $\vec{e}_1$, $\vec{e}_2$, $\vec{e}_3$ and $\vec{e}_4$ be four tetrahedrally directed unit vectors originating from the central atom A in the tetrahedral molecule AB_4 (Fig. 9.1). The point group of a regular tetrahedron (T_d) has symmetry operators $\hat{E}$, $8\hat{C}_3(= 4\hat{C}_3, 4\hat{C}_3^2)$, $3\hat{C}_2$, $6\hat{S}_4(= 3\hat{S}_4, 3\hat{S}_4^3)$, $6\hat{\sigma}_d$.

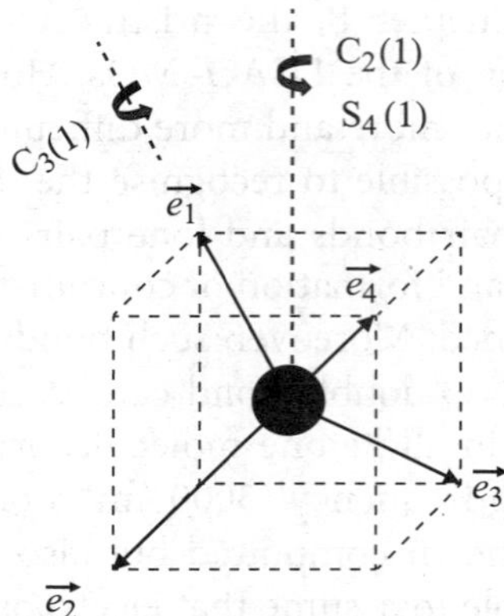

Fig. 9.1 Four tetrahedrally directed basis vectors and some proper and improper axes of rotation for a tetrahedral molecule

Each σ_d plane contains two of the four vectors. Let the symbol $\sigma_d(i, j)$ be used for the symmetry plane containing the vectors $\vec{e}_i$ and $\vec{e}_j$. Then, the six σ_d planes are $\sigma_d(1, 2)$, $\sigma_d(1, 3)$, $\sigma_d(1, 4)$ $\sigma_d(2, 3)$, $\sigma_d(2, 4)$ and $\sigma_d(3, 4)$. With the vectors $\{\vec{e}_i\}$, $i = 1$ to 4 as basis, the characters of a representation Γ of T_d are as given in Table 9.1.

Table 9.1 Characters of a representation of T_d with $\{\vec{e}_i\}$, $i = 1$ to 4 as basis

	χ (characters)				
	$\hat{E}$	$8\hat{C}_3$	$3\hat{C}_2$	$6\hat{S}_4$	$6\hat{\sigma}_d$
Γ	4	1	0	0	2

Let Γ be reducible as

$$\Gamma = a_1A_1 + a_2A_2 + a_3E + a_4T_1 + a_5T_2 \tag{9.1}$$

Using the character table of T_d and the reduction formula, it can be readily verified that $a_1 = 1$, $a_2 = a_3 = a_4 = 0$ and $a_5 = 1$, giving $\Gamma = A_1 + T_2$.

Thus, the atomic orbitals of the central atom must be so combined that there is one orbital of A_1 symmetry and three orbitals of the triply degenerate T_2 symmetry in the combination. The T_d character table shows that atomic orbitals of the central atom transform under the action of the T_d symmetry operators as follows:

$$A_1 : s$$

$$T_2 : (p_x, p_y, p_z); (d_{xy}, d_{yz}, d_{zx})$$

Hence the tetrahedrally directed orbitals are produced by the linear combinations,

s and (p_x, p_y, p_z) orbitals (sp^3 hybridisation)

and s and (d_{xy}, d_{yz}, d_{zx}) orbitals (sd^3 hybridisation)

In the second quantum shell, sd^3 hybridisation is not possible due to the absence of d orbitals; so in the tetrahedral C atom, the hybridisation is of sp^3 type. In transition metal complexes both types are possible, depending on the energy difference between the s and d orbitals. Thus in the MnO_4^- ion, Mn^{7+} exhibits sd^3 hybridisation (three d orbitals of the third quantum shell and the 4s orbital hybridise to give trtrahedrally directed orbitals).

Let us now obtain the explicit forms of the wave functions Ω_1, Ω_2, Ω_3 and Ω_4 of the sp^3 hybrid orbitals which correspond respectively to the vectors $\vec{e}_1, \vec{e}_2, \vec{e}_3$ and $\vec{e}_4$ and are directed accordingly. Each Ω_i, $\{i = 1 \text{ to } 4\}$, is a linear combination of the s, p_x, p_y and p_z wave functions; the reverse is also true. So we may consider the s wave function which is of A_1 symmetry to be obtainable from the set $\{\Omega_i\}$ by applying the projection operator $\hat{P}^{A_1}$. Similarly, the triply degenerate p functions of symmetry type T_2 should be obtainable from the set $\{\Omega_i\}$ by the projection operator $\hat{P}^{T_2}$. Once the s and p_x, p_y, p_z functions are expressed in terms of $\{\Omega_i\}$, a matrix inversion will give each Ω_i in terms of s, p_x, p_y and p_z. From Fig. 9.1, it is easy to verify that the effects of the various T_d symmetry operators on the vectors $\vec{e}_1, \vec{e}_2, \vec{e}_3$ and $\vec{e}_4$ are as in Table 9.2.

Using these results and the correspondence $\Omega_i \leftrightarrow \vec{e}_i$, $i = 1$ to 4, we have :

$$\hat{P}^{(A_1)}\Omega_1 = \frac{1}{g}\sum_R [\chi^j(\hat{R})]^* \hat{R}\Omega_1 \quad (g = 24 \text{ and all } \chi^j(\hat{R}) = 1 \text{ for } j = A_1)$$

$$= \frac{1}{24}[(1)^* \hat{E}\Omega_1 + (1)^* \hat{C}_3(1)\Omega_1 + (1)^* \hat{C}_3^2(1)\Omega_1 + (1)^* \hat{C}_3(2)\Omega_1 + (1)^* \hat{C}_3^2(2)\Omega_1$$

$$+ (1)^* \hat{C}_3(3)\Omega_1 + (1)^* \hat{C}_3^2(3)\Omega_1 + (1)^* \hat{C}_3(4)\Omega_1 + (1)^* \hat{C}_3^2(4)\Omega_1 + (1)^* \hat{C}_2(1)\Omega_1$$

$$+ (1)^* \hat{C}_2(2)\Omega_1 + (1)^* \hat{C}_2(3)\Omega_1 + \text{similar terms for the remaining operations}]$$

$$= \frac{1}{24}[\Omega_1 + (\Omega_1 + \Omega_1) + (\Omega_3 + \Omega_4) + (\Omega_2 + \Omega_4) + (\Omega_2 + \Omega_3) + (\Omega_4 + \Omega_2 + \Omega_3)$$

$$+ (\Omega_2 + \Omega_3) + (\Omega_4 + \Omega_3) + (\Omega_4 + \Omega_2) + (\Omega_1 + \Omega_1 + \Omega_1 + \Omega_4 + \Omega_3 + \Omega_2)]$$

Table 9.2 Effect of the T_d symmetry operators on $\vec{e}_1, \vec{e}_2, \vec{e}_3$ and $\vec{e}_4$

Part A

	$\hat{C}_3(1)$	$\hat{C}_3^2(1)$	$\hat{C}_3(2)$	$\hat{C}_3^2(2)$	$\hat{C}_3(3)$	$\hat{C}_3^2(3)$	$\hat{C}_3(4)$	$\hat{C}_3^2(4)$	$\hat{C}_2(1)$	$\hat{C}_2(2)$	$\hat{C}_2(3)$	$\hat{S}_4(1)$	$\hat{S}_4^3(1)$	$\hat{S}_4(2)$	$\hat{S}_4^3(2)$
$\vec{e}_1$	$\vec{e}_1$	$\vec{e}_1$	$\vec{e}_3$	$\vec{e}_4$	$\vec{e}_2$	$\vec{e}_4$	$\vec{e}_2$	$\vec{e}_3$	$\vec{e}_4$	$\vec{e}_2$	$\vec{e}_3$	$\vec{e}_2$	$\vec{e}_3$	$\vec{e}_4$	$\vec{e}_3$
$\vec{e}_2$	$\vec{e}_3$	$\vec{e}_4$	$\vec{e}_2$	$\vec{e}_2$	$\vec{e}_4$	$\vec{e}_1$	$\vec{e}_3$	$\vec{e}_1$	$\vec{e}_3$	$\vec{e}_1$	$\vec{e}_4$	$\vec{e}_4$	$\vec{e}_1$	$\vec{e}_3$	$\vec{e}_4$
$\vec{e}_3$	$\vec{e}_4$	$\vec{e}_2$	$\vec{e}_4$	$\vec{e}_1$	$\vec{e}_3$	$\vec{e}_3$	$\vec{e}_1$	$\vec{e}_2$	$\vec{e}_2$	$\vec{e}_4$	$\vec{e}_1$	$\vec{e}_1$	$\vec{e}_4$	$\vec{e}_1$	$\vec{e}_2$
$\vec{e}_4$	$\vec{e}_2$	$\vec{e}_3$	$\vec{e}_1$	$\vec{e}_3$	$\vec{e}_1$	$\vec{e}_2$	$\vec{e}_4$	$\vec{e}_4$	$\vec{e}_1$	$\vec{e}_3$	$\vec{e}_2$	$\vec{e}_3$	$\vec{e}_2$	$\vec{e}_2$	$\vec{e}_1$

Part B

	$\hat{S}_4(3)$	$\hat{S}_4^3(3)$	$\hat{\sigma}_d(1, 2)$	$\hat{\sigma}_d(1, 3)$	$\hat{\sigma}_d(1, 4)$	$\hat{\sigma}_d(2, 3)$	$\hat{\sigma}_d(2, 4)$	$\hat{\sigma}_d(3, 4)$
$\vec{e}_1$	$\vec{e}_4$	$\vec{e}_2$	$\vec{e}_1$	$\vec{e}_1$	$\vec{e}_1$	$\vec{e}_4$	$\vec{e}_3$	$\vec{e}_2$
$\vec{e}_2$	$\vec{e}_1$	$\vec{e}_3$	$\vec{e}_2$	$\vec{e}_4$	$\vec{e}_3$	$\vec{e}_2$	$\vec{e}_2$	$\vec{e}_1$
$\vec{e}_3$	$\vec{e}_2$	$\vec{e}_4$	$\vec{e}_4$	$\vec{e}_3$	$\vec{e}_2$	$\vec{e}_3$	$\vec{e}_1$	$\vec{e}_3$
$\vec{e}_4$	$\vec{e}_3$	$\vec{e}_1$	$\vec{e}_3$	$\vec{e}_2$	$\vec{e}_4$	$\vec{e}_1$	$\vec{e}_4$	$\vec{e}_4$

$$= \frac{1}{24}(6\Omega_1 + 6\Omega_2 + 6\Omega_3 + 6\Omega_4)$$

$$\approx \frac{1}{2}(\Omega_1 + \Omega_2 + \Omega_3 + \Omega_4) = \phi_1 \text{ (say)} \quad \text{(after normalisation to unity)}$$

A similar procedure will give

$$\hat{P}^{T_2}\,\Omega_1 = \frac{1}{12}(3\Omega_1 - \Omega_2 - \Omega_3 - \Omega_4);\;\; \hat{P}^{T_2}\,\Omega_2 = \frac{1}{12}(-\Omega_1 + 3\Omega_2 - \Omega_3 - \Omega_4)$$

$$\hat{P}^{T_2}\,\Omega_3 = \frac{1}{12}(-\Omega_1 - \Omega_2 + 3\Omega_3 - \Omega_4);\;\; \hat{P}^{T_2}\,\Omega_4 = \frac{1}{12}(-\Omega_1 - \Omega_2 - \Omega_3 + 3\Omega_4)$$

These four SALCs of T_2 symmetry are linearly dependent because

$$(\hat{P}^{T_2}\,\Omega_1 + \hat{P}^{T_2}\,\Omega_2 + \hat{P}^{T_2}\,\Omega_3 + \hat{P}^{T_2}\,\Omega_4) = 0$$

But we need three linearly independent functions of T_2 symmetry to be constructed from the above four functions. This can be achieved as follows:

$$\hat{P}^{T_2}\,\Omega_1 + \hat{P}^{T_2}\,\Omega_2 = \frac{1}{12}(2\Omega_1 + 2\Omega_2 - 2\Omega_3 - 2\Omega_4)$$

$$\approx \frac{1}{2}(\Omega_1 + \Omega_2 - \Omega_3 - \Omega_4) = \phi_2 \text{ (say)} \quad \text{(after normalisation)}$$

$$\hat{P}^{T_2}\,\Omega_2 + \hat{P}^{T_2}\,\Omega_3 = \frac{1}{12}(-2\Omega_1 + 2\Omega_2 + 2\Omega_3 - 2\Omega_4)$$

$$\approx \frac{1}{2}(\Omega_1 - \Omega_2 - \Omega_3 + \Omega_4) = \phi_3 \text{ (say)}$$

$$\hat{P}^{T_2}\,\Omega_3 + \hat{P}^{T_2}\,\Omega_4 = \frac{1}{12}(-2\Omega_1 - 2\Omega_2 + 2\Omega_3 + 2\Omega_4)$$

$$\approx \frac{1}{2}(\Omega_1 + \Omega_2 - \Omega_3 - \Omega_4) = \phi_2 \quad \text{(obtained earlier)}$$

$$\hat{P}^{T_2}\,\Omega_2 + \hat{P}^{T_2}\,\Omega_4 = \frac{1}{12}(-2\Omega_1 + 2\Omega_2 - 2\Omega_3 + 2\Omega_4)$$

$$\approx \frac{1}{2}(\Omega_1 - \Omega_2 + \Omega_3 - \Omega_4) = \phi_4 \text{ (say)}$$

The orthogonality of ϕ_2, ϕ_3 and ϕ_4 can be easily verified using the condition

$$\langle \Omega_i | \Omega_j \rangle = \delta_{ij}$$

The linear relationships of s, p_x, p_y and p_z with Ω_1, Ω_2, Ω_3 and Ω_4 can now be written in matrix form as

$$[s \quad p_x \quad p_y \quad p_z] = [\Omega_1 \quad \Omega_2 \quad \Omega_3 \quad \Omega_4] \begin{bmatrix} \frac{1}{2} & \frac{1}{2} & \frac{1}{2} & \frac{1}{2} \\ \frac{1}{2} & \frac{1}{2} & -\frac{1}{2} & -\frac{1}{2} \\ \frac{1}{2} & -\frac{1}{2} & -\frac{1}{2} & \frac{1}{2} \\ \frac{1}{2} & -\frac{1}{2} & \frac{1}{2} & -\frac{1}{2} \end{bmatrix}$$

$$\begin{matrix} & \uparrow & \uparrow & \uparrow & \uparrow \\ \text{coefficents of:} & \phi_1 & \phi_2 & \phi_3 & \phi_4 \end{matrix}$$

The transformation matrix is unitary as (i) sum of the modulus-squares of elements in each column = 1 and (ii) inner product of any two different column vectors = 0. Thus the inverse of the matrix is simply its transpose and we have

$$[\Omega_1 \quad \Omega_2 \quad \Omega_3 \quad \Omega_4] = [s \quad p_x \quad p_y \quad p_z] \begin{bmatrix} \frac{1}{2} & \frac{1}{2} & \frac{1}{2} & \frac{1}{2} \\ \frac{1}{2} & \frac{1}{2} & -\frac{1}{2} & -\frac{1}{2} \\ \frac{1}{2} & -\frac{1}{2} & -\frac{1}{2} & \frac{1}{2} \\ \frac{1}{2} & -\frac{1}{2} & \frac{1}{2} & -\frac{1}{2} \end{bmatrix}$$

Hence the expressions for the sp^3 hybrid orbital wave functions are

$$A_1 \text{ symmetry: } \Omega_1 = \frac{1}{2}(s + p_x + p_y + p_z)$$

$$T_2 \text{ symmetry: } \Omega_2 = \frac{1}{2}(s + p_x - p_y - p_z), \Omega_3 = \frac{1}{2}(s - p_x - p_y + p_z)$$

$$\text{and} \qquad \Omega_4 = \frac{1}{2}(s - p_x + p_y - p_z)$$

9.2.2 Trigonal Hybridisation

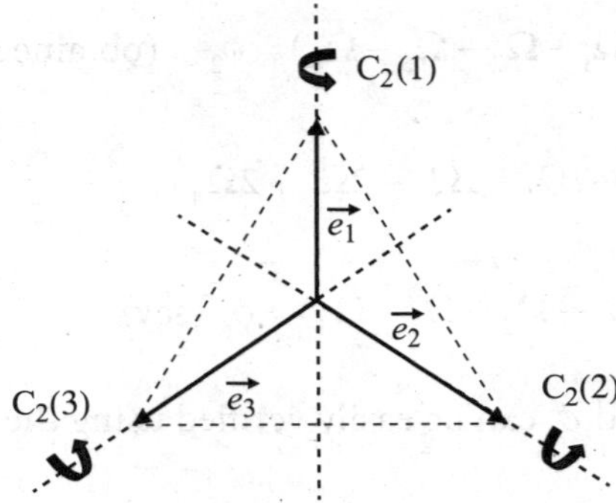

Fig. 9.2 An equilateral triangle and unit vectors $\vec{e}_1, \vec{e}_2, \vec{e}_3$ pointed from the centroid to the vertices. σ_h is the plane of the triangle, the C_3-axis is perpendicular to the σ_h plane. The three C_2-axes are as shown.

In such a hybridisation, the valence orbitals of the central atom combine linearly to form hybrid orbitals directed towards the vertices of an equilateral triangle. Let $\vec{e}_1$, $\vec{e}_2$ and $\vec{e}_3$ be three unit vectors starting from the centroid of the equilateral triangle and pointing towards the vertices (Fig. 9.2). The point group of the molecule under consideration is D_{3h} and the results of the D_{3h} symmetry operations on $\vec{e}_1$, $\vec{e}_2$ and $\vec{e}_3$ are as in Table 9.3.

Table 9.3 Action of D_{3h} symmetry operations on $\vec{e}_1$, $\vec{e}_2$ and $\vec{e}_3$

	$\hat{E}$	$\hat{C}_3$	$\hat{C}_3^2$	$\hat{C}_2(1)$	$\hat{C}_2(2)$	$\hat{C}_2(3)$	$\hat{\sigma}_h$	$\hat{S}_3$	$\hat{S}_3^5$	$\hat{\sigma}_v(1)$	$\hat{\sigma}_v(2)$	$\hat{\sigma}_v(3)$
$\vec{e}_1$	$\vec{e}_1$	$\vec{e}_3$	$\vec{e}_2$	$\vec{e}_1$	$\vec{e}_3$	$\vec{e}_2$	$\vec{e}_1$	$\vec{e}_3$	$\vec{e}_2$	$\vec{e}_1$	$\vec{e}_3$	$\vec{e}_2$
$\vec{e}_2$	$\vec{e}_2$	$\vec{e}_1$	$\vec{e}_3$	$\vec{e}_3$	$\vec{e}_2$	$\vec{e}_1$	$\vec{e}_2$	$\vec{e}_1$	$\vec{e}_3$	$\vec{e}_3$	$\vec{e}_2$	$\vec{e}_1$
$\vec{e}_3$	$\vec{e}_3$	$\vec{e}_2$	$\vec{e}_1$	$\vec{e}_2$	$\vec{e}_1$	$\vec{e}_3$	$\vec{e}_3$	$\vec{e}_2$	$\vec{e}_1$	$\vec{e}_2$	$\vec{e}_1$	$\vec{e}_3$

Thus the characters of a representation (Γ) of D_{3h} with $\{\vec{e}_1, \vec{e}_2, \vec{e}_1\}$ as the basis set are:

χ (character)

	$\hat{E}$	$2\hat{C}_3$	$3\hat{C}_2$	$\hat{\sigma}_h$	$2\hat{S}_3$	$3\hat{\sigma}_v$
Γ	3	0	1	3	0	1

Using the character table of D_{3h} let us reduce Γ as

$$\Gamma = n_1 A_1' + n_2 A_2' + n_3 E' + n_4 A_1'' + n_5 A_2'' + n_6 E''$$

Then by the reduction formula,

$$n_1 = \frac{1}{12}[3\cdot 1^* + 2\cdot 0\cdot 1^* + 3\cdot 1^*\cdot 1 + 1^*\cdot 3 + 2\cdot 1^*\cdot 0 + 3\cdot 1^*\cdot 1] = 1$$

and similarly, $n_3 = 1$, $n_2 = n_4 = n_5 = n_6 = 0$. Hence $\Gamma = A_1' + E'$. So in order to form trigonally directed hybrid orbitals, one valence shell orbital of A_1' symmetry and two valence shell orbitals of the doubly degenerate E' symmetry must combine. The D_{3h} character table also shows that the atomic orbitals which form bases of A_1' and E' representations are:

A_1'	s	d_{z^2}
E'	(p_x, p_y)	$(d_{x^2-y^2}, d_{xy})$

(**Note:** $x^2 + y^2 + z^2 = r^2$ corresponds to the s orbital.)

Hence trigonal hybridisation results from the combinations:

s and (p_x, p_y) orbitals (sp^2 hybridisation)

s and $(d_{x^2-y^2}, d_{xy})$ orbitals (sd^2 hybridisation)

d_{z^2} and (p_x, p_y) orbitals (dp^2 hybridisation)

and d_{z^2} and $(d_{x^2-y^2}, d_{xy})$ orbitals (d^3 hybridisation)

Let the sp^2 hybrid orbital wave functions be Ω_1, Ω_2 and Ω_3. Their expressions in terms of the pure atomic orbitals can be obtained as follows. The hybrid orbitals are stretched along the vectors $\vec{e}_1$, $\vec{e}_2$, $\vec{e}_3$; utilising Table 9.3 and applying the projection operators we find:

$$\hat{P}^{A_1'}\Omega_1 = \frac{1}{12}[1^* \cdot \hat{E}\Omega_1 + 1^* \cdot \hat{C}_3\Omega_1 + 1^* \cdot \hat{C}_3^2\Omega_1 + 1^* \cdot \hat{C}_2(1)\Omega_1 + 1^* \cdot \hat{C}_2(2)\Omega_1 + 1^* \cdot \hat{C}_2(3)\Omega_1$$

$$+ 1^* \cdot \hat{\sigma}_h\Omega_1 + 1^* \cdot \hat{S}_3\Omega_1 + 1^* \cdot \hat{S}_3^5\Omega_1 + 1^* \cdot \hat{\sigma}_v(1)\Omega_1 + 1^* \cdot \hat{\sigma}_v(2)\Omega_1 + 1^* \cdot \hat{\sigma}_v(3)\Omega_1]$$

$$= \frac{1}{12}(4\Omega_1 + 4\Omega_2 + 4\Omega_3) \approx \frac{1}{\sqrt{3}}(\Omega_1 + \Omega_2 + \Omega_3) \quad \text{(after normalisation)}$$

$$\hat{P}^{E'}\Omega_1 = \frac{1}{12}[2^*\hat{E}\Omega_1 + (-1)^*\hat{C}_3\Omega_1 + (-1)^*\hat{C}_3^2\Omega_1 + 0 + 2^*\hat{\sigma}_h\Omega_1 + (-1)^*\hat{S}_3\Omega_1 + (-1)^*\hat{S}_3^5\Omega_1 + 0]$$

$$= \frac{1}{6}[2\Omega_1 - \Omega_2 - \Omega_3] \approx \frac{1}{\sqrt{6}}[2\Omega_1 - \Omega_2 - \Omega_3] \quad \text{(after normalisation)}$$

$$\hat{P}^{E'}\Omega_2 = \frac{1}{\sqrt{6}}[2\Omega_2 - \Omega_3 - \Omega_1] \quad \text{(after normalisation)}$$

$$\hat{P}^{E'}\Omega_3 = \frac{1}{\sqrt{6}}[2\Omega_3 - \Omega_2 - \Omega_1] \quad \text{(after normalisation)}$$

It is easy to verify that $\hat{P}^{E'}\Omega_1 + \hat{P}^{E'}\Omega_2 + \hat{P}^{E'}\Omega_3 = 0$. Hence, these three functions are linearly dependent and are not orthogonal. We can construct two orthogonal functions from them as follows:

(i) Keep $\hat{P}^{E'}\Omega_1$ unchanged.

(ii) $\hat{P}^{E'}\Omega_2 - \hat{P}^{E'}\Omega_3 = \frac{1}{\sqrt{6}}(3\Omega_2 - 3\Omega_3) \approx \frac{1}{\sqrt{2}}(\Omega_2 - \Omega_3)$ (after normalisation).

The functions $\hat{P}^{E'}\Omega_1$ and $\frac{1}{\sqrt{2}}(\Omega_2 - \Omega_3)$ are orthogonal as can be readily verified by recalling that $\langle\Omega_i|\Omega_j\rangle = \delta_{ij}$. As s and (p_x, p_y) are bases of A_1' and E' respectively, we can identify the s orbital as the $\hat{P}^{A_1'}\Omega_1$ function and the (p_x, p_y) set as the set of $\hat{P}^{E'}$ type functions.

$$s = \hat{P}^{A_1'}\Omega_1 = \frac{1}{\sqrt{3}}(\Omega_1 + \Omega_2 + \Omega_3)$$

$$p_x = \hat{P}^{E'}(1) = \frac{1}{\sqrt{6}}[2\Omega_1 - \Omega_2 - \Omega_3]$$

$$p_y = \hat{P}^{E'}(2) = \frac{1}{\sqrt{2}}[\Omega_2 - \Omega_3]$$

In matrix form the relations are,

$$[s \quad p_x \quad p_y] = [\Omega_1 \quad \Omega_2 \quad \Omega_3] \begin{bmatrix} \frac{1}{\sqrt{3}} & \frac{2}{\sqrt{6}} & 0 \\ \frac{1}{\sqrt{3}} & -\frac{1}{\sqrt{6}} & \frac{1}{\sqrt{2}} \\ \frac{1}{\sqrt{3}} & -\frac{1}{\sqrt{6}} & -\frac{1}{\sqrt{2}} \end{bmatrix}$$

It is easy to invert the matrix on the RHS as it is unitary [inverse = conjugate transpose]. Hence expressions for the sp^2 hybridised orbitals in terms of s, p_x and p_y are as follows:

$$[\Omega_1 \quad \Omega_2 \quad \Omega_3] = [s \quad p_x \quad p_y] \begin{bmatrix} \frac{1}{\sqrt{3}} & \frac{2}{\sqrt{6}} & 0 \\ \frac{1}{\sqrt{3}} & -\frac{1}{\sqrt{6}} & \frac{1}{\sqrt{2}} \\ \frac{1}{\sqrt{3}} & -\frac{1}{\sqrt{6}} & -\frac{1}{\sqrt{2}} \end{bmatrix}^{-1} = [s \quad p_x \quad p_y] \begin{bmatrix} \frac{1}{\sqrt{3}} & \frac{1}{\sqrt{3}} & \frac{1}{\sqrt{3}} \\ \frac{2}{\sqrt{6}} & -\frac{1}{\sqrt{6}} & -\frac{1}{\sqrt{6}} \\ 0 & \frac{1}{\sqrt{2}} & -\frac{1}{\sqrt{2}} \end{bmatrix}$$

Written explicitly, $\Omega_1 = \frac{1}{\sqrt{6}}(\sqrt{2}s + 2p_x), \Omega_2 = \frac{1}{\sqrt{6}}(\sqrt{2}s - p_x + \sqrt{3}p_y)$ and

$$\Omega_3 = \frac{1}{\sqrt{6}}(\sqrt{2}s - p_x - \sqrt{3}p_y)$$

The above results may be obtained in an alternative and easier way as follows:

The point group C_3 is a subgroup of D_{3h} and with $\{\vec{e}_1, \vec{e}_2, \vec{e}_3\}$ as basis, a representation (Γ^e) of this subgroup is:

	χ (character)		
	$\hat{E}$	$\hat{C}_3$	$\hat{C}_3^2$
Γ^e	3	0	1

Using the character table of C_3 and the reduction formula, $\Gamma^e = A + E_1 + E_2$. The normalised SALCs corresponding to the component irreducible representations are now constructed as follows:

$$\hat{P}^{A_1}\,\Omega_1 = \frac{1}{3}[1^*.\hat{E}\Omega_1 + 1^*.\hat{C}_3\,\Omega_1 + 1^*.\hat{C}_3^2\,\Omega_1] = \frac{1}{3}(\Omega_1 + \Omega_2 + \Omega_3)$$

$$\approx \frac{1}{\sqrt{3}}(\Omega_1 + \Omega_2 + \Omega_3) \quad \text{(after normalisation)}$$

$$\hat{P}^{E_1}\,\Omega_1 = \frac{1}{3}[1^*\,\hat{E}\Omega_1 + \omega^*\,\hat{C}_3\,\Omega_1 + \omega\hat{C}_3^2\,\Omega_1] = \frac{1}{3}(\Omega_1 + \omega^*\,\Omega_2 + \omega\Omega_3)$$

$$\hat{P}^{E_2}\,\Omega_1 = \frac{1}{3}[1^*\,\hat{E}\Omega_1 + \omega\hat{C}_3\,\Omega_1 + \omega^*\hat{C}_3^2\,\Omega_1] = \frac{1}{3}(\Omega_1 + \omega\Omega_2 + \omega^*\,\Omega_3)$$

The complex coefficients can be removed by addition and subtraction and utilising the fact that $1 + \omega + \omega^* = 0$:

$$\hat{P}^{E_1}\,\Omega_1 + \hat{P}^{E_2}\,\Omega_1 = \frac{1}{3}(2\Omega_1 - \Omega_2 - \Omega_3) \approx \frac{1}{\sqrt{6}}(2\Omega_1 - \Omega_2 - \Omega_3)$$

$$\hat{P}^{E_1}\,\Omega_1 - \hat{P}^{E_2}\,\Omega_1 = \frac{1}{3}[(\omega^* - \omega)\,\Omega_2 - (\omega^* - \omega)\,\Omega_3] \approx \frac{1}{\sqrt{2}}(\Omega_2 - \Omega_3) \text{ (after normalisation)}$$

The s orbital can be identified with $\hat{P}^{A_1}\,\Omega_1 = \frac{1}{\sqrt{3}}(\Omega_1 + \Omega_2 + \Omega_3)$ and (p_x, p_y) with the set $\{\frac{1}{\sqrt{6}}(2\Omega_1 - \Omega_2 - \Omega_3), \frac{1}{\sqrt{2}}(\Omega_2 - \Omega_3)\}$; the rest of the procedure is as before.

9.2.3 Square Planar Hybridisation

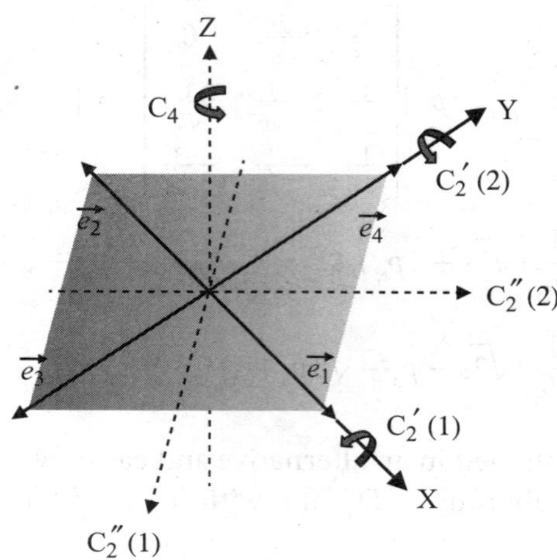

Fig. 9.3 Some symmetry elements of D_{4h}, orientation of coordinate axes and unit vectors $\vec{e}_1$, $\vec{e}_2$, $\vec{e}_3$ and $\vec{e}_4$ along which the square planar hybrid orbitals are stretched.

The hybrid orbitals of the central atom in this case are directed towards the vertices of a square, at the centre of which the central atom is situated. Let $\vec{e}_1$, $\vec{e}_2$, $\vec{e}_3$, $\vec{e}_4$ be four vectors along the diagonals of the square (Fig. 9.3). The hybrid orbitals with wave functions Ω_1, Ω_2, Ω_3, Ω_4 are stretched along these vectors. The X and Y axes are along the diagonals with the origin at the centre of the square. The Z-axis is perpendicular to the XY plane (Fig. 9.3). The principal C_4-axis is along the Z-axis. The point group of a square is D_{4h} and Table 9.4 shows the results of applying D_{4h} symmetry operators on the vectors $\{\vec{e}_i\}$, $i = 1$ to 4.

From these results, the characters of a representation Γ of D_{4h} with $(\vec{e}_1, \vec{e}_2, \vec{e}_3, \vec{e}_4)$ as the basis set are:

χ **(character)**

	$\hat{E}$	$2\hat{C}_4$	$\hat{C}_2$	$2\hat{C}_2'$	$2\,\hat{C}_2''$	$\hat{i}$	$2\hat{S}_4$	$\hat{\sigma}_h$	$2\hat{\sigma}_v$	$2\hat{\sigma}_d$
Γ	4	0	0	2	0	0	0	4	2	0

On reduction, $\Gamma = A_{1g} + B_{1g} + E_u$ (9.3)

The character table of D_{4h} shows that the basis orbitals of the component irreducible representations found in Eq. (9.3) are,

$A_{1g} : s, d_{z^2}$; $\quad B_{1g} : d_{x^2-y^2}$; $\quad E_u : (p_x, p_y)$

Table 9.4 Action of D_{4h} symmetry operators on the vectors $\{\vec{e}_i\}$, $i = 1$ to 4 of Fig. 9.3

Part A

	$\hat{E}$	$\hat{C}_4$	$\hat{C}_4^3$	$\hat{C}_2$	$\hat{C}'_2(1)$	$\hat{C}'_2(2)$	$\hat{C}''_2(1)$	$\hat{C}''_2(2)$
$\vec{e}_1$	$\vec{e}_1$	$\vec{e}_4$	$\vec{e}_3$	$\vec{e}_2$	$\vec{e}_1$	$\vec{e}_2$	$\vec{e}_3$	$\vec{e}_4$
$\vec{e}_2$	$\vec{e}_2$	$\vec{e}_3$	$\vec{e}_4$	$\vec{e}_1$	$\vec{e}_2$	$\vec{e}_1$	$\vec{e}_4$	$\vec{e}_3$
$\vec{e}_3$	$\vec{e}_3$	$\vec{e}_1$	$\vec{e}_2$	$\vec{e}_4$	$\vec{e}_4$	$\vec{e}_3$	$\vec{e}_1$	$\vec{e}_2$
$\vec{e}_4$	$\vec{e}_4$	$\vec{e}_2$	$\vec{e}_1$	$\vec{e}_3$	$\vec{e}_3$	$\vec{e}_4$	$\vec{e}_2$	$\vec{e}_1$

Part B

	$\hat{i}$	$\hat{S}_4$	$\hat{S}_4^3$	$\hat{\sigma}_h$	$\hat{\sigma}_v(1)$	$\hat{\sigma}_v(2)$	$\hat{\sigma}_d(1)$	$\hat{\sigma}_d(2)$
$\vec{e}_1$	$\vec{e}_2$	$\vec{e}_3$	$\vec{e}_4$	$\vec{e}_1$	$\vec{e}_1$	$\vec{e}_2$	$\vec{e}_3$	$\vec{e}_4$
$\vec{e}_2$	$\vec{e}_1$	$\vec{e}_4$	$\vec{e}_3$	$\vec{e}_2$	$\vec{e}_2$	$\vec{e}_1$	$\vec{e}_4$	$\vec{e}_3$
$\vec{e}_3$	$\vec{e}_4$	$\vec{e}_2$	$\vec{e}_1$	$\vec{e}_3$	$\vec{e}_4$	$\vec{e}_3$	$\vec{e}_1$	$\vec{e}_2$
$\vec{e}_4$	$\vec{e}_3$	$\vec{e}_1$	$\vec{e}_2$	$\vec{e}_4$	$\vec{e}_3$	$\vec{e}_4$	$\vec{e}_2$	$\vec{e}_1$

Hence the four hybrid orbitals for square planar geometry are obtained from linear combination of

$d_{x^2-y^2}$, s and (p_x, p_y) orbitals (*dsp*² hybridisation)

and $d_{x^2-y^2}$, d_{z^2} and (p_x, p_y) orbitals (*d*²*p*² hybridisation)

The first case is usually found in transition metal complexes where $3d_{x^2-y^2}$, $4s$ and ($4p_x$, $4p_y$) orbitals are involved.

To find the expressions of the hybrid orbital wave functions $\{\Omega_i\}$, $i = 1$ to 4, in terms of the atomic *s*, *p* and *d* orbitals, we utilise the fact that linear combinations of Ω_i adopted to the symmetries of A_{1g}, B_{1g} and E_u must yield the atomic orbitals *s* or d_{z^2} (for A_{1g}), $d_{x^2-y^2}$ (for B_{1g}) and (p_x, p_y) (for E_u). Hence we first construct the required SALCs of the Ωs.

$$\hat{P}^{A_{1g}}\,\Omega_1 = \frac{1}{16}[4\Omega_1 + 4\Omega_2 + 4\Omega_3 + 4\Omega_4] \approx \frac{1}{2}[\Omega_1 + \Omega_2 + \Omega_3 + \Omega_4]$$

$$\hat{P}^{B_{1g}}\,\Omega_1 = \frac{1}{16}[4\Omega_1 + 4\Omega_2 - 4\Omega_3 - 4\Omega_4] \approx \frac{1}{2}[\Omega_1 + \Omega_2 - \Omega_3 - \Omega_4]$$

$$\hat{P}^{E_u}\,\Omega_1 = \frac{1}{16}[4\Omega_1 - 4\Omega_2] \approx \frac{1}{\sqrt{2}}[\Omega_1 - \Omega_2]$$

$$\hat{P}^{E_u}\,\Omega_3 = \frac{1}{16}[4\Omega_3 - 4\Omega_4] \approx \frac{1}{2}[\Omega_3 - \Omega_4]$$

Hence for *dsp*² hybridisation

$$s = \hat{P}^{A_{1g}}\,\Omega_1 = \frac{1}{2}[\Omega_1 + \Omega_2 + \Omega_3 + \Omega_4];\; d_{x^2-y^2} = \hat{P}^{B_{1g}}\,\Omega_1 = \frac{1}{2}[\Omega_1 + \Omega_2 - \Omega_3 - \Omega_4]$$

$$p_x = \hat{P}^{E_u}\,\Omega_1 = \frac{1}{\sqrt{2}}[\Omega_1 - \Omega_2];\; p_y = \hat{P}^{E_u}\,\Omega_3 = \frac{1}{2}[\Omega_3 - \Omega_4]$$

Using matrix notation and inverting as before,

$$[\Omega_1 \quad \Omega_2 \quad \Omega_3 \quad \Omega_4] = [s \quad d_{x^2-y^2} \quad p_x \quad p_y]\begin{bmatrix} \frac{1}{2} & \frac{1}{2} & \frac{1}{\sqrt{2}} & 0 \\ \frac{1}{2} & \frac{1}{2} & -\frac{1}{\sqrt{2}} & 0 \\ \frac{1}{2} & -\frac{1}{2} & 0 & \frac{1}{\sqrt{2}} \\ \frac{1}{2} & -\frac{1}{2} & 0 & -\frac{1}{\sqrt{2}} \end{bmatrix}^{-1}$$

$$= [s \quad d_{x^2-y^2} \quad p_x \quad p_y]\begin{bmatrix} \frac{1}{2} & \frac{1}{2} & \frac{1}{2} & \frac{1}{2} \\ \frac{1}{2} & \frac{1}{2} & -\frac{1}{2} & -\frac{1}{2} \\ \frac{1}{\sqrt{2}} & -\frac{1}{\sqrt{2}} & 0 & 0 \\ 0 & 0 & \frac{1}{\sqrt{2}} & -\frac{1}{\sqrt{2}} \end{bmatrix}$$

This gives the required expressions for the Ωs in terms of the component atomic orbitals.

9.2.4 Trigonal Bipyramidal Hybridisation

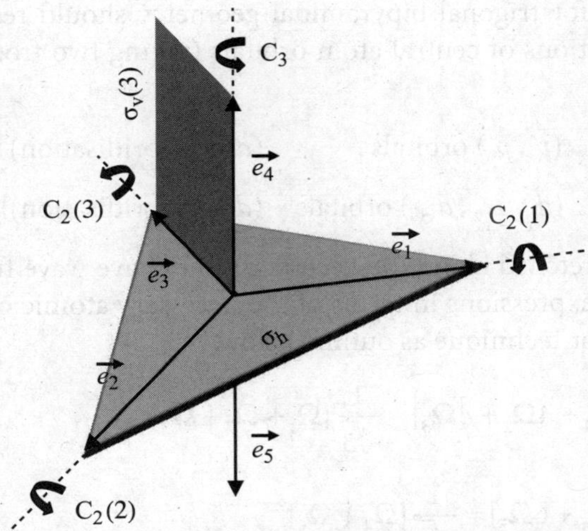

Fig. 9.4 Some symmetry elements of D_{3h} and orientation of axial and equatorial unit vectors along which trigonally hybridised orbitals are stretched

The point group of a molecule or ion having a trigonal bipyramidal shape is D_{3h}. The unit vectors $\vec{e}_1$, $\vec{e}_2$, $\vec{e}_3$ on the equatorial plane and the unit vectors $\vec{e}_4$, $\vec{e}_5$ along the C_3-axis (Fig. 9.4) act as bases of a representation Γ of D_{3h}. The transformation properties of these vectors under the D_{3h} symmetry operators are given in Table 9.5.

Table 9.5 Action of D_{3h} symmetry operators on the basis vectors $\{\vec{e}_i\}$, $i = 1$ to 5 (Fig. 9.4)

	$\hat{E}$	$\hat{C}_3$	$\hat{C}_3^2$	$\hat{C}_2(1)$	$\hat{C}_2(2)$	$\hat{C}_2(3)$	$\hat{\sigma}_h$	$\hat{S}_3$	$\hat{S}_3^5$	$\hat{\sigma}_v(1)$	$\hat{\sigma}_v(2)$	$\hat{\sigma}_v(3)$
$\vec{e}_1$	$\vec{e}_1$	$\vec{e}_3$	$\vec{e}_2$	$\vec{e}_1$	$\vec{e}_3$	$\vec{e}_2$	$\vec{e}_1$	$\vec{e}_3$	$\vec{e}_2$	$\vec{e}_1$	$\vec{e}_3$	$\vec{e}_2$
$\vec{e}_2$	$\vec{e}_2$	$\vec{e}_1$	$\vec{e}_3$	$\vec{e}_3$	$\vec{e}_2$	$\vec{e}_1$	$\vec{e}_2$	$\vec{e}_1$	$\vec{e}_3$	$\vec{e}_3$	$\vec{e}_2$	$\vec{e}_1$
$\vec{e}_3$	$\vec{e}_3$	$\vec{e}_2$	$\vec{e}_1$	$\vec{e}_2$	$\vec{e}_1$	$\vec{e}_3$	$\vec{e}_3$	$\vec{e}_2$	$\vec{e}_1$	$\vec{e}_2$	$\vec{e}_1$	$\vec{e}_3$
$\vec{e}_4$	$\vec{e}_4$	$\vec{e}_4$	$\vec{e}_4$	$\vec{e}_5$	$\vec{e}_5$	$\vec{e}_5$	$\vec{e}_5$	$\vec{e}_5$	$\vec{e}_5$	$\vec{e}_4$	$\vec{e}_4$	$\vec{e}_4$
$\vec{e}_5$	$\vec{e}_5$	$\vec{e}_5$	$\vec{e}_5$	$\vec{e}_4$	$\vec{e}_4$	$\vec{e}_4$	$\vec{e}_4$	$\vec{e}_4$	$\vec{e}_4$	$\vec{e}_5$	$\vec{e}_5$	$\vec{e}_5$

The characters in the representations of Γ with the basis vectors given in Table 9.5 are:

	χ (character)						
	$\hat{E}$	$2\hat{C}_3$	$3\hat{C}_2$	$\hat{\sigma}_h$	$2\hat{S}_3$	$3\hat{\sigma}_v$	
Γ	5	2	1	3	0	3	$\Gamma = 2A_1' + A_2'' + E'$

The result of reduction of Γ is also shown above. Looking up the D_{3h} character table, the component irreducible representations are found to have the following basis functions:

$$A_1': s, d_{z^2}; \qquad A_2'': p_z; \qquad E': (p_x, p_y), (d_{x^2-y^2}, d_{xy})$$

Hence, hybrid orbitals for trigonal bipyramidal geometry should result from the following types of linear combinations of central atom orbitals (taking two from A_1', one from A_2'' and one set from E')

$$d_{z^2}, s, p_z \text{ and } (p_x, p_y) \text{ orbitals} \qquad (dsp^3 \text{ hybridisation})$$

and $$d_{z^2}, s, p_z \text{ and } (d_{x^2-y^2}, d_{xy}) \text{ orbitals} \qquad (d^3sp \text{ hybridisation})$$

If the hybrid orbitals stretched along the vectors $\vec{e}_1$ to $\vec{e}_5$ have wave functions $\Omega_1, \Omega_2, \ldots, \Omega_5$ respectively, then their expressions in terms of the necessary atomic orbitals are found by the usual projection operator technique as outline below.

$$\hat{P}^{A_1'}\Omega_1 = \frac{1}{12}[4\Omega_1 + 4\Omega_2 + 4\Omega_3] \approx \frac{1}{\sqrt{3}}[\Omega_1 + \Omega_2 + \Omega_3]$$

$$\hat{P}^{A_1'}\Omega_4 = \frac{1}{12}[6\Omega_4 + 6\Omega_5] \approx \frac{1}{\sqrt{2}}[\Omega_4 + \Omega_5]$$

$$\hat{P}^{A_2''}\Omega_4 = \frac{1}{12}[6\Omega_4 - 6\Omega_5] \approx \frac{1}{\sqrt{2}}[\Omega_4 - \Omega_5]$$

$$\hat{P}^{E'}\Omega_1 = \frac{1}{12}[4\Omega_1 - 2\Omega_2 - 2\Omega_3] \approx \frac{1}{\sqrt{6}}[2\Omega_1 - \Omega_2 - \Omega_3]$$

$$\hat{P}^{E'}\Omega_2 = \frac{1}{12}[4\Omega_2 - 2\Omega_3 - 2\Omega_1] \approx \frac{1}{\sqrt{6}}[2\Omega_2 - \Omega_3 - \Omega_1]$$

$$\hat{P}^{E'}\Omega_3 = \frac{1}{12}[4\Omega_3 - 2\Omega_1 - 2\Omega_2] \approx \frac{1}{\sqrt{6}}[2\Omega_3 - \Omega_1 - \Omega_2]$$

The three functions corresponding to the E' representation obtained above are not linearly independent as their sum is zero. We can construct two linearly independent functions from them as follows

$$\hat{P}^{E'}\Omega_1 = \frac{1}{\sqrt{6}}[2\Omega_1 - \Omega_2 - \Omega_3] \text{ is kept intact}$$

$$\hat{P}^{E'}\Omega_2 - \hat{P}^{E'}\Omega_3 = \frac{1}{\sqrt{2}}[\Omega_2 - \Omega_3] \text{ (after normalisation)}$$

By the usual matrix inversion procedure described earlier, each Ω_i can now be expressed in terms of the *d*, *s* and *p* functions of the central atom orbitals.

9.2.5 Octahedral Hybridisation

Let us consider the hybridisation of atomic orbitals of the central atom A of an octahedral AB_6 molecule. In order to form bonds directed octahedrally from A to the atoms B at the vertices of an octahedron, the central atom hybrid orbitals $\phi_1, \phi_2, \ldots, \phi_6$ must be stretched along the two parts of each coordinate axis with the origin at A (Fig. 9.5).

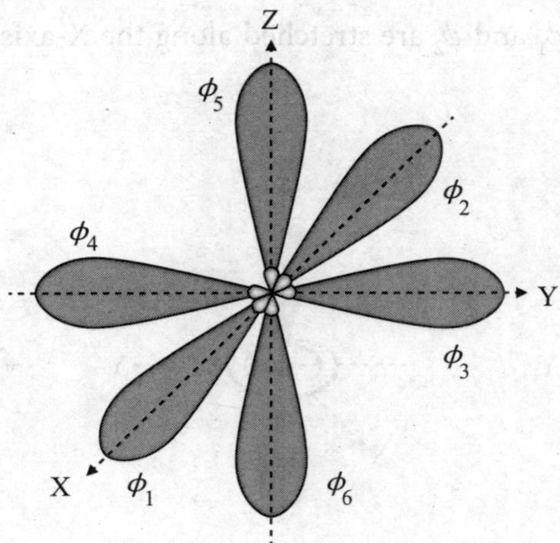

Fig. 9.5 Orientation of hybrid orbitals in octahedral geometry

With these orbitals as bases, the characters of a representations Γ_ϕ of the O_h point group are as follows:

χ (character)

	$\hat{E}$	$8\hat{C}_3$	$6\hat{C}_2'$	$6\hat{C}_4$	$3\hat{C}_2$	$\hat{i}$	$6\hat{S}_4$	$8\hat{S}_6$	$3\hat{\sigma}_h$	$3\hat{\sigma}_d$
Γ_ϕ	6	0	0	2	2	0	0	0	4	2

Applying reduction formula it is found that

$$\Gamma_\phi = A_{1g} + E_g + T_{1u} \tag{9.4}$$

From the character table of O_h it is found that the component irreducible representations found in Eq. (9.4) have basis functions as listed below.

$$A_{1g} : s; \quad T_{1u} : (p_x, p_y, p_z); \quad E_g : (d_{x^2-y^2}, d_{z^2})$$

Hence the required hybrid orbitals (ϕ_1 to ϕ_6) are made up of one *s* orbital, three *p*-orbitals and two *d*-orbitals (that is, the hybridisation is of sp^3d^2 or d^2sp^3 type). The forms of the wave functions ϕ_1 to ϕ_6 in terms of *s*, *p* and *d* functions can be obtained by the usual projection operator method. But in the present case an alternative easier scheme is possible.

Just as the hybrid orbitals ϕ_1 to ϕ_6 are linear combinations of the central atom orbitals $s, p_x, p_y, p_z, d_{x^2-y^2}$ and d_{z^2}, the reverse is also true. Hence the totally symmetric (A_{1g}) orbital *s* should be a symmetric linear combination of hybrid orbitals $\phi_1, \phi_2, \ldots, \phi_6$. This is possible if the ϕ-functions contribute equally to *s* (as the ϕs are all equivalent). Thus for A_{1g} symmetry,

$$s = \frac{1}{\sqrt{6}}(\phi_1 + \phi_2 + \phi_3 + \phi_4 + \phi_5 + \phi_6) \tag{9.5}$$

The normalisation factor $1/\sqrt{6}$ is based on the assumption that the hybrid orbitals form an orthonormal set, $\langle \phi_i | \phi_j \rangle = \delta_{ij}$.

In the E_g set, the atomic orbitals are $d_{x^2-y^2}$ and d_{z^2}. Of these two, $d_{x^2-y^2}$ is stretched along the X- and Y-axes. Again, ϕ_1 and ϕ_2 are stretched along the X-axis while ϕ_3 and ϕ_4 are along the Y-axis.

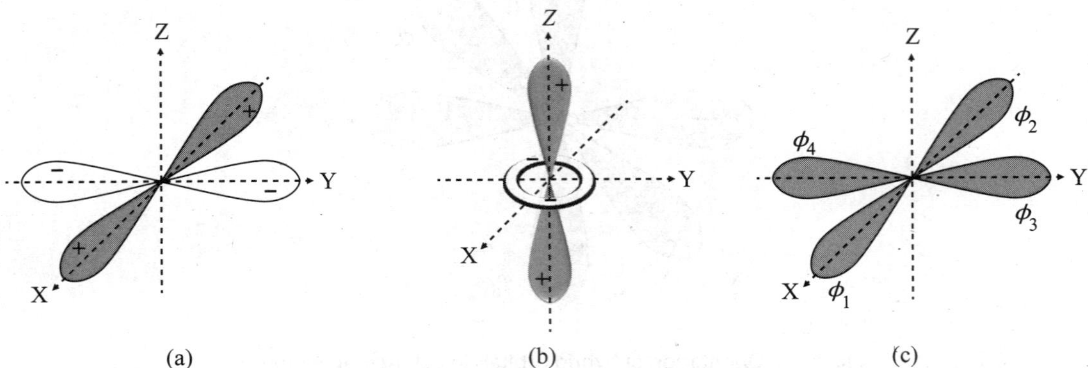

Fig. 9.6 (a) $d_{x^2-y^2}$ orbital (b) d_{z^2} orbital and (c) a hybrid orbital combination of E_g symmetry

Hence looking at Fig. 9.6 and considering the signs of the lobes of $d_{x^2-y^2}$, we have

$$d_{x^2-y^2} = \frac{1}{2}(\phi_1 + \phi_2 - \phi_3 - \phi_4) \tag{9.6}$$

where 1/2 is the normalisation factor.

The d_{z^2} ($\equiv d_{2z^2-x^2-y^2}$) orbital has two parts; one part is stretched along the Z-axis with an amplitude twice as that of the second part, which is a torus on the XY plane with a negative sign [Fig. 9.6(b)]. Hence a comparison of Fig. 9.6(b) with Fig. 9.5 reveals that d_{z^2} must be made up of ϕ_1, ϕ_2, ϕ_3 and ϕ_4 with negative sign and ϕ_5 and ϕ_6 with positive sign; contribution of ϕ_5 and ϕ_6 should be twice as much as that of ϕ_1 to ϕ_4 each. Hence, after normalisation,

$$d_{z^2} = \frac{1}{\sqrt{12}}(2\phi_5 + 2\phi_6 - \phi_1 - \phi_2 - \phi_3 - \phi_4) \tag{9.7}$$

In the same way, the T_{1u} set of atomic orbitals (p_x, p_y, p_z) are easily found to be given by

$$T_{1u}: \begin{cases} p_x = \dfrac{1}{\sqrt{2}}(\phi_1 - \phi_2) \\ p_y = \dfrac{1}{\sqrt{2}}(\phi_3 - \phi_4) \\ p_z = \dfrac{1}{\sqrt{2}}(\phi_5 - \phi_6) \end{cases} \tag{9.8}$$

The relations (9.5) to (9.8) can be written in matrix form as follows:

$$\begin{pmatrix} s \\ p_x \\ p_y \\ p_z \\ d_{x^2-y^2} \\ d_{z^2} \end{pmatrix} = \begin{pmatrix} \frac{1}{\sqrt{6}} & \frac{1}{\sqrt{6}} & \frac{1}{\sqrt{6}} & \frac{1}{\sqrt{6}} & \frac{1}{\sqrt{6}} & \frac{1}{\sqrt{6}} \\ \frac{1}{\sqrt{2}} & -\frac{1}{\sqrt{2}} & 0 & 0 & 0 & 0 \\ 0 & 0 & \frac{1}{\sqrt{2}} & -\frac{1}{\sqrt{2}} & 0 & 0 \\ 0 & 0 & 0 & 0 & \frac{1}{\sqrt{2}} & -\frac{1}{\sqrt{2}} \\ \frac{1}{2} & \frac{1}{2} & -\frac{1}{2} & -\frac{1}{2} & 0 & 0 \\ -\frac{1}{\sqrt{12}} & -\frac{1}{\sqrt{12}} & -\frac{1}{\sqrt{12}} & -\frac{1}{\sqrt{12}} & \frac{1}{\sqrt{3}} & \frac{1}{\sqrt{3}} \end{pmatrix} \begin{pmatrix} \phi_1 \\ \phi_2 \\ \phi_3 \\ \phi_4 \\ \phi_5 \\ \phi_6 \end{pmatrix} \tag{9.9}$$

The square matrix in the RHS of Eq. (9.9) can be easily verified to be unitary. Hence its inverse is equal to its transpose. Matrix inversion will now give, from Eq. (9.9), the following expressions for the hybrid orbitals.

$$\left.\begin{aligned} \phi_1 &= \frac{1}{\sqrt{6}}s + \frac{1}{\sqrt{2}}p_x + \frac{1}{2}d_{x^2-y^2} - \frac{1}{\sqrt{12}}d_{z^2} \\ \phi_2 &= \frac{1}{\sqrt{6}}s - \frac{1}{\sqrt{2}}p_x + \frac{1}{2}d_{x^2-y^2} - \frac{1}{\sqrt{12}}d_{z^2} \\ \phi_3 &= \frac{1}{\sqrt{6}}s + \frac{1}{\sqrt{2}}p_y - \frac{1}{2}d_{x^2-y^2} - \frac{1}{\sqrt{12}}d_{z^2} \\ \phi_4 &= \frac{1}{\sqrt{6}}s - \frac{1}{\sqrt{2}}p_y - \frac{1}{2}d_{x^2-y^2} - \frac{1}{\sqrt{12}}d_{z^2} \\ \phi_5 &= \frac{1}{\sqrt{6}}s + \frac{1}{\sqrt{2}}p_z + \frac{1}{\sqrt{3}}d_{z^2} \\ \phi_6 &= \frac{1}{\sqrt{6}}s - \frac{1}{\sqrt{2}}p_z + \frac{1}{\sqrt{3}}d_{z^2} \end{aligned}\right\} \tag{9.10}$$

Note: Verify that $\langle \phi_i | \phi_j \rangle = 0$ for $i \neq j$ using these expressions.

9.3 EXAMPLES OF MO CALCULATION BY THE LOCALISED MO METHOD

MO calculation for molecules with both σ-and π-bonds is easier in the localised MO method than that followed in Chapter 8 where pristine atomic orbitals were used. Some examples are given below.

9.3.1 MOs of Ethylene Using the Localised MO Method

In the localised MO approach, the C atoms in ethylene are assumed to have sp^2 hybrid orbitals capable of forming σ-bonds. We call these σ-orbitals and in Fig. 9.7, where the C-atoms are labelled 1 and 2, the σ-orbitals are drawn on the XY plane. The π-bonds are formed by overlap of the p_z orbitals of the two C-atoms. The group orbitals of the two C-atoms are first constructed on the basis of the D_{2h} point group of the C_2H_4 molecule. These are then combined with the group orbitals of the four 1s(H) orbitals by the LCAO–MO method.

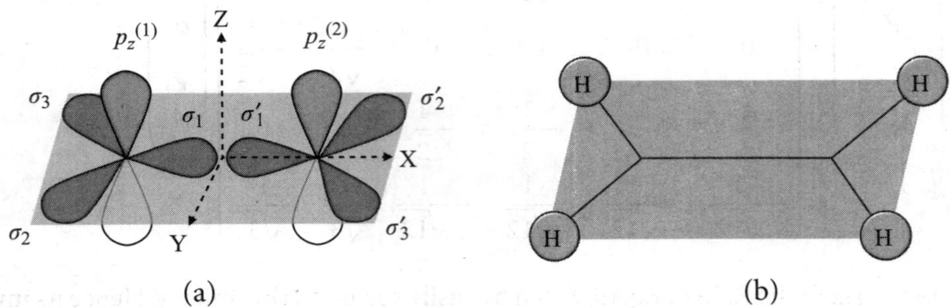

Fig. 9.7 (a) σ and π (that is, p) orbitals of the two C-atoms of ethylene, (b) the 1s orbitals of the four H atoms of ethylene

No symmetry operation of D_{2h} can bring a σ-orbital into a π-orbital and hence the total representation (Γ) of D_{2h} with the entire set of orbitals of the C atom (Fig. 9.7) can be considered as

$$\Gamma = \Gamma_\sigma + \Gamma_\pi \tag{9.11}$$

where Γ_σ is a representation of D_{2h} with the set of σ-orbitals $\{\sigma_1, \sigma_2, \sigma_3, \sigma_1', \sigma_2', \sigma_3'\}$ as basis and Γ_π is a representation with $\{p_z^{(1)}, p_z^{(2)}\}$ as basis. The character of Γ_σ, obtained from the transformation properties of the σ-orbitals are:

χ (character)

	$\hat{E}$	$\hat{C}_2(Z)$	$\hat{C}_2(Y)$	$\hat{C}_2(X)$	$\hat{i}$	$\hat{\sigma}(XY)$	$\hat{\sigma}(XZ)$	$\hat{\sigma}(YZ)$
Γ_σ	6	0	0	2	0	6	2	0

By reduction formula, $\Gamma_\sigma = 2A_g + B_{1g} + B_{2u} + 2B_{3u}$ (9.12)

The carbon p-orbitals, $p_z^{(1)}$ and $p_z^{(2)}$, produce for a representation (Γ_π), the set of characters:

χ (character)

	$\hat{E}$	$\hat{C}_2(Z)$	$\hat{C}_2(Y)$	$\hat{C}_2(X)$	$\hat{i}$	$\hat{\sigma}(XY)$	$\hat{\sigma}(XZ)$	$\hat{\sigma}(YZ)$
Γ_π	2	0	0	- 2	0	- 2	2	0

This reduces as

$$\Gamma_\pi = B_{2g} + B_{1u} \tag{9.13}$$

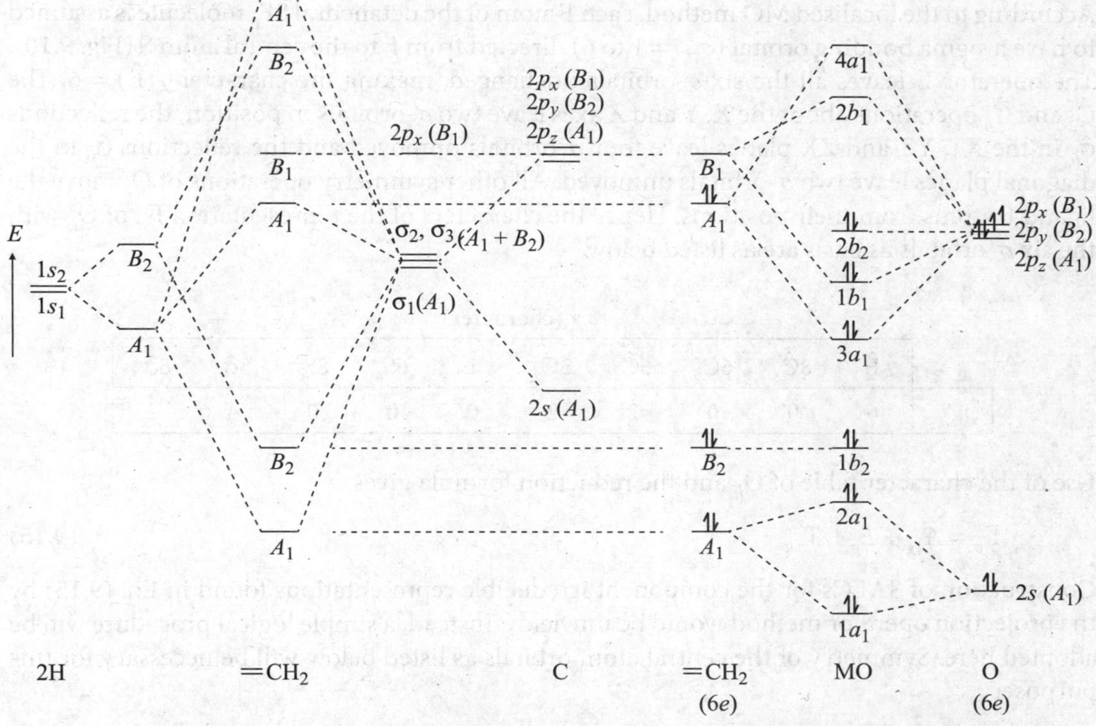

Fig. 9.9 Energy level diagram of the $=CH_2$ fragment and that of formaldehyde placed side by side with the same scale of relative energy

9.3.3 MOs OF SF_6 (OCTAHEDRAL)

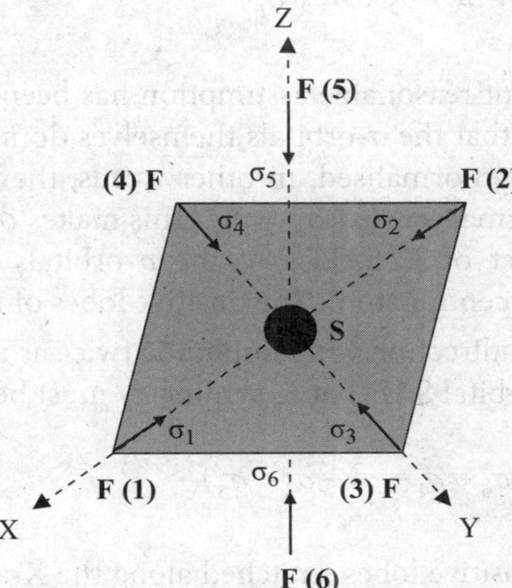

Fig. 9.10 Orientation of the σ orbitals of the F atoms along the coordinate axes in the octahedral SF_6 molecule

According to the localised MO method, each F atom of the octahedral SF_6 molecule is assumed to have a sigma bonding orbital (σ_i, $i = 1$ to 6) directed from F to the central atom S (Fig. 9.10). The operator $\hat{E}$ leaves all the six σ-orbitals unchanged, making the character $\chi(\hat{E}) = 6$. The $\hat{C}_2$ and $\hat{C}_4$ operations about the X, Y and Z axes leave two σ-orbitals in position; the reflections $\hat{\sigma}_h$ in the XY, YZ and ZX planes leave four σ-orbitals unmoved and the reflections $\hat{\sigma}_d$ in the diagonal planes leave two σ-orbitals unmoved. All other symmetry operations of O_h move the ligand F atoms from their positions. Hence the characters of the representation Γ_σ of O_h with the six σ-orbitals as basis are as listed below.

	χ (character)									
	$\hat{E}$	$8\hat{C}_3$	$6\hat{C}_2'$	$6\hat{C}_4$	$3\hat{C}_2$	$\hat{i}$	$6\hat{S}_4$	$8\hat{S}_6$	$3\hat{\sigma}_h$	$6\hat{\sigma}_d$
Γ_σ	6	0	0	2	2	0	0	0	4	2

Use of the character table of O_h and the reduction formula gives

$$\Gamma_\sigma = A_{1g} + E_g + T_{1u} \tag{9.15}$$

Construction of SALCs for the component irreducible representations found in Eq. (9.15) by the projection operator method would be unwieldy. Instead, a simple logical procedure will be adopted here. Symmetry of the central atom orbitals as listed below will be necessary for this purpose:

$$A_{1g} : s; \qquad E_g : (d_{x^2-y^2}, d_{z^2}); \qquad T_{1u} : (p_x, p_y, p_z)$$

The ligand SALC of A_{1g} symmetry has to combine linearly with the spherically symmetric s orbital of A. Hence all the σ-orbitals of B should contribute equally to this SALC. So the required normalised SALC of A_{1g} symmetry is

$$A_{1g} : \phi_1 = \frac{1}{\sqrt{6}}(\sigma_1 + \sigma_2 + \sigma_3 + \sigma_4 + \sigma_5 + \sigma_6) \tag{9.16}$$

At this stage, a simplifying but reasonable assumption has been made. The ligand atoms are so far apart from each other that the σ-orbitals themselves do not overlap significantly; also, the σ-orbitals are individually normalised. In other words, the σ-orbital wave functions are assumed to form an orthonormal set ($\langle \sigma_i | \sigma_j \rangle = \delta_{ij}$). This makes ϕ_1 a normalised SALC.

The doubly degenerate set of E_g-SALCs of the σ-orbitals of B must match with the $d_{x^2-y^2}$ and d_{z^2} orbitals of the central atom. The positive lobes of the d_{z^2} ($\equiv d_{2z^2-x^2-y^2}$) orbital of S are stretched along the Z-direction with amplitude twice as much as the negative torus in the XY plane. Hence one σ-orbital SALC of E_g symmetry must be

$$E_g : \phi_2 = \frac{1}{\sqrt{12}}(2\sigma_5 + 2\sigma_6 - \sigma_1 - \sigma_2 - \sigma_3 - \sigma_4) \tag{9.17}$$

The $d_{x^2-y^2}$ orbital of S has positive lobes stretched along the X-axis and negative lobes along the Y axis, all with equal amplitude. Since σ_1 and σ_2 are along the X-axis while σ_3 and σ_4 are along Y-axis (Fig. 9.10), the second SALC of E_g symmetry is

$$E_g : \phi_3 = \frac{1}{2}(\sigma_1 + \sigma_2 - \sigma_3 - \sigma_4) \tag{9.18}$$

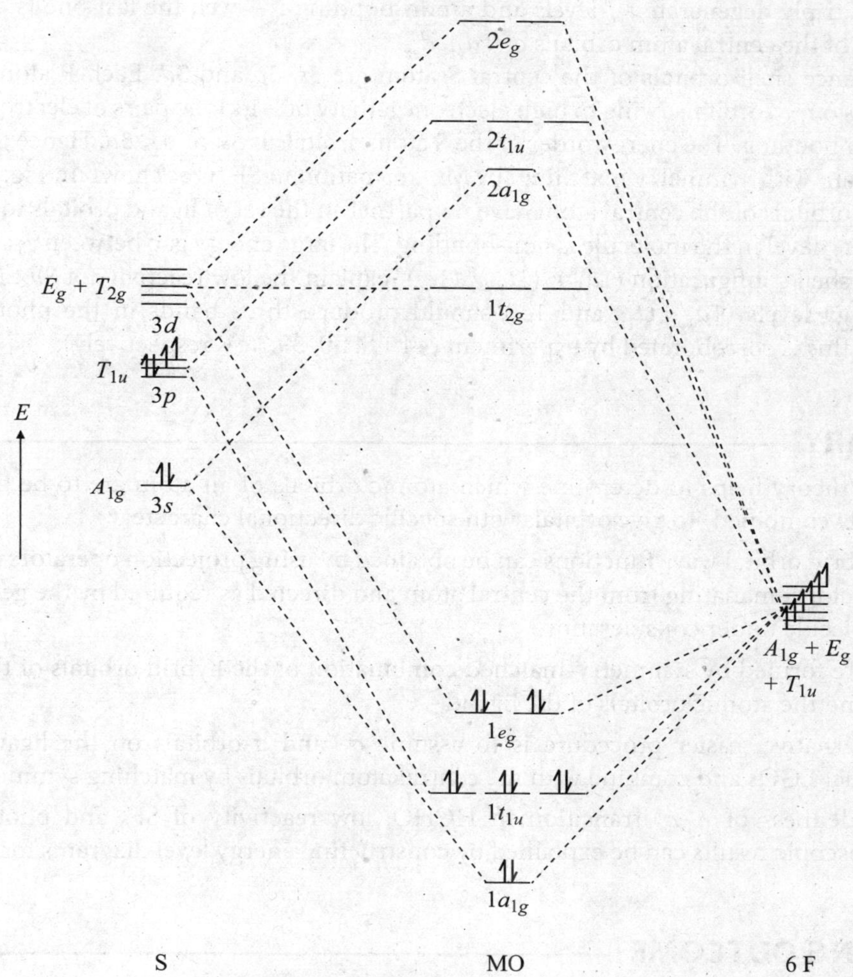

Fig. 9.11 Energy level diagram for SF_6

The p_x, p_y and p_z orbitals of S are stretched respectively along the X, Y and Z axes. Each has positive and negative lobes on the positive and negative sides of the respective axis. Hence according to Fig. 9.10 the T_{1u} SALCs of the ligand σ-orbitals are:

$$T_{1u} : \begin{cases} \phi_4 = \frac{1}{\sqrt{2}}(\sigma_1 - \sigma_2) \\ \phi_5 = \frac{1}{\sqrt{2}}(\sigma_3 - \sigma_4) \\ \phi_6 = \frac{1}{\sqrt{2}}(\sigma_5 - \sigma_6) \end{cases} \tag{9.19}$$

LCAO–MOs from the central atom orbitals and ligand orbitals are now constructed by matching symmetry. There will be two non-degenerate a_{1g} levels, two doubly degenerate e_g levels, two triply degenerate t_{1u} levels and a non-bonding t_{2g} level; the last one is essentially composed of the central atom orbitals d_{xy}, d_{yz}, d_{zx}.

The valence shell orbitals of the central S atom are 3*s*, 3*p* and 3*d*. Each F atom (ligand) contributes one σ-orbital; owing to high electronegativity of F, its lone pairs of electrons are not involved in bonding. The energy order of the S atom orbitals is $3s < 3p < 3d$. Hence the energy level diagram with symmetry matching for MO formation in SF_6 is as shown in Fig. 9.11.

The T_{2g} orbitals of the central atom have no partner in the set of ligand orbitals to combine. Hence the t_{2g} level in the molecule is non-bonding. The large energy gap between e_g and t_{2g} and the closed shell configuration $(1a_{1g})^2\,(1t_{1u})^6\,(1e_g)^4$ explain the low reactivity of SF_6. Moreover, three energy levels ($1a_{1g}$, $1t_{1u}$ and $1e_g$) should produce three bands in the photoelectron spectrum; this is corroborated by experiment (44.1, 41.0, 39.3 eV respectively).

SUMMARY

- Group theory helps to determine which atomic orbitals of an atom are to be hybridised (linearly combined) to give orbitals with specific directional character.
- The hybrid orbital wave functions can be obtained by using projection operators on a set of unit vectors emanating from the central atom and directed as required by the geometry of the molecule under consideration.
- MOs are formed by symmetry-matched combination of the hybrid orbitals of the central atom and the atomic orbitals of the ligand.
- An alternative, easier procedure is to assume σ- and π-orbitals on the ligand atoms, construct LGOs and combine with the central atom orbitals by matching symmetry.
- Forbiddenness of n–π^* transition in HCHO, low reactivity of SF_6 and photoelectron spectroscopic results can be explained by constructing energy level diagrams for the MOs.

LEARNING OUTCOME

After reading this chapter the student will be able to (i) determine which atomic orbitals are to be hybridised to give directed orbitals of specified directions, (ii) find the hybrid orbital wave functions by projection operator–matrix inversion method, (iii) construct MOs of molecules having both σ- and π- bonds, and (iv) explain certain spectroscopic/chemical features of some specific molecules (like HCHO, SF_6).

EXERCISES

1. Pentafluoroantimonate(III) has a square pyramidal geometry with Sb^{+3} ion at the centre of the base of the pyramid. Construct molecular orbitals and the energy level diagram for the system. Examine whether it is paramagnetic or not.
2. Apply the localised MO method to obtain the molecular orbitals and energy level diagram for (a) CH_4 and (b) PF_5 molecules.

3. Construct the LGOs for σ-bond formation in AB_n molecules of the following geometry (a) trigonal prism (AB_6) (b) cubic (AB_8).
4. Show that the HOMO–LUMO transition in formaldehyde is electric dipole forbidden.

Suggested Books and References for Further Reading

1. Pauling L, *J. Am. Chem. Soc.*, 53 (1367) **1931.**
2. Potts A W and Price W C, *Proc. R. Soc. London*, A326 (165) **1972.**
3. Jorgensen W L and Salem L, *The Organic Chemist's Book of Orbitals*, Academic Press, NY, **1973.**
4. Holland D M P *et al.*, *Chem. Phys.*, 192 (333) **1995.**

10 Chemical Bonding III: Hückel Method of π-MO Calculation

Learning objectives

- Assumptions of the HMO theory, α and β parameters
- HMO method of calculation of π-MO energies, eigenvectors, Coulson bond orders, charge densities, delocalisation energy and free valence
- Aromaticity, proof of Hückel's $(4n + 2)$ rule
- π–π^* transitions
- π-systems with heteroatoms

10.1 INTRODUCTION

One approximate method for solving the Schrödinger equation to obtain the wave function and energy of π-type molecular orbitals of conjugated molecules was developed by E Hückel (1931); it utilises the linear variation method and makes the following assumptions:

- σ-electrons and π-electrons can be considered separately; in the conjugated molecule the π-electrons are moving in the field of the nuclei, the core electrons and the σ-bond framework.
- Each of the conjugated atoms (labelled 1, 2, 3, ...) in the molecule contributes one p-type atomic orbital perpendicular to the plane of the σ-bond framework. If these atomic orbital wave functions are $\phi_1, \phi_2, \phi_3, \ldots, \phi_n$ (which are obviously not eigenfunctions of the molecular Hamiltonian, $\hat{H}$) then

$$\begin{aligned}\langle \phi_i | \hat{H} | \phi_j \rangle &= H_{ij} = \beta_{ij} \text{ if } i \text{ and } j \text{ are adjacent atoms}\\ &= \alpha_i \text{ if } i = j\\ &= 0, \text{ if } i \neq j \text{ and } i \text{ and } j \text{ are non-adjacent}\end{aligned} \tag{10.1}$$

$$\begin{aligned}\langle \phi_i | \phi_j \rangle &= S_{ij} = 1 \text{ for } i = j\\ &= 0, \text{ for } i \neq j\end{aligned} \tag{10.2}$$

Here α_i (called *Coulomb integral*) means the energy of a π-electron if it is localised in atom i but feels the field of the other nuclei and electrons (since $\hat{H}$ is the Hamiltonian operator of the whole molecule and not of the atom i) and β_{ij} (called *resonance integral*) means the energy of a π-electron due to its exchange between the atoms i and j.

Starting with a linear combination of the p-type wave functions $\{\phi_i\}$ as a trial function and applying the variation method, one ends up with a secular determinantal equation, the roots of which are the energies of the π-MOs. For small conjugated molecules, application of group theory (GT) is not needed for factoring this determinant but GT assigns symmetry labels to the π-MOs which are needed to judge whether an electronic transition is allowed or not. Apart from this, GT simplifies the calculation of MO energies of relatively large π-systems by block-factoring the secular determinant.

10.2 EXAMPLES WITH SOME SIMPLE SYSTEMS

10.2.1 Ethylene

A trial wave function of a π-MO in this case is

$$\psi = c_1\phi_1 + c_2\phi_2 \qquad (10.3)$$

where c_1 and c_2 are combination coefficients to be determined by the variation method. Functions ϕ_1 and ϕ_2 represent p_z-type atomic orbitals of the conjugated C atoms labelled 1 and 2 (Fig. 10.1).

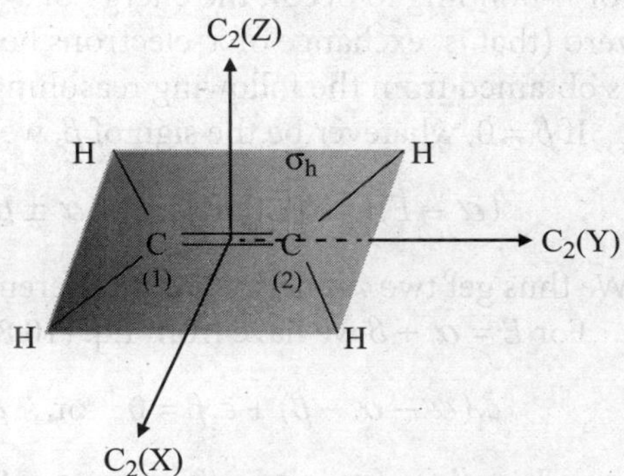

Fig. 10.1 Ethylene molecule with atom labels and symmetry elements (D_{2h})

The π-electronic part of the Schrödinger equation is

$$\hat{\mathrm{H}}\psi = E\psi \qquad (10.4)$$

Multiplying both sides by ψ^*, integrating over whole space and rearranging, we obtain

$$E = \frac{\langle\psi|\hat{\mathrm{H}}|\psi\rangle}{\langle\psi|\psi\rangle} = \frac{c_1^2 H_{11} + 2c_1c_2H_{12} + c_2^2 H_{22}}{c_1^2 S_{11} + 2c_1c_2S_{12} + c_2^2 S_{22}} = \frac{c_1^2\alpha_1 + 2c_1c_2\beta_{12} + c_2^2\alpha_2}{c_1^2 + c_2^2} \qquad (10.5a)$$

where the approximations (10.1) and (10.2) have been used. On simplification this gives

$$c_1^2(\alpha_1 - E) + 2c_1c_2\beta_{12} + c_2^2(\alpha_2 - E) = 0 \qquad (10.5b)$$

The coefficients c_1 and c_2 must be so chosen that E is a minimum, that is, $\dfrac{\partial E}{\partial c_1} = 0$ and $\dfrac{\partial E}{\partial c_2} = 0$.

From Eq. (10.5b), the condition $\dfrac{\partial E}{\partial c_1} = 0$ gives, after dividing by 2,

$$c_1(\alpha_1 - E) + c_2\beta_{12} = 0 \qquad (10.6)$$

Similarly $\dfrac{\partial E}{\partial c_2} = 0$ gives

$$c_1\beta_{12} + c_2(\alpha_2 - E) = 0 \qquad (10.7)$$

In the case of ethylene, $\alpha_1 = \alpha_2 = \alpha$ and $\beta_{12} = \beta_{21} = \beta$ (say). Then the linear equations, Eqs (10.6) and (10.7), give non-trivial solutions for c_1 and c_2 only if the following determinantal equation holds:

$$\begin{vmatrix} \alpha - E & \beta \\ \beta & \alpha - E \end{vmatrix} = 0 \tag{10.8}$$

If β happens to be zero, Eq. (10.8) gives $(\alpha - E)^2 = 0$, that is, $E = \alpha, \alpha$. Thus, if the exchange of electrons between the two C atoms in ethylene were absent, the energy of the π-electron would be the same as that if it was localised on a particular C atom. This is entropically unfavourable for bonding to occur—formation of molecules from atoms leads to a decrease in entropy; hence, for the free energy change to be negative, energy must be released in the process. So, for π-bonding to occur, the energy of a π-electron must be different from α; thus β cannot be zero (that is, exchange of π-electrons between adjacent atoms must be allowed). The sign of β is obtained from the following reasoning.

If $\beta \neq 0$, whatever be the sign of β, we have from Eq. (10.8),

$$(\alpha - E)^2 = \beta^2, \text{ that is, } E = \alpha \pm \beta \tag{10.9}$$

We thus get two π-energy levels different from α. Evaluation of c_1 and c_2 is the next step.

For $E = \alpha + \beta$ we have from Eq. (10.8),

$$\left.\begin{aligned} c_1(\alpha - \alpha - \beta) + c_2\beta = 0 \quad \text{or,} \quad c_1 = c_2 \\ c_1\beta + c_2(\alpha - \alpha - \beta) = 0 \quad \text{or,} \quad c_1 = c_2 \end{aligned}\right\}$$

For normalisation of the wave function, $c_1^2 + c_2^2 = 1$ that is, $c_1 = c_2 = \frac{1}{\sqrt{2}}$.

Thus the π-MO wave function corresponding to $E = \alpha + \beta$ is

$$\psi_{MO} = \tfrac{1}{\sqrt{2}}(\phi_1 + \phi_2) \tag{10.10}$$

Since this is an 'in-phase' combination of the atomic orbitals ϕ_1 and ϕ_2, the resulting MO must be of bonding nature and its energy, $\alpha + \beta$, should be less than α according to the entropy-based argument given earlier. Hence β is negative.

The antibonding MO wave function corresponding to energy $E = \alpha - \beta$ can be similarly found:

$$\psi_{MO} = \tfrac{1}{\sqrt{2}}(\phi_1 - \phi_2) \tag{10.11}$$

The energy level diagram for the bonding (BMO) and anti bonding molecular orbitals (ABMO) of the π-electrons of ethylene are as shown in Fig. 10.2.

Note: Unlike Fig. 8.3(i) the bonding and antibonding MOs are equally lowered and raised; this is because S_{ij} has been assumed to be zero.

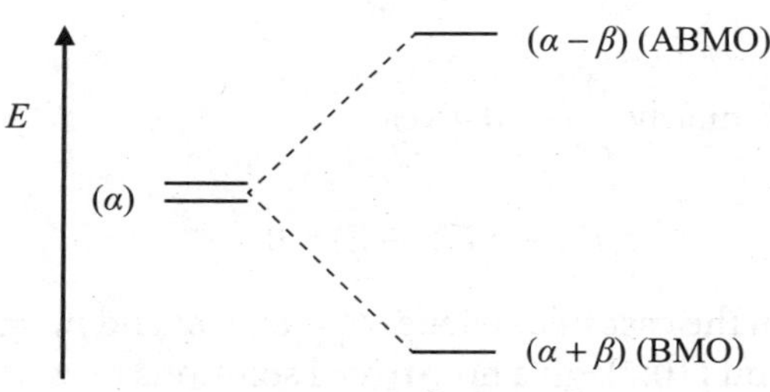

Fig. 10.2 Energy level diagram for ethylene π-MOs

The bonding and autibonding π-MOs of ethylene are usually represented by π and π^* respectively. Thus far, group theory was not needed to obtain their energies and wave functions. We shall now examine whether the π (singlet) $\rightarrow \pi^*$ (singlet) electronic transition is allowed or forbidden. At this stage group theoretical methods will be necessary.

The point group of ethylene molecule is D_{2h} (see Fig. 10.1, where the molecular plane is the XY plane). The action of the various symmetry operators of D_{2h} on the bonding π-MO wave function is as follows. (The atomic orbitals represented by ϕ_1 and ϕ_2 are perpendicular to the XY plane, their positive lobes being directed towards the positive Z-direction.)

$$\hat{E}\left[\frac{1}{\sqrt{2}}(\phi_1+\phi_2)\right]=\frac{1}{\sqrt{2}}(\phi_1+\phi_2) \quad \Rightarrow \chi(E)=1$$

$$\hat{C}_2(Z)\left[\frac{1}{\sqrt{2}}(\phi_1+\phi_2)\right]=\frac{1}{\sqrt{2}}(\phi_1+\phi_2) \quad \Rightarrow \chi(C_2(Z))=1$$

$$\hat{C}_2(Y)\left[\frac{1}{\sqrt{2}}(\phi_1+\phi_2)\right]=\frac{1}{\sqrt{2}}(-\phi_1-\phi_2) \quad \Rightarrow \chi(C_2(Y))=-1$$

$$\hat{C}_2(X)\left[\frac{1}{\sqrt{2}}(\phi_1+\phi_2)\right]=\frac{1}{\sqrt{2}}(-\phi_1-\phi_2) \quad \Rightarrow \chi(C_2(X))=-1$$

$$\hat{i}\left[\frac{1}{\sqrt{2}}(\phi_1+\phi_2)\right]=\frac{1}{\sqrt{2}}(-\phi_2-\phi_1) \quad \Rightarrow \chi(i)=-1$$

$$\hat{\sigma}(XY)\left[\frac{1}{\sqrt{2}}(\phi_1+\phi_2)\right]=\frac{1}{\sqrt{2}}(-\phi_1-\phi_2) \quad \Rightarrow \chi(\sigma(XY))=-1$$

$$\hat{\sigma}(XZ)\left[\frac{1}{\sqrt{2}}(\phi_1+\phi_2)\right]=\frac{1}{\sqrt{2}}(\phi_2+\phi_1) \quad \Rightarrow \chi(\sigma(XZ))=1$$

$$\hat{\sigma}(YZ)\left[\frac{1}{\sqrt{2}}(\phi_1+\phi_2)\right]=\frac{1}{\sqrt{2}}(\phi_1+\phi_2) \quad \Rightarrow \chi(\sigma(YZ))=1$$

From the character table of D_{2h} we find that these characters correspond to the irreducible representation B_{1u}. Thus the bonding π-MO of ethylene acts as a basis of B_{1u}.

In the same way, operation of the D_{2h} symmetry operators on the antibonding π^*-MO wave function yields the following characters.

χ (character)

$\hat{E}$	$\hat{C}_2(Z)$	$\hat{C}_2(Y)$	$\hat{C}_2(X)$	$\hat{i}$	$\hat{\sigma}(XY)$	$\hat{\sigma}(XZ)$	$\hat{\sigma}(YZ)$
1	–1	–1	1	1	–1	–1	1

These are the characters of B_{3g}. It is now possible to examine whether the $\pi \rightarrow \pi^*$ electronic transition in ethylene is allowed or not. The ground state electronic configuration of the system is b_{1u}^2. So, the symmetry of the overall ground state wave function is the direct product $b_{1u} \times b_{1u} = A_g$ (as all the characters will come out be 1 in the direct product).

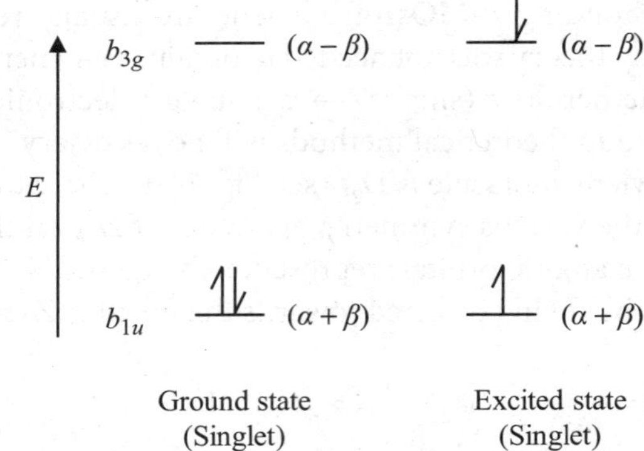

Fig. 10.3 Ground and excited state π-electronic configurations of ethylene

In the excited (singlet) state, the electronic configuration is $b_{1u}^1\ b_{3g}^1$ for which the overall excited state wave function has symmetry $b_{1u} \times b_{3g} = B_{2u}$ as shown below.

D_{2h}	Ê	$\hat{C}_2$(Z)	$\hat{C}_2$(Y)	$\hat{C}_2$(X)	î	σ̂(XY)	σ̂(XZ)	σ̂(YZ)	
	χ (character)								
b_{1u}	1	1	−1	−1	−1	−1	1	1	
b_{3g}	1	−1	−1	1	1	−1	−1	1	
$b_{1u} \times b_{3g}$	1	−1	1	−1	−1	1	−1	1	$= B_{2u}$

Thus, the π (singlet) $\rightarrow \pi^*$(singlet) transition in ethylene amounts to $A_g \rightarrow B_{2u}$. The intensity (I) of this transition is proportional to the square of the transition moment integral, $M = \langle \psi_{ex}|\vec{\mu}|\psi_{gr}\rangle$, where ψ_{ex} and ψ_{gr} are respectively the excited and ground state wave functions and $\vec{\mu}$ is the dipole moment operator given by $\vec{\mu} = \sum_j (-e_j \vec{r}_j)$. Here j is the index for all charged particles (nuclei and electrons) in the system and $\vec{r}_j$ is the position vector of the jth particle with respect to some origin of coordinate system. Each $\vec{r}_j$ contains the Cartesian coordinates x_j , y_j , z_j. Hence the integral M can be decomposed into three components,

$$M_x = \langle \psi_{ex}|\mu_x|\psi_{gr}\rangle, \quad \text{where} \quad \mu_x = \sum_j -e_j\, x_j$$

$$M_y = \langle \psi_{ex}|\mu_y|\psi_{gr}\rangle, \quad \text{where} \quad \mu_y = \sum_j -e_j\, y_j$$

and $$M_z = \langle \psi_{ex}|\mu_z|\psi_{gr}\rangle, \quad \text{where} \quad \mu_z = \sum_j -e_j\, z_j$$

Now, since the coordinate x is a basis of B_{3u}, M_x corresponds to the direct product $B_{2u} \times B_{3u} \times A_g$ = [1 1 −1 −1 1 1 −1 −1] $\neq A_g$. So M_x vanishes. Similarly M_y corresponds to the direct product $B_{2u} \times B_{2u} \times A_g = A_g$. So, $M_y \neq 0$. In the same way, M_z corresponds to

$B_{2u} \times B_{1u} \times A_g = B_{3g} \neq A_g$. Hence M_z also vanishes. Thus the transition moment integral M survives because of the component M_y and so the intensity of absorption $I \neq 0$. The $\pi \to \pi^*$ (singlet–singlet) transition is *electric dipole allowed* and it is polarised in the *y*-direction. This means that as a result of the transition, the electronic charge distribution in the molecule changes in such a way that only the *y*-component of the dipole moment a undergoes major change, while changes in the *x*- and *z*- components are insignificant.

If we consider two dissimilar conjugated atoms, Eq. (10.8) will look like

$$\begin{vmatrix} \alpha_1 - E & \beta_{12} \\ \beta_{21} & \alpha_2 - E \end{vmatrix} = 0 \qquad (10.12)$$

which yields on solving, $E = \dfrac{(\alpha_1 + \alpha_2)}{2} \pm$ something. The energy-level diagram then looks like Fig. 10.4.

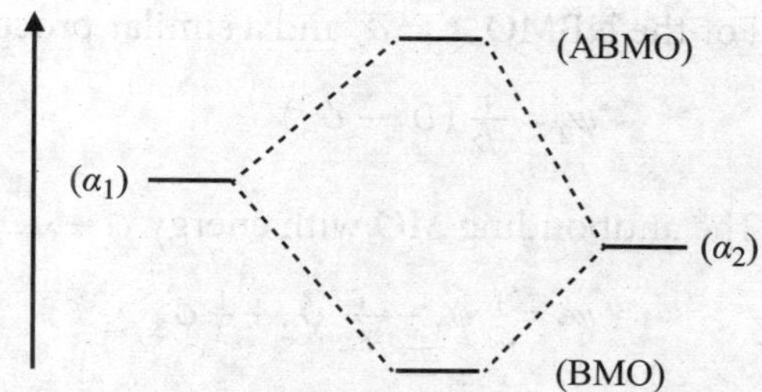

Fig. 10.4 Energy level diagram for π-bonding between two dissimilar conjugated atoms

10.2.2 Allyl Radical, $CH_2 = CH - \dot{C}H_2$

The linear variation method in this case will lead to the following determinantal equation:

$$\begin{vmatrix} \alpha - E & \beta & 0 \\ \beta & \alpha - E & \beta \\ 0 & \beta & \alpha - E \end{vmatrix} = 0 \qquad (10.13)$$

On solving, we get three values of *E*, namely, $\alpha, \alpha \pm \sqrt{2}\beta$. Recalling that β is negative, the π-energy level diagram is:

It is found that one MO has energy α which is the same as that if there were no π-bonding. This is called 'non-bonding molecular orbital' (abbreviated as NBMO). The coefficients of the atomic orbitals in the three MOs are obtained as follows:

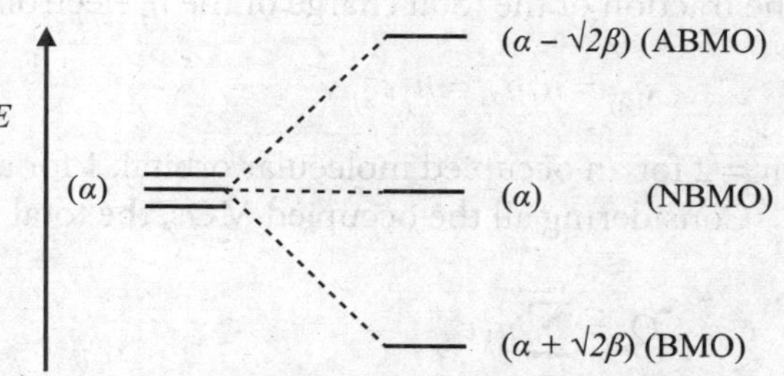

Fig. 10.5 π-MO energy level diagram for the allyl system

When $E = \alpha + \sqrt{2}\beta$, looking at the secular determinant in (10.13) we have:

$$c_1(\alpha - \alpha - \sqrt{2}\beta) + c_2\beta = 0 \quad \text{or,} \quad c_1 = \frac{c_2}{\sqrt{2}}$$

$$c_1\beta + c_2(-\sqrt{2}\beta) + c_3\beta = 0 \quad \text{or,} \quad c_1 - c_2\sqrt{2}c_2 + c_3 = 0$$

$$c_2\beta + c_3(\alpha - \alpha - \sqrt{2}\beta) = 0 \quad \text{or,} \quad c_3 = \frac{c_2}{\sqrt{2}}$$

These, together with the normalisation condition $c_1^2 + c_2^2 + c_3^2 = 1$ give $c_1 = c_3 = \frac{1}{2}$ and $c_2 = \frac{1}{\sqrt{2}}$.

Thus the wave function of the MO with energy $\alpha + \sqrt{2}\beta$ is

$$\psi_1 = \tfrac{1}{2}\,\phi_1 + \tfrac{1}{\sqrt{2}}\,\phi_2 + \tfrac{1}{2}\,\phi_3 \tag{10.14}$$

For the NBMO, $E = \alpha$ and a similar procedure gives the wave function

$$\psi_2 = \tfrac{1}{\sqrt{2}}\,(\phi_1 - \phi_3) \tag{10.15}$$

The antibonding MO with energy $\alpha - \sqrt{2}\beta$ is similarly found to have the wave function

$$\psi_3 = \tfrac{1}{2}\,\phi_1 - \tfrac{1}{\sqrt{2}}\,\phi_2 + \tfrac{1}{2}\,\phi_3 \tag{10.16}$$

10.3 CALCULATION OF CHARGE DENSITIES ON THE CONJUGATED ATOMS

The probability of finding a π-electron of the jth MO on the λth atom is given by

$$p_{\lambda j} = \frac{c_{\lambda j}^2}{\sum_{\lambda} c_{\lambda j}^2} \tag{10.17}$$

But $\sum_{i} c_{\lambda j}^2 = 1$ (normalisation condition) and so, if n_j electrons are present in the λth MO, then the fraction of the total charge of the n_j electrons present on the λth atom is

$$q_{\lambda j} = n_j p_{\lambda j} = n_j\, c_{\lambda j}^2 \tag{10.18}$$

($n_j = 2$ for an occupied molecular orbital, 1 for a half-filled MO and 0 for an unoccupied MO.)

Considering all the occupied MOs, the total electronic charge on the λth atom becomes

$$Q_\lambda = \sum_{j=1}^{\text{occ}} n_j\, c_{\lambda j}^2 \tag{10.19}$$

10.3.1 ETHYLENE

The π-MO wave functions are $\psi_1 = c_{11}\phi_1 + c_{21}\phi_2$ and $\psi_2 = c_{12}\phi_1 + c_{22}\phi_2$; in the ground state of the molecule, two π-electrons occupy the energy level E_1 and the E_2 level is empty.

E (level)	Atom 1	Atom 2	
E_2 —	$\frac{1}{\sqrt{2}}$	$-\frac{1}{\sqrt{2}}$	$Q_1 = 2 \cdot c_{11}^2 + 0 \cdot c_{12}^2 = 2 \times \frac{1}{2} + 0 = 1$
E_1 ↑↓	$\frac{1}{\sqrt{2}}$	$\frac{1}{\sqrt{2}}$	$Q_2 = 2 \cdot c_{21}^2 + 0 \cdot c_{22}^2 = 2 \times \frac{1}{2} + 0 = 1$

Thus each C-atom in the ethylene molecule has one unit of π-electronic charge.

10.3.2 Allyl System

For the allyl cation, $CH_2=CH-\overset{+}{C}H_2$, the number of π-electrons is two and the π-MO wave functions are

$$\psi_1 = c_{11}\phi_1 + c_{21}\phi_2 + c_{31}\phi_3 = \tfrac{1}{2}\phi_1 + \tfrac{1}{\sqrt{2}}\phi_2 + \tfrac{1}{2}\phi_3, \quad \psi_2 = c_{12}\phi_1 + c_{22}\phi_2 + c_{32}\phi_3 = \tfrac{1}{\sqrt{2}}(\phi_1 - \phi_3)$$

and $$\psi_3 = c_{13}\phi_1 + c_{23}\phi_2 + c_{33}\phi_3 = \tfrac{1}{2}\phi_1 - \tfrac{1}{\sqrt{2}}\phi_2 + \tfrac{1}{2}\phi_3$$

Considering the electron occupancies and atomic orbital coefficients of the different π-MOs, the π-electronic charges on the three C-atoms of the allyl cation (with two π-electrons in the E_1 energy level) are calculated as follows:

E		Atom 1	Atom 2	Atom 3	
E_3	—	$\frac{1}{2}$	$-\frac{1}{\sqrt{2}}$	$\frac{1}{2}$	$Q_1 = 2\cdot c_{11}^2 + 0\cdot c_{12}^2 + 0\cdot c_{13}^2 = 2\times\frac{1}{4} + 0 + 0 = 0.5$
E_2	—	$\frac{1}{\sqrt{2}}$	0	$-\frac{1}{\sqrt{2}}$	$Q_2 = 2\cdot c_{21}^2 + 0\cdot c_{22}^2 + 0\cdot c_{23}^3 = 2\times\frac{1}{2} + 0 + 0 = 1.0$
E_1	↑↓	$\frac{1}{2}$	$\frac{1}{\sqrt{2}}$	$\frac{1}{2}$	$Q_3 = 2\cdot c_{31}^2 + 0\cdot c_{32}^2 + 0\cdot c_{33}^2 = 2\times\frac{1}{4} + 0 + 0 = 0.5$

In the case of the allyl radical, there are two π-electrons in the lowest energy level (E_1) and one in the next higher level (E_2). Hence the charges on the atoms are:

$$Q_1 = 2\cdot c_{11}^2 + 1\cdot c_{12}^2 + 0\cdot c_{13}^2 = 2\times\tfrac{1}{4} + 1\times\tfrac{1}{2} + 0 = 1.0$$

$$Q_2 = 2\cdot c_{21}^2 + 1\cdot c_{22}^2 + 0\cdot c_{23}^3 = 2\times\tfrac{1}{2} + 1\times 0 + 0 = 1.0$$

$$Q_3 = 2\cdot c_{31}^2 + 1\cdot c_{32}^2 + 0\cdot c_{33}^2 = 2\times\tfrac{1}{4} + 1\times\tfrac{1}{2} + 0 = 1.0$$

Thus the charges of three π-electrons of the allyl radical are equally distributed over the three carbon atoms.

10.4 CALCULATION OF BOND ORDER AND DELOCALISATION ENERGY

The probability of finding the electron of the jth MO (ψ_j) between the atoms λ and μ is obviously proportional to the product of the contributions of the λth and μth atoms to ψ_j. Hence Coulson defined the *partial bond order* between atoms λ and μ as

$$p_{\lambda\mu}^{(j)} = n_j c_{\lambda j} c_{\mu j} \tag{10.20}$$

Summing the contributions from all the occupied MOs, the total bond order between atoms λ and μ is,

$$P_{\lambda\mu} = \sum_j n_j c_{\lambda j} c_{\mu j} \tag{10.21}$$

where n_j is the number of electrons in the jth MO (= 2, 1 or 0 depending on whether the MO is doubly occupied, half-filled or empty).

Delocalisation energy of a π-conjugated system is defined as the difference between the total π-electron energy of the system and the sum of the energies of the electrons in the π-bonds if such bonds were localised as in isolated double bonds. It is a measure of the stability of the π-system.

Some examples are given below.

10.4.1 Ethylene

		Atom 1	Atom 2
$(\alpha-\beta)$	—	$\frac{1}{\sqrt{2}}$	$-\frac{1}{\sqrt{2}}$
$(\alpha+\beta)$	↑↓	$\frac{1}{\sqrt{2}}$	$\frac{1}{\sqrt{2}}$

$$P_{12} = 2 \cdot \frac{1}{\sqrt{2}} \cdot \frac{1}{\sqrt{2}} + 0 \cdot \frac{1}{\sqrt{2}}\left(-\frac{1}{\sqrt{2}}\right) = 1$$

Thus, there is one π-bond between atoms 1 and 2 in ethylene molecule.

10.4.2 Allyl Radical

This has two π-electrons in the lowest energy level and one in the next higher level.

		Atom 1	Atom 2	Atom 3
$(\alpha-\sqrt{2}\,\beta)$	—	$\frac{1}{2}$	$-\frac{1}{\sqrt{2}}$	$\frac{1}{2}$
(α)	↑	$\frac{1}{\sqrt{2}}$	0	$-\frac{1}{\sqrt{2}}$
$(\alpha+\sqrt{2}\,\beta)$	↑↓	$\frac{1}{2}$	$\frac{1}{\sqrt{2}}$	$\frac{1}{2}$

$$P_{12} = 2 \cdot \frac{1}{2} \cdot \frac{1}{\sqrt{2}} + 1 \cdot \frac{1}{\sqrt{2}} \cdot 0 = 0.708$$

$$P_{23} = 2 \cdot \frac{1}{\sqrt{2}} \cdot \frac{1}{2} + 1 \cdot 0 \cdot \left(-\frac{1}{\sqrt{2}}\right) = 0.708$$

$$P_{13} = 2 \cdot \frac{1}{2} \cdot \frac{1}{2} + 1 \cdot \frac{1}{\sqrt{2}} \cdot \left(-\frac{1}{\sqrt{2}}\right) = 0$$

Thus, in the allyl radical there is a π-bond of order slightly less than unity between each of the atom pairs 1, 2 and 2, 3, and there is no π-bond directly connecting the atoms 1 and 3 (as in $[CH_2 \overset{...}{=} CH \overset{...}{=} CH_2]^{\bullet}$).

Delocalisation energy of allyl radical

$$= (E_\pi)_{\text{total}} - E_\pi \text{ due to one isolated double bond}$$

$$= \{2(\alpha + \sqrt{2}\beta)\} + \alpha - 2(\alpha + \beta)$$

$$= \alpha + 2(\sqrt{2} - 1)\beta$$

Delocalisation energy of allyl carbocation, $[CH_2 \overset{...}{=} CH \overset{...}{=} CH_2]^+$

$$= (E_\pi)_{\text{total}} - E_\pi \text{ due to one isolated double bond}$$

$$= \{2(\alpha + \sqrt{2}\beta)\} - 2(\alpha + \beta)$$

$$= 2(\sqrt{2} - 1)\beta$$

10.4.3 Naphthalene (A More Complicated System)

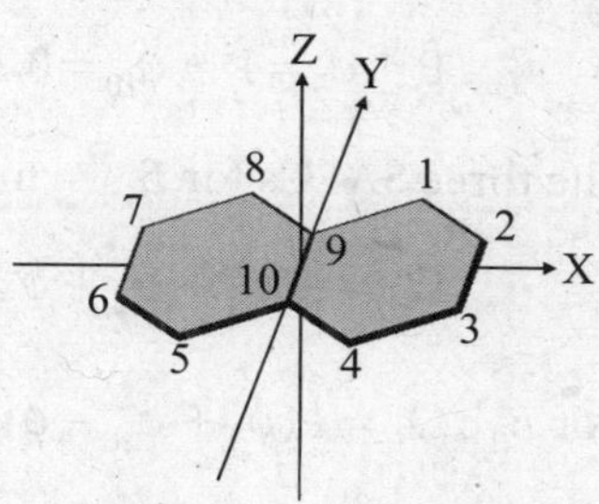

Fig. 10.6 A naphthalene molecule with atom labels

Let $\{\phi_i\}$, i = 1 to 10, be the p_z-type atomic orbital wave functions of the carbon atoms of naphthalene molecule labelled 1 to 10 (Fig. 10.6). With this set as the basis, the characters (χ) of a representation Γ_π, of the molecular point group (D_{2h}) of naphthalene are as tabulated below:

χ (character)

$\hat{E}$	$\hat{C}_2(Z)$	$\hat{C}_2(Y)$	$\hat{C}_2(X)$	$\hat{i}$	$\hat{\sigma}(XY)$	$\hat{\sigma}(XZ)$	$\hat{\sigma}(YZ)$
10	0	0	−2	0	−10	2	0

This is a reducible representation since the sum of the squares of the characters in Γ_π exceeds 8, the order of the group.

Let $$\Gamma_\pi = n_1 A_g + n_2 B_{1g} + n_3 B_{2g} + n_4 B_{3g} + n_5 A_u + n_6 B_{1u} + n_7 B_{2u} + n_8 B_{3u} \tag{10.22}$$

where the coefficients n_1, n_2 ... denote the number of times the corresponding irreducible representations occur in Γ_π. Then by using the character table of D_{2h} (given in Appendix A3) and applying the reduction formula, we find

$$n_1 = \tfrac{1}{8}\{1^* \cdot 10 + 1^* \cdot 0 + 1^* \cdot 0 + 1^* \cdot (-2) + 1^* \cdot 0 + 1^* \cdot (-10) + 1^* \cdot 2 + 1^* \cdot 0\} = 0$$

$$n_2 = \tfrac{1}{8}\{1^* \cdot 10 + 1^* \cdot 0 + (-1)^* \cdot 0 + (-1)^* \cdot (-2) + 1^* \cdot 0 + 1^* \cdot (-10) + (-1)^* \cdot 2 + (-1)^* \cdot 0\} = 0$$

$$n_3 = \tfrac{1}{8}\{1^* \cdot 10 + (-1)^* \cdot 0 + 1^* \cdot 0 + 1^* \cdot (-2) + 1^* \cdot 0 + (-1)^* \cdot (-10) + (-1)^* \cdot 2 + (-1)^* \cdot 0\} = 2$$

and so on. Thus $$\Gamma_\pi = 2B_{2g} + 3B_{3g} + 2A_u + 3B_{1u} \tag{10.23}$$

The next task is to construct the SALCs corresponding to the component irreducible representations of Γ_π using the projection operator technique. For A_u symmetry, two SALCs are needed; they are constructed as follows:

$$\hat{P}^{A_u}\,\phi_1 = \frac{1}{8}\left\{\begin{bmatrix} 1^*\,\hat{E}\,\phi_1 + 1^*\,\hat{C}_2(Z)\,\phi_1 + 1^*\,\hat{C}_2(Y)\,\phi_1 + 1^*\,\hat{C}_2(X)\,\phi_1 + (-1)^*\,\hat{i}\,\phi_1 \\ +(-1)^*\,\hat{\sigma}(XY)\,\phi_1 + (-1)^*\,\hat{\sigma}(XZ)\,\phi_1 + (-1)^*\,\hat{\sigma}(YZ)\,\phi_1 \end{bmatrix}\right.$$

$$= \tfrac{2}{8}(\phi_1 - \phi_4 + \phi_5 - \phi_8) \equiv \tfrac{1}{2}(\phi_1 - \phi_4 + \phi_5 - \phi_8) \quad \text{(after normalisation)}$$

$$= \Omega_1 \text{ (say)} \tag{10.24}$$

Similarly,

$$\hat{\mathrm{P}}^{A_u}\,\phi_2 \equiv \tfrac{1}{2}(\phi_2 - \phi_3 + \phi_6 - \phi_7) = \Omega_2 \text{ (say)} \tag{10.25}$$

$$\hat{\mathrm{P}}^{A_u}\,\phi_9 = \hat{\mathrm{P}}^{A_u}\,\phi_{10} = 0 \text{ (which can be readily verified)}$$

The three SALCs for B_{1u} symmetry are:

$$\Omega_3 = \tfrac{1}{2}(\phi_1 + \phi_4 + \phi_5 + \phi_8),\ \Omega_4 = \tfrac{1}{2}(\phi_2 + \phi_3 + \phi_6 + \phi_7),\ \Omega_5 = \tfrac{1}{\sqrt{2}}(\phi_9 + \phi_{10}) \tag{10.26}$$

$$\text{For } B_{2g}\text{: } \Omega_6 = \tfrac{1}{2}(\phi_1 + \phi_4 - \phi_5 - \phi_8),\ \Omega_7 = \tfrac{1}{2}(\phi_2 + \phi_3 - \phi_6 - \phi_7) \tag{10.27}$$

$$\text{For } B_{3g}\text{: } \Omega_8 = \tfrac{1}{2}(\phi_1 - \phi_4 - \phi_5 + \phi_8),\ \Omega_9 = \tfrac{1}{2}(\phi_2 - \phi_3 - \phi_6 + \phi_7),\ \Omega_{10} = \tfrac{1}{\sqrt{2}}(\phi_9 - \phi_{10}) \tag{10.28}$$

These SALCs $\{\Omega_i\}$, $i = 1$ to 10, are now combined linearly to form the π-MOs of naphthalene:

$$\Psi = \sum_{i=1}^{10} c_i\,\Omega_i \tag{10.29}$$

In constructing the secular determinant we now note that

$$H'_{ij} = \langle \Omega_i | \hat{\mathrm{H}} | \Omega_j \rangle = 0 \quad \text{and} \quad S'_{ij} = \langle \Omega_i | \Omega_j \rangle = 0 \tag{10.30a}$$

when Ω_i and Ω_j are SALCs of two different irreducible representations and that

$$H'_{ij} = \langle \Omega_i | \hat{\mathrm{H}} | \Omega_j \rangle \neq 0 \quad \text{and} \quad S'_{ij} = \langle \Omega_i | \Omega_j \rangle \neq 0 \tag{10.30b}$$

when Ω_i and Ω_j are SALCs of the same irreducible representation. This is because $\hat{\mathrm{H}}$ transforms according to the totally symmetric irreducible representation and the rule for vanishing of integrals operates. The required secular determinant now takes the following block-factored form:

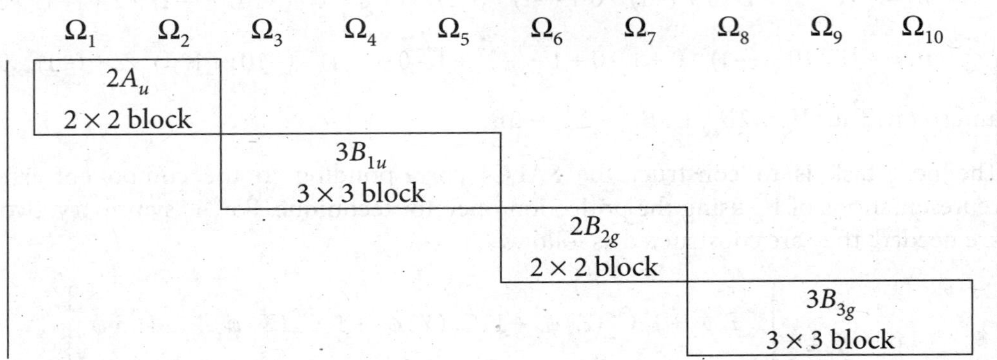

Equating the determinant of the first block (for A_u symmetry) to zero, we have

$$\begin{vmatrix} H'_{11}-ES'_{11} & H'_{12}-ES'_{12} \\ H'_{21}-ES'_{21} & H'_{22}-ES'_{22} \end{vmatrix}=0$$

The required integrals (matrix elements) are now calculated using the Hückel approximations:

$$H'_{11}=\langle\Omega_1|\hat{\mathrm{H}}|\Omega_1\rangle=\langle\tfrac{1}{2}(\phi_1-\phi_4+\phi_5-\phi_8)|\hat{\mathrm{H}}|\tfrac{1}{2}(\phi_1-\phi_4+\phi_5-\phi_8)\rangle$$

$$=\tfrac{1}{4}(\alpha-0+0-0+0+\alpha-0+0+\alpha+\alpha)=\alpha$$

$$S'_{11}=\langle\tfrac{1}{2}(\phi_1-\phi_4+\phi_5-\phi_8)|\tfrac{1}{2}(\phi_1-\phi_4+\phi_5-\phi_8)\rangle=\tfrac{1}{4}(1+1+1+1)=1$$

$$H'_{12}=\langle\tfrac{1}{2}(\phi_1-\phi_4+\phi_5-\phi_8)|\hat{\mathrm{H}}|\tfrac{1}{2}(\phi_2-\phi_3+\phi_6-\phi_7)\rangle$$

$$=\tfrac{1}{4}(\beta+\beta+\beta+\beta)=\beta=H'_{21}$$

In this way, $S'_{12}=S'_{21}=0$, $S'_{22}=1$, $H'_{22}=\alpha-\beta$

Hence the first block gives

$$\begin{vmatrix} \alpha-E & \beta \\ \beta & \alpha-\beta-E \end{vmatrix}=0 \tag{10.31}$$

In the HMO theory it is a usual practice to set the energy scale by taking α as the zero level and β as the unit of energy. Then Eq. (10.31) becomes

$$\begin{vmatrix} -E & 1 \\ 1 & -1-E \end{vmatrix}=0$$

or, $$E^2+E-1=0 \quad \text{or,} \quad E=\frac{-1\pm\sqrt{1+4}}{2}=-1.618,\,0.618 \tag{10.32}$$

The MO coefficients corresponding to the energy eigenvalue −1.618 are obtained as follows:

$$c_1(-E)+c_2\cdot 1=0 \quad \text{or,} \quad c_2=c_1E=-1.618c_1$$

and for normalisation, $c_1^2+c_2^2=1$ which gives $c_1=0.5257$(taking the positive square root). Then $c_2=-1.618c_1=-0.8506$. Hence the wave function of the π-MO with energy −1.618 (in β unit) is

$$\psi_1=c_1\Omega_1+c_2\Omega_2$$

$$=0.526\times\tfrac{1}{2}(\phi_1-\phi_4+\phi_5-\phi_8)-0.851\times\tfrac{1}{2}(\phi_2-\phi_3+\phi_6-\phi_7)$$

$$=0.263(\phi_1-\phi_4+\phi_5-\phi_8)-0.425(\phi_2-\phi_3+\phi_6-\phi_7)$$

Table 10.1 Eigenvalues (in β unit, α=0) and eigenvectors for naphthalene π-MOs

Eigen value	2.303	1.618	1.303	1.000	0.618	−0.618	−1.000	−1.303	−1.618	−2.303
Eigenvector	0.301	0.263	0.400	0	0.425	0.425	0	0.400	0.263	0.301
	0.231	0.425	0.174	0.408	0.263	−0.263	0.408	−0.174	−0.425	−0.231
	0.231	0.425	−0.174	0.408	−0.263	−0.263	−0.408	−0.174	0.425	0.231
	0.301	0.263	−0.400	0	−0.425	0.425	0	0.400	−0.263	−0.301
	0.301	−0.263	−0.400	0	0.425	−0.425	0	0.400	0.263	−0.301
	0.231	−0.425	−0.174	0.408	0.263	0.263	−0.408	−0.174	−0.425	0.231
	0.231	−0.425	0.174	0.408	−0.263	0.263	0.408	−0.174	0.425	−0.231
	0.301	−0.263	0.400	0	−0.425	0.425	0	0.400	−0.263	0.301
	0.461	0	0.347	−0.408	0	0	−0.408	−0.347	0	−0.461
	0.461	0	−0.347	−0.408	0	0	0.408	−0.347	0	0.461

$$= (\phi_1\ \phi_2\ \phi_3\ \phi_4\ \phi_5\ \phi_6\ \phi_7\ \phi_8\ \phi_9\ \phi_{10}) \begin{pmatrix} 0.263 \\ -0.425 \\ 0.425 \\ -0.263 \\ 0.263 \\ -0.425 \\ 0.425 \\ -0.263 \\ 0 \\ 0 \end{pmatrix} \tag{10.33}$$

The column vector thus obtained is called *eigenvector* corresponding to the eigenvalue −1.618. In this way all the ten eigenvectors and eigenvalues can be obtained (Table 10.1).

10.5 CARBOCYCLIC π-SYSTEMS $(CH)_n$, WHERE *n* IS AN INTEGER

Cyclic conjugated hydrocarbons with all *C* atoms in a state of sp^2 hybridisation consist of alternately placed π-bonds in their valence bond structures. Such molecules, ions or radicals are called *carbocyclic π-systems.* The most familiar example is benzene, $(CH)_6$. Calculation of the energy and wave functions of their π-molecular orbitals under the Hückel formalism can be easily performed by applying methods of Group Theory. Some examples are given below.

10.5.1 Benzene

To obtain the π-MO wave functions of benzene, the *C* atoms are labelled 1, 2, ..., 6 (Fig. 10.7). The σ-bonds are formed by the *s*, p_x and p_y orbitals of *C* and the 1*s* orbitals of six *H* atoms. The p_z orbitals of *C* are all perpendicular to the molecular plane, that is, the XY plane. Although the molecular point group is D_{6h} it will be more convenient if the subgroup C_6 of D_{6h} is considered as shown in the following discussions.

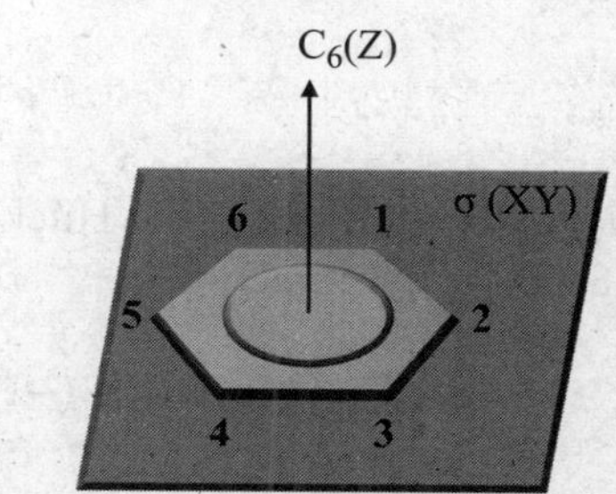

Fig. 10.7 Benzene molecule with atom labels

Let the p_z orbitals of *C* be denoted by the functions $\phi_1, \phi_2, \ldots, \phi_6$ associated respectively with the atoms labelled 1 to 6 (Fig. 10.7). With these functions as basis, a representation Γ_π of the uniaxial rotational subgroup C_6 of D_{6h} has the following set of characters.

χ **(character)**

	$\hat{E}$	$\hat{C}_6$	$\hat{C}_6^2 = \hat{C}_3$	$\hat{C}_6^3 = \hat{C}_2$	$\hat{C}_6^4 = \hat{C}_3^2$	$\hat{C}_6^5$
Γ_π	6	0	0	0	0	0

(The zeros result from the fact that all the p_z-orbitals are shifted from their own positions by the $\hat{C}_6$ rotation operations.) Γ_π is reducible as $\sum_R |\chi(R)|^2 = 36 > 6$, the order of the group C_6. The character table of the C_6 point group [with $\varepsilon = \exp(2\pi i/6), i = \sqrt{-1}$] is given in Appendix A3.

Let Γ_π be reducible as

$$\Gamma_\pi = n_1 A + n_2 B + n_3 E_{1a} + n_4 E_{1b} + n_5 E_{2a} + n_6 E_{2b} \tag{10.34}$$

where E_{1a} and E_{1b} correspond respectively to the first and second rows of E_1. In the same way E_{2a} and E_{2b} are defined. Applying reduction formula we find

$$n_1 = \tfrac{1}{6}\{1^*.6 + 0\} = 1$$

$$n_2 = \tfrac{1}{6}\{1^*.6 + 0\} = 1 \text{ and so on}$$

Thus $\quad \Gamma_\pi = A + B + E_{1a} + E_{1b} + E_{2a} + E_{2b}$ (10.35)

Hence every irreducible representation of C_6 occurs in Γ_π exactly once (Γ_π is a 'regular representation' of C_6; see Appendix A2). We now construct the SALCs corresponding to the component irreducible representations of Γ_π by applying the projection operator $\hat{P}^j(\hat{R}) = \frac{1}{g}\sum_{\hat{R}}[\chi^j(\hat{R})]$ on the functions ϕ_1 to ϕ_6. The SALC of symmetry type A is

$$\Omega_1 = \hat{P}^A\,\phi_1 = \tfrac{1}{6}\{1^*\,\hat{E}\,\phi_1 + 1^*\,\hat{C}_6\,\phi_1 + 1^*\,\hat{C}_3\,\phi_1 + 1^*\,\hat{C}_2\,\phi_1 + 1^*\,\hat{C}_3^2\,\phi_1 + 1^*\,\hat{C}_6^5\,\phi_1\}$$

$$= \tfrac{1}{6}\{\phi_1 + \phi_6 + \phi_5 + \phi_4 + \phi_3 + \phi_2\}$$

$$= \tfrac{1}{\sqrt{6}}\{\phi_1 + \phi_6 + \phi_5 + \phi_4 + \phi_3 + \phi_2\} \tag{10.36}$$

(after normalisation under the Hückel approximation, $\langle\phi_m|\phi_n\rangle = \delta_{mn}$).

The SALC of symmetry B is,

$$\Omega_2 = \hat{P}^B\,\phi_1 = \tfrac{1}{6}\{1^*\,\hat{E}\,\phi_1 + (-1)^*\,\hat{C}_6\,\phi_1 + 1^*\,\hat{C}_3\,\phi_1 + (-1)^*\,\hat{C}_2\,\phi_1 + 1^*\,\hat{C}_3^2\,\phi_1 + (-1)^*\,\hat{C}_6^5\,\phi_1\}$$

$$= \tfrac{1}{6}\{\phi_1 - \phi_6 + \phi_5 - \phi_4 + \phi_3 - \phi_2\}$$

$$= \tfrac{1}{\sqrt{6}}\{\phi_1 - \phi_6 + \phi_5 - \phi_4 + \phi_3 - \phi_2\} \text{ (after normalisation)} \tag{10.37}$$

The SALCs of E_{1a} and E_{1b} symmetry are:

$$\hat{P}^{E_{1a}}\,\phi_1 = \tfrac{1}{6}\{1^*\,\hat{E}\,\phi_1 + (\varepsilon)^*\,\hat{C}_6\,\phi_1 + (-\varepsilon^*)^*\,\hat{C}_3\,\phi_1 + (-1)^*\,\hat{C}_2\,\phi_1 + (-\varepsilon)^*\,\hat{C}_3^2\,\phi_1 + (\varepsilon)^*\,\hat{C}_6^5\,\phi_1\}$$

$$= \tfrac{1}{6}\{\phi_1 + \varepsilon^*\phi_6 - \varepsilon\,\phi_5 - \phi_4 - \varepsilon^*\phi_3 - \varepsilon\,\phi_2\} \tag{10.38}$$

$$\hat{P}^{E_{1b}}\,\phi_1 = \tfrac{1}{6}\{1^*\,\hat{E}\,\phi_1 + (\varepsilon^*)^*\,\hat{C}_6\,\phi_1 + (-\varepsilon)^*\,\hat{C}_3\,\phi_1 + (-1)^*\,\hat{C}_2\,\phi_1 + (-\varepsilon^*)^*\,\hat{C}_3^2\,\phi_1 + (\varepsilon)^*\,\hat{C}_6^5\,\phi_1\}$$

$$= \tfrac{1}{6}\{\phi_1 + \varepsilon\phi_6 - \varepsilon^*\phi_5 - \phi_4 - \varepsilon\phi_3 - \varepsilon^*\phi_2\} \tag{10.39}$$

To remove the complex coefficients we construct additive and subtractive linear combinations, making use of the fact that

$$\varepsilon + \varepsilon^* = 2\cos\tfrac{2\pi}{6} = 1 \quad \text{and} \quad \varepsilon - \varepsilon^* = 2i\sin\tfrac{2\pi}{6} = i\sqrt{3}$$

$$\Omega_3 = \hat{P}^{E_{1a}}\,\phi_1 + \hat{P}^{E_{1b}}\,\phi_1 = \tfrac{1}{6}\{2\phi_1 - \phi_6 - \phi_5 + 2\phi_4 - \phi_3 + \phi_2\}$$

$$= \tfrac{1}{\sqrt{12}}\{2\phi_1 + \phi_2 - \phi_3 + 2\phi_4 - \phi_5 - \phi_6\} \text{ (after normalisation)} \tag{10.40}$$

$$\Omega_4 = \hat{P}^{E_{1a}}\,\phi_1 - \hat{P}^{E_{1b}}\,\phi_1 = \tfrac{i\sqrt{3}}{6}\{-\phi_6 - \phi_5 + \phi_3 + \phi_2\}$$

$$= \tfrac{1}{2}\{\phi_2 + \phi_3 - \phi_5 - \phi_6\} \text{ (after normalisation)} \tag{10.41}$$

The SALCs of E_{2a} and E_{2b} symmetry are similarly determined:

$$\hat{P}^{E_{2a}}\,\phi_1 = \tfrac{1}{6}\{1^*\,\hat{E}\,\phi_1 + (-\varepsilon^*)^*\,\hat{C}_6\,\phi_1 + (-\varepsilon)^*\,\hat{C}_3\,\phi_1 + (1)^*\,\hat{C}_2\,\phi_1 + (-\varepsilon^*)^*\,\hat{C}_3^2\,\phi_1 + (-\varepsilon)^*\,\hat{C}_6^5\,\phi_1\}$$

$$= \tfrac{1}{6}\{\phi_1 - \varepsilon\phi_6 - \varepsilon^*\phi_5 + \phi_4 - \varepsilon\phi_3 - \varepsilon^*\phi_2\}$$

$$= \tfrac{1}{6}\{\phi_1 - \varepsilon^*\phi_2 - \varepsilon\phi_3 + \phi_4 - \varepsilon^*\phi_5 - \varepsilon\phi_6\} \tag{10.42}$$

$$\hat{P}^{E_{2b}}\,\phi_1 = \tfrac{1}{6}\{1^*\,\hat{E}\,\phi_1 + (-\varepsilon)^*\,\hat{C}_6\,\phi_1 + (-\varepsilon^*)^*\,\hat{C}_3\,\phi_1 + (1)^*\,\hat{C}_2\,\phi_1 + (-\varepsilon)^*\,\hat{C}_3^2\,\phi_1 + (-\varepsilon)^*\,\hat{C}_6^5\,\phi_1\}$$

$$= \tfrac{1}{6}\{\phi_1 + \varepsilon^*\phi_6 - \varepsilon\phi_5 + \phi_4 - \varepsilon^*\phi_3 - \varepsilon\phi_2\}$$

$$= \tfrac{1}{6}\{\phi_1 - \varepsilon\phi_2 - \varepsilon^*\phi_3 + \phi_4 - \varepsilon\phi_5 - \varepsilon^*\phi_6\} \tag{10.43}$$

$$\Omega_5 = \hat{P}^{E_{2a}}\,\phi_1 + \hat{P}^{E_{2b}}\,\phi_1 = \tfrac{1}{6}\{2\phi_1 - \phi_6 - \phi_5 + 2\phi_4 - \phi_3 - \phi_2\}$$

$$= \tfrac{1}{\sqrt{12}}\{2\phi_1 - \phi_2 - \phi_3 + 2\phi_4 - \phi_5 - \phi_6\} \text{ (after normalisation)} \tag{10.44}$$

$$\Omega_6 = \hat{P}^{B_{2a}}\,\phi_1 - \hat{P}^{B_{2b}}\,\phi_1 = \tfrac{i\sqrt{3}}{6}\{-\phi_6 + \phi_5 - \phi_3 + \phi_2\}$$

$$= \tfrac{1}{2}\{\phi_2 - \phi_3 + \phi_5 - \phi_6\} \text{ (after normalisation)} \tag{10.45}$$

The π-MOs of benzene are now constructed by linearly combining these SALCs as

$$\Psi_{MO} = \sum_{i=1}^{6} c_i\,\Omega_i \tag{10.46}$$

The resulting secular determinantal equation is obtained in a block-diagonal form:

$$\begin{array}{l|cccccc|}
A: & H'_{11}-ES'_{11} & & & & & \\
B: & & H'_{22}-ES'_{22} & & & & \\
E_1: & & & H'_{33}-ES'_{33} & H'_{34}-ES'_{34} & & \\
 & & & H'_{43}-ES'_{43} & H'_{44}-ES'_{44} & & \\
E_2: & & & & & H'_{55}-ES'_{55} & H'_{56}-ES'_{56} \\
 & & & & & H'_{65}-ES'_{65} & H'_{66}-ES'_{66}
\end{array} = 0 \tag{10.47}$$

The blank spaces (zeros) in the determinant are results of the vanishing integral rule:

$$H'_{ij} = \langle \Omega_i | \hat{\mathrm{H}} | \Omega_j \rangle = 0 \quad \text{and} \quad S'_{ij} = \langle \Omega_i | \Omega_j \rangle = 0$$

whenever Ω_i and Ω_j are bases of different irreducible representations (the Hamiltonian operator belongs to the totally symmetric irreducible representation).

We shall now calculate the non-zero values of H'_{ij} and S'_{ij} occurring in the determinant.

$$\begin{aligned}
H'_{11} &= \langle \Omega_1 | \hat{\mathrm{H}} | \Omega_1 \rangle \\
&= \langle \tfrac{1}{\sqrt{6}} \{\phi_1+\phi_2+\phi_3+\phi_4+\phi_5+\phi_6\} | \hat{\mathrm{H}} | \tfrac{1}{\sqrt{6}} \{\phi_1+\phi_2+\phi_3+\phi_4+\phi_5+\phi_6\} \rangle \\
&= \tfrac{1}{6}[(\alpha+\beta+\beta)\times 6] \quad \text{(using the Hückel approximations)} \\
&= \alpha + 2\beta
\end{aligned}$$

Similarly, $H'_{22} = \alpha - 2\beta, H'_{33} = \alpha + \beta, H'_{44} = \alpha + \beta,$

$$H'_{55} = \alpha - \beta, H'_{66} = \alpha - \beta, H'_{23} = 0$$

$$H'_{34} = 0, H'_{56} = 0, H'_{65} = 0$$

which can be readily verified by using the SALCs. Moreover, each SALC is individually normalised, that is, $S'_{ii} = 1$ and making use of the Hückel approximations, it can be found that $S'_{ij} = 0$ for $i \neq j$ (whatever be the irreducible representation to which Ω_i and Ω_j belong).

Using these two values and equating each block of the secular determinant to zero, it is now easy to find the eigenvalues and eigenvectors for the π-MOs of benzene. For the first block $H'_{11} - ES'_{11} = 0$ gives,

$$\alpha + 2\beta - E \cdot 1 = 0 \quad \text{or} \quad E = \alpha + 2\beta$$

Only one coefficient c_1 is needed for this block as the other elements in the first row of the determinant are each zero. Normalisation of the π-MO with energy $\alpha + 2\beta$ gives $c_1^2 = 1$, that is, $c_1 = 1$ (taking the positive square root only). Thus the π-MO wave function corresponding to energy $\alpha + 2\beta$ is

$$\Psi_1 = 1\cdot\Omega_1 = \frac{1}{\sqrt{6}}\{\phi_1+\phi_2+\phi_3+\phi_4+\phi_5+\phi_6\}$$

$$= (\phi_1 \quad \phi_2 \quad \phi_3 \quad \phi_4 \quad \phi_5 \quad \phi_6)\begin{pmatrix}0.408\\0.408\\0.408\\0.408\\0.408\\0.408\end{pmatrix} \tag{10.48}$$

(Here $\frac{1}{\sqrt{6}} = 0.408$). The symmetry label for this π-MO is A_g (after noting that the wave function is symmetric with respect to the inversion operator of the D_{6h} point group). Similarly the second block for symmetry type B gives $E = \alpha - 2\beta$ and

$$\Psi_2 = 1\cdot\Omega_2 = \frac{1}{\sqrt{6}}\{\phi_1-\phi_2+\phi_3-\phi_4+\phi_5-\phi_6\}$$

$$= (\phi_1 \quad \phi_2 \quad \phi_3 \quad \phi_4 \quad \phi_5 \quad \phi_6)\begin{pmatrix}0.408\\-0.408\\0.408\\-0.408\\0.408\\-0.408\end{pmatrix} \tag{10.49}$$

The third block is a 2 × 2 determinant because it corresponds to the two-dimensional representation, E_1:

$$\begin{vmatrix}\alpha+\beta-E & 0\\ 0 & \alpha+\beta-E\end{vmatrix} = 0 \tag{10.50}$$

So, $(\alpha+\beta-E)^2 = 0$ that is, $E = (\alpha+\beta), (\alpha+\beta)$

Fig. 10.8 Energy order of the benzene π-MOs

The determinant in Eq. (10.50) is in the diagonal form and the zero off-diagonal elements indicate that the SALCs Ω_3 and Ω_4 do not mix. So the π-MO wave functions corresponding to these two degenerate eigenvalues $(\alpha + \beta)$, $(\alpha + \beta)$ are Ω_3 and Ω_4 themselves. In the same way it is found that the fourth block which corresponds to symmetry type E_2 gives $E = (\alpha - \beta)$, $(\alpha - \beta)$ with π-MO wave functions Ω_5 and Ω_6. Figure 10.8 displays schematically the order of energy eigenvalues of the π-MOs of benzene. In the energy scale $(\alpha = 0, \beta = 1)$, these eigenvalues are $\pm 1, \pm 1, \pm 2$.

The results are summarised in Table 10.2.

Table 10.2 Energies and wave functions of the π-MOs of benzene

Symmetry type	Energy	Wave function
A	$\alpha + 2\beta$	$\frac{1}{\sqrt{6}}\{\phi_1 + \phi_2 + \phi_3 + \phi_4 + \phi_5 + \phi_6\}$
E_1	$\alpha + \beta$	$\frac{1}{\sqrt{12}}\{2\phi_1 + \phi_2 - \phi_3 - 2\phi_4 - \phi_5 + \phi_6\}$
	$\alpha + \beta$	$\frac{1}{2}\{\phi_2 + \phi_3 - \phi_5 - \phi_6\}$
E_2	$\alpha - \beta$	$\frac{1}{\sqrt{12}}\{2\phi_1 - \phi_2 - \phi_3 + 2\phi_4 - \phi_5 - \phi_6\}$
	$\alpha - \beta$	$\frac{1}{2}\{\phi_2 - \phi_3 + \phi_5 - \phi_6\}$
B	$\alpha - 2\beta$	$\frac{1}{\sqrt{6}}\{\phi_1 - \phi_2 + \phi_3 - \phi_4 + \phi_5 - \phi_6\}$

The delocalisation energy of benzene is obtained as follows. In the ground state of the molecule, two π-electrons occupy the lowest energy level $(\alpha + 2\beta)$ and the next two higher energy levels (doubly degenerate) are occupied by one pair of π-electrons each. Hence the total π-electron energy is $2(\alpha + 2\beta) + 4(\alpha + \beta) = 6\alpha + 8\beta$. If the π-electrons were localised to form three isolated ethylenic double bonds (as in cyclohexatriene, which is the most stable Kekulè structure in the valence bond model), the total π-electron energy would have been $6(\alpha + \beta)$ (recall that the bonding π-MO of ethylene has energy $\alpha + \beta$). Then the delocalisation energy of benzene, which is the main cause of its aromatic stability, is $(6\alpha + 8\beta) - 6(\alpha + \beta) = 2\beta$.

The charge densities and bond orders can be determined by noting that the A and E_1 orbitals are each doubly occupied. By symmetry, each C atom in benzene should have the same π-electronic charge. This is readily verified:

On atom 1, $Q_1 = 2 \times \left(\frac{1}{\sqrt{6}}\right)^2 + 2 \times \left(\frac{1}{\sqrt{12}}\right)^2 + 2 \times 0$ (collecting coefficients of ϕ_1)

$$= \tfrac{1}{3} + \tfrac{2}{3} = 1$$

On atom 2, $Q_2 = 2 \times \left(\frac{1}{\sqrt{6}}\right)^2 + 2 \times \left(\frac{1}{\sqrt{12}}\right)^2 + 2 \times \left(\frac{1}{2}\right)^2$ (collecting coefficients of ϕ_2)

$$= \tfrac{1}{3} + \tfrac{1}{6} + \tfrac{1}{2} = 1$$

The π-bond orders between all pairs of adjacent C-atoms are also expected to be equal owing to symmetry. This can also be verified easily. Thus,

$$\pi\text{-bond order between atoms 1 and 2, } P_{12} = 2\cdot\left(\frac{1}{\sqrt{6}}\right)\cdot\left(\frac{1}{\sqrt{6}}\right) + 2\cdot\left(\frac{1}{\sqrt{12}}\right)\cdot\left(\frac{2}{\sqrt{12}}\right) + 2.0\cdot\left(\frac{1}{2}\right)$$

$$= \frac{1}{3} + \frac{1}{3} = \frac{2}{3} = 0.667$$

$$\pi\text{-bond order between atoms 2 and 3, } P_{23} = 2\cdot\left(\frac{1}{\sqrt{6}}\right)\cdot\left(\frac{1}{\sqrt{6}}\right) + 2\cdot\left(\frac{1}{\sqrt{12}}\right)\cdot\left(-\frac{1}{\sqrt{12}}\right) + 2\cdot\left(\frac{1}{2}\right)\cdot\left(\frac{1}{2}\right)$$

$$= \frac{1}{3} - \frac{1}{6} + \frac{1}{2} = \frac{2}{3} = 0.667$$

10.6 RELATION BETWEEN THE NUMBER OF NODAL PLANES AND π-MO ENERGIES OF BENZENE

A nodal plane is one in which the lobes of the MOs have opposite signs (out of phase combinations of atomic orbitals) on the two sides of the plane. Looking at the π-MO wave functions of benzene, it is easy to find such planes. In schematic diagrams (Fig.10.9), the + and – signs at the C-sites indicate the signs of the ϕ functions (atomic p_z-functions) of the conjugated atoms; if any ϕ is absent in the expression for a MO wave function, it is indicated by a zero in the diagram. Since β is negative, the energy order in the π-MOs of benzene is,

$$\alpha + 2\beta < \alpha + \beta,\ \alpha + \beta < \alpha - \beta,\ \alpha - \beta < \alpha - 2\beta$$

Thus, as the number of nodal planes increases, the π-MO energy also increases; if two π-MOs have the same number of nodes, they have the same energy.

10.7 HÜCKEL 4*n* + 2 RULE OF AROMATICITY

This rule states that monocyclic coplanar systems of trigonally hybridised atoms which contain $4n + 2$ π-electrons will posses more electronic stability than the corresponding π-systems with isolated double bonds. This property is called *aromaticity*. From the HMO method of calculation on $(CH)_k$ type of carbocyclic systems (k = an integer) it is found that the π-MO energy levels are distributed as in Fig. 10.10.

To fill up the bonding π-MOs in order to get closed shell electronic configuration, the number of electrons required

= 2 (for the A-type MO) + 4 × (an integral number of E_1-type MOs)

= $4n + 2$ (where n is an integer)

It should be mentioned that the $4n + 2$ rule ensures aromatic stability only if the π-system is of D_{nh} point group (that is, planarity of the molecule is a requirement).

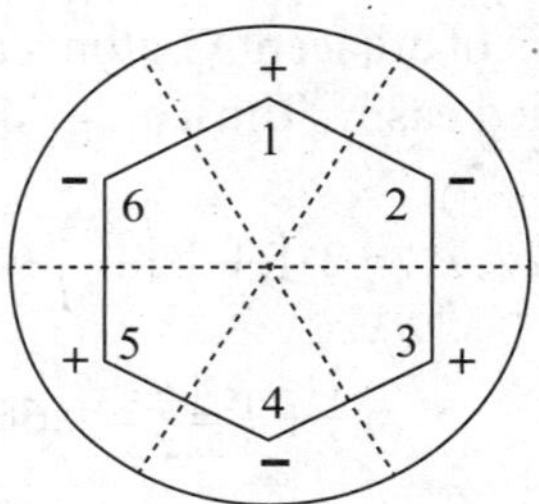

$$\Psi_6(B) = \frac{1}{\sqrt{6}}\{\phi_1 - \phi_2 + \phi_3 - \phi_4 + \phi_5 - \phi_6\}$$

No. of nodal planes = 3, $E = \alpha - 2\beta$

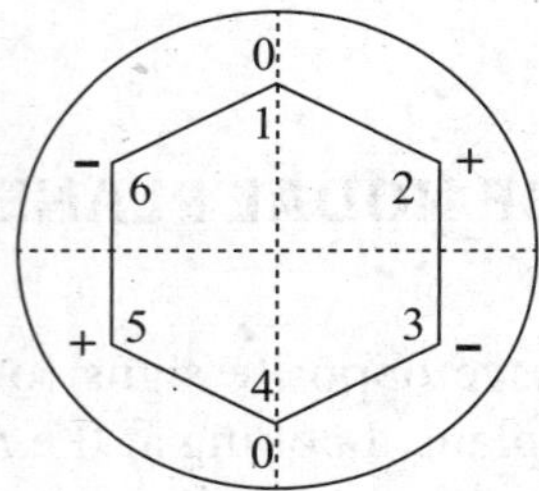

$$\Psi_5(E_2) = \frac{1}{2}\{\phi_2 - \phi_3 + \phi_5 - \phi_6\}$$

No. of nodal planes = 2, $E = \alpha - \beta$

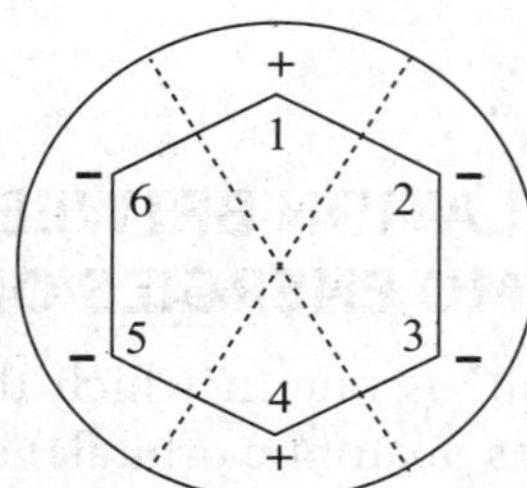

$$\Psi_4(E_2) = \frac{1}{\sqrt{12}}\{2\phi_1 - \phi_2 - \phi_3 + 2\phi_4 - \phi_5 - \phi_6\}$$

No. of nodal planes = 2, $E = \alpha - \beta$

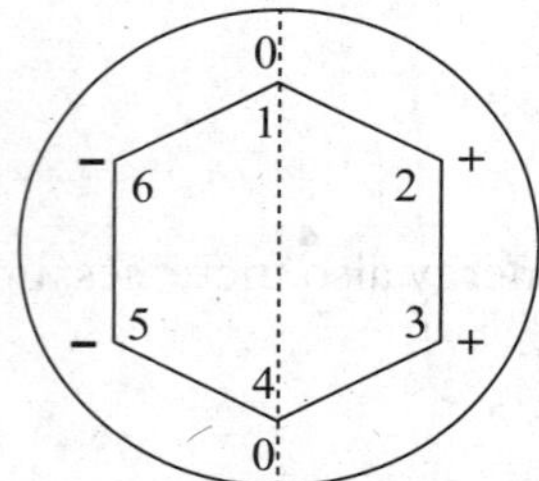

$$\Psi_3(E_1) = \frac{1}{2}\{\phi_2 + \phi_3 - \phi_5 - \phi_6\}$$

No. of nodal planes = 1, $E = \alpha + \beta$

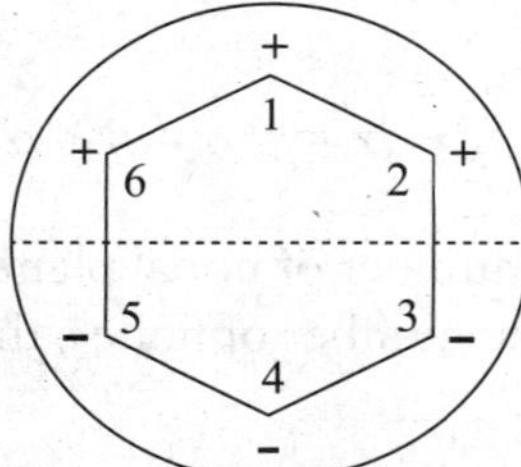

$$\Psi_2(E_1) = \frac{1}{\sqrt{12}}\{2\phi_1 + \phi_2 - \phi_3 - 2\phi_4 - \phi_5 + \phi_6\}$$

No. of nodal planes = 1, $E = \alpha + \beta$

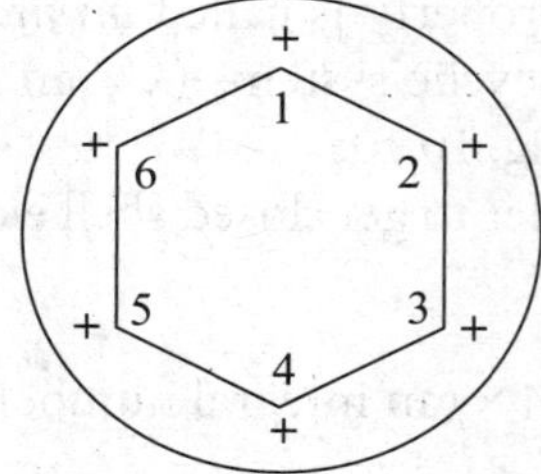

$$\Psi_1(A) = \frac{1}{\sqrt{6}}\{\phi_1 + \phi_2 + \phi_3 + \phi_4 + \phi_5 + \phi_6\}$$

No. of nodal planes = 0, $E = \alpha + 2\beta$

Fig. 10.9 Nodal planes in the benzene π-MOs

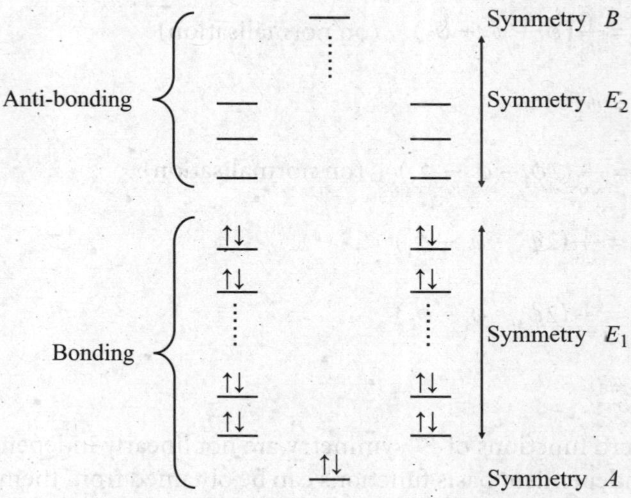

Fig. 10.10 Energy levels of carbocyclic π-systems

10.8 FREE VALENCEY: REACTIVITY OF ATOMIC SITES IN π-CONJUGATED SYSTEMS

The *free valence* (F_r) of the atom labelled r in a π-conjugated system was defined by Coulson as

$$F_r = N_{max} - N_r \tag{10.51}$$

where N_r is the sum of the π-bond orders of all bonds joining atom r to adjacent atoms and N_{max} is the maximum value of any N_r which is generally taken to be $\sqrt{3}$. This value of N_{max} is the sum of π-bond orders of the central C atom of the hypothetical π-system, $(CH_2)_3C$ (trimethylenemethane). On Hückel π-MO calculation this system is found to be a diradical (Fig. 10.11).

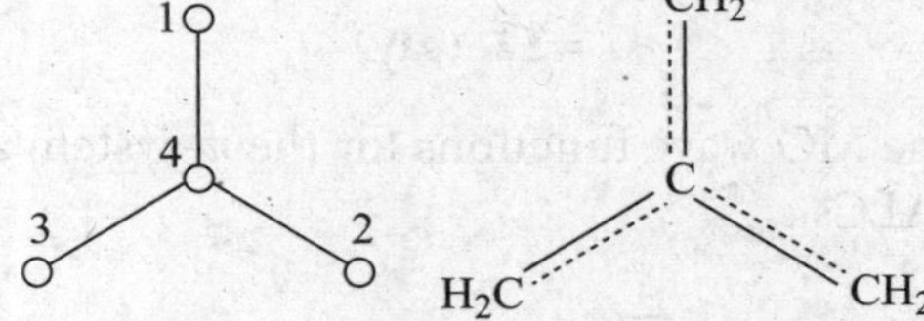

Fig. 10.11 Atom labels in trimethylenemethane

Figure 10.11 shows the labels of the conjugated atoms in trimethylenemethane. Each C atom is taken to be sp^2 hybridised and the p orbitals for π-MO formation are perpendicular to the plane containing the C atoms. The point group of the system is D_{3h}; the C_3 axis passes through the central C atom. The p orbital wave functions are ϕ_1, ϕ_2, ϕ_3 and ϕ_4 for atoms 1 to 4 respectively. With these functions as bases a representation Γ_ϕ of D_{3h} is as follows:

χ **(character)**

	$\hat{E}$	$2\hat{C}_3$	$3\hat{C}_2$	$\hat{\sigma}_h$	$2\hat{S}_3$	$3\hat{\sigma}_v$
Γ_ϕ	4	1	−2	−4	−1	2

On reduction it is found that $\Gamma_\phi = 2A_2'' + E''$.

Hence there should be two non-degenerate π-MOs of A_2'' symmetry and one doubly degenerate π-MO of E'' symmetry. By applying the projection operator method, the required SALCs are found:

$$\Omega_1 = \hat{P}^{A_2''}\,\phi_1 = \tfrac{1}{\sqrt{3}}(\phi_1 + \phi_2 + \phi_3) \quad \text{(on normalisation)}$$

$$\Omega_2 = \hat{P}^{A_2''}\,\phi_4 = \phi_4$$

$$\Omega_3 = \hat{P}^{E''}\,\phi_1 = \tfrac{1}{\sqrt{6}}(2\phi_1 - \phi_2 - \phi_3) \quad \text{(on normalisation)}$$

$$\Omega_4' = \hat{P}^{E''}\,\phi_2 = \tfrac{1}{\sqrt{6}}(2\phi_2 - \phi_3 - \phi_1)$$

$$\Omega_5' = \hat{P}^{E''}\,\phi_3 = \tfrac{1}{\sqrt{6}}(2\phi_3 - \phi_1 - \phi_2)$$

$$\Omega_6 = \hat{P}^{E''}\,\phi_4 = 0$$

But the three non-zero functions of E'' symmetry are not linearly independent as their sum is zero. Two linearly independent basis functions can be obtained from them as follows: keep Ω_3 unchanged; take the sum and difference of the remaining two.

$$\Omega_4' + \Omega_5' = \tfrac{1}{\sqrt{6}}(\phi_2 + \phi_3 - 2\phi_1) = -\tfrac{1}{\sqrt{6}}(2\phi_1 - \phi_2 - \phi_3) \text{ which is the same as } \Omega_3$$

$$\Omega_4' - \Omega_5' = \tfrac{1}{\sqrt{6}}(3\phi_2 - 3\phi_3) \approx \tfrac{1}{\sqrt{2}}(\phi_2 - \phi_3) \quad \text{(on normalisation)}$$

$$= \Omega_4 \text{ (say)}$$

The MO wave functions for the π-system are now obtained from linear combination of these SALCs,

$$\Psi = \sum_{i=1}^{4} c_i\,\Omega_i$$

The secular determinantal equation is block factored as:

$$\begin{matrix} A_2'': \\ A_2'': \\ E'': \left\{ \begin{matrix} \\ \\ \end{matrix} \right. \end{matrix} \begin{vmatrix} H_{11}-E & H_{12} & 0 & 0 \\ H_{21} & H_{22}-E & 0 & 0 \\ 0 & 0 & H_{33}-E & H_{34} \\ 0 & 0 & H_{43} & H_{44}-E \end{vmatrix} = 0 \tag{10.52}$$

On calculating the matrix elements by using the Hückel approximations, Eq. (10.52) reduces to:

$$\begin{vmatrix} \alpha - E & \sqrt{3}\beta \\ \sqrt{3}\beta & \alpha - E \end{vmatrix} = 0 \text{ for the block of } A_2'' \text{ symmetry}$$

and

$$\begin{vmatrix} \alpha - E & 0 \\ 0 & \alpha - E \end{vmatrix} = 0 \text{ for the block of } E'' \text{ symmetry}$$

By the usual procedure, the energy eigenvalues and MO wave functions for the first block are found to be:

$$E = \alpha + \sqrt{3}\beta, \Psi_1(A_2'') = \tfrac{1}{\sqrt{2}}(\Omega_1 + \Omega_2) = \tfrac{1}{\sqrt{6}}(\phi_1 + \phi_2 + \phi_3) + \tfrac{1}{\sqrt{2}}(\phi_4)$$

$$E = \alpha - \sqrt{3}\beta, \Psi_2(A_2'') = \tfrac{1}{\sqrt{2}}(\Omega_1 - \Omega_2) = \tfrac{1}{\sqrt{6}}(\phi_1 + \phi_2 + \phi_3) - \tfrac{1}{\sqrt{2}}(\phi_4)$$

The block corresponding to E'' symmetry is itself in diagonal form. Hence Ω_3 and Ω_4 are themselves the MO wave functions with an energy eigenvalue of α each:

$$E = \alpha, \Psi_3(E'') = \Omega_3 = \tfrac{1}{\sqrt{6}}(2\phi_1 - \phi_2 - \phi_3)$$

$$E = \alpha, \Psi_4(E'') = \Omega_4 = \tfrac{1}{\sqrt{2}}(\phi_2 - \phi_3)$$

The energy level diagram is given in Fig. 10.12, where the distribution of four π-electrons is also shown.

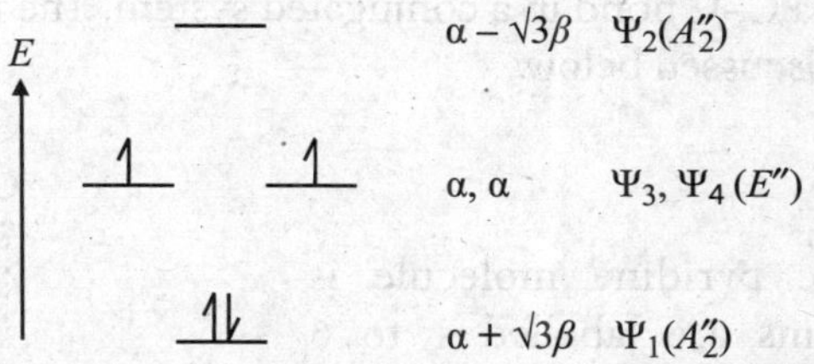

Fig. 10.12 Energy level diagram of trimethylenemethane

It is found that the system is a diradical. The π-bond orders at the central C atom (labelled 4) are as follows:

$$P_{14} = \sum_{i=1}^{occ} \nu_i\, c_{1i}\, c_{4i} = 2 \cdot \tfrac{1}{\sqrt{6}} \cdot \tfrac{1}{\sqrt{2}} + 1 \cdot \tfrac{2}{\sqrt{6}} \cdot 0 + 1 \cdot 0 = \tfrac{1}{\sqrt{3}}$$

Similarly $P_{24} = P_{34} = \tfrac{1}{\sqrt{3}}$

The sum of the bond orders of the bonds connected to the central atom $= 3 \times \tfrac{1}{\sqrt{3}} = \sqrt{3}$. This is taken as N_{max} for the calculation of F_r for the other π-systems. Equation (10.51) indicates that F_r is a measure of the left-over π-bonding power of the atom r; the greater the extent of π-bonding of the position r to others, the lower the value of F_r and the less is the left-over π-bonding ability to an attacking electrophile. Hence sites of higher F_r value will be the preferred sites for electrophilic attack. Some examples are given below.

10.8.1 Butadiene

$$P_{12} = 0.894, F_1 = 1.732 - 0.894 = 0.838$$

$$P_{12} + P_{23} = 0.894 + 0.447 = 1.341, F_2 = 1.732 - (P_{12} + P_{23}) = 0.391$$

Thus, the terminal position of butadiene is more reactive than the 2-position for attack by electrophiles. In this context, the free valence concept was considered as a quantum mechanical version of Thiele's theory of partial valence (or, residual affinity).

10.8.2 Naphthalene

Calculations using results obtained in Section 10.4.3 reveal that $F_1 = 0.453$ and $F_2 = 0.404$. Hence mononitration of naphthalene occurs preferentially at the 1-position.

10.8.3 Anthracene

It will be found by actual calculation that $F_1 = 0.459$, $F_2 = 0.409$ and $F_9 = 0.520$. Hence the 9-position of anthracene is the most susceptible to electrophilic attack.

10.9 HMO CALCULATION FOR HETEROCONJUGATED π-SYSTEMS

In π-conjugated systems with heteroatoms, the coulomb integral α_X of heteroatom (X) and the resonance integral β_{C-X} of the C–X bond differ respectively from α for a carbon atom (sp^2-hybridised) and β for a C–C bond in a conjugated system. The rest of the procedure is the same. Two examples are discussed below.

10.9.1 Pyridine

The point group of the pyridine molecule is C_{2v}. The conjugated atoms are labelled 1 to 6 (Fig. 10.13). The molecular plane is the YZ plane and the p_x-orbitals with wave functions $\{\phi_i\}$, $i = 1$ to 6, combine linearly to produce the π-MOs.

A representation Γ_p of C_{2v}, with the set $\{\phi_i\}$, $i = 1$ to 6 as basis, has the following characters:

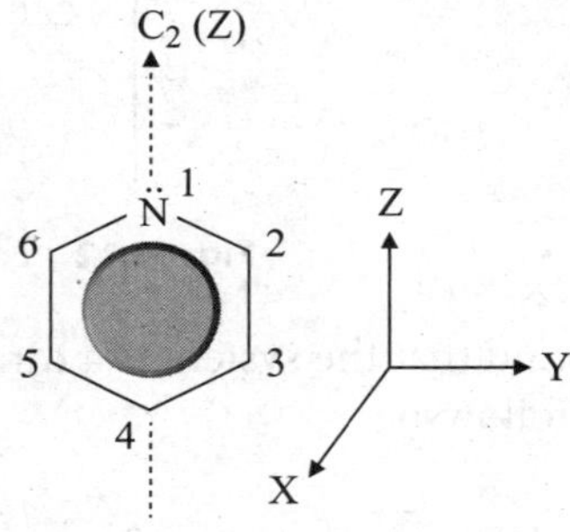

Fig. 10.13 Structure of pyridine with atom labels and orientation of coordinate axes

χ **(character)**

	$\hat{E}$	$\hat{C}_2(Z)$	$\hat{\sigma}_v(XZ)$	$\hat{\sigma}_v(YZ)$
Γ_p	6	−2	2	−6

Applying the reduction formula, we obtain $\Gamma_p = 2A_2 + 4B_1$. Thus two π-MOs are expected to be of A_2 symmetry and four of B_1 symmetry. The required SALCs are now formed by the projection operator method:

$$\hat{P}^{A_2}\,\phi_1 = \tfrac{1}{4}[1\cdot\hat{E}\cdot\phi_1 + 1\cdot\hat{C}_2\cdot\phi_1 + (-1)\cdot\hat{\sigma}_v(XZ)\cdot\phi_1 + (-1)\cdot\hat{\sigma}_v(YZ)\cdot\phi_1]$$

$$= \tfrac{1}{4}[\phi_1 - \phi_1 - \phi_1 + \phi_1] = 0$$

$$\hat{P}^{A_2}\,\phi_2 = \tfrac{1}{4}[1\cdot\hat{E}\cdot\phi_2 + 1\cdot\hat{C}_2\cdot\phi_2 + (-1)\cdot\hat{\sigma}_v(XZ)\cdot\phi_2 + (-1)\cdot\hat{\sigma}_v(YZ)\cdot\phi_2]$$

$$= \tfrac{1}{4}[\phi_2 - \phi_6 - \phi_6 + \phi_2] \approx \tfrac{1}{\sqrt{2}}[\phi_2 - \phi_6] \quad \text{(on normalisation)}$$

$$\hat{P}^{A_2}\,\phi_3 = \tfrac{1}{4}[1\cdot\hat{E}\cdot\phi_3 + 1\cdot\hat{C}_2\cdot\phi_3 + (-1)\cdot\hat{\sigma}_v(XZ)\cdot\phi_3 + (-1)\cdot\hat{\sigma}_v(YZ)\cdot\phi_3]$$

$$= \tfrac{1}{4}[\phi_3 - \phi_5 - \phi_5 + \phi_3] \approx \tfrac{1}{\sqrt{2}}[\phi_3 - \phi_5] \text{ (on normalisation)}$$

$$\hat{P}^{A_2}\,\phi_4 = \tfrac{1}{4}[1\cdot\hat{E}\cdot\phi_4 + 1\cdot\hat{C}_2\cdot\phi_4 + (-1)\cdot\hat{\sigma}_v(XZ)\cdot\phi_4 + (-1)\cdot\hat{\sigma}_v(YZ)\cdot\phi_4]$$

$$= \tfrac{1}{4}[\phi_4 - \phi_4 - \phi_4 + \phi_4] = 0$$

We thus get two normalised SALCs of A_2 symmetry

$$\Omega_1 = \tfrac{1}{\sqrt{2}}[\phi_2 - \phi_6] \text{ and } \Omega_2 = \tfrac{1}{\sqrt{2}}[\phi_3 - \phi_5]$$

The following four normalised SALCs of B_1 symmetry can be similarly found by applying the projection operator method:

$$\hat{P}^{B_1}\,\phi_1 = \phi_1 = \Omega_3 \text{ (say)};\ \hat{P}^{B_1}\,\phi_2 = \tfrac{1}{\sqrt{2}}(\phi_2 + \phi_6) = \Omega_4;\ \hat{P}^{B_1}\,\phi_3 = \tfrac{1}{\sqrt{2}}(\phi_3 + \phi_5) = \Omega_5;$$

$$\hat{P}^{B_1}\,\phi_4 = \phi_4 = \Omega_6$$

Each π-MO is now considered as a linear combination of these SALCs:

$$\Psi_1 = c_1\Omega_1 + c_2\Omega_2 + \cdots + c_6\Omega_6$$

Applying the variation method by the usual procedure, the following secular determinantal equation, with $H_{ij} = \langle\Omega_i|\hat{H}|\Omega_j\rangle$, is arrived at:

$$\begin{array}{c} \overbrace{\Omega_1 \quad \Omega_2}^{A_2} \quad \overbrace{\Omega_3 \quad \Omega_4 \quad \Omega_5 \quad \Omega_6}^{B_1} \end{array}$$

$$\begin{vmatrix} H_{11}-E & H_{12} & & & & \\ H_{21} & H_{22}-E & & & & \\ & & H_{33}-E & H_{34} & H_{35} & H_{36} \\ & & H_{43} & H_{44}-E & H_{45} & H_{46} \\ & & H_{53} & H_{54} & H_{55}-E & H_{56} \\ & & H_{63} & H_{64} & H_{65} & H_{66}-E \end{vmatrix} = 0$$

Because of the vanishing integral rule, $H_{ij} = 0$ whenever Ω_i and Ω_j are SALCs of different irreducible representations and hence the block-factored nature of the secular determinant results.

For the first block, $H_{11} = \langle\Omega_1|\hat{H}|\Omega_1\rangle = \left\langle \tfrac{1}{\sqrt{2}}(\phi_2 - \phi_6)\middle|\hat{H}\middle|\tfrac{1}{\sqrt{2}}(\phi_2 - \phi_6)\right\rangle$

$$= \tfrac{1}{2}[\alpha - 0 - 0 + \alpha] = \alpha$$

$$H_{12} = \langle \Omega_1|\hat{H}|\Omega_1\rangle = \left\langle \tfrac{1}{\sqrt{2}}(\phi_2 - \phi_6)\middle|\hat{H}\middle|\tfrac{1}{\sqrt{2}}(\phi_3 - \phi_5)\right\rangle$$

$$= \tfrac{1}{2}[\beta - 0 - 0 + \beta] = \beta$$

This gives, $\begin{vmatrix} \alpha - E & \beta \\ \beta & \alpha - E \end{vmatrix} = 0$, that is, $E = \alpha \pm \beta$. For the eigenvector corresponding to $E = \alpha + \beta$.

$$c_1(\alpha - E) + c_2\beta = 0$$

or, $\quad c_1(-\beta) + c_2\beta = 0 \quad$ or, $\quad c_1 = c_2$

and $\quad$ normalisation $(c_1^2 + c_2^2 = 1)$ gives $c_1 = c_2 = \frac{1}{\sqrt{2}}$

Thus the MO wave function with energy $E = \alpha + \beta$ is

$$\Psi_1 = c_1\Omega_1 + c_2\Omega_2 = \tfrac{1}{\sqrt{2}}(\Omega_1 + \Omega_2) = \tfrac{1}{\sqrt{2}}(\phi_2 + \phi_3 - \phi_5 - \phi_6)$$

Similarly for $E = \alpha - \beta$ the MO wave function is found to be:

$$\Psi_2 = \tfrac{1}{\sqrt{2}}(\Omega_1 - \Omega_2) = \tfrac{1}{\sqrt{2}}(\phi_2 - \phi_3 + \phi_5 - \phi_6)$$

For the second block two of the necessary integrals are,

$$H_{33} = \langle \Omega_3|\hat{H}|\Omega_3\rangle = \langle \phi_1|\hat{H}|\phi_1\rangle = \alpha_N$$

$$H_{34} = \langle \Omega_3|\hat{H}|\Omega_4\rangle = \left\langle \phi_1\middle|\hat{H}\middle|\tfrac{1}{\sqrt{2}}(\phi_2 + \phi_6)\right\rangle = \tfrac{1}{\sqrt{2}}(\beta_{C-N} + \beta_{C-N}) = \sqrt{2}\beta_{C-N}$$

(where β_{C-N} is the resonance integral for C–N bond). The other integrals are found to be

$$H_{35} = H_{36} = H_{46} = 0$$

$$H_{44} = H_{55} = H_{66} = \alpha$$

$$H_{45} = H_{56} = \beta \text{ (according to the Hückel approximations)}$$

From the second block of the secular determinant we therefore have

$$\begin{vmatrix} \alpha_N - E & \sqrt{2}\beta_{C-N} & 0 & 0 \\ \sqrt{2}\beta_{C-N} & \alpha - E & \beta & 0 \\ 0 & \beta & \alpha - E & \sqrt{2}\beta \\ 0 & 0 & \sqrt{2}\beta & \alpha - E \end{vmatrix} = 0$$

At this stage the usual procedure is to substitute

$$\alpha_N = \alpha + h_N\beta, \beta_{C-N} = k_{C-N}\beta$$

where h_N and k_{C-N} are parameters for the heteroatom N and the bond C–N. Taking α as the zero and β as the unit of energy, this equation becomes

$$\begin{vmatrix} h_N - E & \sqrt{2}k_{C-N} & 0 & 0 \\ \sqrt{2}k_{C-N} & -E & 1 & 0 \\ 0 & 1 & -E & \sqrt{2} \\ 0 & 0 & \sqrt{2} & -E \end{vmatrix} = 0 \tag{10.53}$$

The heteroatom parameters h and k have been estimated for many π-conjugated atoms on the basis of comparison of theoretical results with experimentally observed spectroscopic data [4]. A number of commonly occurring parameters are listed in Table 10.3.

Table 10.3 Some heteroatom parameters for the Hückel method

Element	h (Coulomb integral parameter)	k (Resonance integral parameter)
Nitrogen	$h_N = 0.5$ (*)	$k_{C-N} = 0.8$
	$h_N = 1.5$ (**)	$k_{CN} = 1.0$
Oxygen	$h_O = 1.0$ (*)	$k_{C-O} = 0.8$
	$h_O = 2.0$ (**)	$k_{C=O} = 1.0$

(*) for heteroatom contributing one π-electron and (**) for heteroatom contributing two π-electrons.

Table 10.4 Energies and LCAO coefficients of the π-MOs of pyridine

Eigenvalue	1.9543	1.0618	1.0000	−0.6673	−1.0000	−1.8488
Eigenvector	0.4242	0.6653	0	0.5642	0	0.2431
	0.3856	0.2336	−0.5	−0.4116	−0.5	−0.3568
	0.4142	−0.2842	−0.5	−0.1767	0.5	0.4653
	0.4239	−0.5354	0	0.5295	0	−0.5033
	0.4142	−0.2842	0.5	−0.1767	−0.5	0.4653
	0.3856	0.2336	0.5	−0.4116	0.5	−0.3568

Using the values, $h_N = 0.5$ and $k_{C-N} = 0.8$ in Eq. (10.53), expanding the determinant and solving the biquadratic equation in E, the energy eigenvalues can be found out. The complete results are given in Table 10.4.

10.9.2 1, 4-Benzoquinone

1,4-benzoquinone of point group D_{2h} is shown with atom labels and orientation of coordinate axes in Fig. 10.14. The molecular plane is the YZ plane. Let $\{\phi_i\}$, $i = 1$ to 8, be the wave functions of the p_x-orbitals that take part in π-bond formation. With these functions as basis, a representation Γ_ϕ of D_{2h} is as follows:

χ (character)

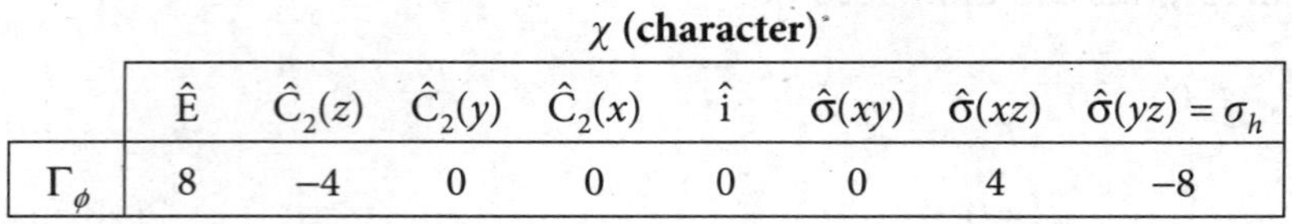

	$\hat{E}$	$\hat{C}_2(z)$	$\hat{C}_2(y)$	$\hat{C}_2(x)$	$\hat{i}$	$\hat{\sigma}(xy)$	$\hat{\sigma}(xz)$	$\hat{\sigma}(yz) = \sigma_h$
Γ_ϕ	8	−4	0	0	0	0	4	−8

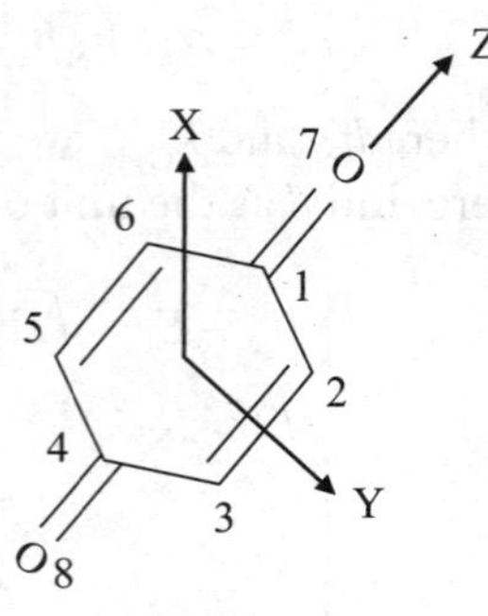

Fig. 10.14 Structure of 1,4-benzoquinone with atom labels and orientation of coordinate axes

By the usual reduction formula it is found that

$$\Gamma_\phi = B_{1g} + 3B_{2g} + A_{1u} + 3B_{3u}$$

The required normalised SALCs are found by the projection operator method to be as follows.

$$B_{1g}: \quad \Omega_1 = \tfrac{1}{2}(\phi_2 + \phi_3 - \phi_5 - \phi_6)$$

$$B_{2g}: \quad \Omega_2 = \tfrac{1}{\sqrt{2}}(\phi_1 - \phi_4); \Omega_3 = \tfrac{1}{\sqrt{2}}(\phi_7 - \phi_8); \Omega_4 = \tfrac{1}{2}(\phi_2 - \phi_3 - \phi_5 + \phi_6)$$

$$A_{1u}: \quad \Omega_5 = \tfrac{1}{2}(\phi_2 - \phi_3 + \phi_5 - \phi_6)$$

$$B_{3u}: \quad \Omega_6 = \tfrac{1}{\sqrt{2}}(\phi_1 + \phi_4); \Omega_7 = \tfrac{1}{\sqrt{2}}(\phi_7 + \phi_8); \Omega_8 = \tfrac{1}{2}(\phi_2 + \phi_3 + \phi_5 + \phi_6)$$

The π-MO wave function, $\Psi = \sum_{i=1}^{8} c_i \phi_i$, on variational treatment yields the secular determinant:

$$\begin{array}{c} \begin{array}{cccccccc} B_{1g} & & B_{2g} & & A_{1u} & & B_{3u} & \\ \Omega_1 & \Omega_2 & \Omega_3 & \Omega_4 & \Omega_5 & \Omega_6 & \Omega_7 & \Omega_8 \end{array} \\ \left| \begin{array}{cccccccc} H_{11}-E & & & & & & & \\ & H_{22}-E & H_{23} & H_{24} & & & & \\ & H_{32} & H_{33}-E & H_{34} & & & & \\ & H_{42} & H_{43} & H_{44}-E & & & & \\ & & & & H_{55}-E & & & \\ & & & & & H_{66}-E & H_{67} & H_{68} \\ & & & & & H_{76} & H_{77}-E & H_{78} \\ & & & & & H_{86} & H_{87} & H_{88}-E \end{array} \right| = 0 \end{array}$$

The matrix elements of the four blocks are found by using the parameters in Table 10.3 to be as follows.

First block: $\quad H_{11} = \langle \Omega_1 | \hat{H} | \Omega_1 \rangle = \alpha + \beta = 1$

Second block: $H_{22} = \langle\Omega_2|\hat{H}|\Omega_2\rangle = \alpha = 0$

$$H_{32} = H_{23} = \langle\Omega_2|\hat{H}|\Omega_3\rangle = \beta_{C=O} = k_{C=O}\beta = k_{C=O} = 1$$

$$H_{42} = H_{24} = \langle\Omega_2|\hat{H}|\Omega_4\rangle = \sqrt{2}\beta = \sqrt{2}$$

$$H_{33} = \langle\Omega_3|\hat{H}|\Omega_3\rangle = \alpha_o = \alpha + h_O\beta = h_O = 1$$

$$H_{43} = H_{34} = \langle\Omega_3|\hat{H}|\Omega_4\rangle = 0$$

$$H_{44} = \langle\Omega_4|\hat{H}|\Omega_4\rangle = \alpha - \beta = -1$$

Third block: $H_{55} = \langle\Omega_5|\hat{H}|\Omega_5\rangle = \alpha - \beta = -1$

Fourth block: $H_{66} = \langle\Omega_6|\hat{H}|\Omega_6\rangle = \alpha = 0$; $H_{76} = H_{67} = \langle\Omega_6|\hat{H}|\Omega_7\rangle = \beta_{C=O} = k_{C=O}\beta = k_{C=O} = 1$

$$H_{86} = H_{68} = \langle\Omega_6|\hat{H}|\Omega_8\rangle = \sqrt{2}\beta = \sqrt{2};$$

$$H_{77} = \langle\Omega_7|\hat{H}|\Omega_7\rangle = \alpha_o = \alpha + h_O\beta = h_O = 1$$

$$H_{87} = H_{78} = \langle\Omega_7|\hat{H}|\Omega_8\rangle = 0; \ H_{88} = \langle\Omega_8|\hat{H}|\Omega_8\rangle = \alpha + \beta = 1$$

The first block is of size 1×1. Hence Ω_1 is itself the eigenfunction, the eigenvalue being $E = H_{11} = 1$. The MO wave function is $\Psi_1 = \Omega_1 = \frac{1}{2}(\phi_2 + \phi_3 - \phi_5 - \phi_6)$.

The second block simplifies to $\begin{vmatrix} -E & 1 & \sqrt{2} \\ 1 & 1-E & 0 \\ \sqrt{2} & 0 & -1-E \end{vmatrix} = 0$

This on expansion of the determinant gives

$$E^3 - 4E + 1 = 0$$

which by the Newton–Raphson method of iteration yields roots: 0.2541, 1.8608, – 2.1149. For $E = 0.2541$, the second block gives

$$c_1(-0.2541) + c_2 + \sqrt{2}c_3 = 0$$

$$c_1 + c_2(1 - 0.2541) + c_3 \cdot 0 = 0$$

which yield $c_1 = -0.7459\ c_2$ and $c_3 = -0.8412\ c_2$. The normalisation condition $c_1^2 + c_2^2 + c_3^2 = 1$ gives

$$[(-0.7459)^2 + 1^2 + (-0.8412)^2]\, c_2^2 = 1$$

from which we find $c_2 = 0.6646$, $c_1 = -0.4957$, $c_3 = -0.5591$. So the MO wave function corresponding to the energy $E = 0.2541$ is $\Psi_2 = c_1\Omega_2 + c_2\Omega_3 + c_3\Omega_4$

$$= (-0.4957) \times \tfrac{1}{\sqrt{2}}(\phi_1 - \phi_4) + (0.6646) \times \tfrac{1}{\sqrt{2}}(\phi_7 - \phi_8) + (-0.5591) \times \tfrac{1}{2}(\phi_2 - \phi_3 - \phi_5 + \phi_6)$$

$$= -0.3505\ (\phi_1 - \phi_4) + 0.4699\ (\phi_7 - \phi_8) - 0.3953\ (\phi_2 - \phi_3 - \phi_5 + \phi_6)$$

In this way, the other two MO wave functions corresponding to $E = 1.8608$ and -2.1149 can be determined for the second block. For the third block, the eigenvalue is -1 and the wave function is $\Psi_5 = \Omega_5 = \frac{1}{2}\ (\phi_2 - \phi_3 + \phi_5 - \phi_6)$. The fourth block simplifies to $\begin{vmatrix} -E & 1 & \sqrt{2} \\ 1 & 1-E & 0 \\ \sqrt{2} & 0 & 1-E \end{vmatrix} = 0$ or, $(1 - E)\ (E^2 - E - 3) = 0$ which gives $E = 1, \dfrac{1 \pm \sqrt{13}}{2} = 1, 2.3028, -1.3028$; the corresponding MO wave functions can be determined as before. (Determination of the full set of eigenfunctions is left as an exercise.)

SUMMARY

- ❑ The Hückel molecular orbital (HMO) method rests on the assumption of σ-π separability.
- ❑ π-MOs are formed by linearly combining p-type atomic orbitals of conjugated atoms and applying the variation method.
- ❑ The Coulomb and resonance integrals, $\alpha = \langle\phi_i|\hat{\mathrm{H}}|\phi_i\rangle$ and $\beta = \langle\phi_i|\hat{\mathrm{H}}|\phi_j\rangle$, are treated as parameters and are not integrated numerically.
- ❑ Roots of the secular equations (the π-MO energies) are obtained conveniently by setting $\alpha = 0$ and $\beta = 1$. The LCAO coefficients give the π-bond orders, charge densities and free valences after filling the π-MOs with electrons according to the *aufbau* principle.
- ❑ HMO method can also be applied to heteroconjugated π-systems and to establish group theoretically the $4n + 2$ rule of aromaticity for carbocyclic π-systems.

LEARNING OUTCOME

After reading this chapter the student will be able to apply the Hückel Molecular Orbital (HMO) method to (i) calculate π-MOs and related quantities like π-bond order, charge density, delocalisation energy and free valence of different sites in a π-conjugated molecule, (ii) prove Hückel's $(4n + 2)$ rule of aromaticity, (iii) judge the allowedness of π–π^* transitions and (iv) perform HMO calculation on π-systems having conjugated hetroatoms.

EXERCISES

1. Find the energy eigenvalues and wave functions of the π-MOs of (a) *cis*-1,3-butadiene and (b) *trans*-1,3-butadiene. Assign symmetry labels to the MOs and arrange them in increasing order of energy. Verify that the number of nodal planes increases with energy in each case.
2. By the Hückel method obtain the π-MO energies, wave functions, delocalisation energy, charge density, bond order and free valence values of important sites in the following molecules, radicals or cations. In each case find the spin multiplicity of the ground state. (The structures are given with sp^2-C atoms denoted by small hollow circles.)
   (a) Methylenecyclopropene (b) trivinylmethyl (c) fulvene and (d) benzyl

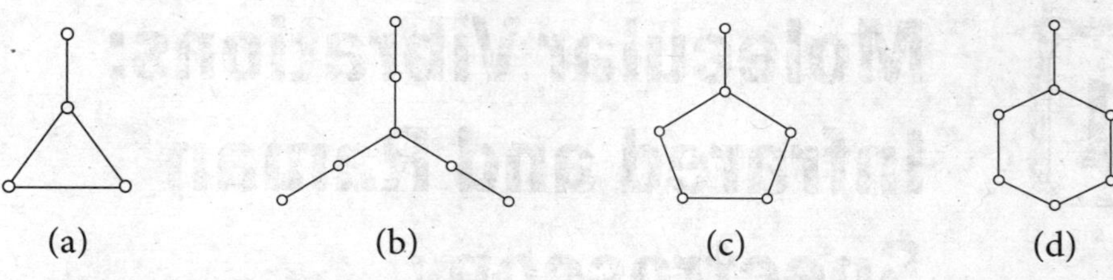

3. Perform Hückel molecular orbital calculation for benzene using (a) the full point group D_{6h} and (b) D_3 subgroup of D_{6h}.

4. Obtain the HMO secular matrix for corranulene ($C_{20}H_{10}$) in a block diagonal form. The molecule has a buckybowl structure with a pentagon surrounded by five hexagonal rings fused together.

5. In the unstable species trimethylene methane, $C(CH_2)_3$, the central carbon atom has three σ-bonds and three π-bonds. By applying the HMO method show that the maximum total binding power of C atom is $3+\sqrt{3}$. Obtain the free valency for each atom.

6. Explain, by π-MO calculation, why a rectangular structure of neutral cyclobutadiene with alternating double and single bonds is more stable than the corresponding square planar structure. (Hint: find delocalisation energy using D_{2h} and D_{4h} point groups for the two structures.)

7. Carry out the complete HMO calculation on *s*-triazene using the following heteroatom parameters for N:

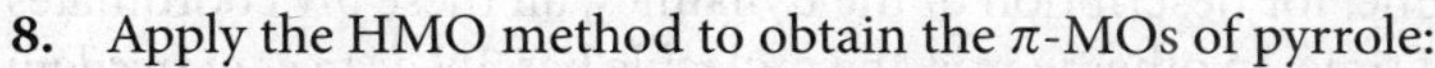

$h_N = 0.5,\ k_{C\text{–}N} = 1.0$

Find the charge densities on each N and C atoms.

8. Apply the HMO method to obtain the π-MOs of pyrrole:

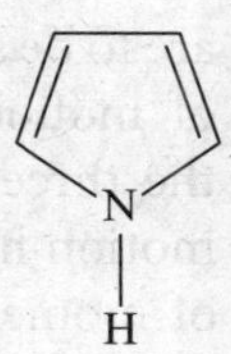

Consider the molecular plane as the YZ plane and take necessary parameters $h_N = 1.5$, $k_{C\text{–}N} = 1.0$. Find the total π-electron energy (E_π) and delocalisation energy (D_π).

Suggested Books and References for Further Reading

1. Hückel E, *Zeitschirift für Physik*, 70 (204) **1931**.
2. Hückel E, *Zeitschirift für Physik*, 72 (310) **1931**.
3. Coulson C A and Streitwieser A, *Dictionary of π-electron Calculations*, W H Freeman, CA, **1965**.
4. Streitwieser A Jr., *Molecular Orbital Theory for Organic Chemists*, John Wiley & Sons. Inc, NY, **1961**.
5. Coulson C A and Longuet-Higgins H C, *Proc. R. Soc. London*, A189 (39) **1947**.

11 Molecular Vibrations: Infrared and Raman Spectroscopy

Learning objectives

- Normal coordinates of molecular vibration—definition, symmetry types and sketch
- Prediction of the number and nature of polarisation of fundamental bands in infrared/Raman spectra of molecules—use of group theoretical selection rules
- Point group–subgroup correlation table for vibrational analysis of linear molecules
- Internal coordinates—assignment of symmetry types to experimental infrared/Raman spectra and extraction of important molecular structural features
- Analysis of observed overtones and combination bands group theoretically

11.1 INTRODUCTION

At ordinary temperatures, molecules execute three types of motion—translation, rotation and vibration. Even at absolute zero, molecular vibration does not stop. Absorption of electromagnetic radiation from the infrared region by matter is interpreted by spectroscopists in terms of vibrational motion of molecules (which is coupled with their rotational motion if the absorbing substance is in gas phase). In this chapter we will discuss molecular vibration only. When a molecule containing N atoms is in motion in free space, the $3N$ position coordinates are changing with time; for description of the dynamics, all these $3N$ coordinates are to be specified as functions of time. In other words, the molecule has $3N$ *degrees of freedom* of motion. The translational motion, however, can be fully described by specifying how the three position coordinates of the centre of mass change with time. Thus, translational motion has three degrees of freedom. Rotational motion of a molecule is described in terms of moments of inertia about three coordinate axes; however, for linear molecules, the line joining the atomic nuclei may be taken as one axis and only two moments of inertia about the other two orthogonal axes are sufficient to describe the rotational motion (moment of inertia about the internuclear axis is zero as the atoms are all situated on this axis). Hence, excluding the translational and rotational degrees of freedom, the number of vibrational degrees of freedom of an N-atomic molecule is $3N - 6$ (or, $3N - 5$ for linear molecules). In other words, an N-atomic molecule has $3N - 6$ (or, $3N - 5$) *modes of vibration*. For example, H_2O has $3 \times 3 - 6 = 3$ and CO_2 has $3 \times 3 - 5 = 4$ modes of vibration. Note that the overall vibrational motion of a molecule is complicated, being the resultant of vibrations involving all the modes taking place simultaneously.

11.2 NORMAL MODES OF VIBRATION OF POLYATOMIC MOLECULES

Let a molecule contain N atoms labelled as 1, 2, 3, ..., N. At the equilibrium configuration, let the coordinates of these atoms with respect to a fixed set of Cartesian axes be (x_{1e}, y_{1e}, z_{1e}), (x_{2e}, y_{2e}, z_{2e}) ..., (x_{Ne}, y_{Ne}, z_{Ne}). As the molecule is executing translational, rotational and vibrational motion, let the Cartesian coordinates of these atoms at any time t be (x_1, y_1, z_1), (x_2, y_2, z_2), ..., (x_N, y_N, z_N). Then the kinetic energy (T) of the molecule is given by

$$T = \tfrac{1}{2}\sum_{i=1}^{N} m_i(\dot{x}_i^2 + \dot{y}_i^2 + \dot{z}_i^2) \tag{11.1}$$

where the dot notation denotes the time-derivative (for example $\dot{x} = \frac{dx}{dt}$).

We can define mass-weighted displacements parallel to the fixed Cartesian axes at time t by:

$$\left.\begin{aligned} &q_1 = \sqrt{m_1}\,(x_1 - x_{1e}),\ q_2 = \sqrt{m_1}\,(y_1 - y_{1e}),\ q_3 = \sqrt{m_1}\,(z_1 - z_{1e}) \\ &q_4 = \sqrt{m_2}\,(x_2 - x_{2e}), \ldots \\ &q_{3N-2} = \sqrt{m_N}\,(x_N - x_{Ne}),\ q_{3N-1} = \sqrt{m_N}\,(y_N - y_{Ne}),\ q_{3N} = \sqrt{m_N}\,(z_N - z_{Ne}) \end{aligned}\right\} \tag{11.2}$$

Then $$\dot{q}_1^2 = \left(\frac{dq_1}{dt}\right)^2 = m_1\dot{x}_1^2,\ \dot{q}_2^2 = m_1\dot{y}_1^2, \ldots, \dot{q}_{3N}^2 = m_N\dot{z}_N^2 \tag{11.3}$$

and Eq. (11.1) takes the form, $$T = \tfrac{1}{2}\sum_{i=1}^{N}\dot{q}_i^2 \tag{11.4}$$

The potential energy V is generally a complicated function of the coordinates but it can be expanded in a Taylor series in terms of the displacements from equilibrium positions of the atoms as follows:

$$V = V\big|_{q_i=0} + \sum_i q_i\left(\frac{\partial V}{\partial q_i}\right)_{q_i=0} + \tfrac{1}{2}\sum_{i,j} q_i\, q_j\left(\frac{\partial^2 V}{\partial q_i\,\partial q_j}\right)_{q_i=0,\,q_j=0} + \cdots \tag{11.5}$$

The potential energy at the equilibrium positions of the atoms $V(q_i = 0)$, may be set to be the zero of energy; also, at such positions of the atoms the molecule has minimum potential energy, that is,

$$\left(\frac{\partial V}{\partial q_i}\right) = 0 \text{ at } q_i = 0 \text{ for } i = 1, 2, 3, \ldots, 3N \tag{11.6}$$

If now it is assumed that the higher order derivatives in Eq. (11.5) are negligible, we have from Eq. (11.5),

$$V = \tfrac{1}{2}\sum_{i,j} f_{ij}\, q_i\, q_j \text{ where } f_{ij} = \left(\frac{\partial^2 V}{\partial q_i\, \partial q_j}\right)_{q_i=0,\, q_j=0} \tag{11.7}$$

[In the one dimensional case, $V = \tfrac{1}{2} f_{ii}\, q_i^2$ and $-\left(\frac{\partial V}{\partial q_i}\right) = -f_{ii}\, q_i$, that is, force $\propto$ – (displacement) which indicates that each f_{ij} term is like a force constant in a simple harmonic motion; for a molecule with N atoms, the matrix with f_{ij} as elements is called *force constant matrix* or *Hessian matrix.*]

The approximation made by neglecting the higher order derivatives in the Taylor series expansion of V [in Eq. (11.5)] is usually called the *harmonic approximation.* The Lagrangian function of the system is

$$L = T - V = \tfrac{1}{2}\sum_{i=1}^{3N} \dot{q}_i^2 - \tfrac{1}{2}\sum_{i,j=1}^{3N} f_{ij}\, q_i\, q_j$$

$$= \tfrac{1}{2}\sum_{i=1}^{3N} \dot{q}_i^2 - \tfrac{1}{2}\sum_{j=1}^{3N}(f_{1j}\, q_1\, q_j + f_{2j}\, q_2\, q_j + \cdots + f_{ij}\, q_i\, q_j + \cdots + f_{3N,j}\, q_{3N}\, q_j)$$

$$= \tfrac{1}{2}\sum_{i=1}^{3N} \dot{q}_i^2 - \tfrac{1}{2}(f_{11}\, q_1^2 + f_{22}\, q_2^2 + \cdots + f_{ii}\, q_i^2 + \cdots + f_{3N,3N}\, q_{3N}^2) - \tfrac{1}{2}(f_{12} + f_{21})\, q_1 q_2$$

$$-\tfrac{1}{2}(f_{13} + f_{31})\, q_1 q_3 - \cdots - \tfrac{1}{2}(f_{3N-1,3N} + f_{3N,3N-1})\, q_{3N-1}\, q_{3N}$$

$$= \tfrac{1}{2}\sum_{i=1}^{3N} \dot{q}_i^2 - \tfrac{1}{2}\sum_{j=1}^{3N} f_{ii}\, q_i^2 - \sum_{i \neq j} f_{ij}\, q_i\, q_j \quad (\text{as } f_{ij} = f_{ji}) \tag{11.8}$$

Hence $$\left(\frac{\partial L}{\partial \dot{q}_i}\right)_{q_i} = \dot{q}_i \text{ and } \left(\frac{\partial L}{\partial q_i}\right)_{\dot{q}_i} = -f_{ii}\, q_i - \sum_{i \neq j} f_{ij}\, q_j = -\sum_{j=1}^{3N} f_{ij}\, q_j,\ i = 1, 2, 3, \cdots, 3N \tag{11.9}$$

Lagrange's equation of motion

$$\frac{d}{dt}\left(\frac{\partial L}{\partial \dot{q}_i}\right)_{q_i} - \left(\frac{\partial L}{\partial q_i}\right)_{\dot{q}_i} = 0 \tag{11.10}$$

gives [from Eq. (11.9)],

$$\ddot{q}_i + \sum_{j=1}^{3N} f_{ij}\, q_j = 0,\ i = 1, 2, 3, \cdots, 3N \tag{11.11}$$

This is equivalent to a set of $3N$ linear equations:

$$\ddot{q}_1 + \sum_{j=1}^{3N} f_{1j}\, q_j = 0; \ddot{q}_2 + \sum_{j=1}^{3N} f_{2j}\, q_j = 0; \ldots; \ddot{q}_{3N} + \sum_{j=1}^{3N} f_{3N,\,j}\, q_j = 0 \tag{11.12}$$

or, in matrix form,

$$\begin{bmatrix} \ddot{q}_1 \\ \ddot{q}_2 \\ \cdot \\ \cdot \\ \ddot{q}_{3N} \end{bmatrix} + \begin{bmatrix} f_{11} & f_{12} & \cdot & \cdot & f_{1,3N} \\ f_{21} & f_{22} & \cdot & \cdot & f_{2,3N} \\ \cdot & \cdot & \cdot & \cdot & \cdot \\ \cdot & \cdot & \cdot & \cdot & \cdot \\ f_{3N,1} & f_{3N,2} & \cdot & \cdot & f_{3N,3N} \end{bmatrix} \begin{bmatrix} q_1 \\ q_2 \\ \cdot \\ \cdot \\ q_{3N} \end{bmatrix} = \mathbf{0} \quad \text{(null matrix)}$$

or, $\quad \ddot{\boldsymbol{q}} + \boldsymbol{f}\boldsymbol{q} = \mathbf{0}$ (11.13)

The matrix $\boldsymbol{f}$ is real and symmetric and can be diagonalised by a suitable unitary matrix $\mathbf{U}$:

$$\mathbf{U}^{-1}\boldsymbol{f}\,\mathbf{U} = \Lambda \tag{11.14}$$

where Λ is a diagonal matrix:

$$\Lambda = \begin{bmatrix} \lambda_1 & 0 & \cdot & \cdot & 0 \\ 0 & \lambda_2 & \cdot & \cdot & 0 \\ \cdot & \cdot & \cdot & \cdot & \cdot \\ \cdot & \cdot & \cdot & \cdot & \cdot \\ 0 & 0 & \cdot & \cdot & \lambda_{3N} \end{bmatrix} \tag{11.15}$$

Here $\lambda_1, \lambda_2, \ldots, \lambda_{3N}$ are the eigenvalues of $\boldsymbol{f}$ and the corresponding eigenvectors are the column vectors of $\mathbf{U}$. From Eq. (11.14) we have

$$\boldsymbol{f} = \mathbf{U}\Lambda\mathbf{U}^{-1} \tag{11.16}$$

and Eq. (11.13) becomes

$$\ddot{\boldsymbol{q}} + \mathbf{U}\Lambda\mathbf{U}^{-1}\,\boldsymbol{q} = \mathbf{0}$$

or, $\quad \mathbf{U}^{-1}\ddot{\boldsymbol{q}} + \Lambda\mathbf{U}^{-1}\,\boldsymbol{q} = \mathbf{0}$

or, $\quad \mathbf{U}^{\dagger}\,\ddot{\boldsymbol{q}} + \Lambda\mathbf{U}^{\dagger}\,\boldsymbol{q} = \mathbf{0} \quad (\text{since } \mathbf{U}^{\dagger} = \mathbf{U}^{-1})$ (11.17)

Let $\quad \mathbf{U}^{\dagger}\,\boldsymbol{q} = \mathbf{Q}$ (11.18)

Then Eq. (11.17) gives $\ddot{\mathbf{Q}} + \Lambda\mathbf{Q} = \mathbf{0}$ (11.19)

Written explicitly, this equation has the form

$$\begin{bmatrix} \ddot{Q}_1 \\ \ddot{Q}_2 \\ \cdot \\ \cdot \\ \ddot{Q}_{3N} \end{bmatrix} + \begin{bmatrix} \lambda_1 & 0 & \cdot & \cdot & 0 \\ 0 & \lambda_2 & \cdot & \cdot & 0 \\ \cdot & \cdot & \cdot & \cdot & \cdot \\ \cdot & \cdot & \cdot & \cdot & \cdot \\ 0 & 0 & \cdot & \cdot & \lambda_{3N} \end{bmatrix} \begin{bmatrix} Q_1 \\ Q_2 \\ \cdot \\ \cdot \\ Q_{3N} \end{bmatrix} = \begin{bmatrix} 0 \\ 0 \\ \cdot \\ \cdot \\ 0 \end{bmatrix} \tag{11.20}$$

This corresponds to a set of linear equations

$$\ddot{Q}_1 + \lambda_1 Q_1 = 0; \quad \ddot{Q}_2 + \lambda_2 Q_2 = 0; \ldots; \quad \ddot{Q}_{3N} + \lambda_{3N} Q_{3N} = 0 \tag{11.21}$$

Now, kinetic energy, $T = \frac{1}{2}\sum_{i=1}^{3N} \dot{q}_i^2 = \frac{1}{2}\begin{bmatrix} \dot{q}_1 & \dot{q}_2 & \cdot & \cdot & \dot{q}_{3N} \end{bmatrix} \cdot \begin{bmatrix} \dot{q}_1 \\ \dot{q}_2 \\ \cdot \\ \cdot \\ \dot{q}_{3N} \end{bmatrix}$

$$= \frac{1}{2}\dot{\boldsymbol{q}}^{\dagger}\dot{\boldsymbol{q}} = \frac{1}{2}(\mathbf{U}\dot{\mathbf{Q}})^{\dagger}(\mathbf{U}\dot{\mathbf{Q}}) \qquad \text{[using Eq. (11.18)]}$$

$$= \frac{1}{2}\dot{\mathbf{Q}}^{\dagger}\mathbf{U}^{\dagger}\mathbf{U}\dot{\mathbf{Q}} = \frac{1}{2}\dot{\mathbf{Q}}^{\dagger}\dot{\mathbf{Q}}$$

$$= \frac{1}{2}\sum_i \dot{Q}_i^2 \tag{11.22a}$$

Potential energy, $V = \frac{1}{2}\sum_{i,j=1}^{3N} f_{ij}\, q_i\, q_j$

$$= \frac{1}{2}\boldsymbol{q}^{\dagger}\boldsymbol{f}\boldsymbol{q} = \frac{1}{2}(\mathbf{UQ})^{\dagger}\boldsymbol{f}(\mathbf{UQ}) \qquad \text{[using Eq. (11.18)]}$$

$$= \frac{1}{2}[\mathbf{Q}^{\dagger}(\mathbf{U}^{\dagger}\boldsymbol{f}\,\mathbf{U})\mathbf{Q}] = \frac{1}{2}[\mathbf{Q}^{\dagger}\Lambda\mathbf{Q}] = \frac{1}{2}\sum_{i=1}^{3N}\lambda_i Q_i^2 \tag{11.22b}$$

Equations (11.22a) and (11.22b) indicate that both kinetic and potential energy of the molecule and consequently the Hamiltonian function ($H = T + V$) of the system are devoid of any cross-term (like $Q_i Q_j$, $i \neq j$) that connects the different Q-type coordinates. Comparing with the standard differential equation $\frac{d^2x}{dt^2} = -\left(\frac{k}{m}\right)x$, we see that each member of the set of equations (11.21) denotes a simple harmonic motion along one of the coordinates $Q_1, Q_2, \ldots, Q_{3N}$ and that these motions are *uncoupled*, that is, non-interfering (owing to absence of cross-terms in H). The coordinates $\{Q_i\}$, $i = 1$ to $3N$, are called *normal coordinates* and the simple harmonic oscillation along each of such coordinates is called a *normal mode of vibration* of the molecule. Note that a normal mode of vibration does not mean vibration of any particular bond or periodic change of any particular bond angle. A normal coordinate Q_i, owing to the transformation (11.18), is a linear combination of mass-weighted coordinates $\{q_j\}$, $j = 1$ to $3N$, as shown below.

$$\mathbf{Q} = \mathbf{U}^{\dagger}\boldsymbol{q} \Rightarrow \begin{bmatrix} Q_1 \\ Q_2 \\ \cdot \\ \cdot \\ Q_{3N} \end{bmatrix} = \begin{bmatrix} c_{11} & c_{12} & \cdot & \cdot & c_{1,3N} \\ c_{21} & c_{22} & \cdot & \cdot & c_{2,3N} \\ \cdot & \cdot & \cdot & \cdot & \cdot \\ \cdot & \cdot & \cdot & \cdot & \cdot \\ c_{3N,1} & c_{3N,2} & \cdot & \cdot & c_{3N,3N} \end{bmatrix} \begin{bmatrix} q_1 \\ q_2 \\ \cdot \\ \cdot \\ q_{3N} \end{bmatrix}$$

that is, $Q_i = \sum_{j=1}^{3N} c_{ij}\, q_j, \quad i = 1 \text{ to } 3N$ (11.23)

where $(c_{i1} \quad c_{i2} \quad \cdot \quad \cdot \quad c_{i,3N})$ is the ith row vector of the matrix $\mathbf{U}^\dagger$ that diagonalises the $\boldsymbol{f}$ matrix. Thus a normal mode involves a number of oscillating atoms of the molecule (in principle, all the N atoms; however, some of the coefficients $\{c_{ij}\}$ may be zero).

Out of the $3N$ eigenvalues $\lambda_1 \quad \lambda_2 \quad \cdot \quad \cdot \quad \lambda_{3N}$ of $\boldsymbol{f}$, six must be zero (five, for linear molecules) since translation and rotation have no restoring force; these motions have zero frequency. The remaining degrees of freedom correspond to vibrational motion with frequencies $\sqrt{\lambda_i}$, $i = 1$ to $3N - 6$ or $3N - 5$ ($\sqrt{\lambda_i}$ is equivalent to $\sqrt{\frac{k}{m}}$ for one-dimensional simple harmonic oscillator, since we have already used mass-weighted displacements instead of Cartesian ones). For stable molecules (unlike transition states in chemical reactions) λ_k must therefore be positive. *Vibrational modes of stable molecules have positive frequencies.*

In brief, a normal mode of vibration of a molecule is an oscillatory motion of a group of constituent atoms such that the centre of mass of the molecule remains in its own position (to disregard translation and rotation). A normal coordinate is a single coordinate along which the progress of a single normal mode of vibration can be followed.

So far we have used the equations of classical mechanics. Once the Hamiltonian function $H = T + V$ is known in terms of the normal coordinates, formulation of the quantum mechanical equations for the vibrational motions becomes easy by the standard substitutions, $\hat{P}_k$ (generalised momenta) $\rightarrow -i\hbar\frac{\partial}{\partial Q_k}$ and $\hat{Q}_k \rightarrow Q_k$. Now, the generalised momentum component P_k associated with the generalised coordinate Q_k is given by

$$P_k = \left(\frac{\partial L}{\partial \dot{Q}_k}\right) = \frac{\partial}{\partial \dot{Q}_k}\left[\tfrac{1}{2}\sum_k \dot{Q}_k^2 - \tfrac{1}{2}\sum_k \lambda_k Q_k^2\right] = \dot{Q}_k \tag{11.24a}$$

Hence the operator forms of generalised momenta and coordinates are:

$$\hat{\dot{Q}}_k = \hat{P}_k \rightarrow -i\hbar\frac{\partial}{\partial Q_k} \quad \text{and} \quad \hat{Q}_k \rightarrow Q_k \tag{11.24b}$$

So the quantum mechanical Hamiltonian operator becomes

$$\hat{H} = \hat{T} + \hat{V} = \tfrac{1}{2}\sum_{k=1}^{s} \hat{\dot{Q}}_k^2 + \tfrac{1}{2}\sum_{k=1}^{s} \lambda_k \hat{Q}_k^2$$

$$= \tfrac{1}{2}\sum_{k=1}^{s} \hat{P}_k^2 + \tfrac{1}{2}\sum_{k=1}^{s} \lambda_k \hat{Q}_k^2$$

$$= -\tfrac{1}{2}\sum_{k=1}^{s} \hbar^2 \frac{\partial^2}{\partial Q_k^2} + \tfrac{1}{2}\sum_{k=1}^{s} \lambda_k Q_k^2 = \sum_{k=1}^{s} \hat{H}_k \tag{11.25}$$

$$\text{where}\quad \hat{H}_k = -\frac{\hbar^2}{2}\frac{\partial^2}{\partial Q_k^2} + \tfrac{1}{2}\lambda_k Q_k^2 \tag{11.26}$$

and $s = 3N - 6$ or $3N - 5$ depending on whether the molecule is non-linear or linear.

The Schrödinger equation for vibrational motion is, then,

$$\hat{H}\Psi = E\Psi \quad \text{or,} \quad \left(\sum_{k=1}^{s}\hat{H}_k\right)\Psi = E\Psi \tag{11.27}$$

As the Hamiltonian is a sum of single normal mode Hamiltonians, we can express the total vibrational wave function Ψ as a product of wave functions of the individual normal modes:

$$\Psi = \psi_1(Q_1)\cdot\psi_2(Q_2)\cdot\ldots\cdot\psi_s(Q_s),\quad s = 3N-6 \text{ or } 3N-5 \tag{11.28}$$

The wave functions $\psi_k(Q_k)$ are solutions of

$$\hat{H}_k\psi_k = E_k\psi_k,\quad k = 1, 2, \ldots, s$$

$$\text{or,}\quad \left(-\frac{\hbar^2}{2}\frac{\partial^2}{\partial Q_k^2} + \tfrac{1}{2}\lambda_k Q_k^2\right)\psi_k = E_k\psi_k$$

or, multiplying both sides by $-\frac{2}{\hbar^2}$ and rearranging,

$$\left[\frac{\partial^2}{\partial Q_k^2} + \left(\frac{2E_k}{\hbar^2} - \frac{\lambda_k}{\hbar^2}Q_k^2\right)\right]\psi_k = 0 \tag{11.29}$$

$$\text{Putting}\quad \frac{2E_k}{\hbar^2} = \varepsilon_k \text{ and } \frac{\lambda_k}{\hbar^2} = \alpha_k^2 \tag{11.30}$$

$$\text{this becomes}\quad \left[\frac{\partial^2}{\partial Q_k^2} + (\varepsilon_k - \alpha_k^2 Q_k^2)\right]\psi_k = 0 \tag{11.31}$$

which is the standard form of the Schrödinger equation for a simple harmonic oscillator and its solution is characterised by a vibrational quantum number v_k which may have any positive integral value, $v_k = 0, 1, 2, 3\ldots$ For the sake of clarity, let us discard the distinguishing label k for the moment (and restore it again when necessary).

For each normal coordinate the solution is

$$\psi_v = N_v \exp\left(-\tfrac{1}{2}\alpha Q^2\right)\cdot H_v(\sqrt{\alpha}\cdot Q) \tag{11.32}$$

Here N_v is a normalisation constant and $H_v(x)$ is the Hermite polynomial of degree v in x. The first two such polynomials, namely,

$$H_0(\sqrt{\alpha}Q)=1 \text{ and } H_1(\sqrt{\alpha}Q)=2\sqrt{\alpha}Q \tag{11.33}$$

will be of interest in the analysis of the fundamental bands of molecular vibrational spectra. The energy corresponding to the v_k level of the normal mode corresponding to coordinate Q_k is

$$E_\mathrm{v}\,(k\text{th mode}) = \left(\mathrm{v}_k + \tfrac{1}{2}\right) h\nu_k \tag{11.34}$$

where ν_k is the frequency of vibration of the kth mode, given by $2\pi\nu_k = \sqrt{\lambda_k}$ and from Eq. (11.30),

$$\nu_k = \frac{\sqrt{\lambda_k}}{2\pi} = \sqrt{\hbar^2\alpha_k^2}\cdot\frac{1}{2\pi} = \frac{\hbar}{2\pi}\alpha_k \tag{11.35}$$

11.3 SYMMETRY OF THE NORMAL COORDINATES Q_k, $k = 1, 2, \ldots, (3N-6)$ or $(3N-5)$

Each normal coordinate (Q_k) *forms a basis of some irreducible representation of the point group of the molecule under consideration.* This important statement can be proved from the expressions for kinetic and potential energies of the vibrating molecule:

Kinetic energy, $T = \frac{1}{2}\sum_{k=1}^{s}\dot{Q}_k^2$

Potential energy, $V = \frac{1}{2}\sum_{k=1}^{s}\lambda_k Q_k^2,\ s = (3N-6) \text{ or } (3N-5)$

where s is the number of vibrational degrees of freedom. By the action of a symmetry operator of the molecular point group, the kinetic and potential energies do not change. Hence the above expressions for T and V suggest that each Q_k must transform by the symmetry operations as follows: if Q_k is non-degenarate, that is, if no other λ has value λ_k, then the action of a symmetry operator $\hat{R}$ must either leave it unchanged or convert it into the negative of itself:

$$\hat{R}Q_k = \pm 1 \cdot Q_k \tag{11.36}$$

Hence in a non-degenerate mode of vibration Q_k forms a basis of a one-dimensional representation (D) of the molecular point group in which the characters are either 1 or –1:

	$\hat{E}$	$\hat{R}_2$	$\hat{R}_2$	…	$\hat{R}_g$	**Basis**
D	1	1 or –1	1 or –1	…	1 or –1	Q_k (non-degenerate)

If Q_k is d-fold degenerate, that is, if there are d number of equal values of λ_k (that is, $\lambda_k = \lambda_{k+1} = \lambda_{k+2} = \ldots = \lambda_{k+d-1}$) corresponding to d number of normal coordinates $Q_k^{(1)}\ (= Q_k)$, $Q_k^{(2)}\ (= Q_{k+1})$, $Q_k^{(3)}\ (= Q_{k+2})$, …, $Q_k^{(d)}\ (= Q_{k+d-1})$ then the operation $\hat{R}$ on any one of such Q_k will, in general, lead to a linear combination of Q_k's:

$$\hat{R}Q_k^{(i)}(1 \le i \le d) = [Q_k^{(1)} \quad Q_k^{(2)} \quad Q_k^{(3)} \quad \ldots \quad Q_k^{(d)}] \cdot [a_{1i} \quad a_{2i} \quad a_{3i} \quad \cdots \quad a_{di}]^T$$

$$= \sum_{l=1}^{d} a_{li}\, Q_k^{(l)} \tag{11.37}$$

If $\hat{S}$ is another symmetry operator of the same point group, then

$$\hat{S}\hat{R}Q_k^{(i)} = \hat{S}\sum_{l=1}^{d} a_{li}Q_k^{(l)} = \sum_{l=1}^{d} a_{li}\,\hat{S}Q_k^{(l)} \quad \text{(as } \hat{S} \text{ is a linear operator)}$$

$$= \sum_{l=1}^{d} a_{li}\sum_{m=1}^{d} b_{ml}Q_k^{(m)} = \sum_{m=1}^{d}\sum_{l=1}^{d} b_{ml}a_{li}Q_k^{(m)} \tag{11.38}$$

(order of summation is immaterial and a_{li} and b_{ml} are numbers).

Owing to the closure property, the combined operator $\hat{S}\hat{R}$ should be equivalent to another symmetry operator, $\hat{T}$ (that is, $\hat{S}\hat{R} = \hat{T}$). Operation of $\hat{T}$ on $Q_k^{(i)}$ should give another linear combination of the normal coordinates,

$$\hat{T}Q_k^{(i)} = \sum_{m=1}^{d} c_{mi}Q_k^{(m)} \tag{11.39a}$$

Comparing Eqs (11.38) and (11.39a), we have

$$c_{mi} = \sum_{l=1}^{d} b_{ml}\, a_{li} \tag{11.39b}$$

But this is the result of the matrix multiplication,

$$\mathbf{C} = \mathbf{BA} \tag{11.39c}$$

where $\mathbf{C} = [c_{mi}]$, $\mathbf{B} = [b_{ml}]$ and $\mathbf{A} = [a_{li}]$ are matrices representing $\hat{T}$, $\hat{S}$ and $\hat{R}$ respectively with $(Q_k^{(1)}, Q_k^{(2)}, Q_k^{(3)} \ldots, Q_k^{(d)})$ as the basis set; these matrices multiply in accordance with symmetry multiplication $\hat{T} = \hat{S}\hat{R}$. Hence a set of d-fold degenerate normal coordinates is a basis set of a d-dimensional irreducible representation of the point group just as a set of d-fold degenerate eigenfunctions of the Hamiltonian operator $\hat{H}$ forms a basis set of a d-dimensional irreducible representation. According to Eqs (11.34) and (11.35), degenerate modes of vibration occur with the same energy and same frequency for a given vibrational quantum number.

11.4 VIBRATIONAL ENERGY STATES OF MOLECULES AND THEIR SYMMETRY

Infrared and Raman spectra of molecules result from transition between their different vibrational energy states. Such a state is described by means of an overall vibrational wavefunction (Ψ_{vib}) of the molecule. In a particular vibrational state of the molecule, the various normal modes vibrate at their own quantum levels designated by the quantum numbers $v_k = 0, 1, 2, 3....$ Although the allowed values of these quantum numbers are the same (0, 1, 2, ...) for all the coordinates $\{Q_i\}$, different modes may vibrate with different quantum numbers in a given state Ψ_{vib}.

11.4.1 THE GROUND STATE

The lowest vibrational state of a molecule occurs when every $v_k = 0$, that is, all the normal modes of vibration are at their respective v = 0 quantum levels. The vibrational wave function of the molecule in its ground state, apart from a normalising factor, is

$$\Psi_0 = \psi_0(Q_1) \cdot \psi_0(Q_2) \cdot \ldots \cdot \psi_0(Q_s), \quad s = 3N - 6 \text{ or } 3N - 5 \tag{11.40}$$

$$= \exp\left(-\tfrac{1}{2}\alpha_1 Q_1^2\right) \cdot \exp\left(-\tfrac{1}{2}\alpha_2 Q_2^2\right) \cdot \ldots \cdot \exp\left(-\tfrac{1}{2}\alpha_s Q_s^2\right)$$

$$= \exp\left(-\tfrac{1}{2}\sum_{k=1}^{s} \alpha_k Q_k^2\right) \tag{11.41}$$

(since the zero-degree Hermite polynomial $H_0(\sqrt{\alpha}Q) = 1$ for each k.)

The exponent in Ψ_0 is of the same form as potential energy V. Hence Ψ_0 is invariant under the action of all symmetry operators of the point group, that is, *Ψ_0 belongs to the totally symmetric irreducible representation of the molecular point group*. The vibrational energy of the ground state is,

$$E_0 = \sum_{k=0}^{s} E_{0_k} = \sum_{k=0}^{s} \tfrac{1}{2} h\nu_k, \quad s = 3N - 6 \text{ or } 3N - 5 \tag{11.42}$$

This energy (E_0) is called the *zero-point energy* of the molecule.

11.4.2 THE FUNDAMENTAL LEVELS

When one particular normal mode corresponding to a coordinate Q_p is in the first excited level ($v_p = 1$) and all the other modes of vibration are in their zero quantum levels ($v_k = 0$, $k \neq p$), the molecule is said to be in one fundamental level of vibration. The vibrational energy of the molecule in this state is

$$E_p = \sum_{k \neq p}^{s} \tfrac{1}{2} h\nu_k + \left(1 + \tfrac{1}{2}\right) h\nu_p = \sum_{k=1}^{s} \tfrac{1}{2} h\nu_k + h\nu_p$$

$$= E_0 + h\nu_p \tag{11.43}$$

which means that excitation of this mode (Q_p) of vibration to its fundamental level requires a photon of energy $h\nu_p$ to be absorbed by the molecule. The vibrational wave function of the molecule in this state is

$$\Psi_p^{(1)} = \psi_0(Q_1)\cdot\psi_0(Q_2)\cdot\ldots\cdot\psi_1(Q_p)\cdot\ldots\cdot\psi_0(Q_s) \tag{11.44}$$

But according to Eq. (11.32),

$$\psi_1(Q_p) = \exp\left(-\tfrac{1}{2}\alpha_p Q_p^2\right)\cdot H_1(\sqrt{\alpha_p}\cdot Q_p) \quad \text{(apart from a normalising factor)}$$

$$= \exp\left(-\tfrac{1}{2}\alpha_p Q_p^2\right)\cdot 2(\sqrt{\alpha_p}\cdot Q_p)$$

$$= \psi_0(Q_p)\cdot 2(\sqrt{\alpha_p}\cdot Q_p)$$

Hence $\Psi_p^{(1)} = 2\sqrt{\alpha_p}[\psi_0(Q_1)\cdot\psi_0(Q_2)\cdot\ldots\cdot\psi_0(Q_p)\cdot\ldots\cdot\psi_0(Q_s)]\cdot Q_p$

$$= \text{constant}\times\Psi_0 Q_p$$

As Ψ_0 belongs to the totally symmetric representation of the point group, $\Psi_p^{(1)}$ transforms under the action of all symmetry operators of the group in the same way as Q_p does. In other words, *symmetry of the fundamental level described by the vibrational wave function* $\Psi_p^{(1)}$ *is the same as that of the normal coordinate* Q_p.

Evidently, the number of possible fundamental levels of a molecule is equal to the number of normal modes of vibration. However, the number of fundamental bands that appear in the infrared or Raman spectrum of a molecule is often less than the number of normal modes because (i) some modes may be degenerate, causing the bands to appear at the same frequency and (ii) excitation from the ground state to many of the fundamental levels of all the modes may be not be allowed by symmetry-based selection rules (will be discussed in Section 11.5).

11.4.3 Overtones and Combination Levels

An overtone level of a molecule is one in which the vibrational quantum number (v) is zero for all the normal modes except one (say, of normal coordinate Q_m), for which the vibrational quantum number (v_m) is greater than unity. If $v_m = 2$, the molecule is said to be in the first overtone level with respect to the *m*th vibrational mode; if $v_m = 3$, it is in the second overtone level with respect to the same normal mode, and so on.

A molecule is said to be in a *combination level* if it is vibrating with non-zero quantum numbers for more than one of the normal modes, for example $v_p \neq 0$, $v_m \neq 0$.

11.5 SELECTION RULES FOR TRANSITION BETWEEN VIBRATIONAL ENERGY STATES: INFRARED AND RAMAN SPECTROSCOPY

Infrared and Raman spectra of molecules are known to result from transition between vibrational levels of molecules (coupled with rotational motion in the gas phase). In the former case, the molecules must have oscillating electric dipoles and in the latter case, oscillating polarisability during molecular vibration. The most general form of the selection rule for any transition, as obtained from time-dependent perturbation theory, is

$$\int \Psi_i^* \, \hat{O} \, \Psi_f \, d\tau \neq 0 \tag{11.45}$$

where Ψ_i and Ψ_f are the wave functions for the initial and final states of the system respectively and $\hat{O}$ is an appropriate operator, for example, the dipole moment operator for electric dipole transition and polarisabity operator for Raman scattering.

11.5.1 Group Theoretical Condition for Appearance of Fundamental Bands in Vibrational Spectroscopy

Such bands appear in the infrared region of the electromagnetic spectrum due to the excitation of a single normal mode of coordinate Q_p from its $v_p = 0$ level to its first higher level ($v_p = 1$), the other normal modes remaining in their respective $v = 0$ levels. The vibrational transition is

$$\psi_0(Q_1)\cdot\psi_0(Q_2)\cdot\ldots\cdot\psi_0(Q_p)\cdot\ldots\cdot\psi_0(Q_s) \rightarrow \psi_0(Q_1)\cdot\psi_0(Q_2)\cdot\ldots\cdot\psi_1(Q_p)\cdot\ldots\cdot\psi_0(Q_s)$$

or,
$$\psi_0(Q_p)\prod_{i\neq p}^{s}\psi_0(Q_i) \rightarrow \psi_1(Q_p)\prod_{i\neq p}^{s}\psi_0(Q_i); \quad s = 3N-5 \text{ or } 3N-6$$

This, in the notation used in the Eq. (11.44), is $\Psi_0 \rightarrow \Psi_p^{(1)}$.

The ground state vibrational wave function is Ψ_0 and it belongs to the totally symmetric irreducible representation D^1 of the molecular point group. The excited state wave function $\Psi_p^{(1)}$ belongs to the irreducible representation D^p for which the normal coordinate Q_p is a basis. For a transition by absorption of infrared radiation, the operator $\hat{O}$ in Eq. (11.45) is the dipole moment operator $\hat{\mu}$, which has three Cartesian components $\hat{\mu}_x$, $\hat{\mu}_y$ and $\hat{\mu}_z$; these components transform respectively as x, y and z. Hence a representation D^μ which has $\hat{\mu}$ as basis must also have x, y and z in the basis set. The integrand in Eq. (11.45) survives only if the direct product representation $D^1(\text{for } \Psi_0) \times D^\mu(\text{for } \hat{\mu}) \times D^p(\text{for } \Psi_p^{(1)})$ contains D^1. But $D^1 \times D^p = D^p$ (since each character in D^1 is unity). Hence $D^1 \times D^\mu \times D^p = D^\mu \times D^p$ and it contains D^1 only if D^p contains x and/or y and/or z as basis.

Thus, the integral in Eq. (11.45) *does not vanish, that is, the transition* $\Psi_0 \rightarrow \Psi_p^{(1)}$ *is electric dipole allowed only if the irreducible representation* D^p*, for which the normal coordinate* Q_p *is a basis, contains any one, two or all of* x, y and z *in the basis set.*

For a Raman transition $\hat{O}$ in Eq. (11.45) is the operator for the molecule's polarisability. When a molecule is exposed to an electric field $\vec{E}$, a dipole moment $\vec{\mu}_{\text{ind}}$ is induced in it, the strength of the induced dipole being proportional to the applied field, $\vec{E}$:

$$\vec{\mu}_{\text{ind}} = \alpha \vec{E} \tag{11.46}$$

The proportionality constant α is called the polarisability of the molecule; it may be alternatively defined as the rate of change of the induced dipole moment with unit change of $\vec{E}$. Both $\vec{\mu}_{\text{ind}}$ (or, briefly $\vec{\mu}$) and $\vec{E}$ are vectors and the rate of change of each component of $\vec{\mu}$ with change in each component of $\vec{E}$ may be arranged in an array of three rows and three columns as follows:

	μ_x	μ_y	μ_z
E_x	$\left(\frac{\partial \mu_x}{\partial E_x}\right)$	$\left(\frac{\partial \mu_y}{\partial E_x}\right)$	$\left(\frac{\partial \mu_z}{\partial E_x}\right)$
E_y	$\left(\frac{\partial \mu_x}{\partial E_y}\right)$	$\left(\frac{\partial \mu_y}{\partial E_y}\right)$	$\left(\frac{\partial \mu_z}{\partial E_y}\right)$
E_z	$\left(\frac{\partial \mu_x}{\partial E_z}\right)$	$\left(\frac{\partial \mu_y}{\partial E_z}\right)$	$\left(\frac{\partial \mu_z}{\partial E_z}\right)$

Accordingly, the polarisability is expressed as a 9-component symmetric matrix (also called *tensor*):

$$\text{Polarisability tensor} = \begin{bmatrix} \alpha_{xx} & \alpha_{yx} & \alpha_{zx} \\ \alpha_{xy} & \alpha_{yy} & \alpha_{zy} \\ \alpha_{xz} & \alpha_{yz} & \alpha_{zz} \end{bmatrix}$$

where $\alpha_{xx} = \left(\frac{\partial \mu_x}{\partial E_x}\right), \alpha_{yx} = \left(\frac{\partial \mu_y}{\partial E_x}\right), \ldots$ Hence to test Raman activity, we have to test six transition moment integrals, namely, $\int \Psi_i^* \, \alpha_{xx} \, \Psi_f \, d\tau, \int \Psi_i^* \, \alpha_{xy} \, \Psi_f \, d\tau \ldots$ (as $\alpha_{xy} = \alpha_{yx}$).

The polarisability components transform (under the action of the symmetry operators) according to the quadratic functions of coordinates, x^2, y^2, z^2, xy, yz, zx (or, as their linear combinations). *Hence, according to the logic followed in the case of electric dipole transitions, a normal mode Q_p is Raman active (Raman transition between the states $\Psi_0 \rightarrow \Psi_p^{(1)}$ is allowed) if D^p contains a representation with any of the quadratic functions x^2, y^2, z^2, xy, yz, zx (or their linear combination) as basis.*

Let ξ be any of the Cartesian coordinates (x, y, z) of a general point. If ξ is a basis of a one-dimensional representation of the group G, then $\hat{R}\xi = +\xi$ or $-\xi$ and $\hat{R}\xi^2 = \xi^2$ for any symmetry operator $\hat{R}$ of G; this means that ξ^2 transforms according to the totally symmetric representation, $D^{(1)}$ of G. Hence $D^{(1)}$ for any group must have the quadratic functions x^2, y^2, z^2 (or their linear combination) in its basis set. Three polarisability tensor components, therefore, transform as the totally symmetric representation, whatever be the molecular point group. We thus arrive at the important general result: *if the normal coordinate of a mode of vibration transforms as the totally symmetric representation of the molecular point group, the mode must be Raman active.*

11.5.2 Infrared–Raman Mutual Exclusion Rule in Vibrational Spectroscopy

If a normal mode (Q_p) of vibration is infrared active, then Q_p must be a basis of the irreducible representation D^p, which contains x and/or y and/or z as basis; for a centrosymmetric species this means that D^p is a u-type representation ($\hat{i}x = -x \ldots$). Consequently D^p cannot have the quadratic functions x^2, $xy \ldots$ as basis and so Q_p cannot be Raman active for a centrosymmetric species. Again, if Q_p of the centrosymmetric species is Raman active, it must be a basis of a representation of g-type $\hat{i}x^2 = x^2$, $\hat{i}xy = xy \ldots$) and so Q_p cannot have x and/or y and/or z as basis and the Q_p mode cannot be infrared active.

In brief, an excited state vibrational wave function $\Psi_p^{(1)}$ cannot be simultaneously of u-type and g-type for a molecule having a centre of symmetry. This is the basis of the infrared–Raman mutual exclusion rule: *for a centrosymmetric molecule (or ion) a normal mode which is infrared active is Raman inactive and vice-versa.*

11.5.3 Overtone, Combination and Hot Bands: Condition for Their Appearance in Infrared/Raman Spectra

Besides the intense fundamental bands, many bands of much lower intensity appear in the infrared and Raman spectra of molecules. These constitute the overtone, combination and hot bands. An overtone band appears when a particular vibrational mode of normal coordinate Q_p is excited from the $v_p = 0$ level to a level beyond $v_p = 1$ by a single photon; thus, the transition $v_p = 0 \rightarrow v_p = 2$ gives the first overtone, $v_p = 0 \rightarrow v_p = 3$ gives the second overtone, and so on. According to the simple harmonic approximation, the first overtone should appear at double the frequency of the fundamental ($v_p = 0 \rightarrow v_p = 1$), the second overtone should appear at three times the frequency of the fundamental and so on. However, for real molecules, the overtones appear at somewhat lower frequencies—anharmonicity reduces the energy gap between successive levels, the effect being more pronounced as the vibrational quantum number increases. A combination band is observed when more than one normal mode is excited by a single photon. For example, the following two transitions give combination bands:

$$\Psi_i(= \psi_0(Q_1)\cdot\psi_0(Q_2)\cdot\ldots\cdot\psi_0(Q_p)\cdot\ldots\cdot\psi_0(Q_q)\cdot\ldots\cdot\psi_0(Q_s)) \rightarrow$$

$$\Psi_f(= \psi_0(Q_1)\cdot\psi_0(Q_2)\cdot\ldots\cdot\psi_1(Q_p)\cdot\ldots\cdot\psi_1(Q_q)\cdot\ldots\cdot\psi_0(Q_s)) : \nu_{\text{obs}} = \nu_p + \nu_q$$

$$\Psi_i(= \psi_0(Q_1)\cdot\psi_0(Q_2)\cdot\ldots\cdot\psi_0(Q_p)\cdot\ldots\cdot\psi_0(Q_q)\cdot\ldots\cdot\psi_0(Q_s)) \rightarrow$$

$$\Psi_f(= \psi_0(Q_1)\cdot\psi_0(Q_2)\cdot\ldots\cdot\psi_2(Q_p)\ldots\psi_1(Q_q)\cdot\ldots\cdot\psi_0(Q_s)) : \nu_{\text{obs}} = 2\nu_p + \nu_q$$

In the above examples ν_p and ν_q are the frequencies of the corresponding fundamentals (that is, ν_p results from $\psi_0(Q_p) \rightarrow \psi_1(Q_p)$ and ν_q from $\psi_0(Q_q) \rightarrow \psi_1(Q_q)$, the other modes $\psi_0(Q_r)$, $r \neq p$ or q, remaining in their respective $v_r = 0$ levels). A *hot band* results when an already excited vibrational mode is further excited. For example, a hot band appears due to the transition

$$\Psi_i(= \psi_0(Q_1)\cdot\psi_0(Q_2)\cdot\ldots\cdot\psi_1(Q_p)\cdot\ldots\cdot\psi_0(Q_s)) \rightarrow$$

$$\Psi_f(= \psi_0(Q_1)\cdot\psi_0(Q_2)\cdot\ldots\cdot\psi_2(Q_p)\cdot\ldots\cdot\psi_0(Q_s))$$

For any vibrational mode of normal coordinate Q_p, the vibrational energy levels [given by $(v_p + \frac{1}{2})h\nu_p$] are equispaced if the simple harmonic approximation is valid, and the hot band resulting from the transition $v_p = 1$ to $v_p = 2$ is expected to be observed at the same frequency as that of the corresponding fundamental transition ($v_p = 0 \rightarrow v_p = 1$). But in reality, the observed frequency is slightly less than that of the fundamental because anharmonicity reduces the energy gap between the successive levels as mentioned earlier. The intensity of a hot band is usually weak due to low thermal population of the initial state. However, the population of the initial state can be increased by raising the temperature. This justifies the name, hot band.

The selection rule for the appearance of overtone, combination and hot bands is the same as Eq. (11.45). In cases where Ψ_f has degenerate vibrational modes, the procedure involving the direct product is not as simple as that for the non-degenerate case, and is outside the scope of this introductory book.

11.6 DETERMINATION OF SYMMETRY TYPES OF NORMAL MODES OF VIBRATION AND PREDICTION OF INFRARED AND RAMAN ACTIVITY OF MOLECULES

Symmetry types of the normal coordinates and infrared/Raman activity of molecules can be ascertained by the use of group character tables without actually calculating the force constants. Although this is a qualitative achievement, the results based on symmetry arguments are significant in the interpretation of spectra.

The normal coordinates are linear combinations of mass-weighted displacement vectors of the atoms in a molecule. The latter, in turn, are resolvable into Cartesian components along axes situated in each individual atom. Hence the collection of individual representations spanned by the $3N$ position coordinates are the same as those spanned by the $3N$ Cartesian components of N displacement vectors of the atoms. The procedure described here for the determination of symmetry of the normal modes of vibration is based on this fact. It involves the following steps:

Step 1: Construct the character set for a reducible representation (Γ_{total}) of the molecular point group, using the Cartesian components of the atomic displacements.

Step 2: Subtract the sum (Γ_{trans}) of the irreducible representations with x, y and z as basis, which correspond to the displacement of the centre of mass due to translational motion of the whole molecule.

Step 3: Subtract the sum of the irreducible representations Γ_{rot}, spanned by the rotation vector components R_x, R_y and R_z.

The result of these three steps is a representation (usually reducible), Γ_{vib}, which on being reduced to the component irreducible representations, gives the number and symmetry types of the normal modes of vibration. These steps will now be illustrated. Conventionally, the principal axis of symmetry of the molecule is taken as the Z-axis.

11.6.1 VIBRATION OF THE H_2O MOLECULE

The O atom of H_2O is labelled 1 and the two H atoms, 2 and 3 (Fig. 11.1). The molecular point group is C_{2v}. At the centre (nucleus) of each atom we imagine a set of Cartesian coordinate axes along which the atomic displacement vectors can be resolved. For the sake of symmetry considerations, keeping aside the actual magnitudes of displacements, we imagine vectors of unit length along these axes, namely, x_1, y_1, z_1 (on the O atom), x_2, y_2, z_2 (on H atom labelled 2)

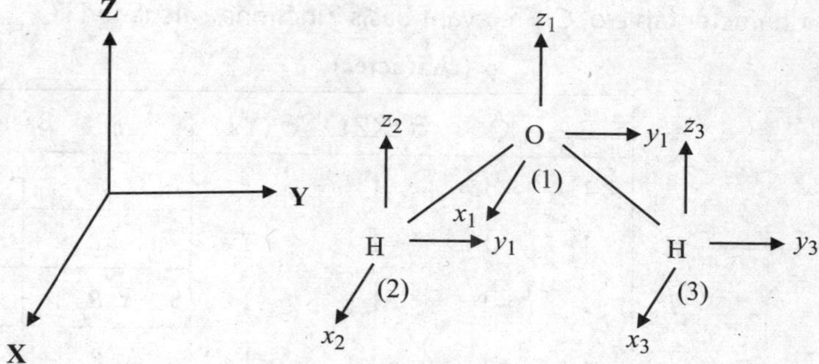

Fig. 11.1 Orientation of Cartesian axes (left) and a set of three unit vectors (x_i, y_i, z_i) on each atom of the H_2O molecule

and x_3, y_3, z_3 (on H atom labelled 3) (Fig. 11.1). [The vector notations (arrow-heads) on x_i, y_i, z_i are omitted in these figures for neatness.] With these nine unit vectors as basis we can construct a representation Γ_{total} of C_{2v} as follows.

The identity operator $\hat{E}$ leaves all the nine vectors unchanged, giving $\chi(\hat{E}) = 9$. By $\hat{C}_2$, the vectors on atoms 2 and 3 are shifted from their positions and so they do not contribute to character; x_1 becomes $-x_1$, y_1 becomes $-y_1$ and z_1 remains unchanged. So, $\chi(\hat{C}_2) = -1 - 1 + 1 = -1$. The $\hat{\sigma}_v(XZ)$ operation interchanges the positions of 2 and 3 (thus contributing zero to character), keeps x_1 and z_1 unchanged and turns y_1 to $-y_1$. Hence $\chi(\hat{\sigma}_v(XZ)) = 1 + 1 - 1 = 1$. $\hat{\sigma}_v(YZ)$ does not shift any atom from its position; y_i and z_i remain unchanged and x_i turns into $-x_i$ for each $i = 1, 2, 3$. Hence $\chi(\hat{\sigma}_v(YZ)) = 3(1 + 1 - 1) = 3$.

Thus, a representation Γ_{total} of C_{2v} is the following:

	χ **(character)**				
	$\hat{E}$	$\hat{C}_2$	$\hat{\sigma}_v(XZ)$	$\hat{\sigma}_v(YZ)$	**Basis**
Γ_{total}	9	–1	1	3	(x_i, y_i, z_i), $i = 1$ to 3

The character table of C_{2v}, relevant basis functions and calculated Γ_{trans} (for translational motion of the centre of mass) and Γ_{rot} (for rotational motion of the molecule) are shown in Table 11.1, where the representation $\Gamma_{vib} = \Gamma_{total} - \Gamma_{trans} - \Gamma_{rot}$ is also shown.

In Table 11.1, Γ_{trans} has been obtained by adding the irreducible representations whose basis functions are x, y and z because the translational motion of the molecule is completely described by the translation of the centre of mass having Cartesian coordinates (x, y, z). Similarly, Γ_{rot} is obtained by adding the irreducible representations whose basis set comprises any one or more of the rotation vector components (R_x, R_y, R_z). Finally, Γ_{vib}, whose basis set consists of the normal coordinates of vibration, is obtained by subtracting ($\Gamma_{trans} + \Gamma_{rot}$) from Γ_{total}.

Application of the reduction formula gives $\Gamma_{vib} = 2A_1 + B_2$, which means that two normal modes of vibration are of A_1 type and one mode is of B_2 type.

Infrared activity: The irreducible representation A_1 has z as basis but not x or y. So the two A_1 modes of vibration of H_2O molecule are infrared active owing to the z-component of the

Table 11.1 Character table of C_{2v}, relevant basis functions, calculated Γ_{trans}, Γ_{rot} and Γ_{vib}

	χ (character)				Basis	
	$\hat{E}$	$\hat{C}_2$	$\hat{\sigma}_v(XZ)$	$\hat{\sigma}_v(YZ)$		
A_1	1	1	1	1	z	x^2, y^2, z^2
A_2	1	1	−1	−1	R_z	xy
B_1	1	−1	1	−1	x, R_y	zx
B_2	1	−1	−1	1	y, R_x	yz
$\Gamma_{trans} = A_1 + B_1 + B_2$	3	−1	1	1	(x, y, z)	
$\Gamma_{rot} = A_2 + B_1 + B_2$	3	−1	−1	−1	(R_x, R_y, R_z)	
$\Gamma_{vib} = \Gamma_{total} - \Gamma_{trans} - \Gamma_{rot}$	3	1	1	3	normal coordinates for vibrational modes	

dipole moment vector. The B_2 representation has y as basis and so the B_2 mode of vibration is also infrared active owing to the y-component of the dipole moment.

Raman activity: The quadratic functions x^2, y^2 and z^2 can individually act as basis of the A_1 representation and the function yz is a basis of B_2. Hence all the modes of vibration are Raman active. [Raman polarisation has been explained in Section 11.10.] We can summarise the results in a tabular form as follows:

Activity	Irreducible representation	Number of fundamental frequencies expected in vibrational spectrum
Infrared	$2A_1, B_2$	3
Raman	$2A_1, B_2$	3
Raman Polarised	$2A_1$	2
Infrared–Raman coincidence	$2A_1, B_2$	3

11.6.2 Vibration of the XeF_4 Molecule

XeF_4 is a square planar molecule with point group D_{4h}. The principal axis C_4, passes through the Xe atom and is perpendicular to the molecular plane which is σ_h. The C_2' axes are along the two diagonals of the square plane. There are two C_2'' axes, each bisecting the pair of opposite ∠F–Xe–F angles. Containing each diagonal there is a σ_v plane and there are two such planes. The centre of symmetry is situated at the centre of the Xe atom. Some of the elements of symmetry are shown in Fig. 11.2 together with the orientation of the Cartesian axes. The Z-axis, taken along the C_4-axis, is pointing above the plane of the paper (not shown).

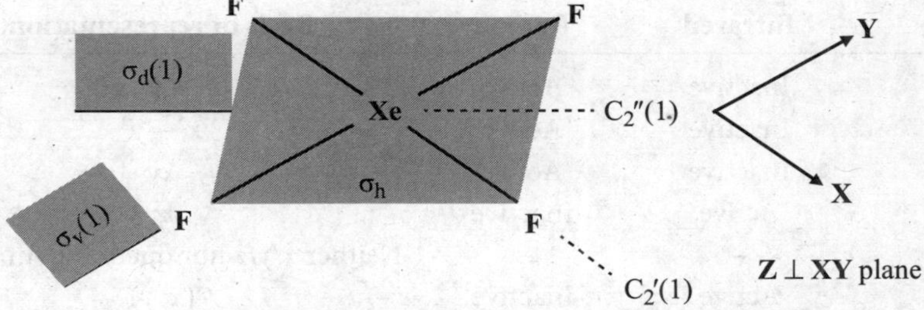

Fig. 11.2 Some symmetry elements of the XeF_4 molecule and orientation of the Cartesian axes

At each atom of XeF_4, we imagine a set of three unit vectors x_i, y_i and z_i parallel to the direction of the Cartesian axes. Using the character table of D_{4h} (Appendix A3) and noting the effect of the symmetry operators on the unit vectors we first construct a representation Γ_{total} for translation, rotation and vibration of the molecule. We need to consider only the unit vectors at the atoms which remain unmoved by the symmetry operators (because those associated with the shifted atoms contribute zero to the character). By the $\hat{\sigma}_v(1)$ operation, for example, three atoms remain unmoved and for each of them x and y exchange their positions while z remains unchanged; thus, character of $\hat{\sigma}_v(1) = 3(0 + 0 + 1) = 3$ and this is the same for $\hat{\sigma}_v(2)$ because these two reflections are in the same conjugate class. Results obtained in this way are shown in Table 11.2 together with Γ_{trans} and Γ_{rot} obtained by adding the irreducible representations whose basis sets contain x, y, z, R_x, R_y and R_z. Γ_{vib} is obtained as $\Gamma_{total} - \Gamma_{trans} - \Gamma_{rot}$.

Table 11.2 Determination of Γ_{total} and Γ_{vib} of XeF_4

	χ (character)									
	$\hat{E}$	$2\hat{C}_4$	$\hat{C}_2$	$2\hat{C}_2'$	$2\hat{C}_2''$	$\hat{i}$	$2\hat{S}_4$	$\hat{\sigma}_h$	$2\hat{\sigma}_v$	$2\hat{\sigma}_d$
No. of unmoved atoms	5	1	1	3	1	1	1	5	3	1
Γ_{total}	15	1	−1	−3	−1	−3	−1	5	3	1
Γ_{trans}	3	1	−1	−1	−1	−3	−1	1	1	1
Γ_{rot}	3	1	−1	−1	−1	3	1	−1	−1	−1
$\Gamma_{vib} = \Gamma_{total} - \Gamma_{trans} - \Gamma_{rot}$	9	−1	1	−1	1	−3	−1	5	3	1

By applying the reduction formula it is found that

$$\Gamma_{vib} = A_{1g} + B_{1g} + B_{2g} + A_{2u} + B_{2u} + 2E_u$$

Thus, out of the $(3 \times 5) - 6 = 9$ modes of vibration of XeF_4, five are non-degenerate and there are two sets of doubly degenerate (E_u) type, each set having its own vibrational energy.

Infrared and Raman activity: The A_{1g} representation does not have x, y and z as basis; so it is infrared inactive. But this mode is Raman active because A_{1g} has the quadratic functions $x^2 + y^2$ and z^2 as basis. In this way the results for all the modes are tabulated below:

Mode	Infrared	Raman	Basis of representation
A_{1g}	Inactive	Active	$x^2 + y^2, z^2$
B_{1g}	Inactive	Active	$x^2 - y^2$
B_{2g}	Inactive	Active	xy
A_{2u}	Active	Inactive	z
B_{2u}	–	–	Neither $x/y/z$ nor quadratic functions
E_u	Active	Inactive	(x, y)

The B_{2u} mode is called a *silent vibration* because it gives no signal either in the infrared or in the Raman spectrum. Thus, for the centrosymmetric XeF_4 molecule, the modes which are active in Raman are inactive in infrared and vice-versa. The results are summarised below.

Activity	Irreducible representation	Number of fundamental frequencies expected in vibrational spectrum
Infrared	A_{2u}, $2E_u$	3
Raman	A_{1g}, B_{1g}, B_{2g}	3
Raman polarised	A_{1g}	1
Silent	B_{2u}	0
Infrared–Raman coincidence	none	0

Some more examples are given below in brief tabular form to be verified by the reader using the necessary character tables.

11.6.3 BF_3 Molecule

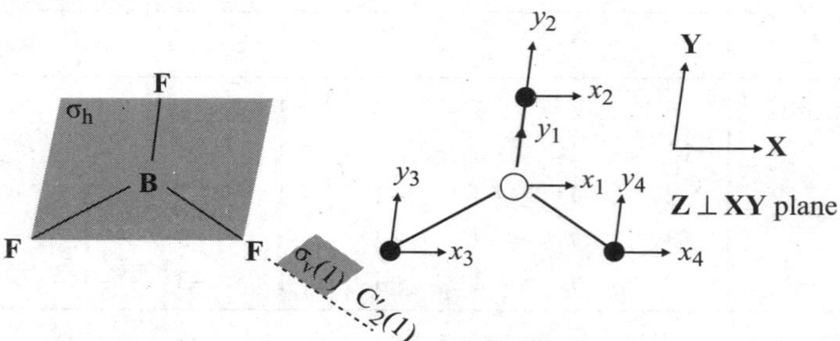

Fig. 11.3 Symmetry elements, structure and orientation of Cartesian axes for the planar BF_3 molecule (point group D_{3h})

	$\hat{E}$	$2\hat{C}_3$	$3\hat{C}_2'$	$\hat{\sigma}_h$	$2\hat{S}_3$	$3\hat{\sigma}_v$
	χ (character)					
No. of unmoved atoms	4	1	2	4	1	2
Γ_{total}	12	0*	−2	4	−2	2
Γ_{trans}	3	0	−1	1	−2	1
Γ_{rot}	3	0	−1	−1	2	−1
$\Gamma_{vib} = \Gamma_{total} - \Gamma_{trans} - \Gamma_{rot}$	6	0	0	4	−2	2

$$\Gamma_{vib} = A_1' + 2E' + A_2''$$

Mode	Infrared	Raman	Basis of representation
A_1'	inactive	active	$x^2 + y^2, z^2$
A_2''	active	inactive	z
$2E'$	active	active	$(x, y), (x^2 - y^2, xy)$

Summarisation table:

Activity	Irreducible representation	Number of fundamental frequencies expected in vibrational spectrum
Infrared	A_2'', $2E'$	3
Raman	A_1', $2E'$	3
Raman polarised	A_1'	1
Infrared–Raman coincidence	$2E'$	2

(A similar case: CO_3^{-2} ion)

[*Note: For a point group where the C_3-axis is the principal axis, $\chi(\hat{C}_3) = 0$ in a representation with (x, y, z) as basis because by a $\hat{C}_3$ rotation, z remains unchanged contributing 1 to χ while x and y transform together to give a matrix with trace $= 2 \cos 120° = -1$.]

11.6.4 NH_3 Molecule

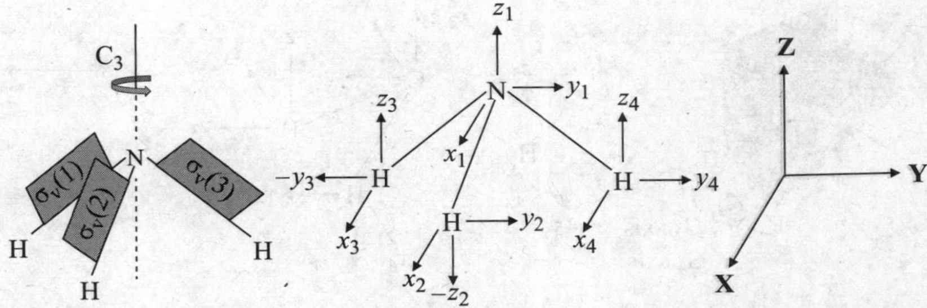

Fig. 11.4 Symmetry elements and orientation of Cartesian axes for the NH_3 molecule (point group C_{3v})

	χ (character)		
	$\hat{E}$	$2\hat{C}_3$	$3\hat{\sigma}_v$
No. of unmoved atoms	4	1	2
Γ_{total}	12	0*	2
Γ_{trans}	3	0	1
Γ_{rot}	3	0	−1
$\Gamma_{vib} = \Gamma_{total} - \Gamma_{trans} - \Gamma_{rot}$	6	0	2

[*Reason for $\chi(\hat{C}_3) = 0$ is given earlier: $\chi(\hat{C}_3) = 1 + 2 \cos 120° = 0$]

$$\Gamma_{vib} = 2A_1 + 2E$$

Mode	Infrared	Raman	Basis of representation
$2A_1$	Active	Active	$z, x^2 + y^2, z^2$
$2E$	Active	Active	$(x, y), (x^2 - y^2, xy)(yz, zx)$

In the infrared spectrum of NH_3, two bands are expected—one band containing two signals due to the two A_1 modes and the other containing two signals due to the two doubly degenerate E modes. The Raman spectrum of NH_3 should also contain two bands coincident with the infrared bands.

Summarisation table:

Activity	Irreducible representation	Number of fundamental frequencies expected in vibrational spectrum
Infrared	$2A_1$, $2E$	4
Raman	$2A_1$, $2E$	4
Raman polarised	$2A_1$	2
Infrared–Raman coincidence	$2A_1$, $2E$	4

11.6.5 Methane (CH_4) Molecule, Point Group T_d

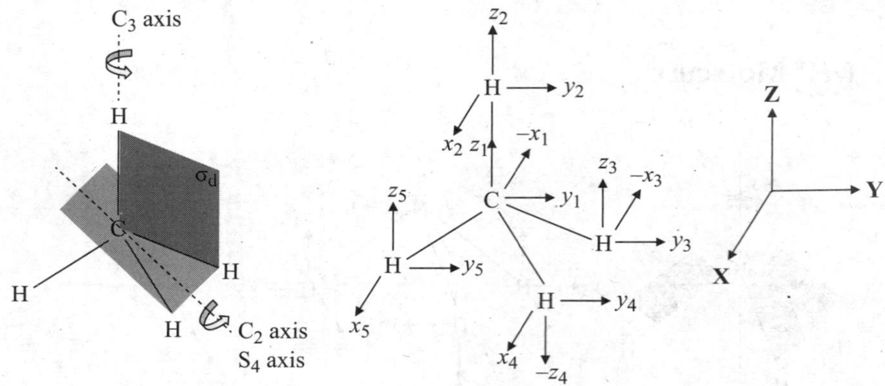

Fig. 11.5 Structure, symmetry elements and orientation of Cartesian axes for CH_4 molecule

	χ (character)				
	$\hat{E}$	$8\hat{C}_3$	$3\hat{C}_2$	$6\hat{S}_4$	$6\hat{\sigma}_d$
No. of unmoved atoms	5	2	1	1	3
Γ_{total}	15	0	−1	−1	3
Γ_{trans}	3	0	−1	−1	1
Γ_{rot}	3	0	−1	1	−1
$\Gamma_{vib} = \Gamma_{total} - \Gamma_{trans} - \Gamma_{rot}$	9	0	1	−1	3

$\Gamma_{vib} = A_1 + E + 2T_2$

Mode	Infrared	Raman	Basis of representation
A_1	Inactive	Active	$x^2 + y^2 + z^2$
E	Inactive	Active	$(2z^2 - x^2 - y^2, x^2 - y^2)$
$2T_2$	Active	Active	$(x, y, z), (xy, yz, zx)$

Summarisation table:

Activity	Irreducible representation	Number of fundamental frequencies expected in vibrational spectrum
Infrared	$2T_2$	2
Raman	$A_1, E, 2T_2$	4
Raman polarised	A_1	1
Infrared–Raman coincidence	$2T_2$	2

11.6.6 CH_3D Molecule, Point Group C_{3v}

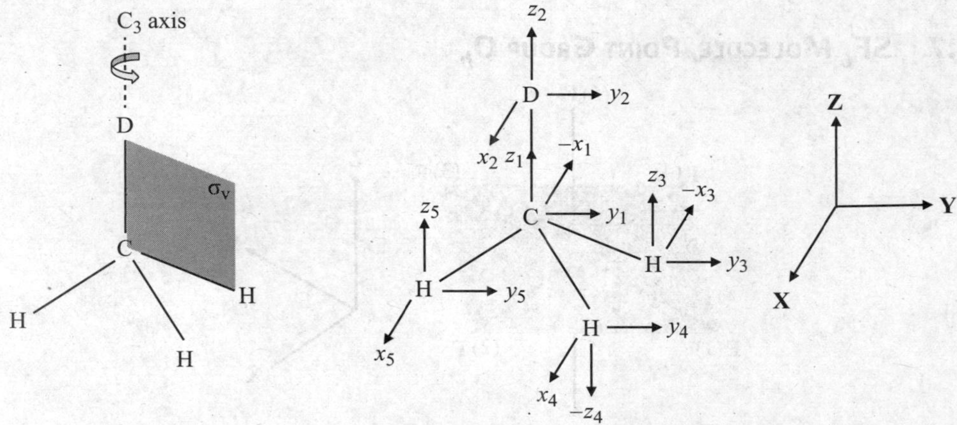

Fig. 11.6 Structure, symmetry elements and orientation of Cartesian axes for CH_3D molecule

	χ (character)		
	$\hat{E}$	$2\hat{C}_3$	$3\hat{\sigma}_v$
No. of unmoved atoms	5	2	3
Γ_{total}	15	0	3
Γ_{trans}	3	0	1
Γ_{rot}	3	0	−1
$\Gamma_{vib} = \Gamma_{total} - \Gamma_{trans} - \Gamma_{rot}$	9	0	3

$\Gamma_{vib} = 3A_1 + 3E$

Mode	Infrared	Raman	Basis of representation
$3A_1$	Active	Active	z, x^2+y^2, z^2
$3E$	Active	Active	$(x, y), (x^2-y^2, xy)(yz, zx)$

Summarisation table:

Activity	Irreducible representation	Number of fundamental frequencies expected in vibrational spectrum
Infrared	$3A_1$, $3E$	6
Raman	$3A_1$, $3E$	6
Raman polarised	$3A_1$	3
Infrared–Raman coincidence	$3A_1$, $3E$	6

Examples 11.6.5 and 11.6.6 show that for CH_3D, the number of fundamental frequencies which appear in the IR spectrum is the same as that in the Raman spectrum. But in the case of CH_4 the number of fundamentals appearing in the infrared differs from that appearing in the Raman spectrum. This fact helps to distinguish between the two molecules.

11.6.7 SF_6 Molecule, Point Group O_h

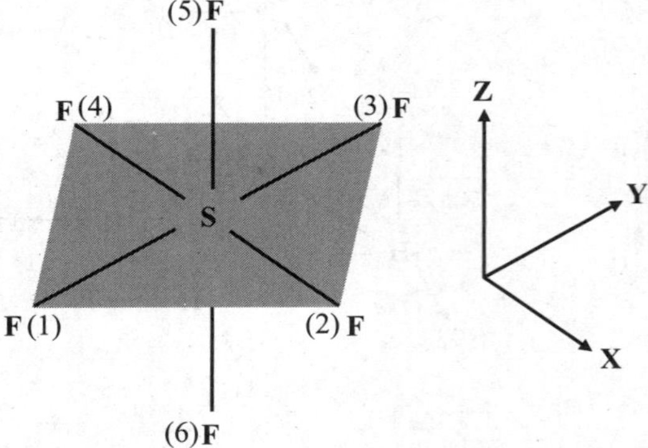

Fig. 11.7 Structure and orientation of Cartesian axes for SF_6 molecule

χ (character)

	$\hat{E}$	$8\hat{C}_3$	$6\hat{C}_2$	$6\hat{C}_4$	$3\hat{C}_2 \equiv 3\hat{C}_4^2$	$\hat{i}$	$6\hat{S}_4$	$8\hat{S}_6$	$3\hat{\sigma}_h$	$3\hat{\sigma}_d$
No. of unmoved atoms	7	0	1	3	3	1	1	1	5	3
Γ_{total}	21	0	−1	3	−3	−3	−1	0	5	3
Γ_{trans}	3	0	−1	1	−1	−3	−1	0	1	1
Γ_{rot}	3	0	−1	1	−1	3	1	0	−1	−1
$\Gamma_{vib} = \Gamma_{total} - \Gamma_{trans} - \Gamma_{rot}$	15	0	1	1	−1	−3	−1	0	5	3

$$\Gamma_{vib} = A_{1g} + E_g + T_{2g} + 2T_{1u} + T_{2u}$$

Mode	Infrared	Raman	Basis of representation
A_{1g}	Inactive	Active	$x^2 + y^2 + z^2$
E_g	Inactive	Active	$(2z^2 - x^2 - y^2, x^2 - y^2)$
T_{2g}	Inactive	Active	(xy, yz, xz)
T_{1u}	Active	Inactive	(x, y, z)
T_{2u}	–	–	–

Summarisation table:

Activity	Irreducible representation	Number of fundamental frequencies expected in vibrational spectrum
Infrared	T_{1u}	1
Raman	A_{1g}, E_g, T_{2g}	3
Raman polarised	A_{1g}	1
Infrared–Raman coincidence	none	0

11.7 VIBRATION OF LINEAR MOLECULES

Linear molecules have point group $C_{\infty v}$ and $D_{\infty h}$ and since the order of such groups is not finite, a straightforward use of the standard reduction formula as in the previous examples is not possible for finding the components of a reducible representation of such groups. However, for small linear molecules, the necessary reduction may be done by inspection of the character tables. Some examples are given here.

11.7.1 NORMAL MODES OF VIBRATION OF THE CO_2 MOLECULE

The molecular point group of CO_2 is $D_{\infty h}$. The symmetry elements and orientation of Cartesian axes are shown in Fig. 11.8. The line joining the nuclei is the C_∞-axis and is taken as the Z-axis. There are an infinite number of σ_v planes, only one of which is shown in Fig. 11.8 for neatness. There is a σ_h plane perpendicular to the C_∞-axis containing the C-atom (shown by a hollow circle). Also, there are an infinite number of C_2-axes lying on the σ_h plane (only one C_2 is shown in Fig. 11.8).

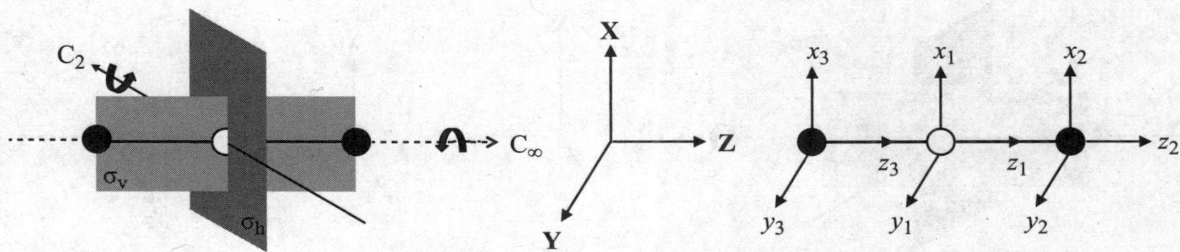

Fig. 11.8 Structure, some relevant symmetry elements and orientation of Cartesian axes for the CO_2 molecule

On each atom of CO_2 we consider unit vectors x_i, y_i and z_i ($i = 1, 2, 3$) parallel to the Cartesian axes. Using these nine vectors as bases we construct a representation Γ_{total}. The symmetry of the vibrational modes is determined as before and the results are given in the following tabular form (where $c = \cos\phi$).

χ **(character)**

	$\hat{E}$	$2\hat{C}_\infty^\phi$	$\infty\hat{\sigma}_v$	$\hat{i}$	$2\hat{S}_\infty^\phi$	$\infty\hat{C}_2$	**Basis**
No. of unmoved atoms	3	3	3	1	1	1	
Γ_{total}	9	$3(1+2c)$	3	-3	$-1+2c$	-1	$(x_i, y_i, z_i), i = 1, 2, 3$
Γ_{trans}	3	$(1+2c)$	1	-3	$-1+2c$	-1	(x, y, z)
Γ_{rot}	2	$2c$	0	2	$-2c$	0	(R_x, R_y)
$\Gamma_{vib} = \Gamma_{total} - \Gamma_{trans} - \Gamma_{rot}$	4	$2+2c$	2	-2	$2c$	0	normal coordinates

Γ_{vib} is reduced by inspection of the $D_{\infty h}$ character table. An important point to note here is that we use only R_x and R_y as basis for Γ_{rot} and disregard R_z since rotation of the molecule about the Z-axis does not effectively contribute to its kinetic energy.

By inspection of the character table of $D_{\infty h}$ we find

$$\Gamma_{vib} = \Pi_u + \Sigma_g^+ + \Sigma_u^+$$

Of the component irreducible representations of Γ_{vib}, Σ_g^+ has basis $x^2 + y^2$ and z^2 but not x or y or z. Σ_u^+ has basis z but no quadratic function of Cartesian coordinates. Π_u has basis set (x, y) but no quadratic function of Cartesian coordinates. Hence the infrared and Raman activity of the fundamentals are as follows:

Mode	**IR activity**	**Raman activity**
Σ_g^+	Inactive	Active
Σ_u^+	Active	Inactive
Π_u	Active	Inactive

The molecule has a centre of symmetry and the infrared–Raman mutual exclusion rule is obeyed.

11.7.2 Vibration of a Homonuclear Diatomic Molecule

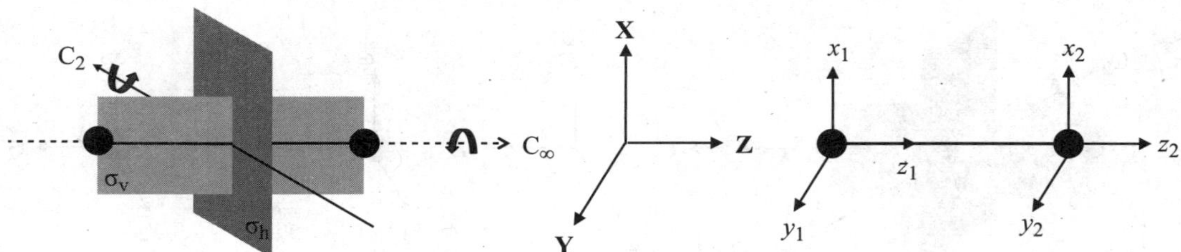

Fig. 11.9 Symmetry elements and orientation of Cartesian axes for a homonuclear diatomic molecule

The description of symmetry elements along with unit vectors (x_1, y_1, z_1 on atom 1 and x_2, y_2, z_2 on atom 2) is the same as in the previous example. The results are summarised in the next table (where $c = \cos\phi$). Thus $\Gamma_{vib} = \Sigma_g^+$ which contains $x^2 + y^2$ and z^2 but not x or y or z. Hence, the only mode of vibration is Raman active but infrared inactive. Physically this is easy to understand. The molecule has only one bond between two similar atoms. This bond can only execute the stretching vibration during which the dipole moment of the molecule does not change (remains zero) but the polarisability varies periodically with time.

	χ (character)						
	$\hat{E}$	$2\hat{C}_\infty^\phi$	$\infty\hat{\sigma}_v$	$\hat{i}$	$2\hat{S}_\infty^\phi$	$\infty\hat{C}_2$	**Basis**
No. of unmoved atoms	2	2	2	0	0	0	
Γ_{total}	6	$2(1+2c)$	2	0	0	0	$(x_i, y_i, z_i), i = 1, 2$
Γ_{trans}	3	$(1+2c)$	1	−3	$-1+2c$	−1	(x, y, z)
Γ_{rot}	2	$2c$	0	2	$-2c$	0	(R_x, R_y)
$\Gamma_{vib} = \Gamma_{total} - \Gamma_{trans} - \Gamma_{rot}$	1	1	1	1	1	1	normal coordinates

In the same way it can be shown that the only normal mode of vibration of a heteronuclear diatomic molecule (of point group $C_{\infty v}$) transforms as Σ^+ whose basis is any of the functions z, $x^2 + y^2$, z^2. Hence the vibration is both infrared and Raman active (which indicates the absence of a centre of symmetry).

11.7.3 Use of Point Group–Subgroup Correlation Table for Vibrational Analysis of Linear Molecules

For linear molecules with more than three atoms, it is difficult to reduce the total representation (Γ_{total}) by inspection. Direct application of the reduction formula, on the other hand, requires a division by infinity (the order of $D_{\infty h}$ or $C_{\infty v}$). To get rid of this difficulty, the method of point group–subgroup correlation is the usual procedure.

Point Group–Subgroup Correlation

Many spectroscopic (and other) properties of molecules belonging to large point groups (G) can be simplified by looking at the problem in terms of a subgroup of G. In such a transition from a group to one of its subgroups, at least one symmetry element (R), other than the identity, must be conserved in order to keep the group-related problem linked to its subgroup. Now, if the characters of $\hat{E}$ and the conserved symmetry operator $\hat{R}$ in one irreducible representation D^i of the subgroup be the same as those in the irreducible representation D^μ of the parent group G, then D^i and D^μ are said to be *correlated*. Sometimes two or more irreducible representations of the subgroup together may be correlated to one irreducible representation of the parent group. After finding the correlation with respect to the conserved symmetry operator $\hat{R}$ it is found that the whole set of characters of the shared symmetry operators of the group and subgroup is the same in the correlated irreducible representations. The following are some examples of such correlations.

Example 11.1 Correlation between the point groups C_{4v} and C_{2v}. The point group C_{2v} is a subgroup of C_{4v} as evident from the list of symmetry operations given below:

$$\begin{array}{llllll} C_{4v}: & \hat{E} & 2\hat{C}_4 & \hat{C}_2 & \overline{\quad 2\hat{\sigma}_v \quad} & 2\hat{\sigma}_d \\ C_{2v}: & \hat{E} & & \hat{C}_2(Z) & \hat{\sigma}(XZ)\ \ \hat{\sigma}(YZ) & \end{array}$$

Let the $\hat{\sigma}(XZ)$ plane of C_{2v} be taken to be the same as one σ_v plane of C_{4v} (that is, the σ_v plane is taken as the conserved symmetry element in the transition from C_{4v} to C_{2v}). In Table 11.3 the character tables of C_{4v} and C_{2v} are arranged side by side, omitting the unshared symmetry operations:

Table 11.3 Demonstration of correlation between C_{4v} and C_{2v}

C_{4v}	$\hat{E}$	~~$2\hat{C}_4$~~	$\hat{C}_2$	$2\hat{\sigma}_v$: $\hat{\sigma}_v(1)$	$2\hat{\sigma}_v$: $\hat{\sigma}_v(2)$	~~$2\hat{\sigma}_d$~~
A_1	1		1	[1]	1	
A_2	1		1	(−1)	−1	
B_1	1		1	[1]	1	
B_2	1		1	(−1)	−1	
E	2		−2	0	0	

C_{2v}	$\hat{E}$	$\hat{C}_2$	$\hat{\sigma}(XZ)$	$\hat{\sigma}(YZ)$
A_1	1	1	[1]	1
A_2	1	1	(−1)	−1
B_1	1	−1	1	−1
B_2	1	−1	−1	1
$B_1 + B_2$	2	−2	0	0

(Arrows connect $A_1(C_{4v}) \leftrightarrow A_1(C_{2v})$, $A_2(C_{4v}) \leftrightarrow A_2(C_{2v})$, $B_1(C_{4v}) \leftrightarrow A_1(C_{2v})$, $B_2(C_{4v}) \leftrightarrow A_2(C_{2v})$, $E(C_{4v}) \leftrightarrow B_1 + B_2(C_{2v})$.)

The character of $\hat{\sigma}(XZ)$ in A_1 of C_{2v} is now enclosed in a square bracket. In the truncated character table of C_{4v} the entries under $\hat{\sigma}_v(1)$ [or $\hat{\sigma}_v(2)$] are searched for the presence of 1 and enclosed in square brackets—they are found in the irreducible representations A_1 and B_1 of C_{4v}. The correlations $A_1(C_{2v}) \leftrightarrow A_1(C_{4v})$ and $A_1(C_{2v}) \leftrightarrow B_1(C_{4v})$ are thus indicated.

It is further found that the character set of A_1 of C_{2v} is the same as the (truncated) character set of each of A_1 and B_1 of C_{4v}. In the same way $A_2(C_{2v})$ is correlated to A_2 and B_2 of C_{4v}. We are now left with the E representation of C_{4v}. In Table 11.3 it is found that the character set of the reducible representation $B_1 + B_2$ of C_{2v} completely coincides with that of E in C_{4v}. Hence $E(C_{4v})$ is correlated to $B_1 + B_2(C_{2v})$. A physical meaning of this particular result is that if the point group of a molecule descends from C_{4v} to C_{2v} by a substitution reaction or by some distortion, a two-fold degenerate energy level belonging to the E-representation splits into two non-degenerate levels of symmetry types B_1 and B_2 in C_{2v}.

If instead of the two σ_v planes of C_{4v}, the two σ_d planes were taken to coincide with the $\hat{\sigma}(XZ)$ and $\hat{\sigma}(YZ)$ planes of C_{2v}, then the correlations should be

$$A_1(C_{2v}) \leftrightarrow A_1(C_{4v}) \quad \text{and} \quad B_2(C_{4v})$$

$$A_2(C_{2v}) \leftrightarrow A_2(C_{4v}) \quad \text{and} \quad B_1(C_{4v})$$

$$B_1 + B_2(C_{2v}) \leftrightarrow E(C_{4v})$$

This is left as an exercise for checking.

Example 11.2 Correlation between D_{2h} and $D_{\infty h}$. D_{2h} is a subgroup of $D_{\infty h}$ as is evident from the following discussion. If the Z-axis is taken as the principal C_2-axis of D_{2h} and the C_∞-axis of $D_{\infty h}$, then the $C_2(X)$ and $C_2(Y)$ axes of D_{2h} are two of the infinite number of C_2-axes of $D_{\infty h}$. Characters of the $\hat{C}_2(Z)$ operation of D_{2h} are obtained from those of $\hat{C}_\infty^\phi$ of $D_{\infty h}$ by putting $\phi = \pi$ [as $\hat{C}_2(Z) \equiv \hat{C}_\infty^\phi$ with $\phi = \pi$]. The $\hat{\sigma}(XY)$ [that is, $\hat{\sigma}_h$] operation of D_{2h} is the $\hat{S}_\infty^\phi$ operation of $D_{\infty h}$ with $\phi = 0$. The $\sigma(YZ)$ and $\sigma(XZ)$ planes of D_{2h} are two of the infinite number of σ_v planes of $D_{\infty h}$ as shown in the following list.

$D_{\infty h}$:	$\hat{E}$	$2\hat{C}_\infty^\phi\ (\phi = \pi)$	$\infty\hat{C}_2$		$\hat{i}$	$2\hat{S}_\infty^\phi\ (\phi = 0)$	$\infty\hat{\sigma}_v$
D_{2h}:	$\hat{E}$	$\hat{C}_2(Z)$	$\hat{C}_2(X)$	$C_2(Y)$		$\hat{\sigma}(XY)$	$\hat{\sigma}(XZ)$ $\hat{\sigma}(YZ)$

Now using the full character table of D_{2h} and $D_{\infty h}$ (from Appendix A3), the characters of the shared operations in their different irreducible representations are arranged side by side as shown below:

$D_{\infty h}$	$\hat{E}$	$2\hat{C}_\infty^\phi$ $(\phi = \pi)$	$\infty\hat{C}_2$: $\hat{C}_2'(1)$	$\infty\hat{C}_2$: $\hat{C}_2'(2)$	$\hat{i}$	$2\hat{S}_\infty^\phi$ $(\phi = 0)$	$\infty\hat{\sigma}_v$: $\hat{\sigma}_v(1)$	$\infty\hat{\sigma}_v$: $\hat{\sigma}_v(2)$
Σ_g^+	1	1	1	1	1	1	1	1
Σ_g^-	1	1	−1	−1	1	1	−1	−1
Π_g	2	−2	0	0	2	−2	0	0

D_{2h}	$\hat{E}$	$\hat{C}_2(Z)$	$\hat{C}_2(Y)$	$\hat{C}_2(X)$	$\hat{i}$	$\hat{\sigma}(XY)$	$\hat{\sigma}(XZ)$	$\hat{\sigma}(YZ)$
A_g	1	1	1	1	1	1	1	1
B_{1g}	1	1	−1	−1	1	1	−1	−1
B_{2g}	1	−1	1	−1	1	−1	1	−1
B_{3g}	1	−1	−1	1	1	−1	−1	1
$B_{2g} + B_{3g}$	2	−2	0	0	2	−2	0	0

It is thus found that A_g and B_{1g} of D_{2h} respectively have the same set of characters as Σ_g^+ and Σ_g^- of $D_{\infty h}$. The set of characters of Π_g in $D_{\infty h}$ is the same as that of $B_{2g} + B_{3g}$ in D_{2h}. Some other correlations can be found in the same way. The results are summarised in Table 11.4.

Table 11.4 Correlation table between D_{2h} and $D_{\infty h}$

$D_{\infty h}$	Σ_g^+	Σ_g^-	Π_g	Δ_g	Σ_u^+	Σ_u^-	Π_u	Δ_u
D_{2h}	A_g	B_{1g}	$B_{2g} + B_{3g}$	$A_g + B_{1g}$	B_{1u}	A_u	$B_{2u} + B_{3u}$	$A_u + B_{1u}$

Note: The correlation in Table 11.4 can be easily checked if we look into the basis sets of the D_{2h} and $D_{\infty h}$ character tables. For example, compare the basis sets of the correlated irreducible representations:

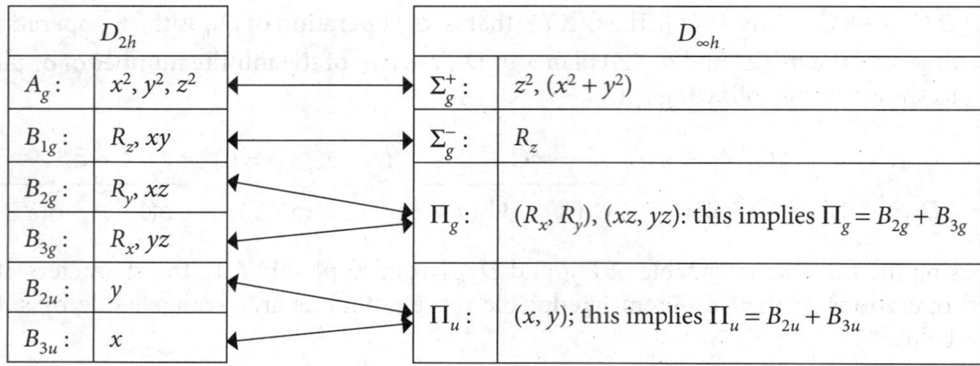

D_{2h}			$D_{\infty h}$	
A_g :	x^2, y^2, z^2	↔	Σ_g^+ :	$z^2, (x^2+y^2)$
B_{1g} :	R_z, xy	↔	Σ_g^- :	R_z
B_{2g} :	R_y, xz	↔	Π_g :	$(R_x, R_y), (xz, yz)$: this implies $\Pi_g = B_{2g} + B_{3g}$
B_{3g} :	R_x, yz	↔		
B_{2u} :	y	↔	Π_u :	(x, y); this implies $\Pi_u = B_{2u} + B_{3u}$
B_{3u} :	x	↔		

Analysis of Normal Modes of Vibration of the Linear Molecule Carbonsuboxide (C_3O_2) Using the ($D_{2h} \leftrightarrow D_{\infty h}$) Correlation Table

Carbonsuboxide (C_3O_2) is a linear molecule with point group $D_{\infty h}$.

Orientation of the coordinate axes and atom labels are shown in Fig. 11.10. The line through the five nuclei is taken as the Z-axis. On each atom (i), unit vectors, $\vec{x}_i, \vec{y}_i, \vec{z}_i$, parallel to the coordinate axes are taken as shown in Fig. 11.10 (where only one set of unit vectors is shown for the sake of clarity).

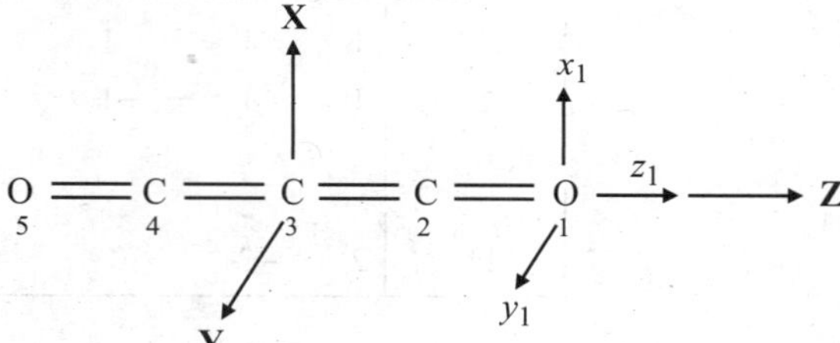

Fig. 11.10 Structure and orientation of Cartesian axes for the C_3O_2 molecule

A representation of $D_{\infty h}$ using all the 15 unit vectors could be constructed, but its reduction by the usual method would be difficult because that would require a division by infinity. The procedure of practical use is the application of correlation tables. $D_{\infty h}$ can be correlated to many subgroups, but a judicious choice is needed for practical application. A C_4 rotation about the C_∞-axis (the Z-axis) turns to X to Y and Y to X; but a C_2 rotation about the same axis turns them into themselves only with a change of sign (X to –X and Y to –Y). Obviously, the subgroup D_{2h} is the better choice.

With the vectors $\vec{x}_i, \vec{y}_i, \vec{z}_i$ ($i = 1$ to 5) as basis a representation (Γ_{total}) of the D_{2h} subgroup is shown in the following tabular form where the representations for translation and rotation about the Cartesian axes (excluding the Z-axis) are also shown. Γ_{trans} is the representation with basis (x, y, z) and Γ_{rot} is the representation spanned by (R_x, R_y). Γ_{vib} is the obtained by subtracting ($\Gamma_{trans} + \Gamma_{rot}$) from Γ_{total}. By reduction formula it is found that

$$\Gamma_{vib} = 2A_g + B_{2g} + B_{3g} + 2B_{1u} + 2B_{2u} + 2B_{3u} \tag{11.47}$$

(Check that the total number of dimensions in the RHS is $3N - 5$.)

χ (character)

	$\hat{E}$	$\hat{C}_2(Z)$	$\hat{C}_2(Y)$	$\hat{C}_2(X)$	$\hat{i}$	$\hat{\sigma}(XY)$	$\hat{\sigma}(XZ)$	$\hat{\sigma}(YZ)$
No. of unmoved atoms	5	5	1	1	1	1	5	5
Γ_{total}	15	−5	−1	−1	−3	1	5	5
Γ_{trans}	3	−1	−1	−1	−3	1	1	1
Γ_{rot}	2	−2	0	0	2	−2	0	0
$\Gamma_{vib} = \Gamma_{total} - \Gamma_{trans} - \Gamma_{rot}$	10	−2	0	0	−2	2	4	4

The task is now to find which irreducible representations of $D_{\infty h}$ are correlated to the component irreducible representation of Γ_{vib} as found in Eq. (11.47). From Table 11.4 we find the following correlations:

D_{2h}:	$2A_g$	$B_{2g} + B_{3g}$	$2B_{1u}$	$2(B_{2u} + B_{3u})$
$D_{\infty h}$:	$2\Sigma_g^+$	Π_g	$2\Sigma_u^+$	$2\Pi_u$

Now looking at the basis sets in the $D_{\infty h}$ character table, the following results are predicted:

Mode	Infrared	Raman	Basis of representation
Σ_g^+	Inactive	Active	$x^2 + y^2, z^2$
Π_g	Inactive	Active	(yz, xz)
Σ_u^+	Active	Inactive	z
Π_u	Active	Inactive	(x, y)

Summarisation table:

Activity	Irreducible representation	Number of fundamental frequencies expected in vibrational spectrum
Infrared	$2\Sigma_u^+, 2\Pi_u$	4
Raman	$2\Sigma_g^+, \Pi_g$	3
Raman polarised	$2\Sigma_g^+$	2
Infrared–Raman coincidence	none	0

11.8 INTERNAL COORDINATES, SYMMETRY COORDINATES AND SKETCHING NORMAL MODES OF VIBRATION USING A PROJECTION OPERATOR

Displacement vectors along the Cartesian coordinate axes have been shown to be useful for determination of symmetry types of the normal modes of vibration, in the previous sections.

From this approach it is difficult to get a clear picture of the direction of movement of the atoms in a particular normal mode. It would be preferable to use specific structural features such as interatomic distances between bonded atoms (called bond length) and interbond angles and to associate their changes to particular vibrational modes. For this purpose, *internal coordinates* are defined as sets of changes (due to vibrations) in bond lengths and in as many mutually independent bond angles as are necessary to provide $3N - 6$ (or, $3N - 5$) internal displacements.

Symmetry coordinates are normalised linear combinations of the internal coordinates adapted to the symmetry of the irreducible representations corresponding to the normal modes of vibration. After constructing the symmetry coordinates from the internal coordinates by applying the projection operator method, it is possible to 'sketch' normal modes of vibration, as will be evident from the following examples.

11.8.1 H_2O Molecule

The internal coordinates for vibration of the H_2O molecule are $\overrightarrow{\Delta r_1}, \overrightarrow{\Delta r_2}$ (for O–H stretching) and $\Delta\theta$ (for bending) (Fig. 11.11). Applying the projection operator for the irreducible representations of the C_{2v} point group on $\overrightarrow{\Delta r_1}$ and $\Delta\theta$ we obtain the symmetry coordinates as shown below.

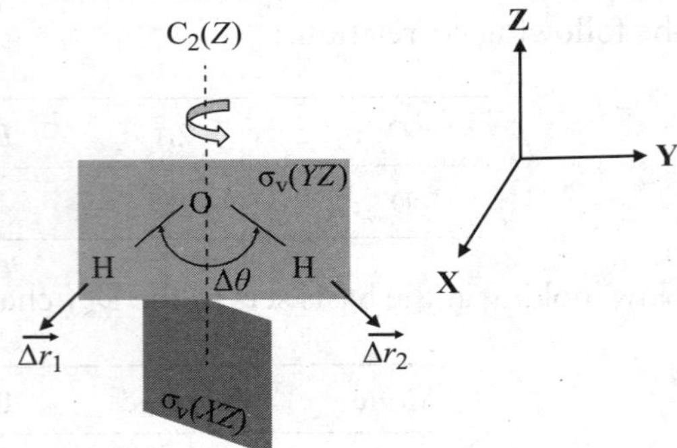

Fig. 11.11 Internal coordinates of the H_2O molecule

By the action of $\hat{E}$, $\hat{C}_2(Z)$, $\hat{\sigma}_v(XZ)$ and $\hat{\sigma}_v(YZ)$, $\overrightarrow{\Delta r_1}$ becomes respectively $\overrightarrow{\Delta r_1}, \overrightarrow{\Delta r_2}, \overrightarrow{\Delta r_2}$ and $\overrightarrow{\Delta r_1}$. Hence

$$\hat{P}^{A_1}(\overrightarrow{\Delta r_1}) = 1^* \cdot \hat{E}(\overrightarrow{\Delta r_1}) + 1^* \cdot \hat{C}_2(Z)(\overrightarrow{\Delta r_1}) + 1^* \cdot \hat{\sigma}_v(XZ)(\overrightarrow{\Delta r_1}) + 1^* \cdot \hat{\sigma}_v(YZ)(\overrightarrow{\Delta r_1})$$

$$= \overrightarrow{\Delta r_1} + \overrightarrow{\Delta r_2} + \overrightarrow{\Delta r_2} + \overrightarrow{\Delta r_1} \approx \frac{1}{\sqrt{2}}(\overrightarrow{\Delta r_1} + \overrightarrow{\Delta r_2}) \quad \text{(on normalisation)}$$

$$= s_1$$

$$\hat{P}^{A_2}(\overrightarrow{\Delta r_1}) = 1^* \cdot \hat{E}(\overrightarrow{\Delta r_1}) + 1^* \cdot \hat{C}_2(Z)(\overrightarrow{\Delta r_1}) + (-1)^* \cdot \hat{\sigma}_v(XZ)(\overrightarrow{\Delta r_1}) + (-1)^* \cdot \hat{\sigma}_v(YZ)(\overrightarrow{\Delta r_1})$$

$$= \overrightarrow{\Delta r_1} + \overrightarrow{\Delta r_2} - \overrightarrow{\Delta r_2} - \overrightarrow{\Delta r_1} = \vec{0} \quad \text{(null vector)}$$

It is similarly found that $\hat{P}^{B_1}(\overrightarrow{\Delta r_1}) = \vec{0}$ and

$$\hat{P}^{B_2}(\overrightarrow{\Delta r_1}) = 1^* \cdot \hat{E}(\overrightarrow{\Delta r_1}) + (-1)^* \cdot \hat{C}_2(Z)(\overrightarrow{\Delta r_1}) + (-1)^* \cdot \hat{\sigma}_v(XZ)(\overrightarrow{\Delta r_1}) + 1^* \cdot \hat{\sigma}_v(YZ)(\overrightarrow{\Delta r_1})$$

$$= \overrightarrow{\Delta r_1} - \overrightarrow{\Delta r_2} - \overrightarrow{\Delta r_2} + \overrightarrow{\Delta r_1} \approx \frac{1}{\sqrt{2}}(\overrightarrow{\Delta r_1} - \overrightarrow{\Delta r_2}) \quad \text{(on normalisation)}$$

$$= s_2$$

We need not apply these symmetry operators on $\overrightarrow{\Delta r_2}$ again because $\overrightarrow{\Delta r_2}$ has already been combined with $\overrightarrow{\Delta r_1}$ and will give the same results.

Next we consider $\Delta\theta$. Figure 11.11 shows that $\Delta\theta$ remains unchanged by the action of all the symmetry operators. Hence,

$$\hat{P}^{A_1}(\Delta\theta) = 1\cdot\hat{E}(\Delta\theta) + 1\cdot\hat{C}_2(Z)(\Delta\theta) + 1\cdot\hat{\sigma}_v(XZ)(\Delta\theta) + 1\cdot\hat{\sigma}_v(YZ)(\Delta\theta)$$

$$= \Delta\theta + \Delta\theta + \Delta\theta + \Delta\theta = 4\Delta\theta \approx \Delta\theta \quad \text{(on normalisation)}$$

$$= s_3$$

$$\hat{P}^{A_2}(\Delta\theta) = \hat{P}^{B_1}(\Delta\theta) = \hat{P}^{B_2}(\Delta\theta) = 0$$

Thus, the symmetry coordinate s_3 is $\Delta\theta$ itself.

We thus find that the symmetry coordinates s_1 and s_3 are of type A_1 and s_2 is of type B_2, which yield the result, $\Gamma_{\text{vib}} = 2A_1 + B_2$ as found earlier. Having found the symmetry coordinates, we can sketch the normal modes.

In s_1, which is the basis of the irreducible representation A_1, the stretching vectors $\overrightarrow{\Delta r_1}$ and $\overrightarrow{\Delta r_2}$ are both positive, meaning that both the bonds are elongating simultaneously (and compressing simultaneously in the next half of the cycle). The resultant of $\overrightarrow{\Delta r_1}$ and $\overrightarrow{\Delta r_2}$ is directed along the Z-axis downward (shown as $\Delta\vec{R}_1$ of Fig. 11.12). Hence, to maintain the centre of gravity of the molecule fixed in position (to avoid an overall translational motion) the O atom should be displaced upward and its displacement vector should be smaller in length than $\overrightarrow{\Delta r_1}$ or $\overrightarrow{\Delta r_2}$ because the O-atom is much heavier than the H-atoms. These three displacement vectors ($\overrightarrow{\Delta r_1}$, $\overrightarrow{\Delta r_2}$ and one for the O atom) comprise the normal mode of vibration for symmetic stretching (v_s) of H_2O (Fig. 11.12).

In s_2 which is a basis of the irreducible representation B_2, $\overrightarrow{\Delta r_1}$ is positive while $\overrightarrow{\Delta r_2}$ has a negative sign; this means that one O–H bond is stretching while the other is being compressed. The resultant ($\vec{R}_2$) of $\overrightarrow{\Delta r_1}$ and $-\overrightarrow{\Delta r_2}$ is directed toward the left of Z-axis and is on the plane of $\overrightarrow{\Delta r_1}$ and $\overrightarrow{\Delta r_2}$. Hence to maintain the centre of gravity of the molecule fixed in position, the O atom must be displaced toward the right and to a lesser extent than $\overrightarrow{\Delta r_1}$ and $\overrightarrow{\Delta r_2}$. These atomic displacements comprise the antisymmetric stretch (v_{as}) (Fig. 11.12).

In order to get a clear picture of the mode of vibration corresponding to s_3 we may consider the increase and decrease of the HÔH bond angle by attaching displacement vectors to the two H-atoms in directions perpendicular to the O–H bonds. For an increase of $\Delta\theta$, these vectors are shown in (a) of Fig. 11.12; constructing their resultant, the displacement vector of O is found to be along the Z-axis downward. For a decrease of $\Delta\theta$, these vectors would be oppositely directed. In any case, however, the 'picture' remains invariant under C_{2v} symmetry operations, as expected for s_3. The in-plane bending mode (s_3) is usually denoted by δ. [For a change of $\Delta\theta$, it is also possible to attach two vectors to the H-atoms keeping them orthogonal to the O–H bonds in a different way as shown in (b) of Fig. 11.12. But in that case, the motion corresponds to a rotation of the molecule and is outside the present discussion of pure vibrations.]

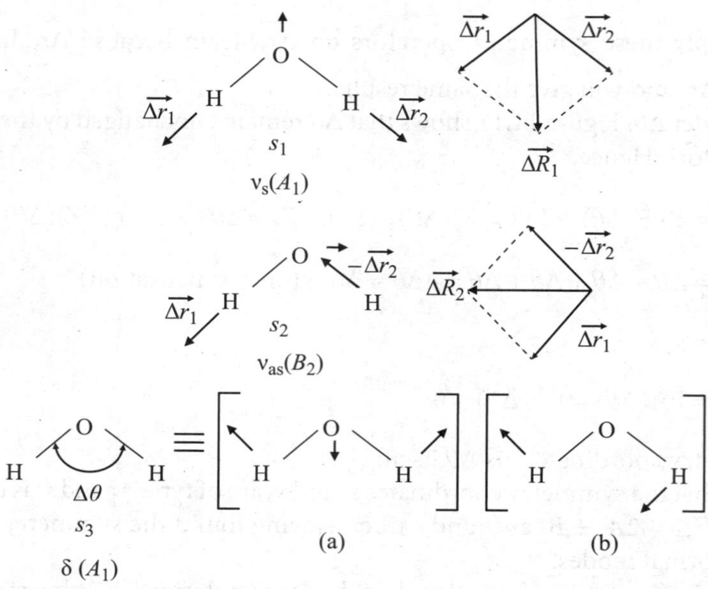

Fig. 11.12 Symmetry coordinates (s_1, s_2, s_3) of the H_2O molecule. Case (b) for s_3 is a pure rotation.

11.8.2 Vibrations of the XeF_4 Molecule

The internal coordinates of this square planar molecule are shown in Fig. 11.13 where four coordinates Δr_i ($i = 1$ to 4) take into account the stretching vibrations, four coordinates $\Delta\beta_i$ (i = 1 to 4) account for out-of-plane bending vibrations, and four other coordinates $\Delta\alpha_i$ ($i = 1$ to 4) account for in-plane bending vibrations. For the sake of clarity only one bending coordinate of each type has been shown.

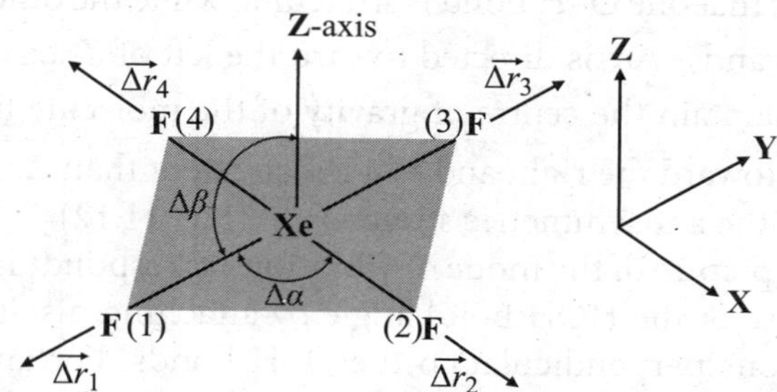

Fig. 11.13 Internal coordinates of XeF_4 and orientation of Cartesian axes

We shall consider the three classes of vibration one by one.

Stretching

With the four coordinates $\overrightarrow{\Delta r_1}$, $\overrightarrow{\Delta r_2}$, $\overrightarrow{\Delta r_3}$ and $\overrightarrow{\Delta r_4}$ as basis a representation of the point group D_{4h} is as follows:

χ (character)

	$\hat{E}$	$2\hat{C}_4$	$\hat{C}_2$	$2\hat{C}_2'$	$2\hat{C}_2''$	$\hat{i}$	$2\hat{S}_4$	$\hat{\sigma}_h$	$2\hat{\sigma}_v$	$2\hat{\sigma}_d$
$\Gamma_{stretch}$	4	0	0	2	0	0	0	4	2	0

This on reduction becomes $A_{1g}+B_{1g}+E_u$. The character table of D_{4h} shows that the bases of A_{1g} and B_{1g} are quadratic functions of coordinates while E_u has basis set (x, y). Hence the former two are Raman active but infrared inactive while the two degenerate vibrations are infrared active and Raman inactive. To obtain symmetry coordinates for the stretching modes, it would be straightforward to use the character table of the whole group. But the labour is much reduced if we use the subgroup C_4 of D_{4h}. The A_{1g}, B_{1g} and E_u representations of D_{4h} correspond to A, B and E representations of C_4. Applying projection operators (for the subgroup C_4), we have the symmetry coordinates s_1, s_2, s_3 and s_4.

$$\hat{P}^A(\overrightarrow{\Delta r_1}) = \tfrac{1}{4}[1^* \cdot \hat{E}(\overrightarrow{\Delta r_1}) + 1^* \cdot \hat{C}_4(\overrightarrow{\Delta r_1}) + 1^* \cdot \hat{C}_2(\overrightarrow{\Delta r_1}) + 1^* \cdot \hat{C}_4^3(\overrightarrow{\Delta r_1})]$$

$$= \tfrac{1}{4}[\overrightarrow{\Delta r_1} + \overrightarrow{\Delta r_2} + \overrightarrow{\Delta r_3} + \overrightarrow{\Delta r_4}]$$

$$= \tfrac{1}{2}[\overrightarrow{\Delta r_1} + \overrightarrow{\Delta r_2} + \overrightarrow{\Delta r_3} + \overrightarrow{\Delta r_4}] \quad \text{(on normalisation)}$$

$$= s_1$$

$$\hat{P}^B(\overrightarrow{\Delta r_1}) = \tfrac{1}{4}[1^* \cdot \hat{E}(\overrightarrow{\Delta r_1}) + (-1)^* \cdot \hat{C}_4(\overrightarrow{\Delta r_1}) + 1^* \cdot \hat{C}_2(\overrightarrow{\Delta r_1}) + (-1)^* \cdot \hat{C}_4^3(\overrightarrow{\Delta r_1})]$$

$$= \tfrac{1}{4}[\overrightarrow{\Delta r_1} - \overrightarrow{\Delta r_2} + \overrightarrow{\Delta r_3} - \overrightarrow{\Delta r_4}]$$

$$= \tfrac{1}{2}[\overrightarrow{\Delta r_1} - \overrightarrow{\Delta r_2} + \overrightarrow{\Delta r_3} - \overrightarrow{\Delta r_4}] \quad \text{(on normalisation)}$$

$$= s_2$$

The E-representation of C_4 contains complex characters. Hence we first construct two sets (E_a and E_b) of real character by adding and subtracting the rows of characters and dividing by 2 and $2i$ respectively:

	$\hat{E}$	$\hat{C}_4$	$\hat{C}_2$	$\hat{C}_4^3$		
E	1	i	−1	$-i$	adding and dividing by 2 →	
	1	$-i$	−1	i	subtracting and dividing by $2i$ →	

	$\hat{E}$	$\hat{C}_4$	$\hat{C}_2$	$\hat{C}_4^3$
E_a	1	0	−1	0
E_b	0	1	0	−1

Using these two sets of real characters and applying the projection operators on $\overrightarrow{\Delta r_1}$ we get two more symmetry coordinates of XeF_4:

$$\hat{P}^{E_a}(\overrightarrow{\Delta r_1}) = \tfrac{1}{4}[1^* \cdot \hat{E}(\overrightarrow{\Delta r_1}) + 0^* \cdot \hat{C}_4(\overrightarrow{\Delta r_1}) + (-1)^* \cdot \hat{C}_2(\overrightarrow{\Delta r_1}) + 0^* \cdot \hat{C}_4^3(\overrightarrow{\Delta r_1})]$$

$$= \tfrac{1}{4}[\overrightarrow{\Delta r_1} - \overrightarrow{\Delta r_3}] \approx \tfrac{1}{\sqrt{2}}[\overrightarrow{\Delta r_1} - \overrightarrow{\Delta r_3}] \quad \text{(on normalisation)}$$

$$= s_3$$

$$\hat{P}^{E_b}(\overrightarrow{\Delta r_1}) = \tfrac{1}{4}[0^* \cdot \hat{E}(\overrightarrow{\Delta r_1}) + 1^* \cdot \hat{C}_4(\overrightarrow{\Delta r_1}) + 0^* \cdot \hat{C}_2(\overrightarrow{\Delta r_1}) + (-1)^* \cdot \hat{C}_4^3(\overrightarrow{\Delta r_1})]$$

$$= \tfrac{1}{4}[\overrightarrow{\Delta r_2} - \overrightarrow{\Delta r_4}] \approx \tfrac{1}{\sqrt{2}}[\overrightarrow{\Delta r_2} - \overrightarrow{\Delta r_4}] \quad \text{(on normalisation)}$$

$$= s_4$$

Restoring the u and g subscripts for the whole group (D_{4h}) the symmetry coordinates s_1, s_2, s_3 and s_4 now become bases of the irreducible representations A_{1g}, B_{1g} and E_u. The stretching modes of vibration of XeF_4 can now be drawn by looking at the expressions for s_1, s_2, s_3 and s_4; a negative sign of any $\overrightarrow{\Delta r_i}$ in such expressions means inversion of direction of the corresponding stretching vectors (Fig. 11.14).

Note that in ν_s and ν_{as}, there is no displacement of the Xe atom because the displacement vectors of four F atoms sum up to zero displacement. But in each of the degenerate stretching modes, the displacement vectors of F have a net resultant and so Xe has an opposite displacement to avoid translation. This is indicated by a double arrow along the appropriate Xe–F bond in Fig. 11.14.

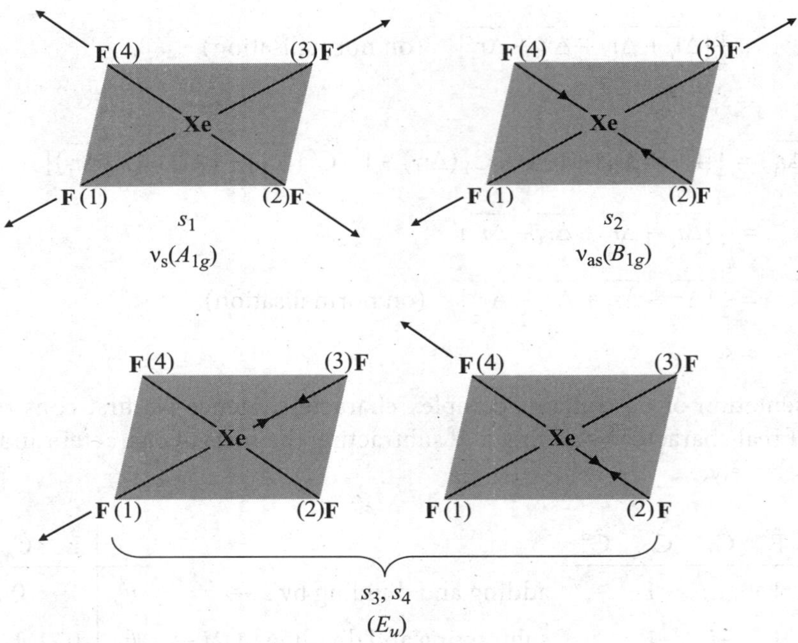

Fig. 11.14 The stretching modes of vibration of XeF_4

Out-of-plane Bending

Next we consider the out-of-plane bending vibrations. The corresponding internal coordinates, $\Delta\beta_i$ ($i = 1$ to 4), are changes in the angles between each Xe–F bond and the Z-axis; they are clearly described by the displacement vectors $\vec{z}_1, \vec{z}_2, \vec{z}_3, \vec{z}_4$ of F-atoms parallel to the Z-axis (Fig. 11.15).

Using the vectors $\vec{z}_1, \vec{z}_2, \vec{z}_3, \vec{z}_4$ as basis a representation Γ_π of the D_{4h} point group is as follows:

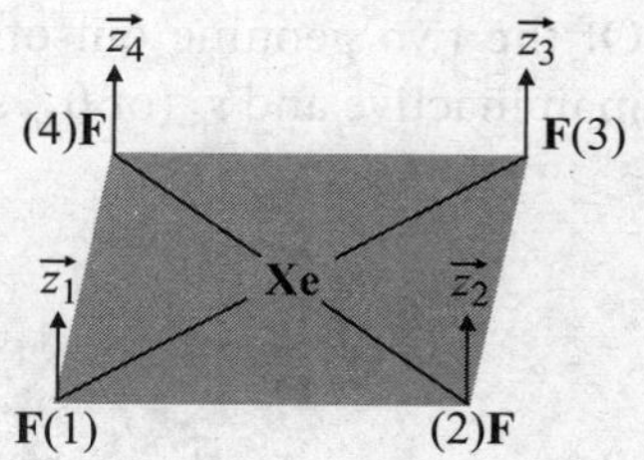

Fig. 11.15 Displacements of F atoms parallel to the Z-axis to take into account out-of-plane bending.

χ **(character)**

	$\hat{E}$	$2\hat{C}_4$	$\hat{C}_2$	$2\hat{C}_2'$	$2\hat{C}_2''$	$\hat{i}$	$2\hat{S}_4$	$\hat{\sigma}_h$	$2\hat{\sigma}_v$	$2\hat{\sigma}_d$
Γ_π	4	0	0	−2	0	0	0	−4	2	0

This reduces to $A_{2u} + B_{2u} + E_g$. We now find the SALCs of A_{2u}, B_{2u} and E_g symmetries by considering, as before, the A, B and E ($\equiv E_a + E_b$) representations of the subgroup C_4 of D_{4h}.

$$\begin{aligned}
\hat{P}^A(\vec{z}_1) &= \tfrac{1}{4}[1^* \cdot \hat{E}(\vec{z}_1) + 1^* \cdot \hat{C}_4(\vec{z}_1) + 1^* \cdot \hat{C}_2(\vec{z}_1) + 1^* \cdot \hat{C}_4^3(\vec{z}_1)] \\
&= \tfrac{1}{4}[\vec{z}_1 + \vec{z}_2 + \vec{z}_3 + \vec{z}_4] \\
&\approx \tfrac{1}{2}[\vec{z}_1 + \vec{z}_2 + \vec{z}_3 + \vec{z}_4] \quad \text{(on normalisation)} \\
&= s_5
\end{aligned}$$

$$\begin{aligned}
\hat{P}^B(\vec{z}_1) &= \tfrac{1}{4}[1^* \cdot \hat{E}(\vec{z}_1) + (-1)^* \cdot \hat{C}_4(\vec{z}_1) + 1^* \cdot \hat{C}_2(\vec{z}_1) + (-1)^* \cdot \hat{C}_4^3(\vec{z}_1)] \\
&= \tfrac{1}{4}[\vec{z}_1 - \vec{z}_2 + \vec{z}_3 - \vec{z}_4] \\
&\approx \tfrac{1}{2}[\vec{z}_1 - \vec{z}_2 + \vec{z}_3 - \vec{z}_4] \quad \text{(on normalisation)} \\
&= s_6
\end{aligned}$$

$$\begin{aligned}
\hat{P}^{E_a}(\vec{z}_1) &= \tfrac{1}{4}[1^* \cdot \hat{E}(\vec{z}_1) + 0^* \cdot \hat{C}_4(\vec{z}_1) + (-1)^* \cdot \hat{C}_2(\vec{z}_1) + 0^* \cdot \hat{C}_4^3(\vec{z}_1)] \\
&= \tfrac{1}{4}[\vec{z}_1 - \vec{z}_3] \\
&\approx \tfrac{1}{\sqrt{2}}[\vec{z}_1 - \vec{z}_3] \quad \text{(on normalisation)} \\
&= s_7
\end{aligned}$$

$$\begin{aligned}
\hat{P}^{E_b}(\vec{z}_1) &= \tfrac{1}{4}[0^* \cdot \hat{E}(\vec{z}_1) + 1^* \cdot \hat{C}_4(\vec{z}_1) + 0^* \cdot \hat{C}_2(\vec{z}_1) + (-1)^* \cdot \hat{C}_4^3(\vec{z}_1)] \\
&= \tfrac{1}{4}[\vec{z}_2 - \vec{z}_4] \\
&\approx \tfrac{1}{\sqrt{2}}[\vec{z}_2 - \vec{z}_4] \quad \text{(on normalisation)} \\
&= s_8
\end{aligned}$$

where s_5, s_6, s_7 and s_8 are the symmetry coordinates for the out-of-plane vibrations. Looking at the signs of the displacement vectors in these expressions we get the following diagrams for the out-of-plane bending vibrations. It is found that two of the modes (s_7 and s_8) are, in fact, rotations, each about a diagonal of the square plane (Fig. 11.16).

Of the two genuine out-of-plane vibrations, s_5 (of A_{2u} symmetry) is infrared active but Raman inactive and s_6 (of B_{2u} symmetry) is 'silent'.

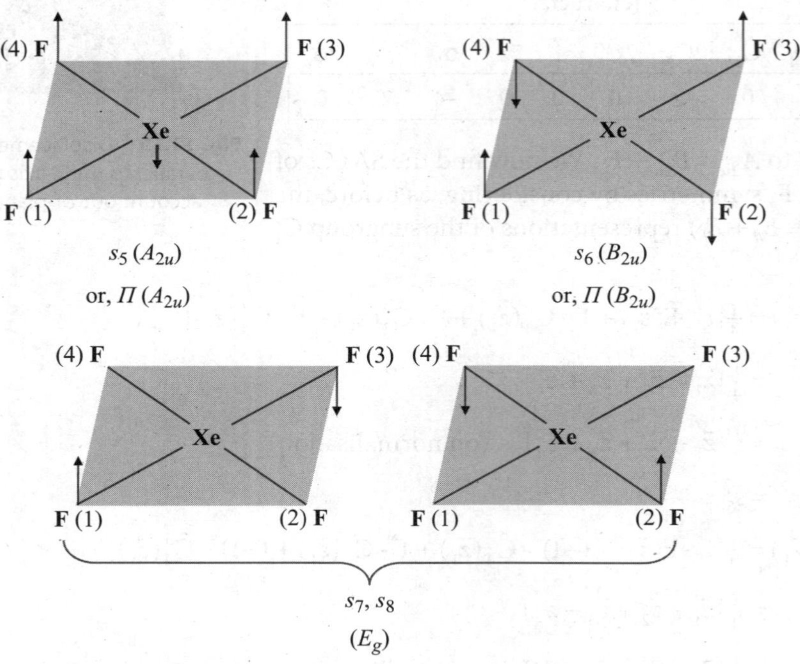

Fig. 11.16 Out-of-plane bending modes of XeF_4; two are proper bending and the other two (of E_g type) are rotations

In-plane Bending

The in-plane vibrations with internal coordinates $\Delta\alpha_1$, $\Delta\alpha_2$, $\Delta\alpha_3$ and $\Delta\alpha_4$ can be indicated by four in-plane displacement vectors $\vec{v}_1, \vec{v}_2, \vec{v}_3$ and $\vec{v}_4$ (Fig. 11.17). This picture is not meant to indicate rotation of the whole molecule about the C_4-axis which could be drawn in the same way. It is supposed that each Xe–F(i) bond moves in the direction of the corresponding $\vec{v}_i$, while the other three remain fixed, to produce the angular displacement $\Delta\alpha_i$ ($i = 1$ to 4). Now with these $\vec{v}_i$ vectors, a representation Γ_δ for the in-plane bending of XeF_4 is the following.

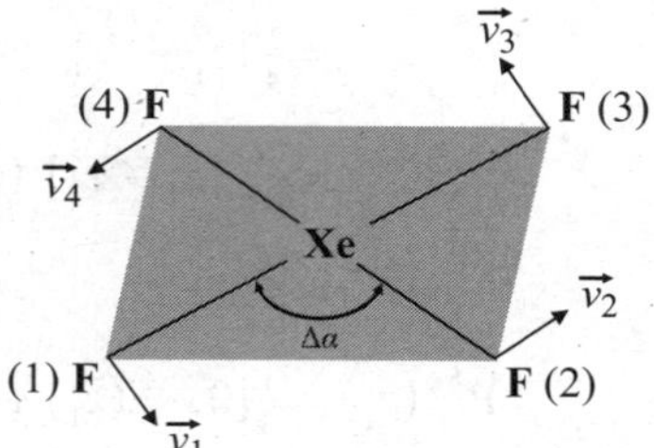

Fig. 11.17 Vectors for in-plane bending of XeF_4

χ **(character)**

	$\hat{E}$	$2\hat{C}_4$	$\hat{C}_2$	$2\hat{C}_2'$	$2\hat{C}_2''$	$\hat{i}$	$2\hat{S}_4$	$\hat{\sigma}_h$	$2\hat{\sigma}_v$	$2\hat{\sigma}_d$
Γ_δ	4	0	0	−2	0	0	0	4	−2	0

This on reduction gives $\Gamma_\delta = A_{2g} + B_{2g} + E_u$

Applying projection operators in a manner analogous to the previous cases, it is found that the symmetry coordinates s_9 to s_{12} are given by:

$$s_9\,(A_{2g}\text{ symmetry}) = \tfrac{1}{2}[\vec{v}_1 + \vec{v}_2 + \vec{v}_3 + \vec{v}_4] \propto (\Delta\alpha_1 + \Delta\alpha_2 + \Delta\alpha_3 + \Delta\alpha_4)$$

$$s_{10}\,(B_{2g}\text{ symmetry}) = \tfrac{1}{2}[\vec{v}_1 - \vec{v}_2 + \vec{v}_3 - \vec{v}_4] \propto (\Delta\alpha_1 - \Delta\alpha_2 + \Delta\alpha_3 - \Delta\alpha_4)$$

$$s_{11}\,(E_{1a}\text{ symmetry}) = \tfrac{1}{\sqrt{2}}[\vec{v}_1 - \vec{v}_3] \propto (\Delta\alpha_1 - \Delta\alpha_3)$$

$$s_{12}\,(E_{1b}\text{ symmetry}) = \tfrac{1}{\sqrt{2}}[\vec{v}_2 - \vec{v}_4] \propto (\Delta\alpha_2 - \Delta\alpha_4)$$

In-plane bending modes can now be drawn by looking at the expressions for $s_i (i = 9$ to $12)$ in terms of $\vec{v}_i$ vectors (Fig. 11.18). When the $\vec{v}_i$ vectors are added as in s_9, the displacements along $\vec{v}_i$ $(i = 1$ to $4)$ occur simultaneously to produce a rotation about the $C_4(Z)$ axis. So this s_9 is not a genuine vibrational mode for XeF_4. Hence there are three genuine in-plane bending modes of XeF_4. Of them, the D_{4h} character table shows that s_{10} of B_{2g} symmetry is Raman active but infrared inactive and the doubly degenerate s_{11}, s_{12} modes of E_u symmetry are infrared active but Raman inactive.

Altogether, the number of modes of genuine vibrations of XeF_4 is found to be 4 (stretching) + 2 (out-of-plane bending) + 3 (in-plane bending) = 9 = $3N - 6$ (with $N = 5$).

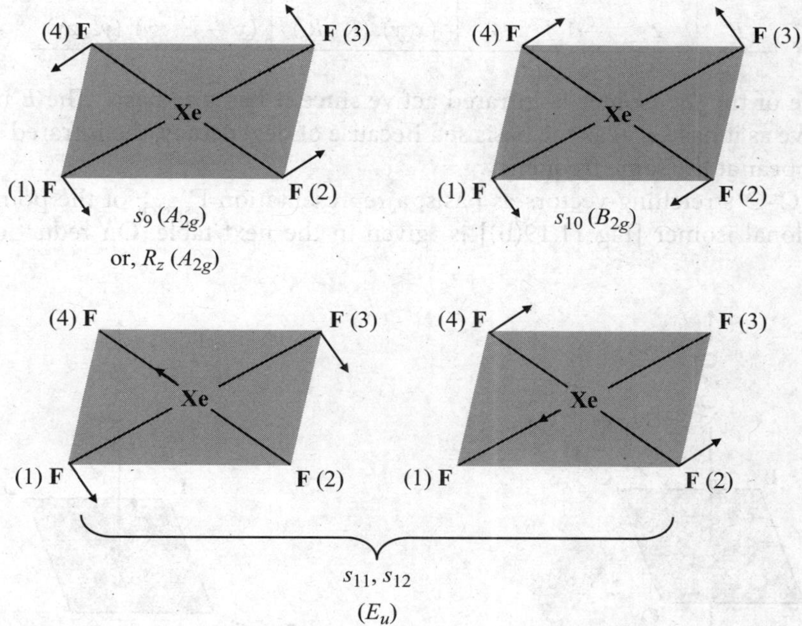

Fig. 11.18 Four in-plane bending modes of XeF_4 (one is pure rotation about the C_4-axis)

11.9 SOME GROUP THEORETICAL APPLICATIONS OF INFRARED AND RAMAN SPECTRA

Infrared and Raman spectra are powerful tools in deducing molecular geometry. Some examples are discussed here.

11.9.1 Distinction Between Facial and Meridonal Isomers of Metal Carbonyls

From the carbonyl stretching region (usually 1700–2100 cm^{-1}) of the infrared spectrum of a transition metal complex of the type $[ML_3(CO)_3]$ it is possible, by a group theoretical analysis, to distinguish between its facial and meridonal isomers. With the C–O stretching vectors in Fig. 11.19(a) as basis, a representation $\Gamma_{stretch}$ of the point group C_{3v} of the facial isomer is as in the adjacent table.

	χ (character)		
	$\hat{E}$	$2\hat{C}_3$	$3\hat{\sigma}_v$
$\Gamma_{stretch}$	3	0	1

In Fig. 11.19(a), all the stretching vectors are shifted from their positions by the $\hat{C}_3$ operation and one vector remains unshifted by each $\hat{\sigma}_v$ operation. By reduction it is found that

$$\Gamma_{stretch} = A_1 + E \quad \text{for the } fac \text{ isomer}$$

The character table of C_{3v} with the necessary basis functions is the following.

C_{3v}	$\hat{E}$	$2\hat{C}_3$	$3\hat{\sigma}_v$		
A_1	1	1	1	z	x^2+y^2, z^2
A_2	1	1	−1	R_z	
E	2	−1	0	$(x, y), (R_x, R_y)$	$(x^2-y^2, xy), (yz, zx)$

The A_1 mode of the *fac* isomer is infrared active since it has z as basis. The E mode is also infrared active as it has (x, y) as the basis set. Because of degeneracy, the infrared signals for E mode will appear at the same frequency.

With the C–O stretching vectors as basis, a representation $\Gamma_{stretch}$ of the point group C_{2v} of the meridonal isomer [Fig 11.19(b)] is given in the next table. On reduction, we find

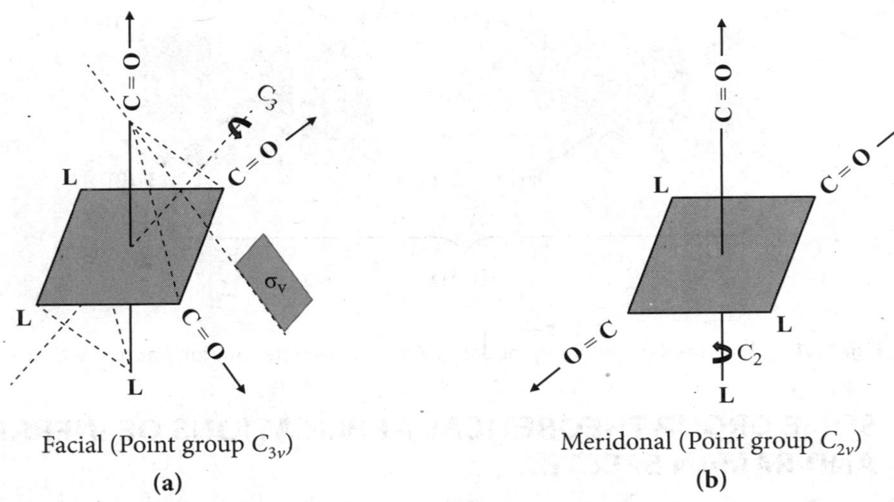

Fig. 11.19 Some symmetry elements and C–O stretching vectors of the facial (a) and meridonal (b) isomers of $[ML_3(CO)_3]$

$\Gamma_{stretch} = 2A_1 + B_2$. All the modes are infrared active, as can be verified by the method already discussed. Hence, in the C–O stretching region of the infrared spectrum of the *fac* isomer, we expect two bands—one for the A_1 mode and the other for the E mode. But in case of the *mer* isomer, stretching should give three bands—two for the A_1 modes and one for the B_2 mode. The expected patterns of the bands are as shown in Fig. 11.20. (Here, the number of absorption peaks is correctly shown but the relative intensities are arbitrary.)

	χ (character)			
	$\hat{E}$	$\hat{C}_2$	$\hat{\sigma}_v(XZ)$	$\hat{\sigma}_v(YZ)$
$\Gamma_{stretch}$	3	1	3	1

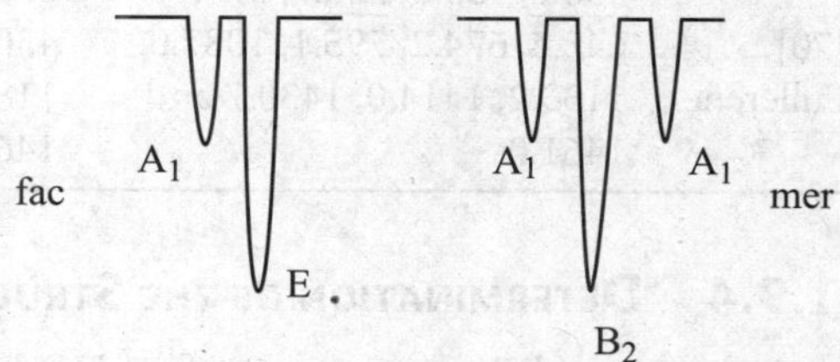

Fig. 11.20 Expected infrared signals in the C–O stretching region of the *fac* and *mer* isomers of $[ML_3(CO)_3]$

11.9.2 Establishing the Square Planar Structure of XeF_4

The structure of the XeF_4 molecule may be square planar or tetrahedral. Considering the difficulty in preparing this inert gas compound, conventional chemical methods for ascertaining the molecular structure is a problem. Infrared and Raman spectroscopy combined with group theory helps at this point. The tetrahedral structure has no centre of symmetry and expects some fundamental vibrations to appear in both infrared and Raman spectra. But if the structure is square planar, there is a centre of symmetry and there should be no infrared–Raman coincidence. Experiment reveals that the latter is the case (Raman: 235 cm^{-1}, 502 cm^{-1} and 543 cm^{-1}; IR: 123 cm^{-1}, 291 cm^{-1} and 586 cm^{-1}). Hence the square planar structure of XeF_4 is established.

11.9.3 Distinction Between [60]- and [70] Fullerene Structures

Fig. 11.21 Top and bottom of (a) [60]fullerene, staggered pentagons and (b) [70] fullerene, eclipsed pentagons

The [60]fullerene molecule has a truncated icosahedral structure. It is spherical in shape, with 12 pentagonal and 20 hexagonal carbon rings embedded on the surface of a sphere. The pentagons are all separated by intervening hexagons. If one imagines three mutually perpendicular diameters, each passing through the centres of a pair of pentagons on opposite sides of a diameter, then the pentagons are found to be arranged in a staggered manner [Fig. 11.21(a)]. But [70]fullerene has 12 pentagons and 25 hexagons fused and embedded on the surface of an ellipsoid. An ellipsoid has one long and a set of short diameters. The pentagons situated on opposite sides of each diameter of the ellipsoid are eclipsed with respect to one another [Fig. 11.21(b)]. Thus [60]fullerene molecule (point group I_h) has a centre of symmetry while [70]fullerene molecule (point group D_{5h}) is not centrosymmetric. Hence there is no infrared–Raman coincidence in the vibrational spectra of [60]fullerene but [70]fullerene exhibits a number of common frequencies in its infrared and Raman spectra. The prominent infrared and Raman bands of the two fullerenes are given below (Kroto et. al., 1991).

	Infrared (cm^{-1})	Raman (cm^{-1})	Coincidence
[60] Fullerene	527.0, 576.5, 1182.5 and 1429.0	272.0, 429.0, 495.0, 570.0, 772.0, 1099.0, 1250.0, 1422.0, 1467.0 and 1573.5	None
[70] Fullerene	458.2, 535.5, 565.7, 578.0, 642.4, 674.2, 795.1, 1087.0, 1133.2, 1414.0, 1430.7 and 1461.0	226.0, 252.0, 259.0, 394.0, 410.0, 417.0, 430.0, 456.0, 566.0, 700.0, 736.0, 1061.0, 1180.0, 1221.0, 1226.0, 1294.0, 1423.0, 1465.0, 1509.0 and 1562.0	456.0, 566.0

11.9.4 Determination of the Structure of $[HF_2]^-$

The infrared and Raman spectra for $[HF_2]^-$ and $[DF_2]^-$ ions, investigated as their potassium salts, show frequencies as given in the following table:

$[HF_2]^-$		$[DF_2]^-$	
Infrared (cm^{-1})	Raman (cm^{-1})	Infrared (cm^{-1})	Raman (cm^{-1})
1550	–	1140	–
1200	–	860	–
–	675	–	675

For the $[HF_2]^-$ ion one can imagine two bent and two linear structures:

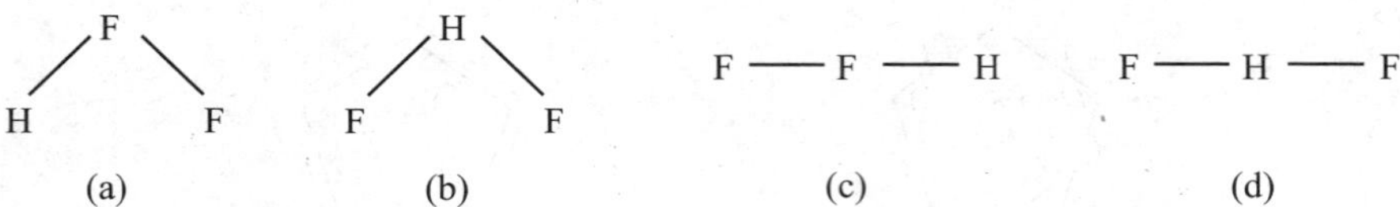

Of these, only structure (d) has a centre of symmetry. The table of infrared and Raman frequencies shows that there is no infrared–Raman coincidence for the ion $[HF_2]^-$. Hence the structure should have a centre of symmetry; this fact suggests the linear structure (d) for the ion and rejects the other structures. This is further supported by the fact that isotopic substitution (H by D) induces no change in one frequency (675 cm^{-1}); this is possible only with one mode of vibration, namely, the symmetric stretching as revealed by the following sketches:

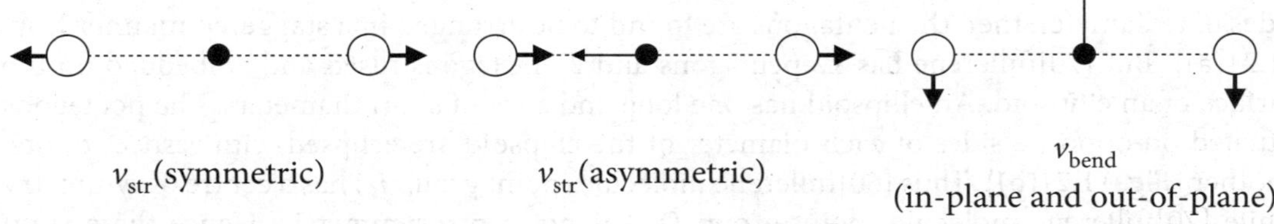

The symmetric stretching involves no displacement of the central atom (H or D), making ν_{str}(symmetric) not responsive to isotopic substitution.

11.10 ASSIGNMENT OF A TOTALLY SYMMETRIC NATURE TO A RAMAN ACTIVE VIBRATIONAL MODE

Whether a given Raman signal arises out of a totally symmetric vibration or not can be determined by noting the polarisation of the signal. The experimental procedure is to use a linearly polarised light beam as incident (exciting) radiation and take the Raman signal intensities, $I_{||}$ and $I_{\perp}$, in two directions, namely perpendicular and parallel, with respect to the incident beam (Fig. 11.22).

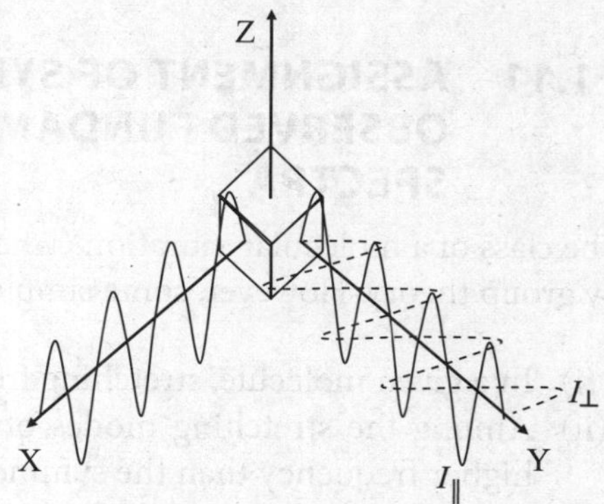

Fig. 11.22 Parallel and perpendicular Raman scattering

The intensity ratio $\rho = \dfrac{I_{\perp}}{I_{||}}$, is called the *depolarisation ratio*. If ρ is close to zero, $I_{\perp} \approx 0$, which means that the Raman scattered light is in the same state of polarisation as that of the incident light, and the vibrational mode responsible for this Raman signal is totally symmetric (A, A_1, A_{1g}, ... as the case may be).

For higher values of ρ, the scattered light is depolarised, and the vibration is not totally symmetric. From the theory of Raman spectra it is found that if $0 \le \rho \le \frac{3}{4}$ the line is polarised (totally symmetric vibration). For $\rho > \frac{3}{4}$, the line is depolarised and the vibration is not totally symmetric. A qualitative explanation is as follows.

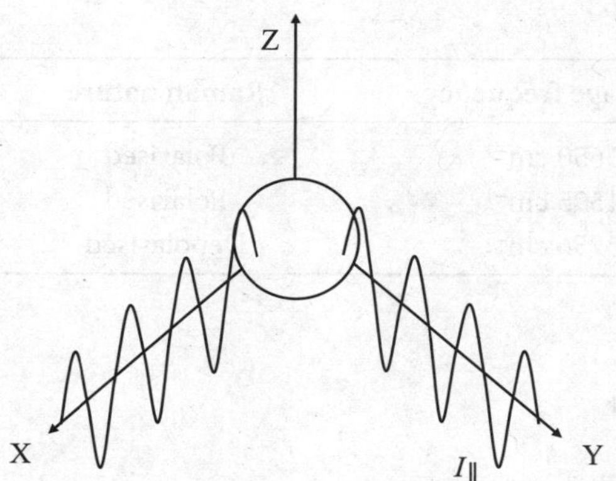

Fig. 11.23 Polarised light scattering by a sphere ($\rho = 0$)

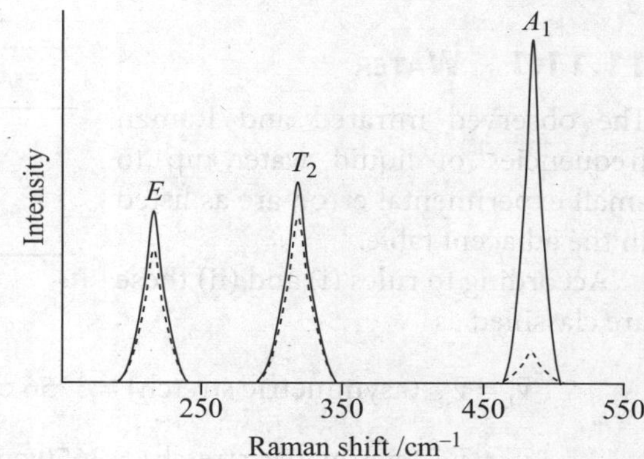

Fig. 11.24 Vibrational Raman spectrum (schematic) of liquid CCl_4 showing polarised and depolarised bands. The solid line corresponds to $I_{||}$ and dotted line to $I_{\perp}$.

If a plane polarised light beam is incident on a spherical top, the plane of the induced dipole in the case of a totally symmetric vibration is always parallel to that of the incident light and the molecule behaves like a tiny sphere. The scattered light is also polarised parallel to the plane of the incident light (Fig. 11.23), making $\rho \approx 0$. Molecules with O_h or T_d symmetry behave in this way, as shown schematically in the vibrational Raman spectrum (Fig. 11.24) of liquid CCl_4. For the E and T_2 bands $\rho = \frac{I_\perp}{I_\parallel} \approx 1$ while for A_1 band $\rho = \frac{I_\perp}{I_\parallel} \approx 0$.

11.11 ASSIGNMENT OF SYMMETRY TYPES TO EXPERIMENTALLY OBSERVED FUNDAMENTAL BANDS IN INFRARED/RAMAN SPECTRA

The class of a molecular vibration (whether it is a stretching or a bending) cannot be revealed by group theory. However, some simple general rules can be stated:

(i) In a given molecule, stretching frequencies are higher than bending frequencies.
(ii) Among the stretching modes of vibration of a molecule, asymmetric stretches are of higher frequency than the symmetric ones.
(iii) For molecules behaving like symmetric tops, the Raman transitions corresponding to the totally symmetric irreducible representation are polarised.
(iv) Out-of-plane bends are generally of higher frequency than in-plane bends.
(v) For a bipyramidal molecule with two sets of bonds, axial and equatorial, the bending vibration involving two equatorial bonds is of lower frequency than the bending vibration involving one axial and one equatorial bond.

Using these rules, symmetry symbols can be assigned to observed frequencies in the infrared and Raman spectra of many relatively simple molecules. Some examples are discussed below.

11.11.1 WATER

The observed infrared and Raman frequencies of liquid water, up to small experimental error, are as listed in the adjacent table.

Average frequency	Raman nature
3650 cm^{-1}	Polarised
1595 cm^{-1}	Polarised
3756 cm^{-1}	Depolarised

According to rules (i) and (ii) these are classified as

$$\bar{\nu}_1 = \bar{\nu}_{str} \text{ (asymmetric stretch)} = 3756 \text{ cm}^{-1}$$

$$\bar{\nu}_2 = \bar{\nu}_{str} \text{ (symmetric stretch)} = 3650 \text{ cm}^{-1}$$

$$\bar{\nu}_3 = \bar{\nu}_{bend} = 1595 \text{ cm}^{-1}$$

According to earlier analysis, $\bar{\nu}_2$ and $\bar{\nu}_3$ are of symmetry type A_1 and $\bar{\nu}_1$ is of type B_2. This is in conformity with rule (iii) and the given polarisation data (namely, $\bar{\nu}_2$ and $\bar{\nu}_3$ are polarised in Raman while $\bar{\nu}_1$ is depolarised in Raman).

11.11.2 BF_3 (Planar)

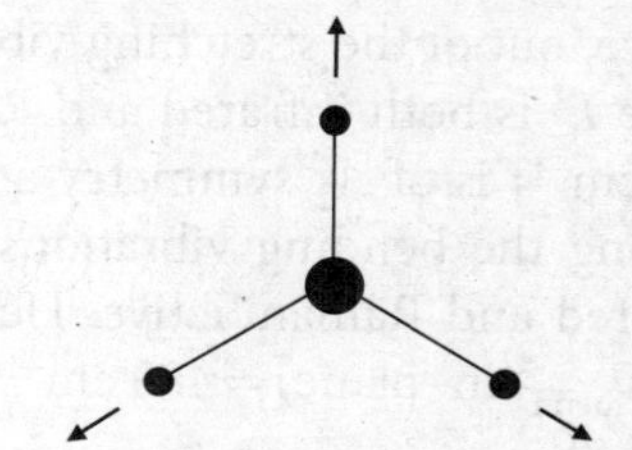

Fig. 11.25 Stretching vibrations of BF_3

The total number of vibrational modes in BF_3 is $3 \times 4 - 6 = 6$; out of these, three are stretching vibrations (due to three bonds) and so there are $6 - 3 = 3$ bending modes. Earlier it was shown that the six vibrational modes correspond to $\Gamma_{vib} = A_1' + 2E' + A_2''$.

If we consider only the stretching vibrations (Fig. 11.25) we get a reducible representation of D_{3h}:

	χ (character)						
	$\hat{E}$	$2\hat{C}_3$	$3\hat{C}_2$	$\hat{\sigma}_h$	$2\hat{S}_3$	$3\hat{\sigma}_v$	
$\Gamma_{stretch}$	3	0	1	3	0	1	$= A_1' + E'$

Hence, the bending vibrations correspond to

$$\Gamma_{bend} = \Gamma_{vib} - \Gamma_{stretch} = E' + A_2''$$

One bending mode is of A_2'' symmetry and the remaining two are of symmetry type E' (doubly degenerate). The infrared and Raman frequencies of BF_3 are given in the adjacent table. The stretching vibrations are of higher frequency than bending; moreover, the asymmetric stretch is of higher frequency than the symmetric stretch and the out-of-plane bending is of a higher frequency than in-plane bending. Hence we classify these results as follows:

	Infrared	Raman
$\bar{\nu}_1$	—	888 cm^{-1}
$\bar{\nu}_2$	692 cm^{-1}	—
$\bar{\nu}_3$	1454 cm^{-1}	1454 cm^{-1}
$\bar{\nu}_4$	480 cm^{-1}	480 cm^{-1}

$\bar{\nu}_{stretch}$	$\bar{\nu}_{bend}$
1454 cm^{-1} (asymmetric)	692 cm^{-1} (out-of-plane)
888 cm^{-1} (symmetric)	480 cm^{-1} (in-plane)

To assign symmetry symbols to this classification we now examine the infrared/Raman activity of the vibrational modes. From the D_{3h} character table we find the linear and quadratic basis functions of the irreducible representations corresponding to the vibrational modes:

Irreducible representation	Basis	
	Linear	Quadratic
A_1'		$x^2 + y^2, z^2$
E'	(x, y)	$(x^2 - y^2, xy)$
A_2''	(z)	

Hence, out of the stretching vibrations, A_1' is Raman active (polarised) and infrared inactive while E' is both infrared and Raman active (depolarised). Hence $\bar{\nu}_1$ [$= \bar{\nu}_{stretch}$(symmetric) = 888 cm^{-1}] is of A_1' symmetry and $\bar{\nu}_3$ [$= \bar{\nu}_{stretch}$(asymmetric) = 1454 cm^{-1}] is of E' symmetry. Among the bending vibrations A_2'' is infrared active but Raman inactive while E' is both infrared and Raman active. Hence $\bar{\nu}_2 = \bar{\nu}_{bend}$ (out-of-plane) = 692 cm^{-1} is of A_2'' type, and $\bar{\nu}_4 = \bar{\nu}_{bend}$ (in–plane) = 480 cm^{-1} is of E' symmetry.

11.11.3 $POCl_3$ (Pyramidal)

The molecular point group of $POCl_3$ is C_{3v} (the P=O bond is along the molecular C_3-axis). No symmetry operation of C_{3v} transforms the P=O stretching vector ($\vec{r}_1$) into the P–Cl stretching vectors ($\vec{r}_2, \vec{r}_3, \vec{r}_4$); but the P–Cl stretching vectors are interconvertible by C_{3v} symmetry operations (Fig. 11.26). Hence we have to consider these two types of vectors separately. The representations with the two sets of vectors as basis and their reduction results are as in the adjacent table:

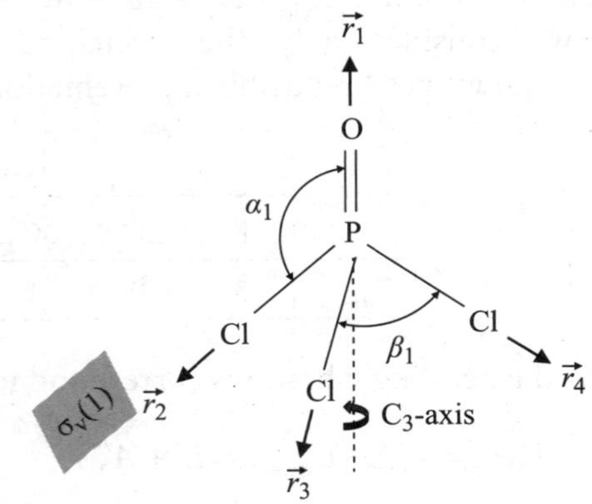

Fig. 11.26 Stretching vectors and bond angles whose changes give the internal coordinates of $POCl_3$. Only one σ_v-plane is shown for the sake of clarity.

A_1 has basis z and also z^2 and $x^2 + y^2$. Hence the A_1 mode is both infrared and Raman active (polarised). E has basis set (x, y) and also $(x^2 - y^2, xy)$ and (xz, yz). Hence the E modes are also both infrared and Raman active (depolarised). The observed experimental data are as follows:

	χ (character)			
	$\hat{E}$	$2\hat{C}_3$	$3\hat{\sigma}_v$	
$\Gamma_{P=O}$	1	1	1	$= A_1$
Γ_{P-Cl}	3	0	1	$= A_1 + E$

	Infrared	Raman
$\bar{\nu}_1$	1291 cm^{-1}	1291 cm^{-1} (polarised)
$\bar{\nu}_2$	581 cm^{-1}	581 cm^{-1} (depolarised)
$\bar{\nu}_3$	487 cm^{-1}	487 cm^{-1} (polarised)
$\bar{\nu}_4$	339 cm^{-1}	339 cm^{-1} (depolarised)
$\bar{\nu}_5$	267 cm^{-1}	267 cm^{-1} (polarised)
$\bar{\nu}_6$	193 cm^{-1}	193 cm^{-1} (depolarised)

Stretching frequencies are higher than the bending ones and also, P=O stretching is expected to be higher than P–Cl stretching (because of lower reduced mass and greater bond force constant of P=O than P–Cl). Moreover, only the vibrational modes of A_1 type have polarised Raman nature. Depending on these facts we assign the symmetry symbols to the stretching vibrations of $POCl_3$ as follows:

1291 cm^{-1} A_1 P=O stretch Both infrared and Raman active (polarised)

581 cm^{-1}	E	P–Cl asymmetric stretch	Both infrared and Raman active (depolarised)
487 cm^{-1}	A_1	P–Cl symmetric stretch	Both infrared and Raman active (polarised)

The internal coordinates which describe the bending vibrations are changes in the Cl–P–O angles α_i, $i = 1, 2, 3$ and the Cl–P–Cl angles β_i, $i = 1, 2, 3$, subject to the condition

$$\beta_1 + \beta_2 + \beta_3 = 2\pi \tag{11.48}$$

For reasons of symmetry, (α_i) and (β_i) are to be treated as separate basis sets because no symmetry operation of C_{3v} can convert any α_i into a β_i. Representations with such basis sets are as in the adjacent table.

χ (character)

	$\hat{E}$	$2\hat{C}_3$	$3\hat{\sigma}_v$	
Γ_α	3	0	1	$= A_1 + E$
Γ_β	3	0	1	$= A_1 + E$

Because of restriction (11.48), all the β_i angles cannot increase or decrease simultaneously. Hence A_1 is not possible with the Cl–P–Cl bendings and hence it is redundant and only two degenerate Cl–P–Cl bendings of type E are possible. There is one Cl–P–O bending of A_1 symmetry and two degenerate Cl–P–O bendings of type E. Because of higher reduced mass, Cl–P–Cl bending is of lower frequency than Cl–P–O bending. Depending on this fact and consulting the C_{3v} character table, the bands resulting from bending are assigned as follows:

267 cm^{-1}	A_1	Cl–P–O bend	Both infrared and Raman active (polarised)
339 cm^{-1}	E	Cl–P–O bend	Both infrared and Raman active (depolarised)
193 cm^{-1}	E	Cl–P–Cl bend	Both infrared and Raman active (depolarised)

11.11.4 PCl_5 (Trigonal Bipyramid)

The molecular point group is D_{3h} and the total number of vibrational modes = 3 × 6 − 6 = 12. There are five bonds and five P–Cl stretching modes of which three are equatorial P–Cl^e and two are axial P–Cl^a stretches. Hence there are 12 − 5 = 7 bending modes. No symmetry operation of D_{3h} converts an axial bond into an equatorial one (Fig. 11.27).

Hence we can consider the axial and equatorial stretches separately. The representations of D_{3h} spanned by the axial and equatorial stretching vectors [$\vec{r}_1^{\,a}$, $\vec{r}_2^{\,a}$ and $\vec{r}_1^{\,e}$, $\vec{r}_2^{\,e}$, $\vec{r}_3^{\,e}$] are as follows.

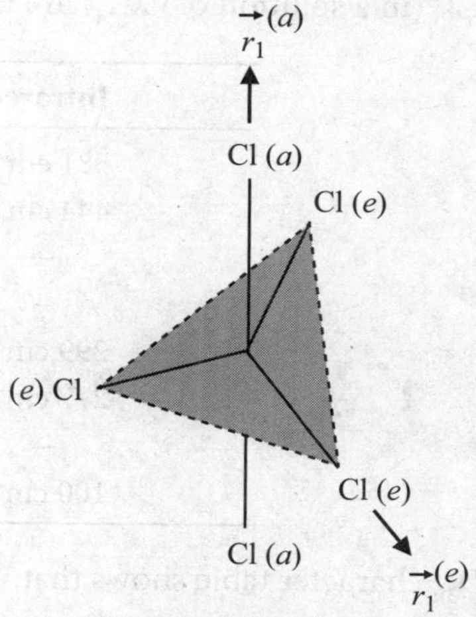

Fig. 11.27 Axial and equatorial bond stretching vectors of PCl_5. Only one stretching vector of each type is shown for clarity.

χ (character)

	$\hat{E}$	$2\hat{C}_3$	$3\hat{C}_2$	$\hat{\sigma}_h$	$2\hat{S}_3$	$3\hat{\sigma}_v$	
$\Gamma^a_{\text{stretch}}$	2	2	0	0	0	2	$= A_1' + A_2''$
$\Gamma^e_{\text{stretch}}$	3	0	1	3	0	1	$= A_1' + E'$

There are two types of angles, Cl^e–P–Cl^e and Cl^e–P–Cl^a, denoted respectively as $\alpha_i^{e,e}$, $i = 1, 2, 3$ and $\alpha_i^{e,a}$, $i = 1, 2, \ldots, 6$ (Fig. 11.28).

Changes in these angles may be taken as internal coordinates with the following restrictions:

$$\alpha_1^{e,e} + \alpha_2^{e,e} + \alpha_3^{e,e} = 2\pi \tag{11.49}$$

$$\alpha_1^{e,a} + \alpha_2^{e,a} + \cdots + \alpha_6^{e,a} = 3\pi \tag{11.50}$$

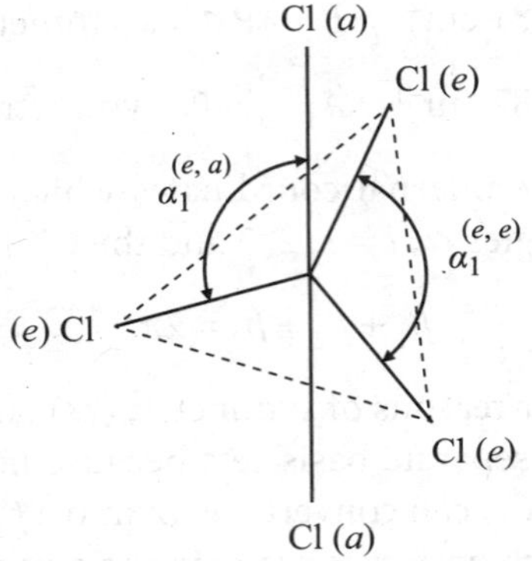

Fig. 11.28 Two types of bond angles whose changes give internal coordinates for the bending vibrations. One of each type is shown for clarity.

No symmetry operation of D_{3h} can convert any $\alpha_i^{e,e}$ into an $\alpha_i^{e,a}$. Hence each type of bond angle should be regarded as forming a separate basis set for representing D_{3h}. Representations with basis set $\{\alpha_i^{e,e}, i = 1, 2, 3\}$ and $\{\alpha_i^{e,a}, i = 1, 2, \cdots, 6\}$ along with their reductions are as follows:

	χ (character)						
	$\hat{E}$	$2\hat{C}_3$	$3\hat{C}_2$	$\hat{\sigma}_h$	$2\hat{S}_3$	$3\hat{\sigma}_v$	
$\Gamma_{\text{bend}}^{e,a}$	6	0	0	0	0	2	$= A_1' + E' + A_2'' + E''$
$\Gamma_{\text{bend}}^{e,e}$	3	0	1	3	0	1	$= A_1' + E'$

(The characters in this table can be easily obtained by noting which angles remain unchanged and which ones are changed by a symmetry operation.)

The experimentally observed infrared and Raman data (within small experimental errors) for PCl_5 (in a solution of CCl_4) are the following:

Infrared	Raman
581 cm^{-1}	579 cm^{-1} (depolarised)
444 cm^{-1}	–
–	393 cm^{-1} (polarised)
–	385 cm^{-1} (polarised)
299 cm^{-1}	–
277 cm^{-1}	279 cm^{-1} (depolarised)
–	261 cm^{-1} (depolarised)
100 cm^{-1}	98 cm^{-1} (depolarised)

The D_{3h} character table shows that:

- A_1' does not have $x/y/z$ as basis but has z^2 and $x^2 + y^2$ as basis. Thus the A_1' modes are infrared inactive and Raman active. Also, they give polarised Raman spectra.
- A_2'' has z as basis but not any quadratic function of coordinates; so this mode is infrared active, Raman inactive.
- E' has basis (x, y) and also $(x^2 - y^2, xy)$; these two (degenerate) modes are both infrared and Raman active.

- E'' has no $x/y/z$ but has (xy, xz) as basis; these two (degenerate) modes are infrared inactive but Raman active.

In the case of bending vibrations, the symmetric A_1' mode is not possible owing to the restrictions (11.49) and (11.50); all the angles (Cl^e–P–Cl^e or Cl^e–P–Cl^a) cannot simultaneously increase or decrease. Hence the A_1' representations for the bending modes found in $\Gamma^{e,e}_{\text{bend}}$ and $\Gamma^{e,a}_{\text{bend}}$ are redundant.

Stretching frequencies are higher than bending frequencies; asymmetric stretches are of higher frequency than symmetric stretches. Also, P–Cl^a (axial) stretching is of a lower frequency than P–Cl^e (equatorial) stretching, because in both cases the reduced mass is the same but the axial bonds are longer and hence have a lower force constant than equatorial bonds. Thus we assign symmetry symbols to the stretching frequencies as follows:

393 cm^{-1}	A_1'	P–Cl^e symmetric stretch	Infrared inactive, Raman active
385 cm^{-1}	A_1'	P–Cl^a symmetric stretch	Infrared inactive, Raman active
444 cm^{-1}	A_2''	P–Cl^a asymmetric stretch	Infrared active, Raman inactive
579/581 cm^{-1}	E'	P–Cl^e stretch	Both infrared and Raman active
299 cm^{-1}	A_2''	Cl^e–P–Cl^a bending	Infrared active, Raman inactive
277/279 cm^{-1}	E'	Cl^e–P–Cl^a bending	Both infrared and Raman active
261 cm^{-1}	E''	Cl^e–P–Cl^a bending	Infrared inactive, Raman active
98/100 cm^{-1}	E'	Cl^e–P–Cl^e bending	Both infrared and Raman active

(Only the modes of A_1' type give polarised Raman spectra.)

11.12 ASSIGNMENT OF SYMMETRY LABELS TO COMBINATION BANDS AND OVERTONES IN VIBRATIONAL SPECTRA

Experimentally obtained infrared and Raman spectra most often contain, besides the fundamental bands predicted from the molecular geometry, many peaks of relatively lower intensity. Assignment of symmetry labels to all of them is difficult (almost impossible) but sometimes quite a few of these additional weak bands can be recognised as overtones and combination bands.

If the kth normal mode is excited to its $\text{v}_k = 2$ level, the resulting absorption band is the first overtone; excitations to the 3rd, 4th, ... vibrational quantum level of this mode result respectively in the 2nd, 3rd, ... overtones. If the fundamental band is denoted by ν_k, the overtone bands are denoted by $2\nu_k$, $3\nu_k$, ... Two points need to be mentioned here:

- If the fundamental band appears at frequency ν_k, the nth overtone appears at a frequency near but somewhat less than $(n + 1)\nu_k$. This is because, for any normal mode, the simple harmonic approximation does not hold good beyond the fundamental level ($\text{v}_k = 1$); the vibrational energy levels become more and more closely spaced as the extent of anharmonicity increases with increase in vibrational quantum number.

- If the simple harmonic approximation holds, the probability of transition to the levels beyond $v_k = 1$ is strictly zero and no overtone band should appear. Some overtones do appear, however, with low intensity; the anharmonic perturbation makes the transitions weakly allowed.

A 'combination band' in infrared/Raman spectra appears if two different modes are simultaneously excited to their fundamental levels; such a band appears at a frequency equal to (somewhat less than) the sum of the fundamental frequencies. Whether a combination band or overtone is symmetry-allowed or not can be judged as follows:

Let ν_a and ν_b be the fundamental frequencies of two different vibrational modes of normal coordinates Q_a and Q_b respectively. Let us consider a combination level composed of these two fundamentals. The wave function for this vibrational state of the molecule is

$$\begin{aligned}\Psi &= [\psi_0(Q_1)\cdot\psi_0(Q_2)\cdot\ldots\cdot\psi_1(Q_a)\cdot\psi_0(Q_{a+1})\ldots\psi_1(Q_b)\cdot\psi_0(Q_{b+1})\ldots\psi_0(Q_s)]\\ &= [\psi_0(Q_1)\cdot\psi_0(Q_2)\cdot\ldots 2\sqrt{\alpha_a}\cdot\psi_0(Q_a)Q_a\cdot\psi_0(Q_{a+1})\ldots\\ &\quad 2\sqrt{\alpha_b}\cdot\psi_0(Q_b)Q_b\cdot\psi_0(Q_{b+1})\ldots\psi_0(Q_s)]\\ &= \text{constant}\times Q_aQ_b\Psi_0\end{aligned}$$

The ground state wave function Ψ_0 transforms according to the totally symmetric representation ($D^{(1)}$) of the group. Hence Ψ transforms according to the direct product representation $D^a \times D^b$ (where D^a and D^b are respectively the irreducible representations according to which the normal coordinates Q_a and Q_b transform). Thus the infrared/Raman activity of a combination level $\nu_a + \nu_b$ is determined by the direct product representation $D^a \times D^b$; if on reduction it is found to contain an irreducible representation with $x/y/z$ as basis, it is infrared active; if any component irreducible representation of $D^a \times D^b$ contains a quadratic function of cartesian coordinates as basis, the combination level is Raman active.

For overtones, the rule is similar. For example, the $2\nu_a$ band is to be treated as $\nu_a + \nu_a$ and the required direct product is $D^a \times D^a$. Since the direct product of any irreducible representation with itself must contain the totally symmetric representation $D^{(1)}$, and since $D^{(1)}$ for any group has quadratic functions of coordinates as basis, *all first overtones are Raman allowed.*

It should be mentioned at this point that fulfilment of the above mentioned conditions does not ensure the appearance of an overtone or combination band because transitions to such excited levels from the ground state are only weakly allowed by anharmonicity. Rather, if a combination or overtone band is suspected in a spectrum, it can be justified by the group theoretical criteria mentioned above. *Group theory can allow or forbid but cannot predict an overtone or a combination band.*

Some examples will now be considered with the possible overtones and combination bands of methane.

It was found in Section 11.6.5 that the tetrahedral AB_4 molecule shows two degenerate fundamentals of symmetry species T_2 in both infrared and Raman. For methane they have been experimentally found at 3020 and 1306 cm^{-1}. To examine whether their first overtones are permitted or not we construct the direct product representation $T_2 \times T_2$ of the point group T_d and reduce:

T_d	$\hat{E}$	$8\hat{C}_3$	$3\hat{C}_2$	$6\hat{S}_4$	$6\hat{\sigma}_d$
T_2	3	0	−1	−1	1
$T_2 \times T_2$	9	0	1	1	1

On reduction it is found that $T_2 \times T_2 = A_1 + E + T_1 + T_2$

From the basis functions of the component irreducible representations of $T_2 \times T_2$ as found from character tables, we get the following results:

Mode	**Infrared**	**Raman**	**Basis of representation**
A_1	Inactive	Active	$x^2 + y^2, z^2$
E	Inactive	Active	$(2z^2 - x^2 - y^2, x^2 - y^2)$
T_1	Inactive	Inactive	No linear or quadric function of coordinates
T_2	Active	Active	$(x, y, z), (xy, yz, xz)$

Hence both the T_2 fundamentals of methane are expected to show their first overtones in the infrared spectrum. Experimentally two such overtones at 6006 cm^{-1} ($\approx 2 \times 3020$ cm^{-1}) and 2600 cm^{-1} ($\approx 2 \times 1306$ cm^{-1}) have been observed.

The only non-degenerate fundamental of A_1 symmetry is found at 2914 cm^{-1}. Since $A_1 \times A_1 = A_1$, which is Raman allowed, an overtone at 5828 cm^{-1} is expected in the Raman spectrum of CH_4, but this has not been observed experimentally. This is an example where group theory is *permissive but not predictive* for overtones.

11.13 FERMI RESONANCE IN VIBRATIONAL SPECTRA

Sometimes in infrared or Raman spectra of molecules, a fundamental band is found to be much less intense, and at the same time, a combination or overtone band is more intense than expected from the usual selection rules. Also, the separation in frequency between the fundamental and the overtone (or, combination) band is found to be somewhat greater than expected. Such a phenomenon was first explained by Enrico Fermi to be caused by a symmetry-induced interaction between a fundamental and a combination (or, overtone) level. It is referred to as *Fermi resonance* and it occurs if two conditions are satisfied: (i) an active fundamental level has energy close to a

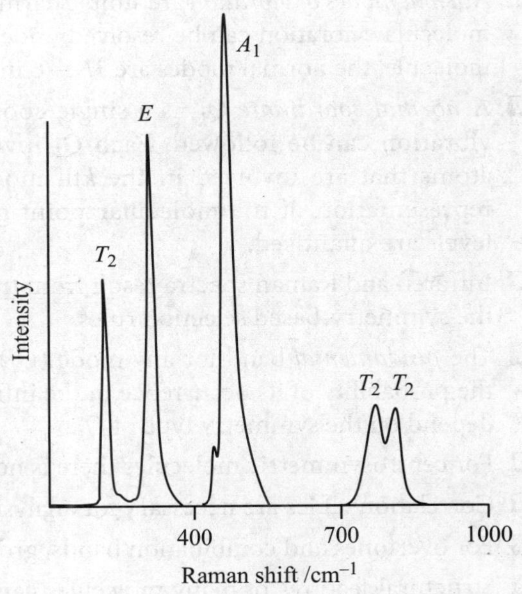

Fig. 11.29 Vibrational Raman spectrum of liquid CCl_4

combination (or overtone) level and (ii) the corresponding excited state wave functions are of the same symmetry species. Under such conditions, the anharmonic terms in the Taylor series expansion of the potential energy V, cause a mixing of the excited states. Such a mixing has the effect mentioned at the beginning (namely, frequency separation and 'borrowing' of intensity by a combination/overtone band from a nearby fundamental).

A familiar example of Fermi resonance is found in the Raman spectrum of liquid CCl_4. The molecular point group is T_d and the vibrational modes are of symmetry type $A_1 + E + 2T_2$ (as shown in the case of CH_4). All the modes are Raman active and four Raman bands are expected. But experimentally five bands are observed (Fig. 11.29). These have been assigned as follows:

A_1	T_2	E	T_2
459 cm^{-1}	315 cm^{-1}	217 cm^{-1}	762 cm^{-1}, 791 cm^{-1}

The first three signals at 459 cm^{-1}, 315 cm^{-1} and 217 cm^{-1} are as expected. The fourth one of symmetry T_2 is predicted to occur at approximately 775 cm^{-1}; but the combination of 459 cm^{-1}(A_1) and 315 cm^{-1} (T_2) occurs at (459 + 315) cm^{-1}, that is, 774 cm^{-1} and its symmetry, the direct product $A_1 \times T_2 = T_2$, is the same as that of the expected fourth band. Hence a Fermi resonance occurs, giving two peaks (762 and 791 cm^{-1}) of equal intensity which are approximately equidistant from the predicted fourth band at 775 cm^{-1}. This *Fermi doublet* has also been found in the infrared spectrum of CCl_4.

SUMMARY

- *Normal modes of vibration* are simple harmonic motions into which the overall complicated molecular vibration can be resolved under a reasonable approximation. For an N-atomic molecule, the normal modes are $3N - 6$ in number ($3N - 5$ for linear molecules).
- A *normal coordinate* Q_k, is a single coordinate along which the kth normal mode of vibration can be followed. Each Q_k involves the mass-weighted displacement of the atoms that are involved in the kth mode and acts as the basis of some irreducible representation of the molecular point group. For each mode the vibrational energy levels are quantised.
- Infrared and Raman spectra result from transition between vibrational levels governed by the symmetry-based selection rules.
- The *fundamental* band for any mode (Q_k) results from the $v = 0 \rightarrow v = 1$ transition and the probability of its occurrence in the infrared/Raman spectra and nature of polarisation depend on the symmetry type of Q_k.
- For centrosymmetric molecules there is no infrared–Raman coincidence.
- Correlation tables are necessary for analysing vibrational spectra of linear molecules.
- For overtones and combination bands, group theory is permissive but not predictive.
- Structural features of many molecules can be obtained by a group theoretical analysis of their infrared and Raman spectra.

LEARNING OUTCOME

After reading this chapter the student will be able to apply group theory to (i) find the symmetry types and infrared/Raman activity of the normal modes of vibration of a molecule, (ii) predict the number of fundamental bands, (iii) construct point group–subgroup correlation tables for application to linear molecules having infinite point groups, (iv) examine the feasibility of a suspected overtone or combination band and (v) interpret infrared/Raman spectra in terms of internal coordinates and thereby extract important structural features.

EXERCISES

1. (a) Consider only the A–X stretching vibrations of the following molecules, find their symmetry types and deduce their infrared and Raman activity using necessary character tables.

(i) AX_3 (planar) (ii) AX_3 (pyramidal) (iii) AX_4 (square planar) (iv) AX_4 (pyramidal) (v) AX_5 (square pyramidal, with A at the centre of the square base) (vi) AX_5 (trigonal bipyramidal) (vi) AX_6 (octahedral)

(b) Determine the symmetry and activity of the normal modes of vibration of the following molecules

(i) CO_3^{2-} (ii) $[PtCl_4]^{2-}$ (iii) *trans*-glyoxal (iv) *trans*-N_2F_2.

2. The symmetry of the octahedral complex $[CoCl_6]^{3-}$ is reduced to D_{4h} by some distortion. Find the symmetries of the infrared and Raman active modes of the complex in the lower symmetry. Explain how these spectra can be used to distinguish this complex from its original octahedral form.

3. Ignoring the H-atoms of the octahedral $[Cr(NH_3)_6]^{3+}$ complex, find the irreducible representations spanned by the vibrational modes of the ion and predict their infrared and Raman activity. In the process, verify that some vibrational modes are silent.

4. Find the symmetry types of the infrared and Raman active vibrational modes of the fumarate ion considering that all the atoms of the ion lie in one plane.

5. Find the number, infrared/Raman activities and Raman polarisation of all the normal modes of vibration of the planar $Pd_2Cl_6^{2-}$ ion having the following structure:

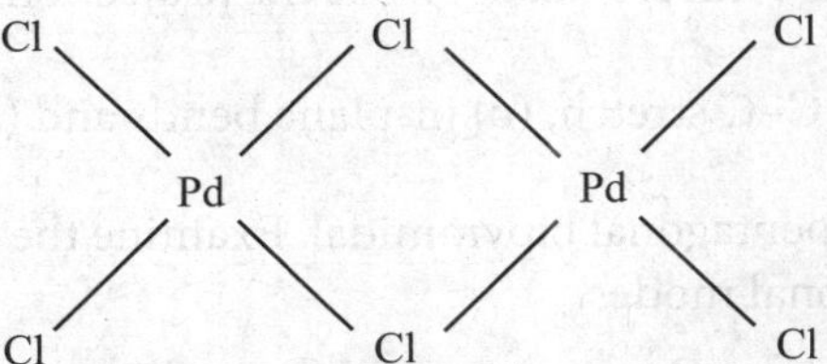

6. Justify the following statements group theoretically:

(a) The Raman spectrum of N_2 shows a band centred at 2331 cm^{-1} but its infrared spectrum shows nothing.

(b) Any hetronuclear diatomic molecule exhibits both infrared and Raman spectra.

7. Explain how you can distinguish between the *cis* and *trans* isomers of the complex $M[L_2(CO)_4]$ by analysing group theoretically the carbonyl stretching region of the infrared and Raman spectra of the complex.

8. The C–O stretching vibration of the compound $Mo(CO)_3[P(OMe)_3]_3$ exhibits bands at 1993, 1919 and 1890 cm^{-1}. Which isomer is it, *fac* or *mer*? Ans. *mer*

9. The complex $Cr(CO)_3(CNCH_3)_3$ shows absorption bands at 1942 and 1860 cm^{-1} for the carbonyl stretch. Which structure, *mer* or *fac*, should be assigned to it? Ans. *fac*

10. For the complex OsO_4N (N = pyridine) the following results were obtained in the Os–O stretching region:

    Infrared (cm^{-1}): 926, 915, 908, 885

    Raman (cm^{-1}): 928 (p), 916(p), 907(p), 886(dp)

    Find which of the following structures is consistent with the given data: (a) trigonal bipyramidal with the N atom at an axial position (b) trigonal bipyramidal with N at one of the equatorial positions (c) square pyramidal with N at the apex of the pyramid (d) square pyramidal with N at a vertex of the square base. Ans: (d)

11. A molecule of the general formula AB_3 may be planar or non-planar. Which vibrations are infrared active and which ones are Raman active in each case?

12. In the Xe–F stretching region (500–700 cm^{-1}) one allowed infrared band and two allowed Raman bands are expected for the XeF_4 molecule. Why?

13. Which vibrational mode of CO_2 will show no change in frequency on substituting ^{12}C by ^{13}C? Is this mode infrared active or Raman active?

14. Apply the projection operator method to sketch the in-plane bending vibrations of XeF_4.

15. Determine the number and symmetry types of the normal modes of vibration of the ethylene molecule. (Choose the *X*-axis to be out of plane and the *Z*-axis along the C=C bond.) Which modes are infrared active? Use the projection operator method to find the symmetry of the
    (a) C–H stretches, (b) C–C stretch, (c) in-plane bends and (d) out-of-plane bends.

16. The structure of IF_7 is pentagonal bipyramidal. Examine the infrared and Raman activity of the possible vibrational modes.

17. 1, 1-dichloroethylene is planar with a C_{2v} symmetry, the C=C coinciding with the C_2-axis. Take this axis as the Z-axis, the YZ plane as the molecular plane and examine the infrared and Raman activity of the possible modes vibration.

18. Methane shows the following fundamental frequencies in infrared and Raman spectra:

Infrared	Raman
3020 cm^{-1}	3020 cm^{-1}
—	2914 cm^{-1}
1526 cm^{-1}	1526 cm^{-1}
1306 cm^{-1}	—

Assign symmetry symbols to the vibrational modes which are responsible for these transitions.

19. Ozone has more than one fundamental band in its infrared spectrum. Show that the molecular configuration of O_3 cannot be an equilateral triangle.
[**Hint:** D_{3h}; $\Gamma_{vib} = A_1' + E'$. Only E' is infrared active—one infrared band is expected if the configuration is an equilateral triangle.]

20. Determine the number and symmetry types of the vibrational modes of *trans*-difluoroethylene molecule. Determine the infrared and Raman activity of each mode. Determine the symmetry of each of the following internal modes:
(a) C–F stretching (b) Out-of-plane bending
(c) H–C–F angle deformation (d) C–C stretching.

21. Determine the number of infrared active CH stretching modes in each of the following cyclic hydrocarbon molecules/ions:

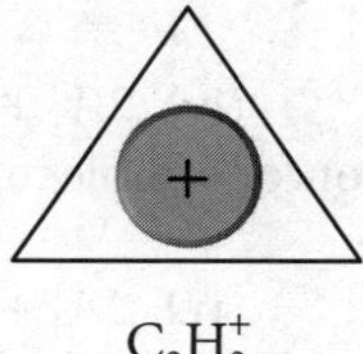

$C_3H_3^+$

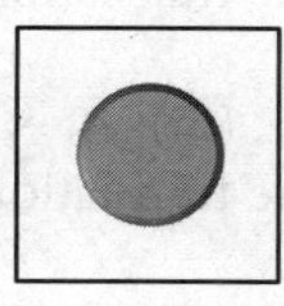
C_4H_4

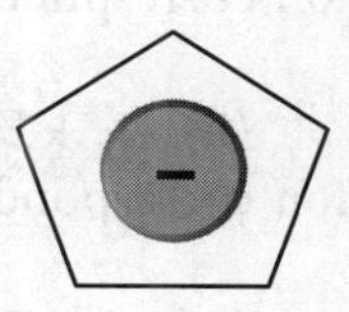

$C_5H_5^-$

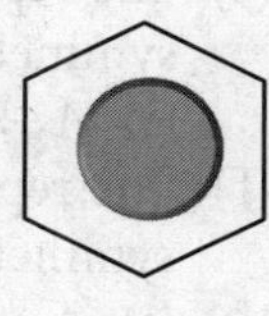
C_6H_6

22. Find the number of infrared and Raman active stretching vibrations of the terminal B–H bonds in diborane.

23. Obtain the correlations between the point groups $C_{\infty v}$ and C_{2v}. Utilise them to find the symmetry types and infrared/Raman activity of the vibrations of H–C≡C–C≡N.

24. Find the symmetry types and infrared/Raman activity of the molecular vibrations of acetylene by the method of point group–subgroup correlation (use Table 11.4).

25. Analyse the stretching and bending vibrations of SO_3^{2-} ion assuming a planar and also a pyramidal structure. Which geometry is consistent with the experimental observation that it shows four fundamental absorption bands in its infrared spectrum?

26. Obtain the correlations between the point groups D_{3h} and C_{3v}. A planar AX_3 molecule is slightly distorted by pulling the A atom above the plane of the three *X* atoms. Utilise the correlations to show how the number of fundamental bands changes in the infrared and Raman spectra by this distortion.

27. Obtain the correlations between the point groups D_{4h} and D_{2d} in the following cases:
(a) $2C_2'$ and $2\sigma_d$ of D_{4h} become $2C_2'$ and $2\sigma_d$ respectively of D_{2d};
(b) $2C_2''$ and $2\sigma_v$ of D_{4h} become $2C_2'$ and $2\sigma_d$ respectively of D_{2d}.

28. In the infrared spectrum of CH_4 two absorption bands at 3020 and 1306 cm^{-1} have been assigned the symmetry label T_2. Two relatively weak bands are observed at 4313 and 1720 cm^{-1}. Examine whether the last two can be combination/difference bands of the former two.

29. Choose the correct option:
(a) For CO_2 the bending vibrations are of symmetry type
(i) A_{1g} (ii) Π_u (iii) Σ_g^+ (iv) A_{2u}
(b) The stretching modes of XeF_4 give a representation which reduces to
(i) $A_1 + B_2 + E$ (ii) $A' + A'' + E'$ (iii) $A_{1g} + B_{1g} + E_u$ (iv) $2A_1 + 2B_2$
(c) A Raman scattering experiment is performed on liquid CCl_4 with plane polarised light. $I_\perp$ and $I_\parallel$ are respectively the intensities of the Raman scattered light polarised in perpendicular and parallel directions with respect to the incident beam. The vibrations which causes $I_\perp < 0.75 I_\parallel$ is of symmetry type
(i) T_2 (ii) T_1 (iii) A_1 (iv) E
(d) Raman polarisation experiment with water shows that there are
(i) two polarised and one depolarised (ii) one polarised and two depolarised
(iii) three depolarised (iv) three polarised vibrational bands.
(e) The representation corresponding to the translational motion of a molecule of C_{3v} symmetry is
(i) $A_2 + E$ (ii) $A_1 + E$ (iii) $A_1 + 2A_2$ (iv) $2A_1 + A_2$
(f) The representation corresponding to the translational motion of a molecule of C_{2v} symmetry is
(i) $A_1 + A_2 + B_2$ (ii) $A_1 + B_1 + B_2$ (iii) $A_2 + B_1 + B_2$ (iv) $2A_1 + B_2$

Suggested Books and References for Further Reading

1. Herzberg G, *Infrared and Raman Spectra of Polyatomic Molecules*, Van Nostrand Reinhold, NY, **1945.**
2. Harris D C and Bertolucci M D, *Symmetry and Spectroscopy: An introduction to Vibrational and Electronic Spectra,* Oxford University Press, Inc., **1978.**
3. Wilson E B, Decius J C and Cross P C, Molecular Vibrations, McGraw-Hill, NY, **1955.**
4. El-Sayed M A and Kaesz H D, *J. Mol. Spec.*, 9 (310) **1962.**
5. Claassen H H, Chernick C L and Malm J G, *J. Am. Chem. Soc.*, 85 (1927) **1963.**
6. Kroto H W, et. al., *Spectrochim. Acta A*, 47 (1289) **1991.**

12 Spin–Orbit Coupling: Term Symbols

Learning objectives

- ➢ Vector atom model for spin–orbit interaction in atoms
- ➢ Atomic term symbols and energy levels from *LS* and *jj* coupling scheme
- ➢ Energy order of the terms in the absence and presence of magnetic field – Hund's rules
- ➢ Landé interval rule for multiplets
- ➢ Obtaining all possible terms for d^n systems with $n \geq 3$ – the hole formalism
- ➢ Molecular term symbols

12.1 INTRODUCTION

Crystal field and ligand field theories for transition metal complexes involve interesting applications of group theory. A preliminary knowledge of spin–orbit interaction in atoms and transition metal ions is required for this purpose. In this chapter the necessary background information is outlined briefly.

The spin angular momentum ($\vec{S}$) of an electron gives rise to a magnetic moment ($\vec{\mu}_s$). The orbital motion of the electron about a positively charged nucleus amounts to a relative motion of the nucleus about the electron, that is, an electric current in a closed loop, which produces a magnetic field ($\vec{H}$) on the electron. The interaction between these two (which results in an interaction energy proportional to $\vec{\mu}_s \cdot \vec{H}$) is called *spin–orbit coupling*. This phenomenon manifests itself in various ways, the most important of which are (i) appearance of fine structures in the high resolution spectra of atoms and (ii) splitting of spectral lines in applied magnetic and electric fields (Zeeman and Stark effect, respectively). Spin–orbit interaction is thus a magnetic interaction.

For the hydrogen atom or a hydrogen-like system, the Schrödinger equation can be solved exactly only if the spin–orbit (S–O) interaction is neglected. The solutions so obtained are the hydrogenic orbitals which act as good starting points for considering the S–O interaction as a perturbation. Such a treatment leads to a number of electronic energy levels (or states) for the atom and these are determined by the combined effect (coupling) of the spin and orbital angular momenta of the electron.

For many-electron atoms, the situation is even more complex. Firstly, the spin angular momentum of an individual electron interacts not only with its own orbital angular momentum but also with those of the others. Secondly, the Schrödinger equation cannot be exactly solved even if the S–O interaction is completely disregarded. The method of dealing with this problem has been so designed that the results of the hydrogen atom problem, where the electrostatic potential energy is of a central field nature,* can be utilised as much as possible. In the final analysis the energy levels of many-electron atoms depend, as in the hydrogen atom case, on the resultant of the orbital and spin angular momenta of all the electrons.

[*If the force derived from a potential is always directed towards a central point, the potential is said to be of a central field nature; an example is the electrostatic potential of the electron in a hydrogen atom.]

12.2 SPIN–ORBIT COUPLING IN ONE-ELECTRON SYSTEMS

Including the spin–orbit interaction (and excluding hyperfine interaction) the Hamiltonian operator for a hydrogen-like system can be written as

$$\hat{H} = \hat{H}_1 + \hat{H}^{SO} \tag{12.1}$$

where the potential energy from spin–orbit interaction is included in $\hat{H}^{SO}$ and is given by

$$\hat{H}^{SO} = \xi(r)\, \hat{L} \cdot \hat{S} \tag{12.2}$$

Here $\xi(r)$ is a quantity which depends on the electron–nucleus distance r and the nuclear charge Z. The operator $\hat{L} \cdot \hat{S}$ corresponds to the vector dot product of the orbital and spin angular momenta, $\vec{L}$ and $\vec{S}$, of the electron. The first part ($\hat{H}_1$) of the Hamiltonian has the well known form,

$$\hat{H}_1 = -\frac{\hbar^2}{2m}\left(\frac{\partial^2}{\partial x^2} + \frac{\partial^2}{\partial y^2} + \frac{\partial^2}{\partial z^2}\right) + \left(\frac{-Ze^2}{4\pi\varepsilon_0 r}\right) \tag{12.3}$$

The electrostatic potential energy in $\hat{H}_1$ is of a central field nature and the Schrödinger equation

$$\hat{H}_1 \Psi_n = E_n^{(0)} \Psi_n \tag{12.4}$$

has exact solutions. The eigenfunctions of $\hat{H}_1$ are functions of the electron coordinates and have parametric dependence on three quantum numbers, n, l and m_l:

$$\Psi_n(r, \theta, \phi) = R_{n,l}(r) Y_l^{m_l}(\theta, \phi) \tag{12.5}$$

Here $R_{n,l}$ is the associated Laguerre function and $Y_l^{m_l}$ is a spherical harmonic function. The latter can be further factorised as

$$Y_l^{m_l}(\theta, \phi) = N \cdot P_l^{m_l}(\cos\theta) \cdot e^{im_l\phi} \tag{12.6}$$

where N is a normalisation factor and $P_l^{m_l}(\cos\theta)$ is the associated Legendre polynomial in $\cos\theta$ of degree l and order m_l. The permissible values, $l = 0, 1, 2, \ldots, (n-1)$ and $m_l = (0, \pm 1, \pm 2, \ldots, \pm l)$ are respectively called *azimuthal* and *magnetic* quantum numbers of the electron and they appear during the mathematical process of solving Eq. (12.4) as necessary integers. The part $Y_l^{m_l}(\theta, \phi)$ of Ψ_n (in Eq. 12.5) is also an eigenfunction of two operators related to the orbital angular momentum of the electron, namely, $\hat{L}^2$ and $\hat{L}_z$ with eigenvalues $l(l+1)\hbar^2$ and $m_l\hbar$ respectively. The operator $\hat{H}_1$ commutes with $\hat{L}^2$ and also with the three Cartesian component operators $\hat{L}_x$, $\hat{L}_y$ and $\hat{L}_z$ of the orbital angular momentum. But $\hat{L}_x$, $\hat{L}_y$ and $\hat{L}_z$ do not commute among themselves (that is, $\hat{L}_x\hat{L}_y - \hat{L}_x\hat{L}_y \neq 0$, and so on). Hence the operators $\hat{H}_1$, $\hat{L}^2$ and any one of $\hat{L}_x$, $\hat{L}_y$, $\hat{L}_z$ should have simultaneous eigenfunctions. It is customary to choose $\hat{L}_z$.

Experiments reveal that in addition to orbital angular momentum, the electron has an intrinsic angular momentum called *spin*. This spin has no 'picture' in conventional (classical) mechanics (like the spinning of a top about its own axis). However, all experimental findings can be explained if it is *postulated* that the intrinsic (spin) angular momentum is describable by means of operators and commutation rules analogous to those of the orbital angular momentum and that an eigenfunction χ_S exists for the spin. Thus, it is postulated that:

- analogous to the orbital angular momentum operator $\hat{L} = \hat{L}_x\vec{e}_1 + \hat{L}_y\vec{e}_2 + \hat{L}_z\vec{e}_3$, there is the spin angular momentum operator $\hat{S} = \hat{S}_x\vec{e}_1 + \hat{S}_y\vec{e}_2 + \hat{S}_z\vec{e}_3$ ($\vec{e}_1$, $\vec{e}_2$ and $\vec{e}_3$ are unit vectors along the coordinate axes),
- analogous to $\hat{L}^2$, there is $\hat{S}^2$: $\hat{L}^2\, Y_l^{m_l}(\theta,\phi) = l(l+1)\hbar^2\; Y_l^{m_l}(\theta,\phi)$

$$\hat{S}^2\,\chi_S = s(s+1)\hbar^2\,\chi_S \text{ where } s = \tfrac{1}{2},$$

- analogous to $\hat{L}_z$, there is $\hat{S}_z$: $\hat{L}_z\, Y_l^{m_l}(\theta,\phi) = m_l\hbar\, Y_l^{m_l}(\theta,\phi)$

$$\hat{S}_z\,\chi_S = m_s\hbar\,\chi_S \text{ where } m_s = \tfrac{1}{2} \text{ or } -\tfrac{1}{2},$$

- analogous to the commutation rules $[\hat{H}, \hat{L}^2] = [\hat{H}, \hat{L}_z] = 0$, $[\hat{L}_x, \hat{L}_y] \neq 0 \neq [\hat{L}_y, \hat{L}_z]$, there exist the rules: $[\hat{H}, \hat{S}^2] = [\hat{H}, \hat{S}_z] = 0$, $[\hat{S}_x, \hat{S}_y] \neq 0 \neq [\hat{S}_y, \hat{S}_z]$.

Hence the operators $\hat{H}_1$, $\hat{L}^2$, $\hat{L}_z$, $\hat{S}^2$ and $\hat{S}_z$ have a simultaneous eigenfunction of the form

$$\Psi = R_{n,l}(r)Y_l^{m_l}(\theta,\phi)\chi_S \tag{12.7}$$

[This can be readily verified: for example

$$\hat{H}_1\Psi = [\hat{H}_1 R_{n,l}(r)Y_l^{m_l}(\theta,\phi)]\chi_S = E_n^{(0)}R_{n,l}(r)Y_l^{m_l}(\theta,\phi)\chi_S$$

$$= E_n^{(0)}\,\Psi \qquad \text{(as } \hat{H}_1 \text{ does not contain any coordinate related to the spin)}$$

$$\hat{L}_z\Psi = R_{n,l}(r)[\hat{L}_z\,Y_l^{m_l}(\theta,\phi)]\chi_S = R_{n,l}(r)N\,P_l^{m_l}(\cos\theta)[\hat{L}_z\,e^{im_l\phi}]\,\chi_S$$

$$= m_l\,\hbar\,R_{n,l}(r)N\,P_l^{m_l}(\cos\theta)e^{im_l\phi}\,\chi_S$$

$$= m_l\hbar\Psi \qquad \left(\text{as } \hat{L}_z = -i\hbar\frac{\partial}{\partial\phi} \text{ contains no coordinate other than } \phi\right)$$

$$\hat{S}_z\Psi = R_{n,l}(r)Y_l^{m_l}(\theta,\phi)\hat{S}_z\,\chi_S = m_s\hbar\,R_{n,l}(r)N\,P_l^{m_l}(\cos\theta)e^{im_l\phi}\,\chi_S = m_s\hbar\Psi\;,\ldots]$$

The function $\Psi_n(r,\theta,\phi) = R_{n,l}(r)Y_l^{m_l}(\theta,\phi)$ is called an *orbital* and it is an exact solution of the truncated Schrödinger equation in which the spin–orbit interaction has been disregarded (along with other magnetic interactions). It is taken as the starting wave function for a perturbation calculation with the operator $\hat{H}^{SO}$ as the perturbing Hamiltonian. The most important point in this perturbation treatment is that the operator $\hat{L}\cdot\hat{S}$ (in $\hat{H}^{SO}$) does not commute with any of $\hat{L}_z$ or $\hat{S}_z$ but it **does** commute with $\hat{L}_z + \hat{S}_z$ and also with the operator $\hat{J}^2$ $(=\hat{J}\cdot\hat{J})$ where $\hat{J}$ is the operator for the resultant angular momentum, $\vec{J} = \vec{L} + \vec{S}$:

$$\hat{L}\cdot\hat{S} = \hat{L}_x\hat{S}_x + \hat{L}_y\hat{S}_y + \hat{L}_z\hat{S}_x \tag{12.8}$$

The spin angular momentum and its components do not have any space coordinate (x, y, z) or (r, θ, φ); so each of the component operators $\hat{S}_x$, $\hat{S}_y$ and $\hat{S}_z$ commutes with the operators $\hat{L}_x$, $\hat{L}_y$, $\hat{L}_z$ and $\hat{L}^2$:

$$\hat{L}_\mu \hat{S}_\nu = \hat{S}_\nu \hat{L}_\mu \quad \text{and} \quad \hat{L}^2 \hat{S}_\nu = \hat{S}_\nu \hat{L}^2 \text{ for } \mu, \nu = x, y, z \tag{12.9}$$

Hence,

$$\hat{L}_z(\hat{L} \cdot \hat{S}) = \hat{L}_z \hat{L}_x \hat{S}_x + \hat{L}_z \hat{L}_y \hat{S}_y + \hat{L}_z^2 \hat{S}_z \tag{12.10}$$

and
$$\begin{aligned}(\hat{L} \cdot \hat{S})\hat{L}_z &= \hat{L}_x \hat{S}_x \hat{L}_z + \hat{L}_y \hat{S}_y \hat{L}_z + \hat{L}_z \hat{S}_z \hat{L}_z \\ &= \hat{L}_x \hat{L}_z \hat{S}_x + \hat{L}_y \hat{L}_z \hat{S}_y + \hat{L}_z^2 \hat{S}_z\end{aligned} \tag{12.11}$$

Subtracting Eq. (12.11) from Eq. (12.10), we have

$$\begin{aligned}[\hat{L}_z, (\hat{L} \cdot \hat{S})] &= (\hat{L}_z \hat{L}_x - \hat{L}_x \hat{L}_z)\hat{S}_x + (\hat{L}_z \hat{L}_y - \hat{L}_y \hat{L}_z)\hat{S}_y \\ &= i\hbar\, \hat{L}_y \hat{S}_x - i\hbar\, \hat{L}_x \hat{S}_y = i\hbar(\hat{L}_y \hat{S}_x - \hat{L}_x \hat{S}_y) \neq 0\end{aligned} \tag{12.12}$$

[Using standard forms of the operators it can be shown that $\hat{L}_z \hat{L}_x - \hat{L}_x \hat{L}_z = i\hbar \hat{L}_y$,... (any textbook on quantum mechanics may be consulted).]

Similarly it can be shown that

$$[\hat{S}_z, (\hat{L} \cdot \hat{S})] = i\hbar(\hat{S}_y \hat{L}_x - \hat{S}_x \hat{L}_y) \neq 0 \tag{12.13}$$

Thus neither $\hat{L}_z$ nor $\hat{S}_z$ commutes with $\hat{L} \cdot \hat{S}$ and hence with $\hat{H}$. However, adding Eqs (12.12) and (12.13) we have

$$[\hat{L}_z + \hat{S}_z, (\hat{L} \cdot \hat{S})] = 0 \tag{12.14}$$

We then define the operator

$$\hat{J}_z = \hat{L}_z + \hat{S}_z \tag{12.15}$$

which, owing to Eq. (12.14), commutes with total Hamiltonian $\hat{H}$:

$$[\hat{J}_z, \hat{H}] = [\hat{J}_z, \hat{H}_1 + \hat{H}^{SO}] = 0 \tag{12.16}$$

The z-components of the total (spin + orbital) angular momentum is, therefore, a constant of motion. After defining the total angular momentum operator as

$$\hat{J} = \hat{J}_x \vec{e}_1 + \hat{J}_y \vec{e}_2 + \hat{J}_z \vec{e}_3$$

it can also be shown that the operator for the scalar square of the total angular momentum, $\hat{J}^2 (= \hat{J} \cdot \hat{J})$, commutes with the total Hamiltonian $\hat{H} (= \hat{H}_1 + \hat{H}^{SO})$. It is therefore justified to take $\vec{J} \cdot \vec{J}$ and the projection J_z as constants of motion instead of the corresponding quantities related to the individual orbital and spin momenta separately. The operators $\hat{J}^2$ and $\hat{J}_z$ have eigenvalues $j(j+1)\hbar^2$ and $m_j\hbar$. From Eq. (12.15) we have

$$m_j \hbar = m_l \hbar + m_s \hbar \quad \text{or,} \quad m_j = m_l + m_s \tag{12.17}$$

A pictorial model that summarises all these quantum mechanical results is the vector atom model discussed below.

In accordance with Heisenberg's uncertainty principle, the orbital angular momentum vector $\vec{L}$ in a given state of the atom should not be exactly specified, some uncertainty must be allowed. Only the magnitude $|\vec{L}| = \sqrt{l(l+1)}\,\hbar$ and the z-component, $L_z = m_l \hbar$ are exactly known but the x- and y-components of $\vec{L}$ are undetermined. For a given n the possible values of l are integers 0, 1, 2,…, $(n-1)$ and for a particular l, the possible values of m_l are $-l$, $-(l-1)$,…, -1, 0, 1,…, $(l-1)$, l. In the vector model this is represented as in Fig. 12.1, where the vector $\vec{L}$ is anywhere on the surface of a cone which is symmetrical about the Z-axis with a semi-vertical angle of $\cos^{-1}\left(\frac{m_l}{\sqrt{l(l+1)}}\right)$ and has a constant height $m_l \hbar$.

If there is no spin–orbit coupling, the statements about the spin angular momentum ($\vec{S}$) are exactly analogous: the magnitude of $\vec{S}$ is $\sqrt{s(s+1)}\,\hbar$ (where the only allowed value of s is 1/2) and the z-component of $\vec{S}$ is $S_z = m_s \hbar$, where the allowed values of m_s are $\pm\frac{1}{2}$ and S_x and S_y are unspecified (Fig. 12.2). In the absence of spin–orbit interaction, the two vectors $\vec{L}$ and $\vec{S}$ should precess independently about the Z-axis (Figs 12.1 and 12.2).

However, spin–orbit interaction *does* exist; the vector $\vec{L}$ affects the precession of $\vec{S}$, and vice versa, by magnetic interaction. But since the interaction is completely internal and no external torque is acting, the overall (resultant) angular momentum is conserved in both magnitude $|\vec{J}|$, and the z-component J_z (the other two components J_x and J_y being non-specified owing to the uncertainty principle). This is possible only if both $\vec{L}$ and $\vec{S}$ precess about their resultant $\vec{J}$, and $\vec{J}$ precesses symmetrically about the Z-axis (Fig. 12.3).

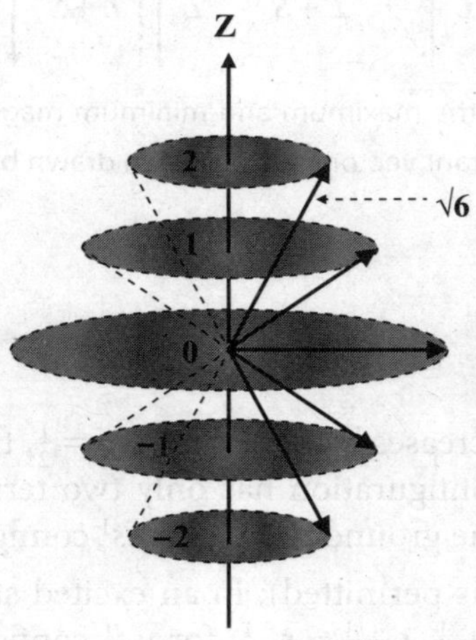

Fig. 12.1 Representation of orbital angular momentum vectors (in units of $\hbar$) for $l = 2$ and $m_l = 2$ to -2. For each m_l, the vector $\vec{L}$ may be anywhere on the surface of a cone maintaining a constant $L_z = (m_l \hbar)$; L_x and L_y are uncertain.

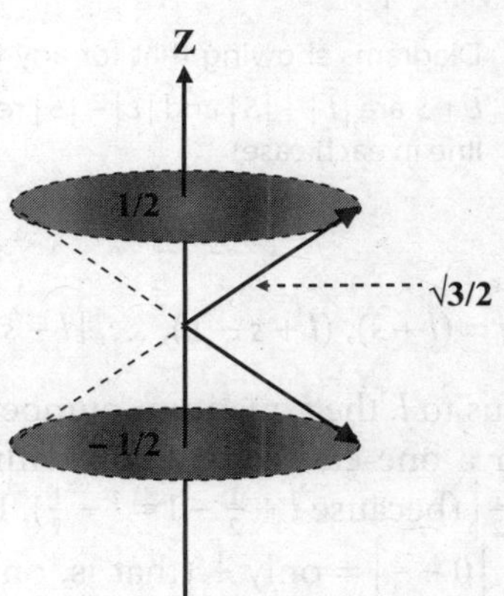

Fig. 12.2 Representation of spin angular momentum vectors (in units of $\hbar$) for $m_s = +1/2$ and $-1/2$. The vector $\vec{S}$ may be anywhere on the surface of a cone maintaining a constant $S_z = m_s \hbar$, S_x and S_y being uncertain.

The possible maximum and minimum values of the magnitude $|\vec{J}| = |\vec{L} + \vec{S}|$ can be easily shown to be $|\vec{L}| + |\vec{S}|$ and $|\vec{L}| - |\vec{S}|$ respectively by drawing vector diagrams (Fig. 12.4).

Consequently, the quantum number j, which gives the magnitude $|\vec{J}| = \sqrt{j(j+1)}\,\hbar$, has values

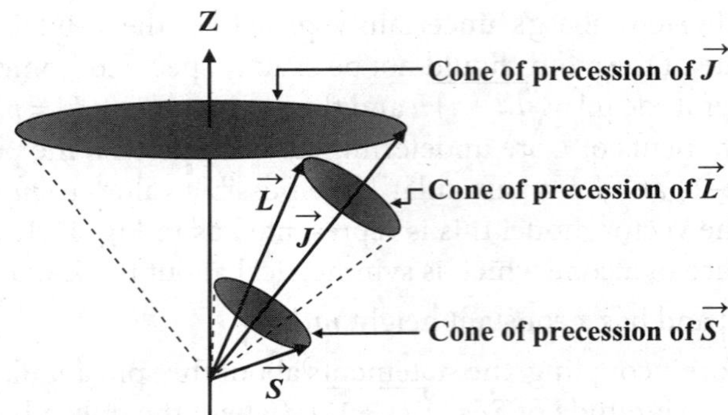

Fig. 12.3 Vector model of spin–orbit coupling in a one-electron system.

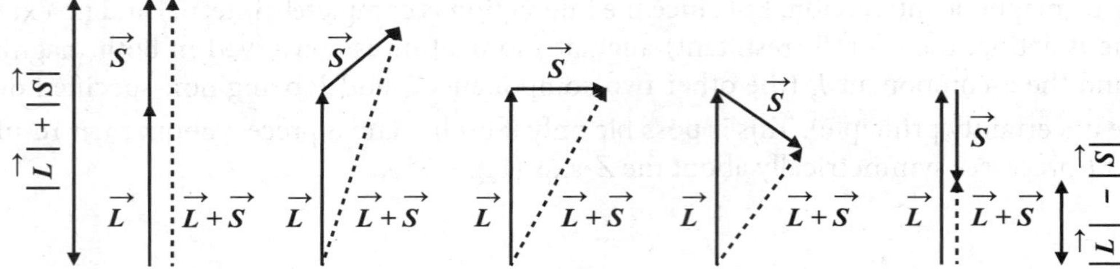

Fig. 12.4 Diagrams showing that for any two vectors $\vec{L}$ and $\vec{S}$ the maximum and minimum magnitudes of $\vec{L}+\vec{S}$ are $|\vec{L}|+|\vec{S}|$ and $|\vec{L}|-|\vec{S}|$ respectively (the resultant vector $\vec{L}+\vec{S}$ has been drawn by a dotted line in each case)

$$j=(l+s),(l+s-1),\ldots,|l-s| \tag{12.18}$$

Analogous to l, the j quantum number in this series decreases by one unit. As $s=\frac{1}{2}$, the above series for a one-electron system with nl^1 electronic configuration has only two terms, $l+\frac{1}{2}$ and $\left|l-\frac{1}{2}\right|$ (because $l+\frac{1}{2}-1=l-\frac{1}{2}$). For a H atom in the ground state with ns^1 configuration, $j=0+\frac{1}{2},\left|0-\frac{1}{2}\right|=$ only $\frac{1}{2}$ (that is, only one value of j is permitted); in an excited state with configuration np^1, two values of j are permitted: $j=1+\frac{1}{2},1-\frac{1}{2}=\frac{3}{2},\frac{1}{2}$; for nd^1 configuration, $j=2+\frac{1}{2},2-\frac{1}{2}=\frac{5}{2},\frac{3}{2}$ and so on. For each j, the permitted maximum and minimum values of m_j can be obtained from the vector model in a simple way:

$$J_z=m_j\hbar=|\vec{J}|\cos\theta=\sqrt{j(j+1)}\,\hbar\cos\theta\approx j\hbar\cos\theta \tag{12.19}$$

which gives $m_j=j\cos\theta$ and since $-1\le\cos\theta\le 1$, we have $-j\le m_j\le j$.

The permitted m_j values give the possible orientations of $\vec{J}$ in space for a particular j (just as m_l gives the possible orientations of the vector $\vec{L}$ for a particular l). In a multiplet caused by spin–orbit coupling, each energy level corresponding to a particular j is $(2j+1)$-fold degenerate (indicated by the m_j values) and such degeneracy is removed by the application of external magnetic or electric fields (Zeeman and Stark effect respectively).

12.3 SPIN–ORBIT COUPLING IN MANY-ELECTRON ATOMS

For an N-electron atom, the Schrödinger equation cannot be solved exactly owing to the inter-electron repulsion term and the complex magnetic interactions (spin–orbit, spin–spin and so on). As an approximation it is necessary to assume that each electron moves independently of the other electrons in an *average* field caused by the nucleus and the other electrons. This assumption leads to the *independent particle model* which provides a suitable starting point for more accurate calculations. The Hamiltonian for an N-electron atom is:

$$\hat{\text{H}} = \sum_{\mu=1}^{N}\left[-\frac{1}{2}\nabla_{\mu}^{2} - \frac{Z}{r_{\mu}}\right] + \sum_{\mu\rangle\nu}^{N}\frac{1}{r_{\mu\nu}} + V_{\text{mag}} \tag{12.20}$$

where r_μ is the distance of the μth electron from the nucleus of charge Z, $r_{\mu\nu}$ is its distance from the νth electron and V_{mag} is the potential from magnetic interactions. In Eq. (12.20), atomic units have been used (and this will be used in the subsequent discussions). If the inter-electron repulsion energy is much stronger than spin–orbit interaction energy, one may start from a Hamiltonian which is free from the $1/r_{\mu\nu}$ terms, then improve the solution by incorporating the $1/r_{\mu\nu}$ terms and finally consider the spin–orbit interaction as a perturbation. On the other hand, if the spin–orbit interactions are much stronger than (or as strong as) the inter-electronic repulsions, a reverse procedure is followed. Accordingly, there are two extreme cases of spin–orbit coupling. The former, called *LS* or Russell–Saunders coupling, is valid for atoms with low or moderate nuclear charge ($Z \leq 30$); the other, called *jj* coupling, is valid for atoms with higher Z and also for phenomena within the nucleus.

[**Note:** In atomic unit (also called the *Hartree unit*),
one unit of mass = the rest mass of the electron,
one unit of distance = the first Bohr radius of the H atom,
one unit of angular momentum = $\hbar$ = the Planck constant divided by 2π, and
one unit of energy = 1 Hartree = twice the ionisation potential of H = 27.2 eV.]

12.3.1 THE *LS* OR RUSSELL–SAUNDERS COUPLING SCHEME

The quantum mechanical rationale for this scheme is briefly outlined here. This will be followed by a summarisation of the scheme in terms of the vector atom model.

Neglecting V_{mag}, the Hamiltonian in Eq. (12.20) can be split up as follows:

$$\hat{\text{H}} = \sum_{\mu=1}^{N}\left[-\frac{1}{2}\nabla_{\mu}^{2} - \frac{Z}{r_{\mu}} + U_{\mu}(r_{\mu})\right] + \sum_{\mu\rangle\nu}^{N}\frac{1}{r_{\mu\nu}} - \sum_{\mu=1}^{N}U_{\mu}(r_{\mu})$$

$$= \hat{\text{H}}_0 + \hat{\text{H}}' \tag{12.21}$$

where $$\hat{\text{H}}_0 = \sum_{\mu=1}^{N}\hat{\text{h}}_0(\mu) = \sum_{\mu=1}^{N}\left[-\frac{1}{2}\nabla_{\mu}^{2} - \frac{Z}{r_{\mu}} + U_{\mu}(r_{\mu})\right] \tag{12.22}$$

and $$\hat{\text{H}}' = \sum_{\mu\rangle\nu}^{N}\frac{1}{r_{\mu\nu}} - \sum_{\mu=1}^{N}U_{\mu}(r_{\mu}) \tag{12.23}$$

This $U_\mu(r_\mu)$ is so chosen that (i) it is a potential of central field nature, (ii) it contains most of the inter-electron repulsions in order that $\hat{H}'$ in Eq. (12.23) is small enough to allow a perturbational treatment and (iii) $U_\mu(r_\mu) \to 0$ as $r_\mu \to 0$ and $U_\mu(r_\mu) \to (N-1)/r_\mu$ as $r_\mu \to \infty$. The last condition means that the μth electron feels the entire nuclear charge Z when it is close to the nucleus and it is screened by $(N-1)$ electrons when it is far apart. The assumption of central field nature for $U_\mu(r_\mu)$ makes the one-electron Hamiltonian $\hat{h}_0(\mu)$ behave like that of the H-atom and the one-electron Schrödinger equation

$$\hat{h}_0\,\psi = \left[-\frac{1}{2}\nabla^2 - \frac{Z}{r} + U(r)\right]\psi = \varepsilon_0\psi \quad \text{(for each } \mu) \tag{12.24}$$

has solutions of the form

$$\psi = R_{n,l}(r)Y_l^{m_l}(\theta, \phi) \tag{12.25}$$

like those of the H-atom. The angular part is the same as that of the H-atom and the radial part $R_{n,l}(r)$ differs in the fact that the electron now feels an effective nuclear charge screened by the other electrons. The functions in Eq. (12.25) are *orbitals* of the many-electron atom and the N electrons occupy them in accordance with the Pauli exclusion principle, giving rise to an *electronic configuration* for the atom. This provides a starting point for further calculation of the wave function and electronic energy of the atom.

The central field potential also makes it possible to express the spin–orbit interaction energy (excluding hyperfine interaction) in the form

$$V_{\text{mag}} = \sum_{\mu=1}^{N} \xi(r_\mu)\vec{L}_\mu \cdot \vec{S}_\mu$$

(like $\hat{H}^{SO}$ in the one-electron case).

Classically, the total orbital angular momentum of a system of N electrons about a point is given by the vector sum,

$$\vec{L} = \sum_{\mu=1}^{N} \vec{L}_\mu = L_x\vec{e}_1 + L_y\vec{e}_2 + L_z\vec{e}_3 \tag{12.26}$$

where $\vec{L}_\mu$ is the orbital angular momentum of the μth electron. The Cartesian components of the total orbital angular momentum are given by

$$L_x = \sum_{\mu=1}^{N} L_{x\mu},\; L_y = \sum_{\mu=1}^{N} L_{y\mu},\; L_z = \sum_{\mu=1}^{N} L_{z\mu} \tag{12.27}$$

and the scalar square of $\vec{L}$ is given by

$$L^2 = \vec{L} \cdot \vec{L} = L_x^2 + L_y^2 + L_z^2 \tag{12.28}$$

These are of the same form as in the one-electron case and the corresponding quantum mechanical operators satisfy analogous equations. The z-component of the total orbital angular momentum is described by the operator

$$\hat{L}_z = \sum_{\mu=1}^{N} \hat{L}_{z\mu} = \frac{1}{i}\sum_{\mu=1}^{N} \frac{\partial}{\partial \phi_\mu} \text{ (in atomic unit, where } \hbar = 1) \tag{12.29}$$

Here ϕ_μ is the polar angle of the μth electron relative to a fixed coordinate system. The operator representing L^2 for an N-electron system is

$$\hat{L}^2 = \hat{L}\cdot\hat{L} = \left[\sum_{\mu=1}^{N}(\hat{L}_{x\mu} + \hat{L}_{y\mu} + \hat{L}_{z\mu})\right]\cdot\left[\sum_{\nu=1}^{N}(\hat{L}_{x\nu} + \hat{L}_{y\nu} + \hat{L}_{z\nu})\right]$$

$$= \sum_{\mu=1}^{N}(\hat{L}^2_{x\mu} + \hat{L}^2_{y\mu} + \hat{L}^2_{z\mu}) + \sum_{\mu\neq\nu}^{N}(\hat{L}_{x\mu}\hat{L}_{x\nu} + \hat{L}_{y\mu}\hat{L}_{y\nu} + \hat{L}_{z\mu}\hat{L}_{z\nu}) \tag{12.30}$$

Let us now examine the commutation relations among the operators $\hat{L}_x$, $\hat{L}_y$, $\hat{L}_z$, $\hat{L}^2$ and $\hat{H}$ for the N-electron case. Since $\hat{H}_0$ of Eq. (12.22) contains a central-field potential, its commutation with $\hat{L}^2$ and $\hat{L}_z$ are obvious (similar to the one-electron case). We, therefore, examine the commutation of $\hat{L}_z$ with the two-electron repulsion operator, $\frac{1}{r_{\mu\nu}}$. For an arbitrary function f of coordinates,

$$\hat{L}_z\left(\frac{1}{r_{\mu\nu}}\right)f = \frac{1}{i}\sum_{\rho=1}^{N}\frac{\partial}{\partial\phi_\rho}\left(\frac{1}{r_{\mu\nu}}\right)f$$

$$= \frac{1}{i}\left[\frac{1}{r_{\mu\nu}}\frac{\partial f}{\partial\phi_1} + \frac{1}{r_{\mu\nu}}\frac{\partial f}{\partial\phi_2} + \cdots + \frac{\partial}{\partial\phi_\mu}\left(\frac{1}{r_{\mu\nu}}\right)f + \frac{\partial}{\partial\phi_\nu}\left(\frac{1}{r_{\mu\nu}}\right)f + \cdots + \frac{1}{r_{\mu\nu}}\frac{\partial f}{\partial\phi_N}\right]$$

$$= \frac{1}{i}\left[\frac{1}{r_{\mu\nu}}\frac{\partial f}{\partial\phi_1} + \frac{1}{r_{\mu\nu}}\frac{\partial f}{\partial\phi_2} + \cdots + \frac{1}{r_{\mu\nu}}\frac{\partial f}{\partial\phi_\mu} + f\frac{\partial\left(\frac{1}{r_{\mu\nu}}\right)}{\partial\phi_\mu} + \frac{1}{r_{\mu\nu}}\frac{\partial f}{\partial\phi_\nu} + f\frac{\partial\left(\frac{1}{r_{\mu\nu}}\right)}{\partial\phi_\nu} + \cdots + \frac{1}{r_{\mu\nu}}\frac{\partial f}{\partial\phi_N}\right]$$

$$= \frac{1}{r_{\mu\nu}}\cdot\frac{1}{i}\sum_{\rho=1}^{N}\frac{\partial f}{\partial\phi_\rho} + \frac{1}{i}\left[\frac{\partial\left(\frac{1}{r_{\mu\nu}}\right)}{\partial\phi_\mu} + \frac{\partial\left(\frac{1}{r_{\mu\nu}}\right)}{\partial\phi_\nu}\right]f$$

Hence

$$\hat{L}_z\left(\frac{1}{r_{\mu\nu}}\right) = \left(\frac{1}{r_{\mu\nu}}\right)\hat{L}_z + \frac{1}{i}\left[\frac{\partial\left(\frac{1}{r_{\mu\nu}}\right)}{\partial\phi_\mu} + \frac{\partial\left(\frac{1}{r_{\mu\nu}}\right)}{\partial\phi_\nu}\right] \tag{12.31}$$

According to Fig. 12.5, $r_{\mu\nu}$ depends only on the angle $\chi = |\phi_\mu - \phi_\nu|$. So, if $\phi_\nu \rangle \phi_\mu$, $\chi = \phi_\nu - \phi_\mu$ and $\frac{\partial\chi}{\partial\phi_\mu} = -1$, $\frac{\partial\chi}{\partial\phi_\nu} = 1$, making

$$\frac{\partial \chi}{\partial \phi_\mu}+\frac{\partial \chi}{\partial \phi_\nu}=0 \tag{12.32}$$

The same is true if $\phi_\mu \rangle \phi_\nu$. Hence,

$$\left[\frac{\partial\left(\frac{1}{r_{\mu\nu}}\right)}{\partial \phi_\mu}+\frac{\partial\left(\frac{1}{r_{\mu\nu}}\right)}{\partial \phi_\nu}\right]=0 \tag{12.33}$$

Equation (12.31) then reduces to

$$\hat{\mathrm{L}}_z\left(\frac{1}{r_{\mu\nu}}\right)=\left(\frac{1}{r_{\mu\nu}}\right)\hat{\mathrm{L}}_z \text{ that is, } \left[\hat{\mathrm{L}}_z, \frac{1}{r_{\mu\nu}}\right]=0 \tag{12.34}$$

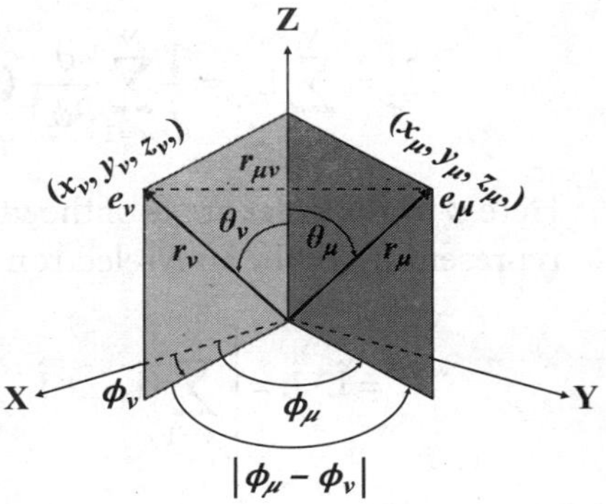

Fig. 12.5 Coordinate system for two electrons e_ν and e_μ

Thus $\hat{\mathrm{L}}_z$ commutes with $\hat{\mathrm{H}}'$ and therefore also with $\hat{\mathrm{H}}$ of Eq. (12.21). In a similar way it can be shown that $\hat{\mathrm{L}}^2$ also commutes with $\hat{\mathrm{H}}$ of Eq. (12.21) for the many-electron system. Thus L^2 and the L_z component of the many-electron system are constants of motion. However, the z-components of the individual orbital angular momenta of electrons are not so, that is, $[\hat{\mathrm{L}}_{z\mu}, \hat{\mathrm{H}}] \neq 0$ as proved below:

Considering the μth electron and an arbitrary function f of coordinates

$$\hat{\mathrm{L}}_{z\mu}\left(\frac{1}{r_{\mu\nu}}\right)f=\frac{1}{i}\frac{\partial}{\partial \phi_\mu}\left(\frac{1}{r_{\mu\nu}}f\right)=\frac{1}{i}\left[\frac{1}{r_{\mu\nu}}\frac{\partial f}{\partial \phi_\mu}+\frac{\partial\left(\frac{1}{r_{\mu\nu}}\right)}{\partial \phi_\mu}f\right] \tag{12.35}$$

Hence $$\hat{\mathrm{L}}_{z\mu}\left(\frac{1}{r_{\mu\nu}}\right)-\left(\frac{1}{r_{\mu\nu}}\right)\hat{\mathrm{L}}_{z\mu}=\left[\hat{\mathrm{L}}_{z\mu}, \left(\frac{1}{r_{\mu\nu}}\right)\right]=\frac{1}{i}\frac{\partial\left(\frac{1}{r_{\mu\nu}}\right)}{\partial \phi_\mu}\neq 0 \tag{12.36}$$

If spin–orbit interaction is totally absent, the above commutation relations of $\hat{\mathrm{L}}^2$ and L_z with $\hat{\mathrm{H}}$ for a many-electron system are valid under the approximation of central field potential. The same is true for the spin operators $\hat{\mathrm{S}}^2$ and $\hat{\mathrm{S}}_z$. But spin–orbit interaction *does* come into play and, as in the one-electron case, it can be shown that the operator $\hat{\mathrm{L}} \cdot \hat{\mathrm{S}}$ (which represents the scalar product $\vec{L} \cdot \vec{S}$) commutes with $\hat{\mathrm{L}}_z + \hat{\mathrm{S}}_z$ instead of $\hat{\mathrm{L}}_z$ and $\hat{\mathrm{S}}_z$ separately ($\hat{\mathrm{S}}_z$ now means the operator for the z-component of the vector sum of the individual spins, that is, $\hat{\mathrm{S}}_z=\sum_{\mu=1}^{N}\hat{\mathrm{S}}_{z\mu}$). We can define the operator

$$\hat{\mathrm{J}}_z=\hat{\mathrm{L}}_z+\hat{\mathrm{S}}_z \tag{12.37}$$

for the many-electron case which now becomes exactly analogous to the one-electron case. The resultant $\hat{\mathrm{J}}$ of $\hat{\mathrm{L}}$ and $\hat{\mathrm{S}}$ has, therefore, a quantised z-component given by

$$J_z=M_J\hbar \tag{12.38}$$

where the quantum number M_J is obtained as

$$M_J=M_L+M_S \tag{12.39}$$

and the quantum numbers M_L and M_S are obtained from the definitions stated earlier: $L_z = \sum_{\mu=1}^{N} L_{z\mu}$ and $S_z = \sum_{\mu=1}^{N} S_{z\mu}$ which give

$$M_L = \sum_{\mu=1}^{N} (m_l)_\mu = (m_l)_1 + (m_l)_2 + \cdots + (m_l)_N \tag{12.40}$$

and $$M_S = \sum_{\mu=1}^{N} (m_s)_\mu = (m_s)_1 + (m_s)_2 + \cdots + (m_s)_N \tag{12.41}$$

In the above equations the individual $(m_l)_\mu$ values may be positive, negative or zero: $(m_l)_\mu = -l_\mu, -(l_\mu - 1), -(l_\mu - 2), \ldots, 0, \ldots, (l_\mu - 2), (l_\mu - 1), l_\mu$; the individual $(m_s)_\mu$ may be $+\frac{1}{2}$ or $-\frac{1}{2}$. Hence M_L will correspond to different series of values incremented by unity and in each such series $(M_L)_{\max} = L$, the resultant orbital angular momentum quantum number. Like the one-electron case, the corresponding spectroscopic state has a letter symbol:

$L =$	0	1	2	3	4	5	…
	S	P	D	F	G	H	…

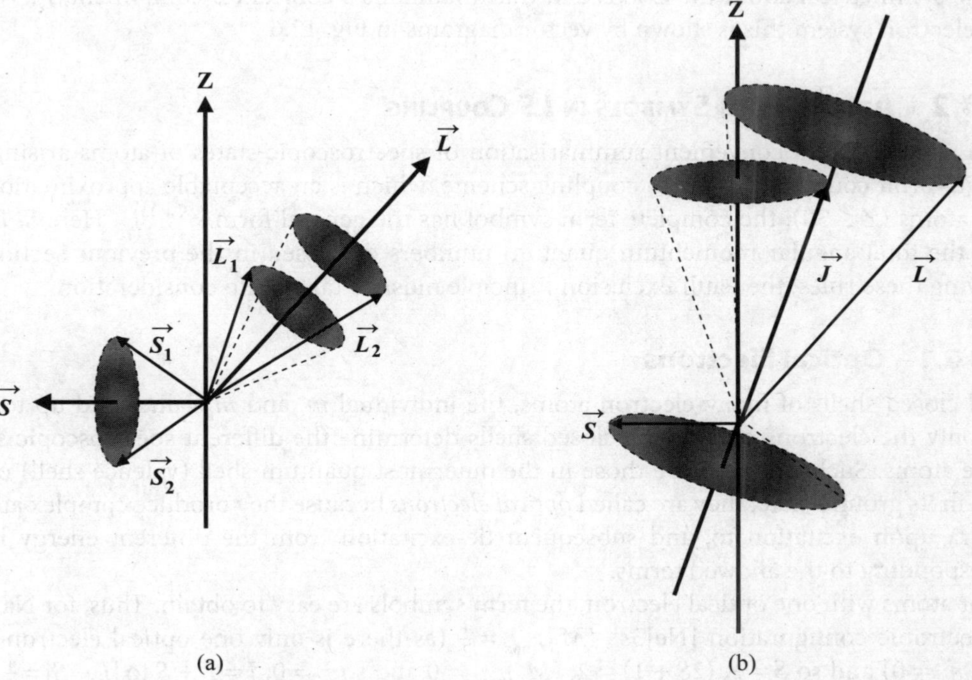

Fig. 12.6 *LS* coupling of angular momenta of two electrons. (a) The orbital angular momenta $\vec{L}_1$ and $\vec{L}_2$ of two electrons precess rapidly about their resultant (vector sum) $\vec{L}$. Similarly their spins $\vec{S}_1$ and $\vec{S}_2$ precess rapidly about their vector sum $\vec{S}$. (b) The total orbital angular momentum $\vec{L}$ and the total spin angular momentum $\vec{S}$ precess slowly about their sum $\vec{J}$. Finally, $\vec{J}$ can be found anywhere on the surface of a cone symmetrical about the Z-axis.

Similarly, different combinations of $(m_s)_\mu$ will produce different series of M_S values, each with an increment of 1 and for each series $(M_S)_{max} = S$ which corresponds to the resultant spin of the many-electron system. The spin multiplicity of the state is $(2S + 1)$; the state is called a spin singlet, doublet, triplet, quartet, … for values of $(2S + 1) = 0, 1, 2, 3, 4, \ldots$ (The symbol S for total spin and S for $L = 0$ should not be confused.) The J quantum number which corresponds to the maximum M_J comes from

$$(M_J)_{max} = (M_L)_{max} + (M_S)_{max} \tag{12.42}$$

or,

$$J = L + S \tag{12.43}$$

For an atomic state with a particular L and S the possible values of J are

$$(L + S), (L + S - 1), \ldots, |L - S| \tag{12.44}$$

which are $(2S + 1)$ in number.

All the above quantum mechanical considerations can be summarised by the vector atom model: the individual orbital angular momenta $\vec{L}_1, \vec{L}_2, \ldots$ of the electrons interact strongly and precess rapidly about their resultant $\vec{L}$, the individual spin angular momenta $\vec{S}_1, \vec{S}_2, \ldots$ interact strongly and precess rapidly about their resultant $\vec{S}$, and the vectors $\vec{L}$ and $\vec{S}$ interact weakly to precess slowly about their resultant $\vec{J} = \vec{L} + \vec{S}$. This $\vec{J}$ may be found anywhere on the surface of a cone symmetrical about the Z-axis and thus maintains a constant z-component, J_z. With a two-electron system this is shown by vector diagrams in Fig. 12.6.

12.3.2 Atomic Term Symbols in *LS* Coupling

Term symbols are a convenient summarisation of spectroscopic states of atoms arising out of spin–orbit coupling. In the *LS* coupling scheme, which is an acceptable approximation for light atoms ($Z \leq 30$), the complete term symbol has the general form ${}^{(2S+1)}L_J$. Here S, L and J are the total angular momentum quantum numbers discussed in the previous section; in applying these rules, the Pauli Exclusion Principle must be taken into consideration.

12.3.2.1 Optical Electrons

In all closed shells of many-electron atoms, the individual m_s and m_l values add up to zero and only the electrons outside the closed shells determine the different spectroscopic states of the atoms. Such electrons are those in the outermost quantum shell (valence shell) of the atom in its ground state; they are called *optical electrons* because they produce complex atomic spectra upon excitation to, and subsequent de-excitation from the different energy levels corresponding to the allowed terms.

For atoms with one optical electron, the term symbols are easy to obtain. Thus, for Na with an electronic configuration [Ne]$3s^1$, $(M_S)_{max} = \frac{1}{2}$ (as there is only one optical electron with $l = 0$, $m_l = 0$) and so $S = \frac{1}{2}$, $(2S + 1) = 2$; $(M_L)_{max} = 0$ and so $L = 0$; $J = L + S$ to $|L - S| = \frac{1}{2}$. The term symbol of the ground state of Na is, therefore, ${}^2\mathrm{S}_{\frac{1}{2}}$ (read as 'doublet ess half'). An excited state of Na has electronic configuration [Ne]$3p^1$. In this case $L = 1$ and $S = \frac{1}{2}$ giving $(2S + 1) = 2$ and $J = L + S$ to $|L - S| = 1 + \frac{1}{2}$ to $1 - \frac{1}{2} = \frac{3}{2}, \frac{1}{2}$. The excited state has terms ${}^2\mathrm{P}_{\frac{3}{2}}$ and ${}^2\mathrm{P}_{\frac{1}{2}}$ which correspond to two microstates differing slightly in energy; the former is $(2\times\frac{3}{2}+1)=4$-fold degenerate given by the M_J values $\frac{3}{2}, \frac{1}{2}, -\frac{1}{2}, -\frac{3}{2}$ and the latter is doubly degenerate with M_J values $\frac{1}{2}$ and $-\frac{1}{2}$.

For systems with more than one optical electron there are many schemes for obtaining the *LS* terms, some of which are discussed in the following sections.

12.3.3 Method of Tabulation of Microstates for Obtaining Atomic Term Symbols

The method is best explained by examples.

Example 1: p^2 electronic configuration

Considering three space functions (for the *p*-orbitals) and two spin functions (up and down spins), there are $3 \times 2 = 6$ spin-orbitals available to *p*-electrons. One spin-orbital can accommodate only one electron according to the Pauli Exclusion Principle. Hence two *p*-electrons (of a p^2 configuration) can occupy the available spin-orbitals in ${}^6C_2 = 15$ ways; this gives rise to 15 possible microstates.

$$(M_L)_{\max} = \max|(m_l)_1| + \max|(m_l)_2| = 1 + 1 = 2$$

$$(M_S)_{\max} = |(m_s)_1| + |(m_s)_2| = \tfrac{1}{2} + \tfrac{1}{2} = 1$$

Possible values of M_L are 2, 1, 0, −1, −2 and those of M_S are 1, 0, −1. The possible m_l and m_s combinations which produce the microstates are shown in Table 12.1. The entries in this table contain pairs of numbers in parenthesis, meaning: (m_l of electron 1, m_l of electron 2). A bar over a value of m_l indicates that the concerned electron has spin $-\frac{1}{2}$; no bar indicates that the m_s value is $+\frac{1}{2}$. Thus, in the first row both the electrons have $m_l = 1$, in the second row one electron has $m_l = 1$ and the other has $m_l = 0$ and so on. Similarly, in the first column both electrons have $m_s = \frac{1}{2}$, in the second column one electron has $m_s = +\frac{1}{2}$ and the other has $m_s = -\frac{1}{2}$ and in the third column, both the electrons have $m_s = -\frac{1}{2}$. In Table 12.1 the combinations that are not permitted by the exclusion principle are crossed (deleted). The number of microstates corresponding to the various M_L and M_S values are shown in Table 12.2.

Table 12.1 Possible combinations of m_l and m_s for p^2 configuration

M_L \ M_S	1	0	−1
2	~~$(1, 1)$~~	$(\bar{1}, 1)$	~~$(\bar{1}, \bar{1})$~~
1	$(1, 0)$	$(\bar{1}, 0), (\bar{0}, 1)$	$(\bar{1}, \bar{0})$
0	~~$(0, 0)$~~, $(1, -1)$	$(\bar{1}, -1), (1, -\bar{1}), (0, \bar{0})$	$(\bar{1}, -\bar{1})$, ~~$(\bar{0}, \bar{0})$~~
−1	$(-1, 0)$	$(-\bar{1}, 0)(-1, \bar{0})$	$(-\bar{1}, \bar{0})$
−2	~~$(-1, -1)$~~	$(-1, -\bar{1})$	~~$(-\bar{1}, -\bar{1})$~~

Table 12.2 Number of microstates for p^2 configuration

M_L \ M_S	1	0	−1
2		1	
1	1	2	1
0	1	3	1
−1	1	2	1
−2		1	
Total number of microstates = 15			

Table 12.3 Number of microstates obtained by subtracting the ^1D microstates from Table 12.2

M_L \ M_S	1	0	−1
1	1	1	1
0	1	2	1
−1	1	1	1

Table 12.4 Microstate counts obtained from Table 12.3 by deleting those of ^{3}P

M_L \ M_S	1	0	−1
1	—	—	—
0	—	1	—
−1	—	—	—

In Table 12.2, only the middle column has entries for all the M_L values. This means that the set of M_L values (2, 1, 0, −1, −2) is possible only with M_S = 0. This corresponds to a ^{1}D state with (2.0 + 1)(2.2 + 1) = 5 microstates. Deleting these microstates, that is, subtracting 1 from the middle column of Table 12.2, we get the microstate counts as shown in Table 12.3.

Table 12.3 shows that the set of M_L values (1, 0, −1) has at least one entry in each column; that is, M_S = (1, 0, −1) is permitted for M_L = (1, 0, −1). Hence we get a ^{3}P state which has (2.1 + 1) (2.1 + 1) = 9 microstates.

Subtracting the microstate counts of ^{3}P from Table12.3 we are left with only $M_L = 0$, $M_S = 0$ (Table 12.4) which gives a ^{1}S state containing (2.0 + 1)(2.0 + 1) = 1 microstate. The terms for the atomic states of a system with p^2 configuration are, therefore, ^{1}D, ^{3}P, ^{1}S (the total number of microstates = 5 + 9 + 1 = 15, as stated earlier).

Example 2: d^2 configuration

Number of spin-orbitals = 5 (d-functions) × 2 (spin-functions) = 10

Number of possible microstates = number of ways in which 2 electrons can fill up these 10 spin-orbitals = $^{10}C_2 = 45$

$$(M_L)_{max} = \max|(m_l)_1| + \max|(m_l)_2| = 2 + 2 = 4$$

$$(M_S)_{max} = |(m_s)_1| + |(m_s)_2| = \frac{1}{2} + \frac{1}{2} = 1$$

Hence the permitted M_L values are ±4, ±3, ±2, ±1, 0 and permitted M_S values are 1, 0, −1. The possible (m_l, m_s) combinations are shown in Table 12.5.

Table 12.5 Microstates for d^2 configuration

M_L \ M_S	1	0	−1
4	~~(2, 2)~~	$(2, \bar{2})$	~~$(\bar{2}, \bar{2})$~~
3	(2, 1)	$(\bar{2}, 1), (2, \bar{1})$	$(\bar{2}, \bar{1})$
2	~~(1, 1)~~, (2, 0)	$(1, \bar{1}), (2, \bar{0}), (\bar{2}, 0)$	~~$(\bar{1}, \bar{1})$~~, $(\bar{2}, \bar{0})$
1	(2, −1), (1, 0)	$(2, -\bar{1}), (\bar{2}, -1), (\bar{1}, 0), (1, \bar{0})$	$(\bar{2}, -\bar{1}), (\bar{1}, \bar{0})$
0	(2, −2), (1, −1), ~~(0, 0)~~	$(2, -\bar{2}), (\bar{2}, -2), (1, -\bar{1}), (\bar{1}, -1), (0, \bar{0})$	$(\bar{2}, -\bar{2}), (\bar{1}, -\bar{1})$, ~~$(\bar{0}, \bar{0})$~~
−1	(−2, 1), (−1, 0)	$(-2, \bar{1}), (-\bar{2}, 1), (-1, \bar{0}), (-\bar{1}, 0)$	$(-\bar{2}, \bar{1}), (-\bar{1}, \bar{0})$
−2	~~(−1, −1)~~, (−2, 0)	$(-1, -\bar{1}), (-2, \bar{0}), (-\bar{2}, 0)$	~~$(-\bar{1}, -\bar{1})$~~, $(-\bar{2}, \bar{0})$
−3	(−2, −1)	$(-\bar{2}, -1), (-2, -\bar{1})$	$(-\bar{2}, -\bar{1})$
−4	~~(−2, −2)~~	$(-2, -\bar{2})$	~~$(-\bar{2}, -\bar{2})$~~

The microstate counts are listed in Table 12.6. The middle column ($M_S = 0$) has at least one entry for all the M_L values from −4 to 4. This gives a ^{1}G state which has (2.0 + 1) (2.4 + 1) = 9 microstates. Subtracting these 9 microstate counts from Table 12.6 (that is, subtracting 1 from the middle column of Table 12.6) we get Table 12.7.

Table 12.6 Microstate counts for d^2 configuration

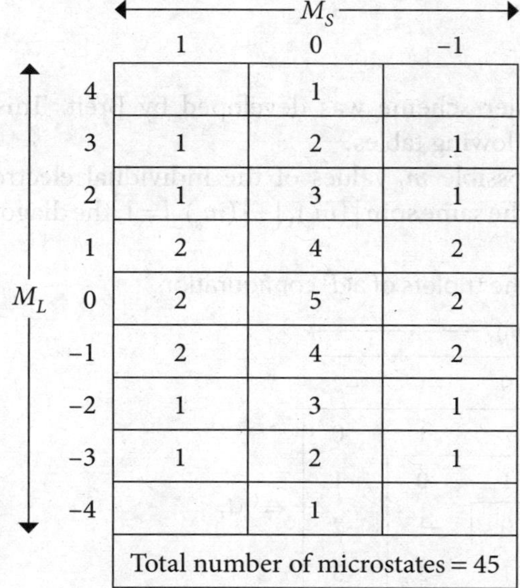

M_L \ M_S	1	0	−1
4		1	
3	1	2	1
2	1	3	1
1	2	4	2
0	2	5	2
−1	2	4	2
−2	1	3	1
−3	1	2	1
−4		1	
Total number of microstates = 45			

Table 12.7 Number of microstates obtained by subtracting the ^{1}G microstate from Table 12.6

M_L \ M_S	1	0	−1
3	1	1	1
2	1	2	1
1	2	3	2
0	2	4	2
−1	2	3	2
−2	1	2	1
−3	1	1	1

Table 12.7 shows that there is at least one entry for the set M_L = (3, 2, 1, 0, −1, −2, −3) in each column; hence the set M_S = (1, 0, −1) is possible for all M_L values from 3 to −3. Thus a ^{3}F

state with (2.1 + 1)(2.3 + 1) = 21 microstates is possible. Subtracting these microstates from Table 12.7 (that is, subtracting 1 from each entry in the columns), we get Table 12.8.

Table 12.8 Microstate count obtained by subtracting the ^{3}F microstates from Table 12.7

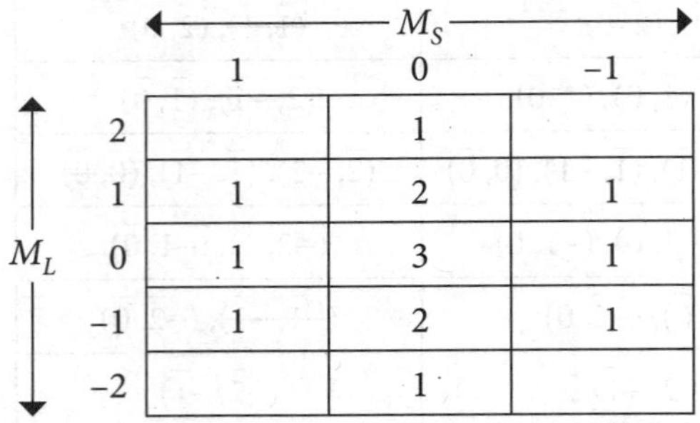

M_L \ M_S	1	0	−1
2		1	
1	1	2	1
0	1	3	1
−1	1	2	1
−2		1	

Table 12.9 Microstate counts obtained by subtracting the ^{1}D microstate counts from Table 12.8

M_L \ M_S	1	0	−1
1	1	1	1
0	1	2	1
−1	1	1	1

In Table 12.8 there is at least one entry for each row in the middle column. Thus one state is possible with $M_S = 0$ and $M_L = (2, 1, 0, -1, -2)$. This corresponds to a ^{1}D state with (2.0 + 1) (2.2 + 1) = 5 microstates. Subtracting the microstate counts of ^{1}D from Table 12.8 we get Table 12.9. Here the set $M_S = (1, 0, -1)$ is possible for all the values of $M_L = (1, 0, -1)$. Hence we get a ^{3}P state with (2.1 + 1)(2.1 + 1) = 9 microstates. When these microstates are subtracted from Table 12.9, we are left with only $M_L = 0$, $M_S = 0$, which corresponds to the ^{1}S term.

Thus the spectroscopic terms for the d^2 configuration are ^{1}G, ^{1}D, ^{1}S, ^{3}F and ^{3}P (the total number of microstates = 9 + 5 + 1 + 21 + 9 = 45 as stated earlier).

12.3.4 Breit's Scheme

For systems with two optical electrons an easier scheme was developed by Breit. This is illustrated for d^2 and pd configurations in the following tables.

The entries in this table are sums of the possible m_l values of the individual electrons $M_L = (m_l)_1 + (m_l)_2$. When the two electrons are of the same spin $|(m_s)_1| + |(m_s)_2| = 1$, the diagonal

Table 12.10 Breit's scheme for the triplets of a d^2 configuration

$(m_l)_2$ \ $(m_l)_1$	2	1	0	−1	−2
2	4	3	2	1	0
1	3	2	1	0	−1
0	2	1	0	−1	−2
−1	1	0	−1	−2	−3
−2	0	1	−2	−3	−4

⇐ M_L

^{3}P ^{3}F

entries of the table are not allowed by the Exclusion principle. $S = (M_S)_{\max} = |(m_s)_1| + |(m_s)_2| = \frac{1}{2} + \frac{1}{2} = 1$ and so spin multiplicity is $(2.S + 1) = (2.1 + 1) = 3$ and the possible values of M_S are 1, 0, –1. The electronic state is therefore a triplet. The set of permitted M_L values are (3, 2, 1, 0, –1, –2, –3) and (1, 0, –1) as obtained from the upper (or lower) triangle of the table. Thus we get the terms ^{3}F and ^{3}P for a d^2 system when the two electrons are of the same spin.

If $(m_s)_1 \neq (m_s)_2$ (that is, the electrons are of opposite spin), $S = (M_S)_{\max} = 0$ and $(2.S + 1) = 0$; the electronic state is a singlet. In this case all combinations of $(m_l)_1$ and $(m_l)_2$ are allowed and one can construct Table 12.11.

Table 12.11 Breit's scheme for the singlets of a d^2 configuration

$(m_l)_2$ \ $(m_l)_1$ →	2	1	0	–1	2
2	4	3	2	1	0
1	3	2	1	0	–1
0	2	1	0	–1	–2
–1	1	0	–1	–2	–3
–2	0	–1	–2	–3	–4
	^{1}S	^{1}P	^{1}D	^{1}F	^{1}G
		(b)		(a)	

⇐ M_L

Here the case of $M_S = 0$ and $M_L = (1, 0, -1)$, which corresponds to a ^{1}P term, has already been considered in forming the ^{3}P term of the above table. Similarly the ^{1}F term is a component of ^{3}F obtained earlier. Hence these two cases, marked (a) and (b), are excluded. Thus the possible singlet states with d^2 configuration are ^{1}G, ^{1}D and ^{1}S. So the atomic terms for d^2 system are ^{3}(P, F) and ^{1}(G, D, S).

Breit's scheme for a *pd* configuration is as follows. The two electrons in this case are non-equivalent. Hence all combinations of m_s may be combined with all m_l combinations as indicated in Tables 12.12 and 12.13.

Table 12.12 Possible m_l combinations

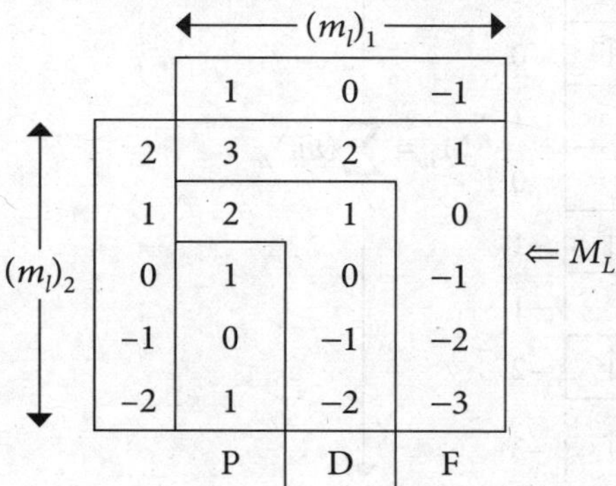

$(m_l)_2$ \ $(m_l)_1$ →	1	0	–1
2	3	2	1
1	2	1	0
0	1	0	–1
–1	0	–1	–2
–2	1	–2	–3
	P	D	F

⇐ M_L

Table 12.13 Possible m_s combinations

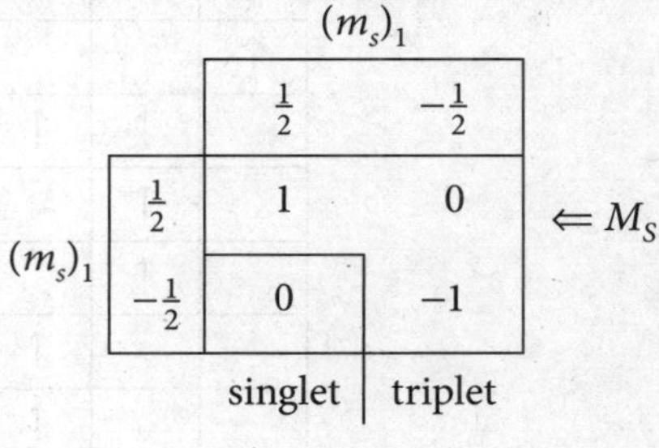

$(m_s)_1$ \ $(m_s)_1$	$\frac{1}{2}$	$-\frac{1}{2}$
$\frac{1}{2}$	1	0
$-\frac{1}{2}$	0	–1
	singlet	triplet

⇐ M_S

The terms for *pd* configuration are, therefore, (^{3}P, ^{3}D, ^{3}F) and (^{1}P, ^{1}D, ^{1}F) (note that the singlets are not components of the corresponding triplets as the two electrons are not equivalent).

12.3.5 Possible Spin Multiplicities for Systems with *n* Equivalent Optical Electrons: Some Examples

Spins of electrons with $m_s = +\frac{1}{2}$ and $-\frac{1}{2}$ are usually called *up* and *down* respectively. For a d^3 system the following combinations are possible:

Three up (or down), giving $S = \frac{3}{2}$, $2S + 1 = 4$.

Two up, one down (≡ one up, two down), giving $S = \frac{1}{2}$, $2S + 1 = 2$.

Thus a d^3 system can have quartet and triplet spin states.

In this way it can be found that a d^4 system can have singlet, triplet and quartet states. A d^5 system can have sextet, quartet and doublet spin states.

12.3.6 Finding Only the High Spin Terms for Systems with *n* Equivalent Optical Electrons ($n \geq 2$)

Obtaining only the high spin (HS) terms for a system with *n* equivalent electrons ($n \geq 2$) is a straightforward process—fill up the available space orbitals, each with one up spin (↑) electron and count the M_L and M_S values. The procedure is illustrated below with some d^n electronic configurations.

High Spin States for a d^2 System

According to Table 12.14 $S = (M_S)_{max} = |\Sigma m_s|$ (for 2 electrons in the system) $= +\frac{1}{2} + \frac{1}{2} = 1$. Spin multiplicity is $(2S + 1) = 3$. For values of M_L the following two sets are possible:

Table 12.14 Table for obtaining high spin terms for d^2 system

$m_l =$ 2	1	0	−1	−2	$M_L = \sum_\mu (m_l)_\mu$
↑	↑				3
↑		↑			2
↑			↑		1
↑				↑	0
	↑	↑			1
	↑		↑		0
	↑			↑	−1
		↑	↑		−1
		↑		↑	−2
			↑	↑	−3

(i) $M_L = \Sigma m_l$ (for two electrons in the system) = 3, 2, 1, 0, –1, –2, –3 (shown in Table 12.14).

$L = (M_L)_{max} = 3$. Term symbol: ^{3}F (triplet F).

(ii) $M_L = 1, 0, -1$ [obtained by excluding from Table 12.14 the rows corresponding to the M_L values considered in case (i)]; $L = (M_L)_{max} = 1$, Term symbol: ^{3}P (Triplet P).

Thus two high-spin states are possible for a d^2 system: ^{3}F, ^{3}P.

d^5 System

$m_l =$	2	1	0	–1	–2
	↑	↑	↑	↑	↑

$S = (M_S)_{max} = |\Sigma m_s|$ (for 5 electrons in the system) $= (+\frac{1}{2}) \times 5 = \frac{5}{2}$, spin multiplicity, $(2S + 1) = 6$, $L = (M_L)_{max} = \Sigma m_l$ (for the five electrons) = 0. Term: ^{6}S.

12.3.7 The Hole Formalism

A *hole* is an absence of an electron from a completely filled subshell and it behaves like a positively charged centre, namely a positron, with the same angular momentum behaviour as that of the electron. Thus all the spectroscopic terms for the configuration p^n are the same as those for p^{6-n}; similarly, d^n(electron) and d^{10-n}(hole) produce the same terms, only the energy order in an applied electric field or in a crystal field is reversed. Thus the *LS* term symbols for two configurations in each of the following pairs are similar:

$$(p^1, p^5), (p^2, p^4), (d^1, d^9), (d^2, d^8), (d^3, d^7), (d^4, d^6)$$

For the high spin states (that is, the states with as many parallel spin electrons as possible) the hole formalism can also be applied by considering the absence of electrons from half-filled subshells; the spin multiplicities, however, must then be carefully taken into account. Thus, a high spin d^3 (electron) system has terms ^{4}F and ^{4}P where the F and P symbols are easily obtained from d^2 (hole). Similarly, d^4 (electron) = d^1 (hole; 5 – 4 = 1); $L = 2$ as for the d^1 (electron) case. Spin multiplicity for the d^4 high spin term = 5 (due to 4 unpaired electrons). So the high spin term for a d^4 (electron) system is ^{5}D.

12.3.8 Energy Order of the Terms Obtained by *LS* Coupling

The electronic configuration of an atom corresponds to occupied orbitals which result from solving the approximate Schrödinger equation, disregarding spin–orbit interactions and much of the electron–electron repulsions. Hence the configuration gives a set of $(2L + 1)$-fold degenerate levels. This degeneracy is removed by taking electron–electron repulsions fully into consideration and some *LS* terms are generated. Each of these terms splits into a set of energy levels when spin–orbit coupling is introduced; every such level is $(2J + 1)$-fold degenerate in the absence of any external magnetic field; these final levels dictate the atomic spectrum of the concerned element. The Zeeman and Stark effects are results of lifting of this $(2J + 1)$-fold degeneracy of the levels.

The terms in the *LS* coupling scheme can be arranged in energy-order by the use of the following rules (called Hund's rules) which are based on experimental observations and theoretical calculations.

Hund's rule 1 When the principal quantum number (n) is the same for a number of terms, the lowest energy term obtained by *LS* coupling is the one with the highest spin multiplicity. This rule is also called *Hund's rule of maximum spin multiplicity.*

Hund's rule 2 For *LS* terms with the same spin multiplicity (and corresponding to the same n), the lowest energy term is the one with the higher value of L.

Thus, in energy, $n^x\mathrm{S} > n^x\mathrm{P} > n^x\mathrm{D}\ldots$ (where x is spin multiplicity)

This rule can be simply rationalised as follows: if electrons during orbital motion stay as much away as possible from each other, electron repulsions are minimised. Hence a pair of electrons labelled μ, v must rotate about the nucleus in the same direction to keep the inter-electron distance $r_{\mu v}$ as large as possible; if they travel in opposite directions, $r_{\mu v}$ decreases. In the former case the individual orbital angular momentum vectors add up to give a large resultant $\vec{L}$ while in the latter case they oppose each other giving a low $\vec{L}$ by vector addition. Thus a large $\vec{L}$ corresponds to a low energy of the atom.

Hund's rule 3 For an electronic configuration with an open shell part nl^x, energy is lower for a term with lower J value if nl^x is less than half-filled; energy is lower for a term with greater J value if nl^x is more than half-filled. (Such energy orders are called 'normal' and 'inverted' *Landé orders* respectively).

Some examples of the applicability of Hund's rules are the following:

Example 1: C($1s^2\ 2s^2\ 2p^2$) and O($1s^2\ 2s^2\ 2p^4$) have the same possible terms $^1\mathrm{S}$, $^1\mathrm{D}$, $^3\mathrm{P}$ according to the hole formalism. The lowest energy state in both cases should correspond to the triplet [Rule 1]; the complete term symbols are $^3\mathrm{P}_2$, $^3\mathrm{P}_1$, $^3\mathrm{P}_0$. Carbon with less than half-filled p subshell has ground state $^3\mathrm{P}_0$ (lowest J) but oxygen with more than half-filled p subshell has ground state $^3\mathrm{P}_2$ (highest J) [Rule 3]. These are summarised in Fig. 12.7.

Example 2: The terms for B ($1s^2\ 2s^2\ 2p^1$) and F ($1s^2\ 2s^2\ 2p^5$) are the same (according to the hole formalism), namely, $^2\mathrm{P}_{\frac{3}{2}}$ and $^2\mathrm{P}_{\frac{1}{2}}$. For B (having less than half-filled p subshell) the lowest energy term is $^2\mathrm{P}_{\frac{1}{2}}$ while for F (with more than half-filled p subshell) it is $^2\mathrm{P}_{\frac{3}{2}}$.

12.3.9 The Energy Separation Between Successive Levels in a Given Spin Multiplet

Symbols of the levels in a given spin multiplet differ in their J values (not in L and S values). Their relative energies can be compared by a simple rule known as the *Landé interval rule* which states that *the energy separation between two successive levels in a given spin multiplet is proportional to the higher J value.* For example, carbon has a triplet $^3\mathrm{P}_2$, $^3\mathrm{P}_1$, $^3\mathrm{P}_0$. From the Landé interval rule we find

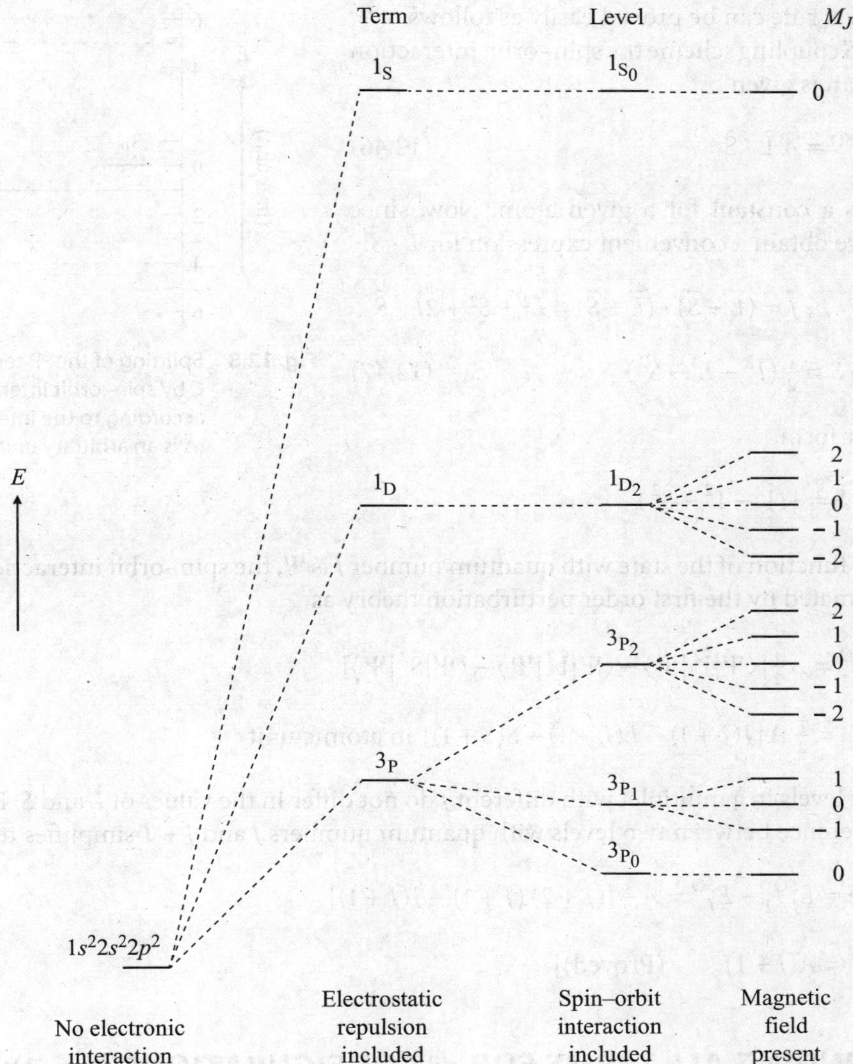

Fig. 12.7 *LS* terms and qualitative energy levels of carbon in the absence and presence of external magnetic field; for oxygen, only the ^{3}P levels are to be arranged in an inverted order

$$\frac{E(^3P_2) - E(^3P_1)}{E(^3P_1) - E(^3P_0)} = \frac{A \cdot 2}{A \cdot 1} = 2 \tag{12.45}$$

This has been experimentally verified for C with electronic configuration $2p^2$ which gives the term ^{3}P. The prediction in Eq. (12.45) by Landé interval rule is shown schematically in Fig. 12.8.

Whether the *LS* coupling scheme is operating in a given system or not can be verified by checking the ratio of predicted energy gaps in a multiplet against the observed experimental values.

[The interval rule can be proved easily as follows:

In the *LS* coupling scheme the spin–orbit interaction Hamiltonian is given by

$$\hat{H}^{SO} = A\,\hat{L}\cdot\hat{S} \tag{12.46}$$

where A is a constant for a given atom. Now, since $\vec{J} = \vec{L} + \vec{S}$ we obtain a convenient expression for $\vec{L}\cdot\vec{S}$:

$$J^2 = \vec{J}\cdot\vec{J} = (\vec{L}+\vec{S})\cdot(\vec{L}+\vec{S}) = L^2 + S^2 + 2\vec{L}\cdot\vec{S}$$

or,
$$\vec{L}\cdot\vec{S} = \tfrac{1}{2}(J^2 - L^2 - S^2) \tag{12.47}$$

In operator form,

Fig. 12.8 Splitting of the ^{3}P term of C by spin–orbit interaction according to the interval rule (Δ is an arbitrary unit of energy)

$$\hat{L}\cdot\hat{S} = \tfrac{1}{2}(\hat{J}^2 - \hat{L}^2 - \hat{S}^2) \tag{12.48}$$

If the wave function of the state with quantum number J is Ψ, the spin–orbit interaction energy E_J^{SO}, is estimated by the first order perturbation theory as

$$\begin{aligned} E_J^{SO} &= A\tfrac{1}{2}[\langle\Psi|\hat{J}^2|\Psi\rangle - \langle\Psi|\hat{L}^2|\Psi\rangle - \langle\Psi|\hat{S}^2|\Psi\rangle] \\ &= \tfrac{1}{2}A\,[J(J+1) - L(L+1) - S(S+1)] \text{ in atomic unit} \end{aligned} \tag{12.49}$$

The energy levels in a multiplet with different J do not differ in the values of L and S. Hence the energy difference between two levels with quantum numbers J and $J+1$ simplifies to

$$\begin{aligned} \Delta E = E_{J+1}^{SO} - E_J^{SO} &= A\,\tfrac{1}{2}[(J+2)(J+1) - J(J+1)] \\ &= A(J+1) \qquad \text{(Proved)]} \end{aligned} \tag{12.50}$$

12.4 FINDING ALL TERMS FOR d^n CONFIGURATIONS ($n > 2$): A MORE CONVENIENT METHOD

Although the method of tabulation of microstates discussed in Section 12.3.3 is quite general, it is laborious for systems with more than 2 electrons. A flowchart method seems to be more convenient and is discussed here for d^3, d^4 and d^5 systems.

12.4.1 d^3 CONFIGURATION

d^3 high spin states are already found to be ^{4}F and ^{4}P. Next we consider the combination of m_l quantum numbers for obtaining the lower spin states of d^3. To do this, we start with d^2 high spin state and add a d-electron as shown in Scheme 12.1.

$d^3 = d^2$(up spin) $+ d^1$(down spin); d^2(up spin) terms are ^{3}F and ^{3}P.

Scheme 12.1

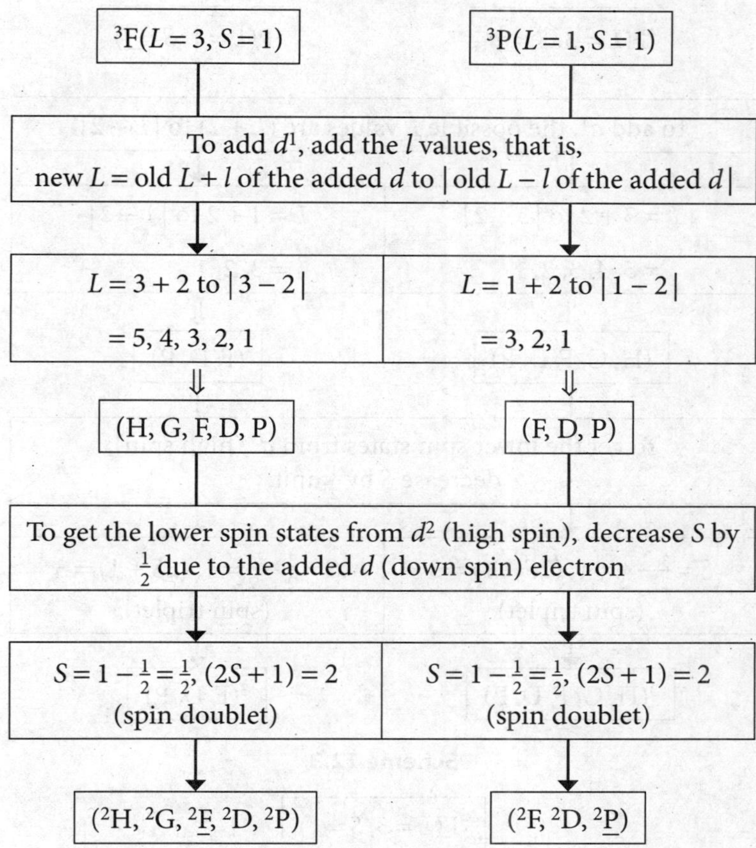

The underlined ^2F and ^2P terms are components of the already found ^4F and ^4P of d^3 (high spin) and hence they are excluded from the d^3 doublets. So the spin doublet terms of a d^3 configuration are

^{2}(H, G, F, D(2), P)

[D(2) means that ^2D term occurs twice]

12.4.2 d^4 Configuration

d^4 high spin states are known to be ^5D (from earlier discussions). For the lower spin states of d^4, two cases may arises:

d^3(up spin) + d^1(down spin) and d^2(up spin) + d^2(down spin)

We consider the former case first. The d^3(up spin) terms are already known to be ^4F and ^4P. Now we follow Scheme 12.2 which is analogous to Scheme 12.1.

Excluding one ^3D (the underlined one in Scheme 12.2) for being a component of ^5D (d^4 up spin), the spin triplets of a d^4 configuration are ^{3}(H, G, F(2), P(2)).

Scheme 12.2

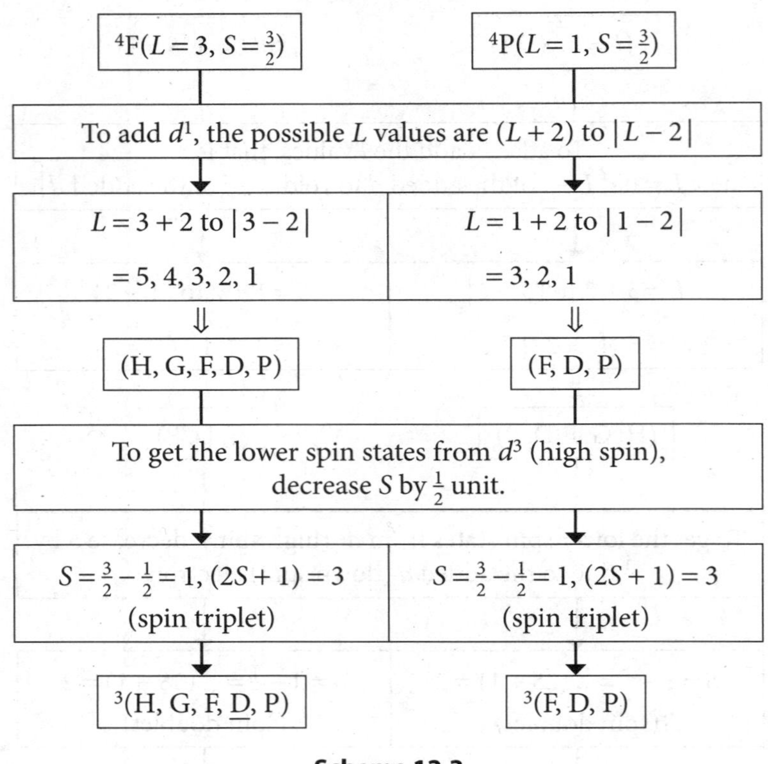

Scheme 12.3

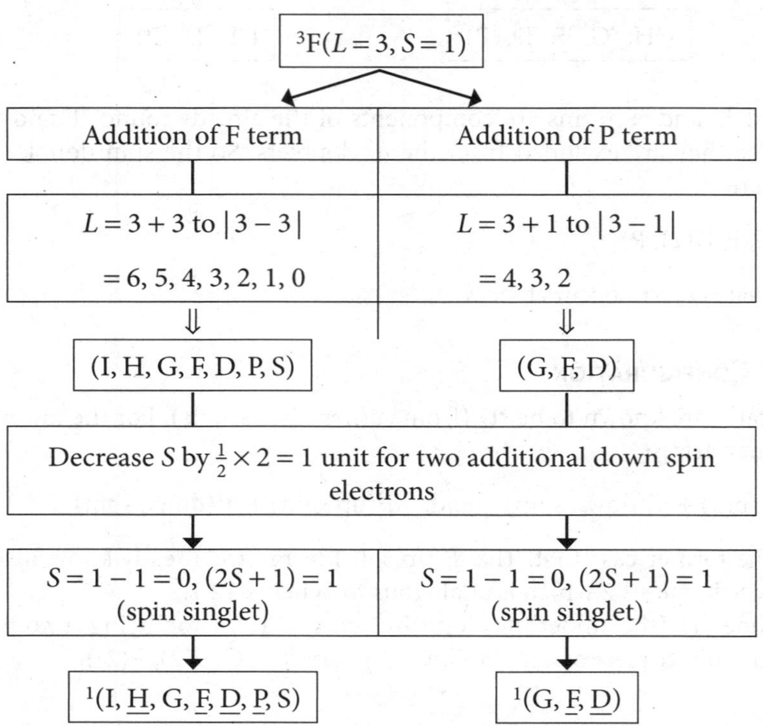

For the next case we start from d^2 (up spin) terms, ^{3}F and ^{3}P. Addition of two down spin d electrons (which also give ^{3}F and ^{3}P) requires addition of F and P terms to each of ^{3}F and ^{3}P (coming from d^2 up spin configuration) and follow a similar procedure (Schemes 12.3 and 12.4).

Scheme 12.4

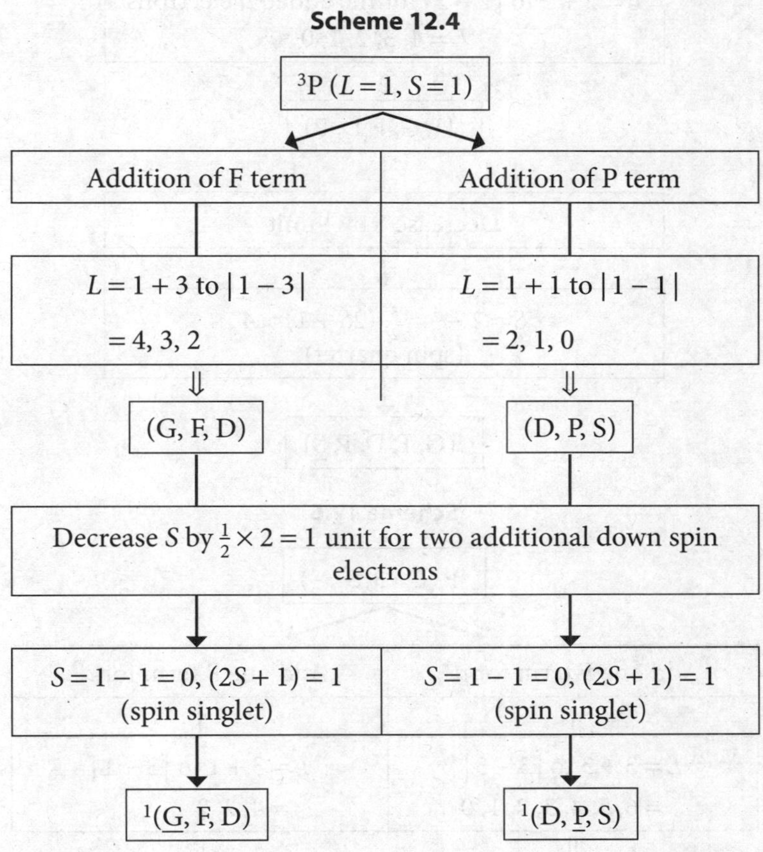

Of these, ^{1}H, one ^{1}G, two ^{1}F, two ^{1}D and two ^{1}P are excluded as they are components ^{3}H, ^{3}G, two ^{3}F, ^{3}D, ^{5}D and two ^{3}P respectively. Hence the list of singlet spin states for d^4 configuration is ^{1}(I, G(2), F, D(2), S(2)).

12.4.3 d^5 Configuration

The d^5 electronic configuration has the high spin term ^{6}S. The lower spin terms may be obtained by two possible spin combinations,

(i) d^4(up spin) + d^1(down spin) and (ii) d^3 (up spin) + d^2(down spin)

d^4 (all spins up) gives ^{5}D. Hence for case (i) the present procedure can be easily applied (Scheme 12.5):

Scheme 12.5

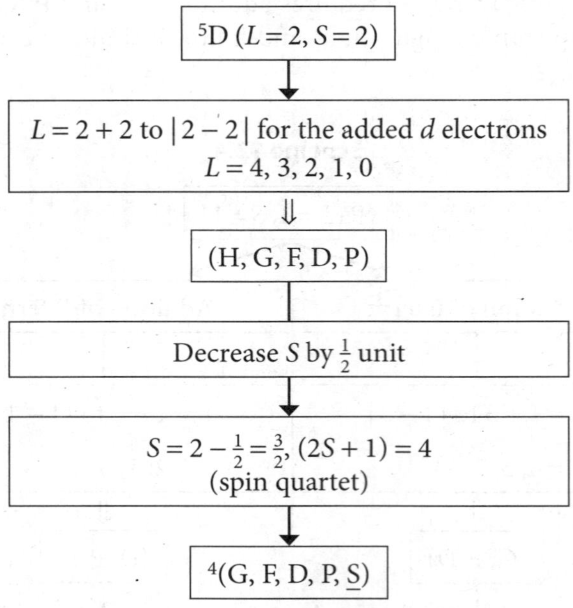

Scheme 12.6

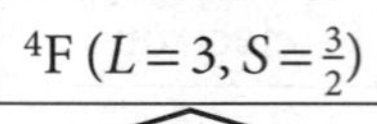

Add one F term for d^2	Add one P term for d^2
$L = 3 + 3$ to $\|3 - 3\|$ $= 6, 5, 4, 3, 2, 1, 0$	$L = 3 + 1$ to $\|3 - 1\|$ $= 4, 3, 2$
⇓ (I, H, G, F, D, P, S)	⇓ (G, F, D)

^4P ($L = 1, S = \frac{3}{2}$)

Add one F term for d^2	Add one P term for d^2
$L = 1 + 3$ to $\|1 - 3\|$ $= 4, 3, 2$	$L = 1 + 1$ to $\|1 - 1\|$ $= 2, 1, 0$
⇓ (G, F, D)	⇓ (D, P, S)

^{4}S is a component of ^{6}S for d^5. Hence the spin quartets of d^5 configuration are ^{4}(G, F, D, P). For case (ii) we note that d^3 (up spin) gives the terms ^{4}F and ^{4}P while d^2 (down spin, which is also a high spin configuration) gives ^{3}F and ^{3}P. So the terms for case (ii) will be generated by adding ^{3}F and ^{3}P terms to each case of ^{4}F and ^{4}P and lowering S by $2 \times \frac{1}{2} = 1$ unit (Scheme 12.6).

For each of these cases S will be $\frac{3}{2} - 1 = \frac{1}{2}$, $(2S + 1) = 2$ (doublet). Hence the doublet terms generated for d^5 are

$$^2(\mathrm{I, H, \underline{G}, \underline{F}, \underline{D}, \underline{P}, \underline{S}}),\ ^2(\mathrm{G, F, D}),\ ^2(\mathrm{G, F, D}),\ ^2(\mathrm{D, P, S})$$

We now exclude one ^{2}S, one ^{2}P, one ^{2}D, one ^{2}F, one ^{2}G as they are respectively components of ^{6}S and the quartet terms already obtained. Thus the spin doublet terms for d^5 are ^{2}(I, H, G(2), F(2), D(3), P, S). The results of the *LS* coupling scheme for some systems with d^n-configuration, which will be useful in the discussion of spectra of transition metal complexes, are summarised in Table 12.15.

Table 12.15 Russell–Saunders terms for d^n-configurations

Configuration	Term
d^1, d^9	^{2}D
d^2, d^8	^{1}G, ^{1}D, ^{1}S, ^{3}F, ^{3}P
d^3, d^7	^{2}H, ^{2}G, ^{2}F, ^{2}D(2), ^{2}P, ^{4}F, ^{4}P
d^4, d^6	^{1}I, ^{1}G(2), ^{1}F, ^{1}D(2), ^{1}S(2), ^{3}H, ^{3}G, ^{3}F, ^{3}D, ^{3}P, ^{5}D
d^5	^{2}I, ^{2}H, ^{2}G(2), ^{2}F(2), ^{2}D(3), ^{2}P, ^{2}S, ^{4}G, ^{4}F, ^{4}D, ^{1}G, ^{6}S
d^{10}	^{1}S

12.5 *jj* COUPLING SCHEME FOR MANY-ELECTRON SYSTEMS

In this scheme, which is valid for atoms with high atomic number ($Z \geq 30$), the orbital and spin angular momenta of each individual electron couple strongly to produce a resultant angular momentum $\vec{j}_\mu (\mu = 1, 2, \ldots, N)$. Then by electron–electron interaction these $\vec{j}_\mu$ vectors couple to produce a resultant angular momentum $\vec{J}$ with quantised magnitudes $\sqrt{J(J+1)}$ and quantised z-component, M_J (in units of $\hbar$). Since the *LS* coupling scheme is sufficient to explain the spectra of transition metal complexes, we shall not elaborate much on the *jj* coupling scheme and only some examples will be given. There is no universally accepted convention for writing *jj* term symbols. It is, however, common practice to write the individual $j_1, j_2, \ldots$ values in a parenthesis and write the resulting J as a subscript.

12.5.1 $6p^17s^1$ Configuration

Here $l_1 = 1, s_1 = \frac{1}{2}, j_1 = l_1 + s_1$ to $|l_1 - s_1| = 1 + \frac{1}{2}$ to $\left|1 - \frac{1}{2}\right| = \frac{3}{2}, \frac{1}{2}$

$l_2 = 0, s_2 = \frac{1}{2}, j_2 = l_2 + s_2$ to $|l_2 - s_2| = 0 + \frac{1}{2}$ to $\left|0 - \frac{1}{2}\right| = \frac{1}{2}$

With $j_1 = \frac{3}{2}$ and $j_2 = \frac{1}{2}$ we have the following values of J:

$$J = j_1 + j_2 \text{ to } |j_1 - j_2| = \tfrac{3}{2} + \tfrac{1}{2} \text{ to } \left|\tfrac{3}{2} - \tfrac{1}{2}\right| = 2, 1$$

With $j_1 = \frac{1}{2}$ and $j_2 = \frac{1}{2}$, we have

$$J = j_1 + j_2 \text{ to } |j_1 - j_2| = \tfrac{1}{2} + \tfrac{1}{2} \text{ to } \left|\tfrac{1}{2} - \tfrac{1}{2}\right| = 1, 0$$

The term symbols for the states arising out of *jj* coupling are written as $(j_1, j_2, \ldots)_J$. Thus we have the following four *jj* term symbols for the $6p^17s^1$ configuration:

$$\left(\tfrac{3}{2}, \tfrac{1}{2}\right)_2, \left(\tfrac{3}{2}, \tfrac{1}{2}\right)_1, \left(\tfrac{1}{2}, \tfrac{1}{2}\right)_1 \text{ and } \left(\tfrac{1}{2}, \tfrac{1}{2}\right)_0$$

[It may be verified that the *LS* scheme also gives four terms with this configuration.]

12.5.2 Configuration with Two Non-Equivalent and Equivalent *p* Electrons

For *jj* coupling of the individual spin and orbital angular momenta we first construct the (m_l, m_s) tables [Tables 12.16 and 12.17].

Table 12.16 Construction of possible $\vec{J}$ vectors for electron 1

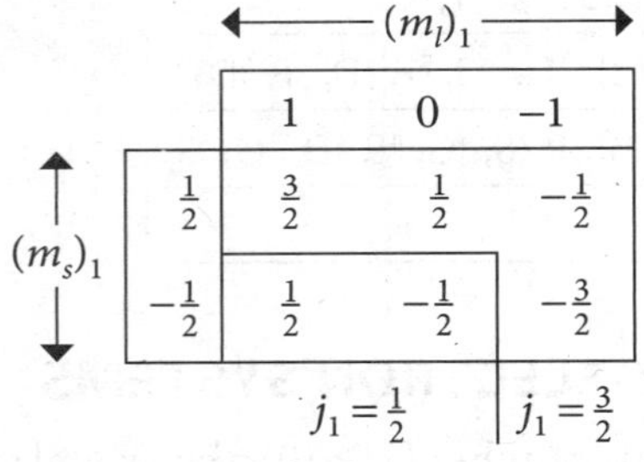

Table 12.17 Construction of possible $\vec{J}$ vectors for electron 2

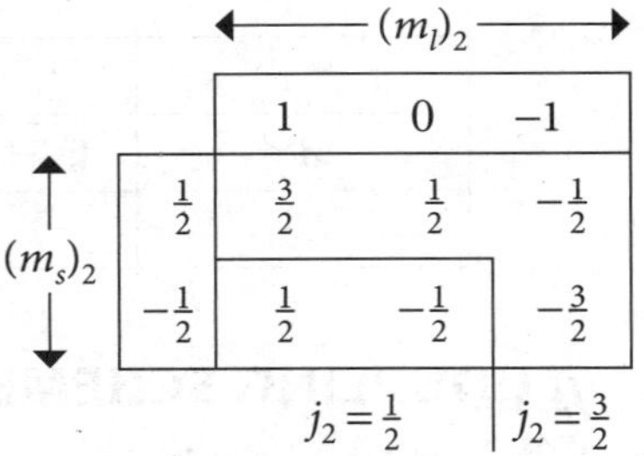

For two non-equivalent *p* electrons (as in the configuration 6*p*7*p*), all j_1, j_2 combinations are possible:

If $j_1 = j_2 = \frac{3}{2}$, $J = j_1 + j_2$ to $|j_1 - j_2| = 3, 2, 1, 0$.

If $j_1 = \frac{3}{2}$, $j_2 = \frac{1}{2}$, $J = j_1 + j_2$ to $|j_1 - j_2| = 2, 1$; the same is true for $j_1 = \frac{1}{2}$ and $j_2 = \frac{3}{2}$.

If $j_1 = j_2 = \frac{1}{2}$, $J = j_1 + j_2$ to $|j_1 - j_2| = 1, 0$.

Thus the *jj* coupling terms for two non-equivalent *p* electrons are

$$\left(\tfrac{3}{2}, \tfrac{3}{2}\right)_{3,2,1,0}, \left(\tfrac{3}{2}, \tfrac{1}{2}\right)_{2,1}, \left(\tfrac{1}{2}, \tfrac{3}{2}\right)_{2,1} \text{ and } \left(\tfrac{1}{2}, \tfrac{1}{2}\right)_{1,0}.$$

The total number of microstates = (7 + 5 + 3 + 1) + (5 + 3) + (5 + 3) + (3 + 1) = 36.

For two equivalent *p* electrons (as in p^2), the exclusion principle must be obeyed.

If $j_1 = j_2 = \frac{3}{2}$, $J = j_1 + j_2$ to $|j_1 - j_2| = \underline{3}, 2, \underline{1}, 0$.

The underlined J values are to be excluded for the following reasons:

The case $J = 3$ corresponds to $(m_l)_1 = (m_l)_2 = 1; (m_s)_1 = (m_s)_2 = \frac{1}{2}$

Similarly the case $J = 1$ corresponds to $(m_l)_1 = (m_l)_2 = 0; (m_s)_1 = (m_s)_2 = \frac{1}{2}$

Thus both these cases violate the exclusion principle.

Hence with $j_1 = j_2 = \frac{3}{2}$ we get the term $\left(\frac{3}{2}, \frac{3}{2}\right)_{2,0}$.

If $j_1 = j_2 = \frac{1}{2}$, $J = j_1 + j_2$ to $|j_1 - j_2| = \underline{1}, 0$

But $J = 1$ is to be disregarded because it corresponds to $(m_l)_1 = (m_l)_2 = 0; (m_s)_1 = (m_s)_2 = \frac{1}{2}$

So in this case the term is $\left(\frac{1}{2}, \frac{1}{2}\right)_0$.

If $j_1 = \frac{3}{2}, j_2 = \frac{1}{2}$ we have $J = j_1 + j_2$ to $|j_1 - j_2| = 2, 1$. The terms are $\left(\frac{3}{2}, \frac{1}{2}\right)_{2,1}$.

[The reverse order, $j_1 = \frac{1}{2}, j_2 = \frac{3}{2}$ is not considered as the electrons are equivalent.]

Thus, a p^2 configuration has the jj coupling terms:

$$\left(\frac{3}{2}, \frac{3}{2}\right)_{2,0}, \left(\frac{3}{2}, \frac{1}{2}\right)_{2,1} \text{ and } \left(\frac{1}{2}, \frac{1}{2}\right)_0$$

The total number of microstates = (5 + 1) + (5 + 3) + (1) = 15, as in the LS coupling scheme.

Note: In the LS coupling scheme, the terms are described by the values of L and S; they are good quantum numbers. In the jj scheme the resultant orbital angular momentum vector does not separately exist nor does the resultant spin; it is the individual j vectors and their resultant J that determine the term and hence the good quantum number in this case is J.

12.6 A SIMPLE EXAMPLE OF INTERPRETATION OF ATOMIC SPECTRA BY *LS* COUPLING

There exists a wealth of literature on interpretation of atomic spectra (White, 1934). Only one very common example is briefly given here.

12.6.1 The Doublet Structure in the Visible Part of the Spectrum of Sodium

The golden yellow flame of Na (or, the golden yellow light emitted by sodium vapour lamps used for illumination of streets) on resolution gives a closely spaced doublet of spectral lines at $\lambda = 589.0$ and 589.6 nm. The ground and some low-lying excited states of sodium have the configurations, [Ne]$3s^1$, [Ne]$3p^1$, [Ne]$3d^1$ and [Ne]$4s^1$. Application of LS coupling scheme gives the following terms from these configurations (only the optical electron should be considered):

$$[\text{Ne}]\,3s^1 : L = 0, S = \tfrac{1}{2}, J = 0 + \tfrac{1}{2} \text{ to } \left|0 - \tfrac{1}{2}\right| = \text{only } \tfrac{1}{2}, (2S+1) = 2; \text{Term} : {}^2\text{S}_{\frac{1}{2}}$$

$$[\text{Ne}]\,3p^1: L = 1, S = \tfrac{1}{2}, J = 1 + \tfrac{1}{2} \text{ to } \left|1 - \tfrac{1}{2}\right| = \tfrac{3}{2}, \tfrac{1}{2}, (2S+1) = 2; \text{Terms: } {}^2\text{P}_{\frac{3}{2}}, {}^2\text{P}_{\frac{1}{2}}$$

$$[\text{Ne}]3d^1: L = 2, S = \tfrac{1}{2}, J = 2 + \tfrac{1}{2} \text{ to } \left|2 - \tfrac{1}{2}\right| = \tfrac{5}{2}, \tfrac{3}{2}, (2S+1) = 2; \text{Terms: } {}^2\text{D}_{\frac{5}{2}}, {}^2\text{D}_{\frac{3}{2}}$$

and so on. These energy levels can be arranged in increasing order, by following Hund's rules (Fig. 12.9).

The selection rules for electronic transitions in atomic spectroscopy are: $\Delta S = 0$, $\Delta L = \pm 1$ and $\Delta J = 0, \pm 1$. Hence both the transitions ${}^2\text{P}_{\frac{3}{2}} \rightarrow {}^2\text{S}_{\frac{1}{2}}$ and ${}^2\text{P}_{\frac{1}{2}} \rightarrow {}^2\text{S}_{\frac{1}{2}}$ are permitted. These two transitions are responsible for the closely spaced sodium-D lines.

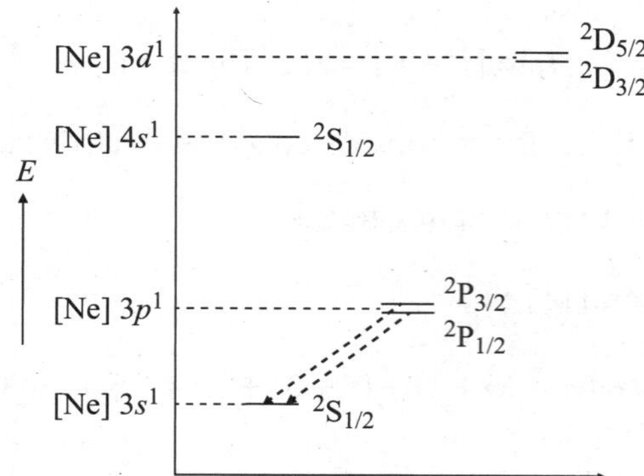

Fig. 12.9 Partial energy level diagram for Na and the spectroscopic transitions responsible for the emission of sodium-D light.

12.7 TERM SYMBOLS FOR ELECTRONIC STATES OF MOLECULES

The electronic configuration of a molecule, which only tells us how many electrons there are in the various MOs, is too crude an approximation to explain the observed UV–vis spectra of molecules. However, by incorporating inter-electron repulsions adequately, it is possible to obtain from a given configuration the possible spectroscopic states of a molecule. Such states are described by *terms* having the general symbol ${}^{2S+1}\Gamma_\mu$ where the left superscript is the spin multiplicity of the state and Γ_μ is the symbol of the irreducible representation of the molecular point group obtained from the direct product of the representations for the occupied MOs. The total degeneracy of a term ${}^{2S+1}\Gamma_\mu$ is $(2S + 1)$ times the dimension of Γ_μ; physically this is the number of independent spin-orbital wave functions required to describe the state. For closed shell configurations, the spin multiplicity is 1 since all MOs are occupied by electron pairs (giving $S = 0$) and the term is always ${}^1A$ (with appropriate subscripts and/or primes) corresponding to the totally symmetric representation. For configurations with only one occupied non-degenerate MO beyond a closed shell, the task is simple. The real difficulty arises when degenerate MOs are partly occupied. The following examples will make the procedure clear.

12.7.1 Terms from the Ground State Electronic Configuration of O_2

Configuration: $(1\sigma_g^+)^2\,(1\sigma_u^{+*})^2\,(2\sigma_g^+)^2\,(2\sigma_u^{+*})^2\,(3\sigma_g^+)^2\,(1\pi_u)^4\,(1\pi_g^*)^2 = [\text{closed shell}]\,(1\pi_g^*)^2$

The possible values of S are 1, 0; $2S + 1 = 3, 1$ (singlets and triplets). The point group of O_2 is $D_{\infty h}$. By inspection of the $D_{\infty h}$ character table the required direct product representation is reduced as

$$\Pi_g \times \Pi_g = \Sigma_g^+ + \Sigma_g^- + \Delta_g$$

The identity $4\cos^2\phi = 2\cos 2\phi + 2$ is required to check this result. The task is now to assign the spin multiplicities (3 or 1). There are two orbitally degenerate MOs (π_g); so four spin-orbitals are possible and each of them can accommodate only one electron according to the Pauli principle. The spin orbitals can be filled in ${}^4C_2 = 6$ ways; thus 6 spin-orbital wave functions, antisymmetric with respect to interchange of electrons, are necessary to describe the possible states corresponding to the given configuration. Let the term symbols be ${}^x\Sigma_g^+$, ${}^y\Sigma_g^-$ and ${}^z\Delta_g$. Then

$$x + y + 2z = 6$$

where the spin multiplicities x, y and z must be either 3 or 1. Obviously z cannot be 3. Hence $z = 1$ and $x + y = 4$. At this point it is necessary to examine the symmetry properties of the spin and orbital wave functions separately.

Spin wave functions of individual electrons are usually denoted by α and β for $m_s = \frac{1}{2}$ and $-\frac{1}{2}$ respectively. The spin part of the π_g MOs occupied by two electrons labelled 1 and 2 may be as follows:

$\alpha(1)\alpha(2)$, $M_S = 1$, symmetric with respect to interchange of electrons

$\beta(1)\beta(2)$, $M_S = -1$, symmetric with respect to interchange of electrons

$\frac{1}{\sqrt{2}}[\alpha(1)\beta(2) + \alpha(2)\beta(1)]$, $M_S = 0$, symmetric with respect to interchange of electrons

$\frac{1}{\sqrt{2}}[\alpha(1)\beta(2) - \alpha(2)\beta(1)]$, $M_S = 0$, antisymmetric with respect to interchange of electrons

The first three functions correspond to the triplet spin state and the fourth one to the singlet state. In brief, triplet spin functions are symmetric and singlets are antisymmetric with respect to interchange of electrons.

The basis function of Σ_g^+ (the totally symmetric representation of $D_{\infty h}$) is invariant under all symmetry operations. Hence the orbital part of the MO wave function which acts as a basis of Σ_g^+ must be symmetric with respect to interchange of electrons; the spin part must therefore be antisymmetric, that is, singlet in order that the overall spin-orbital wave function is antisymmetric with respect to the interchange of electrons. Hence $x = 1$, making $y = 3$. The terms from the configuration $\ldots(1\pi_g^*)^2$ of O_2 are therefore ${}^1\Sigma_g^+$, ${}^3\Sigma_g^-$ and ${}^1\Delta_g$. The energy-order of these terms is obtained from Hund's rules: ${}^3\Sigma_g^- < {}^1\Delta_g < {}^1\Sigma_g^+$ (lower energy for higher spin multiplicity and for terms with the same S, higher z-component of orbital angular momentum).

12.7.2 Terms from the First Excited State Electronic Configuration of O_2

Configuration: $(1\sigma_g^+)^2\,(1\sigma_u^{+*})^2\,(2\sigma_g^+)^2\,(2\sigma_u^{+*})^2\,(3\sigma_g^+)^2\,(1\pi_u)^3\,(1\pi_g^*)^3$ = [closed shell] $(1\pi_u)^3\,(1\pi_g^*)^3$

The hole formalism makes this configuration equivalent to $\ldots 1\pi_u)^1\,(1\pi_g^*)^1$; the required direct product is again $\Pi_u \times \Pi_g = \Sigma_u^+ + \Sigma_u^- + \Delta_u$ and the spin states are singlets and triplets as before. Since the concerned MOs are different, the Pauli principle allows both the spin multiplicities to be assigned to all the orbital symbols and the terms are ${}^1\Sigma_u^+$, ${}^1\Sigma_u^-$, ${}^1\Delta_u$, ${}^3\Sigma_u^+$, ${}^3\Sigma_u^-$ and ${}^3\Delta_u$. Hund's rule states that triplets are lower in energy than singlets and that the Δ states are below the Σ. Any further ordering of energy of the Σ states requires sophisticated calculation or experiment.

SUMMARY

- Magnetic interaction between spin and orbital momenta of electrons in atoms is called *spin–orbit coupling* and it produces the accessible states of atoms. Spectroscopists describe such states by *term symbols* which are determined by the vector atom model.
- For atoms with low or moderate nuclear charge ($Z \le 30$), spin–orbit (SO) interaction energy is much weaker than electron–electron repulsion and in them the *Russell–Saunders* or *LS* coupling scheme operates. In this, the individual orbital angular momenta of the electrons precess *rapidly* about their resultant ($\vec{L}$), the individual spin angular momenta precess *rapidly* about their resultant ($\vec{S}$), and the vectors $\vec{L}$ and $\vec{S}$ precess *slowly* about their resultant $\vec{J}$; this $\vec{J}$ may be anywhere on the surface of a cone symmetrical about the Z-axis maintaining a constant J_Z component.
- For atoms with higher Z, spin–orbit (SO) interaction is stronger and the operating scheme is *jj* coupling. Here, the spin and orbital angular momenta of each electron precess *rapidly* about their resultant $\vec{j}$; these $\vec{j}$ for all electrons precess slowly about their resultant $\vec{J}$ maintaining a quantised J_Z.
- There are several methods for obtaining term symbols for the accessible states of the atoms; some of them are explained.
- Hund's rules are used to determine the relative energy order of the terms and Landé interval rule gives the ratio of energy intervals in a multiplet; these are useful in interpreting complex atomic spectra.
- Electronic states of molecules are given by molecular term symbols obtainable from direct product representations of the occupied molecular orbitals with adequate spin multiplicity.

LEARNING OUTCOME

After reading this chapter the student will be able to (i) understand the quantum mechanical rationale for the vector atom model of spin–orbit coupling in atoms, (ii) apply the model to determine the possible spectroscopic terms in a given electronic configuration using both *LS* and *jj* coupling scheme, (iii) apply Hund's rules to determine the relative energy order of the terms, (iv) apply Landé interval rule to explain the energy orders of levels in a multiplet (v) interpret atomic spectra and (vi) determine molecular term symbols.

EXERCISES

1. Obtain the terms and energy levels by constructing the table of microstates for the electronic configurations (a) p^3 and (b) d^2. What are the lowest energy terms and energy levels in each case?

2. Derive the ground state terms and energy levels for the elements (Sc to Zn) of the third row of the periodic table without using the method of tabulating the microstates.

3. What are the possible terms for the configurations (a) $nsn'p$ (b) $nsn'd$ and (c) $npn'pn''p$? For each term, what are the possible energy levels?

4. The term symbol of the ground state of Na is $^2S_{\frac{1}{2}}$. Is it really a doublet energy level?

5. Draw a qualitative diagram for the terms and energy levels of vanadium, [Ar]$3d^34s^2$ and cobalt, [Ar]$3d^74s^2$. How do their ground states differ and why? (Hint: use the hole formalism)

6. What are the allowed LS terms for the configuration sp^2? Find the J values for each term.

7. Although Si and Pb have similar valence shell configurations, their allowed spectroscopic terms are different. Elucidate.

8. Draw the energy levels of oxygen atom in the presence of an applied magnetic field.

9. Denote the following electronic transitions using spectroscopic term symbols.
   (a) Na: [Ne]$3s^1 \rightarrow$ [Ne]$3p^1$ (b) K: [Ar]$5d^1 \rightarrow$ [Ar]$4p^1$
   (c) Mg: [Ne]$3s^13p^1 \rightarrow$ [Ne]$3s^2$

10. Find the number of microstates for the configurations: $p^3, f^3, d^6, ns^1(n+1)p^1, np^2nd^1, 2p^13p^1$.

11. Choose the correct option(s). (More than one may be correct)
    (a) The spectroscopic terms of a d^2 ion in the free state obeys the energy order:
        (i) $^3P > {}^3F > {}^1S > {}^1D > {}^1G$ (ii) $^1S > {}^1D > {}^1G > {}^3P > {}^3F$
        (iii) $^1G > {}^1D > {}^1S > {}^3P > {}^3F$ (iv) $^3F > {}^3P > {}^1S > {}^1D > {}^1G$
    (b) For the fluorine atom, the energy levels of the ^{2}P terms are in
        (i) normal Landé order (ii) inverted Landé order (iii) no definite order
    (c) The number of microstates for a p^2 configuration in the LS coupling scheme should be (i) different from that in the jj scheme (ii) same as that in the jj scheme (iii) equal to the number of ways of filling 6 boxes with two balls (iv) same as that for a p^3 configuration.

12. Find the ratio of energy separation between successive levels in the ^{4}D multiplet of an atomic spectrum.

13. In the atomic spectrum of an element of intermediate atomic number, the separations between adjacent energy levels of increasing energy are found to be approximately in the

ratio 3:5. Using the Landé interval rule, assign the quantum numbers *L*, *S* and *J* to these levels. [Ans. $L = 1$, $S = 3/2$]

14. Draw the energy levels for the ground state and the excited states corresponding to the first two one-electron promotions for the neutral calcium atom. Predict some possible transitions involving these states. Explain how the validity of *LS* coupling scheme in the case of calcium can be established from the energy separation ratios of the multiplets resulting from these transitions.

15. Find the term symbols for the first excited state configurations of (a) N_2 and (b) benzene.

Suggested Books and References for Further Reading

1. White H E, *Introduction to Atomic Spectra*, McGraw-Hill Book Company Inc., **1934.**
2. Eisberg R and Resnick R, *Quantum Physics of Atoms, Molecules, Solids, Nuclei and Particles*, John Wiley & Sons Inc., NY, **1985.**
3. Pilar F L, *Elementary Quantum Chemistry*, Dover Publications, Inc. NY, **2001.**
4. Russell H N and Saunders F A, *Astrophys. J.*, 61 (38) **1925.**
5. Rakshit S C, *Atoms and d^n-ions in Perturbation Fields*, Sarat Book House, Calcutta, **1998.**

13 Crystal Field Theory and Bonding in Metal Complexes

Learning objectives

- ➢ Splitting of spectroscopic terms of transition metal ions in crystal fields
- ➢ Correlation diagrams, method of descending symmetry, energy level diagram, interpretation of UV–vis spectra of complexes, Jahn–Teller theorem and distortion
- ➢ Inadequacy of CFT: spectrochemical and nephelauxetic series
- ➢ Orgel diagram, vibronic coupling and polarisation
- ➢ Effect of odd number of electrons: concept of double group

13.1 INTRODUCTION

The method of applying group theory to explain how the degeneracy of energy levels of an atom or ion is removed when it is placed in a crystal lattice was shown first by Bethe (1929). Van Vleck extended this method to transition metal complexes to explain their spectral and magnetic properties (1932, 1935). These works are the foundations of the *crystal field theory* (CFT) for transition metal complexes. In this theory, the ligands are treated as point negative charges (or dipoles) surrounding the central metal ion. It is assumed that the negatively charged ligands (or the negative ends of the ligand dipoles) electrostatically repel the metal *d* electrons and split the *d*-orbital energy levels. The splitting patterns are determined by the geometry of the complex (that is, symmetry of the environment surrounding the metal ion). After some initial success the necessity of considering some covalency, that is, some overlap of metal *d*-orbitals and ligand orbitals was soon felt (for example, by the discovery of the *nephelauxetic effect*) and CFT was modified by introducing some adjustable parameters. This adjusted CFT was termed *ligand field theory*. Finally, the molecular orbital theory came in and all properties of transition metal complexes can now be explained by the MO theory. However, the symmetry-related results of the MO theory, ligand field theory and CFT are all the same. It is, therefore, instructive to deal with the CFT in some detail.

CFT is based on a perturbation treatment carried out on the free atom/ion as the starting point (unperturbed system). We begin the discussions by recalling that the point group of a Hamiltonian operator ($\hat{H}$) is the group formed by the symmetry operators with which $\hat{H}$ commutes. The point group of an atom or ion in free space, like that of a sphere, is K_h, of which the three-dimensional pure rotation group (K) is a subgroup. The angular part of the wave function of a many-electron atom, under the *LS* coupling scheme, is given by the spherical harmonic functions $Y_L^{M_L}(\theta, \phi)$. On the one hand, functions of the set $\{Y_L^{M_L}(\theta, \phi)\}$ are eigenfunctions of the Hamiltonian of the free atom, and on the other, they form a basis set of a $(2L+1)$-dimensional irreducible representation of the pure rotation group K.

The central idea in the group theory part of Bethe's original (1929) paper is that if the free atom or ion is perturbed so that the point group of the Hamiltonian changes from K to a finite group G, the $(2L+1)$-fold irreducible representation of K reduces to a number of irreducible representations of the new group G and the original degenerate basis set splits into a number of new basis sets.

A brief discussion of this pure rotation group (K) is therefore necessary.

13.2 THE THREE DIMENSIONAL ROTATION GROUP

The set of all rotations of a sphere about axes passing through its centre has all the four group properties. For example, the result of successive rotations about two different axes is again a rotation about a third axis; this is the closure property. Rotation about a given axis through an angle 2π is the identity and this identity is the same (unique) for all imaginable axes. A rotation through angle θ about a given axis is the inverse of a rotation through angle $(2\pi - \theta)$ about the same axis. The associativity of three different rotations also holds good obviously. All rotations through the same magnitude of angle about different axes are in the same conjugate class because, for any two such axes, one can find a third axis which brings the former two axes into coincidence by rotation (Fig. 13.1) (See Rule 1, Section 1.26).

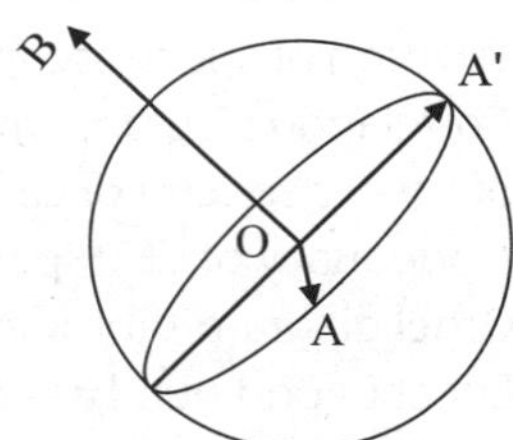

Fig. 13.1 Rotations through the same angle about two different axes, OA and OA′ are equivalent because these two axes can be brought to coincidence by a rotation about OB

13.2.1 CHARACTERS OF REPRESENTATIONS OF THE ROTATION GROUP (*K*)

In terms of spherical polar coordinates (r, θ, ϕ), the function

$$Y_l^m(\theta, \phi) = N \cdot P_l^m(\cos\theta)\, e^{im\phi} \tag{13.1}$$

is called a *spherical harmonic function.* Here N is a normalisation constant and $P_l^m(\cos\theta)$ is the associated Legendre polynomial of degree l and order m. The permissible values of m are the following integers:

$$m = l, (l-1), (l-2), \ldots, 0, \ldots, \\ -(l-2) -(l-1), -l \tag{13.2}$$

Such functions are eigenfunctions of the angular momentum operator, $\hat{L}_Z$ (in other words, angular momenta are quantised with respect to the Z axis). A rotation $\hat{C}_n$ by angle about the Z axis transforms the θ and ϕ coordinates of a general point P to θ' and ϕ' (Fig. 13.2) such that

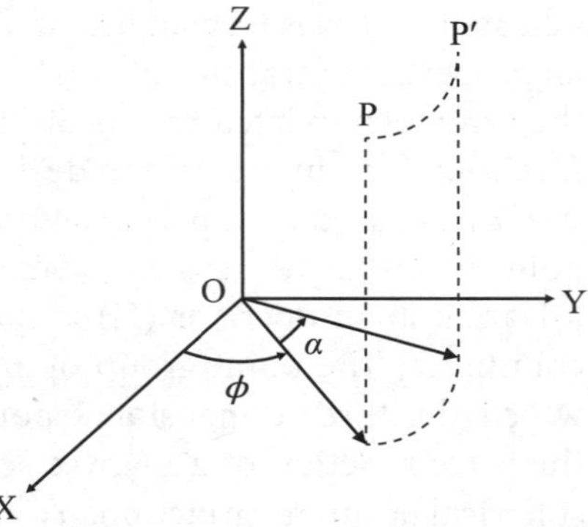

Fig. 13.2 Rotation about the Z-axis through an angle α

$$\theta' = \theta \text{ and } \phi' = \phi + \alpha \tag{13.3}$$

Accordingly, the function in Eq. (13.1) is transformed as

$$\hat{C}_n Y_l^m(\theta, \phi) = N \cdot P_l^m(\cos\theta) e^{im(\phi+\alpha)} = e^{im\alpha} \cdot Y_l^m(\theta, \phi) \tag{13.4}$$

All functions of the type (13.1) thus change, under the action of a $\hat{C}_n$ rotation about the Z-axis through angle $\alpha = 2\pi/n$, according to the following relations:

$$\hat{C}_n(e^{il\phi}) = e^{il\phi} \cdot e^{il\alpha} + 0 \cdot e^{i(l-1)\phi} + 0 \cdot e^{i(l-2)\phi} + \cdots + 0 \cdot e^{il\phi}$$

$$\hat{C}_n(e^{i(l-1)\phi} = 0 \cdot e^{il\phi} + e^{i(l-1)\alpha} \cdot e^{i(l-1)\phi} + 0 \cdot e^{i(l-2)\phi} + \cdots + 0 \cdot e^{il\phi}$$

...

...

$$\hat{C}_n(e^{-il\phi}) = 0 \cdot e^{il\phi} + 0 \cdot e^{i(l-1)\phi} + 0 \cdot e^{i(l-2)\phi} + \cdots + e^{-il\alpha} \cdot e^{il\phi}$$

Thus, with the set of functions $\{e^{il\phi}, e^{i(l-1)\phi}, \cdots, e^{il\phi}\}$ as the basis set, the matrix representation of $\hat{C}_n$ becomes

$$M(\hat{C}_n) = \begin{bmatrix} e^{il\alpha} & 0 & 0 & \cdots & 0 \\ 0 & e^{i(l-1)\alpha} & 0 & \cdots & 0 \\ 0 & 0 & e^{i(l-2)\alpha} & 0 & 0 \\ \cdots & \cdots & \cdots & \cdots & \cdots \\ 0 & 0 & 0 & 0 & e^{-il\alpha} \end{bmatrix} \tag{13.5}$$

Considering all rotations and a particular l, we thus get a set of $(2l+1) \times (2l+1)$ matrices, that is, a $(2l+1)$-dimensional representation of the full rotation group K. The general formula for the character (χ) of $\hat{C}_n$ in such a representation is now obtained as follows:

$$\begin{aligned}
\chi(\hat{C}_n) &= e^{il\alpha} + e^{i(l-1)\alpha} + e^{i(l-2)\alpha} + \cdots + e^{-il\alpha} \\
&= e^{il\alpha}[1 + e^{-i\alpha} + e^{-2i\alpha} + \cdots + e^{-2il\alpha}] \\
&= e^{il\alpha}[1 + b + b^2 + \cdots + b^{2l}] \qquad (\text{where } b = e^{-i\alpha}) \\
&= \frac{e^{il\alpha} \cdot 1 \cdot \{b^{(2l+1)} - 1\}}{b-1} \qquad (\text{sum of a GP series; number of terms} = 2l{+}1) \\
&= \frac{e^{il\alpha} \cdot e^{-2il\alpha} \cdot e^{-i\alpha} - e^{il\alpha}}{e^{-i\alpha} - 1} = \frac{(e^{-il\alpha} \cdot e^{-i\alpha} - e^{il\alpha}) \cdot e^{\frac{i\alpha}{2}}}{(e^{-i\alpha} - 1) \cdot e^{\frac{i\alpha}{2}}} \\
&= \frac{e^{-il\alpha} \cdot e^{\frac{-i\alpha}{2}} - e^{il\alpha} \cdot e^{\frac{i\alpha}{2}}}{e^{\frac{-i\alpha}{2}} - e^{\frac{i\alpha}{2}}} = \frac{e^{-i\left(l+\frac{1}{2}\right)\alpha} - e^{i\left(l+\frac{1}{2}\right)\alpha}}{e^{\frac{-i\alpha}{2}} - e^{\frac{i\alpha}{2}}}
\end{aligned}$$

$$= \frac{-2i \sin\left(l+\frac{1}{2}\right)\alpha}{-2i \sin\frac{\alpha}{2}} = \frac{\sin\left(l+\frac{1}{2}\right)\alpha}{\sin\frac{\alpha}{2}} \tag{13.6}$$

Substituing the value of α (= $2\pi/n$) for the $\hat{C}_n$ rotations we obtain the characters of a $(2l + 1)$-dimensional irreducible representation of the pure rotation group:

$$\left.\begin{aligned}
&\alpha=\pi, \quad \chi^{(l)}(\hat{C}_2) = (-1)^l \\
&\alpha=\tfrac{\pi}{2}, \quad \chi^{(l)}(\hat{C}_4) = \begin{cases} (-1)^{\frac{l}{2}} & (\text{even } l) \\ (-1)^{\frac{(l-1)}{2}} & (\text{odd } l) \end{cases} \\
&\alpha=\tfrac{2\pi}{3}, \quad \chi^{(l)}(\hat{C}_3) = \begin{cases} 1 & (l=3k) \\ 0 & (l=3k+1) \\ -1 & (l=3k+2) \end{cases}
\end{aligned}\right\} \text{with } k \text{ an integer} \tag{13.7}$$

Finally, for $\alpha = 0$ the identity operator has character = $(2l + 1)$ as shown below.

$$\chi^l(\hat{E}) = \lim_{\alpha\to 0} \frac{\sin\left(l+\frac{1}{2}\right)\alpha}{\sin\frac{\alpha}{2}}$$

$$= \lim_{\alpha\to 0} \frac{\left(l+\frac{1}{2}\right)\cos\left(l+\frac{1}{2}\right)\alpha}{\frac{1}{2}\cos\frac{\alpha}{2}} \quad \text{(by the L'Hospital theorem)}$$

$$= \frac{\left(l+\frac{1}{2}\right)}{\frac{1}{2}} = (2l+1) \tag{13.8}$$

In the rotation group K or in any of its subgroups (like the cubic pure rotation group O), two rotations through equal angles (α), irrespective of axis, are of the same conjugate class. Hence one can use the formulae (Eqs 13.6–13.8) to calculate the character of any rotation operation in the point group K or its subgroup O. This fact makes the task easy to show the splitting of atomic energy levels in the octahedral and tetrahedral fields by group theoretical methods, as will be discussed in the following sections.

13.3 EFFECT OF CRYSTAL FIELDS ON ONE-ELECTRON ENERGY LEVEL: SPLITTING

When a free atom or ion is placed in a crystal field, the isotropy of free space is destroyed by the inhomogeneous electric fields within the crystal. The point group of the atom/ion is changed from K_h (= full three-dimensional rotation group, K + centre of inversion and symmetry planes) to some finite group (G) depending on the geometry of the crystalline environment of the atom/ion. The originally irreducible representation of the full rotation group (K) generated by the spherical harmonic functions can then be reduced with respect to the finite group. The basis functions of the component irreducible representations of the latter group (G) must correspond to energy levels of the atom/ion in the crystal field of the symmetry group G. Hence each $(2l + 1)$-fold degenerate energy level of the free atom/ion splits (for $l > 0$) into a set of new energy levels of the atom/ion in the crystal field.

13.3.1 Splitting of Energy Levels of One-electron Orbitals in an Octahedral Field

Let us consider an atom/ion having only one d-electron outside a closed shell configuration. The d orbital energy level of such a system in free space is five-fold degenerate. When it is placed in an octahedral field, the point group changes from K_h to O_h. For investigation of the effect of an octahedral field on the d-orbital energy level, it is convenient to consider the subgroup O of O_h since it is known that all the d-functions are of g-type with respect to the inversion operator $\hat{\imath}$ and that the point group O_h is the direct product group, $O \times C_i$. The characters of the operators in the pure rotation subgroup O of O_h with the d-orbitals as basis are as follows [obtained by putting $l = 2$ in Eqs (13.6)–(13.8)].

χ (character)

O	$\hat{E}$	$8\hat{C}_3$	$3\hat{C}_2\ (=\hat{C}_4^2)$	$6\hat{C}_4$	$6\hat{C}_2'$
Γ_d	5	−1	1	−1	1

From the character table of O given in Appendix A3, we find (by inspection or by applying the reduction formula) that

$$\Gamma_d = e + t_2 \tag{13.9}$$

(According to convention, lower case symbols will be used for irreducible representations with one-electron wave functions as basis; for states arising out of LS coupling upper case symbols E, T … will be used.)

Now by restoring the g-subscript (for the symmetric nature of d-orbitals with respect to $\hat{\imath}$) we find that the five-fold degenerate d-orbital energy level of the free ion is split into a set of doubly degenerate e_g and a set of triply degenerate t_{2g} levels when the free atom/ion is placed in an octahedral field. The O_h character table also shows that the set of basis functions of e_g is $\{d_{x^2-y^2}, d_{z^2}\}$ and that of t_{2g} is $\{d_{xy}, d_{yz}, d_{zx}\}$.

In the same way, the set of characters of O with the s-orbital as basis is {1, 1, 1, 1, 1} (obtained by putting $l = 0$ in Eqs 13.6–13.8). The s-orbital is non-degenerate and there is no possibility of further splitting. The representation {1, 1, 1, 1, 1} of O is also in agreement with this expectation; being one dimensional it cannot be reduced further. When the point group O changes to O_h, the symmetry label changes from a_1 to a_{1g} (the s-function being symmetric with respect to the inversion operator).

With p- and f-orbitals ($l = 1$ and 3 respectively), the set of characters of the representation Γ_p and Γ_f of O are as given in Table 13.1. For Γ_p, the sum of the squares of the characters = 24 = order of the group O. Hence Γ_p is irreducible and inspection of the character table of O reveals that $\Gamma_p = t_1$. Considering the inversion operation of O_h, the p-functions (being one-degree functions of coordinates) are of u-type and hence $\Gamma_p = t_{1u}$. Thus, in an octahedral crystal field, the p-orbitals of the central ion or atom behave as bases of a triply degenerate irreducible representation t_{1u}, and hence the p-level also does not split.

Table 13.1 Characters of O for the representations Γ_s, Γ_p and Γ_f

O	$\hat{E}$	$8\hat{C}_3$	$3\hat{C}_2\ (=\hat{C}_4^2)$	$6\hat{C}_4$	$6\hat{C}_2'$	Basis
			χ (character)			
Γ_s	1	1	1	1	1	s-orbital
Γ_p	3	0	−1	1	1	p-orbitals
Γ_f	7	1	−1	−1	−1	f-orbitals

Γ_f is reducible because the sum of the squares of the characters in $\Gamma_f = 1.7^2 + 8.(1)^2 + 3.(-1)^2 + 6.(-1)^2 + 6.(-1)^2 = 72$, which is greater than the order (24) of the group O. By the usual reduction procedure and using the character table of O, it is found that

$$\Gamma_f = a_2 + t_1 + t_2 \qquad (13.10)$$

Hence, after restoring the u-subscript for O_h symmetry (as f-functions are cubic functions of coordinates),

$$\Gamma_f = a_{2u} + t_{1u} + t_{2u} \qquad (13.11)$$

The seven-fold degenerate energy level of the f-orbitals, therefore, splits into one non-degenerate a_{2u} level and two triply degenerate levels t_{1u} and t_{2u} in an octahedral crystal field. In this way the splitting of energy levels for all types of orbitals with higher l values can be found and the results are given in Table 13.2.

Table 13.2 Splitting of one-electron levels in an octahedral field

One-electron orbital	l	Splitting in an O_h field
s	0	a_{1g}
p	1	t_{1u}
d	2	$e_g + t_{2g}$
f	3	$a_{2u} + t_{1u} + t_{2u}$
g	4	$a_{1g} + e_g + t_{1g} + t_{2g}$
h	5	$e_u + 2t_{1u} + t_{2u}$
i	6	$a_{1g} + a_{2g} + e_g + t_{1g} + 2t_{2g}$

(Note that the sum of the dimensions of the component irreducible representations in the third column of Table 13.2 is $2l + 1$ in each case.)

13.3.2 Splitting of Energy Levels of One-electron Orbitals in Tetrahedral and Other Fields of Lower Symmetry

The point group T_d has the symmetry operators $\hat{E}$, $8\hat{C}_3$, $3\hat{C}_2$, $6\hat{S}_4$, $6\hat{\sigma}_d$. With the angular parts of the one-electron atomic orbitals as basis, the characters of the rotation operators $\hat{C}_3$ and $\hat{C}_2$ and of the identity operator $\hat{E}$ can be obtained by using Eqs (13.6–13.8) as before:

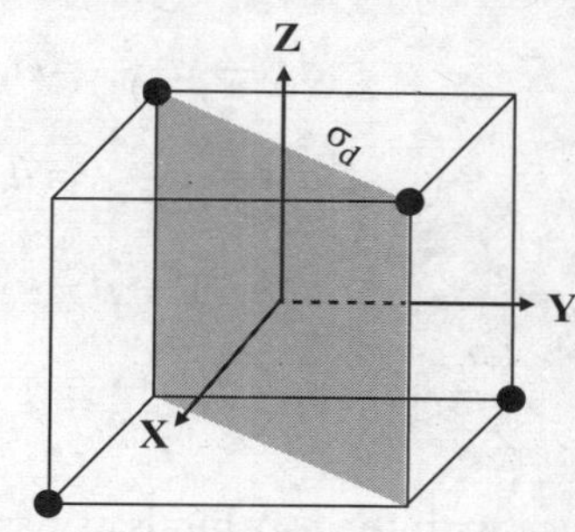

Fig. 13.3 The vertices (•) of a terahedron embedded in a cube and orientation of coordinate axes.

For *s* orbital, $l=0: \quad \chi(\hat{C}_3)=1, \quad \chi(\hat{C}_2)=1, \quad \chi(\hat{E})=1$

For *p* orbitals, $l=1: \quad \chi(\hat{C}_3)=0, \quad \chi(\hat{C}_2)=-1, \chi(\hat{E})=3$

For *d* orbitals, $l=2: \quad \chi(\hat{C}_3)=-1, \chi(\hat{C}_2)=1, \quad \chi(\hat{E})=5$

For *f* orbitals, $l=3: \quad \chi(\hat{C}_3)=1, \quad \chi(\hat{C}_2)=-1, \chi(\hat{E})=7$

The characters of the remaining two operators $\hat{S}_4(z)$ and $\hat{\sigma}_d$ can be obtained by noting their effects on the coordinates (x, y, z) of a general point. A tetrahedron embedded in a cube and the coordinate axes are shown in Fig. 13.3. If a point (x, y, z) transforms to (x', y', z') by $\hat{S}_4(Z)$, then

$$x' = x\cos 90° - y\sin 90° = -y$$

$$y' = x\sin 90° + y\cos 90° = x$$

$$z' = -z$$

One of the six $\hat{\sigma}_d$ planes is shown in Fig. 13.3. It contains the Z-axis and bisects the angle between the X and Y axes. If by this $\hat{\sigma}_d$ operation, the point (x, y, z) transforms to (x', y', z')then

$$x' = x\cos(2\times 45°) + y\sin(2\times 45°) = y$$

$$y' = x\sin(2\times 45°) - y\cos(2\times 45°) = x$$

$$z' = z$$

Hence the *s*-orbital $(\cong x^2+y^2+z^2)$ remains unchanged by both $\hat{S}_4(Z)$ and $\hat{\sigma}_d$, giving characters $\chi(\hat{S}_4)=1$ and $\chi(\hat{\sigma}_d)=1$. By $\hat{S}_4(Z)$, the functions p_x, p_y and p_z change respectively to $-p_y$, p_x and $-p_z$ giving $\chi\hat{S}_4 = 0+0+(-1) = -1$ for a representation with the *p*-orbitals as basis. Similarly by $\hat{\sigma}_d$, the orbitals p_x, p_y and p_z transform to p_y, p_x and p_z respectively giving $\chi(\hat{\sigma}_d) = 0+0+1 = 1$.

The five *d*-orbital wave functions transform as follows:

By $\hat{S}_4(Z)$: $d_{xy} \rightarrow d_{(-y)(x)} = -d_{xy}$

$$d_{yz} \rightarrow d_{(x)(-z)} = -d_{zx}$$

$$d_{zx} \rightarrow d_{(-z)(-y)} = d_{yz}$$

$$d_{x^2-y^2} \rightarrow d_{(-y)^2-(x)^2} = -d_{x^2-y^2}$$

$$d_{z^2} \rightarrow d_{(-z)^2} = d_{z^2}$$

By $\hat{\sigma}_d$: $d_{xy} \rightarrow d_{(y)(x)} = d_{xy}$

$$d_{yz} \rightarrow d_{(x)(z)} = d_{zx}$$

$$d_{zx} \rightarrow d_{(z)(y)} = d_{yz}$$

$$d_{x^2-y^2} \rightarrow d_{(y)^2-(x)^2} = -d_{x^2-y^2}$$

$$d_{z^2} \rightarrow d_{(z)^2} = d_{z^2}$$

Hence with the d-orbitals as basis, the characters are $\chi\hat{S}_4(Z) = (-1) + 0 + 0 + (-1) + 1 = -1$ and $\chi(\hat{\sigma}_d) = 1 + 0 + 0 + (-1) + 1 = 1$

The seven f-orbital wave functions transform as follows:

By $\hat{S}_4(Z)$: $f_{xyz} \rightarrow f_{(-y)(x)(-z)} = f_{xyz}$

$$f_{z(x^2-y^2)} \rightarrow f_{(-z)\{(-y)^2-(x)^2\}} = f_{z(x^2-y^2)}$$

$$f_{y(z^2-x^2)} \rightarrow f_{(-x)\{(-z)^2-(-y)^2\}} = -f_{x(z^2-y^2)}$$

$$f_{x(z^2-y^2)} \rightarrow f_{(-y)\{(-z)^2-(x)^2\}} = -f_{y(z^2-x^2)}$$

$$f_{x(5x^2-3r^2)} \rightarrow f_{(-y)\{5(-y)^2-3r^2\}} = -f_{y(5y^2-3r^2)}$$

$$f_{y(5y^2-3r^2)} \rightarrow f_{(x)\{5(x)^2-3r^2\}} = f_{x(5x^2-3r^2)}$$

$$f_{z(5z^2-3r^2)} \rightarrow f_{(-z)\{5(-z)^2-3r^2\}} = -f_{z(5z^2-3r^2)}$$

By $\hat{\sigma}_d$: $f_{xyz} \rightarrow f_{(y)(x)(z)} = f_{xyz}$

$$f_{z(x^2-y^2)} \rightarrow f_{(z)\{(y)^2-(x)^2\}} = -f_{z(x^2-y^2)}$$

$$f_{y(z^2-x^2)} \rightarrow f_{(x)\{(z)^2-(y)^2\}} = f_{x(z^2-y^2)}$$

$$f_{x(z^2-y^2)} \rightarrow f_{(y)\{(z)^2-(x)^2\}} = f_{y(z^2-x^2)}$$

$$f_{x(5x^2-3r^2)} \rightarrow f_{(y)\{5(y)^2-3r^2\}} = f_{y(5y^2-3r^2)}$$

$$f_{y(5y^2-3r^2)} \rightarrow f_{(x)\{5(x)^2-3r^2\}} = f_{x(5x^2-3r^2)}$$

$$f_{z(5z^2-3r^2)} \rightarrow f_{(z)\{5(z)^2-3r^2\}} = f_{z(5z^2-3r^2)}$$

Thus with the f-orbitals as basis, $\chi(\hat{S}_4) = 1 + 1 + 0 + 0 + 0 + 0 + (-1) = 1$ and $\chi(\hat{\sigma}_d) = 1 + (-1) + 0 + 0 + 0 + 0 + 1 = 1$.

The complete representations of T_d with the s, p, d and f-orbitals as basis are, therefore, as follows:

T_d	$\hat{E}$	$8\hat{C}_3$	$3\hat{C}_2$	$6\hat{S}_4$	$6\hat{\sigma}_d$
Γ_s	1	1	1	1	1
Γ_p	3	0	−1	−1	1
Γ_d	5	−1	1	−1	1
Γ_f	7	1	−1	1	1

By inspection of character table of T_d or by applying the standard reduction formula it is found that

$$\Gamma_s = a_1, \Gamma_p = t_2, \Gamma_d = e + t_2, \Gamma_f = a_1 + t_1 + t_2 \quad (13.12)$$

In a tetrahedral crystal field (T_d), therefore, the energy levels of the *s* and *p*-orbitals do not split, that of *d*-orbitals splits into a doubly degenerate (e) and a triply degenerate (t_2) level and the *f*-orbital energy level splits into one non-degenerate a_1 and two triply degenerate t_1 and t_2 levels.

By the same procedure, the splitting of one-electron energy levels in other fields of lower symmetry such as D_{4h}, D_3, D_{2d} ... can be determined. Some results are given in Table 13.3 which can be readily verified.

Table 13.3 Splitting of one-electron energy levels in tetrahedral and some other fields

One-electron level	*l*	Symmetry of crystal field			
		T_d	D_{4h}	D_3	D_{2d}
s	0	a_1	a_{1g}	a_1	a_1
(*) *p*	1	t_2	$a_{2u} + e_u$	$a_2 + e$	$b_2 + e$
d	2	$e + t_2$	$a_{1g} + b_{1g} + b_{2g} + e_g$	$a_1 + 2e$	$a_1 + b_1 + b_2 + e$
(*) *f*	3	$a_1 + t_1 + t_2$	$a_{2u} + b_{1u} + b_{2u} + 2e_u$	$a_1 + 2a_2 + 2e$	$a_1 + a_2 + b_2 + 2e$
g	4	$a_1 + e + t_1 + t_2$	$2a_{1g} + a_{2g} + b_{1g} + b_{2g} + 2e_g$	$2a_1 + a_2 + 32e$	$2a_1 + a_2 + b_1 + b_2 + 2e$
(*) *h*	5	$e + t_1 + 2t_2$	$a_{1u} + 2a_{2u} + b_{1u} + b_{2u} + 3e_u$	$a_1 + 2a_2 + 4e$	$a_1 + a_2 + b_1 + 2b_2 + 3e$

Splitting patterns under O_h and T_d fields are closely related as expected, since both an octahedron and a tetrahedron can be generated from the same cube. Tables 13.2 and 13.3 indicate that the results in a T_d environment can be obtained from those in an O_h environment by the following simple manipulations:

- for orbitals with even values of *l* (such as *s, d, g,*....) remove the *g*-subscript, and
- for orbitals with odd values of *l* (such as *p, f, h,*....) remove the *u*-subscript and replace the subscript 1 by 2 and vice-versa.

The rows marked with an asterisk (*) in Table 13.3 may be compared with the corresponding rows in Table 13.2 to verify the manipulations stated above. The case of *p*-orbitals may be easily verified from the character tables; thus (x, y, z), and consequently, (p_x, p_y, p_z), are basis sets of T_{1u} for O_h and of T_2 for T_d.

13.4 THE EXTENT OF SPLITTING OF ENERGY LEVELS OF ONE-ELECTRON *d*-ORBITALS IN O_h AND T_d FIELDS

From symmetry considerations alone it is not possible to estimate the energy separation of the split up levels; actual computation is necessary for the purpose. Group theory says that the wave functions of the $d_{x^2-y^2}$ and d_{z^2} orbitals form bases of E_g representation of the O_h point group while the d_{xy}, d_{yz} and d_{zx} orbitals from bases of the T_{2g} representation; hence the two sets of orbitals will correspond to different energy levels but the actual energies can only be found by calculation. However, crystal field theory considers only the electrostatic effect of the surrounding point charges or point dipoles on the central atom orbitals and it is possible to determine qualitatively the relative energy order of the split up levels by considering the average distances of the orbitals from the point charges/dipoles. As a first order electrostatic effect the total energy of the one-electron *d*-orbitals increases, as if the central ion were surrounded by a spherical distribution of negative charges. In the octahedral field, this total charge is kept the same as in the sphere but it is now redistributed to the vertices of an octahedron, maintaining the same average ligand-to-charge distance as that in the sphere; this maintains the *barycentre*

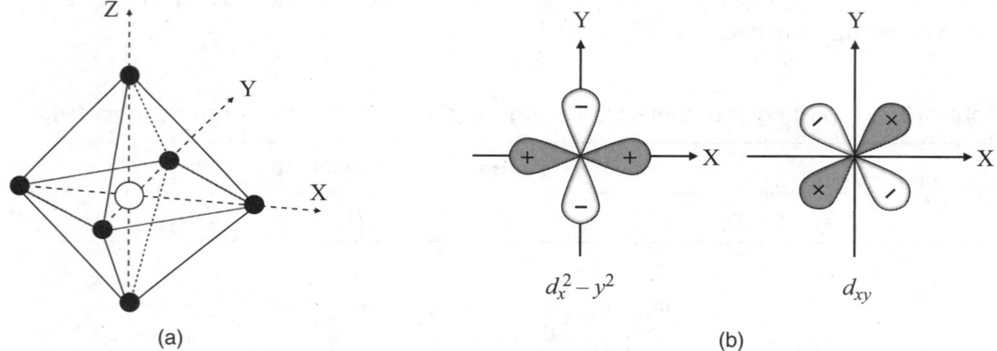

Fig. 13.4 (a) Position of metal ion (○), ligands (•) and coordinate axes in O_h symmetry. (b) Orientation of $d_{x^2-y^2}$ and d_{xy} orbitals with respect to coordinate axes.

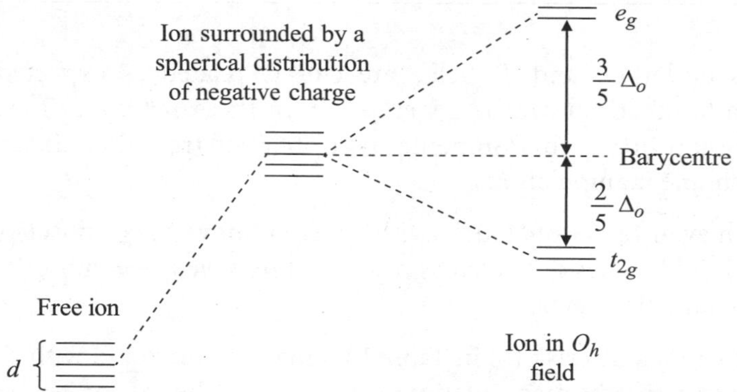

Fig. 13.5 Splitting of energy levels of one-electron *d*-orbitals of a transition metal ion in a crystal field of O_h symmetry

of the energy levels fixed. But in the crystal field of O_h symmetry, the $d_{x^2-y^2}$ and d_{z^2} orbitals, which are stretched along the coordinate axes (Fig. 13.4), face the ligand negative charges directly while the d_{xy}, d_{yz} and d_{zx} orbitals are away from the ligands since they are stretched in a region between the axes (Fig. 13.4). Hence the energy of the $d_{x^2-y^2}$ and d_{z^2} orbitals is raised from the barycentre while that of the d_{xy}, d_{yz} and d_{zx} orbitals is lowered (Fig. 13.5).

The energy separation $\varepsilon(e_g) - \varepsilon(t_{2g})$ is designated by Δ_O (or, in many applications, by 10 Dq). If with respect to the barycentre, the energy of each e_g-orbital is raised by a fraction λ of Δ_O and that of each t_{2g}-orbital is lowered by a fraction $(1 - \lambda)$ of Δ_O, then, since there are two e_g orbitals and three t_{2g} orbitals, we have

$$2\lambda\Delta_O = 3(1-\lambda)\Delta_O \quad \text{that is,} \quad \lambda = \tfrac{3}{5} \text{ and } (1-\lambda) = \tfrac{2}{5} \tag{13.13}$$

Thus each e_g orbital is higher in energy by $(3/5)\Delta_O$ and each t_{2g} orbital is lower in energy by $(2/5)\Delta_O$ with respect to the barycentre (Fig. 13.5).

In a tetrahedral crystal field, the energy order of the split up levels is just opposite to that in an O_h field. The T_d character table reveals that the d_{xy}, d_{yz} and d_{zx} orbitals form the bases of the T_2 representation and $d_{x^2-y^2}$ and d_{z^2} orbitals form the bases of the E representation of the T_d point group. But, by inscribing a tetrahedron inside a cube, it can be shown (Fig. 13.6) that the $d_{x^2-y^2}$ and d_{z^2} orbitals are at a distance of $(a\sqrt{2}/2)$ from the nearest ligand charge while

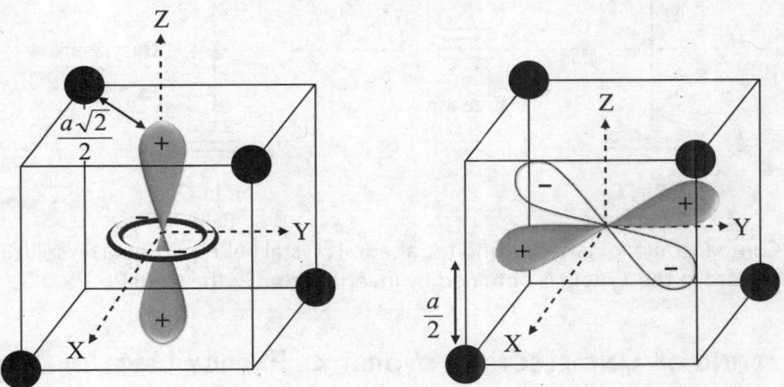

Fig. 13.6 A tetrahedron inscribed in a cube and orientation of d_{z^2} and d_{xy} orbitals of the central metal ion.

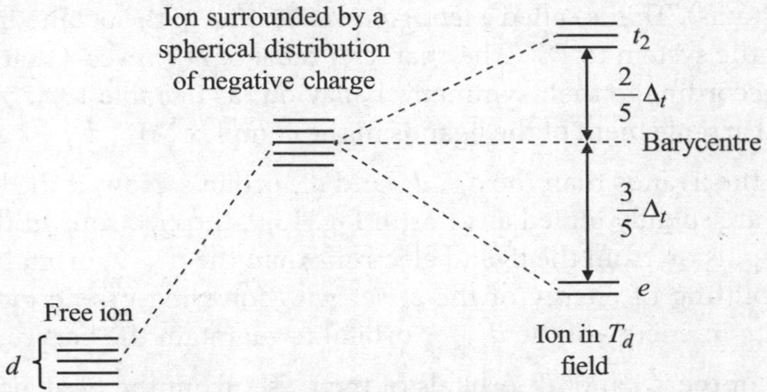

Fig. 13.7 Splitting of one-electron d-orbitals of a transition metal ion in a T_d crystal field

each of the d_{xy}, d_{yz} and d_{zx} orbitals is at a distance of $(a/2)$ from the corresponding nearest ligand charge, where a is the length of the cube in which the tetrahedron is inscribed. Thus, the $d_{x^2-y^2}$, d_{z^2} orbitals feel less electrostatic repulsion than d_{xy}, d_{yz}, d_{zx} orbitals. The energy order of e and t_2 sets is therefore as shown in Fig. 13.7 where the extent of tetrahedral crystal field splitting is Δ_t.

The number of ligands in a tetrahedral complex is (4/6) times that in an octahedral complex. Another factor of (2/3) comes from a geometrical calculation of electronic repulsions, making

$$|\Delta_t| : |\Delta_o| = \tfrac{4}{6} \times \tfrac{2}{3} = 4:9$$

or, $$\Delta_t = -\tfrac{4}{9}\Delta_o \tag{13.14}$$

for the same ligands, same metal ion and same average metal–ligand distance. Here the modulus sign indicates the magnitude of splitting and the negative sign in Eq. (13.14) is due to the fact that the energy order is reversed in going from O_h symmetry to T_d or vice versa. These are represented in Fig. 13.8, where the barycentre for O_h and T_d symmetries are shown (artificially) to be at the same level.

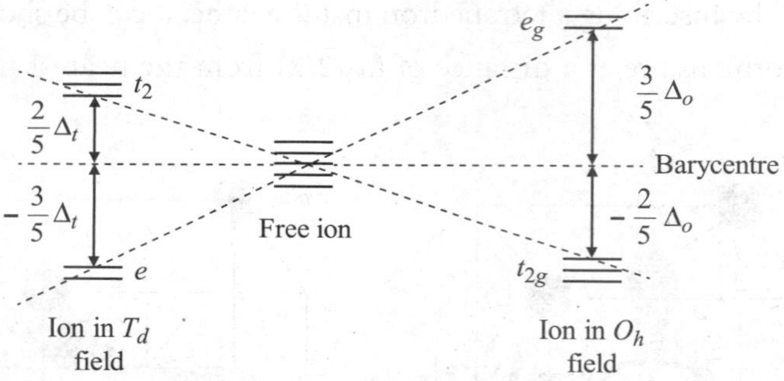

Fig. 13.8 Comparison of octahedral and tetrahedral crystal field splitting. The splitting pattern in the T_d field is obtained by inverting that in the O_h field

13.4.1 Splitting of One-electron *d*-orbital Energy Level in Distorted Octahedral and Square Planar Fields

Let the molecule/ion of an octahedral complex be somewhat elongated (or compressed) along one axis (say, the Z-axis). This is called a *tetragonal distortion* of the octahedron and it converts the point group of the system to D_{4h}. The character table of D_{4h} reveals that the d-orbitals can now be classified according to their symmetry behaviour as in Table 13.4.

In an octahedral arrangement of the ligands in the complex ML_6, $d_{x^2-y^2}$ and d_{z^2} orbitals are (equally) closer to the ligands than the d_{xy}, d_{yz} and d_{zx} orbitals. Now, if the ligands on the two ends of the Z-axis are slightly pulled away as in Fig. 13.9, the electrons in the d_{z^2} orbitals feel less electrostatic repulsion from the ligand electrons than the $d_{x^2-y^2}$ orbital.

This causes a splitting of energy of the e_g set with lowering of energy of the d_{z^2} orbital and an equal raising in energy of the $d_{x^2-y^2}$ orbital to maintain the barycentre. On the other hand, the distance of the d_{yz} and d_{zx} orbitals of the t_{2g} set from the two ligands on the Z-axis increases, which lowers the energy of the d_{yz} and d_{zx} orbitals; to maintain the barycentre, the

Table 13.4 Irreducible representations of D_{4h} with d-orbitals as basis

Irreducible representation of D_{4h}	Basis orbital
A_{1g}	d_{z^2}
B_{1g}	$d_{x^2-y^2}$
B_{2g}	d_{xy}
E_g	(d_{yz}, d_{zx})

energy of the d_{xy} orbital increases. Thus the t_{2g} set also splits by the tetragonal elongation and the ratio *increase*: *decrease* of energy for this set is 2:1 (the energy of one orbital rises while the energies of two orbitals are lowered) (Fig. 13.10).

For square planar complexes, the point group is again D_{4h}. But now the ligands along the Z-axis are absent and hence the energies of the d_{yz}, d_{zx} and d_{z^2} orbitals are still lower. The splitting patterns under tetragonally distorted

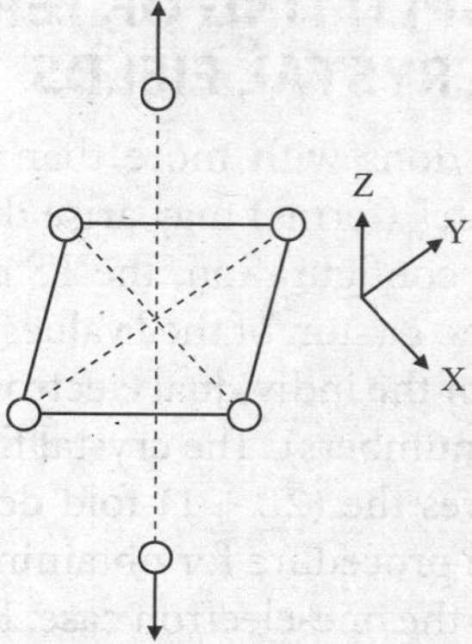

Fig. 13.9 Tetragonal elongation of an octahedral complex along the Z-axis; ligands are denoted by (O). In the case of tetragonal compression, the splitting is similar but the energy order is reversed in the individual t_{2g} and e_g sets.

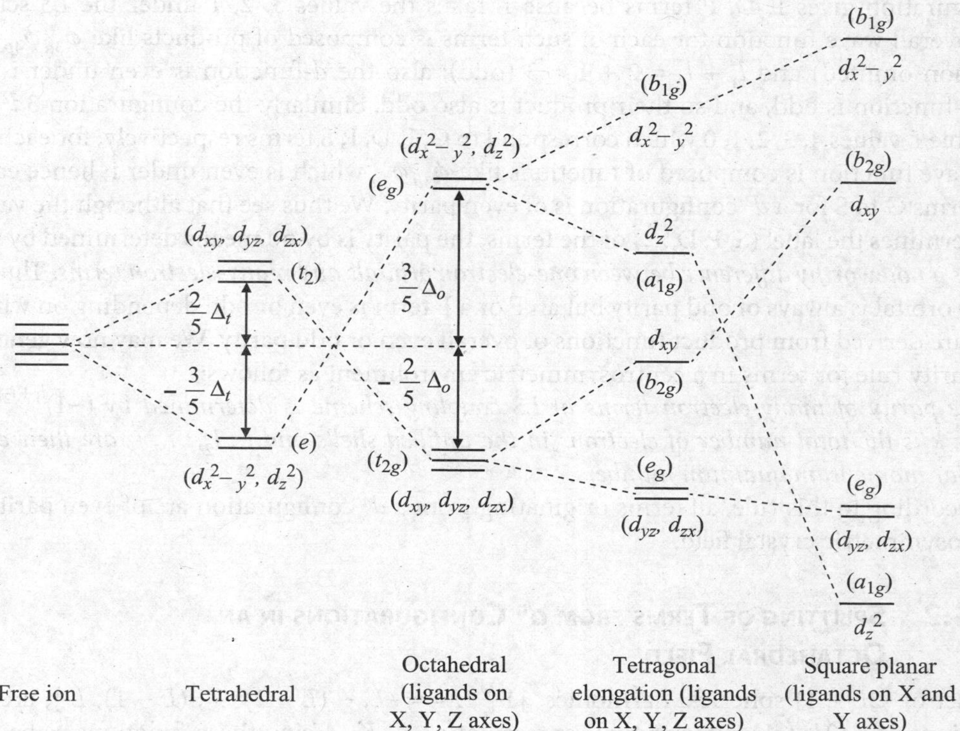

Fig. 13.10 Splitting of one-electron d-orbital energy levels in crystal fields of different symmetry

octahedral and square planar crystal fields are best displayed by a comparison with that in an octahedral field (Fig. 13.10).

13.5 SPLITTING OF TERMS OF MULTI-ELECTRON SYSTEMS IN CRYSTAL FIELDS

For atoms/ions with more than one electron outside a closed shell configuration, several energy states (terms) may arise due to spin–orbit coupling. For transition metal ions with d^n electronic configuration, the *LS* coupling scheme works well and will be employed here. In this scheme L (sum of the l values of the individual electrons) and S (sum of the spin quantum numbers of the individual electrons) are sufficient to describe the atomic states (they are good quantum numbers). The crystal field being electrostatic, does not affect the spin multiplicities but removes the $(2L + 1)$-fold degeneracy of the energy level of a term designated by the L value. The procedure for obtaining the pattern of splitting of a many-electron term is similar to that for the one-electron case. However, the parity of the many-electron wave function with respect to inversion must be adequately taken care of while applying this procedure.

13.5.1 The Parity Rule

The wave function of a multi-electron system is expressed in terms of the product of wave functions of the individual electrons. Hence in a centrosymmetric field, the parity of a term should depend on the sum of the l values of the individual electrons. For example, a $3d^1 4p^1$ configuration gives F, D, P terms because L takes the values 3, 2, 1 under the *LS* scheme. The overall wave function for each of such terms is composed of products like $\phi_{3d}\phi_{4p}$ (spin function omitted) and $l_1 + l_2 = 2 + 1 = 3$ (odd); also the d-function is even under $\hat{i}$ while the p-function is odd, and so their product is also odd. Similarly the configuration $3d^2$ gives possible L values 4, 3, 2, 1, 0 which correspond to G, F, D, P, S terms respectively; for each term the wave function is composed of functions like $\phi_{3d}\phi_{3d}$ which is even under $\hat{i}$; hence each of the terms G to S for a d^2 configuration is of even parity. We thus see that although the value of L determines the label (S, P, D,...) of the terms, the parity is by no means determined by $(-1)^L$. *This is a noteworthy difference between one-electron orbitals and many-electron terms.* Thus, an f or a p orbital is always of odd parity but an F or a P term is even or odd depending on whether they are derived from product functions of overall even or odd parity. We may now generalise the parity rule for terms in a centrosymmetric environment as follows:

The parity of many-electron terms in LS coupling scheme is determined by $(-1)^{l_1+l_2+\cdots+l_k}$ *where k is the total number of electrons in the unfilled shells and* $l_1, l_2, \ldots, l_k$ *are their orbital angular momentum quantum numbers.*

According to this rule, all terms originating from a d^n configuration are of even parity in a centrosymmetric crystal field.

13.5.2 Splitting of Terms from d^n Configurations in an Octahedral Field

The set of $(2L + 1)$ spherical harmonics, $\{Y_L^M, M = -L, -(L-1) \cdots, (L-1), L\}$, are basis functions of a $(2L + 1)$-dimensional representation of K_h. Using these functions as basis the characters of the rotation operators are:

$$\alpha=\pi,\quad \chi^{(L)}(\hat{C}_2)=\ (-1)^L$$

$$\left.\begin{array}{l}\alpha=\frac{2\pi}{3},\quad \chi^{(L)}(\hat{C}_3)=\begin{cases}1, & L=0,3,6,\ldots\\ 0, & L=1,4,7,\ldots\\ -1, & L=2,5,8,\ldots\end{cases}\\ \alpha=\frac{\pi}{2},\quad \chi^{(L)}(\hat{C}_4)=\begin{cases}1 & L=0,1,4,5,\ldots\\ -1, & L=2,3,6,7,\ldots\end{cases}\end{array}\right\} \quad (13.15)$$

These are obtained by replacing l by L in Eq. (13.7). Since the Y_L^M functions span the representation of all rotations, they must also act as bases of representations (D_L) of the finite rotation group O. The first few character sets are listed below:

O	χ (character)				
	$\hat{E}$	$6\hat{C}_4$	$3\hat{C}_2=(\hat{C}_4^2)$	$8\hat{C}_3$	$6\hat{C}_2'$
$L=0$: D_0	1	1	1	1	1
$L=1$: D_1	3	1	−1	0	−1
$L=2$: D_2	5	−1	1	−1	1
$L=3$: D_3	7	−1	−1	1	−1
$L=4$: D_4	9	1	1	0	1

The task is now to find which irreducible representations of O add up to give these D_L's. The use of standard reduction formula (or inspection of the character table of O) yields the following results:

$D_0=A_1$, which, being one-dimensional, cannot be reduced further.

$D_1=T_1$, by comparison of characters. Since T_1 is a single irreducible representation, D_1 also cannot be reduced further.

Hence the terms S ($L=0$) and P ($L=1$) do not split in a cubic O field.

$D_2=E+T_2$. This means that the D term ($L=2$) must split in a cubic O field into a doubly degenerate E and a triply degenerate T_2 term.

$D_3=A_2+T_1+T_2$. This means that the F term ($L=3$) must split in a cubic O field into one non-degenerate A_2, one triply degenerate T_1 and one triply degenerate T_2 term.

$D_4=A_1+E+T_1+T_2$. Thus a G term ($L=4$) of the free ion must split in a cubic O field into a non-degenerate A_1, a doubly degenerate E, one triply degenerate T_1 and one triply degenerate T_2 term.

The terms S to G are all results of spin–orbit coupling for a d^n configuration. Hence they are all of even parity. The splitting of the terms in an O_h environment are, therefore, obtained by just inserting the g-subscript.

13.6 ENERGY LEVELS OF ELECTRONIC STATES (TERMS) OF TRANSITION METAL IONS IN CRYSTAL FIELDS—CORRELATION DIAGRAMS

UV–vis spectra of transition metal complexes are the result of electronic transitions between different spectroscopic terms of d^n metal ions in crystal fields. The relative energy order of the states depends not only on the symmetry but also on the strength of the field which is measured by the magnitude $|\Delta|$, of splitting; the latter, in turn, depends on the nature of the ligand (for example, the F^- ion produces a weak field while the CN^- produces a strong field). Interpretation of electronic spectra, therefore, requires a knowledge of how the energies of the spectroscopic terms vary with gradual change of $|\Delta|$ from weak to strong crystal fields under different environments (octahedral, tetrahedral, ...). The actual energy of a term can be obtained only from rigorous quantum mechanical calculations. Group theory, however, is powerful enough to give the relative energies (energy order) of the states from symmetry considerations alone and provide appropriate selection rules for electronic transitions; these provide an almost complete interpretation of the spectra of transition metal complexes.

A *correlation diagram* shows how the free ion terms split in a weak field and how their energy changes from the weak field to the strong field limit as the magnitude of crystal field is increased under a given symmetry environment. The diagram has $|\Delta|$ as abscissa and energy as the ordinate. On the extreme left hand side of the diagram $|\Delta| = 0$ and the energies of the free ion terms are shown. On the extreme right are shown the energy levels of the states which result from interaction with a very strong crystal field—so strong that the inter-electron interactions (mainly repulsions) become negligible. In the following sections the method of construction of correlation diagrams for O_h and T_d symmetry is explained.

13.6.1 Correlation Diagram for a d^2 System in O_h Symmetry

An example of a d^2 system with octahedral geometry is $[V(H_2O)_6]^{+3}$. The free ion terms for d^2 configuration are ^1S, ^1D, ^1G, ^3P and ^3F. The states for very strong interaction with the octahedral field are obtained from the possible configurations, t_{2g}^2, $t_{2g}^1e_g^1$ and e_g^2 in increasing order of energy. The states corresponding to such configurations are obtained by reducing the direct product representations $T_{2g} \times T_{2g}$, $T_{2g} \times E_g$ and $E_g \times E_g$, of the O_h point group (small letter *t*, *e* ... will be used for orbitals and capital letters *T*, *E* ... for states). As before, the task is simplified by considering the *O* subgroup of O_h and then restoring the *g* subscripts. Recall that all the resulting states will be of *g*-type under the O_h environment since they all originate from *d*-orbitals. Using the character table of *O* we find the following results for the necessary direct products.

O	$\hat{E}$	$8\hat{C}_3$	$3\hat{C}_2$	$6\hat{C}_4$	$6\hat{C}_2'$	
$T_2 \times T_2$	9	0	1	1	1	$= A_1 + E + T_1 + T_2$
$T_2 \times E$	6	0	−2	0	0	$= T_1 + T_2$
$E \times E$	4	1	4	0	0	$= A_1 + A_2 + E$

Hence, restoring the *g*-subscript, the results for an O_h environment are:

$$T_{2g} \times T_{2g} = A_{1g} + E_g + T_{1g} + T_{2g}$$

$$T_{2g} \times E_g = T_{1g} + T_2$$

$$E_g \times E_g = A_{1g} + A_{2g} + E_g$$

Thus, by using the results in Section 13.5.2 for the left side and the above results for the right hand side, we construct a partial correlation diagram (Fig. 13.11), where the asterisk (*)

indicates spin-multiplicities to be determined later. The occurrence of four levels from the $t_{2g}^1 e_g^1$ configuration in Fig. 13.11 instead of two, as expected from the direct product results, has been explained later in connection with spin multiplicities.

Spin multiplicity does not change in going from the free ion state to the weak field interaction. Hence the terms on the left hand side of Fig. 13.11 have been completely specified. In order to assign spin multiplicities to the terms on the right, we may adopt any of the following two methods.

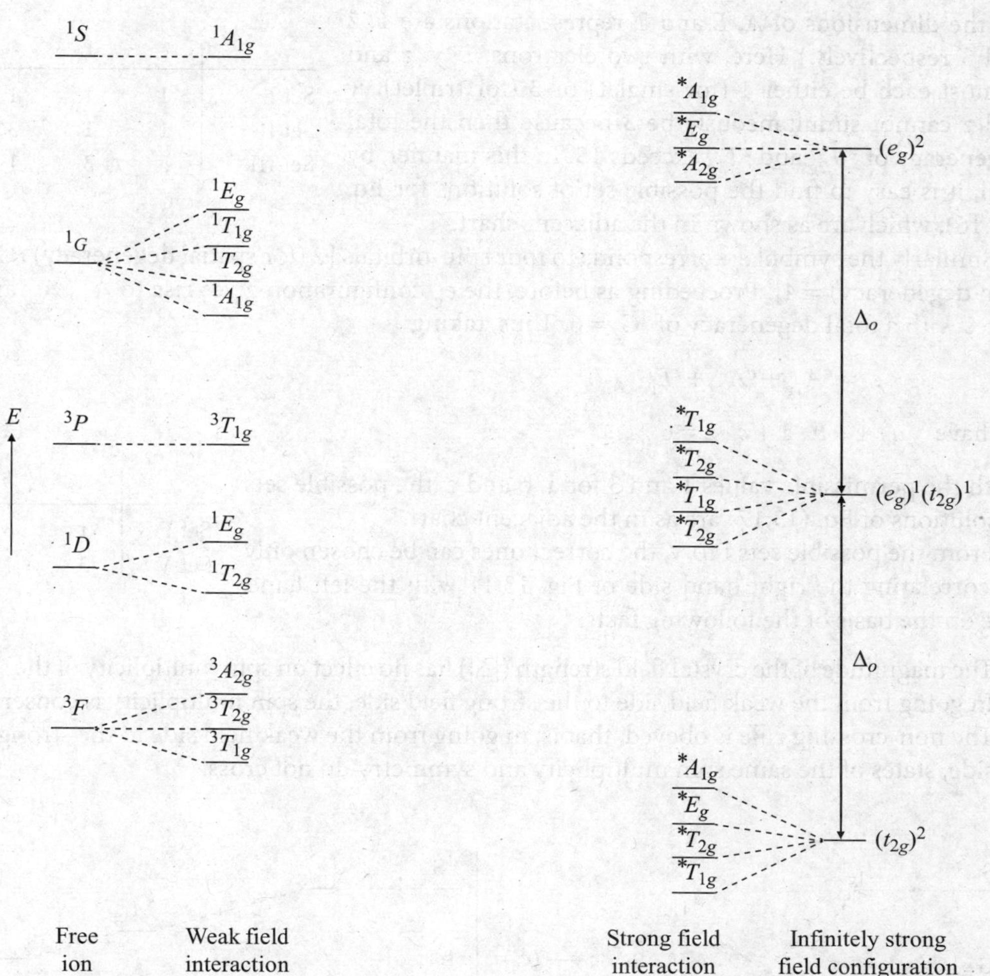

Fig. 13.11 Partial correlation diagram for a d^2 system in an O_h environment [the asterisk (*) indicates spin multiplicity to be determined later]

13.6.1.1 Method I

The symbol t_{2g} refers to a set of 3-fold degenerate orbitals having different orientations in space. Again, an electron in such an orbital may have spin $+\frac{1}{2}$ and $-\frac{1}{2}$. In other words, t_{2g} corresponds to a set of $3 \times 2 = 6$ spin-orbitals. According to the Pauli Exclusion Principle, a spin-orbital can be occupied by one and only one electron. The six spin-orbitals can be occupied by two electrons in ${}^6C_2 = 15$ ways. Thus the t_{2g}^2 configuration has a total degeneracy of 15. When the infinite crystal field strength is relaxed to a finite but strong field the t_{2g}^2 configuration splits into separate states A_{1g}, E_g, T_{1g} and T_{2g} but the total degeneracy must remain the same. Hence, if x, y, z and w be the respective spin degenaracies, that is, if

$$t_{2g} \times t_{2g} = {}^xA_{1g} + {}^yE_g + {}^zT_{1g} + {}^wT_{2g}$$

then, $$x \cdot 1 + y \cdot 2 + z \cdot 3 + w \cdot 3 = 15 \tag{13.16}$$

(as the dimensions of A, E and T representations are 1, 2 and 3 respectively.) Here, with two electrons, x, y, z and w must each be either 1 (for singlet) or 3 (for triplet). w and z cannot simultaneously be 3 because then the total degeneracy of ${}^wT_{2g}$ and ${}^zT_{1g}$ exceeds 15. In this manner, by trial, it is easy to find the possible set of solutions for Eq. (13.16), which are as shown in the adjacent chart.

	x	y	z	w
Set I	1	1	1	3
Set II	1	1	3	1
Set III	3	3	1	1

Similarly the symbol e_g corresponds to four spin-orbitals [2 (for spatial degeneracy) × 2 (for spin degeneracy) = 4]. Proceeding as before, the e_g^2 configuration gives rise to A_{1g}, A_{2g} and E_g states with a total degeneracy of ${}^4C_2 = 6$. Thus, taking

$$e_g \times e_g = {}^aA_{1g} + {}^bA_{2g} + {}^cE_g$$

we have $$a \cdot 1 + b \cdot 1 + c \cdot 2 = 6 \tag{13.17}$$

With the permissible values 1 and 3 for a, b and c, the possible sets of solutions of Eq. (13.17) are as in the adjacent chart.

	a	b	c
Set IV	1	3	1
Set V	3	1	1

From the possible sets I to V, the correct ones can be chosen only by correlating the right hand side of Fig. 13.11 with the left hand side on the basis of the following facts:

- The magnitude of the crystal field strength ($|\Delta|$) has no effect on spin multiplicity of the states. In going from the weak field side to the strong field side, the spin multiplicity is conserved.
- The non-crossing rule is obeyed, that is, in going from the weak field side to the strong field side, states of the same spin multiplicity and symmetry do not cross.

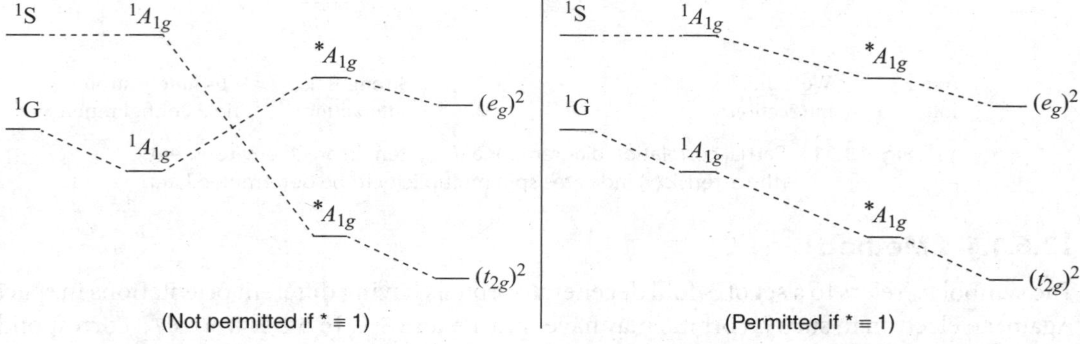

Fig. 13.12 Allowed and forbidden correlations of ${}^1A_{1g}$ on the left with ${}^*A_{1g}$ on the right

- There is a one-to-one correspondence between the states of a particular multiplicity and symmetry on the two sides of the correlation diagram (in order that the total spin degeneracy is conserved).

On the left of Fig. 13.11, there are two singlet A_{1g} levels and no triplet A_{1g}. These $^1A_{1g}$ levels on the left can be correlated to the $^*A_{1g}$ levels on the right, avoiding crossing, in only one way (Fig. 13.12).

Hence both the A_{1g} levels on the right must be singlets. Thus the possibilities III and V, for the t_{2g}^2 and e_g^2 configurations respectively, are discarded. Accepting Set IV for e_g^2 we find:

$$e_g \times e_g = {}^1A_{1g} + {}^3A_{2g} + {}^1E_g$$

The $^3A_{2g}$ level thus found from e_g^2 has to be correlated with the only $^3A_{2g}$ level on the left (from ^{3}F of the free ion). The ^1E_g level found from e_g^2 has to be correlated with ^1E_g from ^{1}G, and not with ^1E_g from ^{1}D, in order to avoid crossing. After these, the only E_g levels remaining are ^1E_g from ^{1}D on the left and $^*E_g$ from t_{2g}^2 on the right. Hence this $^*E_g$ on the right must also be a singlet.

We are now left with only T_{1g} and T_{2g} levels on both sides. The $t_{2g}^1e_g^1$ configuration admits of both singlet and triplet terms because the two electrons with different spatial wave functions can have both $+\frac{1}{2}$ spins, both $-\frac{1}{2}$ spins or one spin $+\frac{1}{2}$ and the other $-\frac{1}{2}$. Hence the T_{1g} and T_{2g} levels resulting from $t_{2g}^1e_g^1$ are each of two types, singlets and triplets, the latter type being of lower energy than the former as shown in the complete correlation diagram (Fig. 13.13). The low energy $^3T_{1g}$ (from ^{3}F) on the left must be connected to the $^*T_{1g}$ from the low-energy configuration t_{2g}^2; the next higher energy $^3T_{1g}$ (from ^{3}P) is to be connected to the $^*T_{1g}$ from the excited state configuration $t_{2g}^1e_g^1$ to avoid crossing. The remaining connections can be drawn only in the way shown in Fig. 13.13.

13.6.1.2 Method II: Method of Descending Symmetry

(When a molecule or ion is subjected to a perturbation, the point group usually changes to one of its subgroups. The utility of point group–subgroup correlation tables was discussed in connection with molecular vibrations in Chapter 11. Another interesting application of such tables will be found in the determination of spin multiplicities of perturbed atomic terms by the method of descending symmetry which is now being discussed.)

The algebraic method I, although very instructive, is impractical for d^n systems with $n > 2$.

However, there is a general method (due to Bethe), known as the *method of descending symmetry*. In this method, it is assumed that if a system is perturbed and its point group changes to a subgroup, the spin multiplicities do not change. The states of the parent system, which are bases of the various irreducible representations of the parent group, evolve into states which are bases of irreducible representations of the subgroup, without change in multiplicity. If the subgroup is so chosen that all its irreducible representations are one-dimensional, then spin multiplicities can be assigned to states corresponding to the lower symmetry subgroup without ambiguity. By reverting to the parent group, the multiplicities of the parent states can be assigned. The procedure is illustrated here for a d^2-configuration in O_h symmetry under an infinitely strong field. The configurations are t_{2g}^2, $t_{2g}^1e_g^1$ and e_g^2 in the increasing order of energy. Looking up the correlation table of O_h

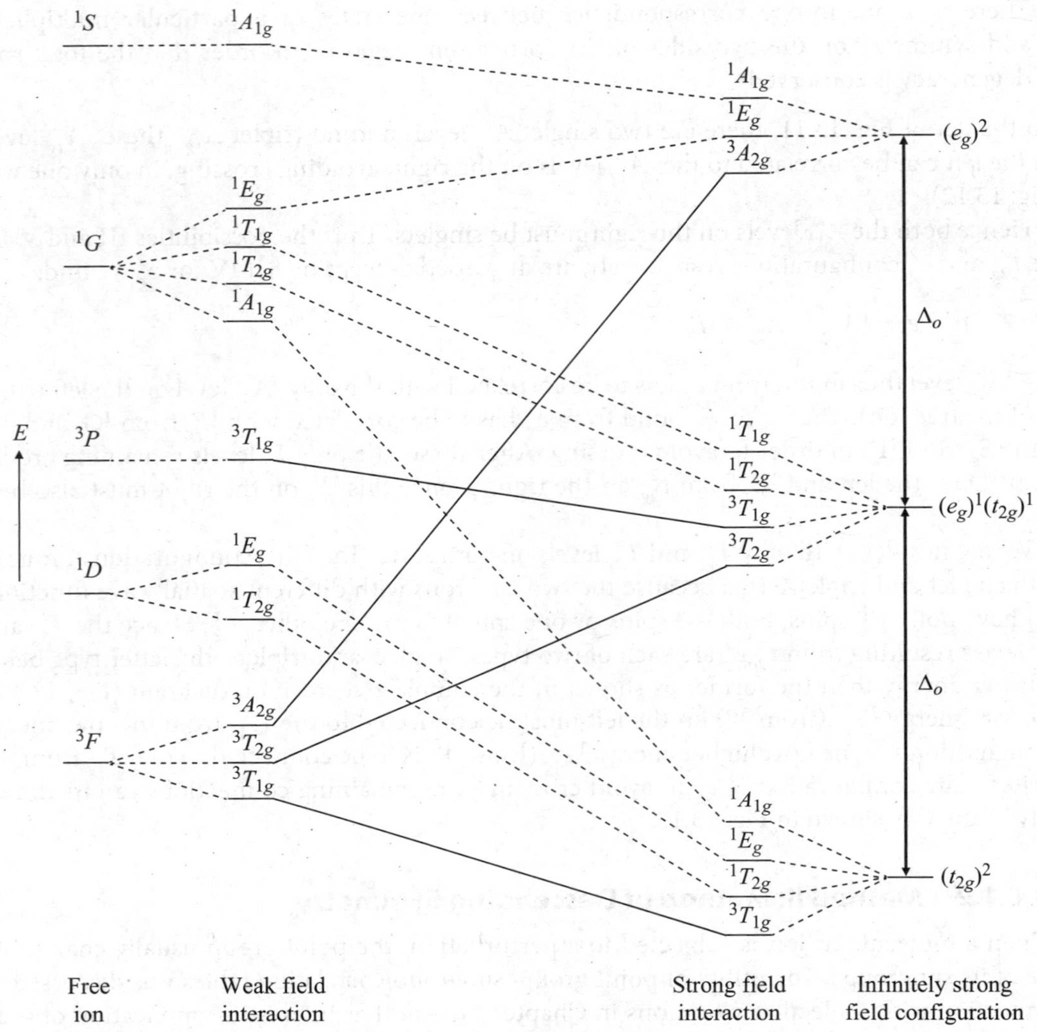

Fig. 13.13 Complete correlation diagram for a d^2 ion in a crystal field of octahedral symmetry

given in Appendix A4, it is found that in the rows corresponding to t_{2g} and e_g, only the subgroups C_{2v} and C_{2h} have one dimensional irreducible representations. Hence either of these two may be used. The D_{4h} subgroup is not suitable for the present purpose because it has the two-dimensional E representation in the row corresponding to t_{2g}. We show here the procedure with the C_{2v} point group.

O_h correlates to C_{2v}

t_{2g} correlates to $a_1 + b_1 + b_2$

Hence t_{2g}^2 correlates to $(a_1 + b_1 + b_2) \times (a_1 + b_1 + b_2)$ (direct product)

Table 13.5 Result of decomposition of the direct product $(a_1 + b_1 + b_2) \times (a_1 + b_1 + b_2)$

State (in C_{2v}) from direct product	Orbital configuration	Spin multiplicity	Allowed state in C_{2v}
$a_1 \times a_1 = A_1$	a_1^2	only singlet	${}^1A_1$
$a_1 \times b_1 = b_1 \times a_1 = B_1$	$a_1^1 b_1^1$	both singlet and triplet	${}^1B_1, {}^3B_1$
$a_1 \times b_2 = b_2 \times a_1 = B_2$	$a_1^1 b_2^1$	both singlet and triplet	${}^1B_2, {}^3B_2$
$b_1 \times b_2 = b_2 \times b_1 = A_2$	$b_1^1 b_2^1$	both singlet and triplet	${}^1A_2, {}^3A_2$
$b_1 \times b_1 = A_1$	b_1^2	only singlet	${}^1A_1$
$b_2 \times b_2 = A_1$	b_2^2	only singlet	${}^1A_1$

By decomposing the direct product we have the states, orbital designations and spin multiplicities as in Table 13.5. It is found that under C_{2v}, A_1 occurs three times, A_2 two times and B_1, B_2 each two times. The results of Table 13.5 can be physically realised if we distribute the two electrons of the t_{2g}^2 configuration among the a_1, b_1 and b_2 levels which are due to lowering of symmetry from O_h to C_{2v}. The following distributions are possible:

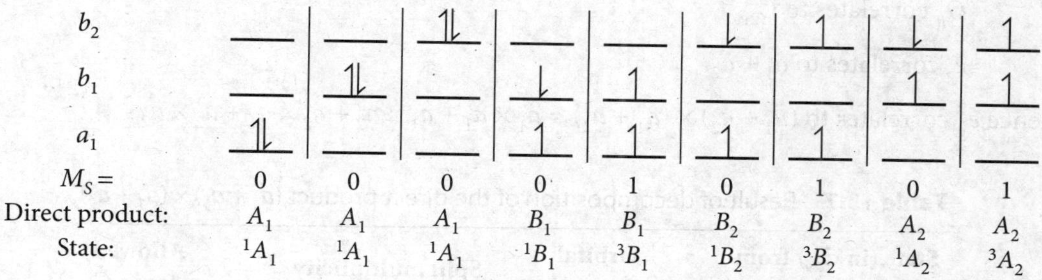

Table 13.6 Relevant correlation results for O_h and C_{2v}: irreducible representations of O_h to which those of C_{2v} (obtained by the direct products in Table 13.5) revert

O_h		C_{2v}
A_{1g}	⟵	A_1
E_g	⟵	$A_1 + A_2$
T_{2g}	⟵	$A_1 + B_1 + B_2$
T_{1g}	⟵	$A_2 + B_1 + B_2$

(Taking A_1 three times, A_2 two times and B_1, B_2 two times each)

We look up the correlation table (Appendix A4) again and find the irreducible representations of O_h to which the representations obtained under C_{2v} (from direct product) revert. The result is shown in Table 13.6.

Under C_{2v}, all the allowed A_1 states are singlets (Table 13.5). Hence in all the combinations under C_{2v} which contain A_1 (the first three of Table 13.6) are singlets, and the resulting states to which they correlate under O_h must also be singlets:

C_{2v}		O_h
$^1A_1$	$\longrightarrow$	$^1A_{1g}$
$^1A_1 + {}^1A_2$	$\longrightarrow$	^1E_g
$^1A_1 + {}^1B_1 + {}^1B_2$	$\longrightarrow$	$^1T_{2g}$

We are now left with only the triplets, $^3A_2$, $^3B_1$ and $^3B_2$ and their combination gives $^3T_{1g}$.

C_{2v}		O_h
$^3A_2 + {}^3B_1 + {}^3B_2$	$\longrightarrow$	$^3T_{1g}$

Hence the t_{2g}^2 configuration splits into $^1A_{1g}$, ^1E_g, $^1T_{2g}$ and $^3T_{1g}$ as shown in the final correlation diagram (Fig. 13.12). For the e_g^2 configuration the procedure is similar. The steps are outlined in the following:

O_h correlates to C_{2v}

e_g correlates to $a_1 + a_2$

Hence e_g^2 correlates to $(a_1 + a_2) \times (a_1 + a_2) = a_1 \times a_1 + a_1 \times a_2 + a_2 \times a_1 + a_2 \times a_2$

Table 13.7 Result of decomposition of the direct product $(a_1 + a_2) \times (a_1 + a_2)$

State (in C_{2v}) from direct product	**Orbital configuration**	**Spin multiplicity**	**Allowed state in C_{2v}**
$a_1 \times a_1 = A_1$	a_1^2	only singlet	$^1A_1$
$a_1 \times a_2 = a_2 \times a_1 = A_2$	$a_1^1 a_2^1$	both singlet and triplet	$^1A_2$, $^3A_2$
$a_2 \times a_2 = A_1$	a_2^2	only singlet	$^1A_1$

Under C_{2v}, A_1 occurs two times and A_2 once.

Going back to O_h, using correlation table,

O_h		C_{2v}
A_{1g}	$\longleftarrow$	A_1
A_{2g}	$\longleftarrow$	A_2
E_g	$\longleftarrow$	$A_1 + A_2$

Table 13.7 reveals that for A_1 only singlet is allowed. So in the sum $A_1 + A_2$, A_2 must be a singlet. Hence we obtain the following result:

C_{2v}		O_h
$^1A_1$	$\longrightarrow$	$^1A_{1g}$
$^1A_1 + {}^1A_2$	$\longrightarrow$	^1E_g

Only A_2 is left with a multiplicity of 3. Hence the inverse correlation $C_{2v} \rightarrow O_h$ gives,

O_h		C_{2v}
$^3A_{2g}$	$\longleftarrow$	$^3A_2$

Thus the e_g^2 configuration splits into $^1A_{1g}$, ^1E_g and $^3A_{2g}$ in the O_h field as shown in the final correlation diagram (Fig. 13.13). The $t_{2g}^1 e_g^1$ configuration gives states for which both singlets and triplets are allowed by the Pauli principle. No special method for assigning spin multiplicity is, therefore, needed.

13.6.2 Energy Level Diagram for a d^2 Ion in a Tetrahedral Environment

The free ion terms of a d^2 ion are ^{1}(S, D, G) and ^{3}(P, F), all of even parity. In a weak tetrahedral field they split just as in an O_h field, only the g-subscript is not necessary. Maintaining spin multiplicity conservation, the splitting is as follows:

$$^1\text{S} \rightarrow {}^1A_1$$

$$^1\text{G} \rightarrow {}^1A_1 + {}^1E + {}^1T_1 + {}^1T_2$$

$$^3\text{P} \rightarrow {}^3T_1$$

$$^1\text{D} \rightarrow {}^1E + {}^1T_2$$

$$^3\text{F} \rightarrow {}^3A_2 + {}^3T_1 + {}^3T_2$$

The left hand side of the required correlation diagram is thus known. Regarding the right hand side, the configurations under an infinitely strong tetrahedral environment are e^2, $e^1t_2^1$ and t_2^2 in increasing order of energy (here the set of d-orbitals splits into e and t_2 sets, the former being lower in energy than the latter by an amount Δ_t).

Using the correlation table for T_d, it is again found that the C_{2v} subgroup is suitable for applying the method of symmetry descent, since C_{2v} has all one-dimensional representations in common with the E and T_2 representations of T_d. The remaining task is just the same as in the case of O_h symmetry, the only difference being that the levels corresponding to e_2^2 configuration should be placed lower and those corresponding to t_2^2 at a higher position. The correlation lines are then drawn, avoiding crossing as before (Fig. 13.14).

13.7 EXPLANATION OF SPECTRAL PROPERTIES OF TRANSITION METAL COMPLEXES USING ENERGY LEVEL DIAGRAMS

The energy level diagrams (that is, the weak field–strong field correlation diagrams) are useful for explaining the spectral absorption bands of transition metal complexes. Such bands, which

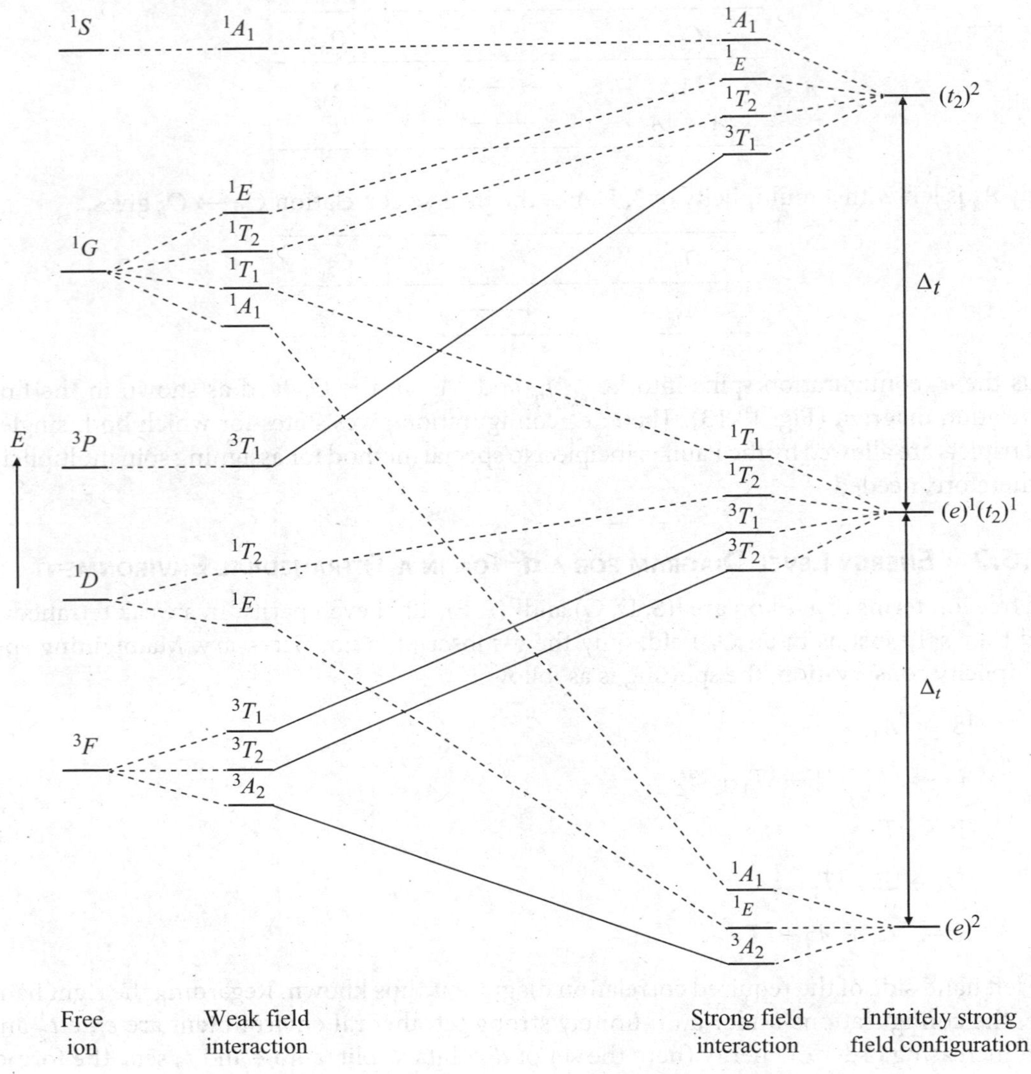

Fig. 13.14 Correlation diagram for a d^2 ion in a tetrahedral field

usually appear in the UV–vis range, are the results of electronic transitions between states obeying two selection rules:

(i) $\Delta S = 0$. This rule (called *spin selection rule*) means that transition between two states of different spin multiplicity is not allowed.

(ii) For centrosymmetric molecules (or ions) transitions between states of the same parity ($u \leftrightarrow u$ and $g \leftrightarrow g$) are not allowed. This parity selection rule is also called the *Laporte rule*.

Crystal field theory assumes that the *d*-orbitals are perturbed but their *d*-character is retained when the metal ion forms a complex or is placed at a site within a crystal. (Ligand Field Theory or Molecular Orbital Theory allows for combination of *d*-orbitals with ligand orbitals but then the metal *d*-orbitals have a dominating contribution, so that the results obtained with Crystal Field Theory are found to be qualitatively valid.)

Spectroscopic states of the complexes are derived from the d^n configurations of the metal ions and accordingly, the spectroscopic transitions in metal complexes are called *d–d transitions*. As the *d*-orbitals are all of *g*-type, such transitions in octahedral complexes (which have an inversion centre) should not be observed; however, there are certain modes of vibration which destroy the regular octahedral geometry and the inversion centre is lost; such vibrations are coupled with the electronic motion and the *d–d* transition becomes weakly allowed. This phenomenon is called *vibronic coupling* (coupling of vibrational and electronic motions). Tetrahedral complexes are not centrosymmetric and hence the spin allowed *d–d* transitions in such complexes are usually more intense (about 10^3 times) than the octahedral complexes with the same ligands.

Apart from *d–d* transitions, there are *charge transfer*, and intra- and inter-ligand electron transfer transitions. In a charge transfer transition, electrons are transferred from the metal to ligand (or in the reverse direction) by absorption of light; such transitions are both spin and symmetry allowed and are therefore more intense than *d–d* transitions. These two types of transitions are not included in the present discussion because they are outside the domain of crystal field theory. Interpretation of electronic spectra of transition metal complexes using energy level diagrams is illustrated below.

13.7.1 UV–VIS SPECTRA OF THE COMPLEX ION $[V(H_2O)_6]^{3+}$

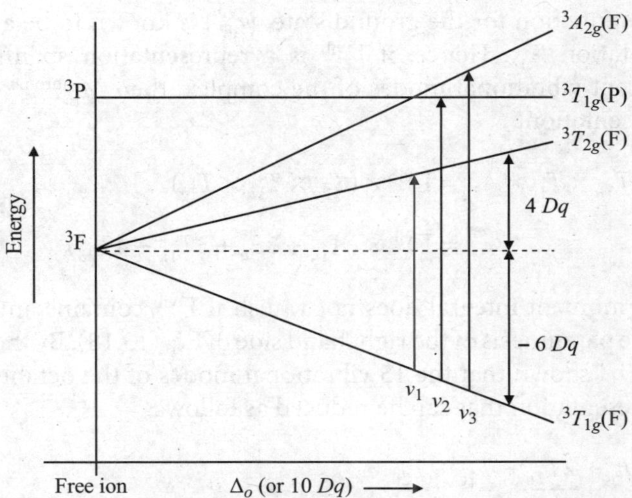

Fig. 13.15 Relevant part of the correlation diagram for a d^2-ion in an octahedral field and spin-allowed transitions for the $[V(H_2O)_6]^{3+}$ ion

The V^{3+} ion in the complex is a d^2 system in an octahedral environment. The appropriate correlation diagram (Fig. 13.15) shows that the possible spin allowed transitions are from the ground state $^3T_{1g}$(F) to the nearby excited states $^3T_{2g}$(F), $^3T_{1g}$(P) and $^3A_{2g}$(F) where the symbols in parenthesis indicate the parent states in the free ion. Experimentally, two absorption bands at 17200 and 25600 cm^{-1} are observed which impart green colour to an aqueous solution of V^{3+} salts and these are ascribed to the transitions $^3T_{1g}$ (F) → $^3T_{2g}$ (F) and $^3T_{1g}$ (F) → $^3T_{1g}$ (P); the third transition to the $^3A_{2g}$ (F) state occurs at a much higher energy and is obscured by high energy charge transfer transitions.

If we consider only electronic wave functions of the ground and excited states, then none of the transitions indicated in Fig. 13.15 is allowed as revealed by the direct products; the dipole moment operator $\hat{\mu}$ (with three Cartesian components $\hat{\mu}_x, \hat{\mu}_y, \hat{\mu}_z$) transforms according to the T_{1u} representation of O_h. For $^3T_{1g}$ (F) → $^3T_{2g}$ (F), the transition moment integral $\int (\psi^e_{\text{gr}})^* \hat{\mu} (\psi^e_{\text{ex}})\, d\tau$ corresponds to the direct product

$$T_{2g} \times T_{1u} \times T_{1g} = A_{1u} + A_{2u} + 2E_u + 3T_{1u} + 4T_{2u}$$

which can be verified using the O_h character table. None of the irreducible representations in the result of reduction is the totally symmetric representation A_{1g}. Hence the transition moment integral vanishes and the transition is not allowed. The transition is spin allowed but orbitally forbidden. The explanation for the appearance of the corresponding absorption band with weak intensity in terms of vibronic coupling was provided by Van Vleck. According to this mechanism, certain vibrational modes are also excited along with the electronic transition and the intensity integral (transition moment integral) should be

$$M^{\text{vibronic}} = \int (\psi^e_{\text{ex}}\ \psi^{\text{vib}}_{\text{ex}})^* \hat{\mu} (\psi^e_{\text{gr}}\ \psi^{\text{vib}}_{\text{gr}})\, d\tau$$

The vibrational wave function for the ground state, $\psi^{\text{vib}}_{\text{gr}}$, is known to be a basis of the totally symmetric representation A_{1g}. Hence, if Γ^{vib} is a representation spanned by the normal coordinates of different vibrational modes of the complex, then M^{vibronic} corresponds to the direct product representation,

$$\begin{aligned} T_{2g} \times \Gamma^{\text{vib}} \times T_{1u} \times T_{1g} \times A_{1g} &= \Gamma^{\text{vib}} \times (T_{2g} \times T_{1u} \times T_{1g}) \\ &= \Gamma^{\text{vib}} \times (A_{1u} + 2E_u + 3T_{1u} + 4T_{2u}) \end{aligned} \tag{13.18}$$

Thus, the transition moment integral does not vanish if Γ^{vib} contains any of the irreducible representations in the parenthesis of the right hand side of Eq. (13.18). By the method discussed in Chapter 11 it can be shown that the 15 vibrational modes of the octahedral AB_6 molecule correspond to a representation that can be reduced as follows:

$$\Gamma^{\text{vib}} = A_{1g} + E_g + \underline{2T_{1u}} + T_{2g} + \underline{T_{2u}}$$

The underlined representations are also present the parenthesis of the RHS of Eq. (13.18). Hence the direct product representation on the left hand side of Eq. (13.18) *does* have a totally symmetric component (A_{1g}). In other words, the intensity integral becomes non-zero because at least one of the two T_{1u} modes or the T_{2u} mode of vibration is excited simultaneously with the electronic excitation. Physically this means that the atomic displacement vectors in the two T_{1u} modes and the T_{2u} mode of vibration are such that they destroy the centre of symmetry of the molecule during such vibrations.

13.7.2 A d^2-SYSTEM IN A TETRAHEDRAL ENVIRONMENT

A portion of the correlation diagram relevant to the spin allowed transitions from the ground state of a d^2-ion in a tetrahedral environment is shown in Fig. 13.16. This figure indicates that the possible spin-allowed transitions are

$$^3A_2 \rightarrow {}^3T_2,\ {}^3A_2 \rightarrow {}^3T_1\,(\text{F}) \text{ and } {}^3A_2 \rightarrow {}^3T_1\,(\text{P})$$

The set of Cartesian coordinates (x, y, z) is a basis set of the irreducible representation T_2 of the T_d point group. Hence the components of the dipole moment operator $\hat{\mu}$ also transform according to T_2. Considering only the electronic (not simultaneously vibrational) transition, the intensity integral corresponds to direct product representations which can be reduced as shown below:

$$^3A_2 \rightarrow {}^3T_2 \quad : \quad A_2 \times T_2 \times T_2 = A_2 + E + T_1 + T_2$$

$$^3A_2 \rightarrow {}^3T_1\,(\text{F}) \quad : \quad A_2 \times T_2 \times T_1 = A_1 + E + T_1 + T_2$$

$$^3A_2 \rightarrow {}^3T_1\,(\text{P}) \quad : \quad A_2 \times T_2 \times T_1 = A_1 + E + T_1 + T_2$$

The totally symmetric representation A_1 is contained only in the reduction results for the last two transitions; hence the transition moment integrals for them do not vanish. These two are, therefore, symmetry allowed and should give rise to strong absorption bands in the electronic spectrum. But the reduction result for the transition $^3A_2 \rightarrow {}^3T_2$ (ν_1 in Fig. 13.16) does not contain A_1 and hence its intensity integral vanishes. Consequently this transition is not allowed without consideration of vibronic coupling. To test whether this is vibronically allowed, we need to consider the vibronic transition moment integral:

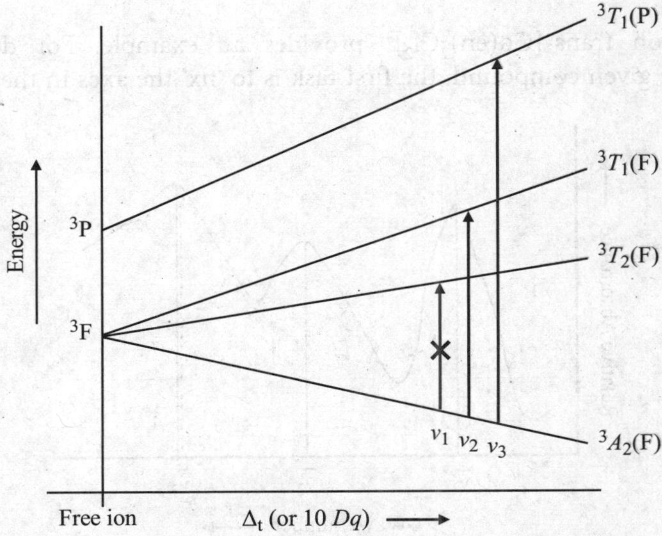

Fig. 13.16 Energy level diagram for a d^2-ion in a tetrahedral field (relevant portion only) and possible spin-allowed transitions. ν_2 and ν_3 are symmetry allowed and ν_1 is symmetry forbidden if vibronic coupling is not considered.

$$M^{\text{vibronic}} = \int (\psi^{e}_{\text{ex}}\ \psi^{\text{vib}}_{\text{ex}})^* \hat{\mu} (\psi^{e}_{\text{gr}}\ \psi^{\text{vib}}_{\text{gr}})\, d\tau \text{ corresponds to } T_2 \times \Gamma^{\text{vib}} \times T_2 \times A_2 \times A_1$$

(as the ground vibrational wave function transforms as A_1)

$$= \Gamma^{\text{vib}} \times (T_2 \times T_2 \times A_2)$$

$$= \Gamma^{\text{vib}} \times (A_2 + E + T_1 + T_2) \qquad (13.19)$$

For a tetrahedral AB_4 molecule, the normal coordinates of nine vibrational modes are bases of a representation Γ^{vib} which reduces as follows (see Chapter 11):

$$\Gamma^{\text{vib}} = A_1 + \underline{E} + \underline{2T_2}$$

As $E \times E$ and $T_2 \times T_2$ must contain the totally symmetric irreducible representation, A_1 (see Section 7.5.1), the integral in Eq. (13.19) does not vanish. Hence the transition ${}^3A_2 \rightarrow {}^3T_2$ becomes vibronically allowed if the E or T_2 mode of vibration is excited simultaneously with the electronic transition.

13.8 VIBRONIC POLARISATION IN *d–d* TRANSITIONS

For a perfectly octahedral complex, the coordinate (X-, Y- and Z-) axes are indistinguishable and there is no question of preferential absorption of a plane polarised light along any one of the axial directions. (For a perfectly tetrahedral molecule also, the situation is similar because a tetrahedron can be inscribed in a cube.) However, if the symmetry is lowered by substitution, the molecule may show different extents of absorption of plane-polarised electromagnetic radiation polarised along different axes. When a *d–d* transition which is allowed due to vibronic coupling shows different extents of absorption of polarised light along different coordinate axes depending on the symmetry of the relevant vibrational modes, the phenomenon is called *vibronic polarisation.*

The complex ion trans-$[Co(en)_2Cl_2]^+$ provides an example. For detecting vibronic polarisation with a given compound, the first task is to 'fix' the axes in the molecule. This is

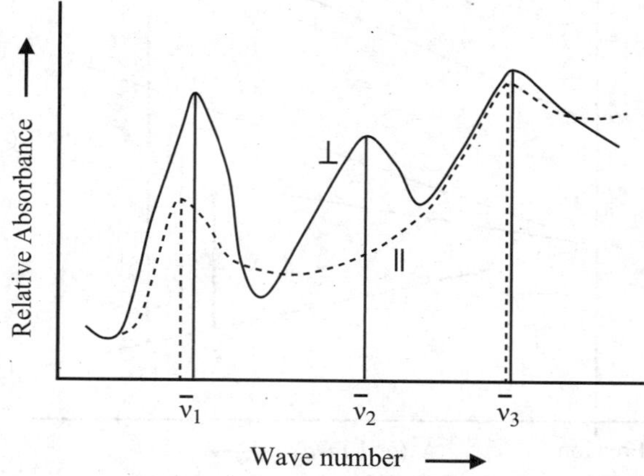

Fig. 13.17 Relative position and intensity of absorption of polarised light by *trans*-$[Co(en)_2Cl_2]Cl \cdot HCl \cdot 2H_2O$ single crystal. (||): incident light polarised parallel to the Cl–Co–Cl axis and (⊥): incident light polarised perpendicular to this axis.

possible by preparing single crystals of the compound because in such crystals all the molecules of the sample are held in space in a regular manner giving rise to the possible crystallographic axes; these axes may be identified with or related to the coordinate axes for each molecule of the sample. Figure 13.17 is a schematic diagram for the absorption spectrum of *trans*-$[Co(en)_2Cl_2]Cl \bullet HCl \bullet 2H_2O$ single crystal with light polarised parallel and perpendicular to the Cl–Co–Cl axis (taken as the Z-axis).

Light from an ordinary source is allowed to pass through a polarising prism and the emerging polarised light is incident on the single crystal sample mounted on a turntable. By rotating the table, adequate crystal planes can be chosen so that the electric vector of the incident light beam may be parallel or perpendicular to the required axis (here, the Cl–Co–Cl axis). From Fig. 13.17 it is observed that the absorption band in the region of about 27000–40000 cm^{-1} (peaked at $\bar{\nu}_3$) is not significantly polarised, that is, separate light beams polarised parallel and perpendicular to the Cl–Co–Cl axis (Z-axis) are absorbed equally. At about 22000 cm^{-1} (= $\bar{\nu}_2$) there is a strongly polarised band: the light beam polarised perpendicular to the Z-axis is strongly absorbed but the one polarised parallel to Z-axis is not absorbed. At about 16000 cm^{-1} (= $\bar{\nu}_1$) there is a significant difference between the absorption intensities for the two directions of polarisation, but the absorption bands are clearly present for both. These experimental observations can be explained in terms of vibronic coupling with the help of group theory as follows.

The molecular structure of *trans*-$[Co(en)_2Cl_2]^+$ ion is shown in Fig. 13.18 along with the coordinate axes. Here, 'en' means ethylene diamine. However, regarding the crystal field perturbation of the *d*-orbitals and vibrational modes affecting the *d*–*d* transitions, only the skeleton part $[CoN_4Cl_2]^+$ (Fig. 13.18) is relevant. The point group is D_{4h}. The centre of inversion is present in D_{4h} and hence any observed *d*–*d* transition in the complex ion must be only vibronically allowed.

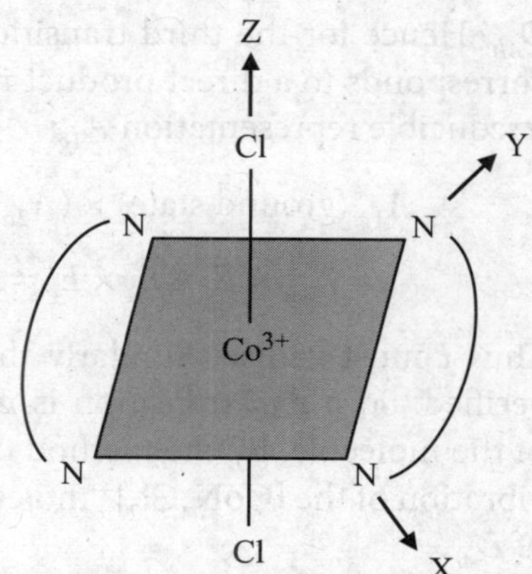

Fig. 13.18 Structure of *trans*-$[Co(en)_2Cl_2]^+$ ion and coordinate axes

Co^{3+} is a d^6 system and the ligands here produce a moderately strong crystal field making the complex low-spin; the ground and excited state electronic configurations are t_{2g}^6 and $t_{2g}^5 e_g^1$ respectively. This is supported by the fact that the complex in the ground state is experimentally found to be diamagnetic. (In fact all Co^{3+} complexes except $[CoF_6]^{3-}$ and $[CoF_3(H_2O)_3]$ are diamagnetic, supporting the t_{2g}^6 configuration for the ground state; even $[Co(H_2O)_6]^{3+}$ is diamagnetic.) If the crystal field is purely of O_h symmetry, the ground state is $^1A_{1g}$ and the excited state configuration $t_{2g}^5 e_g^1 \equiv t_{2g}^1$ (hole) e_g^1, corresponds to the following direct product representation of O_h:

$$T_{2g} \times E_g = T_{1g} + T_{2g}$$

Regarding spin multiplicity, both singlets and triplets are possible for the excited states T_{1g} and T_{2g}. For a spin-allowed transition from the ground $^1A_{1g}$ state however, the accessible excited states are singlets ($^1T_{1g}$ and $^1T_{2g}$). Let us see what happens to these states when the symmetry of the crystal field is lowered from O_h to D_{4h}. The use of the correlation table (Appendix A4) gives the following results:

O_h		D_{4h}
A_{1g}	correlates to	A_{1g}
T_{1g}	correlates to	$A_{2g} + E_g$
T_{2g}	correlates to	$B_{2g} + E_g$

Hence for the Co^{3+} complex of D_{4h} symmetry, the following d–d transitions are expected.

List I: $^1A_{1g} \rightarrow {}^1A_{2g}$, $^1A_{1g} \rightarrow {}^1B_{2g}$ and $^1A_{1g} \rightarrow {}^1E_g$

It is easy to show that without consideration of molecular vibration, the transition moment integral vanishes in each of the above cases. Among the Cartesian components of the dipole moment vectors, μ_z transforms as A_{2u} and (μ_x, μ_y) transform as the E_u representation of D_{4h}. Hence for the third transition in list I, for example, the integral $I = \int \psi^*(A_{1g})\,\hat{\mu}\,\psi(E_g)\,d\tau$ corresponds to a direct product representation which does not contain the totally symmetric irreducible representation A_{1g}:

$$A_{1g}\text{ (ground state)} \times (A_{2u} + E_u \text{ for components of } \hat{\mu}) \times E_g \text{ (excited state)}$$
$$= A_{2u} \times E_g + E_u \times E_g \neq A_{1g} \quad (\text{since } g \times u \text{ cannot be } g)$$

Thus I must vanish. Similarly the null results for the other two transitions of list I can be verified. So a d–d transition is not possible without the involvement of vibrational motion of the molecule. By the method discussed in Chapter 11, the $3 \times 7 - 6 = 15$ normal modes of vibration of the $[CoN_4Cl_2]^+$ moiety are found to span the following irreducible representations of D_{4h}.

List II: $2A_{1g}$, B_{1g}, B_{2g}, E_g, $2A_{2u}$, B_{1u}, $3E_u$ (total dimension = 15)

The transition $^1A_{1g} \rightarrow {}^1E_g$ has an electronic transition moment integral which corresponds to the direct product representation $A_{2u} \times E_g + E_u \times E_g$ as already shown. When $A_{2u} \times E_g$ and $E_u \times E_g$ are reduced by using the D_{4h} character table, the result is as shown in Table 13.8.

Table 13.8 Direct product representations for the electronic part of the transition moment integral for the transition $^1A_{1g} \rightarrow {}^1E_g$ and their reduction.

	χ (character)										
D_{4h}	$\hat{E}$	$2\hat{C}_4$	$\hat{C}_2$	$2\hat{C}_2'$	$2\hat{C}_2''$	$\hat{i}$	$2\hat{S}_4$	$\hat{\sigma}_h$	$2\hat{\sigma}_v$	$2\hat{\sigma}_d$	Reduction result
$A_{2u} \times E_g$	2	0	−2	0	0	−2	0	2	0	0	$= \underline{E_u}$
$E_u \times E_g$	4	0	4	0	0	−4	0	−4	0	0	$= A_{1u} + \underline{B_{1u}} + \underline{A_{2u}} + B_{2u}$

The irreducible representations which have been underlined in the reduction results in Table 13.8 are present in List II of irreducible representations spanned by the vibrational modes.

Table 13.9 Vibronically allowed *d–d* transitions of *trans*-$[Co(en)_2Cl_2]^+$

Transition	Polarisation with vibronic coupling	
	z	(x, y)
${}^1A_{1g} \rightarrow {}^1A_{2g}$	Forbidden	Allowed
${}^1A_{1g} \rightarrow {}^1B_{2g}$	Allowed	Allowed
${}^1A_{1g} \rightarrow {}^1E_g$	Allowed	Allowed

Hence the vibronic ${}^1A_{1g} \rightarrow {}^1E_g$ transition is polarised in the Z-direction (owing to the reduction result A_{2u} which has z as basis) and also in the perpendicular direction, that is, in the XY plane [owing to the reduction result E_u, which has (x, y) as basis set]. In this way the remaining two *d–d* transitions can be examined for their vibronic polarisation and the results are summarised in Table 13.9.

13.9 UTILISING THE HOLE FORMALISM FOR CONSTRUCTION OF CORRELATION DIAGRAMS

The difficulty in constructing correlation diagrams for d^n systems should obviously increase with an increase in the number of electrons (n). However, much labour and time can be saved by utilising the hole formalism discussed in Section 12.3.7. It states that the d^{10-n} electronic configuration behaves in the same way as the corresponding d^n electronic configuration at each point on the abscissa of the correlation diagram except that the interactions with the ligand environments are of opposite sign (that is, the energy order of states is reversed). Physically, the states arising out of the d^{10-n} electronic configuration are equivalent to states from n holes in the d^{10} electronic (closed shell) configuration. A hole, that is, the absence of an electron, may be considered as a positron. The angular momentum behaviour of a positron is the same as that of an electron but the ligands electrostatically attract positrons and repel electrons—this leads to a reversal of energy order of the states in going from d^n to d^{10-n} electronic configuration under a given environment. For example, the correlation diagram of d^8 configuration in an octahedral field is obtainable from that of the d^2 configuration in the same field simply by inverting the energy order of the t_{2g}^2, $t_{2g}^1e_g^1$ and e_g^2 configurations along the vertical axis; similarly the correlation diagram of d^7 can be obtained from that of d^3. Again, for geometric reasons, the energy order of states derived from a d^n configuration under a T_d field is just the reverse of that under an O_h field. Hence the hole formalism leads to the following general relationship:

$$d^n\,(\text{oct}) \equiv d^{10-n}\,(\text{tet}) \leftarrow \text{energy order reversed} \rightarrow d^n\,(\text{tet}) \equiv d^{10-n}\,(\text{oct})$$

13.10 ORGEL DIAGRAM: A PARTIAL CORRELATION DIAGRAM FOR HIGH SPIN COMPLEXES

Partial energy level diagrams, suitable for analysis of spectra of high spin complexes of transition metal ions in weak octahedral and tetrahedral fields, were given by Orgel. These diagrams correlate the free ion states with those which result from splitting under weak fields; the correlation lines show the variation of energy of the states with strength (Δ_t or Δ_o) of the field.

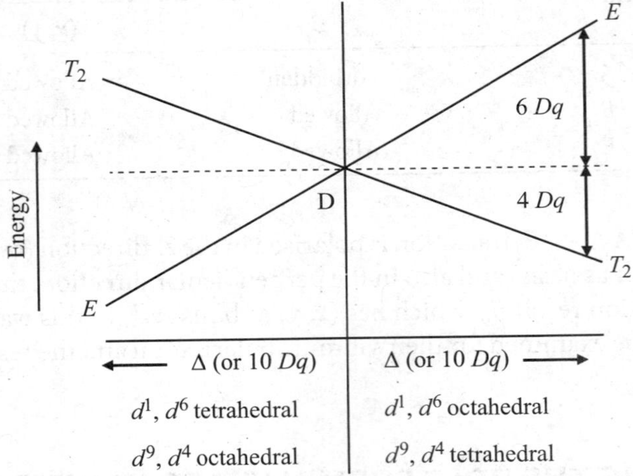

Fig. 13.19. Orgel (splitting) diagram for a D term

The configurations d^1 and its hole counterpart d^9, give rise to a ^{2}D state which splits into $^2T_{2g}$ and ^2E_g under an octahedral field and into $^2T_2$ and ^2E under a tetrahedral field. The energy orders in these two types of fields are just opposite to each other for a particular electronic configuration and so the required correlation can be shown on two sides of a single diagram (Fig. 13.19). The highest spin terms of d^n configuration can be summarised (from Table 12.15) as follows.

Configuration	High spin term*	Hole equivalent
d^1, d^6	D	d^9, d^4
d^2, d^7	P, F	d^8, d^3
d^3, d^8	P, F	d^7, d^2

(* spin multiplicity is not mentioned for the time being)

From this compilation it can be noted that besides the hole formalism, another relationship, applicable only to high spin states, exists between the d^x and d^{5+x} configurations (for $x = 1$, 2, 3). This is because the m_l values of five electrons in the half-filled d^5 configuration add up to zero and the remaining x electrons occupy the available d orbitals of highest m_l (Hund's rule). Thus only two correlation diagrams, one for D terms (Fig. 13.19) and another for the F and P terms (Fig. 13.20) will be needed for all the d^n configurations. The noteworthy features of these diagrams are the following:

- The F term splits into T_{1g}, T_{2g} and A_{2g} in an octahedral field and into T_1, T_2 and A_2 levels in a tetrahedral field [Fig. 13.20, shown on two sides of the energy axis (ordinate): the g-subscripts are dropped].

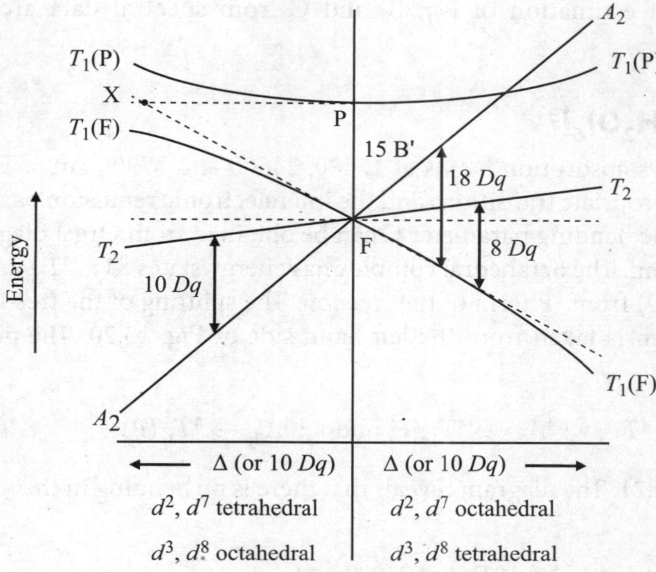

Fig. 13.20 Orgel (splitting) diagram for P and F terms

- The P term does not split—it becomes T_1 in an octahedral/tetrahedral field. The parentage F or P is mentioned in parenthesis and the correlation lines T_1(F) ↔ T_1(F) and T_1(P) ↔ T_1(P) are drawn. Up to a certain magnitude of Δ (or 10 *Dq*) there is no crossing, but beyond this the two lines intersect at X. This crossing of correlation lines of the same symmetry species (here, T_1) is not allowed by the non-crossing rule and so the lines bend and stretch away from one another (Fig. 13.20).

The energy separations between the lines are results of quantum mechanical calculations and do not have much to do with the applications of group theory being presented in this book. However, they are needed to assign spectral absorption bands of transition metal complexes. Such assignments yield the splitting parameter (*Dq*) and the extent of bending (C) of the same-symmetry correlation lines from experimentally observed UV–vis spectra of transition metal complexes. The low-energy, spin-allowed transitions can be assigned with the help of Orgel diagrams if the following points are kept in consideration:

- The difference in energy between the F and P terms resulting from inter-electron repulsions in *LS* coupling is 15B for a free ion which is modified to 15B′ in the complex due to the electronic environment produced by the ligands. This B is called a *Racah parameter* and its modification to B′ by the effect of ligands is called the *nephelauxetic effect* which is discussed in a subsequent section.
- The magnitude of splitting, that is, the energy difference $\varepsilon(E) \sim \varepsilon(T_2)$ for a D term splitting or $\varepsilon(T_2) \sim \varepsilon(A_2)$ for an F term splitting is 10*Dq*. The energy difference $\varepsilon(T_2) \sim \varepsilon(T_1)$ without bending is 8*Dq*.
- The correlation line T_1(F) ↔ T_1(F) is bent downward and T_1(P) ↔ T_1(P) is bent upward. Hence for calculating ν for any transition involving these levels, the bending parameter (C) is to be judiciously added or subtracted.

Some examples of estimation of Dq, B′ and C from spectral data are discussed in the following.

13.10.1 $[Cr(H_2O)_6]^{3+}$

$[Cr(H_2O)_6]^{3+}$ shows absorption bands at 17400, 24600 and 37900 cm^{-1}. These bands can be assigned to the appropriate transitions and the interelectronic repulsion parameter B′, the field strength Dq and the bending parameter C can be obtained from Orgel diagram as follows.

Cr^{3+} is a d^3 system. The octahedral complex has energy states $^4A_{2g}$, $^4T_{2g}$ and $^4T_{1g}$(F) from the ^{4}F term, and $^4T_{1g}$(P) from ^{4}P term of the free ion. The splitting of the free ion terms is shown in Fig. 13.21, which is taken from the left hand side of Fig. 13.20. The possible low-energy transitions are

$$\nu_1\text{: } {}^4A_{2g} \rightarrow {}^4T_{2g},\ \nu_2\text{: } {}^4A_{2g} \rightarrow {}^4T_{1g}(\text{F}) \text{ and } \nu_3\text{: } {}^4A_{2g} \rightarrow {}^4T_{1g}(\text{P})$$

as shown in Fig. 13.21. The diagram reveals that there is no bending in the required correlation lines for ν_1. Hence

$$\nu_1 = \varepsilon(T_{2g}) - \varepsilon(A_{2g}) = 10Dq = 17400 \text{ cm}^{-1}$$

As the line for T_{1g}(F) is bent downward,

$$\begin{aligned}\nu_2 &= \varepsilon[T_{1g}(\text{F})] - C - \varepsilon(A_{2g}) = \varepsilon[T_{1g}(\text{F})] - \varepsilon(T_{2g}) - C + \varepsilon(T_{2g}) - \varepsilon(A_{2g}) \\ &= 8Dq - C + 10Dq = 18Dq - C = 24600 \text{ cm}^{-1}\end{aligned}$$

The line for T_{1g}(P) is bent upward. Hence

$$\nu_3 = \varepsilon\{T_{1g}(\text{P})\} + C - \varepsilon(A_{2g}) = \varepsilon(\text{P}) + C - \varepsilon(A_{2g})$$

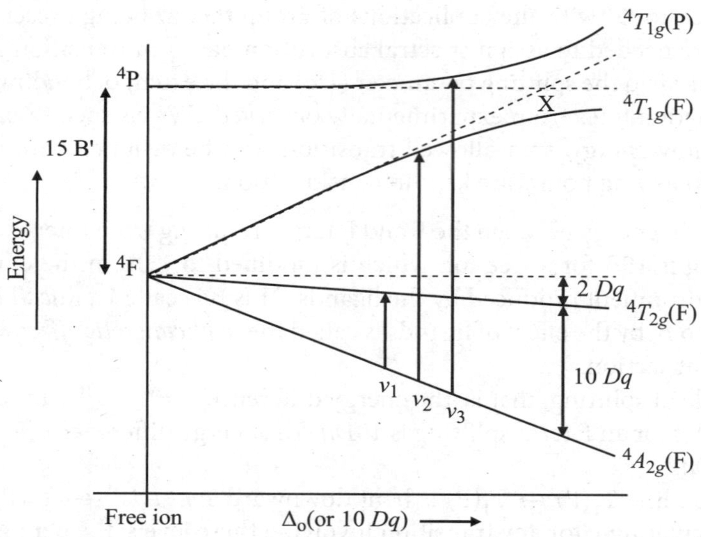

Fig. 13.21 Orgel diagram showing the three low-energy, spin allowed transitions of $[Cr(H_2O)_6]^{3+}$

$$= \varepsilon(\text{P}) - \varepsilon(\text{F}) + C + \varepsilon(\text{F}) - \varepsilon(T_{2g}) + \varepsilon(T_{2g}) - \varepsilon(A_{2g})$$

$$= 15\text{B}' + C + 2Dq + 10Dq$$

$$= 15\text{B}' + 12Dq + C = 37900 \text{ cm}^{-1}$$

Here $\varepsilon\{T_{1g}(\text{P})\} = \varepsilon(\text{P})$ = energy of the ^{4}P level of the free ion and $\varepsilon(\text{F})$ is the energy of the ^{4}F level of the free ion.

On solving these equations it is found that $Dq = 1740 \text{ cm}^{-1}$, $C = 6720 \text{ cm}^{-1}$ and $\text{B}' = 686.6 \text{ cm}^{-1}$ [B for Cr^{3+} (free ion) = 106 cm^{-1}].

13.10.2 $[CoF_6]^{4-}$ Ion in the Crystal of $KCoF_3$

In crystalline $KCoF_3$, the Co^{2+} ion is octahedrally surrounded by six F^- ions. Experimentally observed absorption bands are centered at 7150, 15200 and 19200 cm^{-1}. Co^{2+} is a d^7 ion and its high-spin complex is expected to show absorption bands, according to the right hand side of Fig. 13.20, due to the following transitions:

$$\nu_1\text{: } {}^3T_{1g}(\text{F}) \rightarrow {}^3T_{2g}(\text{F}),\ \nu_2\text{: } {}^3T_{1g}(\text{F}) \rightarrow {}^3A_{2g}(\text{F}) \text{ and } \nu_3\text{: } {}^3T_{1g}(\text{F}) \rightarrow {}^3T_{1g}(\text{P})$$

These are indicated in Fig. 13.22. The parameters B′ and C are obtained from the assignments,

$$\nu_1 = \varepsilon\{T_{2g}(\text{F})\} - [\varepsilon\{T_{1g}(\text{F})\} - C] = 8Dq + C = 7150 \text{ m}^{-1}$$

$$\nu_2 = \varepsilon\{A_{2g}(\text{F})\} - [\varepsilon\{T_{1g}(\text{F})\} - C]$$

$$= \varepsilon\{A_{2g}(\text{F})\} - \varepsilon\{T_{2g}(\text{F})\} + [\varepsilon\{T_{2g}(\text{F})\} - \varepsilon\{T_{1g}(\text{F})\}] + C$$

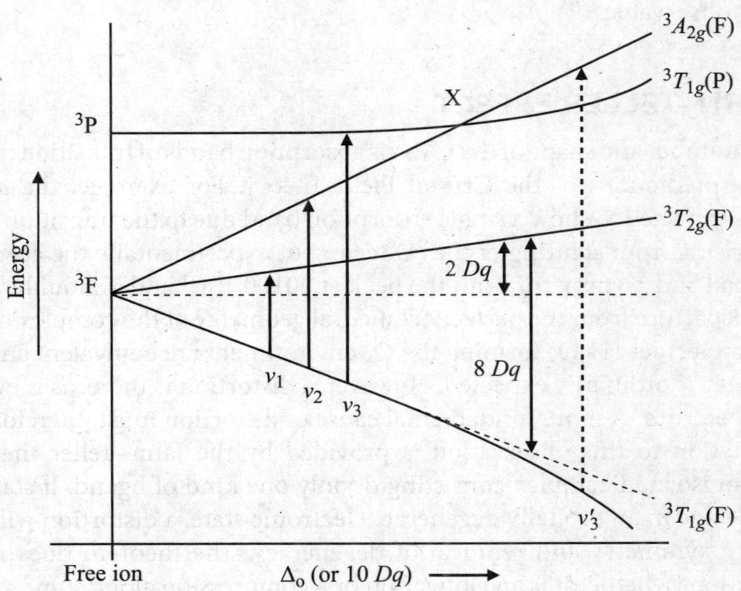

Fig. 13.22 Three spin allowed transitions for $[CoF_6]^{4-}$ ion in the crystal of $KCoF_3$. The dotted vertical line represents a transition for $[V(H_2O)_6]^{3+}$.

$$= 10Dq + 8Dq + C = 18Dq + C = 15200 \text{ cm}^{-1}$$

$$\nu_3 = [\varepsilon\{T_{1g}(\text{P})\} + C] - [\varepsilon\{T_{1g}(\text{F})\} - C]$$

But $\varepsilon\{T_{1g}(\text{P})\} = \varepsilon(\text{P})$ as the P term does not split. Hence

$$\nu_3 = [\varepsilon(\text{P}) - \varepsilon\{T_{1g}(\text{F})\} + 2C = 15\text{B}' + (8-2)\,Dq + 2C = 19200 \text{ cm}^{-1}$$

On solving, $Dq = 815$ cm^{-1}, $C = 530$ cm^{-1} and B$' = 876.6$ cm^{-1} [B for Co^{2+} (free ion) $= 970$ cm^{-1}].

13.10.3 Vanadium Salts

Aqueous solutions of trivalent vanadium salts show two absorption bands at 17200 and 25600 cm^{-1}, which impart green colour to the solutions. In $[V(H_2O)_6]^{3+}$, the d^2 ion V^{3+}, is in an octahedral environment. The necessary Orgel diagram is Fig. 13.22 which is taken from the right hand side of Fig. 13.20. There should be three spin-allowed transitions, namely,

$$\nu_1: {}^3T_{1g}(\text{F}) \rightarrow {}^3T_{2g}(\text{F}),\ \nu_2: {}^3T_{1g}(\text{F}) \rightarrow {}^3T_{1g}(\text{P}) \text{ and } \nu_3': {}^3T_{1g}(\text{F}) \rightarrow {}^3A_{2g}(\text{F})$$

Since one transition (ν_3') is not experimentally observed, the value of Δ_o must be greater than that corresponding to the $T_{1g}(\text{P}) - A_{2g}(\text{F})$ cross-over point (X) as shown by a dotted vertical line in Fig. 13.22. This transition occurs at a high energy where charge transfer transition also occurs and obscures the measurement. However, the third transition (ν_3') has been observed in the solid state, for example, in Al_2O_3 doped with V^{3+}. Here O atoms surround the V^{3+} ion in an octahedral manner and produce nearly the same Δ_o as the H_2O ligands. Charge transfer transition in the solid occurs at a much higher energy, making ν_3' (*d–d* transition) experimentally observable.

13.11 JAHN–TELLER EFFECT

Sometimes the number and shape of the UV–vis absorption bands of transition metal complexes differ from the predictions of the Crystal Field Theory. For example, the d^1 ion complex $[Ti(H_2O)_6]^{3+}$, is expected to show a single absorption band due to the transition ${}^2T_{2g} \rightarrow {}^2E_g$, the energy levels arising from splitting of the ^{2}D term; but experimentally the absorption band is found to be broad and asymmetric with the peak at 20100 cm^{-1} and a shoulder at about 18000 cm^{-1}. A slight departure from the perfect octahedral geometry of the complex can explain this; but the ligand molecules (H_2O) forming the O_h environment are equivalent and no distortion from O_h geometry is ordinarily expected. However, a distortion is there, as is evident from the experimental spectrum. A more fundamental cause of distortion must, therefore, be sought.

An interpretation to this observation is provided by the Jahn–Teller theorem which is valid even for an isolated complex consisting of only one kind of ligand. It states that for any non-linear molecule in an orbitally degenerate electronic state, a distortion will occur leading to a lowering of symmetry and removal of degeneracy. The theorem does not predict the nature of distortion (whether it is an elongation or a compression along some axis, or a change of bond angle, and so on). However, for a centrosymmetric species, the centre of symmetry is preserved in Jahn–Teller distortion.

The asymmetric spectrum of $[Ti(H_2O)_6]^{3+}$ can be explained in terms of Jahn–Teller distortion as follows. The set of *d* orbitals of a d^1 system (Ti^{3+}) splits up in an octahedral environment into a t_{2g} set (d_{xy}, d_{yz}, d_{zx}) and an e_g set ($d_{x^2-y^2}$, d_{z^2}). The electronic configuration t_{2g}^1 of the ground state of the complex is orbitally degenerate since the single electron can occupy any of the t_{2g} orbitals (d_{xy}, d_{yz}, d_{zx}). Hence Jahn–Teller distortion is possible. Now, maintaining the inversion centre, O_h symmetry can be lowered to D_{4h} symmetry by elongation or compression along one axis, say the Z-axis. In the case of compression, the ligands on the Z-axis of the O_h complex come closer to the metal ion. This increases the energy of any *d*-orbital which has electron density on or near the Z-axis. Thus, the energy of the d_{z^2} orbital of the e_g set will be increased and that of the $d_{x^2-y^2}$ orbital of the same set (which has no *z* in the wave function) will be lowered. The extent of this raising and lowering of energy will be the same ($\frac{1}{2}\delta_1$) in order to maintain the barycentre. Again, in the t_{2g} set the d_{yz} and d_{zx} orbitals are equally close to the Z-axis compared to d_{xy}. Hence the energy of the d_{yz} and d_{zx} orbitals will be increased by an amount $\frac{1}{3}\delta_2$ while that of d_{xy} will be lowered by $\frac{2}{3}\delta_2$ in order to maintain the barycentre. From the D_{4h} character table we find that d_{z^2}, $d_{x^2-y^2}$, (d_{yz}, d_{zx}) and d_{xy} are the basis sets of the irreducible representations a_{1g}, b_{1g}, e_g and b_{2g} respectively. Thus the degeneracy of the t_{2g} and e_g sets will be removed by Jahn–Teller distortion (Fig. 13.23).

When these split up orbitals in the D_{4h} environment are occupied by one electron, the resulting (spin doublet) states are $^2B_{2g}$, ^2E_g, $^2B_{1g}$, $^2A_{1g}$ in increasing order of energy. Thus, three spin-allowed transitions are expected, namely, $^2B_{2g} \rightarrow {}^2A_{1g}$, $^2B_{2g} \rightarrow {}^2B_{1g}$ and $^2B_{2g} \rightarrow {}^2E_g$ (Fig. 13.23). Of them, the first two are in the visible range and the third one is a low energy transition in the infrared region which is not detected by the UV–vis spectrophotometer. The two visible absorption bands overlap significantly to produce the asymmetric band with one peak and a shoulder close to it.

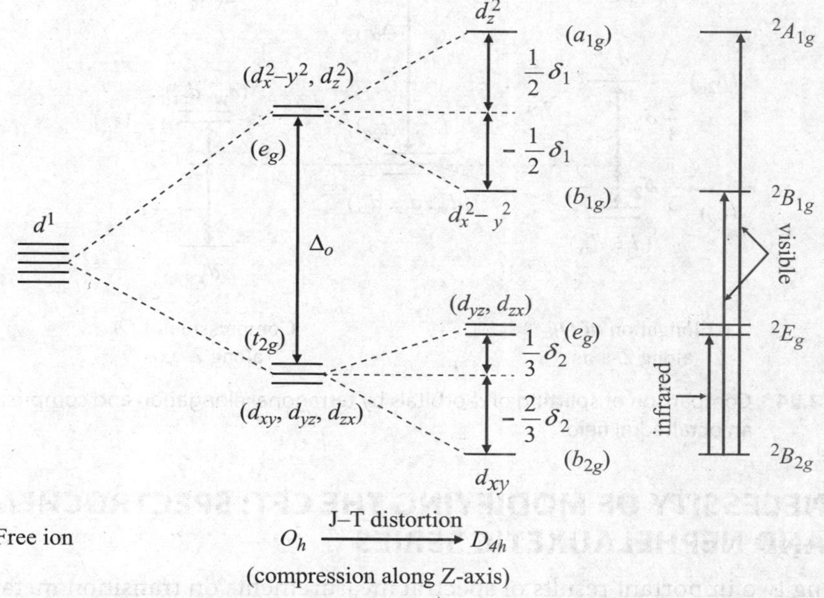

Fig. 13.23 Removal of degeneracy by Jahn–Teller distortion from O_h to D_{4h} and the possible low energy transitions of the distorted $[Ti(H_2O)_6]^{3+}$ complex

The distortion by compression also leads to some overall stabilisation of the d^1-complex by an amount $\frac{2}{3}\delta_2$ due to a change of configuration from $t_{2g}^1(O_h)$ to $b_{2g}^1(D_{4h})$.

It should be mentioned at this point that an elongation along the Z-axis produces the same spectral pattern for the d^1 ion complex. But this is not energetically favoured. Such an elongation lowers the energy of the d_{z^2} orbital of the e_g set and the (d_{yz}, d_{zx}) orbitals of the t_{2g} set. In Fig. 13.24, the two splitting patterns by Jahn–Teller distortion are shown side by side for a comparison.

With one d electron in configuration t_{2g}^1, elongation produces a stabilisation of the complex by an amount of $\frac{1}{3}\delta_2$ while compression stabilises by an amount $\frac{2}{3}\delta_2$. Thus compression is the preferred distortion in this case.

Complexes of the following types are immune to Jahn–Teller distortion due to lack of orbital degeneracy:

$$d^3\ (t_{2g}^3);\quad d^5\ (\text{high spin},\ t_{2g}^3e_g^2);\quad d^6\ (\text{low spin},\ t_{2g}^6);\quad d^8\ (t_{2g}^6e_g^2);\quad d^{10}\ (t_{2g}^6e_g^4)$$

Octahedral complexes of other d^n ion [d^1, d^2, d^4, d^5 (low spin), d^6 (high spin), d^7, d^9] undergo Jahn–Teller distortion and their description as octahedral, tetrahedral and so on, is often really an approximation to their true structures.

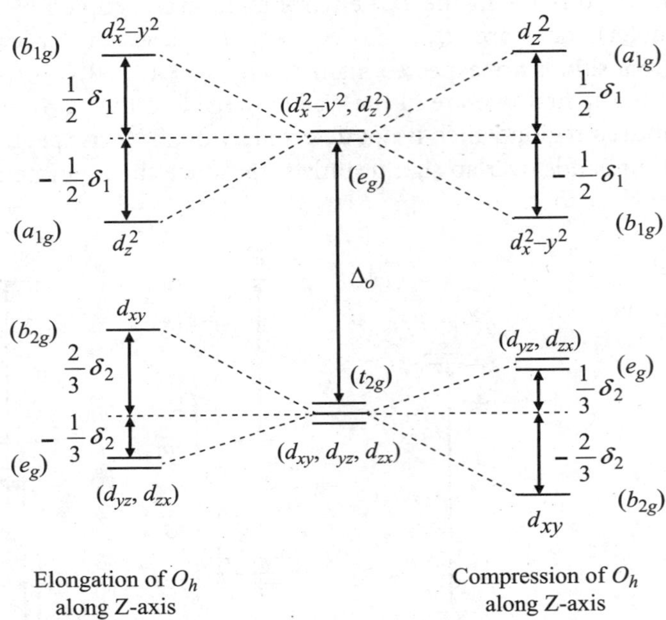

Fig. 13.24 Comparison of splitting of d-orbitals by tetragonal elongation and compression of an octahedral field

13.12 NECESSITY OF MODIFYING THE CFT: SPECTROCHEMICAL AND NEPHELAUXETIC SERIES

The following two important results of spectral measurements on transition metal complexes indicated that the pure ionic model of the CFT needs to be modified by incorporating some degree of covalence in the metal–ligand linkage:

(i) For a particular transition metal ion, the extent of splitting of *d*-orbital energies (10*Dq* or Δ) varies from ligand to ligand; with some common ligands 10*Dq* increases in the (experimentally determined) order

$$I^- < Br^- < S^{2-} < SCN^- < Cl^- < NO_3^- < F^- < OH^- < ox^{2-} < H_2O < NCS^- < CH_3CN$$

$$< NH_3 < en < dipy < phen < NO_2^- < CN < CO$$

This series is called the *spectrochemical series*. The most noteworthy fact is that the neutral ligand CO causes the largest splitting (even larger than the negatively charged ligands). This cannot be rationalised by the purely ionic CFT.

(ii) The magnitudes of energy separation between the Russell–Saunders terms are smaller in the complexed ion than in the free ion. The free ion values are available from a compilation by C E Moore. A decrease in term-energy separation in complexes suggests that inter-electron repulsion in metal *d* orbitals decreases by complex formation. This is possible if *d* electrons are delocalised into ligand orbitals. Jørgensen suggested that in the complex, the metal's *d*-electron cloud 'expands' into the region of the ligands, and termed such expansion *nephelauxetic effect*. This amounts to a degree of covalency, that is, a certain extent of overlap of *d*-orbitals with ligand orbitals. In terms of the Racah parameter B, Jørgensen defined the nephelauxetic ratio (β) in terms of the energy (ε) of the *F* and *P* terms as follows:

$$\beta = \frac{B'}{B} = \frac{[\varepsilon(^xF) - \varepsilon(^xP)] \text{ for complex}}{[\varepsilon(^xF) - \varepsilon(^xP)] \text{ for free ion}} \quad \text{(where } x \text{ is spin multiplicity)}$$

This ratio depends on the metal ion and also on the ligand strength. For a given metal ion and given geometry of the complex, as the ligand strength increases, the overlap of electron clouds also increases; this causes greater and greater deviation of β from unity. Depending on this fact, some common ligands have been arranged in order of their power of making covalent bond with metals, in the following series:

$$F^- < H_2O < urea < NH_3 < en < ox^{2-} < NCS^- < Cl^- < CN^- < N_3^- < Br^- < I^-$$

Here β increases from left to right. This series is called the *nephelauxetic series*.

Electron spin density measurements also support the view that metal *d*-orbitals overlap with ligand orbitals. These facts led to the modification of the CFT by incorporation of the adjustable parameters (B′, C ...) and actual calculation of the metal–ligand orbital overlap. The improved form is known as Ligand Field Theory (LFT). Finally, the molecular orbital method came in. In Section 13.15 it will be shown that the MO procedure applied to an octahedral complex leads to the same splitting of the set of metal *d*-orbitals as does the CFT method—LFT retains the symmetry aspects of CFT and modifies the energy aspects for analysis of the spectral measurements.

13.13 THE EFFECT OF ELECTRON SPIN IN CRYSTAL FIELD SPLITTING OF TERMS: DOUBLE GROUPS

So far the effect of electron spin on crystal field splitting of terms was completely neglected; the crystal field only perturbed the energy levels resulting from orbital motion of the electrons and it was assumed that the electron spin does not 'feel' the crystal field. This is a satisfactorily valid assumption if the Russell–Saunders coupling scheme operates. However, for systems in which the j–j coupling scheme operates, the good quantum number is $J = L + S$ instead of L and the effect of electron spin on crystal field splitting of terms should be included by taking the $(2J + 1)$-fold degenerate representations of the full rotation group (K) instead of the $(2L + 1)$-fold degenerate representations; this means that the angular momentum eigenfunctions which form bases of the irreducible representations of K are Y_J^M instead of Y_L^M as taken before.

For a multi-electron system, L has only integral values while $J = L + S$ can have both integral and half-integral values. A peculiar situation arises with J when the characters of rotation operators are calculated by means of Eq. (13.6), using J instead of l—the rotations $\hat{C}_n(\alpha)$ and $\hat{C}_n(\alpha + 2\pi)$ about an axis do not always give the same character although the angle α and $(\alpha + 2\pi)$ physically mean the same position in space:

$$\chi^J[\hat{C}_n(\alpha+2\pi)] = \frac{\sin\left[\left(J+\frac{1}{2}\right)(\alpha+2\pi)\right]}{\sin\frac{1}{2}(\alpha+2\pi)} = \frac{\sin\left[\left(J+\frac{1}{2}\right)\alpha+(2J+1)\pi\right]}{\sin\left(\pi+\frac{\alpha}{2}\right)}$$

$$= \begin{cases} \dfrac{\sin\left[\left(J+\frac{1}{2}\right)\alpha+(\text{even integer})\cdot\pi\right]}{-\sin\frac{\alpha}{2}} & \text{for half integral } J \\[2ex] \dfrac{\sin\left[\left(J+\frac{1}{2}\right)\alpha+(\text{odd integer})\cdot\pi\right]}{-\sin\frac{\alpha}{2}} & \text{for integral } J \end{cases}$$

$$= \begin{cases} \dfrac{\sin\left(J+\frac{1}{2}\right)\alpha}{-\sin\frac{\alpha}{2}} & \text{for half integral } J \\[2ex] \dfrac{-\sin\left(J+\frac{1}{2}\right)\alpha}{-\sin\frac{\alpha}{2}} & \text{for integral } J \end{cases}$$

$$= \begin{cases} -\chi^J[\hat{C}_n(\alpha)] & \text{for half integral } J \\ \chi^J[\hat{C}_n(\alpha)] & \text{for integral } J \end{cases} \tag{13.20}$$

Thus, for integral J, the same value of $\chi^J[\hat{C}_n(\alpha)]$ is returned on a further rotation through 2π, as is expected in a real physical world; but with half-integral J, $\chi^J[\hat{C}_n(\alpha + 2\pi)]$ is not the same as $\chi^J[\hat{C}_n(\alpha)]$. Hence the character for a $\hat{C}_n(\alpha)$ rotation is 'double valued' for half-integral J. However, even for half-integral J,

$$\chi^J[\hat{C}_n(\alpha+4\pi)] = \frac{\sin\left[\left(J+\frac{1}{2}\right)(\alpha+4\pi)\right]}{\sin\frac{1}{2}(\alpha+4\pi)} = \frac{\sin\left[\left(J+\frac{1}{2}\right)\alpha+(2J+1)\,2\pi\right]}{\sin\left(2\pi+\frac{\alpha}{2}\right)}$$

$$= \frac{\sin\left[\left(J+\frac{1}{2}\right)\alpha + (\text{even integer})\cdot\pi\right]}{+\sin\frac{\alpha}{2}}$$

$$= \chi^J[\hat{C}_n(\alpha)] \tag{13.21}$$

Thus $\hat{C}_n(4\pi) = \hat{E}$ (the identity operator) but $\hat{C}_n(2\pi) \neq \hat{E}$ in the representation with basis functions $\{Y_J^M, J \text{ half-integral}\}$. We are thus led to a world of fiction—a single 2π-rotation does not bring one to an identical position but two successive 2π-rotations do. To overcome this strange situation Bethe proposed to use an operator $\hat{Q}$ such that

$$\hat{Q} \neq \hat{E} \text{ but } \hat{Q}^2 = \hat{E} \equiv \text{a } 4\pi\text{-rotation} \tag{13.22}$$

This $\hat{Q}$ amounts to a 2π-rotation $\hat{C}_n^n$, about a hypothetical axis which has no analogue in the familiar three dimensional space. Using this $\hat{Q}$ as a new symmetry operator it is possible to construct, from a group $G = \{\hat{E}, \hat{A}, \hat{B}, \ldots\}$, a new group $G' = \{\hat{E}, \hat{A}, \hat{B}, \ldots \hat{Q}, \hat{Q}\hat{A}, \hat{Q}\hat{B}, \ldots\}$. This augmented group G' is called *double group* of the original group G. The order of the double group is obviously twice that of the original group. The number of conjugate classes in G' is however not twice that in G, and is determined on the basis of the following rules:

(i) $\hat{Q}$ being a rotation operator commutes with all other operators of a rotation group. Hence $\hat{Q}$ forms a class by itself in the rotation group.
(ii) $\hat{C}_2$ and $\hat{Q}\hat{C}_2$ are in the same class if and only if there is a C_2-axis perpendicular to the one in question. [This cannot be proved easily; for details the reader is directed to the original paper by Bethe (1929) and also the books by Hamermesh (1964) and Landau and Lifshtz (1977).]
(iii) For *ordinary* groups, $\hat{C}_n^n = \hat{E}$ and the rotation $\hat{C}_n^p$ is in the same class as $\hat{C}_n^{n-p}$. But in the case of double groups, $\hat{C}_n^n \neq \hat{E}$; $\hat{C}_n^{2n} = \hat{E}$ and the rotation $\hat{C}_n^p$ is conjugate to $\hat{C}_n^{2n-p} \equiv \hat{C}_n^n \cdot \hat{C}_n^{n-p} \equiv \hat{Q} \cdot \hat{C}_n^{n-p}$.

The conjugate classes in the double group of a rotation group will be as follows:

$(\hat{E})$, $(\hat{Q})$, $(\hat{C}_2 + \hat{Q}\hat{C}_2)$, $(\hat{C}_n^p + \hat{Q} \cdot \hat{C}_n^{n-p}$ for the individual values of the numbers p and n)

For constructing the character table of a double group, at first the group elements are classified according to the above observations and then the dimensions of the possible irreducible representations are determined by the usual dimensionality theorem,

$$\sum l_i^2 = \text{order of the group} \tag{13.23}$$

where l_i is the dimension of the ith irreducible representation.

Half the elements of the double group G' come from the ordinary group G and the remaining ones result from multiplication of these by $\hat{Q}$. Hence the character table of G' has two parts. The first part contains the set of characters of the ordinary (or normal, that is, single-valued) representations (Γ_1, Γ_2, … Γ_c in Bethe's notation) and the characters in the augmented part are those of the additional irreducible representations Γ_{c+1}, Γ_{c+2}, … (c = number of conjugate classes in the ordinary group G); they are evaluated according to Eqs (13.20) and (13.21). In particular, we note the following:

(a) $\chi[\hat{Q}\cdot\hat{C}_n(\alpha)] = \chi[\hat{C}_n(\alpha+2\pi)]$ (since $\hat{Q}$ means a 2π rotation)

$$= \begin{cases} \chi[\hat{C}_n(\alpha)] & \text{for integal } J\text{, that is, for the single valued representations } \Gamma_1, \Gamma_2, \ldots, \Gamma_c \\ -\chi[\hat{C}_n(\alpha)] & \text{for half integal } J\text{, that is, for the double valued representations } \Gamma_{c+1}, \Gamma_{c+2}, \ldots \end{cases}$$

(b) $\chi(\hat{Q}\cdot\hat{C}_2) = \chi(\hat{C}_2)$ [by rule (ii)]

(c) $\chi(\hat{Q}) = \chi[\hat{C}_n(2\pi)]$

$$= \begin{cases} \chi(\hat{E}) & \text{for } \Gamma_1, \Gamma_2, \ldots, \Gamma_c \text{ (for the single valued representations)} \\ -\chi(\hat{E}) & \text{for } \Gamma_{c+1}, \Gamma_{c+2} \ldots \text{ (for the double valued representations)} \end{cases}$$

For the augmented part of the double group $\chi(\hat{Q}\cdot\hat{C}_2) = \chi(\hat{C}_2) = 0$ because a combination of (a) and (b) gives $\chi(\hat{Q}\cdot\hat{C}_2) = \chi(\hat{C}_2) = -\chi(\hat{C}_2)$, whatever be the two-fold rotation axis, if there exists another C_2-axis perpendicular to the one in question. The procedure is illustrated below.

13.13.1 The Double Group *O′* of the Cubic Pure Rotation Group *O*

The cubic *O* group has 24 operations divided into five distinct classes: $\hat{E}, 8\hat{C}_3$ (that is, $4\hat{C}_3$ and $4\hat{C}_3^2$), $3\hat{C}_2 (\equiv 3\hat{C}_4^2)$, $6\hat{C}_4$, $6\hat{C}_2'$ and five irreducible representations Γ_1, Γ_2, Γ_2, Γ_4 and Γ_5 are possible for *O*. Multiplying by $\hat{Q}$ and obeying the classification rules we have, for the double group *O′*, the following 48 operations divided into 8 conjugate classes (I to VIII):

$$\underset{\text{I}}{(\hat{E})}\ \underset{\text{II}}{(\hat{Q})}\ \underset{\text{III}}{(4\hat{C}_3 + 4\hat{Q}\hat{C}_3^2)}\ \underset{\text{IV}}{(4\hat{C}_3^2 + 4\hat{Q}\hat{C}_3)}\ \underset{\text{V}}{(3\hat{C}_2 + 3\hat{Q}\hat{C}_2)}\ \underset{\text{VI}}{(3\hat{C}_4 + 3\hat{Q}\hat{C}_4^3)}\ \underset{\text{VII}}{(3\hat{C}_4^3 + 3\hat{Q}\hat{C}_4)}\ \underset{\text{VIII}}{(6\hat{C}_2' + 6\hat{Q}\hat{C}_2')}$$

Thus *O′* has 8 irreducible representations—the five normal single-valued representations Γ_1, Γ_2, Γ_2, Γ_4 and Γ_5 and three double-valued representations, Γ_6, Γ_7 and Γ_8. If l_i is the dimension of the *i*th representation,

$$\sum_{i=1}^{8} l_i^2 = 48 \tag{13.24}$$

which is satisfied only by the set of l_i values: {1, 1, 2, 3, 3, 2, 2, 4}. The first five l_i values are the dimensions of the irreducible representations Γ_1 to Γ_5 of *O*. Hence for *O′* the representations Γ_6, Γ_7 and Γ_8 have dimensions 2, 2 and 4 respectively which are the characters $\chi(\hat{E})$ for the double group *O′*. The second column, which lists the value of $\chi(\hat{Q})$, can be written down by using the relation (c) stated above [$\chi(\hat{Q}) = \chi(\hat{E})$ for Γ_1 to Γ_5 and $\chi(\hat{Q}) = -\chi(\hat{E})$ for Γ_6, Γ_7 and Γ_8], and the orthogonality of the first two character-columns can be immediately verified (Table 13.10). The columns under $(\hat{C}_2 + \hat{Q}\hat{C}_2)$ and $(\hat{C}_2' + \hat{Q}\hat{C}_2')$ (that is, classes V and VIII of the above list) contain 0 (zero) in the entries for the double-valued representations (Table 13.10) according to the result (c). For the remaining part of the required character table we proceed as follows: For any representation $\chi(\hat{E})$ = dimension of the representation. Hence,

$$2J+1=2 \text{ or } J=\tfrac{1}{2} \text{ for } \Gamma_6 \text{ and } \Gamma_7;\ 2J+1=4 \text{ or } J=\tfrac{3}{2} \text{ for } \Gamma_8$$

For $$J=\tfrac{1}{2},\ \chi^J(\hat{C}_3)=\frac{\sin\left[\left(\frac{1}{2}+\frac{1}{2}\right)\frac{2\pi}{3}\right]}{\sin\left(\frac{1}{2}\cdot\frac{2\pi}{3}\right)}=1=-\chi^J(\hat{Q}\hat{C}_3)$$

$$\chi^J(\hat{C}_4)=\frac{\sin\left[\left(\frac{1}{2}+\frac{1}{2}\right)\frac{\pi}{2}\right]}{\sin\left(\frac{1}{2}\cdot\frac{\pi}{2}\right)}=\sqrt{2}=-\chi^J(\hat{Q}\hat{C}_4)$$

For $$J=\tfrac{3}{2},\ \chi^J(\hat{C}_3)=\frac{\sin\left[\left(\frac{3}{2}+\frac{1}{2}\right)\frac{2\pi}{3}\right]}{\sin\left(\frac{1}{2}\cdot\frac{2\pi}{3}\right)}=-1=-\chi^J(\hat{Q}\hat{C}_3)$$

$$\chi^J(\hat{C}_4)=\frac{\sin\left[\left(\frac{3}{2}+\frac{1}{2}\right)\frac{\pi}{2}\right]}{\sin\left(\frac{1}{2}\cdot\frac{\pi}{2}\right)}=0=-\chi^J(\hat{Q}\hat{C}_4)$$

We can now immediately fill up the columns under $(\hat{C}_3+\hat{Q}\hat{C}_3^2)$, $(\hat{C}_4+\hat{Q}\hat{C}_4^3)$ and $(\hat{C}_4^3+\hat{Q}\hat{C}_4)$ for any one of Γ_6, Γ_7 $(J=\frac{1}{2})$ and Γ_8 $(J=\frac{3}{2})$. One cannot use the same entries for Γ_6 and Γ_7; the complete array of entries is then found from orthogonality of Γ_6 and Γ_7 (Table 13.10).

Table 13.10 Character table of the double group *O′* for cubic group *O*

Mulliken symbol	Bethe notation	χ (character)							
		$\hat{E}$	$\hat{Q}$	$4\hat{C}_3+4\hat{Q}\hat{C}_3^2$	$4\hat{C}_3^2+4\hat{Q}\hat{C}_3$	$3\hat{C}_2+3\hat{Q}\hat{C}_2$	$6\hat{C}_2'+6\hat{Q}\hat{C}_2'$	$3\hat{C}_4+3\hat{Q}\hat{C}_4^3$	$3\hat{C}_4^3+3\hat{Q}\hat{C}_4$
A_1	Γ_1	1	1	1	1	1	1	1	1
A_2	Γ_2	1	1	1	1	1	−1	−1	1
E	Γ_3	2	2	−1	−1	2	0	0	0
T_1	Γ_4	3	3	0	0	−1	−1	1	1
T_2	Γ_5	3	3	0	0	−1	1	−1	−1
	Γ_6	2	−2	1	−1	0	0	$\sqrt{2}$	$-\sqrt{2}$
	Γ_7	2	−2	1	−1	0	0	$-\sqrt{2}$ (*)	$\sqrt{2}$ (*)
	Γ_8	4	−4	−1	1	0	0	0	0

(*) these entries are obtained by making the representations Γ_6 and Γ_7 orthogonal.

13.13.2 Construction of Character Table of the Double Group D_4' of the Pure Rotational Point Group D_4

The point group D_4 has eight symmetry operations divided into five classes: $(\hat{E})$, $(2\hat{C}_4)$ (that is, $\hat{C}_4$ and $\hat{C}_4^3$), $(\hat{C}_2)$ (that is, $\hat{C}_4^2$), $(2\hat{C}_2')$ and $(2\hat{C}_2'')$. For a ready reference the character table of D_4 is given in Table 13.11.

Multiplying the symmetry operators of D_4 by $\hat{Q}$ and classifying in accordance with rules (a) to (c), we find the following seven classes for the double group D_4':

$$(\hat{E}), (\hat{Q}), (\hat{C}_4 + \hat{Q}\hat{C}_4^3), (\hat{C}_4^3 + \hat{Q}\hat{C}_4),$$
$$(\hat{C}_2 + \hat{Q}\hat{C}_2), 2(\hat{C}_2' + \hat{Q}\hat{C}_2'),$$
$$2(\hat{C}_2'' + \hat{Q}\hat{C}_2'')$$

Table 13.11 Character table of the D_4 point group

D_4	$\hat{E}$	$2\hat{C}_4$	$\hat{C}_2$	$2\hat{C}_2'$	$2\hat{C}_2''$
A_1	1	1	1	1	1
A_2	1	1	1	−1	−1
B_1	1	−1	1	1	−1
B_2	1	−1	1	−1	1
E	2	0	−2	0	0

Sixteen symmetry operations of D_4' are divided into seven classes. Hence there are seven irreducible representations of which five are single-valued corresponding to the five irreducible representations (A_1 to E) of D_4; the remaining two, Γ_6 and Γ_7 (according to Bethe's notation), are double-valued representations. If $l_1, l_2, \ldots, l_7$ are the dimensions of the irreducible representations of D_4', then

$$l_1^2 + l_2^2 + l_3^2 + l_4^2 + l_5^2 + l_6^2 + l_7^2 = 16$$

With integral non-zero values of l_i, this is satisfied by

$$\{l_i\} = 1, 1, 1, 1, 2, 2, 2$$

Table 13.12 Character table of the double group D_4' of the point group D_4

Mulliken symbol	Bethe notation	$\hat{E}$	$\hat{Q}$	$\hat{C}_4 + \hat{Q}\hat{C}_4^3$	$\hat{C}_4^3 + \hat{Q}\hat{C}_4$	$\hat{C}_2 + \hat{Q}\hat{C}_2$	$2(\hat{C}_2' + \hat{Q}\hat{C}_2')$	$2(\hat{C}_2'' + \hat{Q}\hat{C}_2'')$
		χ (character)						
A_1'	Γ_1	1	1	1	1	1	1	1
A_2'	Γ_2	1	1	1	1	1	−1	−1
B_1'	Γ_3	1	1	−1	−1	1	1	−1
B_2'	Γ_4	1	1	−1	−1	1	−1	1
E'	Γ_5	2	2	0	0	−2	0	0
	Γ_6	2	−2	$\sqrt{2}$	$-\sqrt{2}$	0	0	0
	Γ_7	2	−2	$-\sqrt{2}$	$\sqrt{2}$	0	0	0

The l_i values are the characters of $\hat{E}$; the first five are also the characters of $\hat{Q}$ in Γ_1 to Γ_5 and the negative of the remaining two are $\chi(\hat{Q})$ in Γ_6 and Γ_7. The characters in the other columns can be easily determined by the rules stated earlier. The resulting character table is given in Table 13.12.

The application of the double group is best explained with the help of an example. Let us consider a d^1 or d^9 ion in a D_{4h} environment [a square planar Cu(II) or Ag(II) complex is an example]. The ground state of the free ion is ^{2}D. We can analyse its term splitting in the following two ways.

(i) Spin–orbit (S–O) interaction followed by crystal field effect: If the S–O coupling is stronger than crystal field effect, we first consider the splitting by S–O interaction into the fine structure levels, ${}^2D_{5/2}$ and ${}^2D_{3/2}$. With one d electron (or one hole), $J = 2 + \frac{1}{2}$ to $|2 - \frac{1}{2}|$ that is, 5/2 and 3/2, and the fine structure levels are as shown in Fig. 13.25. The D_{4h} environment now splits each of the levels corresponding to the individual J values. Using the two half-integral J values, the characters of the representations $\Gamma_{5/2}$ and $\Gamma_{3/2}$ of the double group D_4' (pure rotation group) corresponding to D_{4h} are as in Table 13.13. The characters have been calculated using the relations discussed earlier.

Table 13.13 Characters of representations of D_4' with $J = 5/2$ and 3/2

	χ (character)						
	$\hat{E}$	$\hat{Q}$	$\hat{C}_4 + \hat{Q}\hat{C}_4^3$	$\hat{C}_4^3 + \hat{Q}\hat{C}_4$	$\hat{C}_2 + \hat{Q}\hat{C}_2$	$2(\hat{C}_2' + \hat{Q}\hat{C}_2')$	$2(\hat{C}_2'' + \hat{Q}\hat{C}_2'')$
$\Gamma_{5/2}$	6	−6	$-\sqrt{2}$	$\sqrt{2}$	0	0	0
$\Gamma_{3/2}$	4	−4	0	0	0	0	0

By using the standard reduction formula or by inspection it can be found that

$$\Gamma_{5/2} = \Gamma_6 + 2\Gamma_7 \text{ and } \Gamma_{3/2} = \Gamma_6 + \Gamma_7$$

where Γ_6 and Γ_7 are the double-valued irreducible representations of D_4' (Table 13.12). Hence the fine structure level corresponding to $J = 3/2$ further splits in the crystal field into Γ_6 and Γ_7 levels and the one corresponding to $J = 5/2$ splits into one Γ_6 and two Γ_7 levels (Fig. 13.25). The ordering of energy-levels in Fig. 13.25 is arbitrary; the true order can be obtained only by actual calculation.

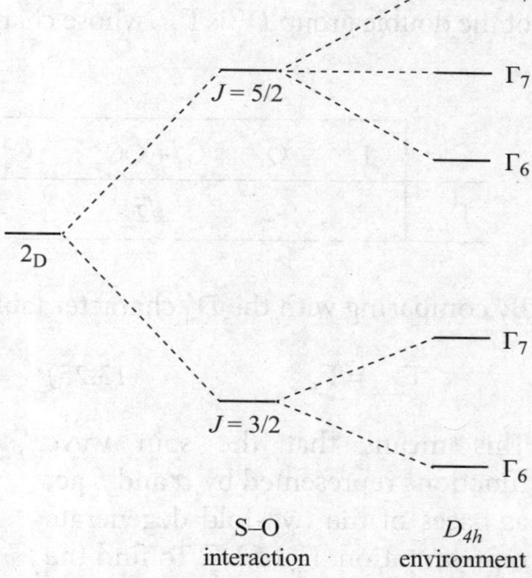

Fig. 13.25 Splitting of fine structure levels of the ${}^2D$ term in a D_{4h} environment

(ii) Crystal field effect followed by S–O interaction: If the S–O interaction is weaker than the crystal field effect, the procedure should be adopted in the reverse order. First, the crystal field splitting of the ${}^2D$ terms ($L = 2$) in the D_{4h} field has to be considered by the usual procedure. The characters of the operations of the rotation group D_4 are as follows:

$$\chi(\hat{E}) = 2L + 1 = 5,\ \chi(\hat{C}_4) = \frac{\sin\left[\left(L + \frac{1}{2}\right)\frac{\pi}{2}\right]}{\sin\left(\frac{1}{2}\cdot\frac{\pi}{4}\right)} = \frac{\sin\left[\left(2 + \frac{1}{2}\right)\frac{\pi}{2}\right]}{\sin\left(\frac{\pi}{4}\right)} = -1$$

$$\chi(\hat{C}_2) = \frac{\sin\left[\left(L + \frac{1}{2}\right)\pi\right]}{\sin\left(\frac{1}{2}\cdot\pi\right)} = \frac{\sin\left[\left(2 + \frac{1}{2}\right)\pi\right]}{\sin\left(\frac{\pi}{2}\right)} = 1$$

Hence a representation (Γ_L) of D_4 with the five orbital wave functions (corresponding to $M_L = 2, 1, 0, -1, -2$) as basis are:

D_4	$\hat{E}$	$2\hat{C}_4$	$\hat{C}_2$	$2\hat{C}_2'$	$2\hat{C}_2''$
Γ_L	5	−1	1	1	1

The usual reduction procedure gives

$$\Gamma_L = A_1 + B_1 + B_2 + E$$

Hence by the crystal field effect in a D_{4h} environment, the ²D level splits into A_{1g}, B_{1g}, B_{2g} and E_g levels (in one-to-one correspondence with the component irreducible representations of Γ_L found above). Each of these is now to be subjected to S–O coupling. The S–O wave function is of the type $\psi\alpha$ or $\psi\beta$ where ψ is the orbital part of the wave function and α and β are the spin wave functions corresponding to $m_s = \frac{1}{2}$ and $-\frac{1}{2}$ respectively. In group theory, the effect of the crystal field of D_{4h} symmetry can be found by considering the direct product of the representations Γ^{ψ} and Γ^{s}, where Γ^{ψ} is a representation with orbital wave function as basis and Γ^{s} is a representation with the spin wave function as basis. Here Γ^{ψ} is any of the representations A_1, B_1, B_2, E and Γ^{s} is to be constructed. With one unpaired electron (or hole), $S = \frac{1}{2}$ and the corresponding representation of the double group D_4' is $\Gamma_{1/2}$ whose characters are as tabulated below:

	χ (character)						
	$\hat{E}$	$\hat{Q}$	$\hat{C}_4 + \hat{Q}\hat{C}_4^3$	$\hat{C}_4^3 + \hat{Q}\hat{C}_4$	$\hat{C}_2 + \hat{Q}\hat{C}_2$	$2(\hat{C}_2' + \hat{Q}\hat{C}_2')$	$2(\hat{C}_2'' + \hat{Q}\hat{C}_2'')$
$\Gamma_{1/2}$	2	−2	$\sqrt{2}$	$-\sqrt{2}$	0	0	0

By comparing with the D_4' character table (Table 13.12) it is found that

$$\Gamma_{1/2} = \Gamma_6 \tag{13.25}$$

This means that the spin wave functions represented by α and β act as bases of the two-fold degenerate representation, Γ_6 of D_4'. To find the complete pattern of term splitting in a D_{4h} crystal field we have to find the direct product of Γ_6 with each of the component irreducible representations of Γ_L. By actual calculation of the direct products followed by reduction, it will be found that

$$\left.\begin{aligned} A_1 \times \Gamma_6 &= \Gamma_6 \\ B_1 \times \Gamma_6 &= \Gamma_7 \\ B_2 \times \Gamma_6 &= \Gamma_7 \\ E \times \Gamma_6 &= \Gamma_6 + \Gamma_7 \end{aligned}\right\} \tag{13.26}$$

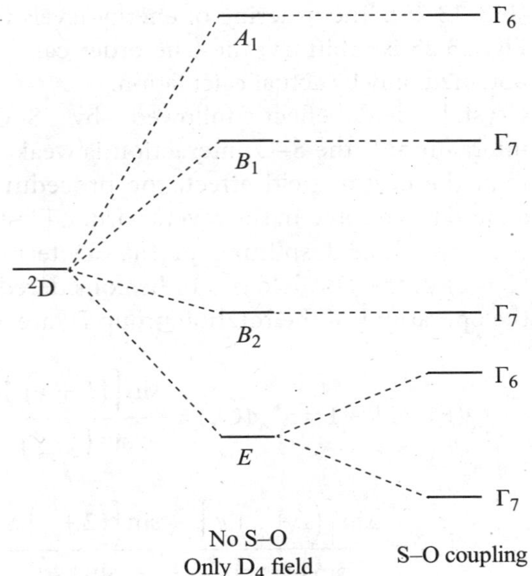

Fig. 13.26 Fine structure (spin–orbit) splitting of a ²D level in a D_{4h} crystal field when the crystal field effect is stronger than the spin–orbit coupling

Thus finally the ^{2}D term is found to split into two Γ_6 and three Γ_7 levels in the D_{4h} environment. The energy level diagram is somewhat like Fig. 13.26.

Figures 13.25 and 13.26 reveal that in the ultimate analysis, five levels, two of Γ_6 type and three of Γ_7 type are obtained in both cases. Note that Γ_6 and Γ_7 are both two dimensional. This means that the final split up levels are all two-fold degenerate. This is a manifestation of an important theorem called *Kramers' theorem* which states that *in the absence of externally applied magnetic fields, the energy levels of systems with an odd number of electrons must be at least two-fold degenerate.* Speaking more specifically, the spin degeneracy persists even after lifting orbital degeneracy by lowering of symmetry (from *K* to finite groups).

13.14 SPLITTING OF TERMS IN O_h ENVIRONMENT FOR SYSTEMS WITH HALF-INTEGRAL *J* AND KRAMERS' DEGENERACY

For half-integral *J*, the characters of a double-valued representation D_J of the double group O' (for the pure rotation group *O*) are as follows:

$$\chi(\hat{E}) = 2J+1,\ \chi(\hat{Q}) = -\chi(\hat{E})$$

$$\chi(\hat{C}_2) = \chi(\hat{Q}\hat{C}_2) = 0$$

$$\chi(\hat{C}_n^p) = \chi(\hat{Q}\hat{C}_n^{n-p}) = \frac{\sin\left[\left(J+\frac{1}{2}\right)\frac{2p\pi}{n}\right]}{\sin\left(\frac{p\pi}{n}\right)}$$

Results for the first few values of *J* are given below:

χ (character)

	$\hat{E}$	$\hat{Q}$	$4\hat{C}_3+4\hat{Q}\hat{C}_3^2$	$4\hat{C}_3^2+4\hat{Q}\hat{C}_3$	$3\hat{C}_2+3\hat{Q}\hat{C}_2$	$6\hat{C}_2'+6\hat{Q}\hat{C}_2'$	$3\hat{C}_4+3\hat{Q}\hat{C}_4^3$	$3\hat{C}_4^3+3\hat{Q}\hat{C}_4$
$D_{1/2}$	2	−2	1	−1	0	0	$\sqrt{2}$	$-\sqrt{2}$
$D_{3/2}$	4	−4	−1	1	0	0	0	0
$D_{5/2}$	6	−6	0	0	0	0	$-\sqrt{2}$	$\sqrt{2}$
$D_{7/2}$	8	−8	1	−1	0	0	0	0
$D_{9/2}$	10	−10	−1	1	0	0	$\sqrt{2}$	$-\sqrt{2}$

Using the character table of the double group O' and the standard reduction formula (or by inspection) it is found that

$$\left.\begin{aligned} D_{1/2} &= \Gamma_6 \\ D_{3/2} &= \Gamma_8 \\ D_{5/2} &= \Gamma_7+\Gamma_8 \\ D_{7/2} &= \Gamma_6+\Gamma_7+\Gamma_8 \\ D_{9/2} &= \Gamma_6+2\Gamma_8 \end{aligned}\right\} \tag{13.27}$$

[A check for the correctness of the results is the equality of the sum of dimensions on the two sides; for example, dim. $(D_{9/2}) = 10 = \text{dim. } \Gamma_6 + (\text{dim. } \Gamma_8) \times 2 = 2 + 4 \times 2$.]

All the double-valued representations, irrespective of the value of L or J, reduce to Γ_6, Γ_7, Γ_8 or their linear sum. Each Γ_i ($i = 6, 7, 8$) is at least two-fold degenerate. Hence the spin-degeneracy is retained even after lowering of symmetry from K (the free ion) to O' (the crystal field). Such residual degeneracy is called *Kramers' degeneracy*. It is a manifestation of Kramers' theorem (stated earlier) and it can be proved in a generalised form on the basis of time reversal symmetry, which is outside the scope of this elementary book.

This section is concluded with a specific example: splitting of ^{4}F terms in an octahedral crystal field. If the spin effects are neglected, the orbital degeneracies of a ^{4}F term are lifted by crystal O_h field into $^4A_{2u}$, $^4T_{1u}$ and $^4T_{2u}$ levels as shown earlier. Each of these levels is further split by spin–orbit interaction. To find these fine structure levels, we have to combine the $^4A_2$, $^4T_1$ and $^4T_2$ representations of O with the representation $D_{3/2}$ (for $S = 3/2$ which gives the quartet-term) of O'. Taking the appropriate direct product and reducing as before, we find

$$A_2 \times \Gamma_8 = \Gamma_8 \quad (\text{for } D_{3/2})$$

$$T_1 \times \Gamma_8 = \Gamma_6 + \Gamma_7 + 2\Gamma_8$$

$$T_2 \times \Gamma_8 = \Gamma_6 + \Gamma_7 + 2\Gamma_8$$

Thus the $^4A_2$ level keeps its four-fold spin degeneracy (dim. $\Gamma_8 = 4$), but each of the four $^4T_1$ and $^4T_2$ levels splits into four distinct levels of degeneracy 2, 2, 4 and 4 (Fig. 13.27).

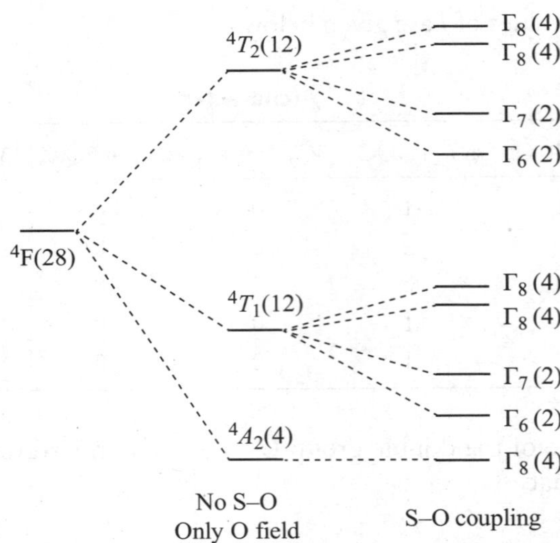

Fig. 13.27 Splitting of a ^{4}F term into fine structure levels in an octahedral field

13.15 MO THEORY FOR BONDING IN TRANSITION METAL COMPLEXES

The procedure for constructing MOs of transition metal complexes is the same as that for simple molecules discussed in Chapter 9. The orbitals of the ligands are first combined linearly with symmetry adaptation to the point group of the complex to form LGOs. The metal orbitals are classified according to their symmetry types. The LGOs are then combined linearly with the metal orbitals by matching symmetry. The procedure automatically brings out the splitting of the metal *d*-orbital energy levels without necessitating any further assumption as in the CFT or LFT. Some examples are discussed in the following sections.

13.15.1 Octahedral Complexes of the Type ML_6

According to the localised MO method, each ligand L of the octahedral ML_6 molecule is assumed to have a sigma bonding orbital (σ_i, $i = 1$ to 6) directed from L to the central atom M. Each L is also associated with two mutually orthogonal pi-bonding orbitals π_i and π_i' which are both orthogonal to the corresponding σ_i. Such orbitals are represented in Fig. 13.28 by arrows pointing towards the positive lobes of the respective orbitals. Chemical bonding occurs by overlap of these ligand orbitals with the valence shell orbitals of M.

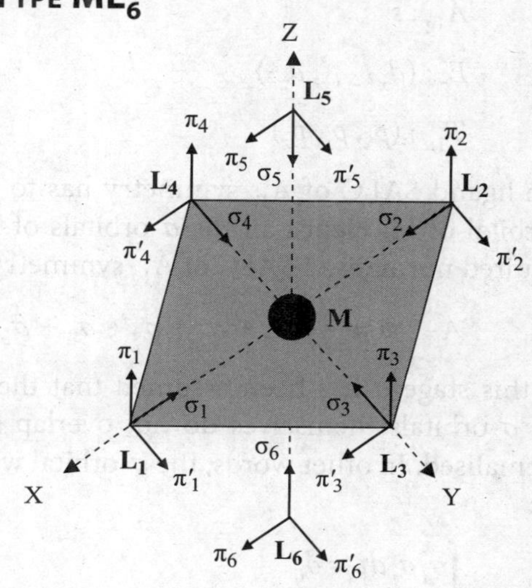

Fig. 13.28 Orientation of the σ and π type orbitals of the ligands along the coordinate axes in the octahedral ML_6 molecule (central atom = M; ligand atoms = L_i, $i = 1$ to 6)

No symmetry operation of the molecular point group O_h can transform the orbitals of M to those of L and vice versa. Hence the full representation Γ of O_h with all the orbitals of L and M as basis can be reduced as

$$\Gamma = \Gamma_M + \Gamma_L \qquad (13.28)$$

where Γ_M is a representation of O_h with the valence shell orbitals of M as basis, and Γ_L is a representation of O_h with σ, π and π' orbitals as basis. Again, no symmetry operation of O_h transforms the σ-orbitals into the π or π'- orbitals and we can straightaway write

$$\Gamma_L = \Gamma_\sigma + \Gamma_\pi \qquad (13.29)$$

where Γ_σ is a representation of O_h with the six σ-orbitals ($\sigma_1, \sigma_2, \ldots, \sigma_6$) as basis and Γ_π is a representation with the twelve π-type orbitals ($\pi_1, \pi_1', \pi_2, \pi_2', \ldots, \pi_6'$) as basis.

First we consider the reduction of Γ_σ. With reference to Fig. 13.28, it can be readily verified that the operator $\hat{E}$ leaves all the six σ-orbitals unchanged, making the character $\chi(\hat{E}) = 6$, the $\hat{C}_2$ and $\hat{C}_4$ operations about the X, Y and Z axes leave two σ-orbitals in position; the reflections $\hat{\sigma}_h$ in the XY, YZ and ZX planes leave four σ-orbitals unmoved and the reflections $\hat{\sigma}_d$ in the dihedral planes leave two σ-orbitals unmoved. All other symmetry operations of O_h move the ligand atoms from their positions. Hence the characters of the representation Γ_σ of O_h are as listed below.

χ **(character)**

	$\hat{E}$	$8\hat{C}_3$	$6\hat{C}_2'$	$6\hat{C}_4$	$3\hat{C}_2$	$\hat{i}$	$6\hat{S}_4$	$8\hat{S}_6$	$3\hat{\sigma}_h$	$6\hat{\sigma}_d$
Γ_6:	6	0	0	2	2	0	0	0	4	2

Using the character table of O_h and the reduction formula we find

$$\Gamma_\sigma = A_{1g} + E_g + T_{1u} \tag{13.30}$$

Instead of the projection operator method we shall use the simple procedure discussed in Section 9.3.3 to construct the SALCs for the irreducible representations found in Eq. (13.30). Symmetry of the central atom orbitals as listed below will be necessary for this purpose:

$$A_{1g} : s$$

$$E_g : (d_{x^2-y^2}, d_{z^2})$$

$$T_{1u} : (p_x, p_y, p_z)$$

The ligand SALC of A_{1g} symmetry has to combine linearly with the spherically symmetric s orbital of M. Hence all the σ-orbitals of L should contribute equally to this SALC. So the required normalised SALC of A_{1g} symmetry is

$$A_{1g} : \phi_1 = \frac{1}{\sqrt{6}}(\sigma_1 + \sigma_2 + \sigma_3 + \sigma_4 + \sigma_5 + \sigma_6) \tag{13.31}$$

At this stage it has been assumed that the ligand atoms are far apart from each other and the σ-orbitals themselves do not overlap significantly; also, the σ-orbitals are individually normalised. In other words, the σ-orbital wavefunctions are assumed to form an orthonormal set:

$$\int \sigma_i^* \sigma_j \, d\tau = \delta_{ij} \tag{13.32}$$

This makes ϕ_1 a normalised SALC. The doubly degenerate set of E_g-SALCs of the σ-orbitals of L must match with the $d_{x^2-y^2}$ and d_{z^2} orbitals of the central atom. The positive lobes of the d_{z^2} ($\equiv d_{2z^2-x^2-y^2}$) orbital of M are stretched along the Z-direction with an amplitude twice as much as the negative torus in the XY plane. Hence one σ-orbital SALC of E_g symmetry must be

$$E_g : \phi_2 = \frac{1}{\sqrt{12}}(2\sigma_5 + 2\sigma_6 - \sigma_1 - \sigma_2 - \sigma_3 - \sigma_4) \tag{13.33}$$

The $d_{x^2-y^2}$ orbital of M has positive lobes stretched along the X-axis and negative lobes along the Y axis, all with equal amplitude. Since σ_1 and σ_2 are along the X-axis while σ_3 and σ_4 are along Y-axis (Figure 13.28), the second SALC of E_g symmetry is

$$E_g : \phi_3 = \frac{1}{2}(\sigma_1 + \sigma_2 - \sigma_3 - \sigma_4) \tag{13.34}$$

The p_x, p_y and p_z orbitals of M are stretched respectively along the X, Y and Z axes. Each has positive and negative lobes on the positive and negative sides of the respective axis. Hence according to Fig. 13.28 the T_{1u} SALCs of the ligand σ-orbitals are:

$$T_{1u} : \begin{cases} \phi_4 = \frac{1}{\sqrt{2}}(\sigma_1 - \sigma_2) \\ \phi_5 = \frac{1}{\sqrt{2}}(\sigma_3 - \sigma_4) \\ \phi_6 = \frac{1}{\sqrt{2}}(\sigma_5 - \sigma_6) \end{cases} \tag{13.35}$$

In the next step we consider the reduction of Γ_π. The M–L π-bonds are formed by the twelve π-orbitals of the ligands, denoted by $\{\pi_1, \pi_1', \ldots, \pi_6, \pi_6'\}$ and represented by arrows in Fig. 13.28. The effect of the O_h symmetry operations on these orbitals are summarised below:

Operation	Result	Character
$\hat{E}$	All π orbitals unmoved	$\chi(\hat{E}) = 12$
$\hat{C}_4$ (about X-axis)	$\pi_1 \to \pi_1'$, $\pi_1' \to \pi_1$, $\pi_2 \to -\pi_2'$, $\pi_2' \to \pi_2$; all other π orbitals are shifted from their position	$\chi(\hat{C}_4) = 0$
$\hat{C}_2$ (about X-axis)	π_1, π_1', π_2, π_2' remain in position but are reversed in sign; all other π orbitals are shifted	$\chi(\hat{C}_2) = -4$
$\hat{\sigma}_h$ (XY plane)	$\pi_i \to -\pi_i$, $\pi_i' \to \pi_i'$, $i = 1, 2, 3, 4$; all other π-orbitals are shifted	$\chi(\hat{\sigma}_h) = (1-1)\times 4 + 0 = 0$
$\hat{\sigma}_d$ (in a plane bisecting the edges L_1L_3 and L_4L_2 and containing the Z-axis)	$\pi_5 \to \pi_5'$, $\pi_6 \to \pi_6'$; all other π orbitals are shifted from their position	$\chi(\hat{\sigma}_d) = 0$
Symmetry operations of other classes of O_h	All L atoms are shifted	$\chi(\hat{R}) = 0$ $\hat{R} = \hat{C}_3, \hat{C}_2', \hat{S}_4, \hat{i}, \hat{S}_6)$

Hence the characters of the representation Γ_π with the ligand π-orbitals as basis are as follows:

χ (character)

	$\hat{E}$	$8\hat{C}_3$	$6\hat{C}_2'$	$6\hat{C}_4$	$3\hat{C}_2$	$\hat{i}$	$6\hat{S}_4$	$8\hat{S}_6$	$3\hat{\sigma}_h$	$6\hat{\sigma}_d$
Γ_π:	12	0	0	0	−4	0	0	0	0	0

By usual procedure Γ_π reduces as

$$\Gamma_\pi = T_{1g} + T_{2g} + T_{1u} + T_{2u} \tag{13.36}$$

To construct the ligand SALCs of the component irreducible representations, we need the central atom orbitals of adequate symmetry. They are

$$T_{2g}: (d_{xy}, d_{yz}, d_{zx})$$

$$T_{1u}: (p_x, p_y, p_z)$$

No valence shell orbital of the central atom acts directly as basis of the T_{1g} and T_{2u} representations which occur in Eq. (13.36). So we have to resort to an alternative procedure. First we consider the cases of T_{2u} and T_{2g} following the logic presented for the σ-orbitals.

The d_{xy} orbital of the central atom must match with one T_{2g}-SALC of the π-orbitals. In Fig. 13.28, the vectors representing the π-orbitals on the XY plane are π_1', π_2', π_3' and π_4'. For adequate matching of these orbitals with the d_{xy} orbital of M, the positive and negative lobes of the π-orbitals must be oriented as shown in Fig. 13.29.

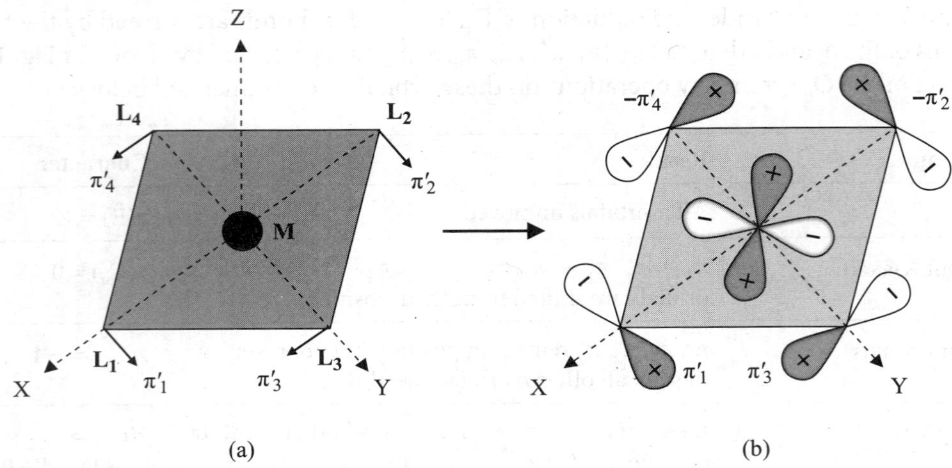

Fig. 13.29 (a) Necessary vectors taken out of Fig. 13.28 to denote the lines of ligand π type orbitals on the XY plane (b) Orientation of ligand π-orbital lobes for matching with d_{xy} of M.

(The π-orbital at L_2 is parallel to the Y axis but in order to match with the nearest lobe of the central d_{xy} orbital, it has a negative lobe in the positive direction of Y; hence it is $-\pi_2'$. The signs of the other three π-orbitals are assigned in a similar way.) Thus a T_{2g} SALC of π-orbitals, matching with the symmetry of d_{xy}, is

$$d_{xy}, T_{2g}(1): \psi_1 = \tfrac{1}{2}(\pi_1' - \pi_2' + \pi_3' - \pi_4') \tag{13.37}$$

By the same way two other SALCs of T_{2g} symmetry are found:

$$\left.\begin{aligned} d_{yz}, T_{2g}(2): \psi_2 &= \tfrac{1}{2}(\pi_3 - \pi_4 + \pi_5' - \pi_6') \\ d_{zx}, T_{2g}(3): \psi_3 &= \tfrac{1}{2}(\pi_1 - \pi_2 + \pi_5 - \pi_6) \end{aligned}\right\} \tag{13.38}$$

We now need to find the SALCs for T_{1g} and T_{2u}. In the character table of O_h there are no central atom orbitals of these types; instead, we find the rotation vector components R_x, R_y and R_z as bases of T_{1g}. We, therefore, need to find which π-vectors correspond to R_x, R_y and R_z from Fig. 13.28. To consider the rotation of the octahedron about the Z-axis, we imagine infinitesimal displacement vectors v_1, v_2, v_3, v_4 of the vertices of the square on the XY plane at the positions of L_1, L_2, L_3 and L_4. These vectors lie on the XY plane and indicate rotation about the Z-axis [Fig. 13.30(a)].

Comparing with the vectors π_1', π_2', π_3' and π_4' which represent the π-orbitals of ligands in Fig. 13.28, we find that v_1, v_2, v_3 and v_4 correspond respectively to π_1', $-\pi_2'$, $-\pi_3'$ and π_4' in so far as symmetry properties are concerned. Thus one π-orbital SALC of T_{1g} symmetry is

$$R_z, T_{1g}(1): \psi_4 = \tfrac{1}{2}(\pi_1' - \pi_2' - \pi_3' + \pi_4') \tag{13.39}$$

In the same way, two other π-orbital SALCs of T_{1g} symmetry are obtained by considering rotations about the X and Y axes:

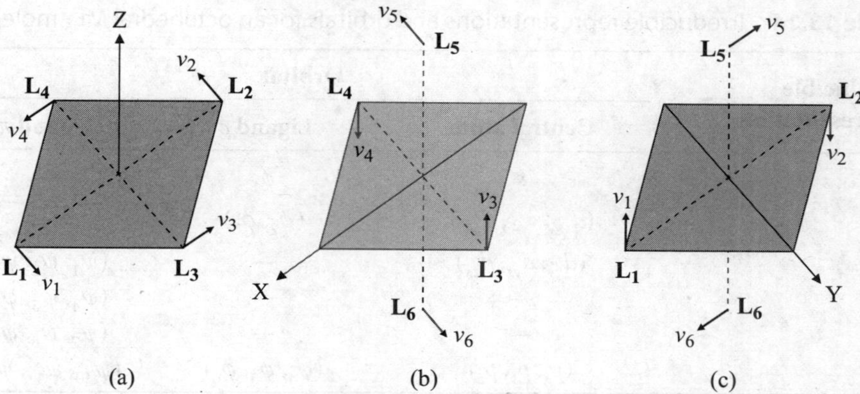

Fig. 13.30 (a) Rotation of the octahedron about the Z-axis, as represented by the basis R_z of T_{1g}, (b) Rotation about the X-axis ($\nu_3 \cong \pi_3$, $\nu_4 \cong -\pi_4$, $\nu_5 \cong -\pi_5'$, $\nu_6 \cong \pi_6'$), (c) Rotation about the Y-axis ($\nu_1 \cong \pi_1$, $\nu_2 \cong -\pi_2$, $\nu_5 \cong -\pi_5$, $\nu_6 \cong \pi_6$)

$$\left.\begin{aligned} R_x, T_{1g}(2): \psi_5 &= \tfrac{1}{2}(\pi_3 - \pi_4 - \pi_5' + \pi_6') \\ R_y, T_{1g}(3): \psi_6 &= \tfrac{1}{2}(\pi_1 - \pi_2 - \pi_5 + \pi_6) \end{aligned}\right\} \tag{13.40}$$

For the SALCs of T_{2u} symmetry, this diagrammatic procedure cannot be applied since the s, p or d functions are not found as basis of T_{2u} in the O_h character table. The projection operator method can, however, be applied and the result is:

$$T_{2u}: \begin{cases} \psi_7 = \tfrac{1}{2}(\pi_3' + \pi_4' - \pi_5 - \pi_6) \\ \psi_8 = \tfrac{1}{2}(\pi_1' + \pi_2' - \pi_5' - \pi_6') \\ \psi_9 = \tfrac{1}{2}(\pi_1 + \pi_2 - \pi_3 - \pi_4) \end{cases} \tag{13.41}$$

To find the ligand π-orbitals of T_{1u} symmetry, which will match with the metal p-orbitals, we can adopt a similar procedure by choosing vectors from Fig. 13.28. The result will be as follows:

$$\left.\begin{aligned} x, T_{1u}(1): \psi_{10} &= \tfrac{1}{\sqrt{2}}(\pi_3' + \pi_4') \\ y, T_{1u}(2): \psi_{11} &= \tfrac{1}{\sqrt{2}}(\pi_1' + \pi_2') \\ z, T_{1u}(3): \psi_{12} &= \tfrac{1}{2}(\pi_1 + \pi_2 + \pi_3 + \pi_4) \end{aligned}\right\} \tag{13.42}$$

13.15.2 The Energy Level Diagram

The central atom orbitals and the ligand orbitals for σ and π bonding are now arranged according to their symmetry in Table 13.14 where each horizontal row corresponds to one component irreducible representation of Γ_M and/or Γ_L.

Table 13.14 Irreducible representations and orbitals for an octahedral ML_6 molecule

Irreducible Representation	Orbital		
	Central atom	Ligand σ	Ligand π
A_{1g}	s	ϕ_1	—
E_g	$(d_{x^2-y^2}, d_{z^2})$	(ϕ_2, ϕ_3)	—
T_{2g}	(d_{xy}, d_{yz}, d_{zx})	—	(ψ_1, ψ_2, ψ_3)
T_{1g}	—	—	(ψ_4, ψ_5, ψ_6)
T_{2u}	—	—	(ψ_7, ψ_8, ψ_9)
T_{1u}	(p_x, p_y, p_z)	(ϕ_4, ϕ_5, ϕ_6)	$(\psi_{10}, \psi_{11}, \psi_{12})$

It is now evident that LCAO-MOs can be constructed from the central atom orbitals and ligand orbitals by matching symmetry. Two cases may arise.

13.15.2.1 Case I: ML_6 Molecules Having Only σ-Bonds

$[Co(H_2O)_6]^{3+}$ is an example of this type. In this case we expect from Table 13.14 that there will be two non-degenerate a_{1g} levels, two doubly degenerate e_g levels, two triply degenerate t_{1u} levels and a non-bonding t_{2g} level; the last one is essentially composed of the central atom orbitals d_{xy}, d_{yz}, d_{zx}. For $[Co(H_2O)_6]^{3+}$, the energy order of the central atom orbitals is $3d < 4s < 4p$ and hence the energy level diagram for the MOs is as shown in Fig. 13.31. Co^{3+} is a d^6 system and from H_2O molecules 12 electrons are available for σ bonding. With these 18 electrons the electronic configuration is $(1a_{1g})^2\ (1e_g)^4\ (1t_{1u})^6\ (1t_{2g})^6$. The most commonly observed electronic transition is $1t_{2g} \rightarrow 2e_g$. The energy separation between $1t_{2g}$ and $2e_g$ is called crystal field splitting energy Δ_0, and its magnitude varies with the ligands.

13.15.2.2 Case II: Octahedral ML_6 Molecules Having Both σ and π Bonds

In this case the σ-MOs will be modified and also some new MOs will appear due to participation of the ligand π-orbitals. Thus, from Table 13.14, we expect two non-degenerate a_{1g} levels, two doubly degenerate e_g levels, three triply degenerate t_{1u} levels, two triply degenerate t_{2g} levels, and two non-bonding levels essentially coming from the ligand π-orbitals—one triply degenerate t_{1g} and one triply degenerate t_{2u}. The set of central atom orbitals (d_{xy}, d_{yz}, d_{zx}) which comprises the non-bonding t_{2g} level in the absence of π-bonding, now forms bonding and antibonding levels (denoted by $1t_{2g}$ and $2t_{2g}$ in Fig. 13.32) by combination with the ligand π-orbitals of T_{2g} symmetry. The splitting energy Δ_0 is therefore affected by π-bonding ligands and it is either increased or decreased, depending on the number of d electrons of the central atom. In the general energy level diagram (Fig. 13.32), the correlations are shown by solid lines on both sides which are strong. For π-bonding, such lines are solid on one side and dotted on the other side; this is to mean that π-type interactions are dominated either by M or by L. If the π-MO has a dominant contribution from M, the correlation line is solid on the M side and broken on the L side. Similarly, if the ligand π-orbital dominates, the correlation line is solid on the L side and broken on the M side. Note that the relative positions of the $1t_{2g}$

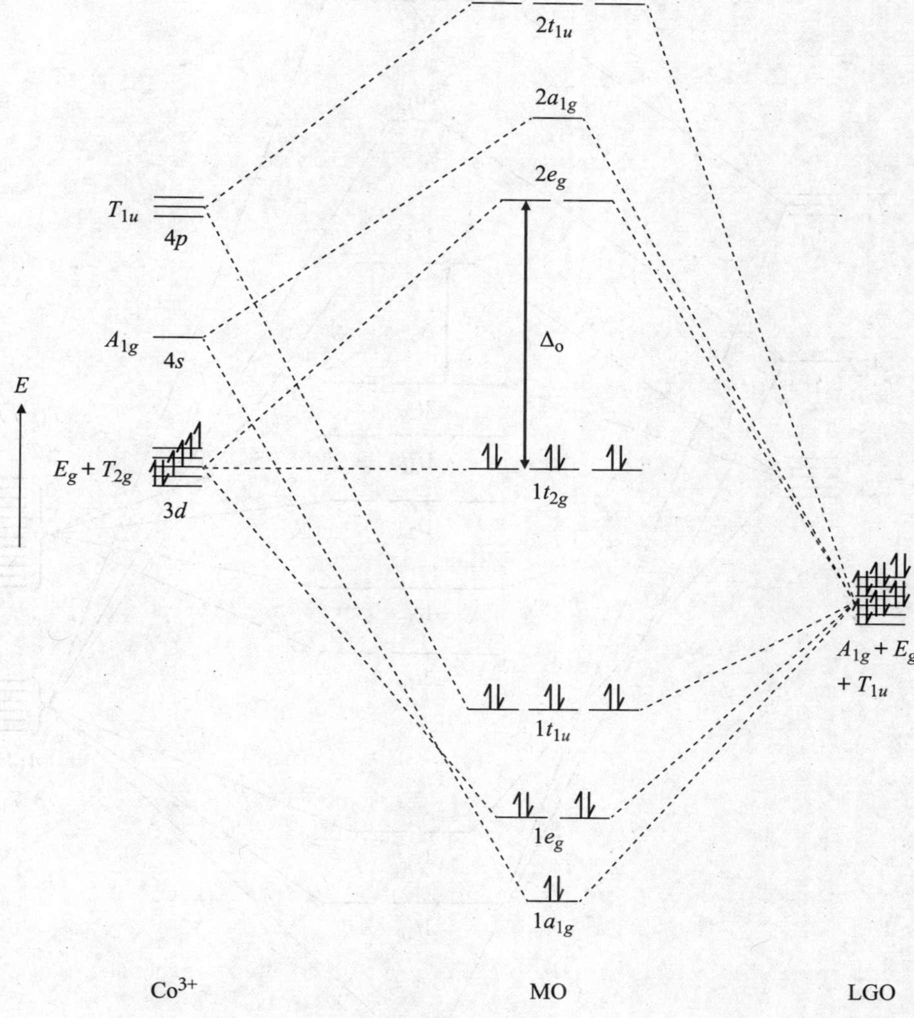

Fig. 13.31 Energy level diagram for $[Co(H_2O)_6]^{3+}$

and $2e_g$ levels depend on the number of *d* electrons in M and the strength of interaction with the ligand.

13.15.3 Metal Sandwich Compounds

Organometallic compounds of the general formula $(C_nH_n)_2M$, where a metal atom M is sandwiched between two parallel carbocyclic rings (C_nH_n) and is equivalently bonded to all the carbon atoms of each ring, are called *metal sandwich compounds*. Important examples are ferrocene, $[(C_5H_5)_2\,Fe]$ and dibenzene chromium, $[(C_6H_6)_2\,Cr]$.

Symmetry considerations lead to a convincing explanation of the structure, bonding and magnetic/spectroscopic properties of such compounds. Let us consider ferrocene. The two

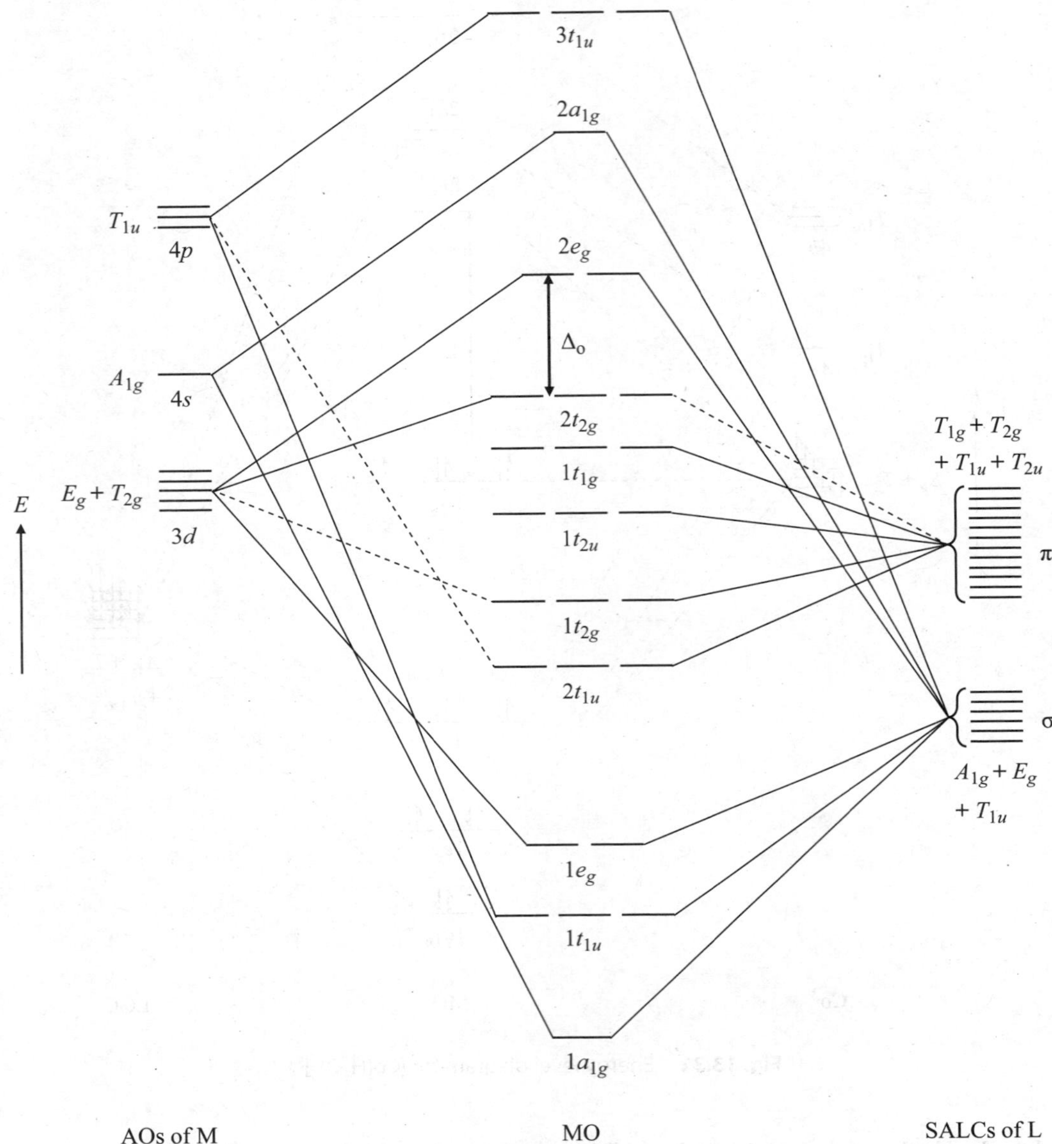

Fig. 13.32 Qualitative energy level diagram for ML_6 molecule; the σ- and π-orbitals of L have the same energy according to the assumptions made in the text—they are drawn separately only for the sake of clarity.

cyclopentadienyl (Cp) rings in this molecule are parallel and have a staggered conformation in the solid state as revealed by X-ray crystallography, and the Fe atom is situated at the centre of inversion. In the vapour phase, it has a pentagonal prism configuration with the Fe atom at a point on the common C_5-axis and is equidistant from the rings [Figs. 13.33 (a) and (b)]. This means that the energy barrier for rotation of one ring (about the C_5-axis) with respect to the other is not very high (about 5 kJ.mol^{-1}).

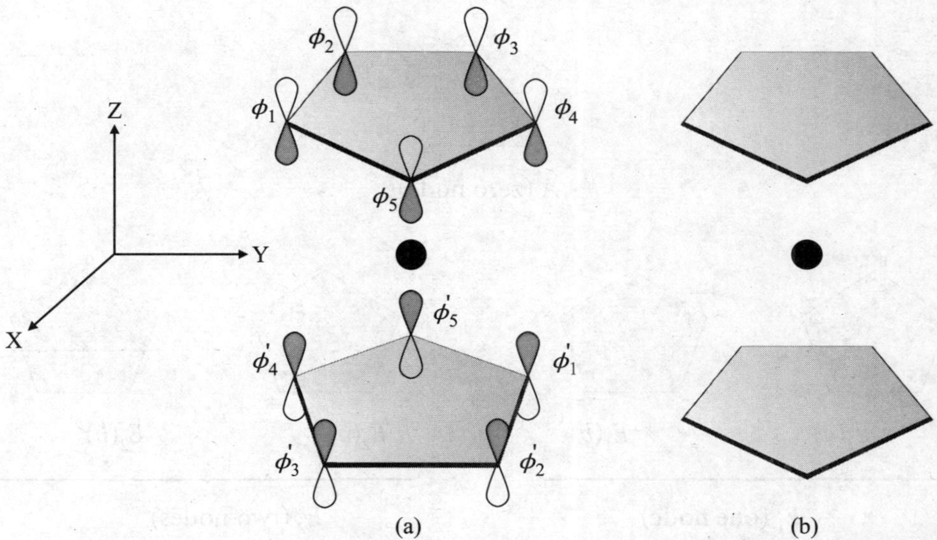

Fig. 13.33 Two conformational isomers of metallocene: (a) staggered and (b) eclipsed. Orientation of coordinate axes, interacting *p*-orbitals of C atoms in the Cp rings and position (•) of M in a staggered metallocene.

The point group of a metallocene $(C_nH_n)_2M$, with staggered orientation of the Cp rings is D_{5d}. Bonding in such molecules may be considered as due to interaction of the π-type orbitals of the Cp rings with the metal valence shell orbitals 3*d*, 4*s* and 4*p*. The two rings in staggered conformation, orientation of the coordinate axes and the set of interacting *p*-orbitals $\{\phi_1, \phi_2, \ldots, \phi_5, \phi'_1, \phi'_2, \ldots, \phi'_5\}$ are shown in Fig. 13.33(a).

The first task is to construct the group orbitals of the Cp rings. With set of orbitals $\{\phi_1, \ldots, \phi'_5\}$ as basis, a representation Γ_ϕ of the D_{5d} point group is as follows:

χ **(character)**

	$\hat{E}$	$2\hat{C}_5$	$2\hat{C}_5^2$	$5\hat{C}_2$	$\hat{i}$	$2\hat{S}_{10}^3$	$2\hat{S}_{10}$	$5\hat{\sigma}_v$
Γ_ϕ	10	0	0	0	0	0	0	2

This reduces as

$$\Gamma_\phi = A_{1g} + A_{2u} + E_{1g} + E_{1u} + E_{2g} + E_{2u} \tag{13.43}$$

The D_{5d} character table also shows that the available valence shell orbitals of the metal which are bases of the component irreducible representations found in Eq. (13.43) are as follows:

$$A_{1g} : s, d_{z^2};\ A_{2u} : p_z;\ E_{1g} : (d_{xz}, d_{yz});\ E_{2g} : (d_{x^2-y^2}, d_{xy});\ E_{1u} : (p_x, p_y)$$

The five π-MOs of the cyclopentadienyl anion $(C_5H_5^-)$, as obtained by the Hückel MO theory, are represented in Fig. 13.34 in the increasing order of energy, depending on the number of nodes. In the notation of cyclic C_5 group, these are (i) one non-degenerate MO of

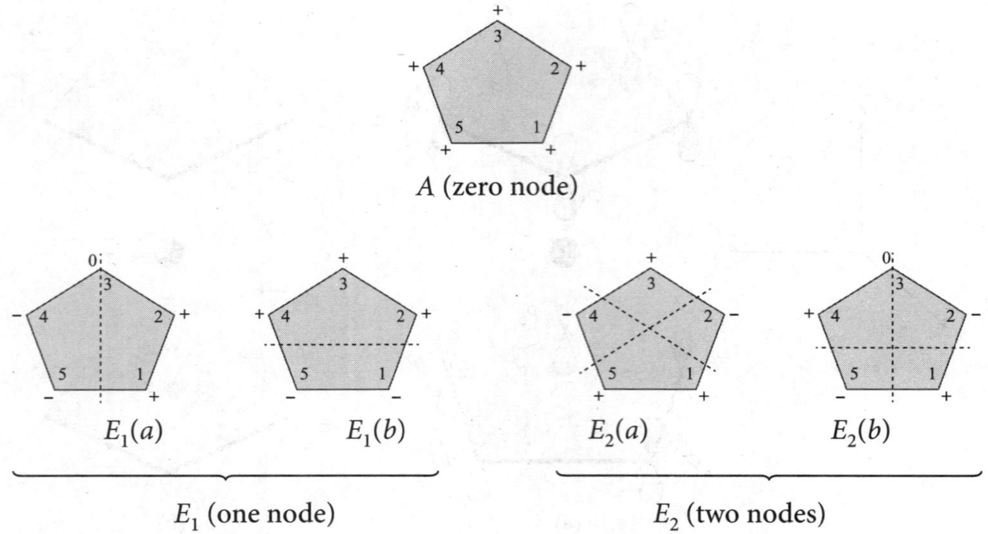

Fig. 13.34 Schematic representation of five π-MOs of Cp^-; the signs + and – represent the signs of the coefficients $\{c_r\}$ in the LCAO-MO expressions $\sum_r c_r \phi_r$. Zero (0) indicates that ϕ_3 is absent in the corresponding MO expression.

symmetry A, having the lowest π-MO energy, (ii) two doubly degenerate MOs of symmetry E_1 with the next higher energy and (iii) two doubly degenerate MOs of symmetry E_2 with the highest π-MO energy for $C_5H_5^-$.

Two A-type MOs, one for each $C_5H_5^-$ ring, can be placed one above the other with staggered orientation in two ways (Fig. 13.35).

The in-phase stacking of the A-MOs of Cp^- is obviously of the g-type, and so of lower energy, than the out-of-phase stacking which is of the u-type. It can be easily verified that no symmetry operation of D_{5d} can change the g-type combination of Fig. 13.35; hence this a

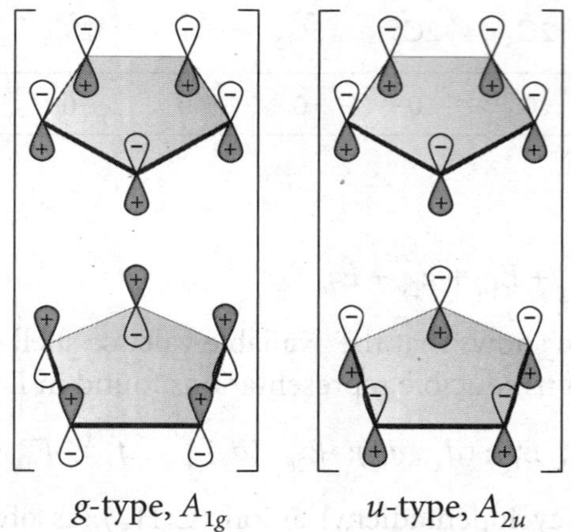

Fig. 13.35 Combination of the A-type MOs of two Cp^- rings of a metallocene in two ways, in-phase and out-of-phase

ligand group orbital of symmetry type A_{1g}. Symmetry operations on the u-type combination produce the characters of the A_{2u} representation [for example, rotation about a C_2-axis which passes through the inversion centre brings the figure into the negative of itself and yields character $\chi(\hat{C}_2') = -1$].

Each MO of the doubly degenerate set E_1 will produce two combinations on stacking—E_{1g} and E_{1u}; in this case the former is of lower energy. Thus, with the two E_1-type MOs of Fig. 13.34, the possible stackings are as in (a) to (d) of Fig. 13.36.

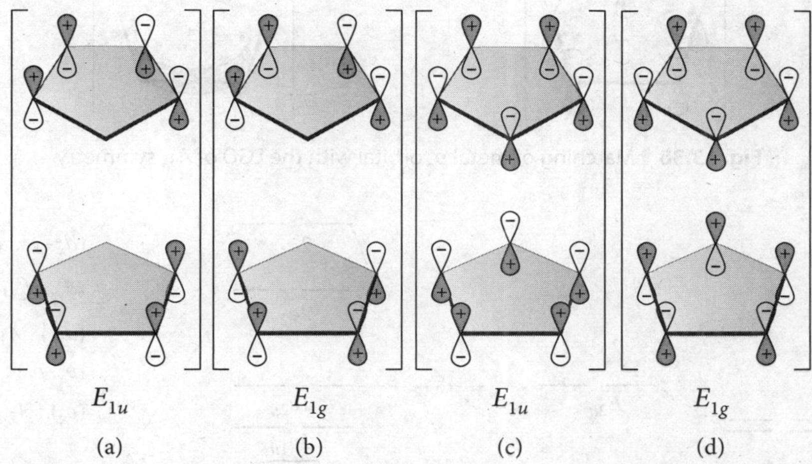

Fig. 13.36 (a) In-phase stacking of $E_1(a)$ MOs of the two Cp^- ions, with resulting symmetry E_{1u}
(b) Out-of-phase stacking of $E_1(a)$ MOs of Cp^-, with resulting symmetry E_{1g}
(c) In-phase stacking of $E_1(b)$ MOs of Cp^-, with resulting symmetry E_{1u}
(d) Out-of-phase stacking of $E_1(b)$ MOs of Cp^-, with resulting symmetry E_{1g}

Similarly, the higher energy $E_2(a)$ and $E_2(b)$ MOs of Cp^- will produce four LGOs, two of g-type and two of u-type. Drawing schematic diagrams as in Fig. 13.36, it can be shown that in this case the g-type LGOs are of lower energy than the u-type ones (as in the case of A_{1g} and A_{2u}). Thus in the D_{5d} environment the LGOs have the qualitative energy order as in Fig. 13.37.

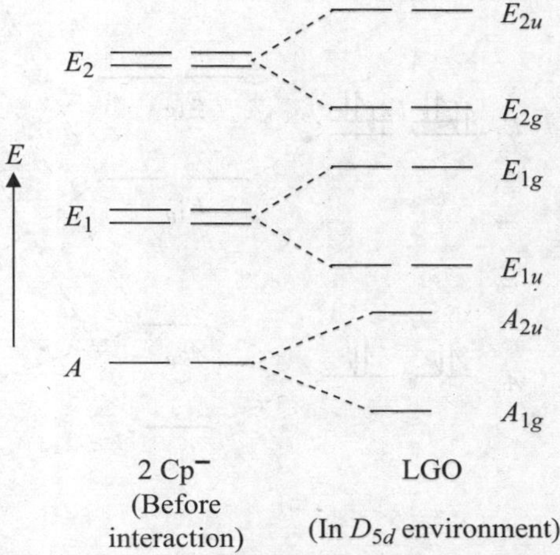

Fig. 13.37 Energy order of $(Cp^-)_2$ LGOs in a D_{5d} environment

The LGOs now combine linearly with the metal orbitals of matching symmetry to produce the molecular orbitals of a metallocene. Symmetry matching for one typical case is shown in Fig. 13.38. The energy level diagram of a metallocene is shown in Fig. 13.39, where the electron occupancy is shown for ferrocene. It is found that all the bonding MOs in ferrocene are occupied (that is, a closed shell configuration results). Moreover, this configuration conforms to the

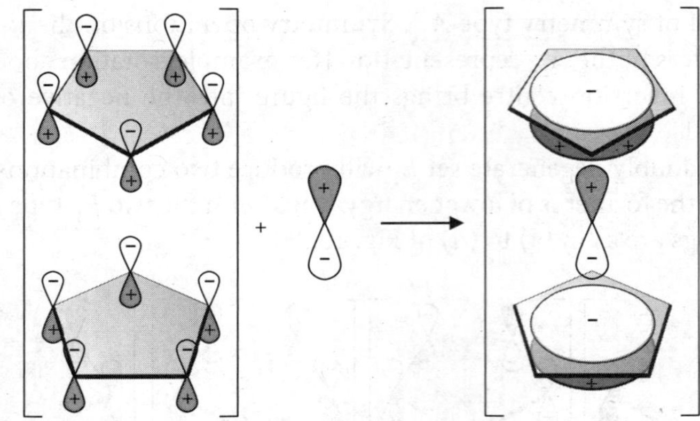

Fig. 13.38 Matching of metal p_z orbital with the LGO of A_{2u} symmetry

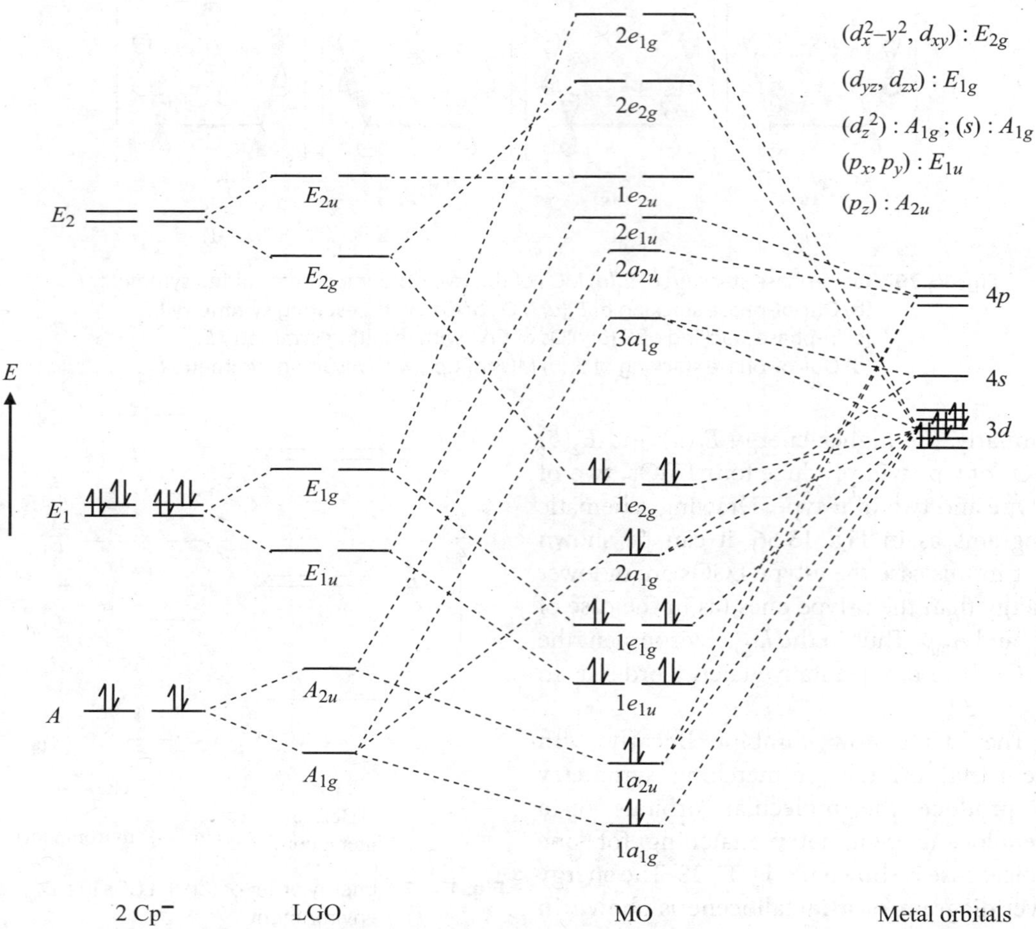

$(d_{x^2-y^2}, d_{xy}) : E_{2g}$

$(d_{yz}, d_{zx}) : E_{1g}$

$(d_{z^2}) : A_{1g}$; $(s) : A_{1g}$

$(p_x, p_y) : E_{1u}$

$(p_z) : A_{2u}$

E

E_2 — E_{2u}, E_{2g}

E_1 — E_{1g}, E_{1u}

A — A_{2u}, A_{1g}

MO: $2e_{1g}$, $2e_{2g}$, $1e_{2u}$, $2e_{1u}$, $2a_{2u}$, $3a_{1g}$, $1e_{2g}$, $2a_{1g}$, $1e_{1g}$, $1e_{1u}$, $1a_{2u}$, $1a_{1g}$

Metal orbitals: $4p$, $4s$, $3d$

2 Cp⁻   LGO   MO   Metal orbitals

Fig. 13.39 The energy level diagram of metallocene; electron occupancy shown for ferrocene

18 electron rule for coordination complexes. Hence this compound is stable; it is diamagnetic since all electrons are paired. $[(C_5H_5)_2\,Co]$, having 19 (= 18 + 1) electrons, has the extra electron in the $3a_{1g}$ level; hence it is paramagnetic corresponding to one unpaired electron and it is easily oxidised to $[(C_5H_5)_2\,Co]^+$ in order to gain the stable 18 electron configuration.

SUMMARY

- ❑ For a free atom, the spherical harmonic functions $Y_l^m(\theta, \phi)$ are eigenfunctions of the angular momentum operator $\hat{L}_Z$ and are bases of a $(2l + 1)$-fold degenerate irreducible representation of the pure rotation group *K*.
- ❑ When the atom/ion is placed in a crystal lattice of point group G, the reduction of symmetry from *K* to *G* causes the $(2l + 1)$-fold degenerate energy levels to split into a number of energy levels of lower degeneracy.
- ❑ Spectroscopic 'terms' associated with the energy levels split accordingly due to perturbation by the crystal field.
- ❑ In the purely ionic model of crystal field theory (CFT) the ligands are treated as point negative charges (or, dipoles) around the metal ion; they split the free ion terms by electrostatic perturbation; the *d*-character of the metal ion orbitals is retained. The splitting pattern depends entirely on the symmetry (point group) of the complex and the extent of splitting depends on the field strength produced by the ligands.
- ❑ Weak to infinite field correlation diagrams, drawn by obeying the non-crossing rule and spin multiplicity conservation, are used to interpret UV–vis absorption spectra of transition metal complexes. The allowedness of an electronic transition is judged by the vanishing or survival of the transition moment integral.
- ❑ Small deviations from the predicted spectral band are explained in terms of the Jahn–Teller theorem.
- ❑ Partial correlation diagrams (Orgel diagrams) are useful for assignment of spin-allowed transitions in high-spin complexes. Orbitally forbidden transitions may occur with low intensity because of vibronic coupling.
- ❑ For complexes of the first row-transition metals the *LS* coupling scheme and character tables of ordinary point groups are adequate. However, when the *jj* coupling scheme is needed, systems with odd number of electrons (half-integral *J*) require character tables of double groups.
- ❑ The extent of splitting (Δ or 10 *Dq*) in a given geometry varies from ligand to ligand; also, the energy separation between the Russell–Saunders terms (called *Racah parameter*, B) is smaller in the complex ion than in the free ion and this separation varies with the nature of the ligand.
- ❑ The above two observations lead to the *spectrochemical* and *nephelauxetic* series and suggest some overlap of ligand orbitals with metal *d*-orbitals. Some adjustments along this line in the CFT led to the *ligand field theory* (LFT).
- ❑ Finally, the MO theory, which includes all the symmetry aspects of CFT/LFT, came in.

LEARNING OUTCOME

After reading this chapter the student will be able to (i) construct characters of irreducible representations of pure rotation groups with the spherical harmonic functions $Y_l^m(\theta, \phi)$ as basis, (ii) find the splitting pattern of terms of a d^n-ion in crystal fields of different geometry, (iii) draw correlation diagrams and energy levels of the d^n-ions in different fields, (iv) predict electronic transitions using full and partial (Orgel) correlation diagrams (v) understand the necessity of incorporating some metal–ligand orbital overlap in the purely ionic CFT and (vi) construct the character tables of double groups.

EXERCISES

1. Explain how a one-electron *f* level splits in (a) a cubic O field, (b) O_h field, (c) D_3 field, (d) T_d field and (e) D_{2d} field. Check your result with Table 13.3.
2. Explain how (a) a G term from a d^n configuration and (b) a G term from a p^3s configuration split in an octahedral field.
3. Although *d–d* transitions are not allowed for octahedral complexes, $[Ti(H_2O)_6]^{3+}$ shows an absorption band of low intensity in the visible region (peak approximately at 500 nm). Draw the necessary Orgel diagram and explain how this is possible.
4. Show that a single crystal of the complex $[Cr(urea)_3]^{3+}$ will exhibit polarised UV–vis absorption spectrum. (Hint: Take urea to be a weak ligand, construct an Orgel diagram, find the spin-allowed transitions and then proceed.)
5. Explain, by drawing an Orgel diagram and assigning the possible electronic transitions, why aqueous HCl solution of $CoCl_2$ is intensely coloured. Show that one spin-allowed transition of the absorbing species is only vibronically allowed.
6. Construct the correlation diagram for a d^2 ion in an O_h environment by the method of descending symmetry using the C_{2h} subgroup.
7. Use the correlation table for T_d given in Appendix A4 and construct a weak-field strong-field correlation diagram for a d^2 system in a T_d environment by the method of descending symmetry.
8. A configuration t_{2g}^2 in an infinitely strong O_h field is somewhat relaxed. Find the possible spin multiplicities of the resulting states.
9. Find the states (including multiplicity) into which the configuration e_g^2 in an infinitely strong O_h field evolves, when the field is slightly relaxed.
10. Absorption bands are observed for the complex $[Cr(H_2O)_6]^{3+}$ at 17400, 24600 and 37900 cm^{-1}. Explain their occurrence. Which of them are in the visible range?
11. Find which transitions are responsible for the colour of aqueous $NiSO_4$ solution.
12. Explain why $[FeF_6]^{3-}$ is colourless whereas $[CoF_6]^{3-}$ and $[Fe(CN)_6]^{3-}$ are coloured.
13. Show the crystal field splitting of the energy levels of one-electron *d*-orbitals in a square planar field. Explain why virtually all square planar ML_4 complexes are immune to Jahn–Teller distortion.

14. Distribute the operations of the double group D_3' into classes and show that two 1-dimensional and one 2-dimensional double-valued representations are possible for D_3'.

15. Obtain the characters of the double-valued representations of D_4' and O' for $J = \frac{7}{2}$ and $\frac{9}{2}$.

16. Check the orthogonality of the single-valued and double-valued representations in Tables 13.10 and 13.12.

17. Construct the ligand group orbitals for σ bond formation with metal ion in a square planar complex of type AB_4 {like $[Ni(CN)_4]^{2-}$, $[Pt(Cl)_4]^{2-}$...}. Draw energy level diagrams and indicate the splitting of the metal valence orbitals.

18. Assuming a tetrahedral geometry for MnO_4^- construct energy level diagram for the σ- and π-MOs and explain the intense colour of the MnO_4^- ion.

19. Explain the intense colour of the tetrahedral complex ion $[Co(Cl)_4]^{2-}$ using molecular orbital theory.

20. Draw the in-phase and out-of-phase stacking of the E_1 MOs of two cyclopentadienyl rings in a D_{5d} environment, find their symmetry types and determine which orbitals of the valence shell of a transition metal atom can combine with such stacking to form a metallocene complex.

21. Explain the formation of MOs and draw the energy level diagram for dibenzene chromium with point group D_{6h}.

22. Choose the correct option:

(a) Cobaltocene, $[(C_5H_5)_2Co]$, is

(i) diamagnetic (ii) diamagnetic and easily oxidisable
(iii) paramagnetic and easily oxidisable (iv) paramagnetic but stable.

(b) Dipole moment of ferrocene is

(i) directed from one Cp ring to the other through the Fe atom
(ii) zero, because the molecule has a centre of inversion
(iii) zero, because the overall complex is neutral
(iv) non-zero, because the Fe atom is closer to one ring than the other.

(c) In a benzene dimer with the two rings parallel in a D_{6h} geometry, the A_{1g} MOs of the individual benzene molecules combine in phase to produce group orbitals of (i) A_{1u} symmetry (ii) A_{1g} symmetry (iii) E_u symmetry (iv) B_{1g} symmetry.

(d) For a cyclopentadienyl dimer with the two rings parallel in a D_{5d} geometry, the individual $E_1(a)$ MOs of Cp^- will produce LGOs of (i) g-type (ii) u-type (iii) neither g nor u type (iv) A_1 type. Given:

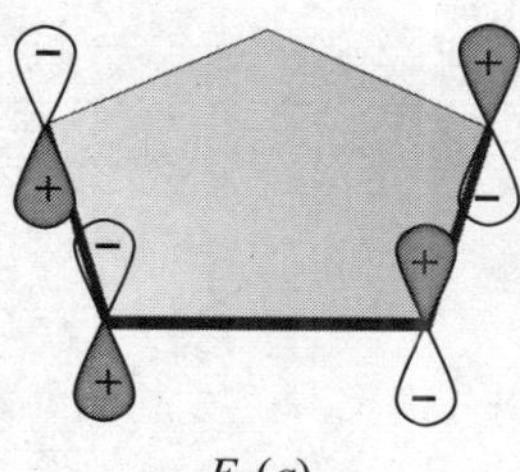

$E_1(a)$

Suggested Books and References for Further Reading

1. Bethe H, *Ann Physik*, 3 (135) **1929**.
2. Van Vleck J H, *Phys. Rev.*, 41 (208) **1932**.
3. Van Vleck J H, *J. Chem. Phys.*, 3 (803 and 807) **1935**.
4. Yamada et al, *Bull. Chem. Soc. Japan*, 25 (172) **1952**.
5. Yamada et al, *Bull. Chem. Soc. Japan*, 28 (222) **1955**.
6. Ballhausen C J and Moffitt W, *J. Inorg. Nucl. Chem.*, 3 (178) **1956**.
7. Orgel L E, *J. Chem. Phys.*, 23 (1004) **1955**.
8. Jahn H A and Teller E, *Proc. R. Soc.*, A161 (220) **1937**.
9. Moore C E, *Atomic Energy Levels, Nat. Bur. Std. Circ.*, 2 (467) **1952**.
10. Jørgensen C K, The nephelauxetic series, *Prog. Inorg. Chem.*, John Wiley, NY, 4 (73) **1962**.
11. Kramers H A, *Proc. Amsterdam Acad.*, 33 (959) **1930**.
12. Wilkinson G et al, *J. Am. Chem. Soc.*, 74 (2125) **1952**.
13. Dunitz J D and Orgel L E, *Nature*, 171 (121) **1953**.
14. Hamermesh M, Group Theory and its Application to Physical Problems, Addison-Wesley, Reading, MA, **1964**.
15. Landau L D and Lifshitz E M, Quantum Mechanics: Non-Relativistic Theory, Pergamon, Oxford, **1977**.

14 Orbital Symmetry in Pericyclic Reactions

Learning objectives

- Pericyclic reactions, concerted mechanism
- The principle of conservation of orbital symmetry
- Woodward–Hoffmann rules for prediction of thermal and photochemical paths

14.1 INTRODUCTION

A chemical reaction is said to involve a *concerted mechanism* if all bond breaking and bond making occur in a single step without the appearance of any reaction intermediate. There is an important group of organic reactions, called *pericyclic reactions*, which proceed by a concerted mechanism and are stereospecific. Some of them occur thermally and some photochemically. Well known examples are the Diels–Alder reaction (thermal) and synthesis of Vitamin D from ergosterol in the human body by exposure to sunlight. It was shown by Woodward and Hoffmann that the MO theory can be used to predict which path, thermal or photochemical, should be followed by such a reaction simply by considering orbital symmetry properties alone (without requiring a single numerical calculation of energy or wave function). Stereospecificity results from such symmetry control of the reaction path.

14.2 CONSERVATION OF ORBITAL SYMMETRY AND PERICYCLIC REACTIONS

A chemical reaction is called *pericyclic* if it is essentially a transformation of σ-orbitals into π-orbitals (or the reverse) through a cyclic transition state in a concerted mechanism. The σ–π transformation in a pericyclic reaction either starts from a cyclic reactant or ends up with a cyclic product. A symmetry-based explanation of the mechanism and stereospecificity of pericyclic reactions was given by R B Woodward and R Hoffmann on the basis of a general principle:

The symmetry of the participating orbitals in a pericyclic reaction must be conserved throughout the reaction. In other words, *if the set of participating orbitals have an axis or plane of symmetry in the reactant state, it (that is, the axis or the plane) has to be present in the product as well as in the transition state.*

The question that arises in applying this principle is, which orbitals are the participating orbitals? Just as the valence shell atomic orbitals take part in bonding, it is the highest occupied molecular orbitals (HOMO) and the lowest unoccupied molecular orbitals (LUMO) of the reacting molecules that take part in chemical reactions; these are called *frontier orbitals* (after K Fukui).

There are several types of pericyclic reactions—cycloaddition, electrocyclic reaction, sigmatropic reaction, and so on. Certain group transfer reactions and eliminations also have been found to belong to this category. We shall discuss the first two types, which will be sufficient to demonstrate the role of symmetry in chemical reactions.

14.3 ELECTROCYCLIC REACTIONS

An electrocyclic reaction is one in which a linear conjugated system (I) containing $k\ \pi$-electrons is cyclised into a conjugated system (II) containing $(k-2)\ \pi$-electrons by the formation of a σ-bond between the termini of (I); the reverse of this reaction is also considered *electrocyclic.* The general scheme for such reactions is shown in Fig. 14.1.

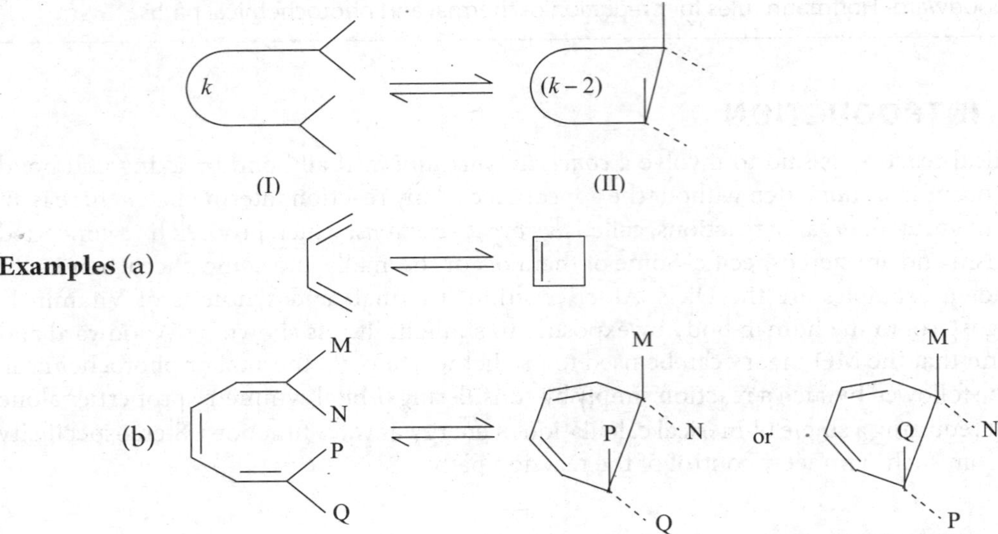

Fig. 14.1 General scheme for electrocyclic reactions. (a) *cis*-buta-1,3diene to cyclobutene interconversion without substituents and (b) the same reaction with substiuents.

14.3.1 Conrotatory and Disrotatory Transformation of Orbitals

The formation of a σ-bond between the termini of a linear polyene in an electrocyclic reaction involves the breaking of two π-bonds at the termini and rotation of the relevant p-orbitals, accompanied by a change of hybridisation of the terminal C atoms from sp^2 to sp^3. Similarly the reverse reaction (ring opening) involves the breaking of a σ-bond, rotation of the σ-orbitals of the related C atoms and a change of hybridisation from sp^3 to sp^2 in a concerted manner. However, the rotation of the two orbitals of the terminal C atoms may be either in the same sense (both clockwise or both anticlockwise) or in the opposite sense (one clockwise and the other anticlockwise). The former rotation is called *conrotatory*; during this only a C_2-axis of symmetry is conserved throughout the reaction; this means that the symmetry type of the set of rotating orbitals with respect to a C_2-axis is the same in the reactant, product and transition state. The other type of rotation is called *disrotatory*; during this a symmetry plane (σ) perpendicular to the plane of the relevant part of the polyene molecule is conserved. Figure 14.2 shows some conrotatory and disrotatory transformations:

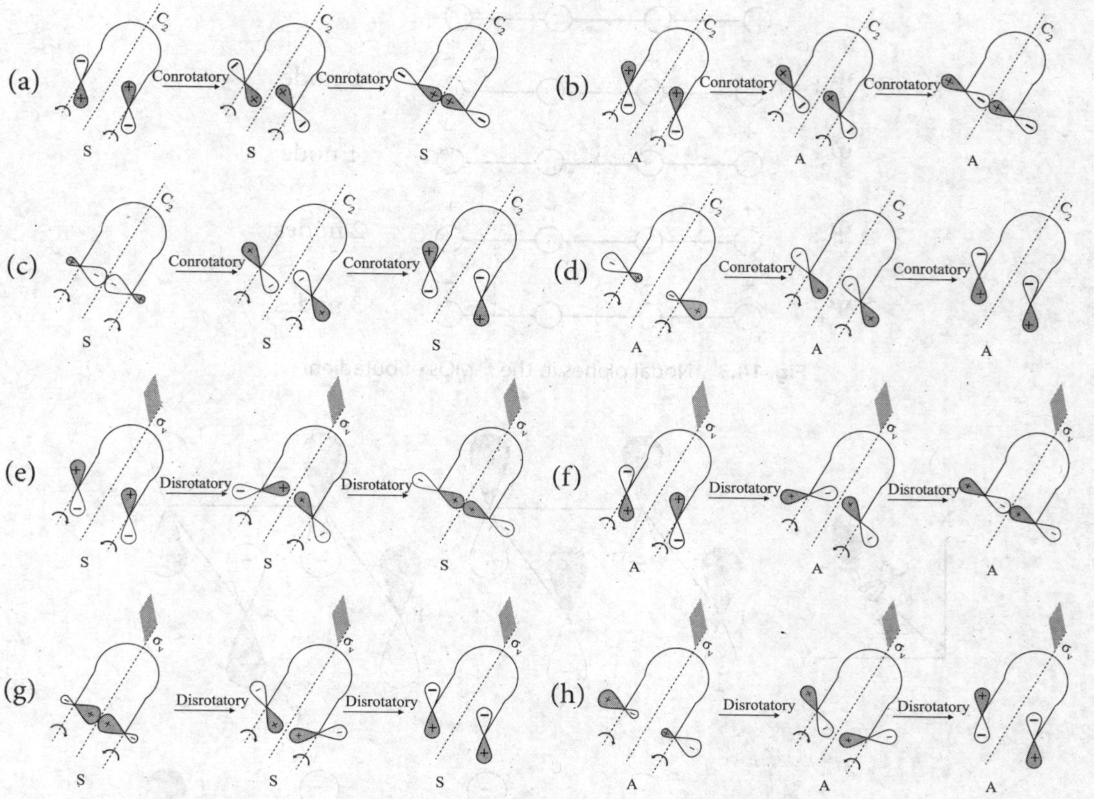

Fig. 14.2 Conservation of orbital symmetry in some conrotatory and disrotatory transformations (A = antisymmetric and S = symmetric with respect to the C_2-axis or σ-plane)

14.3.2 Explanation of the *cis*-Butadiene–Cyclobutene Interconversion

To explain this concerted reaction, a knowledge of the π-MOs of the reactants and products is needed. Let us consider *cis*-1,3-butadiene of point group C_{2v} first. The four sp^2 hybridised C atoms are labelled 1, 2, 3 and 4, the C_2-axis is taken as the Z-axis and the plane containing the C atoms as the σ_v(YZ) plane. The atomic orbitals that take part in π-MO formation are the p_x-orbitals of carbon atoms. Denoting these orbitals as $\phi_i\{i = 1 \text{ to } 4\}$ the π-MO wave functions can be obtained by the Hückel method. Since this method considers only the adjacencies of the conjugated carbon atoms, it does not distinguish between *cis*- and *trans*-butadienes; the π-MOs can be drawn without much calculation simply by utilising the fact that the nodal planes separating the opposite phases of the atomic p-orbitals are to be symmetrically placed in the σ-bond framework of the molecule (Figs 14.3 and 14.4). The results are summarised in Table 14.1 (where N is the normalising factor). Orientation of the component p-orbitals (just before overlap) in the π-MOs (Fig. 14.4), can be drawn by noting the sign-change patterns in Fig. 14.3. The symmetry types of the π-MOs with respect to the $\hat{C}_2$(Z) and $\hat{\sigma}_v$(XZ) operations are shown in Table 14.1 where A and S denote *antisymmetric* and *symmetric* respectively.

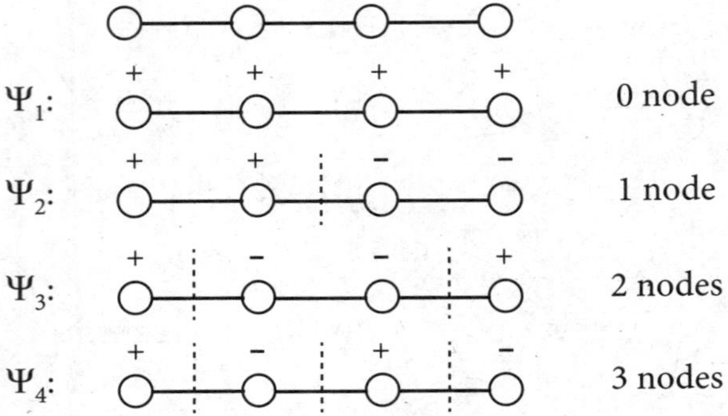

Fig. 14.3 Nodal planes in the π-MOs of butadiene

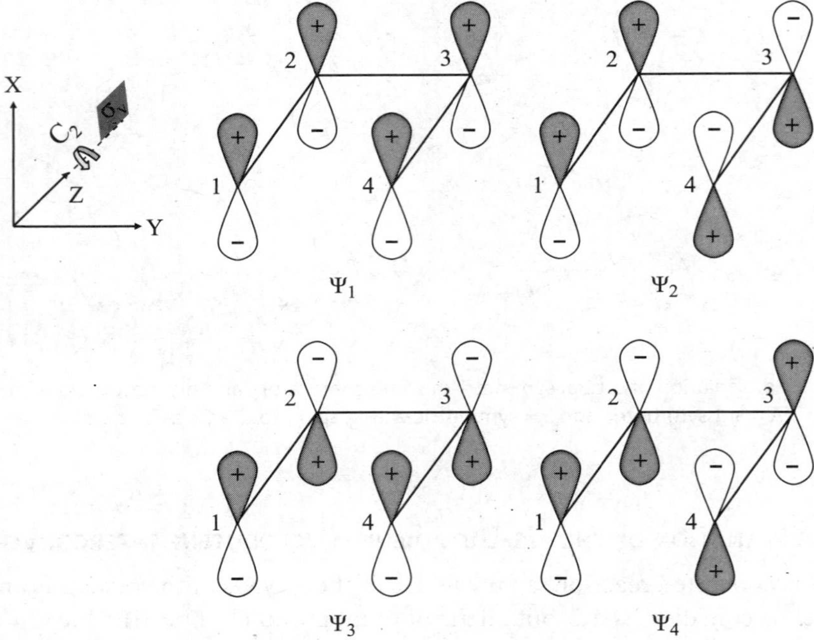

Fig. 14.4 Orientation of p_x orbitals forming the four π-MOs of *cis*-1,3-butadiene

Table14.1 Wave functions and symmetry properties of the π-MOs of *cis*-1,3-butadiene

Wave function	No. of nodal planes*	Symmetry with respect to C_2-axis	Symmetry with respect to σ_v-plane
$\Psi_1 = N(\phi_1 + \phi_2 + \phi_3 + \phi_4)$	0	A	S
$\Psi_2 = N(\phi_1 + \phi_2 - \phi_3 - \phi_4)$	1	S	A
$\Psi_3 = N(\phi_1 - \phi_2 - \phi_3 + \phi_4)$	2	A	S
$\Psi_4 = N(\phi_1 - \phi_2 + \phi_3 - \phi_4)$	3	S	A

(* perpendicular to the plane of the C atom nuclei)

Next, let us consider the four MOs of cyclobutadiene into which the four *p*-orbitals of *cis*-butadiene transform by the cyclisation reaction. Two of them are σ-bonding and antibonding MOs between the terminal C-atoms 1 and 4 and the other two are π-bonding and antibonding MOs between atoms 2 and 3. The component σ- and π-orbitals, (σ_1, σ_4) and (π_2, π_3) respectively, on the C-atoms just before forming these MOs are shown in Fig. 14.5, where the same set of coordinate axes as in Fig. 14.4 are used.

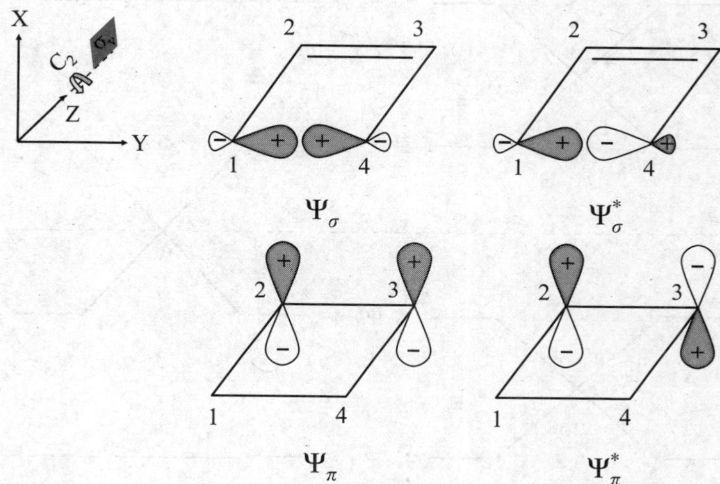

Fig. 14.5 σ- and π-orbitals of C-atoms in cyclobutene just before forming the (1, 4) σ-bond and (2, 3) π-bond, orientation of coordinate axes and some symmetry elements

LCAO-MO wave functions of these MOs and their symmetry types with respect to the $C_2(Z)$ axis and $\sigma_v(XZ)$ plane are as follows:

MO	Symmetry with respect to C_2(Z)	Symmetry with respect to σ_v(XZ)
$\Psi_\sigma = N_1(\sigma_1 + \sigma_2)$	S	S
$\Psi^*_\sigma = N^*_1(\sigma_1 - \sigma_2)$	A	A
$\Psi_\pi = N_2(\pi_1 + \pi_2)$	A	S
$\Psi^*_\pi = N^*_2(\pi_1 - \pi_2)$	S	A

(The Ns are normalisation constants)

As usual, the σ-orbitals are much more bonding (and antibonding) than the π-orbitals and hence the energy-order of the MOs is

$$E(\Psi_\sigma) < E(\Psi_\pi) < E(\Psi^*_\pi) < E(\Psi^*_\sigma)$$

When the π-MOs of *cis*-butadiene are converted into σ-MOs of cyclobutene (or vice versa), one element of symmetry must be conserved throughout the course of the reaction, according to the principle of conservation of orbital symmetry. Cyclobutene and *cis*-butadiene have two elements of symmetry in common—a C_2-axis and a σ_v-plane. Hence two correlation diagrams, one conserving the C_2-axis and the other conserving the σ_v-plane, are possible for the transformation. The former is conrotatory and the latter disrotatory, as explained in the earlier section. In Fig. 14.6 the two correlation diagrams are merged.

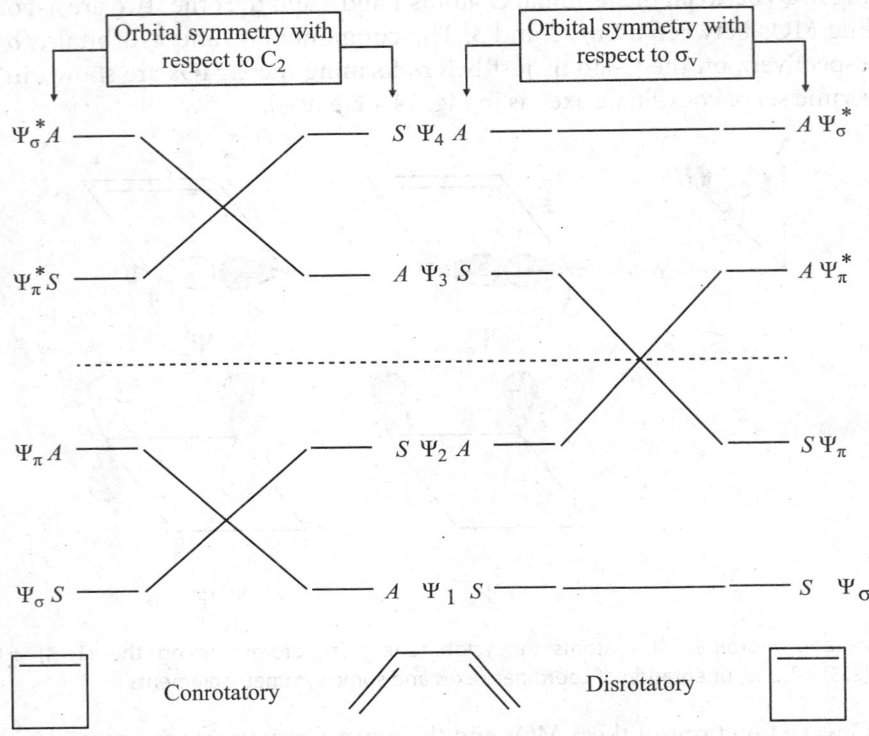

Fig. 14.6 Orbital correlation diagrams for conrotatory and disrotatory transformation of *cis*-butadiene into cyclobutene (and the opposite of it)

The orbital correlation diagram can now be utilised to predict which type of transformation will be allowed by symmetry under thermal and photochemical conditions.

The ground state electronic configurations of *cis*-butadiene and cyclobutene are $(\Psi_1)^2\,(\Psi_2)^2$ and $(\Psi_\sigma)^2\,(\Psi_\pi)^2$. Figure 14.6 shows that in the conrotatory mode of transformation these two configurations are correlated. But in the disrotatory mode, $(\Psi_1)^2\,(\Psi_2)^2$ is correlated to $(\Psi_\sigma)^2\,(\Psi_\pi^*)^2$ which is a very highly excited state of cyclobutene. Hence under thermal conditions, the conrotatory transformation is symmetry-allowed while the disrotatory one is forbidden.

Figure 14.6 also shows that the first excited state of *cis*-butadiene with electronic configuration $(\Psi_1)^2\,(\Psi_2)^1\,(\Psi_3)^1$ is correlated, by conrotatory transformation to $(\Psi_\sigma)^1\,(\Psi_\pi)^2\,(\Psi_\sigma^*)^1$; this is a highly excited state of cyclobutene. But by the disrotatory transformation, $(\Psi_1)^2\,(\Psi_2)^1\,(\Psi_3)^1$ is correlated to $(\Psi_\sigma)^2\,(\Psi_\pi)^1\,(\Psi_\pi^*)^1$ which is the first excited state of cyclobutene. Photochemically, therefore, the first excited state of the reactant easily evolves into the first excited state of the product. Hence the disrotatory transformation is symmetry-allowed photochemically and forbidden thermally.

The same conclusions can be drawn if state (configuration) correlation diagrams are constructed with the help of orbital correlation diagrams. Symmetry type of a state (configuration) can be obtained from the direct product of the irreducible representations corresponding to the occupied orbitals. In the case of conrotatory transformation the subgroup C_2, that is ($\hat{E}$, $\hat{C}_2$), and in the case of disrotatory transformation the subgroup C_s, that is ($\hat{E}$, $\hat{\sigma}$), of the full point group of the reactant/product molecule should be considered. Whatever be the operation, $\hat{C}_2$ or $\hat{\sigma}$, the character is 1 for a symmetric orbital and −1 for an antisymmetric orbital. Hence all the symmetric MOs of *cis*-butadiene and cyclobutene are bases of the *A* representation and all the antisymmetric MOs are bases of the *B* representation of the C_2 point group. Similarly, all the symmetric MOs are bases of A' representation and antisymmetric MOs are bases of A'' representation of the C_s point group. Irreducible representations spanned by the ground state and one excited state of *cis*-butadiene and cyclobutene are as follows:

	Configuration	Irreducible representation of C_2	Irreducible representation of C_s
cis-Butadiene	$(\Psi_1)^2\,(\Psi_2)^2$	$B \times B \times A \times A = A$	$A' \times A' \times A'' \times A'' = A'$
	$(\Psi_1)^2\,(\Psi_2)^1\,(\Psi_3)^1$	$B \times B \times A \times B = B$	$A' \times A' \times A'' \times A' = A''$
Cyclobutene	$(\Psi_\sigma)^2\,(\Psi_\pi^*)^2$	$B \times B \times A \times A = A$	$A' \times A' \times A'' \times A'' = A'$
	$(\Psi_\sigma)^2\,(\Psi_\pi)^1\,(\Psi_\pi^*)^1$	$A \times A \times B \times A = B$	$A' \times A' \times A' \times A'' = A''$

By using Fig. 14.6 it can be found that the configurations $(\Psi_\sigma)^2\,(\Psi_\pi)^1\,(\Psi_\pi^*)^1$ and $(\Psi_\sigma)^1\,(\Psi_\pi)^2\,(\Psi_\sigma^*)^1$ of cyclobutene correspond respectively to $(\Psi_2)^2\,(\Psi_1)^1\,(\Psi_4)^1$ and $(\Psi_2)^1\,(\Psi_1)^2\,(\Psi_3)^1$ of *cis*-butadiene in the conrotatory mode. But the energy order of these configurations in cyclobutene is opposite to that in *cis*-butadiene. Hence a state correlation diagram for the

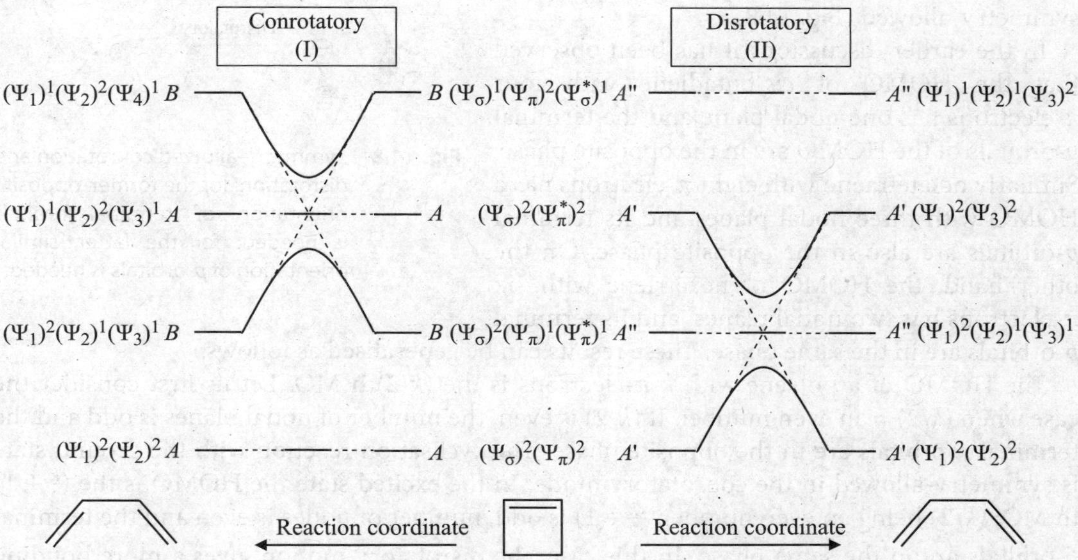

Fig. 14.7 State (configuration) correlation diagram for (I) conrotatory and (II) disrotatory ring opening of cyclobutene to form *cis*-butadiene

conrotatory ring opening can be shown as in Fig. 14.7 (I). States of the same symmetry and the same number of unpaired electrons are correlated. The broken lines in Fig. 14.7 (I) indicate correspondence between similar states but such correlations are not permitted by the non-crossing rule. Hence in the conrotatory process, the lower energy B state of cyclobutene correlates to the lower energy B state of *cis*-butadiene via an energy barrier. The ground states A of the two, however, directly correlate to one another. The conrotatory process, therefore, is symmetry-allowed under thermal conditions while the photochemical conrotatory process is symmetry-forbidden.

For disrotatory ring opening, the cyclobutene ground state $(\Psi_\sigma)^2\,(\Psi_\pi)^2 \equiv (A')$ correlates to $(\Psi_1)^2\,(\Psi_3)^2 \equiv (A')$ according to Fig. 14.6 and the *cis*-butadiene ground state $(\Psi_1)^2\,(\Psi_2)^2 \equiv (A')$ correlates to $(\Psi_\sigma)^2\,(\Psi_\pi^*)^2 \equiv (A')$; but the non-crossing rule does not permit such correlations and so the ground states of the two compounds correlate only via an energy barrier [Fig. 14.7 (II)]. On the other hand, the excited states $(\Psi_\sigma)^2\,(\Psi_\pi)^1\,(\Psi_\pi^*)^1$, (A'') and $(\Psi_1)^2\,(\Psi_2)^1\,(\Psi_3)^1$, (A'') correlate directly. Hence, in the disrotatory mode photochemical reaction (excited state to excited state) is symmetry-allowed but the thermal reaction (ground state to ground state) is symmetry-forbidden.

14.3.3 Derivation of a General Rule for Electrocyclic Reactions

We can predict whether an electrocyclic reaction will occur thermally or photochemically by noting the orientation of the terminal p-orbitals in the highest occupied molecular orbital (HOMO) of the reactant polyene. If the terminal p-orbitals are in opposite phase, conrotation is symmetry-allowed and if they are in the same phase, disrotation is symmetry-allowed (Fig. 14.8).

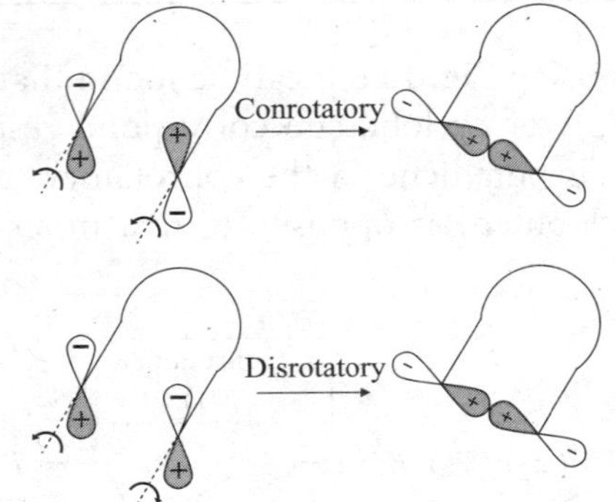

Fig. 14.8 Symmetry-allowed conrotation and disrotation: for the former, opposite orientation of terminal orbitals is needed; for the latter, similar orientation of p-orbitals is needed.

In the earlier discussions it has been observed that the HOMO of *cis*-butadiene with four π-electrons has one nodal plane and the terminal p-orbitals of the HOMO are in the opposite phase. Similarly octatetraene with eight π-electrons has a HOMO with three nodal planes and its terminal p-orbitals are also in the opposite phase. On the other hand, the HOMO of hexatriene with six π-electrons has two nodal planes, and its terminal p-orbitals are in the same phase. These results can be generalised as follows:

The HOMO of a polyene with k π-electrons is the $(k/2)$th MO. Let us first consider the case when $(k/2)$ = an even number. If $(k/2)$ is even, the number of nodal planes is odd and the terminal p-orbitals are in the opposite phase. The cyclisation reaction with the ground state is symmetry-allowed in the conrotatory mode. In the excited state the HOMO is the $(\frac{k}{2}+1)$ th MO; $(k/2)$ being an even number, $(\frac{k}{2}+1)$ is odd, number of nodes is even and the terminal p-orbitals are in the same phase. In this case, the disrotatory motion gives a more bonding σ-MO than the conrotatory mode. Thus the disrotatory conversion is symmetry-allowed photochemically. The opposite is the event for odd values of $k/2$.

Now, for any non-zero integer n, $k/2$ is even for $k/2 = 2n$, that is, $k = 4n$ and $k/2$ is odd for $k/2 = 2n + 1$, that is, $k = 4n + 2$. Hence the general rule can be tabulated as follows:

No. of π-electrons in the polyene	Symmetry allowed thermal reaction	Symmetry allowed photochemical reaction
$4n$	Conrotatory	Disrotatory
$4n + 2$	Disrotatory	Conrotatory

These are the Woodward–Hoffmann rules for electrocyclic reactions. Some examples are given below.

π-system	Reaction	Thermal (Ground state)	Photochemical (Excited state)
4π :	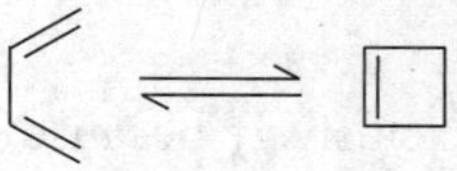	Conrotatory	Disrotatory
$(4 + 2)\pi$:	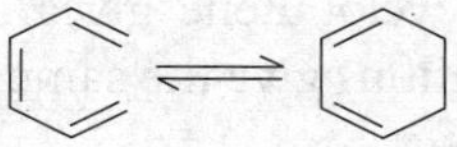	Disrotatory	Conrotatory
8π :	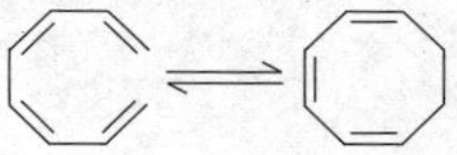	Conrotatory	Disrotatory
2π :	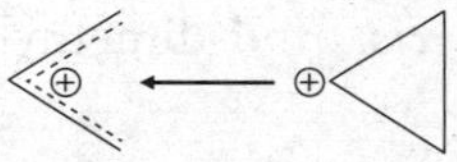 ⊕ ⟵ ⊕	Disrotatory	Conrotatory
4π :	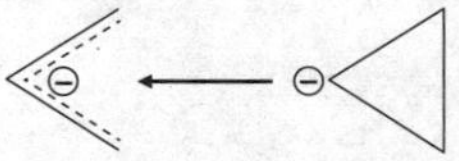 ⊖ ⟵ ⊖	Conrotatory	Disrotatory
4π :	 ⊕ ⟶ ⊕	Conrotatory	Disrotatory
6π :	 ⊖ ⟶ ⊖	Disrotatory	Conrotatory

Note: Owing to high angle strain, the cyclopropenyl system is highly unstable and only its transformation to the linear system is experimentally observed; the reverse reaction, that is, cyclisation is not observed. On the other hand, owing to aromatic delocalisation, the cyclopentadienyl system is stable and cyclisation from the corresponding linear system is almost exclusively observed.

14.3.4 Stereospecificity of Electrocyclic Reactions

As expected from the preceding discussions, the reactions should be highly stereospecific. This has been verified by a large number of experiments with substituted cyclobutenes and *cis*-butadienes. Some examples are given in the following reaction schemes:

(i) *trans*-dimethylcyclobutene is expected to give two possible products in the thermal ring opening because there are two possible conrotatory motions:

cis, cis-dimethyl-1,3-butadiene

trans, trans-dimethyl-1,3-butadiene

However, the *trans, trans* product is formed exclusively because of unfavourable steric situation in the *cis, cis* product. *cis*-dimethylcyclobutene gives *cis,trans*-dimethyl-1,3-butadiene in the thermal reaction (two conrotations give the same product):

cis, trans-dimethyl-1,3-butadiene

trans, cis-dimethyl-1,3-butadiene

(ii) Under thermal conditions cyclisation reactions of the following trienes are disrotatory and the products as mentioned are formed exclusively.

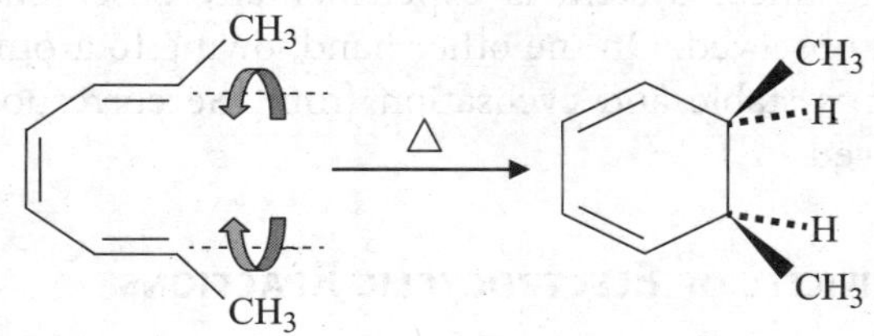

trans, trans-dimethyl hexa-1,3,5-triene

cis-1,2-dimethylcyclohexa-2,4-diene

CH_3 / CH_3 —Δ→ CH_3, H, CH_3, H

trans, cis-dimethyl hexa-1,3,5-triene *trans*-1,2-dimethyl cyclohexa-2,4-diene

14.4 CYCLOADDITION REACTIONS

When two olefin molecules react in a concerted manner to form a cyclic molecule, the reaction is called *cycloaddition*. They are intermolecular analogues of electrocyclic reactions (which are intramolecular); cycloadditions can be explained on the basis of orbital symmetry just like electrocyclic reactions. The most common cycloaddition reactions are—dimerisation of ethylene to cyclobutane and the Diels–Alder reaction.

14.4.1 Dimerisation of Ethylene

Experimentally it has been observed that ethylene undergoes dimerisation to cyclobutane when irradiated with UV radiation but not by heating:

|| + || ⇌ (hν) □

Two π-bonds of the olefin molecules are converted into two σ-bonds of the product molecule. The most favourable way in which the ethylene molecules approach each other to form the cyclic adduct is one in which the planes of the molecules are parallel (Fig. 14.9).

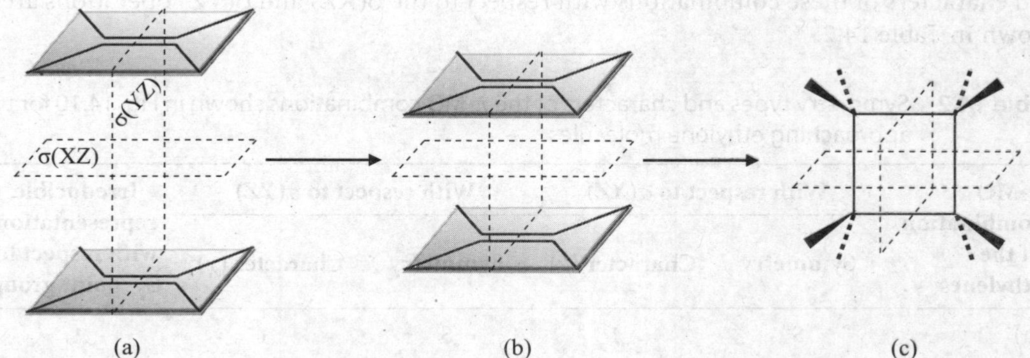

Fig. 14.9 (a) Two isolated ethylene molecules approaching one another
(b) Two ethylene molecules sufficiently close for the reaction to occur
(c) Product cyclobutane; two planes of symmetry

When the reactant molecules are in parallel planes but are far apart, the point group of the combined system is D_{2h}; when the planar cyclobutane molecule is formed, the point group is

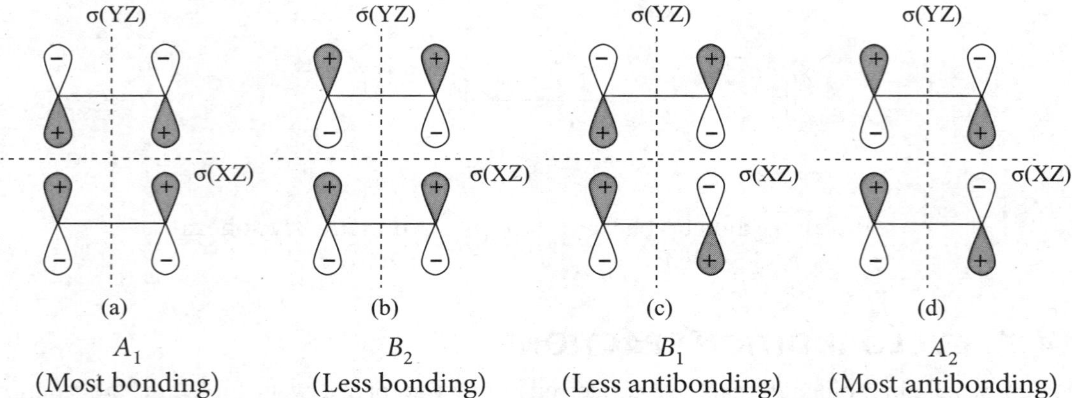

Fig. 14.10 Possible orientations of the carbon *p*-orbitals of the π-MOs as two ethylene molecules approach one another.

D_{4h}. The C_{2v} point group is a subgroup of both D_{2h} and D_{4h}. Hence according to the principle of conservation of orbital symmetry, the reacting orbitals should conserve their symmetry type throughout the reaction path with respect to at least one of the symmetry elements of C_{2v}. In the present case, symmetry with respect to the two σ_v planes are conserved (Fig. 14.9).

One of these planes $\sigma(XZ)$ is parallel to the plane of the ethylene molecules, the other $\sigma(YZ)$ is perpendicular to it and bisects the two π-bonds of the reacting molecules. The orientation of the component *p*-orbitals of the π-MOs, as the molecules approach one another, is shown in Fig. 14.10.

The first two, (a) and (b), result from bonding π-MOs of the two ethylene molecules approaching one another in an in-phase and out-of-phase manner respectively. Similarly, the third and the fourth ones, (c) and (d), result from the antibonding π*-MOs of the individual olefin molecules approaching in-phase and out-of-phase. The symmetry types and characters of these combinations with respect to the $\hat{\sigma}(XZ)$ and $\hat{\sigma}(YZ)$ operations are as shown in Table 14.2.

Table 14.2 Symmetry types and characters of the π-MO combinations shown in Fig. 14.10 for two approaching ethylene molecules

π-MO combination of the ethylenes	**With respect to $\sigma(XZ)$**		**With respect to $\sigma(YZ)$**		**Irreducible representation with respect to C_{2v} point group**
	Symmetry	**Character (χ)**	**Symmetry**	**Character (χ)**	
(a)	S	1	S	1	A_1
(b)	A	−1	S	1	B_2
(c)	S	1	A	−1	B_1
(d)	A	−1	A	−1	A_2

When the distance between the ethylene molecules is large, there is no intermolecular interaction and the bonding MOs of the two molecules have the same energy irrespective of their in-phase and out-of-phase orientation for approach. Hence in the left of the orbital correlation diagram (Fig. 14.11) the energy levels of (a) and (b) (that is, A_1 and B_2 MOs of the dimer) are shown to be degenerate; for the same reason, the π^*-MO combinations, B_1 and A_2 are degenerate. As the molecules come closer and interaction starts, this degeneracy is destroyed and the energy levels split; the orbital energy increases with an increasing number of nodal planes in the order: $A_1 < B_2 < B_1 < A_2$. This explains the left hand side of Fig. 14.11.To get the orbital energy levels on the right hand side of the correlation diagram, we need the σ-orbitals of cyclobutane into which the π-MOs of ethylene transform. The orientation of the component σ-orbitals in forming the σ-MOs resulting from cyclisation are as shown schematically in Fig. 14.12.

Considering the σ(YZ) and σ(XZ) planes, the symmetry types are found to be A_1, B_1, B_2 and A_2 respectively for (a), (b), (c) and (d) of Fig. 14.12. The right hand side of the orbital correlation diagram can now be drawn in Fig. 14.11.

With four π-electrons the occupied orbitals of the ethylene dimer in the ground state are A_1 and B_2. They correlate respectively to the lowest A_1-MO and the B_2-MO of high energy. Hence the conversion is not possible under thermal conditions. However, if one ethylene molecule is excited such that one electron is promoted from the B_2-orbital to the B_1-orbital, then the cyclisation reaction is allowed because the reactant MOs (A_1, B_2, B_1) correlate to the same set of orbitals of the product according to Fig. 14.11; thus the photochemical conversion of the ethylene dimer from an excited state to cyclobutane (also in an excited state) is symmetry-allowed.

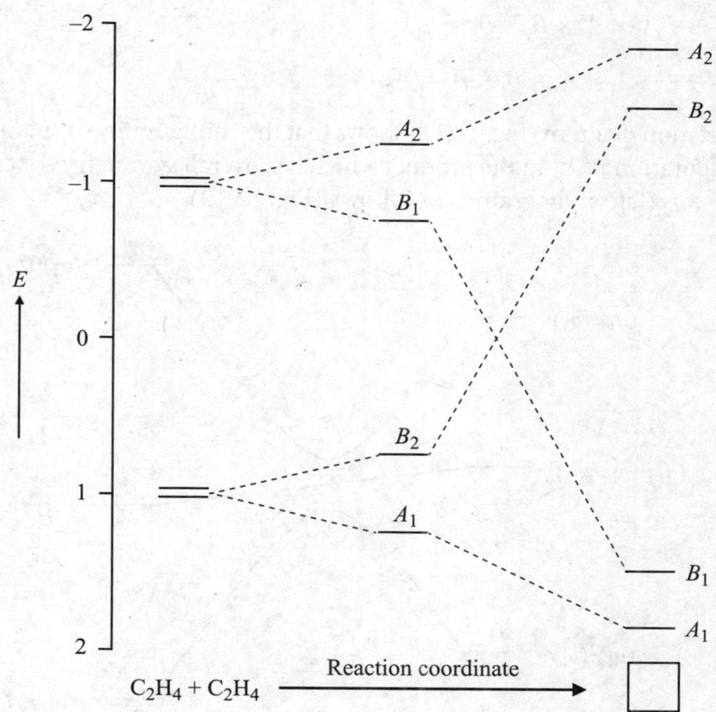

Fig. 14.11 Orbital correlation diagram for the dimerisation of ethylene. Left: ethylene dimer with no interaction, Centre: two ethylene molecules close enough for significant interaction, Right: cyclobutene σ-MOs

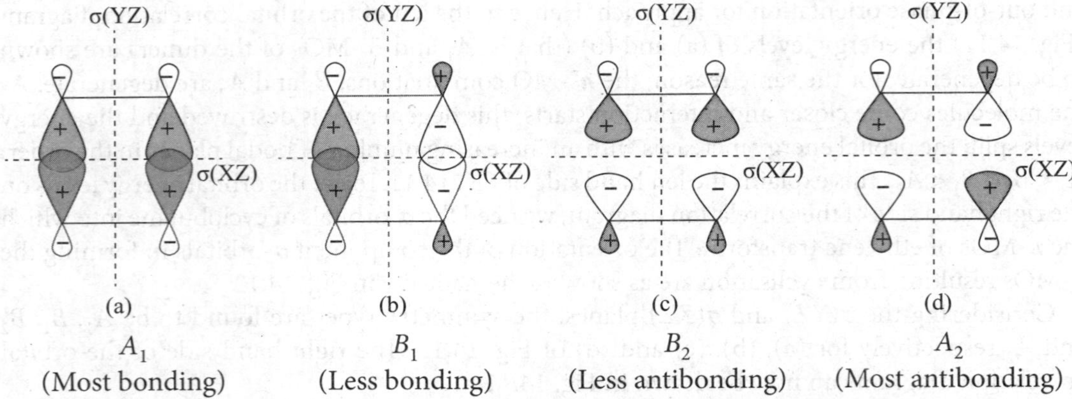

Fig. 14.12 Symmetry and orientation of σ-orbitals in the cyclobutene σ-MOs resulting from cyclisation

A more satisfactory explanation comes from the state correlation diagram. The ground state electronic configuration of ethylene dimer is $a_1^2\, b_2^2$; the symmetry of the ground state is obtained from the direct product $A_1 \times A_1 \times B_2 \times B_2 = A_1$. Similarly, the lowest and the next higher energy state of the product (cyclobutane) have electronic configuration and symmetry as specified below:

$$a_1^2\, b_1^2 \longrightarrow A_1 \times A_1 \times B_1 \times B_1 = A_1$$

$$a_1^2\, b_1^1\, b_2^1 \longrightarrow A_1 \times A_1 \times B_1 \times B_2 = A_2$$

The orbital correlation diagram (Fig. 14.11) shows that the configuration $a_1^2\, b_2^2$ of the reactant goes over to the configuration $a_1^2\, b_2^2$ in the product which has overall symmetry $A_1 \times A_1 \times B_2 \times B_2 = A_1$. Hence the state correlation diagram is as follows (Fig. 14.13).

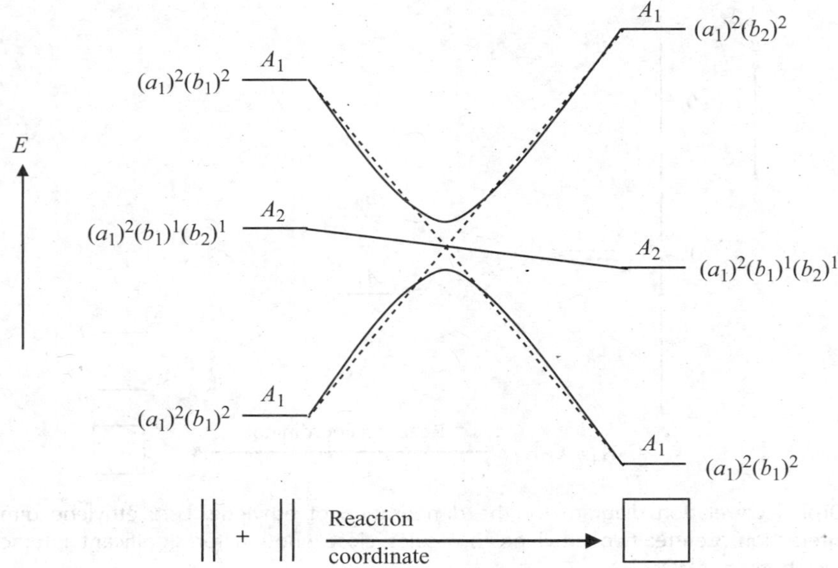

Fig. 14.13 State correlation diagram for the dimerisation of ethylene to cyclobutane

In Fig. 14.13, correlation of the configuration $a_1^2\, b_2^2$ of the reactant to the same configuration of the product (shown by dotted line) is not allowed by the non-crossing rule. The conversion of the ground state of the reactant into that ($a_1^2\, b_1^2$) of the product may occur only via a high energy barrier; the reaction is, therefore, thermally forbidden. On the other hand, if the ethylene dimer is excited to its first excited state, $a_1^2\, b_1^1\, b_2^1$ of A_2 symmetry, it can smoothly convert into the product A_2 state (of configuration $a_1^2\, b_1^1\, b_2^1$) without requiring further energy. Hence the photochemical reaction is symmetry allowed.

14.4.2 The Diels–Alder Reaction

The reaction between a diene and an olefin, in which an electron-withdrawing group (such as cyano, carbonyl, and so on) is attached, is found to occur under thermal conditions to form a cyclic addition product. Such reactions are called *Diels–Alder reactions*. An example is

100°C

Butadiene + Acraldehyde → Tetrahydrobenzaldehyde

The mechanism of such reactions has been proposed to be a concerted one, guided by the principle of conservation of orbital symmetry. Schematically the simplest Diels–Alder reaction is

(diene C1–C4) + (ethylene C5–C6) → cyclohexene (C1–C6)

For a concerted mechanism, the reactant molecules must approach one another with their molecular planes parallel (Fig. 14.14).

The only symmetry element that is conserved throughout the reaction path is a plane of symmetry (σ) perpendicular to the planes of the reactant molecules, and the orbitals are to be classified as symmetric (A) or antisymmetric (B) with respect to this σ.

The orientation of p-orbitals of the π-MOs of butadiene, in increasing order of energy, ($\Psi_1 < \Psi_2 < \Psi_3 < \Psi_4$) is shown in Fig. 14.4. HMO calculation shows that their energies are respectively 1.618, 0.618, −0.618 and −1.618, while the energies of the π-MOs of ethylene are $\pm$ 1. Hence, in the parallel stacking of one

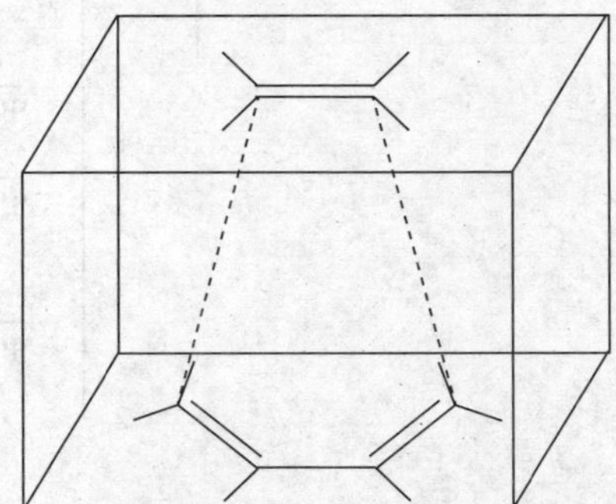

Fig. 14.14 Suitable configuration for the approach of two reactant molecules in a concerted Diels–Alder reaction

ethylene and one butadiene molecule (as in Fig. 14.14), the energy levels of the combined system in the absence of any interaction are in the order shown on the left of Fig. 14.15. The product molecule has two newly formed σ-bonds between the atom pairs (4, 5) and (1, 6) and one π-bond between atoms 2 and 3. The corresponding MOs of the product are formed as follows:

Two low energy π-MOs of butadienes, with 0 and 1 node respectively are symbolised as in Fig. 14.16. They combine with the p-orbitals of the two π-MOs of ethylene in a bonding and antibonding manner to produce σ-MOs (Fig. 14.16). Their symmetry types with respect to a reflection plane (σ) which is conserved in the reaction are also indicated in this figure. There may be more combinations to produce higher energy σ-MOs, but the ones in Fig. 14.16 are of relevance to the discussion of Diels–Alder reactions. The newly formed π-bond between atoms 2 and 3 involve π and π^*-MOs; they are of symmetry A and B respectively with respect to the σ-plane which is conserved in the reaction. Moreover, the σ-MOs are far more bonding (and antibonding) than the π-MOs. Hence the energy levels of the product MOs are as shown on the right hand side of Fig. 14.15. The correlation diagram (Fig. 14.15) can now be completed by symmetry matching. It is found from the diagram that all the filled orbitals of the reactant correlate with the filled MOs of the product in the ground state. Hence the Diels–Alder reaction is symmetry allowed under thermal conditions, as observed in experiments.

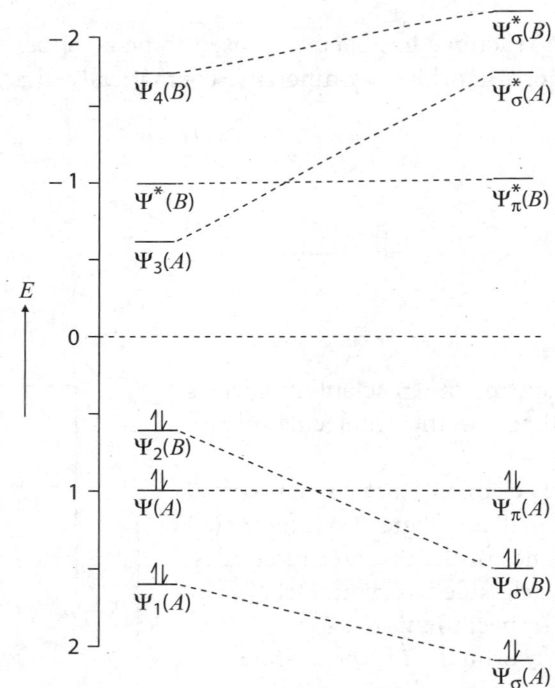

Fig. 14.15 Orbital correlation diagram for the ethylene + butadiene Diels–Alder reaction. Left: (ethylene + butadiene) with no interaction. Right: product MOs.

14.4.3 General Rule for Cycloaddition Reactions

The ethylene + ethylene cycloaddition, where 4π-electrons of the reactant molecules are involved, is photochemically allowed. The ethylene + butadiene cycloaddition, where 6π-electrons are involved, is thermally allowed by symmetry. In both the cases the bonds broken or formed are on the same face of each of the reactant molecules; such additions are called *supra–supra* cycloadditions. The results obtained can be generalised as follows:

For supra–supra cycloaddition of one olefin having m π-electrons to another having n π-electrons to form a cyclic product with $(m + n - 4)$ π-electrons,

- the reaction is photochemically allowed if $m + n = 4k$
- the reaction is thermally allowed if $m + n = 4k + 2$ (k = an integer)

The derivation of these rules for all types of cycloaddition, namely supra–supra, supra–antara, and so on, may be found in the original works of Woodward and Hoffmann (1965–1971).

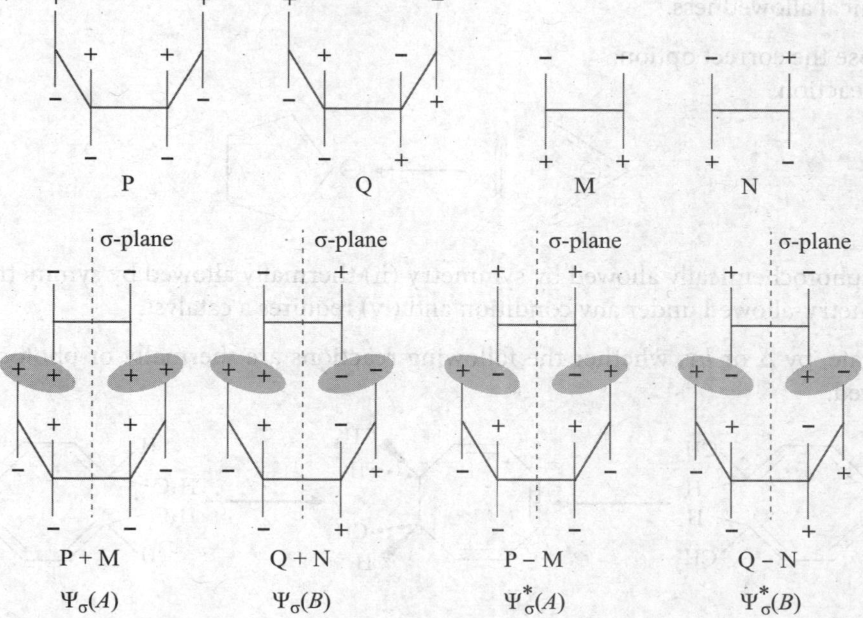

Fig. 14.16 The lowest energy π-MOs of butadiene (P and Q), two π-MOs of ethylene (M and N); formation of σ-MOs by combination of the π-MOs of butadiene and ethylene

SUMMARY

- ❑ Organic pericyclic reactions, which involve interconversion of σ- and π-bonds through a cyclic transition state, are either photochemically (P) or thermally (Th) allowed.
- ❑ Woodward and Hoffmann gave a set of rules which predict the actual course (P/Th) taken by such reactions and their stereospecificity from the count of σ/π-electrons participating in the reaction.

- These rules are based on a general principle which states that the symmetry of the relevant orbitals with respect to a C_2-axis or a σ-plane must be conserved throughout the reaction path.

LEARNING OUTCOME

After reading this chapter the student will be able to (i) apply Woodward–Hoffmann rules to predict which path – thermal or photochemical – is adopted by given reactant(s) for an electrocyclic reaction or cycloaddition, (ii) to predict which stereoisomer is produced specifically in such a reaction and (iii) draw orbital and state correlation diagrams to justify such predictions.

EXERCISES

1. Construct the orbital and state (configuration) correlation diagrams for conrotatory and disrotatory ring opening of cyclohexadiene to *cis*-hexatriene. Discuss its thermal/photochemical allowedness.

2. Choose the correct option:
   The reaction

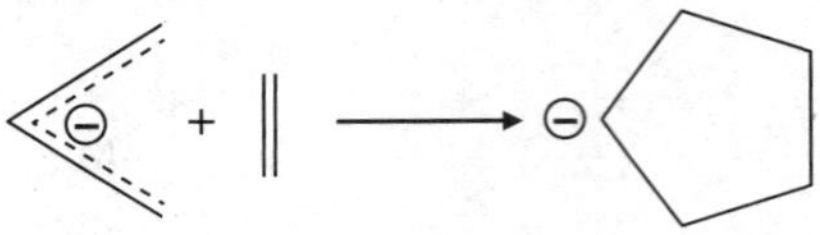

   is (i) photochemically allowed by symmetry (ii) thermally allowed by symmetry (iii) not symmetry-allowed under any condition and (iv) requires a catalyst.

3. Indicate, by Δ or $h\nu$, whether the following reactions are thermally or photochemically allowed:

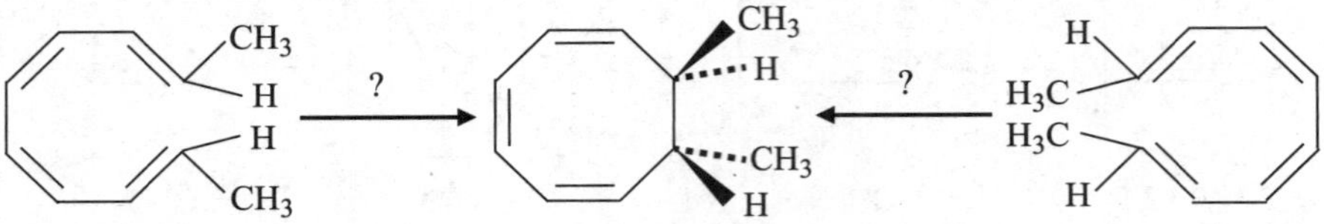

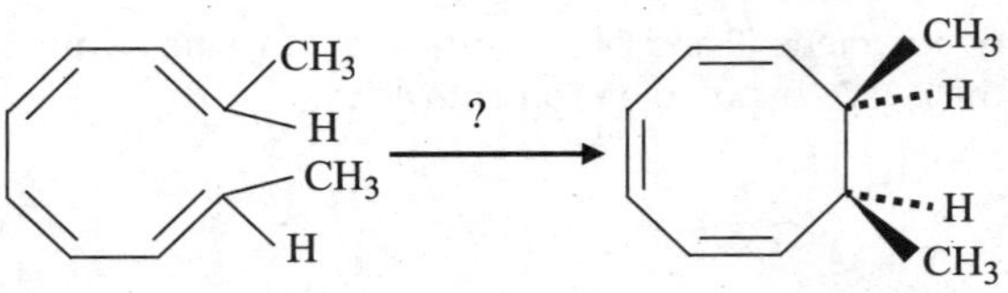

Suggested Books and References for Further Reading

1. Woodward R B and Hoffmann R, *The conservation of orbital symmetry*, Verlag Chemie GmbH, Weinheim/Academic Press Inc., **1971**.
2. Woodward R B and Hoffmann R, *J. Am. Chem. Soc.*, 87 (395, 2045 and 2511) **1965**.
3. Winter R E K, *Tetrahedron Letters*, (1207) **1965**.

Proof of the Great Orthogonality Theorem

The proof of the Great orthogonality theorem is based on some other theorems related to group representations which are stated and proved first.

Theorem 1: Any representation of a group by matrices with non-vanishing determinants is equivalent through a similarity transformation to a representation by unitary matrices.

Proof: Let the set of operators $\{\hat{E}\ (=\hat{R}_1), \hat{R}_2, \hat{R}_3, \ldots, \hat{R}_g\}$ of a group G of order g be represented by the set of non-singular matrices $D = \{\mathbf{D}(\hat{R}_1), \mathbf{D}(\hat{R}_2), \mathbf{D}(\hat{R}_3), \ldots, \mathbf{D}(\hat{R}_g)\}$. For simplicity we shall denote the matrix $\mathbf{D}(\hat{R}_i)$ by $\mathbf{D}_i$ for each $\hat{R}_i$. Using these matrices we can define a new matrix

$$\mathbf{H} = \sum_{i=1}^{g} \mathbf{D}_i \mathbf{D}_i^{\dagger} \tag{A1.1}$$

Each matrix $\mathbf{D}_i \mathbf{D}_i^{\dagger}$ is hermitian (that is, self-adjoint) because

$$(\mathbf{D}_i \mathbf{D}_i^{\dagger})^{\dagger} = (\mathbf{D}_i^{\dagger})^{\dagger} \mathbf{D}_i^{\dagger} = \mathbf{D}_i \mathbf{D}_i^{\dagger}$$

Hence $\mathbf{H}$ is also hermitian. So we can diagonalise $\mathbf{H}$ by solving the associated secular equations. Let the diagonal matrix containing the eigenvalues of $\mathbf{H}$ be $\mathbf{d}$ and the unitary matrix containing the orthogonal normalised eigenvectors be $\mathbf{U}$:

$$\mathbf{d} = \mathbf{U}^{-1}\,\mathbf{H}\mathbf{U} \tag{A1.2}$$

Using this $\mathbf{U}$ one can construct new matrices $\mathbf{D}_i'$ as

$$\mathbf{D}_i' = \mathbf{U}^{-1}\,\mathbf{D}_i\mathbf{U} \text{ for all } \hat{R}_i \text{ in } G \tag{A1.3}$$

$\mathbf{d}$ can be expressed in terms of these transformed matrices as follows

$$\mathbf{d} = \mathbf{U}^{-1}\mathbf{H}\mathbf{U} = \mathbf{U}^{-1}\left(\sum_{i=1}^{g} \mathbf{D}_i \mathbf{D}_i^{\dagger}\right)\mathbf{U} = \mathbf{U}^{-1}\sum_{i=1}^{g} \mathbf{D}_i \mathbf{D}_i^{\dagger}\,\mathbf{U}$$

$$= \sum_{i=1}^{g} \mathbf{U}^{-1}\mathbf{D}_i\mathbf{U}\mathbf{U}^{-1}\mathbf{D}_i^{\dagger}\,\mathbf{U}$$

$$= \sum_{i=1}^{g} (\mathbf{U}^{-1}\mathbf{D}_i\mathbf{U})(\mathbf{U}^{-1}\mathbf{D}_i\mathbf{U})^{\dagger}$$

$$= \sum_{i=1}^{g} \mathbf{D}_i'\mathbf{D}_i'^{\dagger} \tag{A1.4}$$

From this it can be shown that each diagonal element of **d** is non-negative:

A general diagonal element in each term of the summation in equation (A1.4) is

$$(\mathbf{D}_i'\mathbf{D}_i'^{\dagger})_{\lambda\lambda} = \sum_{\mu}(\mathbf{D}_i')_{\lambda\mu}(\mathbf{D}_i'^{\dagger})_{\mu\lambda} \text{ (by the rule of matrix multiplication)}$$

$$= \sum_{\mu}(\mathbf{D}_i')_{\lambda\mu}(\mathbf{D}_i')^{*}_{\lambda\mu} \text{ (as † means conjugate transpose)}$$

$$= \sum_{\mu}\left|(\mathbf{D}_i')_{\lambda\mu}\right|^2 \geq 0$$

Hence each element in **d** is non-negative and real. This enables one to define and construct the matrices $\mathbf{d}^{\frac{1}{2}}$ and $\mathbf{d}^{-\frac{1}{2}}$ by simply taking the appropriate powers of the diagonal elements of **d**. Now taking advantage of the commutation of diagonal matrices we have

$$\mathbf{d}^{-\frac{1}{2}}\mathbf{d}\mathbf{d}^{-\frac{1}{2}} = \mathbf{I} \text{ (the unit matrix)}$$

or, $$\mathbf{d}^{-\frac{1}{2}}\left(\sum_i \mathbf{D}_i'\mathbf{D}_i'^{\dagger}\right)\mathbf{d}^{-\frac{1}{2}} = \mathbf{I} \text{ (using equation A1.4)} \tag{A1.5}$$

We now define a new set of doubly transformed matrices $\mathbf{D}_j''$ from the matrices $\mathbf{D}_j'$ $[= \mathbf{D}'(\hat{\mathrm{R}}_j)]$:

$$\mathbf{D}_j'' = \mathbf{d}^{-\frac{1}{2}}\mathbf{D}_j'\mathbf{d}^{\frac{1}{2}} \text{ for all } \hat{\mathrm{R}}_j \text{ in } G. \tag{A1.6}$$

It can be shown that the matrices $\mathbf{D}_j''$ are unitary:

$$\mathbf{D}_j''\mathbf{D}_j''^{\dagger} = \mathbf{d}^{-\frac{1}{2}}\mathbf{D}_j'\mathbf{d}^{\frac{1}{2}}\left(\mathbf{d}^{-\frac{1}{2}}\mathbf{D}_j'\mathbf{d}^{\frac{1}{2}}\right)^{\dagger}$$

$$= \mathbf{d}^{-\frac{1}{2}}\mathbf{D}_j'\mathbf{d}^{\frac{1}{2}}\mathbf{d}^{\frac{1}{2}}\mathbf{D}_j'^{\dagger}\mathbf{d}^{-\frac{1}{2}} \text{ (as } \left(\mathbf{d}^{\frac{1}{2}}\right)^{\dagger} = \mathbf{d}^{\frac{1}{2}}, \mathbf{d} \text{ is being real and symmetric)}$$

$$= \mathbf{d}^{-\frac{1}{2}}\mathbf{D}_j'\mathbf{d}^{\frac{1}{2}}\mathbf{I}\mathbf{d}^{\frac{1}{2}}\mathbf{D}_j'^{\dagger}\mathbf{d}^{-\frac{1}{2}} \text{ (inserting I does not change the result)}$$

$$= \mathbf{d}^{-\frac{1}{2}}\mathbf{D}_j'\mathbf{d}^{\frac{1}{2}}\left\{\mathbf{d}^{-\frac{1}{2}}\left(\sum_i \mathbf{D}_i'\mathbf{D}_i'^{\dagger}\right)\mathbf{d}^{-\frac{1}{2}}\right\}\mathbf{d}^{\frac{1}{2}}\mathbf{D}_j'^{\dagger}\mathbf{d}^{-\frac{1}{2}} \text{ (using Eq. A1.5)}$$

$$= \mathbf{d}^{-\frac{1}{2}}\mathbf{D}_j'\left(\sum_i \mathbf{D}_i'\mathbf{D}_i'^{\dagger}\right)\mathbf{D}_j'^{\dagger}\mathbf{d}^{-\frac{1}{2}}$$

$$= \sum_i \mathbf{d}^{-\frac{1}{2}}\mathbf{D}_j'\mathbf{D}_i'\mathbf{D}_i'^{\dagger}\mathbf{D}_j'^{\dagger}\mathbf{d}^{-\frac{1}{2}} \tag{A1.7}$$

If $\hat{R}_j \hat{R}_i = \hat{R}_k$ then $\mathbf{D}'_j \mathbf{D}'_i = \mathbf{D}'_k$ and $\mathbf{D}'^{\dagger}_i \mathbf{D}'^{\dagger}_j = \mathbf{D}'^{\dagger}_k$ and the sum over i can be replaced by a sum over k because according to the rearrangement theorem, $\hat{R}_j \hat{R}_i$ for a fixed $\hat{R}_j$ and varying $\hat{R}_i$ (from $i = 1$ to g) will produce all the $\hat{R}_k$, $k = 1$ to g, each $\hat{R}_k$ occurring only once. Hence from Eq. (A1.7),

$$\mathbf{D}''_j \mathbf{D}''^{\dagger}_j = \sum_k \mathbf{d}^{-\frac{1}{2}} \mathbf{D}'_k \mathbf{D}'^{\dagger}_k \mathbf{d}^{-\frac{1}{2}} = \mathbf{I} \text{ (using Eq. A1.5)} \tag{A1.8}$$

Thus $\mathbf{D}''_j$ is unitary for all $\hat{R}_j$ in G. (Proved)

Theorem 2 (also called Schur's Lemma 1): If a hermitian matrix **A** commutes with all the matrices $\mathbf{D}(\hat{R}_i)$ of an irreducible representation of a group, then **A** must be a constant matrix.

Proof: Let $G = \{\hat{E} (= \hat{R}_1), \hat{R}_2, \hat{R}_3, \ldots, \hat{R}_g\}$ be a group of order g and let $\{\mathbf{D}(\hat{R}_i), i = 1 \text{ to } g\}$ be the set of matrices of an irreducible representation D of G. For simplicity such matrices will be henceforth denoted by $\mathbf{D}(\hat{R})$. If there exists a matrix **A** such that

$$\mathbf{AD}(\hat{R}) = \mathbf{D}(\hat{R})\,\mathbf{A} \text{ for all } \hat{R} \text{ in } G$$

then it is required to prove that **A** must be a constant matrix (that is, a constant scalar times the unit matrix of the size of each $\mathbf{D}(\hat{R})$).

The matrix **A** can be diagonalised by a unitary transformation:

$$\mathbf{U}^{-1}\mathbf{AU} = \mathbf{\Lambda} = \text{a diagonal matrix} \tag{A1.9}$$

Using this **U** one can carry out a similarity transformation on each matrix $\mathbf{D}(\hat{R})$ and define a new matrix $\mathbf{D}'(\hat{R})$ for each $\hat{R}$:

$$\mathbf{U}^{-1}\mathbf{D}(\hat{R})\,\mathbf{U} = \mathbf{D}'(\hat{R}) \text{ (for all } \hat{R}) \tag{A1.10}$$

We now examine the commutability of these new matrices with $\mathbf{\Lambda}$:

$$\begin{aligned}
\mathbf{\Lambda D}'(\hat{R}) - \mathbf{D}'(\hat{R})\,\mathbf{\Lambda} &= \mathbf{U}^{-1}\mathbf{AUU}^{-1}\mathbf{D}(\hat{R})\,\mathbf{U} - \mathbf{U}^{-1}\mathbf{D}(\hat{R})\,\mathbf{UU}^{-1}\mathbf{AU} \\
&= \mathbf{U}^{-1}\mathbf{AD}(\hat{R})\,\mathbf{U} - \mathbf{U}^{-1}\mathbf{D}(\hat{R})\,\mathbf{AU} \\
&= \mathbf{U}^{-1}\{\mathbf{AD}(\hat{R}) - \mathbf{D}(\hat{R})\,\mathbf{A}\}\,\mathbf{U} \\
&= \mathbf{U}^{-1}[\mathbf{A}, \mathbf{D}(\hat{R})]\,\mathbf{U} \\
&= \mathbf{0} \text{ as } \mathbf{A} \text{ commutes with } \mathbf{D}(\hat{R})
\end{aligned}$$

Hence each $\mathbf{D}'(\hat{R})$ commutes with $\mathbf{\Lambda}$:

$$\mathbf{\Lambda D}'(\hat{R}) = \mathbf{D}'(\hat{R})\,\mathbf{\Lambda} \text{ for all } \hat{R}_i,\ i = 1 \text{ to } g \tag{A1.11}$$

Comparing the (p, q) elements of both sides,

$$\sum_{r=1}^{n} \mathbf{\Lambda}_{pr}\,[\mathbf{D}'(\hat{R})]_{rq} = \sum_{r=1}^{n} [\mathbf{D}'(\hat{R})]_{pr}\,\mathbf{\Lambda}_{rq} \tag{A1.12}$$

where n is the dimension of the representation D. As $\mathbf{\Lambda}$ is a diagonal matrix,

$$\Lambda_{pr} = 0 \text{ for } p \neq r$$

and $$\Lambda_{rq} = 0 \text{ for } r \neq q$$

Putting $r = p$ in the LHS and $r = q$ in the RHS of Eq. (A1.12) we have

$$\Lambda_{pp} [\mathbf{D}'(\hat{R})]_{pq} = [\mathbf{D}'(\hat{R})]_{pq} \Lambda_{qq}$$

or, $$[\mathbf{D}'(\hat{R})]_{pq} (\Lambda_{pp} - \Lambda_{qq}) = 0 \tag{A1.13}$$

Let us now consider a single diagonal element of $\mathbf{\Lambda}$, say Λ_{11}. If it is different from all the other diagonal elements of $\mathbf{\Lambda}$, then from Eq. (A1.13)

$$p = 1 \text{ gives } [\mathbf{D}'(\hat{R})]_{1q} \text{ (as } \Lambda_{11} \neq \Lambda_{qq}, q = 2, 3, \ldots, n)$$

and $$q = 1 \text{ gives } [\mathbf{D}'(\hat{R})]_{p1} = 0 \text{ (as } \Lambda_{11} \neq \Lambda_{pp}, p = 2, 3, \ldots, n)$$

Thus excepting the (1, 1) element of $\mathbf{D}(\hat{R})$ all elements in the first row (for $q = 2$ to n) and first column (for $p = 2$ to n) of $\mathbf{D}(\hat{R})$ are zero. The matrix $\mathbf{D}(\hat{R})$ is, therefore, block-factored as:

$$\mathbf{D}(\hat{R}) = \left[\begin{array}{c|cccc} \mathbf{D}(\hat{R})_{11} & 0 & 0 & \ldots & 0 \\ \hline 0 & & & & \\ 0 & & \text{block of size} & & \\ \vdots & & (n-1)\times(n-1) & & \\ 0 & & & & \end{array}\right] \quad \text{for all } \hat{R} \text{ in group G}$$

Hence the representation D is reducible, contrary to the starting hypothesis. So Λ_{11} cannot be different from the other diagonal elements of $\mathbf{\Lambda}$. In this way it can be shown that each diagonal element of $\mathbf{\Lambda}$ must be the same as the others:

$$\Lambda_{11} = \Lambda_{22} = \Lambda_{33} = \ldots = \Lambda_{nn} = c \text{ (say)}$$

$$\text{Hence } \mathbf{\Lambda} = \begin{bmatrix} c & 0 & 0 & \ldots & 0 \\ 0 & c & 0 & \ldots & 0 \\ 0 & 0 & c & \ldots & 0 \\ \ldots & \ldots & \ldots & \ldots & \ldots \\ 0 & 0 & 0 & 0 & c \end{bmatrix} = c\mathbf{I}$$

where $\mathbf{I}$ is the unit matrix of size $n \times n$, and c is a constant. (Proved)

Theorem 3 (Schur's Lemma 2): If a group has two different irreducible representations $D^{\nu} = \{\mathbf{D}^{\nu}(\hat{R}_i), i = 1 \text{ to } g\}$ and $D^{\mu} = \{\mathbf{D}^{\mu}(\hat{R}_i), i = 1 \text{ to } g\}$ of dimension n_{ν} and n_{μ} respectively and a rectangular matrix $\mathbf{A}$ exists such that

$$\mathbf{A}\mathbf{D}^{\nu}(\hat{R}) = \mathbf{D}^{\mu}(\hat{R})\,\mathbf{A} \text{ for all } \hat{R} \text{ in G}$$

then—

Case (a): for $n_\nu = n_\mu$, **A** is either non-singular and D^ν and D^μ are equivalent representations or else **A** is the null matrix;

Case (b): for $n_\nu \neq n_\mu$, $\mathbf{A} = \mathbf{0}$ (null matrix)

Proof: All representations by non-singular matrices can be made unitary and so without loss of generality we may assume that the matrices $\mathbf{D}^\nu$ and $\mathbf{D}^\mu$ are unitary.

It is given that $\mathbf{A}\mathbf{D}^\nu(\hat{R}) = \mathbf{D}^\mu(\hat{R})\,\mathbf{A}$ for all $\hat{R}$ in G (A1.14)

Taking adjoints of both sides,

$$[\mathbf{D}^\nu(\hat{R})]^\dagger\,\mathbf{A}^\dagger = \mathbf{A}^\dagger\,[\mathbf{D}^\mu(\hat{R})]^\dagger \tag{A1.15}$$

As the matrices $\mathbf{D}^\nu$ and $\mathbf{D}^\mu$ are unitary

$$[\mathbf{D}^\alpha(\hat{R})]^\dagger = [\mathbf{D}^\alpha(\hat{R})]^{-1} = \mathbf{D}^\alpha(\hat{R}^{-1}) \text{ for both } \alpha = \nu \text{ and } \alpha = \mu$$

Hence from (A1.14)

$$\mathbf{D}^\nu(\hat{R}^{-1})\,\mathbf{A}^\dagger = \mathbf{A}^\dagger\,\mathbf{D}^\mu(\hat{R}^{-1})$$

and right-multiplying both sides by **A**,

$$\mathbf{D}^\nu(\hat{R}^{-1})\,\mathbf{A}^\dagger\mathbf{A} = \mathbf{A}^\dagger\,\mathbf{D}^\mu(\hat{R}^{-1})\,\mathbf{A} \tag{A1.16}$$

Now, $\hat{R}^{-1}$ are all members of G since each $\hat{R}$ in G has a unique inverse. Equation (A1.14) therefore holds if $\hat{R}$ is replaced by the respective $(\hat{R}^{-1})$ and we have,

$$\mathbf{A}\mathbf{D}^\nu(\hat{R}^{-1}) = \mathbf{D}^\mu(\hat{R}^{-1})\,\mathbf{A}$$

Left-multiplying both sides by $\mathbf{A}^\dagger$ we have

$$\mathbf{A}^\dagger\,\mathbf{A}\mathbf{D}^\nu(\hat{R}^{-1}) = \mathbf{A}^\dagger\,\mathbf{D}^\mu(\hat{R}^{-1})\,\mathbf{A} \tag{A1.17}$$

Comparing Eqs (A1.15) and (A1.16),

$$\mathbf{D}^\nu(\hat{R}^{-1})\,\mathbf{A}^\dagger\mathbf{A} = \mathbf{A}^\dagger\,\mathbf{A}\mathbf{D}^\nu(\hat{R}^{-1}) \text{ for all } (\hat{R}^{-1}) \text{ in G}$$

and replacing all $(\hat{R}^{-1})$ by $\hat{R}$ as before,

$$[\mathbf{D}^\nu(\hat{R})]\,\mathbf{A}^\dagger\mathbf{A} = \mathbf{A}^\dagger\mathbf{A}\,[\mathbf{D}^\nu(\hat{R})] \tag{A1.18}$$

Hence by Lemma 1, $\mathbf{A}^\dagger\mathbf{A}$ is a constant matrix:

$$\mathbf{A}^\dagger\mathbf{A} = c\mathbf{I} \tag{A1.19}$$

where c is a scalar constant and **I** is the unit matrix of the size of each $\mathbf{D}^\nu(\hat{R})$.

Now consider case (a): $n_\nu = n_\mu$. In this case **A** and $\mathbf{A}^\dagger$ are square matrices and so also is $\mathbf{A}^\dagger\mathbf{A}$; they have corresponding determinants.

det ($\mathbf{A}^\dagger\mathbf{A}$) = det $\mathbf{A}^\dagger$ · det $\mathbf{A}$ = (det $\mathbf{A}$)^2 (as a determinant remains unchanged by transposition)

or, $\quad$ det ($c\mathbf{I}$) = (det $\mathbf{A}$)^2

or, $\quad$ c^{n_ν} = (det $\mathbf{A}$)^2

So if $c \neq 0$, det $\mathbf{A} \neq 0$ and $\mathbf{A}$ has an inverse, and from Eq. (A1.14) we have

$$\mathbf{A}^{-1}\,\mathbf{A}\mathbf{D}^{\nu}(\hat{R}) = \mathbf{A}^{-1}\,\mathbf{D}^{\mu}(\hat{R})\,\mathbf{A}$$

or, $\quad$ $\mathbf{D}^{\nu}(\hat{R}) = \mathbf{A}^{-1}\,\mathbf{D}^{\mu}(\hat{R})\,\mathbf{A}$ for all $\hat{R}$ in G.

Thus the representation D^ν and D^μ are equivalent.

Or else, if $c = 0$, $\mathbf{A}^\dagger\mathbf{A} = \mathbf{0}$ from Eq. (A1.19)

or, $$\sum_{j=1}^{n_\nu} (A^\dagger)_{ij}\, A_{jk} = 0$$

or, $$\sum_{j=1}^{n_\nu} A^*_{ji}\, A_{jk} = 0, \quad i = 1, 2, \ldots, n_\nu \;\text{ and }\; j = 1, 2, \ldots, n_\nu$$

If $i = k$ this becomes

$$\sum_{j=1}^{n_\nu} A^*_{ji}\, A_{ji} = 0 \;\text{ i.e., }\; \sum_{j=1}^{n_\nu} |A_{ji}|^2 = 0$$

This is possible only if each A_{ji} is individually zero, that is, $\mathbf{A} = \mathbf{0}$.

For case (b), $n_\nu \neq n_\mu$ and without loss of generality we may take $n_\nu > n_\mu$. Then $\mathbf{A}$ is a rectangular matrix; for the multiplication $\mathbf{A}\mathbf{D}^\nu(\hat{R})$, $\mathbf{A}$ has to have n_ν rows and for the multiplication $\mathbf{D}^\mu(\hat{R})\,\mathbf{A}$, there should be n_μ columns in $\mathbf{A}$; we can convert it into a $n_\nu \times n_\nu$ square matrix $\mathbf{B}$ by filling out $(n_\nu - n_\nu)$ columns by zero. Clearly, $\mathbf{A}^\dagger\mathbf{A} = \mathbf{B}^\dagger\mathbf{B}$ = a square matrix of size $n_\nu \times n_\nu$.

Then $\quad$ det ($\mathbf{B}^\dagger\mathbf{B}$) = det $\mathbf{B}^\dagger$ · det $\mathbf{B}$ = 0

or, $\quad$ det ($\mathbf{A}^\dagger\mathbf{A}$) = 0

and therefore, as shown in case (a), $\mathbf{A} = \mathbf{0}$ (null matrix).

Proof of the Great Orthogonality Theorem

Let D^i be an l_i-dimensional irreducible representation of the group G. The operators $\hat{S}$ of G are represented by non-singular $l_i \times l_i$ matrices $\mathbf{D}^i(\hat{S})$. According to Theorem 1 we may take these matrices to be unitary.

Using a non-singular arbitrary matrix $\mathbf{X}$ of the size of each $\mathbf{D}^i(\hat{S})$ we can define a new matrix $\mathbf{A}$ as follows:

$$\mathbf{A} = \sum_{\hat{S}} \mathbf{D}^i(\hat{S})\,\mathbf{X}\mathbf{D}^i(\hat{S}^{-1}) \tag{A1.20}$$

The commutability of **A** with each matrix $\mathbf{D}^i(\hat{R})$ of the representation can be readily tested: For any operation $\hat{R}$ in G,

$$\mathbf{D}^i(\hat{R})\,\mathbf{A} = \mathbf{D}^i(\hat{R}) \sum_{\hat{S}} \mathbf{D}^i(\hat{S})\,\mathbf{X}\mathbf{D}^i(\hat{S}^{-1})$$

$$= \mathbf{D}^i(\hat{R}) \sum_{\hat{S}} \mathbf{D}^i(\hat{S})\,\mathbf{X}\mathbf{D}^i(\hat{S}^{-1})\,\mathbf{I} \quad \text{(where } \mathbf{I} \text{ is the unit matrix of size } l_i \times l_i)$$

$$= \sum_{\hat{S}} \mathbf{D}^i(\hat{R})\,\mathbf{D}^i(\hat{S})\,\mathbf{X}\mathbf{D}^i(\hat{S}^{-1})\,\mathbf{D}^i(\hat{R}^{-1})\,\mathbf{D}^i(\hat{R})$$

$$(\text{as } \mathbf{D}^i(\hat{R}^{-1})\,\mathbf{D}^i(\hat{R}) = \mathbf{D}^i(\hat{R}^{-1}\hat{R}) = \mathbf{D}^i(\hat{E}) = \mathbf{I})$$

$$= \sum_{\hat{S}} \mathbf{D}^i(\hat{R}\,\hat{S})\,\mathbf{X}\mathbf{D}^i(\hat{R}\,\hat{S})^{-1}\,\mathbf{D}^i(\hat{R}) \tag{A1.21}$$

Let $\hat{R}\,\hat{S} = \hat{T}$. For a fixed $\hat{R}$ and varying $\hat{S}$, the list of operations $\hat{R}\,\hat{S}$ in the above summation will contain all the operations exactly once according to the rearrangement theorem; the index of summation can therefore be replaced by $\hat{T}$. Hence from Eq. (A1.21),

$$\mathbf{D}^i(\hat{R})\,\mathbf{A} = \sum_{\hat{T}} \mathbf{D}^i(\hat{T})\,\mathbf{X}\mathbf{D}^i(\hat{T}^{-1})\,\mathbf{D}^i(\hat{R})$$

$$= \mathbf{A}\mathbf{D}^i(\hat{R}) \tag{A1.22}$$

which is true for all $\hat{R}$ in G. Thus **A** commutes with all the matrices $\mathbf{D}^i(\hat{R})$ of the irreducible representation. So according to Schur's Lemma 1, **A** should be a constant matrix:

$$\mathbf{A} = c\mathbf{I} \text{ (c is a constant)} \tag{A1.23}$$

or,

$$\sum_{\hat{R}} \mathbf{D}^i(\hat{R})\,\mathbf{X}\mathbf{D}^i(\hat{R}^{-1}) = c\mathbf{I}$$

Taking the (m, n) element of both sides,

$$\sum_{\hat{R}} \sum_{p=1}^{l_i} \sum_{q=1}^{l_i} [\mathbf{D}^i(\hat{R})]_{mp}\,\mathbf{X}_{pq}\,[\mathbf{D}^i(\hat{R}^{-1})]_{qn} = c(\mathbf{I})_{mn} = c\delta_{mn} \tag{A1.24}$$

The value of c depends on the choice of **X**. As **X** is arbitrary we may take all $\mathbf{X}_{pq} = 0$ except $\mathbf{X}_{m'n'} = 1$. In the summations over p and q only the term with $p = m'$ and $q = n'$ will survive. Then we have from Eq. (A1.24)

$$\sum_{\hat{R}} [\mathbf{D}^i(\hat{R})]_{mm'} \cdot 1 \cdot [\mathbf{D}^i(\hat{R}^{-1})]_{n'n} = c_{m'n'}\,\delta_{mn} \tag{A1.25}$$

The subscripts m' and n' with c have been used to indicate its dependence on the particular choice of **X**. The value of $c_{m'n'}$ is obtained from Eq. (A1.25) as follows:

Choosing $m = n$ and summing over all m, we have

$$\sum_m \sum_{\hat{R}} [\mathbf{D}^i(\hat{R})]_{mm'} \cdot 1 \cdot [\mathbf{D}^i(\hat{R}^{-1})]_{n'm} = c_{m'n'} \sum_{m=1}^{l_i} \delta_{mm} = l_i \, c_{m'n'} \quad \text{(A1.26)}$$

The LHS of Eq. (A1.26) is

$$\sum_{\hat{R}} \sum_m [\mathbf{D}^i(\hat{R}^{-1})]_{n'm} [\mathbf{D}^i(\hat{R})]_{mm'} = \sum_{\hat{R}} [\mathbf{D}^i(\hat{E})]_{n'm'} = g\delta_{n'm'}$$

Hence from Eq. (A1.26),

$$c_{m'n'} = \frac{g}{l_i} \delta_{n'm'}$$

Using this value of $c_{m'n'}$ in Eq. (A1.25)

$$\sum_{\hat{R}} [\mathbf{D}^i(\hat{R})]_{mm'} [\mathbf{D}^i(\hat{R}^{-1})]_{n'n} = \frac{g}{l_i} \delta_{n'm'} \delta_{mn}$$

As $\mathbf{D}^i(\hat{R}^{-1})$ is unitary, this can be written as:

$$\sum_{\hat{R}} [\mathbf{D}^i(\hat{R})]^*_{nn'} [\mathbf{D}^i(\hat{R})]_{mm'} = \frac{g}{\sqrt{l_i l_i}} \delta_{nm} \delta_{n'm'} \ (\text{as } \delta_{nm} = \delta_{mn}) \quad \text{(A1.27)}$$

Next let us consider the case of two non-equivalent irreducible representations, D^i and D^j ($i \neq j$). As before, we can define a matrix **A**, using an arbitrary matrix **X**, as:

$$\mathbf{A} = \sum_{\hat{S}} \mathbf{D}^i(\hat{S}) \, \mathbf{X} \mathbf{D}^j(\hat{S}^{-1}) \quad \text{(A1.28)}$$

where **X** is a rectangular matrix having l_i columns and l_j rows (l_i and l_j being the dimensions of D^i and D^j respectively).

$$\begin{aligned}
\mathbf{D}^i(\hat{R})\,\mathbf{A} &= \mathbf{D}^i(\hat{R}) \sum_{\hat{S}} \mathbf{D}^i(\hat{S}) \, \mathbf{X}\mathbf{D}^j(\hat{S}^{-1}) \\
&= \sum_{\hat{S}} \mathbf{D}^i(\hat{R}) \, \mathbf{D}^i(\hat{S}) \, \mathbf{X}\mathbf{D}^j(\hat{S}^{-1}) \\
&= \sum_{\hat{S}} \mathbf{D}^i(\hat{R}) \, \mathbf{D}^i(\hat{S}) \, \mathbf{X}\mathbf{D}^j(\hat{S}^{-1}) \, \mathbf{I} \\
&= \sum_{\hat{S}} \mathbf{D}^i(\hat{R}) \, \mathbf{D}^i(\hat{S}) \, \mathbf{X}\mathbf{D}^j(\hat{S}^{-1}) \, \mathbf{D}^j(\hat{R}^{-1}) \, \mathbf{D}^j(\hat{R}) \\
&= \sum_{\hat{S}} \mathbf{D}^i(\hat{R}\,\hat{S}) \, \mathbf{X}\mathbf{D}^j(\hat{R}\,\hat{S})^{-1} \, \mathbf{D}^j(\hat{R})
\end{aligned}$$

$$= \sum_{\hat{T}} \mathbf{D}^i(\hat{T})\, \mathbf{X}\mathbf{D}^j(\hat{T}^{-1})\, \mathbf{D}^j(\hat{R})$$

$$= \mathbf{A}\mathbf{D}^j(\hat{R}) \text{ for all } \hat{R} \text{ in } G \qquad \text{(A1.29)}$$

where $\hat{R}\,\hat{S} = \hat{T}$. The index of summation has been changed following the argument given earlier.

Hence according to Schur's Lemma 2, $\mathbf{A} = \mathbf{0}$ which gives

$$\mathbf{A} = \sum_{\hat{S}} \mathbf{D}^i(\hat{S})\, \mathbf{X}\mathbf{D}^j(\hat{S}^{-1}) = \mathbf{0} \qquad \text{for } i \neq j$$

or, $$A_{mn} = \sum_{\hat{S}} \sum_{p=1}^{l_i} \sum_{q=1}^{l_j} [\mathbf{D}^i(\hat{S})]_{mp} [\mathbf{X}]_{pq} [\mathbf{D}^j(\hat{S}^{-1})]_{qn} = 0 \text{ (scalar)} \qquad \text{(A1.30)}$$

As $\mathbf{X}$ is arbitrary we may take all $\mathbf{X}_{pq} = 0$ except $\mathbf{X}_{m'n'} = 1$. Then, as before,

$$\sum_{\hat{S}} [\mathbf{D}^i(\hat{S})]_{mm'} [\mathbf{X}]_{m'n'} [\mathbf{D}^j(\hat{S}^{-1})]_{n'n} = 0$$

or, $$\sum_{\hat{S}} [\mathbf{D}^i(\hat{S})]_{mm'} [\mathbf{D}^j(\hat{S}^{-1})]_{n'n} = 0$$

Using the unitary nature of matrices,

$$\sum_{\hat{S}} [\mathbf{D}^j(\hat{S})]^*_{nn'} [\mathbf{D}^i(\hat{S})]_{mm'} = 0 \qquad \text{(A1.31)}$$

Equation (A1.27) involves the different row and column indices of the matrices in the same irreducible representation D^i, while Eq. (A1.31) combines the matrix elements of different irreducible representations, D^i and D^j. These two equations can be combined as

$$\sum_{\hat{R}} [\mathbf{D}^j(\hat{R})]^*_{nn'} [\mathbf{D}^i(\hat{R})]_{mm'} = \frac{g}{\sqrt{l_i\, l_i}}\, \delta_{ij}\, \delta_{nm}\, \delta_{n'm'}$$

which is the *Great Orthogonality Theorem.*

A2 Regular Representation of Point Groups and the Celebrated Theorem

If the multiplication table of a group is known, a reducible representation can be constructed in the following way. We shall consider the C_{3v} point group for example. Its multiplication table is given below with the symmetry operators as column headings and the corresponding inverses as the row labels.

Second operation ↓ / First operation →	$\hat{E}$	$\hat{C}_3$	$\hat{C}_3^2$	$\hat{\sigma}_v(1)$	$\hat{\sigma}_v(2)$	$\hat{\sigma}_v(3)$
$\hat{E}^{-1} \equiv \hat{E}$	$\hat{E}$	$\hat{C}_3$	$\hat{C}_3^2$	$\hat{\sigma}_v(1)$	$\hat{\sigma}_v(2)$	$\hat{\sigma}_v(3)$
$\hat{C}_3^{-1} \equiv \hat{C}_3^2$	$\hat{C}_3^2$	$\hat{E}$	$\hat{C}_3$	$\hat{\sigma}_v(2)$	$\hat{\sigma}_v(3)$	$\hat{\sigma}_v(1)$
$(\hat{C}_3^2)^{-1} = \hat{C}_3$	$\hat{C}_3$	$\hat{C}_3^2$	$\hat{E}$	$\hat{\sigma}_v(3)$	$\hat{\sigma}_v(1)$	$\hat{\sigma}_v(2)$
$[\hat{\sigma}_v(1)]^{-1} = \hat{\sigma}_v(1)$	$\hat{\sigma}_v(1)$	$\hat{\sigma}_v(2)$	$\hat{\sigma}_v(3)$	$\hat{E}$	$\hat{C}_3$	$\hat{C}_3^2$
$[\hat{\sigma}_v(2)]^{-1} = \hat{\sigma}_v(2)$	$\hat{\sigma}_v(2)$	$\hat{\sigma}_v(3)$	$\hat{\sigma}_v(1)$	$\hat{C}_3^2$	$\hat{E}$	$\hat{C}_3$
$[\hat{\sigma}_v(3)]^{-1} = \hat{\sigma}_v(3)$	$\hat{\sigma}_v(3)$	$\hat{\sigma}_v(1)$	$\hat{\sigma}_v(2)$	$\hat{C}_3$	$\hat{C}_3^2$	$\hat{E}$

A representation (Γ) called *regular representation* of C_{3v} is obtained as follows. For the matrix of a symmetry operator $\hat{R}$, replace $\hat{R}$ by 1 (one) and all other elements by 0 (zero) in the above multiplication table. Thus

$$\mathbf{M}(\hat{E}) = \begin{bmatrix} 1 & 0 & 0 & 0 & 0 & 0 \\ 0 & 1 & 0 & 0 & 0 & 0 \\ 0 & 0 & 1 & 0 & 0 & 0 \\ 0 & 0 & 0 & 1 & 0 & 0 \\ 0 & 0 & 0 & 0 & 1 & 0 \\ 0 & 0 & 0 & 0 & 0 & 1 \end{bmatrix}, \quad \mathbf{M}(\hat{C}_3) = \begin{bmatrix} 0 & 1 & 0 & 0 & 0 & 0 \\ 0 & 0 & 1 & 0 & 0 & 0 \\ 1 & 0 & 0 & 0 & 0 & 0 \\ 0 & 0 & 0 & 0 & 1 & 0 \\ 0 & 0 & 0 & 0 & 0 & 1 \\ 0 & 0 & 0 & 1 & 0 & 0 \end{bmatrix}$$

$$\mathbf{M}(\hat{C}_3^2) = \begin{bmatrix} 0 & 0 & 1 & 0 & 0 & 0 \\ 1 & 0 & 0 & 0 & 0 & 0 \\ 0 & 1 & 0 & 0 & 0 & 0 \\ 0 & 0 & 0 & 0 & 0 & 1 \\ 0 & 0 & 0 & 1 & 0 & 0 \\ 0 & 0 & 0 & 0 & 1 & 0 \end{bmatrix}, \quad \mathbf{M}\{\sigma_v(1)\} = \begin{bmatrix} 0 & 0 & 0 & 1 & 0 & 0 \\ 0 & 0 & 0 & 0 & 0 & 1 \\ 0 & 0 & 0 & 0 & 1 & 0 \\ 1 & 0 & 0 & 0 & 0 & 0 \\ 0 & 0 & 1 & 0 & 0 & 0 \\ 0 & 1 & 0 & 0 & 0 & 0 \end{bmatrix}$$

$$\mathbf{M}\{\sigma_v(2)\} = \begin{bmatrix} 0 & 0 & 0 & 0 & 1 & 0 \\ 0 & 0 & 0 & 1 & 0 & 0 \\ 0 & 0 & 0 & 0 & 0 & 1 \\ 0 & 1 & 0 & 0 & 0 & 0 \\ 1 & 0 & 0 & 0 & 0 & 0 \\ 0 & 0 & 1 & 0 & 0 & 0 \end{bmatrix}, \quad \mathbf{M}\{\sigma_v(3)\} = \begin{bmatrix} 0 & 0 & 0 & 0 & 0 & 1 \\ 0 & 0 & 0 & 0 & 1 & 0 \\ 0 & 0 & 0 & 1 & 0 & 0 \\ 0 & 0 & 1 & 0 & 0 & 0 \\ 0 & 1 & 0 & 0 & 0 & 0 \\ 1 & 0 & 0 & 0 & 0 & 0 \end{bmatrix}$$

It can be verified that these matrices multiply according to the symmetry multiplication table of C_{3v}. Thus, $\hat{\sigma}_v(1)\hat{C}_3 = \hat{\sigma}_v(2)$ and

$$\mathbf{M}\{\hat{\sigma}_v(1)\}\,\mathbf{M}(\hat{C}_3) = \begin{bmatrix} 0 & 0 & 0 & 1 & 0 & 0 \\ 0 & 0 & 0 & 0 & 0 & 1 \\ 0 & 0 & 0 & 0 & 1 & 0 \\ 1 & 0 & 0 & 0 & 0 & 0 \\ 0 & 0 & 1 & 0 & 0 & 0 \\ 0 & 1 & 0 & 0 & 0 & 0 \end{bmatrix} \begin{bmatrix} 0 & 1 & 0 & 0 & 0 & 0 \\ 0 & 0 & 1 & 0 & 0 & 0 \\ 1 & 0 & 0 & 0 & 0 & 0 \\ 0 & 0 & 0 & 0 & 1 & 0 \\ 0 & 0 & 0 & 0 & 0 & 1 \\ 0 & 0 & 0 & 1 & 0 & 0 \end{bmatrix}$$

$$= \begin{bmatrix} 0 & 0 & 0 & 0 & 1 & 0 \\ 0 & 0 & 0 & 1 & 0 & 0 \\ 0 & 0 & 0 & 0 & 0 & 1 \\ 0 & 1 & 0 & 0 & 0 & 0 \\ 1 & 0 & 0 & 0 & 0 & 0 \\ 0 & 0 & 1 & 0 & 0 & 0 \end{bmatrix} = \mathbf{M}\{\hat{\sigma}_v(2)\}$$

Hence the above set of matrices really forms a representation of C_{3v}. The dimension of Γ is 6, which is also equal to the character of $\hat{E}$ in Γ. It is also evident that the characters of all the other symmetry operators in Γ is zero, that is, $\chi(\hat{R}) = 0$ for all $\hat{R} \neq \hat{E}$. The set of characters in Γ is as follows

	$\hat{E}$	$2\hat{C}_3$[that is, $\hat{C}_3$, $\hat{C}_3^2$]	$3\,\hat{\sigma}_v$[that is, $\hat{\sigma}_v(1)$, $\hat{\sigma}_v(2)$, $\hat{\sigma}_v(3)$]
Γ	6	0	0

Let $\Gamma = n_1A_1 + n_2A_2 + n_3E$. Using the reduction formula we find that $n_1 = 1$, $n_2 = 1$, $n_3 = 2$. Thus Γ contains each irreducible representation a number of times equal to the dimensionality of the irreducible representation. This result can be generalised for any point group as discussed below.

Let $G = \{\hat{E}, \hat{R}_2, \hat{R}_3, \ldots, \hat{R}_g\}$ be a group of order g. Its multiplication table (T) is constructed according to the rearrangement theorem, by taking the operator $\hat{R}_i$ as the first operator and $\hat{R}_i^{-1}$ as the second:

First operation ⟶

Second operation ↓

T =		$\hat{E}$	$\hat{R}_2$	$\hat{R}_3$	...	...	$\hat{R}_g$
	$\hat{E}^{-1} \equiv \hat{E}$	$\hat{E}$	$\hat{R}_1$	$\hat{R}_2$	...	...	$\hat{R}_g$
	$\hat{R}_2^{-1}$	$\hat{R}_2^{-1}$	$\hat{E}$	...	...	...	...
	$\hat{R}_3^{-1}$	$\hat{R}_3^{-1}$	...	$\hat{E}$	...	...	...
	...	...	...	...	...	...	...
	...	...	...	...	...	...	...
	$\hat{R}_g^{-1}$	$\hat{R}_g^{-1}$	...	...	...	...	$\hat{E}$

(The blank places are filled up according to the rearrangement theorem.)

The representative matrix of any operator $\hat{R}_i$ is now constructed by putting 1 at places where $\hat{R}_i$ appears and 0 at all other places in the multiplication table (T). The set of matrices so constructed is a representation of G if

$$\mathbf{M}(\hat{R}_i)\mathbf{M}(\hat{R}_j) = \mathbf{M}(\hat{R}_k) \quad \text{whenever } \hat{R}_i\hat{R}_j = \hat{R}_k$$

for any $\hat{R}_i$, $\hat{R}_j$, $\hat{R}_k$ belonging to G.

The (α, β) element of the product matrix in the LHS is

$$[\mathbf{M}(\hat{R}_i)\mathbf{M}(\hat{R}_j)]_{\alpha\beta} = \sum_{\lambda}[\mathbf{M}(\hat{R}_k)]_{\alpha\lambda}[\mathbf{M}(\hat{R}_k)]_{\lambda\beta} \tag{A2.1}$$

Now, the λth column of the group multiplication table (T) contains $\hat{R}_i$ only at one position according to the rearrangement theorem; let this position be the intersection of the αth row and λth column. Then according to the method of construction of the matrices, the element at the (α, λ) position of $\mathbf{M}(\hat{R}_i)$ is unity and zero at the other places in the λth column.

$$[\mathbf{M}(\hat{R}_i)]_{\alpha\lambda} = \begin{cases} 1, \text{ if the } (\alpha, \lambda) \text{ position of T contains } \hat{R}_i \\ 0, \text{ otherwise} \end{cases} \tag{A2.2}$$

$$\text{Similarly,} \quad [\mathbf{M}(\hat{R}_j)]_{\lambda\beta} = \begin{cases} 1, \text{ if the } (\lambda, \beta) \text{ position of T contains } \hat{R}_j \\ 0, \text{ otherwise} \end{cases} \tag{A2.3}$$

Then from Eq. (A2.1)

$$[\mathbf{M}(\hat{R}_i)\,\mathbf{M}(\hat{R}_j)]_{\alpha\beta} = [\mathbf{M}(\hat{R}_k)]_{\alpha 1}[\mathbf{M}(\hat{R}_k)]_{1\beta} + [\mathbf{M}(\hat{R}_k)]_{\alpha 2}[\mathbf{M}(\hat{R}_k)]_{2\beta} + \cdots + [\mathbf{M}(\hat{R}_k)]_{\alpha\lambda}[\mathbf{M}(\hat{R}_i)]_{\lambda\beta} + \cdots$$

$$= \begin{cases} 0 + \cdots + 0 + 1 + 0 + \cdots, \text{ if T contains } \hat{R}_i \text{ at the } (\alpha, \lambda) \\ \qquad \text{position and simultaneously} \\ \qquad \hat{R}_j \text{ at the } (\lambda, \beta) \text{ position} \\ = 0, \text{ otherwise} \end{cases} \tag{A2.4}$$

The column headings of T contain the operators $\hat{R}_i$ and the rows of T are labelled by the inverses $\hat{R}_i^{-1}$. Hence the result (A2.4) means that $[\mathbf{M}(\hat{R}_i)\,\mathbf{M}(\hat{R}_j)]_{\alpha\beta} = 1$ is possible only if

$$\hat{R}_\alpha^{-1}\hat{R}_\lambda = \hat{R}_i \text{ and } \hat{R}_\lambda^{-1}\hat{R}_\beta = \hat{R}_j \text{ simultaneously,}$$

That is, $\hat{R}_i\hat{R}_j = \hat{R}_\alpha^{-1}\hat{R}_\lambda\hat{R}_\lambda^{-1}\hat{R}_\beta = \hat{R}_\alpha^{-1}\hat{R}_\beta$

or, $\hat{R}_k = \hat{R}_\alpha^{-1}\hat{R}_\beta$ that is, the (α, β) position of T contains $\hat{R}_k$

So $[\mathbf{M}(\hat{R}_k)]_{\alpha\beta} = 1$ [if the first conditions of Eqs (A2.2) and (A2.3) are satisfied]

$= 0$ otherwise

Hence $\mathbf{M}(\hat{R}_i)\,\mathbf{M}(\hat{R}_j) = \mathbf{M}(\hat{R}_k)$ implies $\hat{R}_i\hat{R}_j = \hat{R}_k$ (Proved).

According to construction, the character set of the regular representation of a general point group $G = \{\hat{E}, \hat{R}_2, \hat{R}_3, \ldots, \hat{R}_g\}$ of order g is:

	$\hat{E}$	$\hat{R}_2$	$\hat{R}_3$	...	$\hat{R}_g$
Γ	g	0	0	...	0

Let the possible irreducible representations of G be D^1 of dimension l_1, D^2 of dimension l_2, …, D^c of dimension l_c, where c is the possible number of conjugate classes in G. The dimension of each irreducible representation is equal to the character of $\hat{E}$ in that representation:

$$\chi^i(\hat{E}) = l_i \text{ for each } i \text{ from 1 to } c$$

Let $\Gamma = n_1D^1 + n_2D^2 + \cdots + n_cD^c$

Then by the reduction formula,

$$n_i = \frac{1}{g}\sum_{\hat{R}}[\chi^j(\hat{R})]^*\chi^\Gamma(\hat{R})$$

$$= \frac{1}{g}\{[\chi^j(\hat{E})]^*\chi^\Gamma(\hat{E}) + [\chi^j(\hat{R}_2)]^*\chi^\Gamma(\hat{R}_2) + \cdots + [\chi^j(\hat{R}_g)]^*\chi^\Gamma(\hat{R}_g)\}$$

$$= \frac{1}{g}\{g\cdot l_i + 0 + \cdots + 0\} = l_i$$

Thus, in the representation Γ each irreducible representation (D^i) occurs, and the number of times it occurs is equal to its dimension (l_i). This result is known as the *Celebrated theorem* and justifies the name *regular representation* for Γ.

Then the matrix for the identity operator in Γ must contain as block factors the matrix for $\hat{E}$ in D^1 occurring l_1 times, matrix for $\hat{E}$ in D^2 occurring l_2 times and so on.

$$\mathbf{M}^\Gamma(\hat{E}) = \mathbf{M}^1(\hat{E}) \oplus \mathbf{M}^1(\hat{E}) \oplus \cdots l_1 \text{ times} \oplus \mathbf{M}^2(\hat{E}) \oplus \mathbf{M}^2(\hat{E}) \oplus \cdots l_2 \text{ times}$$

$$\cdots \oplus \mathbf{M}^c(\hat{E}) \oplus \mathbf{M}^c(\hat{E}) \oplus \cdots l_c \text{ times}$$

Taking traces of both sides,

$$\chi^{\Gamma}(\hat{E}) = \chi^{1}(\hat{E}) + \chi^{1}(\hat{E}) + \cdots\ l_1 \text{ times} + \chi^{2}(\hat{E}) + \chi^{2}(\hat{E}) + \cdots\ l_2 \text{ times}$$

$$\cdots + \chi^{c}(\hat{E}) + \chi^{c}(\hat{E}) + \cdots\ l_c \text{ times}$$

or, $$\chi^{\Gamma}(\hat{E}) = l_1\chi^{1}(\hat{E}) + l_2\chi^{2}(\hat{E}) + \cdots + l_c\chi^{c}(\hat{E})$$

or, $$g = l_1^2 + l_2^2 + \cdots + l_c^2$$

which completes the proof of the result of Section 5.4.1.

Character Tables of Some Important Point Groups

1. The Nonaxial Groups

C_1	$\hat{E}$
A	1

C_s	$\hat{E}$	$\hat{\sigma}_h$		
A'	1	1	x, y, R_z	x^2, y^2, z^2, xy
A''	1	−1	z, R_x, R_y	xz, yz

C_i	$\hat{E}$	$\hat{i}$		
A_g	1	1	R_x, R_y, R_z	$x^2, y^2, z^2, xy, xz, yz$
A_u	1	−1	x, y, z	

2. The C_n Groups

C_2	$\hat{E}$	$\hat{C}_2$		
A	1	1	z, R_z	x^2, y^2, z^2, xy
B	1	−1	x, y, R_x, R_y	xz, yz

C_3	$\hat{E}$	$\hat{C}_2$	$\hat{C}_3^2$		$\varepsilon = \exp(2\pi i/3)$
A	1	1	1	z, R_z	$x^2 + y^2, z^2$
E	$\begin{Bmatrix} 1 \\ 1 \end{Bmatrix}$	$\begin{matrix} \varepsilon \\ \varepsilon^* \end{matrix}$	$\begin{matrix} \varepsilon^* \\ \varepsilon \end{matrix}$	$(x, y)(R_x, R_y)$	$(x^2 - y^2, xy)(xz, yz)$

C_4	$\hat{E}$	$\hat{C}_4$	$\hat{C}_2$	$\hat{C}_4^3$		
A	1	1	1	1	z, R_z	$x^2 + y^2, z^2$
B	1	−1	1	−1		$x^2 - y^2, xy$
E	$\begin{Bmatrix} 1 \\ 1 \end{Bmatrix}$	$\begin{matrix} i \\ -i \end{matrix}$	$\begin{matrix} -1 \\ -1 \end{matrix}$	$\begin{matrix} -i \\ i \end{matrix}$	$(x, y)(R_x, R_y)$	(xz, yz)

C_5	$\hat{E}$	$\hat{C}_5$	$\hat{C}_5^2$	$\hat{C}_5^3$	$\hat{C}_5^4$		$\varepsilon = \exp(2\pi i/5)$
A	1	1	1	1	1	z, R_z	$x^2 + y^2, z^2$
E_1	$\begin{Bmatrix} 1 \\ 1 \end{Bmatrix}$	$\begin{matrix} \varepsilon \\ \varepsilon^* \end{matrix}$	$\begin{matrix} \varepsilon^2 \\ \varepsilon^{2*} \end{matrix}$	$\begin{matrix} \varepsilon^{2*} \\ \varepsilon^2 \end{matrix}$	$\begin{matrix} \varepsilon^* \\ \varepsilon \end{matrix}$	$(x, y)(R_x, R_y)$	(xz, yz)
E_2	$\begin{Bmatrix} 1 \\ 1 \end{Bmatrix}$	$\begin{matrix} \varepsilon^2 \\ \varepsilon^{2*} \end{matrix}$	$\begin{matrix} \varepsilon^* \\ \varepsilon \end{matrix}$	$\begin{matrix} \varepsilon \\ \varepsilon^* \end{matrix}$	$\begin{matrix} \varepsilon^{2*} \\ \varepsilon^2 \end{matrix}$		$(x^2 - y^2, xy)$

C_6	$\hat{E}$	$\hat{C}_6$	$\hat{C}_3$	$\hat{C}_2$	$\hat{C}_3^2$	$\hat{C}_6^5$		$\varepsilon = \exp(2\pi i/6)$
A	1	1	1	1	1	1	z, R_z	$x^2 + y^2, z^2$
B	1	−1	1	−1	1	−1		
E_1	$\begin{Bmatrix} 1 \\ 1 \end{Bmatrix}$	$\begin{matrix} \varepsilon \\ \varepsilon^* \end{matrix}$	$\begin{matrix} -\varepsilon^* \\ -\varepsilon \end{matrix}$	$\begin{matrix} -1 \\ -1 \end{matrix}$	$\begin{matrix} -\varepsilon \\ -\varepsilon^* \end{matrix}$	$\begin{matrix} \varepsilon^* \\ \varepsilon \end{matrix}$	$(x, y)(R_x, R_y)$	(xz, yz)
E_2	$\begin{Bmatrix} 1 \\ 1 \end{Bmatrix}$	$\begin{matrix} -\varepsilon^* \\ -\varepsilon \end{matrix}$	$\begin{matrix} -\varepsilon \\ -\varepsilon^* \end{matrix}$	$\begin{matrix} 1 \\ 1 \end{matrix}$	$\begin{matrix} -\varepsilon^* \\ -\varepsilon \end{matrix}$	$\begin{matrix} -\varepsilon \\ -\varepsilon^* \end{matrix}$		$(x^2 - y^2, xy)$

C_7	$\hat{E}$	$\hat{C}_7$	$\hat{C}_7^2$	$\hat{C}_7^3$	$\hat{C}_7^4$	$\hat{C}_7^5$	$\hat{C}_7^6$		$\varepsilon = \exp(2\pi i/7)$
A	1	1	1	1	1	1	1	z, R_z	x^2+y^2, z^2
E_1	1	ε	ε^2	ε^3	ε^{3*}	ε^{2*}	ε^*	$(x, y)(R_x, R_y)$	(xz, yz)
	1	ε^*	ε^{2*}	ε^{3*}	ε^3	ε^2	ε		
E_2	1	ε^2	ε^{3*}	ε^*	ε	ε^3	ε^{2*}		x^2-y^2, xy
	1	ε^{2*}	ε^3	ε	ε^*	ε^{3*}	ε^2		
E_3	1	ε^3	ε^*	ε^2	ε^{2*}	ε	ε^{3*}		
	1	ε^{3*}	ε	ε^{2*}	ε^2	ε^*	ε^3		

C_8	$\hat{E}$	$\hat{C}_8$	$\hat{C}_4$	$\hat{C}_8^3$	$\hat{C}_2$	$\hat{C}_8^5$	$\hat{C}_4^3$	$\hat{C}_8^7$		$\varepsilon = \exp(2\pi i/8)$
A	1	1	1	1	1	1	1	1	z, R_z	x^2+y^2, z^2
B	1	−1	1	−1	1	−1	1	−1		
E_1	1	ε	i	$-\varepsilon^*$	−1	$-\varepsilon$	$-i$	ε^*	$(x, y)(R_x, R_y)$	(xz, yz)
	1	ε^*	$-i$	$-\varepsilon$	−1	$-\varepsilon^*$	i	ε		
E_2	1	i	−1	$-i$	1	i	−1	$-i$		x^2-y^2, xy
	1	$-i$	−1	i	1	$-i$	−1	i		
E_3	1	$-\varepsilon$	i	ε^*	−1	ε	$-i$	$-\varepsilon^*$		
	1	$-\varepsilon^*$	$-i$	ε	−1	ε^*	i	$-\varepsilon$		

3. The D_n Groups

D_2	$\hat{E}$	$\hat{C}_2(z)$	$\hat{C}_2(y)$	$\hat{C}_2(x)$		
A	1	1	1	1		x^2, y^2, z^2
B_1	1	1	−1	−1	z, R_z	xy
B_2	1	−1	1	−1	y, R_y	xz
B_3	1	−1	−1	1	x, R_x	yz

D_3	$\hat{E}$	$2\hat{C}_3$	$3\hat{C}_2$		
A_1	1	1	1		x^2+y^2, z^2
A_2	1	1	−1	z, R_z	
E	2	−1	0	$(x, y)(R_x, R_y)$	$(x^2-y^2, xy)(xz, yz)$

D_4	$\hat{E}$	$2\hat{C}_4$	$\hat{C}_2 = \hat{C}_4^2$	$2\hat{C}_2'$	$2\hat{C}_2''$		
A_1	1	1	1	1	1		x^2+y^2, z^2
A_2	1	1	1	−1	−1	z, R_z	
B_1	1	−1	1	1	−1		x^2-y^2
B_2	1	−1	1	−1	1		xy
E	2	0	−2	0	0	$(x, y)(R_x, R_y)$	(xz, yz)

D_5	$\hat{E}$	$2\hat{C}_5$	$2\hat{C}_5^2$	$5\hat{C}_2$		
A_1	1	1	1	1		x^2+y^2, z^2
A_2	1	1	1	−1	z, R_z	
E_1	2	2cos 72°	2cos 144°	0	$(x, y)(R_x, R_y)$	(xz, yz)
E_2	2	2cos 144°	2cos 72°	0		(x^2-y^2, xy)

D_6	$\hat{E}$	$2\hat{C}_6$	$2\hat{C}_3$	$\hat{C}_2$	$3\hat{C}_2'$	$3\hat{C}_2''$		
A_1	1	1	1	1	1	1		x^2+y^2, z^2
A_2	1	1	1	1	−1	−1	z, R_z	
B_1	1	−1	1	−1	1	−1		
B_2	1	−1	1	−1	−1	1		
E_1	2	1	−1	−2	0	0	$(x, y)(R_x, R_y)$	(xz, yz)
E_2	2	−1	−1	2	0	0		(x^2-y^2, xy)

4. The C_{nv} Groups

C_{2v}	$\hat{E}$	$\hat{C}_2$	$\hat{\sigma}_v(xz)$	$\hat{\sigma}_v(yz)$		
A_1	1	1	1	1	z	x^2+y^2, z^2
A_2	1	1	−1	−1	R_z	xy
B_1	1	−1	1	−1	x, R_y	xz
B_2	1	−1	−1	1	y, R_x	yz

C_{3v}	$\hat{E}$	$2\hat{C}_3$	$3\hat{\sigma}_v$		
A_1	1	1	1	z	x^2+y^2, z^2
A_2	1	1	−1	R_z	
E	2	−1	0	$(x, y)(R_x, R_y)$	$(x^2-y^2, xy)(xz, yz)$

C_{4v}	$\hat{E}$	$2\hat{C}_4$	$\hat{C}_2$	$2\hat{\sigma}_v$	$2\hat{\sigma}_d$		
A_1	1	1	1	1	1	z	x^2+y^2, z^2
A_2	1	1	1	−1	−1	R_z	
B_1	1	−1	1	1	−1		x^2-y^2
B_2	1	−1	1	−1	1		xy
E	2	0	−2	0	0	$(x, y)(R_x, R_y)$	(xz, yz)

C_{5v}	$\hat{E}$	$2\hat{C}_5$	$2\hat{C}_5^2$	$5\hat{\sigma}_v$		
A_1	1	1	1	1	z	x^2+y^2, z^2
A_2	1	1	1	−1	R_z	
E_1	2	2cos 72°	2cos 144°	0	$(x, y)(R_x, R_y)$	(xz, yz)
E_2	2	2cos 144°	2cos 72°	0		(x^2-y^2, xy)

C_{6v}	$\hat{E}$	$2\hat{C}_6$	$2\hat{C}_3$	$\hat{C}_2$	$3\hat{\sigma}_v$	$3\hat{\sigma}_d$		
A_1	1	1	1	1	1	1	z	x^2+y^2, z^2
A_2	1	1	1	1	−1	−1	R_z	
B_1	1	−1	1	−1	1	−1		
B_2	1	−1	1	−1	−1	1		
E_1	2	1	−1	−2	0	0	$(x, y)(R_x, R_y)$	(xz, yz)
E_2	2	−1	−1	2	0	0		(x^2-y^2, xy)

5. The C_{nh} Groups

C_{2h}	$\hat{E}$	$\hat{C}_2$	$\hat{i}$	$\hat{\sigma}_h$		
A_g	1	1	1	1	R_z	x^2, y^2, z^2, xy
B_g	1	−1	1	−1	R_x, R_y	xz, yz
A_u	1	1	−1	−1	z	
B_u	1	−1	−1	1	x, y	

C_{3h}	$\hat{E}$	$\hat{C}_3$	$\hat{C}_3^2$	$\hat{\sigma}_h$	$\hat{S}_3$	$\hat{S}_3^5$		$\varepsilon=\exp(2\pi i/3)$
A'	1	1	1	1	1	1	R_z	x^2+y^2, z^2
E'	1	ε	ε^*	1	ε	ε^*	(x, y)	(x^2-y^2, xy)
	1	ε^*	ε	1	ε^*	ε		
A''	1	1	1	−1	−1	−1	z	
E''	1	ε	ε^*	−1	$-\varepsilon$	$-\varepsilon^*$	(R_x, R_y)	(xz, yz)
	1	ε^*	ε	−1	$-\varepsilon^*$	$-\varepsilon$		

C_{4h}	$\hat{E}$	$\hat{C}_4$	$\hat{C}_2$	$\hat{C}_4^3$	$\hat{i}$	$\hat{S}_4^3$	$\hat{\sigma}_h$	$\hat{S}_4$		
A_g	1	1	1	1	1	1	1	1	R_z	x^2+y^2, z^2
B_g	1	−1	1	−1	1	−1	1	−1		(x^2-y^2, xy)
E_g	1	i	−1	$-i$	1	i	−1	$-i$	(R_x, R_y)	(xz, yz)
	1	$-i$	−1	i	1	$-i$	−1	i		
A_u	1	1	1	1	−1	−1	−1	−1	z	
B_u	1	−1	1	−1	−1	1	−1	1		
E_u	1	i	−1	$-i$	−1	$-i$	1	i	(x, y)	
	1	$-i$	−1	i	−1	i	1	$-i$		

C_{5h}	$\hat{E}$	$\hat{C}_5$	$\hat{C}_5^2$	$\hat{C}_5^3$	$\hat{C}_5^4$	$\hat{\sigma}_h$	$\hat{S}_5$	$\hat{S}_5^7$	$\hat{S}_5^3$	$\hat{S}_5^9$		$\varepsilon=\exp(2\pi i/5)$
A'	1	1	1	1	1	1	1	1	1	1	R_z	x^2+y^2, z^2
E_1'	1	ε	ε^2	ε^{2*}	ε^*	1	ε	ε^2	ε^{2*}	ε^*	(x, y)	
	1	ε^*	ε^{2*}	ε^2	ε	1	ε^*	ε^{2*}	ε^2	ε		
E_2'	1	ε^2	ε^*	ε	ε^{2*}	1	ε^2	ε^*	ε	ε^{2*}		(x^2-y^2, xy)
	1	ε^{2*}	ε	ε^*	ε^2	1	ε^{2*}	ε	ε^*	ε^2		

(Continued)

Continued

C_{5h}	$\hat{E}$	$\hat{C}_5$	$\hat{C}_5^2$	$\hat{C}_5^3$	$\hat{C}_5^4$	$\hat{\sigma}_h$	$\hat{S}_5$	$\hat{S}_5^7$	$\hat{S}_5^3$	$\hat{S}_5^9$		$\varepsilon = \exp(2\pi i/5)$
A''	1	1	1	1	1	−1	−1	−1	−1	−1	z	
E_1''	1	ε	ε^2	ε^{2*}	ε^*	−1	$-\varepsilon$	$-\varepsilon^2$	$-\varepsilon^{2*}$	$-\varepsilon^*$	(R_x, R_y)	(xz, yz)
	1	ε^*	ε^{2*}	ε^2	ε	−1	$-\varepsilon^*$	$-\varepsilon^{2*}$	$-\varepsilon^2$	$-\varepsilon$		
E_2''	1	ε^2	ε^*	ε	ε^{2*}	−1	$-\varepsilon^2$	$-\varepsilon^*$	$-\varepsilon$	$-\varepsilon^{2*}$		
	1	ε^{2*}	ε	ε^*	ε^2	−1	$-\varepsilon^{2*}$	$-\varepsilon$	$-\varepsilon^*$	$-\varepsilon^2$		

C_{6h}	$\hat{E}$	$\hat{C}_6$	$\hat{C}_3$	$\hat{C}_2$	$\hat{C}_3^2$	$\hat{C}_6^5$	$\hat{i}$	$\hat{S}_3^5$	$\hat{S}_6^5$	$\hat{\sigma}_h$	$\hat{S}_6$	$\hat{S}_3$		$\varepsilon = \exp(2\pi i/6)$
A_g	1	1	1	1	1	1	1	1	1	1	1	1	R_z	$x^2 + y^2, z^2$
B_g	1	−1	1	−1	1	−1	1	−1	1	−1	1	−1		
E_{1g}	1	ε	$-\varepsilon^*$	−1	$-\varepsilon$	ε^*	1	ε	$-\varepsilon^*$	−1	$-\varepsilon$	ε^*	(R_x, R_y)	(xz, yz)
	1	ε^*	$-\varepsilon$	−1	$-\varepsilon^*$	ε	1	ε^*	$-\varepsilon$	−1	$-\varepsilon^*$	ε		
E_{2g}	1	$-\varepsilon^*$	$-\varepsilon$	1	$-\varepsilon^*$	$-\varepsilon$	1	ε^*	$-\varepsilon$	1	$-\varepsilon^*$	$-\varepsilon$		$(x^2 - y^2, xy)$
	1	$-\varepsilon$	$-\varepsilon^*$	1	$-\varepsilon$	$-\varepsilon^*$	1	$-\varepsilon$	$-\varepsilon^*$	1	$-\varepsilon$	$-\varepsilon^*$		
A_u	1	1	1	1	1	1	−1	−1	−1	−1	−1	−1	z	
B_u	1	−1	1	−1	1	−1	−1	1	−1	1	−1	1		
E_{1u}	1	ε	$-\varepsilon^*$	−1	$-\varepsilon$	ε^*	−1	$-\varepsilon$	ε^*	1	ε	$-\varepsilon^*$	(x, y)	
	1	ε^*	$-\varepsilon$	−1	$-\varepsilon^*$	ε	−1	$-\varepsilon^*$	ε	1	ε^*	$-\varepsilon$		
E_{2u}	1	$-\varepsilon^*$	$-\varepsilon$	1	$-\varepsilon^*$	$-\varepsilon$	−1	ε^*	ε	−1	ε^*	ε		
	1	$-\varepsilon$	$-\varepsilon^*$	1	$-\varepsilon$	$-\varepsilon^*$	−1	ε	ε^*	−1	ε	ε^*		

6. The D_{nh} Groups

D_{2h}	$\hat{E}$	$\hat{C}_2(z)$	$\hat{C}_2(y)$	$\hat{C}_2(x)$	$\hat{i}$	$\hat{\sigma}(xy)$	$\hat{\sigma}(zx)$	$\hat{\sigma}(yz)$		
A_g	1	1	1	1	1	1	1	1		x^2, y^2, z^2
B_{1g}	1	1	−1	−1	1	1	−1	−1	R_z	xy
B_{2g}	1	−1	1	−1	1	−1	1	−1	R_y	xz
B_{3g}	1	−1	−1	1	1	−1	−1	1	R_x	yz
A_u	1	1	1	1	−1	−1	−1	−1		
B_{1u}	1	1	−1	−1	−1	−1	1	1	z	
B_{2u}	1	−1	1	−1	−1	1	−1	1	y	
B_{3u}	1	−1	−1	1	−1	1	1	−1	x	

D_{3h}	$\hat{E}$	$2\hat{C}_3$	$3\hat{C}_2$	$\hat{\sigma}_h$	$2\hat{S}_3$	$3\hat{\sigma}_v$		
A_1'	1	1	1	1	1	1		$x^2 + y^2, z^2$
A_2'	1	1	−1	1	1	−1	R_z	
E'	2	−1	0	2	−1	0	(x, y)	$(x^2 - y^2, xy)$
A_1''	1	1	1	−1	−1	−1		
A_2''	1	1	−1	−1	−1	1	z	
E''	2	−1	0	−2	1	0	(R_x, R_y)	(xz, yz)

D_{4h}	$\hat{E}$	$2\hat{C}_4$	$\hat{C}_2$	$2\hat{C}_2'$	$2\hat{C}_2''$	$\hat{i}$	$2\hat{S}_4$	$\hat{\sigma}_h$	$2\hat{\sigma}_v$	$2\hat{\sigma}_d$		
A_{1g}	1	1	1	1	1	1	1	1	1	1		x^2+y^2, z^2
A_{2g}	1	1	1	−1	−1	1	1	1	−1	−1	R_z	
B_{1g}	1	−1	1	1	−1	1	−1	1	1	−1		x^2-y^2
B_{2g}	1	−1	1	−1	1	1	−1	1	−1	1		xy
E_g	2	0	−2	0	0	2	0	−2	0	0	(R_x, R_y)	(xz, yz)
A_{1u}	1	1	1	1	1	−1	−1	−1	−1	−1		
A_{2u}	1	1	1	−1	−1	−1	−1	−1	1	1	z	
B_{1u}	1	−1	1	1	−1	−1	1	−1	−1	1		
B_{2u}	1	−1	1	−1	1	−1	1	−1	1	−1		
E_u	2	0	−2	0	0	−2	0	2	0	0	(x, y)	

D_{5h}	$\hat{E}$	$2\hat{C}_5$	$2\hat{C}_5^2$	$5\hat{C}_2$	$\hat{\sigma}_h$	$2\hat{S}_5$	$2\hat{S}_5^3$	$5\hat{\sigma}_v$		
A_1'	1	1	1	1	1	1	1	1		x^2+y^2, z^2
A_2'	1	1	1	−1	1	1	1	−1	R_z	
E_1'	2	2cos 72°	2cos 144°	0	2	2cos 72°	2cos 144°	0	(x, y)	
E_2'	2	2cos 144°	2cos 72°	0	2	2cos 144°	2cos 72°	0		(x^2-y^2, xy)
A_1''	1	1	1	1	−1	−1	−1	−1		
A_2''	1	1	1	−1	−1	−1	−1	1	z	
E_1''	2	2cos 72°	2cos 144°	0	−2	−2cos 72°	−2cos 144°	0	(R_x, R_y)	(xz, yz)
E_2''	2	2cos 144°	2cos 72°	0	−2	−2cos 144°	−2cos 72°	0		

D_{6h}	$\hat{E}$	$2\hat{C}_6$	$2\hat{C}_3$	$\hat{C}_2$	$3\hat{C}_2'$	$3\hat{C}_2''$	$\hat{i}$	$2\hat{S}_3$	$2\hat{S}_6$	$\hat{\sigma}_h$	$3\hat{\sigma}_d$	$3\hat{\sigma}_v$		
A_{1g}	1	1	1	1	1	1	1	1	1	1	1	1		x^2+y^2, z^2
A_{2g}	1	1	1	1	−1	−1	1	1	1	1	−1	−1	R_z	
B_{1g}	1	−1	1	−1	1	−1	1	−1	1	−1	1	−1		
B_{2g}	1	−1	1	−1	−1	1	1	−1	1	−1	−1	1		
E_{1g}	2	1	−1	−2	0	0	2	1	−1	−2	0	0	(R_x, R_y)	(xz, yz)
E_{2g}	2	−1	−1	2	0	0	2	−1	−1	2	0	0		(x^2-y^2, xy)
A_{1u}	1	1	1	1	1	1	−1	−1	−1	−1	−1	−1		
A_{2u}	1	1	1	1	−1	−1	−1	−1	−1	−1	1	1	z	
B_{1u}	1	−1	1	−1	1	−1	−1	1	−1	1	−1	1		
B_{2u}	1	−1	1	−1	−1	1	−1	1	−1	1	1	−1		
E_{1u}	2	1	−1	−2	0	0	−2	−1	1	2	0	0	(x, y)	
E_{2u}	2	−1	−1	2	0	0	−2	1	1	−2	0	0		

D_{8h}	$\hat{E}$	$2\hat{C}_8$	$2\hat{C}_8^3$	$2\hat{C}_4$	$\hat{C}_2$	$4\hat{C}_2'$	$4\hat{C}_2''$	$\hat{i}$	$2\hat{S}_8$	$2\hat{S}_8^3$	$2\hat{S}_4$	$\hat{\sigma}_h$	$4\hat{\sigma}_d$	$4\hat{\sigma}_v$		
A_{1g}	1	1	1	1	1	1	1	1	1	1	1	1	1	1		x^2+y^2, z^2
A_{2g}	1	1	1	1	1	−1	−1	1	1	1	1	1	−1	−1	R_z	

(Continued)

Continued

D_{8h}	$\hat{E}$	$2\hat{C}_8$	$2\hat{C}_8^3$	$2\hat{C}_4$	$\hat{C}_2$	$4\hat{C}_2'$	$4\hat{C}_2''$	$\hat{i}$	$2\hat{S}_8$	$2\hat{S}_8^3$	$2\hat{S}_4$	$\hat{\sigma}_h$	$4\hat{\sigma}_d$	$4\hat{\sigma}_v$		
B_{1g}	1	−1	−1	1	1	1	−1	1	−1	−1	1	1	1	−1		
B_{2g}	1	−1	−1	1	1	−1	1	1	−1	−1	1	1	−1	1		
E_{1g}	2	$\sqrt{2}$	$-\sqrt{2}$	0	−2	0	0	2	$\sqrt{2}$	$-\sqrt{2}$	0	−2	0	0	(R_x, R_y)	(xz, yz)
E_{2g}	2	0	0	−2	2	0	0	2	0	0	−2	2	0	0		(x^2-y^2, xy)
E_{3g}	2	$-\sqrt{2}$	$\sqrt{2}$	0	−2	0	0	2	$-\sqrt{2}$	$\sqrt{2}$	0	−2	0	0		
A_{1u}	1	1	1	1	1	1	1	−1	−1	−1	−1	−1	−1	−1		
A_{2u}	1	1	1	1	1	−1	−1	−1	−1	−1	−1	−1	1	1	z	
B_{1u}	1	−1	−1	1	1	1	−1	−1	1	1	−1	−1	−1	1		
B_{2u}	1	−1	−1	1	1	−1	1	−1	1	1	−1	−1	1	−1		
E_{1u}	2	$\sqrt{2}$	$-\sqrt{2}$	0	−2	0	0	−2	$-\sqrt{2}$	$\sqrt{2}$	0	−2	0	0	(x, y)	
E_{2u}	2	0	0	−2	2	0	0	−2	0	0	2	2	0	0		
E_{3u}	2	$-\sqrt{2}$	$\sqrt{2}$	0	−2	0	0	−2	$\sqrt{2}$	$-\sqrt{2}$	0	−2	0	0		

7. The D_{nd} Groups

D_{2d}	$\hat{E}$	$\hat{C}_2$	$2\hat{C}_2'$	$2\hat{S}_4$	$2\hat{\sigma}_d$		
A_1	1	1	1	1	1		x^2+y^2, z^2
A_2	1	1	−1	1	−1	R_z	
B_1	1	1	1	−1	−1		x^2-y^2
B_2	1	1	−1	−1	1	z	xy
E	2	−2	0	0	0	$(x, y); (R_x, R_y)$	(xz, yz)

D_{3d}	$\hat{E}$	$2\hat{C}_3$	$3\hat{C}_2$	$\hat{i}$	$2\hat{S}_6$	$3\hat{\sigma}_d$		
A_{1g}	1	1	1	1	1	1		x^2+y^2, z^2
A_{2g}	1	1	−1	1	1	−1	R_z	
E_g	2	−1	0	2	−1	0	(R_x, R_y)	$(x^2-y^2, xy), (xz, yz)$
A_{1u}	1	1	1	−1	−1	−1		
A_{2u}	1	1	−1	−1	−1	1	z	
E_u	2	−1	0	−2	1	0	(x, y)	

D_{4d}	$\hat{E}$	$2\hat{S}_8$	$2\hat{C}_4$	$\hat{C}_2$	$4\hat{C}_2'$	$2\hat{S}_8^3$	$4\hat{\sigma}_d$		
A_1	1	1	1	1	1	1	1		x^2+y^2, z^2
A_2	1	1	1	1	−1	1	−1	R_z	
B_1	1	−1	1	1	1	−1	−1		
B_2	1	−1	1	1	−1	−1	1	z	
E_1	2	$\sqrt{2}$	0	−2	0	$-\sqrt{2}$	0	(x, y)	
E_2	2	0	−2	2	0	0	0		(x^2-y^2, xy)
E_3	2	$-\sqrt{2}$	0	−2	0	$\sqrt{2}$	0	(R_x, R_y)	(xz, yz)

D_{5d}	$\hat{E}$	$2\hat{C}_5$	$2\hat{C}_5^2$	$5\hat{C}_2$	$\hat{i}$	$2\hat{S}_{10}^3$	$2\hat{S}_{10}$	$5\hat{\sigma}_d$		
A_{1g}	1	1	1	1	1	1	1	1		x^2+y^2, z^2
A_{2g}	1	1	1	−1	1	1	1	−1	R_z	
E_{1g}	2	2cos 72°	2cos 144°	0	2	2cos 72°	2cos 144°	0	(R_x, R_y)	(xz, yz)
E_{2g}	2	2cos 144°	2cos 72°	0	2	2cos 144°	2cos 72°	0		(x^2-y^2, xy)
A_{1u}	1	1	1	1	−1	−1	−1	−1		
A_{2u}	1	1	1	−1	−1	−1	−1	1	z	
E_{1u}	2	2cos 72°	2cos 144°	0	−2	−2cos 72°	−2cos 144°	0	(x, y)	
E_{2u}	2	2cos 144°	2cos 72°	0	−2	−2cos 144°	−2cos 72°	0		

D_{6d}	$\hat{E}$	$2\hat{C}_6$	$2\hat{C}_3$	$\hat{C}_2$	$6\hat{C}_2'$	$2\hat{S}_4$	$2\hat{S}_{12}$	$2\hat{S}_{12}^5$	$6\hat{\sigma}_d$		
A_1	1	1	1	1	1	1	1	1	1		x^2+y^2, z^2
A_2	1	1	1	1	−1	1	1	1	−1	R_z	
B_1	1	1	1	1	1	−1	−1	−1	−1		
B_2	1	1	1	1	−1	−1	−1	−1	1	z	
E_1	2	1	−1	−2	0	0	$\sqrt{3}$	$-\sqrt{3}$	0	(x, y)	
E_2	2	−1	−1	2	0	−2	1	1	0		(x^2-y^2, xy)
E_3	2	−2	2	−2	0	0	0	0	0		
E_4	2	−1	−1	2	0	2	−1	−1	0		
E_5	2	1	−1	−2	0	0	$-\sqrt{3}$	$\sqrt{3}$	0	(R_x, R_y)	(xz, yz)

8. The S_n Groups

S_4	$\hat{E}$	$\hat{S}_4$	$\hat{C}_2$	$\hat{S}_4^3$		
A	1	1	1	1	R_z	x^2+y^2, z^2
B	1	−1	1	−1	z	x^2-y^2, xy
E	{1	i	−1	$-i$}	$(x, y), (R_x, R_y)$	(xz, yz)
	{1	$-i$	−1	i}		

S_6	$\hat{E}$	$\hat{C}_3$	$\hat{C}_3^2$	$\hat{i}$	$\hat{S}_6^5$	$\hat{S}_6$		$\varepsilon = \exp(2\pi i/3)$
A_g	1	1	1	1	1	1	R_z	x^2+y^2, z^2
E_g	{1	ε	ε^*	1	ε	ε^*}	(R_x, R_y)	$(x^2-y^2, xy), (xz, yz)$
	{1	ε^*	ε	1	ε^*	ε}		
A_u	1	1	1	−1	−1	−1	z	
E_u	{1	ε	ε^*	−1	$-\varepsilon$	$-\varepsilon^*$}	(x, y)	
	{1	ε^*	ε	−1	$-\varepsilon^*$	$-\varepsilon$}		

S_8	$\hat{E}$	$\hat{S}_8$	$\hat{C}_4$	$\hat{S}_8^3$	$\hat{C}_2$	$\hat{S}_8^5$	$\hat{C}_4^3$	$\hat{S}_8^7$		$\varepsilon = \exp(2\pi i/8)$
A	1	1	1	1	1	1	1	1	R_z	x^2+y^2, z^2
B	1	−1	1	−1	1	−1	1	−1	z	
E_1	$\begin{cases}1\\1\end{cases}$	$\begin{matrix}\varepsilon\\\varepsilon^*\end{matrix}$	$\begin{matrix}i\\-i\end{matrix}$	$\begin{matrix}-\varepsilon^*\\-\varepsilon\end{matrix}$	$\begin{matrix}-1\\-1\end{matrix}$	$\begin{matrix}-\varepsilon\\-\varepsilon^*\end{matrix}$	$\begin{matrix}-i\\i\end{matrix}$	$\begin{matrix}\varepsilon^*\\\varepsilon\end{matrix}$	$(x, y), (R_x, R_y)$	
E_2	$\begin{cases}1\\1\end{cases}$	$\begin{matrix}i\\-i\end{matrix}$	$\begin{matrix}-1\\-1\end{matrix}$	$\begin{matrix}-i\\i\end{matrix}$	$\begin{matrix}1\\1\end{matrix}$	$\begin{matrix}i\\-i\end{matrix}$	$\begin{matrix}-1\\-1\end{matrix}$	$\begin{matrix}-i\\i\end{matrix}$		(x^2-y^2, xy)
E_3	$\begin{cases}1\\1\end{cases}$	$\begin{matrix}-\varepsilon\\-\varepsilon^*\end{matrix}$	$\begin{matrix}i\\-i\end{matrix}$	$\begin{matrix}\varepsilon^*\\\varepsilon\end{matrix}$	$\begin{matrix}-1\\-1\end{matrix}$	$\begin{matrix}\varepsilon\\\varepsilon^*\end{matrix}$	$\begin{matrix}-i\\i\end{matrix}$	$\begin{matrix}-\varepsilon^*\\-\varepsilon\end{matrix}$		(xz, yz)

9. The Cubic Groups

T	$\hat{E}$	$4\hat{C}_3$	$4\hat{C}_3^2$	$3\hat{C}_2$		$\varepsilon = \exp(2\pi i/3)$
A	1	1	1	1		$x^2+y^2+z^2$
E	$\begin{cases}1\\1\end{cases}$	$\begin{matrix}\varepsilon\\\varepsilon^*\end{matrix}$	$\begin{matrix}\varepsilon^*\\\varepsilon\end{matrix}$	$\begin{matrix}1\\1\end{matrix}$		$(2z^2-x^2-y^2, x^2-y^2)$
T	3	0	0	−1	$(R_x, R_y, R_z); (x, y, z)$	(xy, xz, yz)

T_h	$\hat{E}$	$4\hat{C}_3$	$4\hat{C}_3^2$	$3\hat{C}_2$	$\hat{i}$	$4\hat{S}_6$	$4\hat{S}_6^5$	$3\hat{\sigma}_h$		$\varepsilon = \exp(2\pi i/3)$
A_g	1	1	1	1	1	1	1	1		$x^2+y^2+z^2$
E_g	$\begin{cases}1\\1\end{cases}$	$\begin{matrix}\varepsilon\\\varepsilon^*\end{matrix}$	$\begin{matrix}\varepsilon^*\\\varepsilon\end{matrix}$	$\begin{matrix}1\\1\end{matrix}$	$\begin{matrix}1\\1\end{matrix}$	$\begin{matrix}\varepsilon\\\varepsilon^*\end{matrix}$	$\begin{matrix}\varepsilon^*\\\varepsilon\end{matrix}$	$\begin{matrix}1\\1\end{matrix}$		$(2z^2-x^2-y^2, x^2-y^2)$
T_g	3	0	0	−1	1	0	0	1	(R_x, R_y, R_z)	(xy, xz, yz)
A_u	1	1	1	1	−1	−1	−1	−1		
E_u	$\begin{cases}1\\1\end{cases}$	$\begin{matrix}\varepsilon\\\varepsilon^*\end{matrix}$	$\begin{matrix}\varepsilon^*\\\varepsilon\end{matrix}$	$\begin{matrix}1\\1\end{matrix}$	$\begin{matrix}-1\\-1\end{matrix}$	$\begin{matrix}-\varepsilon\\-\varepsilon^*\end{matrix}$	$\begin{matrix}-\varepsilon^*\\-\varepsilon\end{matrix}$	$\begin{matrix}-1\\-1\end{matrix}$		
T_u	3	0	0	−1	−1	0	0	1	(x, y, z)	

T_d	$\hat{E}$	$8\hat{C}_3$	$3\hat{C}_2$	$6\hat{S}_4$	$6\hat{\sigma}_d$		
A_1	1	1	1	1	1		$x^2+y^2+z^2$
A_2	1	1	1	−1	−1		
E	2	−1	2	0	0		$(2z^2-x^2-y^2, x^2-y^2)$
T_1	3	0	−1	1	−1	(R_x, R_y, R_z)	
T_2	3	0	−1	−1	1	(x, y, z)	(xy, xz, yz)

O	$\hat{E}$	$8\hat{C}_3$	$6\hat{C}_2$	$6\hat{C}_4$	$3\hat{C}_2(=\hat{C}_4^2)$		
A_1	1	1	1	1	1		$x^2+y^2+z^2$
A_2	1	1	−1	−1	1		

(Continued)

Continued

O	$\hat{E}$	$8\hat{C}_3$	$6\hat{C}_2$	$6\hat{C}_4$	$3\hat{C}_2(=\hat{C}_4^2)$		
E	2	−1	0	0	2		$(2z^2 - x^2 - y^2, x^2 - y^2)$
T_1	3	0	−1	1	−1	(R_x, R_y, R_z); (x, y, z)	
T_2	3	0	1	−1	−1		(xy, xz, yz)

O_h	$\hat{E}$	$8\hat{C}_3$	$6\hat{C}_2$	$6\hat{C}_4$	$3\hat{C}_2(=\hat{C}_4^2)$	$\hat{i}$	$6\hat{S}_4$	$8\hat{S}_6$	$3\hat{\sigma}_h$	$3\hat{\sigma}_d$		
A_{1g}	1	1	1	1	1	1	1	1	1	1		$x^2 + y^2 + z^2$
A_{2g}	1	1	−1	−1	1	1	−1	1	1	−1		
E_g	2	−1	0	0	2	2	0	−1	2	0		$(2z^2 - x^2 - y^2, x^2 - y^2)$
T_{1g}	3	0	−1	1	−1	3	1	0	−1	−1	(R_x, R_y, R_z)	
T_{2g}	3	0	1	−1	−1	3	−1	0	−1	1		(xy, xz, yz)
A_{1u}	1	1	1	1	1	−1	−1	−1	−1	−1		
A_{2u}	1	1	−1	−1	1	−1	1	−1	−1	1		
E_u	2	−1	0	0	2	−2	0	1	−2	0		
T_{1u}	3	0	−1	1	−1	−3	−1	0	1	1	(x, y, z)	
T_{2u}	3	0	1	−1	−1	−3	1	0	1	−1		

10. The Groups for Linear Molecules

$C_{\infty v}$	$\hat{E}$	$2\hat{C}_\infty^\phi$	...	$\infty\hat{\sigma}_v$		
$A_1 \equiv \Sigma^+$	1	1	...	1	z	$x^2 + y^2, z^2$
$A_2 \equiv \Sigma^-$	1	1	...	−1	R_z	
$E_1 \equiv \Pi$	2	$2\cos\phi$	...	0	$(x, y), (R_x, R_y)$	(xz, yz)
$E_2 \equiv \Delta$	2	$2\cos 2\phi$	...	0		$(x^2 - y^2, xy)$
$E_3 \equiv \Phi$	2	$2\cos 3\phi$	...	0		
...	...	...	...	...		

$D_{\infty h}$	$\hat{E}$	$2\hat{C}_\infty^\phi$	...	$\infty\hat{\sigma}_v$	$\hat{i}$	$2\hat{S}_\infty^\phi$	...	$\infty\hat{C}_2$		
Σ_g^+	1	1	...	1	1	1	...	1		$x^2 + y^2, z^2$
Σ_g^-	1	1	...	−1	1	1	...	−1	R_z	
Π_g	2	$2\cos\phi$	...	0	2	$-2\cos\phi$	...	0	(R_x, R_y)	(xz, yz)
Δ_g	2	$2\cos 2\phi$	...	0	2	$2\cos 2\phi$	...	0		$(x^2 - y^2, xy)$
...	...	...	...	...	...	...	...	...		
Σ_u^+	1	1	...	1	−1	−1	...	−1	z	
Σ_u^-	1	1	...	−1	−1	−1	...	1		
Π_u	2	$2\cos\phi$	...	0	−2	$2\cos\phi$	...	0	(x, y)	
Δ_u	2	$2\cos 2\phi$	...	0	−2	$-2\cos 2\phi$	...	0		
...	...	...	...	...	...	...	...	...		

11. The Icosahedral Groups (see note below)

I_h	$\hat{E}$	$12\hat{C}_5$	$12\hat{C}_5^2$	$20\hat{C}_3$	$15\hat{C}_2$	$\hat{i}$	$12\hat{S}_{10}$	$12\hat{S}_{10}^3$	$20\hat{S}_6$	$15\hat{\sigma}$		
A_g	1	1	1	1	1	1	1	1	1	1		$x^2 + y^2 + z^2$
T_{1g}	3	$\frac{1}{2}(1+\sqrt{5})$	$\frac{1}{2}(1-\sqrt{5})$	0	−1	3	$\frac{1}{2}(1-\sqrt{5})$	$\frac{1}{2}(1+\sqrt{5})$	0	−1	(R_x, R_y, R_z)	
T_{2g}	3	$\frac{1}{2}(1-\sqrt{5})$	$\frac{1}{2}(1+\sqrt{5})$	0	−1	3	$\frac{1}{2}(1+\sqrt{5})$	$\frac{1}{2}(1-\sqrt{5})$	0	−1		
G_g	4	−1	−1	1	0	4	−1	−1	1	0		
H_g	5	0	0	−1	1	5	0	0	−1	1		$(2z^2 - x^2 - y^2,$ $x^2 - y^2, xy, yz, zx)$
A_u	1	1	1	1	1	−1	−1	−1	−1	−1		
T_{1u}	3	$\frac{1}{2}(1+\sqrt{5})$	$\frac{1}{2}(1-\sqrt{5})$	0	−1	−3	$-\frac{1}{2}(1-\sqrt{5})$	$-\frac{1}{2}(1+\sqrt{5})$	0	1	(x, y, z)	
T_{2u}	3	$\frac{1}{2}(1-\sqrt{5})$	$\frac{1}{2}(1+\sqrt{5})$	0	−1	−3	$-\frac{1}{2}(1+\sqrt{5})$	$-\frac{1}{2}(1-\sqrt{5})$	0	1		
G_u	4	−1	−1	1	0	−4	1	1	−1	0		(xz, yz)
H_u	5	0	0	−1	1	−5	0	0	1	−1		

Note: The blocked portion is the character table of the pure rotational icosahedral group (I) with the g subscript omitted for the Mulliken symbols; the basis set (x, y, z) corresponds to the irreducible representation T_1 in I.

Correlation Tables for O_h and T_d Groups

These tables show correlations of the octahedral and tetrahedral groups with their subgroups which occur commonly in the study of transition metal complexes.

O_h	O	T_d	D_{4h}	D_{2d}	C_{4v}	C_{2v}	D_{3d}	D_3	C_{2h}
A_{1g}	A_1	A_1	A_{1g}	A_1	A_1	A_1	A_{1g}	A_1	A_g
A_{2g}	A_2	A_2	B_{1g}	B_1	B_1	A_2	A_{2g}	A_2	B_g
E_g	E	E	$A_{1g}+B_{1g}$	A_1+B_1	A_1+B_2	A_1+A_2	E_g	E	A_g+B_g
T_{1g}	T_1	T_1	$A_{2g}+E_g$	A_2+E	A_2+E	$A_2+B_1+B_2$	$A_{2g}+E_g$	A_2+E	A_g+2B_g
T_{2g}	T_2	T_2	$B_{2g}+E_g$	B_2+E	B_2+E	$A_1+B_1+B_2$	$A_{1g}+E_g$	A_1+E	$2A_g+B_g$
A_{1u}	A_1	A_2	A_{1u}	B_1	A_2	A_2	A_{1u}	A_1	A_u
A_{2u}	A_2	A_1	B_{1u}	A_1	B_2	A_1	A_{2u}	A_2	B_u
E_u	E	E	$A_{1u}+B_{1u}$	A_1+B_1	A_2+B_2	A_1+A_2	E_u	E	A_u+B_u
T_{1u}	T_1	T_2	$A_{2u}+E_u$	B_2+E	A_1+E	$A_1+B_1+B_2$	$A_{2u}+E_u$	A_2+E	A_u+2B_u
T_{2u}	T_2	T_1	$B_{2u}+E_u$	A_2+E	B_1+E	$A_2+B_1+B_2$	$A_{1u}+E_u$	A_1+E	$2A_u+B_u$

T_d	T	C_{3v}	C_{2v}	D_{2d}
A_1	A	A_1	A_1	A_1
A_2	A	A_2	A_2	B_1
E	E	E	A_1+A_2	A_1+B_1
T_1	T	A_2+E	$A_2+B_1+B_2$	A_2+E
T_2	T	A_1+E	$A_1+B_1+B_2$	B_2+E

K	O	D_4	D_3
S	A_1	A_1	A_1
P	T_1	A_2+E	A_2+E
D	$E+T_2$	$A_1+B_1+B_2+E$	A_1+2E
F	$A_2+T_1+T_2$	$A_2+B_1+B_2+2E$	A_1+2A_2+2E
G	$A_1+E+T_1+T_2$	$2A_1+A_2+B_1+B_2+2E$	$2A_1+A_2+3E$
H	$E+2T_1+T_2$	$A_1+2A_2+B_1+B_2+3E$	A_1+2A_2+4E

Index

E

F

G

H

I

J

K

L